THE PHILOSOPHY OF MATHEMATICS TODAY

The Philosophy of Mathematics Today

edited by

MATTHIAS SCHIRN

CLARENDON PRESS · OXFORD
1998

Oxford University Press, Great Clarendon Street, Oxford OX2 6DP
Oxford New York
Athens Auckland Bangkok Bogotá Buenos Aires
Calcutta Cape Town Chennai Dar es Salaam Delhi
Florence Hong Kong Istanbul Karachi Kuala Lumpur
Madras Melbourne Mexico City Mumbai Nairobi Paris
São Paolo Singapore Taipei Tokyo Toronto Warsaw
and associated companies in
Berlin Ibadan

Oxford is a trade mark of Oxford University Press

Published in the United States by
Oxford University Press Inc., New York

British Library Cataloguing in Publication Data
Data available

Library of Congress Cataloging in Publication Data

The philosophy of mathematics today / edited by Matthias Schirn.
Papers from a conference held in Munich from June 28 to July 4, 1993.
Includes bibliographical references and index.
1. Mathematics—Philosophy—History—20th century—Congresses.
I. Schirn, Matthias.
QA8.P45 1997 510′.1—dc21 97-38028

ISBN 0–19–823654–9

1 3 5 7 9 10 8 6 4 2

Typeset by Best-set Typesetter Ltd., Hong Kong
Printed in Great Britain by
Bookcraft (Bath) Ltd
Midsomer Norton, Avon

ACKNOWLEDGEMENTS

Most of the papers collected here derive from the conference 'Philosophy of Mathematics Today', held in Munich from 28 June to 4 July 1993 and organized by the editor of this volume. The conference was generously sponsored by the Deutsche Forschungsgemeinschaft and received additional financial support from the Mind Association. I wish to express my gratitude to both institutions. I am also grateful to Crispin Wright for his proposal to apply for a grant from the Mind Association and to Peter Clark, in his capacity as Secretary of the Association, for strongly supporting my application. Many thanks are due to Uwe Lück for his help during the conference. I would also like to thank all the participants for creating a most pleasant atmosphere, stimulating and yet very relaxed, indeed conducive to an intense discussion of the philosophy of mathematics over nearly a week. Finally, it is a pleasure to thank Peter Momtchiloff of Oxford University Press for his interest in this volume, his collaboration, and his encouragement and friendly advice.

M. S.

Mexico City, April 1996

ACKNOWLEDGEMENTS

CONTENTS

NOTES ON THE CONTRIBUTORS

Paul Benacerraf is Stuart Professor of Philosophy at Princeton University. He has been a John Simon Guggenheim Fellow and a Fellow of the Center for Advanced Study in the Behavioral Sciences. His publications include *Logicism: Some Considerations* (1960), *Philosophy of Mathematics* (ed. with H. Putnam) (1964; second revised edition 1983); 'What Numbers Could Not Be' (1965), 'God, the Devil, and Gödel' (1967), 'Mathematical Truth' (1973), 'Frege: The Last Logicist' (1981), and 'Skolem and the Skeptic' (1985).

George Boolos was Professor of Philosophy at the Massachusetts Institute of Technology. He was Vice-President of the Association for Symbolic Logic and an editor of the *Journal of Symbolic Logic*. His publications include *Computability and Logic* (with R. C. Jeffrey) (1974), *The Unprovability of Consistency: An Essay in Modal Logic* (1979), *Meaning and Method: Essays in Honor of Hilary Putnam* (ed.) (1990), and *The Logic of Provability* (1993). He is also the author of numerous articles on logic and the philosophy of mathematics and on Frege. He died on 27 May 1996.

John P. Burgess is Professor of Philosophy at Princeton University. He has been an editor of the *Journal of Symbolic Logic*. His many publications on logic and the philosophy of mathematics include 'Forcing' (1977), 'Why I Am Not a Nominalist' (1983), 'Synthetic Mechanics' (1984), 'The Truth is Never Simple' (1986), 'Non-Classical Logic and Ontological Non-Commitment' (1995). His book (with G. Rosen) *A Subject with No Object: Strategies for Nominalistic Reconstrual of Mathematics* was published in 1997 by Oxford University Press.

Charles S. Chihara is Professor of Philosophy at the University of California at Berkeley. His chief research interests are in the philosophies of mathematics and logic, philosophical logic, and metaphysics. He is the author of *Ontology and the Vicious Circle Principle* (1973), *Constructibility and Mathematical Existence* (1990), and *Worlds of Possibility* (1998). He has also published many articles in the areas of his research interests.

Michael Detlefsen is Professor of Philosophy at the University of Notre Dame, where he is also editor-in-chief of the *Notre Dame Journal for Formal Logic*. He is subject-editor in logic and philosophy of mathematics for the forthcoming encyclopaedia of philosophy by Routledge. He has visited at the philosophical faculty of the University of Split and the department of philosophy of the University of Konstanz. He is the author of *Hilbert's Program* (1986), and of various papers in the philosophy of mathematics and logic. Currently, he is working on two books, one on constructivism, the other on Gödel's theorems.

Michael Dummett is Fellow of New College, Oxford and Emeritus Wykeham Professor of Logic at the University of Oxford. In 1976 he delivered the William James Lectures at Harvard University. Among his many publications are *Frege: Philosophy of Language* (1973, 1981), *Elements of Intuitionism* (1976), *Truth and Other Enigmas* (1978), *The Interpretation of Frege's Philosophy* (1981), *Frege: Philosophy of Mathematics* (1991), *Frege and Other Philosophers* (1991), *The Logical Basis of Metaphysics* (1991), and *The Seas of Language* (1993).

Hartry Field is Professor of Philosophy of New York University. His research interests are in metaphysics, epistemology, and the philosophies of language, mind, logic, mathematics, and science. His publications include *Science without Numbers: A Defence of Nominalism* (1980) (for which he won the Lakatos Prize 1986) and *Realism, Mathematics and Modality* (1989). He has been a Visiting Fellow at All Souls College, Oxford and a Visiting Professor at the Massachusetts Institute of Technology and at the University of Arizona.

Kit Fine is Professor of Philosophy at New York University. His principal research interests are in metaphysics, logic, and the philosophy of language. He is the author of *Worlds, Times and Selves* (with A. N. Prior) (1979) and *Reasoning with Arbitrary Objects* (1985). He is also the author of numerous articles in the areas of his research interests. He has been a Visiting Fellow at All Souls College, Oxford and a visitor at the universities of Salzburg, Siena, Toronto, and Arizona.

Bob Hale is Professor of Metaphysical Philosophy at the University of Glasgow. His principal research interests lie in the epistemology and ontology of mathematics and related areas of philosophy of language and metaphysics. His book *Abstract Objects* was published in 1987 by Blackwell. He is currently editing, jointly with Crispin Wright, the Blackwell companion to the Philosophy of Language, to which he is also contributing various chapters.

Richard G. Heck, Jnr. graduated from the Massachusetts Institute of Technology in 1991 and is currently Associate Professor of Philosophy at Harvard University. His chief interests lie in the philosophies of language, logic, and mathematics. He has published articles on Frege in the *Journal of Symbolic Logic, Noûs, Mind, History and Philosophy of Logic*, and in several collections.

Geoffrey Hellman is Professor of Philosophy at the University of Minnesota, Minneapolis. His main research interests lie in the philosophy of mathematics, philosophy of quantum mechanics and natural science, and philosophy of logic. He is the author of *Mathematics without Numbers* (1989). In addition, his most significant publications include 'Physicalism: Ontology, Determination, and Reduction' (with F. Thompson, 1975), 'Realist Principles' (1983), 'Bell-Type Inequalities in the Non-Deal Case: Proof of a Conjecture of Bell' (1992), and 'Gleason's Theorem is Not Constructively Provable' (1993).

Penelope Maddy is Professor of Philosophy at the University of California, Irvine. Her principal research interest is the philosophy of set theory. She is the author of *Realism in Mathematics* (1990) and *Naturalism in Mathematics* (1997). Her most important articles include 'Perception and Mathematical Intuition' (1980), 'Proper Classes' (1983), 'Believing the Axioms: I and II' (1988), 'Indispensability and Practice' (1992), 'Does V equal L?' (1993), and 'Set Theoretic Naturalism' (1996).

Karl-Georg Niebergall graduated from the University of Munich in 1995 with a thesis on the metamathematics of non-axiomatizable theories. He is currently a research fellow of the Deutsche Forschungsgemeinschaft and teaches at the University of Munich.

Charles D. Parsons is Edgar Pierce Professor of Philosophy at Harvard University. He is the author of *Mathematics in Philosophy* (1983), and editor, with Solomon Feferman and others, of Kurt Gödel, *Collected Works*, iii, *Unpublished Essays and Lectures* (1995). He has also published numerous articles on logic and the philosophy of mathematics. He has been a Visiting Fellow at All Souls College and a Fellow of the Netherlands Institute for Advanced Study.

Michael D. Resnik is Distinguished Professor of Philosophy at the University of North Carolina at Chapel Hill. His principal research interests lie in the philosophy of mathematics, the philosophy of logic, and the theory of rationality. Among his most important publications are: *Frege and the Philosophy of Mathematics* (1980), *Choices* (1987), and *Mathematics as a Science of Patterns* (1997).

Matthias Schirn is Professor of Philosophy at the University of Munich. He has been a Visiting Lecturer in Oxford and Cambridge, and a Visiting Professor in Minneapolis, Campinas, São Paulo, Buenos Aires, Mexico City, Santiago de Compostela, and other universities in Europe and North and South America. He has published numerous articles on topics in the philosophies of language, logic and mathematics, and in epistemology. His book *Gottlob Frege: Grundlagen der Logik und Mathematik* is forthcoming from Walter de Gruyter.

Stewart Shapiro is Professor of Philosophy at the Ohio State University at Newark and the University of St Andrews. He specializes in logic and the philosophies of logic and mathematics. His publications include *Foundations Without Foundationalism: A Case for Second-Order Logic* (1991), 'Second-Order Languages and Mathematical Practice' (1985), 'Principles of Reflection and Second-Order Logic' (1987), 'Second-Order Logic, Foundations, and Rules' (1990), and 'Modality and Ontology' (1993).

Peter Simons is Professor of Philosophy at the University of Leeds. He has also taught in Austria, Switzerland, and the USA. He is currently the editor of *History and Philosophy of Logic.* He is the author of *Parts: A Study in Ontology* (1987), *Philosophy and Logic in Central Europe from Bolzano to Tarski: Selected Essays* (1992), and many articles.

William W. Tait is Professor of Philosophy at the University of Chicago. He has been a member of the Institute for Advanced Study in Princeton and a Visiting Professor of Mathematics at the University of Aarhus, Denmark. His most important articles on logic and the philosophy of mathematics include 'Intensional Interpretations of Functionals of Finite Type I' (1967), 'Normal Derivability in Classical Logic' (1968), 'Finitism' (1981), 'Truth and Proof: The Platonism of Mathematics' (1986), and 'The Law of Excluded Middle and the Axiom of Choice' (1994).

Crispin Wright is Professor of Logic and Metaphysics at the University of St Andrews, and formerly Nelson Professor of Philosophy at the University of Michigan. He has held visiting positions at the University of Pennsylvania, Princeton University, and other institutions. He is the author of *Wittgenstein on the Foundations of Mathematics* (1980), *Frege's Conception of Numbers as Objects* (1983), *Realism, Meaning and Truth* (1986), and *Truth and Objectivity* (1992). He is also the author of numerous articles on the philosophies of language, mathematics and mind, and on the theory of knowledge.

Introduction

MATTHIAS SCHIRN

The Philosophy of Mathematics Today—the title of this collection is meant to convey a twofold objective: first, to document important approaches, tendencies, and results in current philosophy of mathematics, and second, to provide fruitful impulses for future work in several of its areas. In my view, the volume reflects, to a considerable extent, the different methods and viewpoints of current research in the field. If one were to attempt to characterize the essays collected here in terms of a feature they all share, one would probably appeal to their being essentially concerned with foundational issues. Questions regarding the historical development of mathematics or the growth of mathematical knowledge play little role in these essays.[1]

This volume is divided into five parts: I. Ontology, Models, and Indeterminacy; II. Mathematics, Science, and Method; III. Finitism and Intuitionism; IV. Frege and the Foundations of Arithmetic; and V. Sets, Structure, and Abstraction. Classical positions in the philosophy of mathematics are brought into focus and are critically discussed in Parts III and IV and also, though to a lesser extent, in Part V. The philosophical legacy of Frege and Hilbert predominates in these parts. I welcome this; yet what pleases me more is the fact that the collection maintains a reasonable balance between the treatment of some central historical views and the analysis of topics which have emerged recently in the discussion of the philosophy of mathematics and have come to dominate a large part of it.

I am very grateful to my friend Daniel Isaacson for his helpful comments on an earlier version of this Introduction. I only regret that Daniel did not finish for publication the paper 'On the Need for New Concepts in Mathematics' which he presented at the conference. My thanks go also to Robert Bublak and Karl-Georg Niebergall for discussing several issues of this volume.

[1] Perhaps the most important approach to the history of mathematics in recent years is Philip Kitcher's study *The Nature of Mathematical Knowledge*, OUP, New York, 1984. It offers a novel account of how mathematical knowledge evolves, ascribing to the present mathematical community and to previous communities an epistemological significance with which they are not usually credited. The doctrine that mathematical knowledge is a priori is expressly rejected. Furthermore, Kitcher holds that the growth of mathematical knowledge is far more akin to the growth of scientific knowledge than is usually assumed.

I. ONTOLOGY, MODELS, AND INDETERMINACY

Indeterminacy and *logical consequence* are two key issues of Part I. Both are discussed in the context of model theory. Hartry Field is concerned with the indeterminacy which we encounter in our use of certain fundamental mathematical notions such as the notion of set, and scrutinizes in this connection Hilary Putnam's model-theoretic argument in 'Models and Reality'.[2] Stewart Shapiro and Charles Chihara argue in defence of a model-theoretic account of the notion of logical consequence, and reject some objections raised by John Etchemendy in his book *The Concept of Logical Consequence*.[3] There is also a link tying Paul Benacerraf's essay 'What Mathematical Truth Could Not Be—I' to Field's paper 'Do We Have a Determinate Conception of Finiteness and Natural Number?' In the final section of his essay, Benacerraf comments on Putnam's and Crispin Wright's reflections on the Löwenheim–Skolem Theorem. In what follows, I shall first comment on Benacerraf's twofold challenge to Platonism and shall then briefly summarize Etchemendy's critique of Tarski's model-theoretic account of the concept of logical consequence.

Mathematical Platonism is the doctrine that numbers, functions, sets, and the like are abstract (that is, non-spatial, non-temporal, causally inert) entities existing independently of the human mind. The advocate of this doctrine, the Platonist, characteristically holds that mathematical statements are either true or false, regardless of whether or not we are capable of determining their truth-value.

In his influential paper 'What Numbers Could Not Be',[4] Benacerraf questioned the legitimacy of mathematical Platonism on ontological grounds. He argued that numbers cannot be sets, because there is no good reason to hold that any particular number is some particular set. Plainly, there is no unique set-theoretic specification of, for example, the natural or the real numbers. The best we can achieve is to determine the numbers of either kind up to isomorphism. Not only does the identification of numbers with sets appear to be undermined by the arbitrariness of our actual choices, but also by the arbitrariness of our taking sets to be the basic objects of mathematics. It is possible, for instance, to regard ordinal numbers as basic and to define sets in terms of them. By way of extending his argument from sets to objects in general, Benacerraf concludes that numbers cannot be objects at all. Only by appeal to non-structural properties of numbers could we pick out a number as a determinate object. Yet in stating the properties of numbers which are necessary and jointly sufficient for the study of arithmetic, we merely characterize an abstract structure. According to this view,

[2] *Journal of Symbolic Logic*, 45 (1980), 464–82.
[3] Harvard University Press, Cambridge, Mass., 1990.
[4] *Philosophical Review*, 74 (1965), 47–73.

then, any object can play the role of, say 3, that is, any object can be the third element in some progression or ω-sequence.

Benacerraf's view that we cannot encounter objects which *are* natural numbers owes its credibility to the assumption that the structural characterization he suggests in effect provides all necessary and sufficient properties of the natural numbers. However, I think that he fails to establish this assumption conclusively. Its plausibility is obviously due to the fact that hitherto no further features of the numbers could be found. Benacerraf's ontological argument merely shows that we cannot single out the numbers as determinate objects by appeal to the conditions which could be specified up to now. On the other hand, it is hard to see what could be said about the nature of the numbers beyond their structural characterization. So it seems to me that unless someone has succeeded in refuting Benacerraf's ontological argument—which, to the best of my knowledge, so far nobody has—the conception of numbers as objects remains a dogma bequeathed to us by Frege.

Almost every philosophical doctrine has its Achilles' heel, and mathematical Platonism is no different. Among the many problems it faces the most pressing is often regarded as one belonging not to ontology, but rather to epistemology: the Platonist must explain how we can have cognitive access to the remote realm of abstract mathematical objects, that is, he must explain how we can attain mathematical knowledge at all. Frege believed that we conceive the abstract objects of arithmetic by virtue of a special logical faculty, namely by virtue of what I shall later introduce as logical or Fregean abstraction. About half a century later, Gödel claimed that it is a faculty of mathematical intuition which allows us cognitive access to the mathematical realm, a claim which Frege probably would have rejected out of hand. Gödel argues that 'despite their remoteness from sense experience, we do have something like a perception also of the objects of set theory, as is seen from the fact that the axioms force themselves upon us as being true.'[5] However, I shall refrain here from trying to analyse the loose parallel Gödel proposes between our ordinary perception of physical objects and the (mysterious) quasi-perceptual intuition of mathematical objects he is postulating. In my view, his conception of how we can come into contact with these objects suffers from considerable vagueness.

In his likewise influential paper 'Mathematical Truth',[6] Benacerraf challenged mathematical Platonism also on epistemological grounds. Roughly speaking, the challenge is that we must be able to combine the standard

[5] K. Gödel, 'What is Cantor's Continuum Problem?', *American Mathematical Monthly*, 54: 515–25; rev. and expanded version in P. Benacerraf and H. Putnam (eds.), *Philosophy of Mathematics: Selected Readings*, CUP, Cambridge, 1983, 470–85, here 483f.

[6] *Journal of Philosophy*, 70 (1973), 661–79.

Platonistic account of mathematical truth with a causal account of mathematical knowledge without offending against the plausible presupposition that we do have some mathematical knowledge. But how can abstract entities enter into any physical interaction? Given that we possess some mathematical knowledge, it follows that either our best theory of mathematical truth at hand (i.e. the standard Platonistic view) or the best theory of knowledge (and here Benacerraf favours the causal theory) comes to naught. It is true that the causal theory is now widely disavowed. All the same, it remains for the Platonist to explain how we come to know anything at all about the subject-matter of mathematics.

Philosophers of mathematics have responded in various ways to the epistemological challenge just laid out. Of course, I cannot and need not review the many facets of the discussion. Suffice it for me to do two things: first, to sketch one prominent response to Benacerraf's epistemological argument, and second, to present a reformulated version of the argument which plays an essential role in Bob Hale's paper 'Is Platonism Epistemologically Bankrupt?' and which I likewise consider important.

On the face of it, the epistemological challenge might loosen its grip for the Platonist if he or she could allow that mathematical objects be construed not as full-fledged abstract entities, but rather as entities to which we may have access on the strength of some pedestrian cognitive faculty we undoubtedly possess. However, a strategy along these lines would, I think, be bound to fail if it were to involve the advocacy of a brand of Platonism not deserving of its name.

Penelope Maddy, in her defence of set-theoretic realism (i.e. of a version of so-called compromise Platonism) seems to have few if any qualms about the idea of abandoning the traditional Platonist's view that mathematical objects are not part of the physical world. Although in this respect she dissociates herself from Gödel, she sees herself joining him closely by taking perception-like intuition to be the faculty that brings us into contact with the mathematical realm. And like Gödel's Platonism, her own naturalized Platonism rests on an analogy between mathematics and natural science.

According to Maddy, sets of physical objects have location in space and time; and we perceive such sets in much the same way as we perceive the physical objects themselves. In the second chapter of her book *Realism in Mathematics*,[7] she describes how we gradually develop the ability of perceiving sets as well as the process leading us from the perception of sets to mathematical intuitions about sets, and, finally, how her epistemological approach can account for the general beliefs that underlie the most elementary set-theoretic axioms. Like Gödel, Maddy believes that not all

[7] OUP, Oxford, 1990; cf. also Maddy, 'Perception and Mathematical Intuition', *Philosophical Review*, 89 (1980), 163–96.

axioms of set theory can be justified on intuitive grounds. The more theoretical part of set theory is left for 'extrinsic' justification.[8] Maddy regards the question of how the axioms of mathematics are justified as the proper successor to Benacerraf's challenge.[9]

Maddy's set-theoretic realism has provoked a number of forceful objections which, for reasons of space, I shall not review here.[10] Instead I would now like to return to Benacerraf's epistemological challenge itself and present Field's reformulation of it.

Both in the Introduction to his collection *Realism, Mathematics and Modality*[11] and in its title essay,[12] Field rephrases Benacerraf's epistemological argument in such a way that it does not depend on any assumption about necessary and sufficient conditions for knowledge. Field suggests we understand the argument not as a challenge to our ability to justify our mathematical beliefs, but rather as a challenge to our ability to explain the reliability of these beliefs. The challenge, so construed, is then to furnish an account of the 'mechanisms' which explain how our beliefs about mathematical objects, conceived of as mind-independent abstract entities, can reflect the facts about them so well.

> The idea is that *if it appears in principle impossible to explain this*, then that tends to *undermine* the belief in mathematical entities, *despite* whatever reason we might have for believing in them. Of course, the reasons for believing in mathematical entities (in particular the indispensability arguments) still need to be addressed, but the role of the Benacerrafian challenge (as I see it) is to raise the cost of thinking that the postulation of mathematical entities is a proper solution, and to thereby increase the motivation for showing that mathematics is not really indispensable after all. (Field 1989, 26)

In Field's view, fleshing out this sketch would require four things. Firstly, one would have to formulate in more precise terms the claim that our mathematical beliefs are reliable. Secondly, one would need to show that the Platonist must accept this claim. The third thing one would have to do is to argue that the Platonist must commit himself to the possibility of explaining the reliability in question. Finally, to render it plausible that Benacerraf's challenge is insuperable, one would need to argue that the relevant facts about the Platonist's conception of mathematical

[8] Cf. Maddy, 'The Roots of Contemporary Platonism', *Journal of Symbolic Logic*, 54 (1989), 1121–44, at 1140f., ch. 4 of her 1990 book, and her recent paper 'Set Theoretic Naturalism', *Journal of Symbolic Logic*, 61 (1996), 490–514.

[9] Cf. 'Philosophy of Mathematics: Prospects for the 1990s', *Synthese*, 88 (1991), 155–66.

[10] See e.g. C. Chihara, 'A Gödelian Thesis Regarding Mathematical Objects: Do They Exist? And Can We Conceive Them?', *Philosophical Review*, 91 (1982), 211–27 and Maddy's reply in her 1990 book, 150ff. The most recent extensive critique of Maddy's position is M. Balaguer's 'Against (Maddian) Naturalized Platonism', *Philosophia Mathematica*, 3 (1994), 97–108.

[11] Blackwell, Oxford, 1989. [12] See Field (1989, essay 7).

objects makes it impossible to provide an explanation of the reliability claim.[13]

How well Field's reformulation of the epistemological challenge accords with Benacerraf's intention in 'Mathematical Truth' I find hard to say. However, we might know more about this, and shall certainly know more about mathematical truth and mathematical knowledge, once Benacerraf's projected paper 'What Mathematical Truth Could Not Be—II' has seen the light of day.

I now turn briefly to Etchemendy's critique of the model-theoretic account of logical consequence. In his article 'On the Concept of Logical Consequence' (1936),[14] Tarski aimed at framing a precise definition of the notion of logical consequence that was faithful to the pre-analytical, intuitive notion. In fact, he made it clear that the acceptability of the envisaged definition depended crucially on how closely it corresponded 'in essentials to the common concept' (413). The proposed definition is as follows: *The sentence X follows logically from the sentences of the class K if and only if every model of the class K is also a model of the sentence X* (417).[15]

Etchemendy's principal thesis is that the model-theoretic account of logical truth and logical consequence fails to capture our pre-theoretic understanding of these notions. The salient feature of our ordinary under-

[13] Cf. Field's ensuing remarks in the Introduction to his 1989 book, 27 ff.—M. Balaguer, in his paper 'A Platonist Epistemology' (*Synthese*, 103 (1995), 303–25), attempts to meet Benacerraf's epistemological challenge 'head on' by constructing an epistemology of abstract mathematical objects. He claims that what he calls *full-blooded Platonism* (or FBP) enables us to explain how we can acquire knowledge of the mathematical realm. Intuitively expressed, FBP is the view that all possible mathematical objects exist, or, rendered more precisely, that all possible mathematical objects which possibly *could* exist actually *do* exist. 'Possible' is here used in the sense of 'logically possible'. Balaguer argues that all consistent purely mathematical theories truly describe some part of the mathematical realm. He calls a sentence or theory *purely mathematical* if it speaks of nothing but the mathematical realm. In Section 4, dealing with FBP and the notion of mathematical truth, Balaguer observes: 'According to FBP, both ZFC and ZF+not-C truly describe parts of the mathematical realm; but there is nothing wrong with this, because they describe *different* parts of that realm' (315). Let me make just one comment. In n. 7 of his paper (322), Balaguer gives a formal characterization of FBP within second-order logic. Let 'Mx' mean 'x is a mathematical object' and let x be a first-order variable and Y a second-order variable. FBP can then be expressed as follows: $\forall Y(\Diamond\exists x(Mx \wedge Yx) \rightarrow \exists x(Mx \wedge Yx))$. Now, this formal explication hardly squares with Balaguer's informal remarks about FBP. Here is an example. (a) Choose $\forall y(y = x)$ for Yx. Since $\exists x(Mx \wedge \forall y (y = x))$ ('There is exactly one mathematical object') is consistent, it follows with FBP: $\exists x(Mx \wedge \forall y(y = x))$. (b) Take now $\exists z(Mz \wedge z \neq x \wedge \forall y(y = x \vee y = z))$ for Yx. Since $\exists x(Mx \wedge \exists z(Mz \wedge z \neq x \wedge \forall y(y = x \vee y = z)))$ ('There are exactly two mathematical objects') is likewise consistent, it follows with FBP: $\exists xz(Mx \wedge Mz \wedge z \neq x \wedge \forall y(y = x \vee y = z))$. In other words: FBP is inconsistent.

[14] In A. Tarski, *Logic, Semantics, Metamathematics*, 2nd edn., ed. J. Corcoran, Hackett Publishing Co., Indianapolis, 1990.

[15] Vann McGee ('Two Problems with Tarski's Theory of Consequence', *Proceedings of the Aristotelian Society*, 92 (1992), 273–92, at 273) has pointed out that the concept of a model which Tarski employs in his 1936 article differs in significant ways from the concept of a model we use today.

standing of logical consequence is clearly a modal one: the premisses of a logically valid argument *cannot* be true if the conclusion is false. Etchemendy claims that the technical and mathematical attraction of Tarski's quantificational account of logical truth derives directly from the following principle: If a universally quantified sentence is true, then all its instances are logically true (98). Etchemendy calls this principle the *reduction principle* and presents it as the 'linchpin' of the model-theoretic account of consequence. It is plain enough that the principle as it stands is wrong. The mere truth of a universal generalization can only guarantee that its instances are true. Yet it cannot warrant that they have any of the features which are usually regarded as separating logical truths from non-logical truths. Etchemendy considers two modifications of the reduction principle (cf. 101, 110) which might resolve this difficulty, but rejects these as well. The second modification is as follows: If a universally quantified sentence is true, and the constant expressions appearing in its matrix are of a distinctively logical sort, then all of its instances are logically true. According to Etchemendy, this version suffers from the same defect as the original principle. Thus he concludes that the reduction principle, no matter how it is stated, is irremediably flawed.

The quintessence of Etchemendy's detailed analysis appears to be this: Equating logical consequence with model-theoretic consequence is as mistaken as identifying it with derivability. The question as to whether one sentence follows logically from another does not amount to asking whether there are interpretations that make the latter true and the former false: logically valid arguments can fail this test, while invalid arguments can escape unscathed by it. Hence, we have no guarantee that an application of the model-theoretic account of logical consequence to any particular language will be extensionally correct. There may indeed be cases in which this account determines the extension correctly, but this is not because it captures the pre-analytic notion of logical consequence.[16]

So much to my introductory remarks about Platonism and logical consequence. It is time now to comment on the essays themselves in Part I.

Paul Benacerraf's paper 'What Mathematical Truth Could Not Be—I' opens the collection. It was written in response to an invitation to review developments concerning the issues discussed in his articles 'What Numbers Could Not Be' (henceforth WNCNB) and 'Mathematical Truth' (henceforth MT)—whence comes its title. However, the discussion of MT is left for a sequel, WMTCNB—II. The paper starts with a prologue followed by

[16] McGee (1992, 275) endorses Etchemendy's view that it is not logically necessary that a sentence true in every model be valid. McGee argues that what he calls the *contingency problem* (saying that what sentences are valid ought not to be a matter of contingent fact) can be overcome by making certain metaphysical assumptions. For a defence of Tarski's concept of logical consequence see G. Ray, 'Logical Consequence. A Defense of Tarski', *Journal of Philosophical Logic*, 25 (1996), 617–77.

three sections. The first describes the philosophical atmosphere in which both papers were written. The second section examines the challenge raised in WNCNB to arithmetical Platonism, seeks to remove actual or possible misunderstandings and gestures towards the directions one might take to fill in what Benacerraf considers to be the missing steps in his argument. It is pointed out that in WNCNB Parts I 'The education' and II 'The dilemma' are really independent of Part III 'Way out'; further that the conclusions of the latter part were drawn with much less conviction than those of the first two. Nonetheless, Benacerraf is still convinced that there is something right about the argument laid out in Part III, saying that not only could numbers not be sets, they could not even be numbers. The advocate of Platonistic reductionism, so we are told, may have three options to escape the attack launched in WNCNB. Option (a) is that he contrives a satisfactory way of defending the Frege–Russell analysis of number. Option (b) is that he (as Benacerraf puts it) 'bites the bullet, sticks to [his] realist guns and insists that the impossibility of establishing any answer as the right answer *does not undermine the thought that there is a right answer*'. Finally, option (c) is to advocate 'unbridled Occamism in a holist setting'. Benacerraf considers the pros and cons of these options. He claims that what is most appealing about favouring option (a), especially as compared with (b) and (c), is that it seems sensitive to arithmetical practice. In the third and final section, Benacerraf discusses the uses of metamathematical results in philosophy, focusing especially on Putnam's and Wright's appeals to the Löwenheim–Skolem Theorems to impose certain constraints on our conceptual abilities. One point made is that in these, as well as in the similar cases of appeals to Gödel's Incompleteness Theorems, the desired philosophical result cannot be achieved without considerable philosophical presupposition.

Bob Hale, in his paper 'Is Platonism Epistemologically Bankrupt?', argues against the widespread belief in contemporary philosophy of mathematics that mathematical Platonism faces a special epistemological problem. Hale concentrates mainly on a discussion of Field's version of the challenge which he considers to be more threatening than Benacerraf's original version. Recall that according to Field's version the acausality of abstract objects (and more generally, their lack of spatio-temporal location) precludes any explanation of how we can reliably form true mathematical beliefs. Hale holds that if we have a satisfactory epistemology of necessary truth in general, there is good reason to believe that this challenge can be met. He aims to establish, in particular, that existential commitments of mathematical statements, Platonistically understood, pose no insurmountable obstacle to their being necessarily true.

At the outset of his paper 'Do We Have a Determinate Conception of Finiteness and Natural Number?', Hartry Field discusses two arguments for the indeterminacy inherent in the notion of set: the 'size of the continuum argument' and the 'Skolem–Putnam argument'. In his influential article 'Models and Reality', Putnam had raised a challenge that in Field's opinion seems very difficult to meet: explain by virtue of what we 'intend' a standard as opposed to a non-standard interpretation of 'set'. Surely, if the challenge cannot be met, it seems that we must conclude that 'set' and '∈' are referentially indeterminate. Assuming the correctness of Putnam's argument for first-order set theory, Field claims that it is also valid for second-order set theory. Hence, second-order quantification would appear to be highly indeterminate, too. Furthermore, there seem to be compelling reasons to maintain that the arguments for the indeterminacy of 'set' and second-order quantification carry over to our notion of finiteness and our notion of natural number. Yet, as Field stresses, it seems especially hard to accept that we lack a determinate conception of finiteness or a conception of the natural numbers that is determinate even up to isomorphism. According to Field, the notion of finitude seems much more basic than the notion of set or second-order quantification and seems presupposed even by our notion of inferential procedure.

Field examines three possible reactions to this situation. The first is that the arguments for the indeterminacy of 'set' and second-order quantification are fallacious. Field rejects the prima facie promising view that the disquotational properties of truth render the indeterminacy concerns incoherent. The second possible reaction is to accept that even 'finite' (and 'there are only finitely many') is highly indeterminate. This radical indeterminist position is said to involve the view that at any time our notion of finitude is exhausted by the maximal mathematical theory of finiteness implicitly accepted by us at that time. Field argues that this view can be defended by pointing out that it is only apparently incoherent, and further, that even a radical version of the position which identifies truth with provability is not undermined by Gödel's Theorem. The third possible reaction is that the cases of second-order quantification and of finiteness differ from one another significantly. By way of scrutinizing Putnam's argument in 'Models and Reality', Field shows how this view might be advocated.

Stewart Shapiro, in his essay 'Logical Consequence: Models and Modality', seeks to vindicate the model-theoretic formulation as a viable account of the notion of logical consequence. The defence is mainly directed against the attack Etchemendy has mounted on the model-theoretic characterization of this notion. Shapiro begins by sketching the historical and philosophical background of the notion of logical consequence. In what follows, he attempts to answer the question 'What is logical theory about?' In order to shed light on the success of model theory, he examines the relationship between ordinary languages used in mathematics, and formal logic, deductive systems, and model theory. The orientation is that a formal language is a mathematical model of a natural language of mathematics, in much the same way that point-mass systems are mathematical models of physical systems. It is argued that with a suitable choice for the border between logical and non-logical terminology, and with a sufficiently rich set-theoretic metatheory, model-theoretic consequence is a reasonable mathematical model of correct reasoning, one that answers to the modal, semantic, and epistemic intuitions which underlie the pre-theoretic notion of logical consequence.

Charles Chihara's paper 'Tarski's Thesis and the Ontology of Mathematics' is closely linked to the topic of Shapiro's study. Like Shapiro, he is concerned with criticisms which have recently been levelled against the widely held view that model theory yields a precise and satisfactory account of the pre-theoretic concept of logical consequence. Chihara canvasses several of the objections raised by Etchemendy and McGee to what the latter calls *Tarski's Thesis*, that is the thesis that a sentence of a formalized language is logically valid (or logically true) just in case it is true in every model. Chihara attaches particular importance to examining Etchemendy's so-called 'contingency objection' which raises questions about the ontology of mathematics. In the concluding section of his paper, Chihara offers a justification of Tarski's Thesis.

II. MATHEMATICS, SCIENCE, AND METHOD

This part brings together papers dealing with different topics and pursuing different aims. It is less homogeneous than, say, Parts III and IV; nonetheless, it seems to me that the authors of Part II share a central concern: roughly speaking, they all deal more or less with questions of mathematical methodology, and some touch on issues relating to both mathematical and scientific theories. The latter applies especially to the papers by Burgess and Resnik.

The purpose of Penelope Maddy's essay 'Naturalizing Mathematical Methodology' is to provide a methodological characterization of naturalism traceable to sources in Quine and Wittgenstein. Put succinctly, she takes naturalism to be an answer to a metaphilosophical question, namely the question that concerns the proper relations between philosophy and the sciences. Maddy begins with a discussion of naturalism itself and then considers a case-study in contemporary set theory: the Lebesgue measurability (or not) of Σ^1_2 sets of real numbers. One allegedly naturalistic approach to this case from the theory of sets is criticized. Later in her paper, Maddy attempts to gain some perspective on the problem of finding a genuinely naturalistic approach to mathematical methodology by discussing Wittgenstein's views on the nature of philosophy. She argues that Wittgenstein was wrong in supposing that the mathematician's interest in pure mathematics arises solely from a fog of misguided Platonistic prose, that he missed the many 'intra-mathematical' sources of interest. Maddy concludes that a properly naturalized mathematical methodology requires that whatever extensions of mathematical practice we might suggest must make sense from a purely mathematical point of view.

In his essay 'Occam's Razor and Scientific Method', John Burgess addresses the question whether the maxim that abstract entities are not to be multiplied beyond necessity should be accepted as a rule of scientific method. According to him, this question was implicit in the controversy between Carnap and Quine over the viability of nominalism. It has recently become explicit, because nominalists have advocated alternatives to current scientific theories, claiming for them the advantage of ontological parsimony regarding abstract entities. Critics have doubted whether, from a scientific point of view, this is an advantage. Burgess considers a number of historical cases from this angle, beginning with the debate among mathematicians about intuitionism. He points out that intuitionism has been interpreted as ontologically motivated, by a mentalistic conception of mathematical objects, but has also been interpreted as non-ontologically motivated, by a verificationist view of mathematical truth. Burgess surmises that the precept of minimizing commitments to the ontology of abstract entities failed to serve as a rule of scientific method. He speculates further that science has in an important sense been indifferent to ontology.

Geoffrey Hellman, in his essay 'Beyond Definitionism—But Not Too Far Beyond', discusses the definitionist programme of predicativist mathematics originating with Henri Poincaré and Hermann Weyl. By 'definitionism' Hellman refers to the view developed by Solomon Feferman that all mathematical objects are presentable by finite symbolic expressions. One of the most recent systems of variable types over the natural numbers is Feferman's W, suitable for carrying out predicative analysis, and which, in spite of its expressive power, is a conservative extension of Peano Arithmetic (PA). Focusing on systems such as W, Hellman lists and contrasts the principal advantages and disadvantages of the definitionist programme. He then suggests liberalizing this foundational programme to eliminate its disadvantages, while most of its advantages are preserved. It is claimed that the resulting system is endowed with a number of virtues: its proof-theoretic strength exceeds that of classical analysis and it avoids certain limitations of predicativist analysis.

In his paper 'Holistic Mathematics', Michael Resnik seeks to adduce strong arguments in favour of what he calls epistemic or confirmational holism. He characterizes this doctrine by the thesis that no claim of theoretical science can be confirmed or refuted in isolation, but only as part of a system of hypotheses. Resnik examines the arguments for Duhem's and Quine's versions of holism and discusses several objections which have recently been raised to Quine's holism and his views on confirming mathematical principles. As to the advocacy of holism, it is suggested that scientific and mathematical theories, together with their attendant methodologies, are capable of being partially ordered as more or less global. Resnik maintains that holistic considerations may license the introduction of new entities, new methods, and new fundamental laws into a theory, but that they do not extend to the more global theories, where mathematics is singled out as the most global theory we possess. He expresses the belief that the proposed view of science will serve to meet some of the objections directed against Quine's holism.

III. FINITISM AND INTUITIONISM

Part III comprises three papers, two of which (those of Parsons, and Niebergall and Schirn) are almost wholly devoted to Hilbert's conception of finitist metamathematics. Detlefsen's essay is concerned not only with Hilbert's finitist programme, but also with the intuitionist programme of Brouwer and Weyl.

Hilbert regarded classical mathematics as the paradigm of 'unchallengeable truth' and had a twofold motive for vindicating all of it by means of a finitist metamathematical consistency proof for its formalized counterpart. On the one hand, Russell's and Zermelo's discovery of the logical paradoxes lurking in naïve set theory with its unrestricted principle of comprehension shook the logical foundations of number theory and analysis. On the other hand, Hilbert was worried about the vigorous attack which Brouwer and Weyl had launched on classical analysis. Hilbert felt especially called upon to meet the intuitionistic challenge concerning the validity of the principle of excluded middle. He tried to make it clear that the paradoxes of set theory are in no way due to the employment of that principle, but rather to the use of illicit and meaningless concept formations which are supposed to be banned once and for all from his proof theory.

This is familiar ground. What is presumably less well known is the fact that there are remarks in the first volume of *Grundlagen der Mathematik* by Hilbert and Bernays[17] which, on the face of it, suggest that Hilbert took the methodological standpoint of Brouwer's intuitionism to embody a certain extension of his own finitist point of view. As Hilbert and Bernays point out, Brouwer admits that an assumption is introduced about the presence of an inference or a proof, without the inference or the proof being determined according to intuitive characteristics. From an epistemological

[17] Springer-Verlag, Berlin, 1934; 2nd edn., 1968, 43.

point of view, such an extension of the finitist attitude is said to amount to the method of adjoining to the intuitive insights (gained in finitist reasoning) reflections of a generally logical character. Hilbert and Bernays observe that the extension so conceived is requisite if by means of finitist considerations one wishes to go beyond a certain elementary domain.

Now, it would probably be too hasty to conclude from these isolated remarks that in the early 1930s Hilbert in effect considered his finitism of the 1920s to be a restricted form of intuitionism. And even if he did, it is by no means clear whether, on closer examination of the epistemological underpinnings of the two doctrines, such a view could be sustained. It is true that Gödel[18] described both what he took to be the kinship and the difference between finitism and intuitionism in a way which comes relatively close to Hilbert's and Bernays's brief characterization. According to Gödel, the shared component of finitism and intuitionism is the idea that we are entitled to speak of mathematical objects only to the extent that we can exhibit them. However, Detlefsen argues—to my mind, convincingly—that what separates intuitionism from finitism is not only the nature of the objects exhibited, but the very nature of exhibition itself.

Hilbert's finitist metamathematics in the 1920s *qua* contentual theory of formalized proofs (henceforth referred to as 'M') is a non-formalized, non-axiomatized, and surveyable 'theory' with a non-mathematical vocabulary. Designed as a collection of sentences dealing with concrete and surveyable figures (in particular with formulas and proofs) it wears its mark of distinction on its sleeve: intuitive evidence. Even if M were clad in formal garb (i.e. couched in a formal language), the resulting version (call it MM) need not be recursively enumerable.[19] Furthermore, there is ample reason to suppose that MM may be considerably weaker than Primitive Recursive Arithmetic (PRA).

As far as Hilbert's extension of the finitist point of view in the 1930s is concerned, it was guided by two central, though possibly conflicting, ideas: first, to make sure that it preserved at least the core of M; second, to conduct indeed, within the extended proof-theoretic bounds, a consistency proof for a large part of mathematics, in particular for second-order arithmetic (Z_2). In the first volume of *Grundlagen der Mathematik*, metamathematics is not yet formalized. If it were to appear in the formal garment of a mathematical theory, it would be approximately as strong as PRA, but we have no reason to assume that it would go beyond PRA + $\mathrm{Con}_{\mathrm{pra}}$. In the second

[18] 'Über eine bisher noch nicht benützte Erweiterung des finiten Standpunktes', *Dialectica*, 12 (1958), 280–87.

[19] Notice that the question whether M or MM is recursively enumerable, i.e. axiomatizable, concerns only the complexity of the set of Gödel numbers of sentences belonging to M or MM within the arithmetical hierarchy. It asks whether we *could* give a recursive set of axioms for M or MM.

volume,[20] which, unlike the first, was written in the face of Gödel's Incompleteness Theorems probably from beginning to end, non-axiomatized metamathematics is replaced with an axiomatized version, and the intuitive down-to-earth 'modes of inference' of M give way to the loftier ones of a formalized metatheory. The range of what Hilbert and Bernays still claim to be finitistically admissible now at least includes PA, but the authors seem to suggest that it could be even as wide as PA + TI[α], where TI[α] is the schema of transfinite induction up to α, for a possibly rather large countable α. In short, in Hilbert and Bernays (1939), the new finitist parameters are kept flexible (too flexible I think) to make allowances for special proof-theoretic needs. The question as to how the new 'finitist' perspective relates to the old one is almost completely passed over in silence. But I think there can be no doubt that the canons of M, such as its informal character, its surveyability, its intuitive evidence, and its unquestionable soundness, are sacrificed for the sole purpose of strengthening and refining the proof-theoretic repertoire.

To conclude my comment on finitism, a few remarks on what has come to be known as partial realizations of Hilbert's programme may be in order. Stephen G. Simpson[21] has argued that Gödel's Second Incompleteness Theorem does not rule out significant partial realizations of Hilbert's programme. Simpson regards Hilbert's programme as consisting of three steps. The first is to isolate the finitist part of mathematics; the second step is to reconstitute infinitistic mathematics as a fairly strong formal theory, Z_2, for instance. The third step is to carry out a finitist consistency proof for that theory. Since Simpson endorses W. W. Tait's thesis that finitist reasoning is essentially primitive recursive reasoning in the sense of Skolem[22] the final step in Hilbert's programme is then to show that the consistency of Z_2 can be proved in the formal system PRA. If this could be achieved, Z_2 would be conservative over PRA with respect to Π^0_1-sentences.

Of course, Gödel's Second Incompleteness Theorem blocks any such realization of the final step in Hilbert's programme. According to Simpson, the main question is then: How much of infinitistic mathematics can be developed in subtheories of Z_2 which are conservative over PRA with respect to Π^0_1-sentences? Investigations by Harvey Friedman[23] and by Simpson himself[24] on what has been termed *reverse mathematics* have dis-

[20] Springer-Verlag, Berlin, 1939; 2nd edn., 1970.

[21] 'Partial Realizations of Hilbert's Program', *Journal of Symbolic Logic*, 53 (1988), 349–63.

[22] Cf. Tait, 'Finitism', *Journal of Philosophy*, 78 (1981), 524–46.

[23] Cf. 'Systems of Second Order Arithmetic with Restricted Induction', I, II (abstracts), *Journal of Symbolic Logic*, 41 (1976), 557–9; H. Friedman, S. G. Simpson, and R. L. Smith, 'Countable Algebra and Set Existence Axioms', *Annals of Pure and Applied Logic*, 25 (1983), 141–81.

[24] Cf. e.g. 'Subsystems of Z_2 and Reverse Mathematics', appendix in G. Takeuti, *Proof Theory*, North-Holland, Amsterdam, 2nd enlarged edn., 1987.

closed that quite a substantial portion of infinitistic mathematics can be constructed in subsystems of Z_2 meeting the requirements just mentioned. In other words, reverse mathematics is taken to serve as a heuristic guideline for discovering subtheories of Z_2 which prove to be relevant for partial realizations of Hilbert's programme.

However valuable and fruitful this branch of research appears in its own right, it does not square with Hilbert's original programme in two important respects. Firstly, I disagree with Tait, and consequently with Simpson, that the core of M is adequately captured by the formal theory PRA. Recall my claim that M is, by its very nature, an informal 'theory' and that MM is possibly much weaker than PRA. However, since partial realizations of Hilbert's programme along the lines of Simpson's proposal rest crucially on the identification of M with PRA, I think that they fail to do justice to Hilbert's pre-Gödelian conception of metamathematics. Secondly, even if we granted that M (or MM) is equal to PRA, partial realizations of Hilbert's programme would still be far removed from Hilbert's idea of carrying out finitist consistency proofs for formalized mathematical theories. To put it succinctly, from the fact that a recursively enumerable theory T is Π^0_1-conservative over PRA, it follows precisely that the consistency of T cannot be proved in PRA. Admittedly, it is quite true that a proof of the Π^0_1-conservativeness of a theory T whose consistency does not appear to be secured may reinforce our confidence that T is indeed consistent. Nevertheless, such a relative consistency proof is a matter quite distinct from the direct or absolute consistency proofs Hilbert sketched or merely envisaged in his classical papers on proof theory.

Charles Parsons, in his essay 'Finitism and Intuitive Knowledge', deals with the relation between finitist proof and intuitive knowledge. The thesis that a proof of a sentence S according to the finitary method yields intuitive knowledge of S is called *Hilbert's Thesis*. Parsons argues that the elementary axioms of PRA are intuitively known when they are interpreted relative to Hilbert's intuitive model of strings. He further maintains that the logical inferences employed in PRA preserve intuitive knowledge and that induction can reasonably be held to do this. Primitive recursion proves to be an obstacle for a defence of Hilbert's Thesis. Parsons holds that addition and multiplication can be seen intuitively to be well defined. Some arguments of Hilbert and Bernays to extend this to exponentiation and to generalize it are examined. The upshot is that Hilbert and Bernays have failed to advance a non-question-begging argument for the claim that all primitive recursive functions can be considered intuitively to be well defined. Yet Parsons maintains that there is no clear case against Hilbert's Thesis either. In a final section, Parsons examines the consequences stemming from the assumption that every string can be intuited.

While in his discussion of finitism Parsons accounts for Hilbert's and Bernays's view in the early 1930s, Karl-Georg Niebergall and Matthias Schirn, in their essay 'Hilbert's Finitism and the Notion of Infinity', focus entirely on the finitist point of view which Hilbert developed in his papers on proof theory between 1922 and 1928. In the two central sections, they examine Hilbert's assumptions of infinity and consider the pros and cons of alternatives to the standard formalization of consistency assertions. It is argued that assumptions of infinity do not occur in formalized mathematics, but, rather unexpectedly, they do underlie Hilbert's finitist metamathematics M. The fact that in advocating M Hilbert did not dispense with assumptions of infinity is diagnosed as the vulnerable spot of his approach. In the concluding section of their essay, Niebergall and Schirn argue that it is only for such a restricted case in which the consistency of PA ought to be provable in PA that Hilbert's programme can be carried out, provided that a consistency proof along the lines of contemporary proof theory has been replaced with what they call an 'approximative consistency proof'. They define this notion as follows (for axiomatizable theories S and T with representation τ): S proves the approximative consistency of T $:\Leftrightarrow \forall n S \vdash \neg \mathrm{Proof}_{\tau}(n, \bot)$. Approximative consistency proofs are taken to be in the spirit of Hilbert's finitist point of view. Niebergall and Schirn concede, though, that Hilbert's programme, when it is understood in the standard way, cannot be salvaged by conducting approximative instead of ordinary consistency proofs.

According to Michael Detlefsen, the requirement that existence claims must be supported by an 'exhibition' of an object of the kind claimed to exist is one of the characteristic marks of the constructivist tradition in the philosophy of mathematics. Detlefsen terms this the *exhibition requirement*. In his essay 'Constructive Existence Claims', he examines two versions of this requirement, the version advocated by Brouwer and Weyl and the version inherent in Hilbert's finitism. Detlefsen argues that the two versions are significantly different and that it is a mistake to construe the exhibition requirement as a condition embodied in all forms of constructivism. Moreover, he rejects the view, held by Gödel and others, that finitism amounts to a kind of restricted intuitionism.

IV. FREGE AND THE FOUNDATIONS OF ARITHMETIC

It is perhaps a striking feature of this volume that it includes a relatively large number of papers on Frege's philosophy of arithmetic. In addition to the essays of Part IV, those by Kit Fine and Peter Simons, in Part V, deal partly with Frege's method of logical abstraction. I take this to reflect the renewed interest in Frege's philosophy of mathematics that has been flourishing for at least a decade, thanks to the work of the late George Boolos, Michael Dummett, Richard Heck, Crispin Wright, and others.[25]

It is well known that Frege's principal scientific aim was to show that arithmetic (number theory and analysis) is a branch of logic. His *Begriffsschrift*[26] had provided the essential logical techniques for carrying out the logicist programme. Unlike the *Begriffsschrift* and Frege's *opus magnum Grundgesetze der Arithmetik*,[27] his second book, *Die Grundlagen der Arithmetik*,[28] was written entirely in natural language (German) and relied only on a few technical devices such as the use of bound variables. *Grundlagen* was designed to supply not only a mathematical analysis of the concept of number, but also a philosophical justification of the claim that the truths of arithmetic are analytic, that is, that they can be derived exclusively from primitive truths of logic and definitions. While Frege thought that the proof-sketches in *Grundlagen* made this claim no more than plausible, in *Grundgesetze* he set out to establish the validity of logicism beyond doubt by providing gapless proofs for the fundamental theorems of arithmetic.

In *Grundlagen*, Frege defines the number belonging to the concept F (symbolically: $N_xF(x)$), after two failed attempts at definition, as the extension of the second-level concept *equinumerous with the concept F.* (Henceforth, 'E' is to abbreviate 'equinumerous'.) However, this explicit definition (call it (D)) rests on the questionable assumption that we intuitively know what the extension of a concept in general is and, hence, falls short of solving what has come to be known as 'the Julius Caesar problem'. The problem is that the criterion of identity embodied in the tentative contextual definition (C) '$N_xF(x) = N_xG(x) := E_x(F(x),G(x))$' is powerless to decide whether Julius Caesar is identical with, say, the number of planets. We are thus left with the problem that (D), like (C), fails to fix uniquely the reference of the cardinality operator '$N_x\varphi(x)$'.

It is worth noting that Frege uses (D) only in his sketchy proof of the equivalence '$N_xF(x) = N_xG(x) \equiv E_x(F(x),G(x))$'; let us call this equivalence (T). The theory obtained by adjoining (T) to second-order logic is consistent.[29] Having derived (T) from (D), Frege never appeals to (D) again. Instead, he derives familiar theorems of cardinal arithmetic, in particular, the fundamental proposition that every finite number is succeeded by a

[25] See e.g. M. Dummett, *Frege: Philosophy of Mathematics*, Duckworth, London, 1991; W. Demopoulos (ed.), *Frege's Philosophy of Mathematics*, Harvard University Press, Cambridge, Mass. 1995; M. Schirn (ed.), *Frege: Importance and Legacy*, Walter de Gruyter, Berlin, 1996.

[26] L. Nebert, Halle a/S, 1879.

[27] Vol. I, H. Pohle, Jena, 1893; vol. II, H. Pohle, Jena, 1903.

[28] W. Koebner, Breslau, 1884.

[29] Cf. J. P. Burgess, Review of C. Wright, *Frege's Conception of Numbers as Objects*, Aberdeen University Press, Aberdeen, 1983, *Philosophical Review*, 93 (1984), 638–40; G. Boolos, 'The Consistency of Frege's *Foundations of Arithmetic*', in J. J. Thomson (ed.), *On Being and Saying. Essays for Richard Cartwright*, MIT Press, Cambridge, Mass. 1987, 3–20.

finite number, from (T) together with his other definitions. In the literature on Frege, the derivability of number theory from (T) within second-order logic was first noted by Charles Parsons in his essay 'Frege's Theory of Number'.[30] It has been given due attention in Crispin Wright's study *Frege's Conception of Numbers as Objects* and in several articles by George Boolos.[31]

In *Grundgesetze*, when it comes to the construction of arithmetic, Frege proceeds in much the same way as in *Grundlagen*. Here, too, he first derives (T) from his explicit definition of the cardinality operator (which now appears as a function-name not of second, but of first level) and subsequently deduces the basic laws of number theory from (T) within axiomatic second-order logic. In *Grundgesetze*, (T) is presented in a somewhat different fashion, but the new version is otherwise equivalent to the old one in *Grundlagen*. Thanks to Richard Heck,[32] it has now come to be known that in *Grundgesetze* Frege in fact provided an axiomatization of arithmetic. Although Frege proves each of the Dedekind–Peano axioms, he is primarily concerned to prove his own axioms for arithmetic. The latter are, of course, equivalent to the former. In all likelihood, Frege was well aware that the axioms of arithmetic are derivable, in second-order logic, from (T). Thus the main theorem of *Grundgesetze*, which Boolos has justly called *Frege's Theorem*, is that (T) implies the axioms of second-order arithmetic.

One of the most important approaches to Frege's foundational programme is, I think, Boolos's reconstruction of Frege's arithmetic as well as his thoroughgoing analysis of the issue of consistency with respect to both *Grundlagen* and *Grundgesetze*. In one of his papers (1987; see n. 29), Boolos presents a formal theory FA ('Frege's arithmetic') which comprises the real content of Frege's programme and for which a simple consistency proof can be given. The logic underlying FA is standard axiomatic second-order logic. Boolos shows that the definitions and theorems of sections 68–83 of *Grundlagen* can be formulated and proved in FA, along Fregean lines.[33] Certainly, if analysis is consistent, so is FA. Boolos (1987, 10; 1990, 269; see n. 31) demonstrates that the converse also holds. The equiconsistency of analysis and FA can be proved in PRA. As far as the formal system of *Grundgesetze* is concerned, Boolos (1986–7, 145 ff.; see n. 31) has suggested

[30] In M. Black (ed.), *Philosophy in America*, Cornell University Press, Ithaca, NY, 1965, 180–203.

[31] 'The Consistency of Frege's *Foundations of Arithmetic*', op. cit.; 'Saving Frege from Contradiction', *Proceedings of the Aristotelian Society*, 1986–7, 137–51; 'The Standard of Equality of Numbers', in Boolos (ed.), *Meaning and Method: Essays in Honor of Hilary Putnam*, CUP, Cambridge, 1990, 261–77.

[32] 'The Development of Arithmetic in Frege's *Grundgesetze der Arithmetik*', *Journal of Symbolic Logic*, 58 (1993), 579–601.

[33] Cf. Boolos (1987, 10–14) and the appendix in Boolos (1990, 275 f.).

a repair that saves it from contradiction. In his later controversy with Dummett on the question as to what caused the inconsistency of the system of *Grundgesetze*, Boolos has argued against Dummett that the culprit is what Frege took it to be, namely Axiom V, and not any deficiency in his stipulations concerning the second-order quantifier.[34]

The two papers by Crispin Wright 'On the Harmless Impredicativity of $N^=$ ("Hume's Principle")' and 'Response to Dummett' and the one by Michael Dummett 'Neo-Fregeans: In Bad Company?' are a continuation of their debate on the viability or failure of Frege's introduction of logical objects (or of a range of abstract singular terms respectively) by means of abstraction. The debate actually began with Dummett's critical assessment in his book *Frege: Philosophy of Mathematics* of Wright's neo-Fregean project. Roughly speaking, that project is based upon the thesis that our knowledge of number theory may be construed as deriving a priori from (T) (also referred to by Wright as '$N^=$' and by Boolos as 'Hume's principle'). Wright has argued that a form of logicism, too, can survive as a programme, at least for number theory. In his book, Dummett complains that Wright's advocacy of Frege's proposed method of bestowing a reference upon abstract singular terms by appeal to an abstraction principle fails to pay due attention to *Grundgesetze*, where courses-of-values are introduced in a fashion quite analogous to Frege's mooted (but rejected) introduction of cardinal numbers via a contextual definition in *Grundlagen*. The introduction of courses-of-values by means of Axiom V led to a contradiction, however. For Dummett, the inconsistency of Axiom V discloses a fundamental flaw in Frege's conception of how logical objects may be given to us. This flaw is said to be also present in a case like $N^=$, which arguably does not fall prey to a contradiction. In Dummett's view, we face a vicious circularity whenever we attempt to define a singular term-forming operator Σ (e.g. the cardinality operator 'the number of Fs') by reference to a domain of quantification which is intended to include the referents of the very terms which that operator enables us to form. As the title of Wright's paper suggests, its principal objective is to show that the 'explanatory' principle $N^=$, although indeed irremediably impredicative as conceived of by Frege, does not give rise to a vicious circularity. He first reviews why that principle does not generate a contradiction after the fashion of Axiom V, and then canvasses two lines of argument which might be taken to support Dummett's view that, even if consistent, $N^=$ is viciously circular. He then sets about constructing an inductive argument that $N^=$ can supply the basis for an

[34] Cf. G. Boolos, 'Whence the Contradiction?', *Proceedings of the Aristotelian Society*, Suppl. Vol., 67 (1993), 213–33, repr. in M. Schirn (ed.), *Frege: Importance and Legacy*, op. cit., 234–52; M. Dummett, 'Basic Law V', *Proceedings of the Aristotelian Society*, 94 (1994), 243–9. A shorter version of Dummett's paper (without the response to Peter Clark) appeared in *Frege: Importance and Legacy*, 253–60.

understanding of the specific kinds of use of the cardinality operator which Frege's reconstruction of arithmetic requires.

Dummett begins his reply with a kind of stage-setting and then examines what he presents as 'Wright's abstraction thesis': We are entitled to *stipulate* outright that the abstraction principle under discussion shall be taken as holding, and that this stipulation is a legitimate way of introducing the operator Σ as forming genuine singular terms. In Dummett's view, this thesis would be defensible only if it were subject to a restriction that would exclude any abstraction principle leading to inconsistency. He adds that even in that case Wright would not have a thesis for which he had offered any justification. An appeal to Frege's context principle would justify the thesis only if it could be shown that a truth-value had been secured for every statement involving Σ; but this claim cannot be sustained. Dummett then objects to Wright's sundering the specification of the senses of Σ-terms from the determination of the domain containing the objects that Σ-terms denote. Not only is this unfaithful to Frege, but it leaves us with no strategy for determining that domain—the problem that Frege believed he had solved. Wright's inductive argument that his imaginary subject Hero can use $N^=$ to fix the truth-values of essential arithmetical statements suffers from the same ill-foundedness as Frege's.

In his 'Response to Dummett', Wright attempts to clarify some crucial aspects of his position and argues that Dummett's view that the inconsistency of certain abstraction principles like Frege's Axiom V somehow creates a 'rotten-apple effect, permeating the whole barrel of abstractions' (as Wright puts it) is without compelling grounds. He focuses on Dummett's contention that if $N^=$ is to serve as an explanation of the meaning of number-theoretic statements, a specification must be provided of a domain over which its individual variables are to range. According to Dummett, the neo-Fregean has just two options to meet this requirement. He may provide the required specification in advance and independently of $N^=$, in which case the infinity of the domain will have to be recognized independently. Or he may try to construe $N^=$ as specifying simultaneously both the meaning of the cardinality operator and the domain over which its individual variables range, a ploy which Dummett argues is doomed to failure. Against this, Wright suggests that there is another possibility: we can attain a conception of the infinity of the sequence of natural numbers if we may legitimately add to what he calls a 'range-unspecific' understanding of $N^=$ the interpretation of the range of its first-order quantifiers as entirely unrestricted. Wright acknowledges that this proposal rests on several assumptions; for one thing, that there is a justifiable use of wholly unrestricted first-order quantification. But he contends that cogent objections against these assumptions remain to be disclosed.

Frege's outline of his logicist programme in *Grundlagen* culminates in §§82–3, in the sketch of a proof that every finite cardinal number has a successor, from which, by appeal to propositions whose proofs have been outlined in earlier sections of his book, the infinity of the natural sequence of numbers follows. In their paper '*Die Grundlagen der Arithmetik*, §§82–3', George Boolos and Richard Heck discuss two proofs of the statement that every finite cardinal number has a successor which might be extracted from §§82–3. The first is said to be very much akin to a proof of this statement which Frege gives in *Grundgesetze*, while the second is taken to correspond to the proof-sketch given in §§82–3 more closely than does the first. Boolos and Heck argue that there is scarcely any evidence suggesting that it was the first 'beautiful' proof that Frege intended to provide in *Grundlagen*. They arrive at the conclusion that the proof as presented in §§82–3 is marred by an oversight, that there are several modes of remedying it and that Frege's way of repairing it cannot be improved on.

The chief purpose of Richard Heck's paper 'The Finite and the Infinite in Frege's *Grundgesetze der Arithmetik*' is to present Frege's proofs, in *Grundgesetze*, of theorems concerning finitude and infinity in a form in which they will be more accessible to potential readers. Heck discusses Frege's characterization of the concept of finitude by commenting on the proof of Theorem 327 of *Grundgesetze*, which states that if the objects falling under a concept F can be ordered as a simple series which *ends*, then the number belonging to F is finite. Special attention is paid to assessing the significance of Theorems 327 and 348 (the converse of Theorem 327). One point made is that in these two theorems Frege shows how the use of natural numbers as finite ordinals can be reconstructed from their use as finite cardinals. Heck emphasizes, however, that there is no obvious reason to regard either the ordinals or the cardinals as being more fundamental. He further points out that the purely second-order characterization of finitude which Frege provides in *Grundgesetze* is equivalent to Cantor's, not to Dedekind's. Heck ventures to surmise that Frege discovered, around 1892, at least the axiom of countable choice, if not the full axiom of choice, and that he was reflecting both on its truth-value and on its epistemological status around 1897. Heck concludes with considerations on the value of formalization and Frege's conception of logical truth.

V. SETS, STRUCTURE, AND ABSTRACTION

The so-called structuralist view of mathematical objects is undoubtedly one of the most influential doctrines in contemporary philosophy of mathematics. Philosophers such as Hellman, Resnik, Shapiro, and others subscribe to it wholeheartedly, though what each of them understands by structuralism must be differentiated in one or the other respect. Advocates of mathematical structuralism of whatever brand would, however, unanimously agree that number theory, for instance, is not about a particular collection of abstract objects (the natural numbers), but rather about the natural number *structure*. Similarly, analysis is construed as being about the real number structure, set theory as being about the cumulative hierarchy structure (for example), etc. A structure can be explained as a domain of objects together with certain functions and relations on the domain fulfilling certain conditions. If we call a structure in this sense a relational system, we may give a different explanation: a structure is the class of all relational systems isomorphic to a given relational system. Typical examples of structures are algebraic structures, topological structures, order structures, etc. Dedekind's analysis of the concept of number in *Was sind und was sollen die Zahlen*?[35] can be seen as the germ-cell of the varieties of mathematical structuralism that have emerged during approximately the last three decades. If I am right, not only is Dedekind the founder of structuralism, but he is also the first to wed structure to abstraction. The question as to whether the wedding he suggests is a lucky one or rather a *mésalliance*, I shall set aside here. In any case, it will be useful to cast a glance at Dedekind's analysis before I say some words about today's structuralism.

Dedekind's introduction of the natural numbers proceeds via a characterization of the natural number structure, that is, the form of any set of objects that has a distinguished initial object and a successor relation which satisfies the induction principle. (The natural number structure is exemplified by the Zermelo numerals, the finite (von Neumann) ordinals, etc.) As is well known, Dedekind holds that we arrive at the natural numbers by means of abstraction, more specifically: that we create the natural numbers if, in considering a simply infinite system N, ordered by a mapping φ, we abstract completely from the specific nature of the elements, maintain only their distinguishability, and consider only the relations into which they are placed to one another by φ.[36] According to this view, then, the natural numbers are objects which have no properties except those they possess by virtue of having their position in the (abstract) simply infinite system.[37] As has been pointed out by Charles Parsons,[38] Dedekind's account of the

[35] Vieweg & Sohn, Braunschweig, 1888.

[36] Frege probably would have stigmatized this kind of abstraction as *psychological*, had he commented on it.

[37] Recall that Benacerraf dismisses the claim that there are abstract objects possessing none but structural properties. Dummett (*Frege: Philosophy of Mathematics*, 295 f.) labels Dedekind's version of structuralism 'mystical structuralism' and Benacerraf's version 'hard-headed structuralism'.

[38] 'The Structuralist View of Mathematical Objects', *Synthese*, 84 (1990), 303–46, at 306.

natural numbers is meant to rest on two substantive claims: the claim that simply infinite systems exist (Section 72), and the famous categoricity theorem (Section 132), stating that all simply infinite systems are isomorphic.[39] Since for Dedekind the process of abstraction requires that there be a non-abstract system from which to start, he felt the need to prove the existence of an infinite system, which had to be a non-mathematical one (Section 66).[40]

Few if any structuralists would deny that Michael Resnik's study 'Mathematics as a Science of Patterns: Ontology and Reference'[41] is the *locus classicus* of contemporary mathematical structuralism. His approach is intended to resolve the 'plight of the Platonist', from the point of view of both ontology and epistemology. According to Resnik, we have only structures in mathematics. The objects to which the terms of a mathematical theory refer are structureless points or positions in structures. As positions in structures, they have no identity or features outside of a structure. For epistemological purposes, Resnik prefers to speak of patterns instead of structures. He construes both patterns and their positions as abstract entities.

The most important relationships among patterns are, according to Resnik, structural similarity (congruence and equivalence) and structural containment (occurrence of patterns within patterns). Congruence is said to be the strongest of the equivalence relations that can occur between patterns and mutual occurrence the weakest. If patterns come into play as models of formal theories, then congruence is the relation which holds between isomorphic models. In this context, the question under which conditions a pattern A is identical with a pattern B appears to force itself upon us. Resnik admits that he has failed to suggest any identity conditions for patterns. Could congruence serve as a criterion of identity? Resnik's answer is no. The natural number sequence, for example, is congruent to the even number sequence. Now if we take their congruence as sufficient for their identity, then their corresponding positions should count as identical too. But then 1 would be 2. Resnik believes that the exclusion of patterns from the field of the relation of identity tallies with mathematical practice. Number theory quantifies over the numbers (positions), but not over the number-theoretic structure, set theory quantifies over sets, but not over the set-theoretic iterative hierarchy.

[39] Parsons (1990; see n. 38) ascribes to Dedekind what he calls 'eliminative structuralism'. See his arguments against eliminative structuralism as well as his discussion of the limits of non-eliminative structuralism in his 1990 paper. Although Parsons presents what he considers to be a defensible version of structuralism, he concludes that structuralism is after all not the whole truth about mathematical objects.

[40] Cf. Dummett, *Frege*: *Philosophy of Mathematics*, 296.

[41] *Noûs*, 15 (1981), 529–50; see also Resnik, 'Mathematics as a Science of Patterns: Epistemology', *Noûs*, 16 (1982), 95–105.

Resnik's argument that whenever we are given two structures, there is no fact of the matter as to whether they are the same or different I take to be open to question. Shapiro[42] even goes so far as to regard it as absurd, when it is taken to its logical consequence. He stresses that a criterion of structure-identity is needed if structuralism itself is to be framed with any rigour.

Another much discussed version of structuralism is Hellman's modal approach.[43] In contrast to Shapiro's and Resnik's structuralism, Hellman's modal version does not presuppose the existence of structures. In a sense, this feature brings it into the vicinity of Field's modal fictionalism.[44] According to Hellman, the modal part of the second-order language enables him to avoid ontological burden. Instead of asserting, for instance, that there is a natural or real number structure, he only commits himself to asserting that it is possible that there be such a structure. Shapiro (1993, 466; see n. 42) has observed that modal structuralism (if it is to be viable) requires that an epistemology of abstract objects be replaced with an epistemology of possible abstract objects. However, there is prima facie no reason to believe that the latter should fare better than the former. Shapiro is willing to concede, though, that in set theory Hellman's approach can claim a number of advantages.[45]

To conclude these introductory remarks, let me comment on abstraction. I shall focus on Frege's method of endowing abstract singular terms with a reference by fixing the sense of a 'recognition-sentence'. There are two reasons for doing this. Firstly, Frege's method plays an essential role in several essays in Parts IV and V. Secondly, I believe that Frege was the first to give an account of logical abstraction in more or less precise terms. As a matter of fact, abstraction was and remained for him the means *par excellence* for introducing logical objects. In *Grundgesetze*, the question 'How do we conceive logical objects, in particular numbers?' is highlighted as the prime problem of arithmetic, and the answer along the lines suggested by Frege is that we conceive them through logical abstraction. Although his foundational project foundered on Russell's Paradox it is, as Kit Fine shows, worth exploring the question as to what extent a Fregean theory of abstraction could be developed and used as a foundation for both number theory and analysis.

Frege uses the word 'abstraction' for the most part in a psychological sense, especially when severely taking to task certain rival theories of arithmetic. In contrast to psychological abstraction, which he regarded as a thorn

42 Shapiro, 'Modality and Ontology', *Mind*, 102 (1993), 455–81, at 476.

43 G. Hellman, *Mathematics Without Numbers*, OUP, Oxford, 1989.

44 Cf. H. Field, *Realism, Mathematics and Modality*, op. cit.

45 See also Shapiro's review of Hellman's 1989 book in *Noûs*, 27 (1993), 522–5.

in the flesh and combated vigorously, his own contextual method of introducing abstract objects may be called *logical* abstraction.

A schema for a Fregean abstraction principle can be stated as follows:

$$\text{(S)}\quad Q(\alpha) = Q(\beta) \equiv R_{eq}(\alpha,\beta).$$

Here Q is a singular term-forming operator, α and β are free variables of the appropriate type, ranging over the members of a given domain, and R_{eq} is an equivalence relation holding between the values of α and β. In *Grundlagen*, Frege's paradigms for logical abstraction are

$$\text{(1)}\quad D(a) = D(b) \equiv a \parallel b$$

and

$$\text{(2)}\quad N_xF(x) = N_xG(x) \equiv E_x(F(x),G(x)).$$

In *Grundgesetze*, his paradigm is

$$\text{(3)}\quad \grave{\varepsilon}f(\varepsilon) = \grave{\alpha}g(\alpha) \equiv \forall x(f(x) \equiv g(x)),$$

that is, Axiom V, the structural analogue of (2).[46] In (1) the direction operator 'D(x)' operates on singular terms, in (2) '$N_x\varphi(x)$' operates on one-place predicates of first level, and in (3) the course-of-values operator '$\grave{\varepsilon}\varphi(\varepsilon)$' acts on monadic function-names of first level. The right-hand branch of each biconditional (1), (2), and (3) embodies a specific criterion of identity for the abstracta of the corresponding type: parallelism, equinumerosity, and coextensiveness. A Fregean abstraction principle thus satisfies what Frege considers to be the cardinal prerequisite for any methodologically sound introduction of logical objects.[47] It is true enough, though, that he rejects (1) and (2) as contextual definitions of 'D(a)' and '$N_xF(x)$' respectively,[48] precisely because he holds that in either case the criterion of identity fails to do what it ought to do, namely to cover all conceivable cases—whence comes his Julius Caesar problem, a version of which he had also to face in *Grundgesetze* regarding (3).

In *Grundlagen*, Frege maintains that one and the same (judgeable) content can be carved up in distinct ways and so emerge in sentences of different logical form. He describes this process as follows:

[46] Axiom V has a Janus-faced nature. It is designed to fix (at least partially) the references of certain terms, although in Frege's view this can never be the task of an axiom, but only that of a definition or an elucidation.

[47] For reasons of space, I must ignore the important role that the context principle plays for Fregean abstraction.

[48] Notice that, according to Frege, (1) and (2) figure only as *partial* contextual definitions, to be supplemented by explanations of all other relevant statements about directions or numbers.

The judgement 'The straight line *a* is parallel to the straight line *b*', in symbols *a* || *b*, can be regarded as an identity-statement. If we do this, we obtain the concept of direction and say: 'the direction of the straight line *a* is identical with the direction of the straight line *b*'. We thus replace the symbol || with the more general symbol = by distributing the particular content of the former symbol to *a* and to *b*. We split up the content in a way different from the original way, and thereby obtain a new concept. (*Grundlagen*, §64)

What we are certainly given here is the transition from one mode of speaking to another, designed to fix the reference of a new term, the direction operator. Both sides of (1) are considered to have the same content, that is, to express the same thought. And it seems that for Frege our grasp of the thought of the new sentence, that is, the identity-statement (henceforth (B)), is mediated by our recognition of its synonymy with the old one, that is, the equivalence statement (henceforth (A)).[49] Yet contrary to what Frege contends, (A) cannot be regarded as an identity-statement, at least not in a strict sense. Even if we were to accept his stipulation that (A) and (B) express the same thought, we could only grant that (A) can be transformed into (B).

As far as the second half of the quoted passage is concerned, I find it hard to follow. How are we to understand Frege's talk of carving up the same content in a way different from the original way, thereby obtaining a new concept? On the face of it, it might suggest that what he has in mind here is the purely syntactic device, always endorsed by him, of removing some or all occurrences of an expression from a sentence, and of marking the resulting gap(s) as an argument-place of the appropriate type, leaving a function-name (concept-expression or relation-expression).[50] This operation (call it *gap formation*) is supposed to go hand in hand with the process of analysing or dividing a thought into thought-components (call this *decomposition*). Closer examination, however, reveals that the 'recarving' of the content of (A) does not involve gap formation. First, the transition from (A) to (B), as described by Frege, is not a purely syntactic operation. It is rather a mixed operation, carried out both on the level of signs and on the level of content.[51] In particular, replacing the symbol '||' with the symbol '=' has nothing to do with gap formation. Second, we can neither extract

[49] In *Grundgesetze*, Frege neither claims nor denies explicitly that the two sides of Axiom V express the same sense. He stipulates only that they are to have the same reference. For a detailed discussion of what Dummett calls 'the synonymy thesis', that is, the thesis that the two sides of (1) (and also those of (2)) express the same thought, see his book *Frege: Philosophy of Mathematics*, 168 ff.

[50] The procedure of gap formation is described in precise and more general terms in §26 of *Grundgesetze*, vol. I.

[51] What is it to mean precisely that the particular content of '||' is distributed to *a* and to *b*? Frege hardly provides a clue for answering this question.

from (A) the direction operator nor can we discern in (A) the equality sign.[52]

Clearly, the division of a thought into thought-components is inextricably intertwined with the analysis of a sentence which expresses it; and it is in terms of gap formation that the latter operation proceeds. Or is there any device distinct from decomposition which Frege could justifiably call a *dissection* of a thought? I do not think so. In short, instead of characterizing the transition from (A) to (B) as resting on a dissection of a content in a way different from the original one, Frege should really have said that one and the same content is presented in two different ways.

Frege uses (1) only for the sake of illustration. The case with which he is really concerned in *Grundlagen* is, of course, (2). Following Fine, we may call (1) an *objectual* principle of abstraction and (2) and (3) *conceptual* principles of abstraction. Unlike the objectual principles, the conceptual ones involve a 'projection' from the larger domain of concepts (or functions) into the smaller domain of abstract objects of a certain kind; and the latter may, of course, fall under the former (if they are of first level). As Fine observes, it is this 'reflexive' feature of the conceptual principles that makes them so powerful, but also so vulnerable. While (1) does not require the existence of any more of abstracta (directions) than there are lines, (2) requires the existence of n + 1 abstracta (cardinal numbers), given n objects of the original kind. In fact, the introduction of the cardinality operator '$N_x\varphi(x)$' relies crucially on the assumption that the domain of the first-order variables is infinite. As far as (3) is concerned, Frege had to learn that the demand it makes on the size of the domain is unrealizable.[53] So much to Fregean abstraction.

In his essay 'Zermelo's Conception of Set Theory and Reflection Principles', W. W. Tait discusses the development of set theory along the lines of Zermelo's conception in 'Über Grenzzahlen und Mengenbereiche' (1930). Part of the paper is concerned with explaining how the reflection principle

[52] What is the original way of splitting up the content that (A) and (B) have in common supposed to be? The thought expressed by the original sentence (A) could be decomposed, for example, into the senses of '*a*', '*b*', and '||' by analysing (A) into just these three expressions. It would be awkward, however, to see this operation as bringing it about that the thought in question emerges in (A).

[53] Richard Heck ('On the Consistency of Second-Order Contextual Definitions', *Noûs*, 26 (1992), 491–4) points out that it is recursively undecidable whether an arbitrary second-order contextual definition (henceforth C) is consistent. He contends that the contextual definition '$Q_xF(x) = Q_xG(x) \equiv \forall x(F(x) \equiv G(x)) \vee A$' is consistent if and only if A is satisfiable, where A is an arbitrary second-order sentence. If we assume that A is satisfiable, then there is some model for second-order logic with respect to which A is true. Heck claims that if we make the opposite assumption, we have '$Q_xF(x) = Q_xG(x) \equiv \forall x(F(x) \equiv G(x))$', since A is logically false. But this is just the inconsistent Axiom V. Thus, if we could decide the consistency of C, we could decide the satisfiability of an arbitrary second-order sentence. Since we cannot do this, Heck concludes that the consistency of C is at least as undecidable as the satisfiability of an arbitrary second-order sentence.

for generating cardinal numbers, implicit in Zermelo's paper, may be understood following Zermelo's conception. Up to now, reflection has been applied only to obtain cardinals indescribable relative to some class of formulas. The other part of Tait's paper indicates how reflection may be extended well beyond this.

How do structure and abstraction in mathematics go together? As we have seen, Dedekind offers an analysis of the natural numbers that combines structure and abstraction in a certain way. In his paper 'Structure and Abstraction', Peter Simons pays tribute to Dedekind, but takes a different route. He proposes that the most attractive Platonist ontology for mathematics links Dedekind's structuralism to Frege's account of contextual definition in *Grundlagen*. Following Frege, it is claimed that predicates applying to abstracta differ in sense from their cognates applying to concreta. The problem of the sense, if any, in which abstracta may be said to exist is examined, but no definitive answer is suggested. Simons points out that in modern discussion 'structuralism' serves as a label for two different doctrines: quantificational structuralism, in which abstract structures are not referred to, but occur as models of formalized theories, and referential structuralism, in which structures are referred to by terms. In Simons's view, the latter accords slightly better with abstraction than the former. Resnik, the foremost proponent of referential structuralism, dismisses structures of structures on the grounds that all structures consist of structureless points. Contrary to Resnik, Simons argues that structures of structures make sense and offer an enlightening structuralist understanding of set theory and its role in mathematics. He thinks that if mathematics is taken to be the theory of pure structure, an account must be given of what purity is. Simons offers a specification in terms of the relations among the structural elements all having only logical properties. The question as to how the recursion is to be extended into the transfinite he considers to be an open problem, however.

Kit Fine, in his essay 'The Limits of Abstraction', sets himself the task of determining both the scope and the limits of abstraction by investigating the philosophical and logical foundations of abstraction theory. The paper is divided into three parts. The first provides the philosophical background and motivation for the ensuing formal parts. Much attention is devoted to Frege's abstraction principles, especially in connection with his introduction of directions and cardinal numbers in *Grundlagen*. In this respect, the first part of Fine's essay appears to be intimately related to the debate between Wright and Dummett. The investigation in the first part is guided by three central questions: (i) What are the correct principles of abstraction? (ii) In what sense do they serve to define the abstracts which they are supposed to introduce? (iii) To what extent can they serve as a foundation of mathematics? In the second part, Fine deals with several issues concerning the acceptability of a principle of abstraction. The basic languages and systems of interest are developed, some basic proof- and model-theoretic results established and various requirements for acceptability discussed. The topic of invariance plays an important role. In the final part, Fine devises a general theory of abstraction which satisfies the formal limitations set out in the second part. The resulting theory is shown to provide a foundation for both number theory and analysis.

I. Ontology, Models, and Indeterminacy

1

What Mathematical Truth Could Not Be—I

PAUL BENACERRAF

'[I]t can be said that the history of Western thought can be regarded as a history of our relationship to the diagonal argument.'

David King, 'From Gödel to Derrida: Undecidability, Indeterminacy, and Infinity', Ph.D. Dissertation, Murdoch University, Murdoch, W. Australia

A PARABLE

The Cohens were very exacting in their requirements on just the right match for their only son Abie, and Hyman Lapidus, the shatchen *to whom they had entrusted the search, at wits' end after having proposed match after match to the Cohens only to have them all turned down, came by one day with a proposal, his 'final offer':*

Lapidus Mrs Cohen, I have just the right girl for Abie. She is beautiful, from an old and wealthy family—gentle, loving, in short, just the kind of girl you would want to welcome into your family as the mother of your grandchildren, . . . Fine country homes . . .

Cohens So what's the problem? Always, there is a problem!

Lapidus Well, there *is* just one little thing I thought I should mention: She isn't Jewish, but trust me, she is an angel and . . .

Cohens [Consternation. Outrage. etc.] 'Trust you!' 'Angel!' Get out! Of course we can't consider anyone who is not Jewish for our Abie. You want our grandchildren should be goyem!?!

Lapidus Wait! Not so quick; this is an *exceptional* individual, generous, . . .

Cohens Absolutely not! Get Out!

Lapidus At least let me say who she is: It is Princess Margaret! You know, from England, from where the Rothschilds are . . . You can imagine—(God forbid anything should happen!), but just think, if something *should* happen to her—and these days, one never knows, does one?—your Abie, King from England,[1] and your grandchildren, heirs to the throne, . . . And she is *such* a fine person, etc.

After several backs and forths. . . .

Cohens All right. Enough already! Done!

Later, outside the Cohen home (wiping his brow)

Lapidus Whew! Well, that's *half* the job!

(Anonymous)

[1] I am grateful to an anonymous referee for pointing out that the calamity envisaged in this parable would *not* place Abie on the throne of England. I am, however, constrained to report Lapidus's arguments as he offered them.

I. PROLOGUE

The philosophy of mathematics is philosophy in a pure state, stripped of all worldly appendages; austere; philosophy without the sugar-coating of a pretence to Relevance to Life; but also heady, perhaps best taken in short sips. Much of the present paper was written for a conference entitled 'Philosophy of Mathematics Today', for which I was invited by Matthias Schirn to reflect on two papers that I published decades ago[2] and to comment on how the issues discussed in them had fared in the intervening years.

Rereading them at this distance, I was surprised to see how much they were the product of a particular time and soon found myself plunged into the philosophical atmosphere from which they had sprung. The papers in question have chequered histories. 'What Numbers Could Not Be' (henceforth 'WNCNB') was written at the invitation of Max Black for possible inclusion in his volume *Philosophers in America* (1965). The book-jacket describes the essays selected for the volume as 'representing the best of the lively and penetrating work of the most talented younger American philosophers', an ambition that the wily Max had managed to convey in his letter of invitation, making it very hard to refuse to submit an entry. But, a representative of 'the best of the lively and penetrating work of the most talented younger American philosophers' turned out to be what WNCNB could not be—it didn't make the cut (happily for it, I think, since a number of excellent papers were long entombed there, including Charles Parsons' 'Frege's Theory of Number').

'Mathematical Truth' (henceforth MT) was written in December–January of 1967–8, sent to a couple of trusted friends, and put away in a drawer. Although persuaded of the thrust of the paper, I was very unhappy with its detail. Despite that, having nothing else in my drawer, it was what I read when invited to give talks. It was a lead balloon, meeting with extraordinary and uniform silence on the part of every audience. Through those trusted (but not trustworthy) friends, it also received a wide underground circulation in manuscript form, which is how it remained until I was invited to make it the topic of a symposium of the American Philosophical Association in December 1973, roughly six years after writing it. Despite

The present paper is a slightly revised version of a paper that first appeared under the same title in *Benacerraf and His Critics*, A. Morton and S. Stich (eds.), Blackwell, Oxford, 1996. The two versions differ slightly in a number of places and in that *Recantation* (see n. 14 below) appears as an Appendix of the earlier version, but was omitted from the present one.

[2] 'What Numbers Could Not Be', *Philosophical Review*, 74 (1965)—page references will be to the original printing—and 'Mathematical Truth', *Journal of Philosophy*, 70(19). Both are repr. in *Philosophy of Mathematics: Selected Readings*, 2nd edn., P. Benacerraf and H. Putnam (eds.), Cambridge University Press 1983 (hereafter, B&P). They will be called WNCNB and MT for the balance of the present paper.

my dissatisfaction with it, I agreed, since, given its wide circulation, by then I felt an obligation to publish at least a version of the much longer work.

As its title suggests, the present paper began as a commentary on how the issues that are taken up in these two papers had fared in the intervening years. However, it had already grown beyond decent limits before broaching those specifically addressed in MT. Regrettably, that discussion is left for a sequel. The present paper begins with a bit of amateur history (perhaps more reminiscence than credible history) that aims at providing the reader who may not have sampled the philosophical atmosphere of those times with at least some sense of context—whether *the* context or not it is not for me to say. Section III follows that up with an analysis of the argument of WNCNB, including my best present take on where the discussion of these last decades has left the issues.

Section IV is devoted to a discussion of some important arguments, currently in the air, all of which are, like the arguments of WNCNB, attempts to extract philosophical juice from metamathematical results. The burden of that section will be a reminder that in such cases, e.g. the philosophical 'implications' of the Gödel Incompleteness and the Löwenheim–Skolem theorems, the philosophical juice can be validly extracted only if a substantial portion of the yield is simultaneously supplied through the back door.

These happen all to be 'limitative' theorems—theorems that establish bounds on what can be done ('proven', 'expressed', etc.) using the resources of certain formal languages, or what can be mirrored or modelled if we limit the modelling apparatus to what is available in those languages or 'proof' mechanisms. In some cases, the conclusion drawn is that if some concept cannot be exactly modelled using those resources, we simply do not possess it, whatever we may think about the matter; in others, it is inferred *from the mere fact that we can prove the limitative theorems*, that, in our mathematical practice, we are not subject to the limits that they demonstrate exist for formal languages/systems of proof. My modest point, in all of these cases, is that to have any weight, the argument must always contain two pieces: (a) the metamathematical result itself—the 'easy part'—'That's *half* the job!'; and (b) the supply of philosophical 'juice'—which I will call 'the Princess Margaret premiss', for the obvious reasons. Both must be there, and the quality of the overall argument will depend crucially on the quality of what has to be an *independent* argument for the Princess Margaret premiss. Therefore, presented with an argument purporting to draw an important philosophical conclusion from this or that metamathematical result, we should immediately beware—especially if the conclusion is supposed to be that the limits (legitimately) established to hold for the formalism, are *thereby* established to hold for the unformalized practice of mathematics.

To be fully explicit: This is not to say that we couldn't *in fact* be subject to the same limitations that have been proved to hold of the formal languages and systems that we study in metamathematics (or, in other cases, free from them)—it is merely to point out that, if it is to be at all probative, the argument that purports to show this must include a convincing demonstration of the relevant isomorphism (or lack thereof) between our own powers and the relevant features of the systems. The point is modest because it addresses only the *dialectical* position, never the substantive matter at issue. Much of this ground is well ploughed and familiar. My excuse for returning to it is the continuing interest and complexity of the central case of 'Skolemite' arguments—indeterminacy arguments based on the Löwenheim–Skolem theorems. Although they began life almost as puns—arguments that no philosopher would take seriously once the source of the claimed 'indeterminacy' had been pointed out—they have returned in recent years, revived single-handedly by Putnam in 'Models and Reality', his Presidential Address to the Association of Symbolic Logic,[3] and continue to garner adherents. They are considerably more fetching in their most recent garb than they were at birth, and as such, deserve our renewed attention.

II. 'PHILOSOPHY IN AMERICA' IN THE 1950s AND 1960s

Rereading them today, I am struck with how 'negative' both papers seem in both their tone and outlook. Both take aim at a picture or set of views, philosophical and metaphilosophical, that were in the air at the time and that had even attracted me to philosophy. Fresh, clean, unproblematic, methodologically secure, these views were the form and the accomplishments of a philosophy that had entered a new age, in which genuine progress was possible—philosophy on the Leibnizian ideal, where collaborative calculation replaced endless ideological squabbling.

The soldiers in this war against darkness were united against the common metaphysical enemy; but they were not all of one mind. There were rifts, the biggest of which (in the Anglo-American Group, at least) followed the natural geographical barrier provided by the Atlantic Ocean. The American side exhibited the most pronounced 'scientific'—indeed *scientistic*—attitude, according to which the philosophy of science (broadly conceived to include mathematics) was philosophy enough. Our leaders—members and descendants of the Vienna Circle, other intellectual offspring of Frege and Hilbert, Russell—were a typically Anglo-American bunch: the most prominent ones who come immediately to mind are Rudolf Carnap, Carl Gustaf Hempel, Herbert Feigl, Hans Reichenbach, and Alfred Tarski. There were others, but these were the ones I knew, either directly or, in

[3] 'Models and Reality', repr. in B&P, pp. 421–44.

Reichenbach's case, only indirectly, through his student and my teacher, Hilary Putnam.

Princeton was a very exciting philosophical place to be in the 1950s and 1960s—in some obvious ways (it was the early stomping ground of the young Hilary Putnam burgeoning with ideas and an irrepressible need to communicate them), but also in others that my own philosophical myopia kept hidden from me at the time. For example, Alonzo Church was a local fixture, exemplifying his essence—large, quietly sowing the seeds of dissent. His role in the development of philosophy in the mid-twentieth century would be a fascinating topic for another time.

Another local figure played a central role in the philosophical drama of the time. Slight as Church was large, a shadowy figure, as he had been in Vienna as well (according to Kreisel, although Kreisel's word on such matters may be less than sure), listening quietly, saying little unless pressed. Still, shadowy or not, he occupied a special place. He was revered as a god, his name not to be sullied in the day-to-day struggle of the reconstruction of philosophy. But he made the orthodox uncomfortable. Like Church, he followed no one; and like Church, he was an embarrassment to the movement, since he was not ideologically pure but insisted on propounding views that sounded suspiciously like the metaphysics that had been marked for extinction.

That was Gödel, of course. In a less tolerant group, either his sins—the most famous was his scuttling of the Hilbert Programme by boring from within (even as he struggled faithfully to advance it?)—or, quite simply, his overtly platonistic and counter-revolutionary philosophical pronouncements, would have led to his removal, not just from the altar or even the inner circle, but from the party roll altogether. That he remained a revered figure despite his subversive views and acts of aggression against the new order is a testimonial to the extreme tolerance of its apostles. Most of them were saints—who more so than Hempel or Carnap? But even a saint can't very well expel God from Heaven. So you accommodate; and they accommodated, albeit uneasily. We will return to Gödel, because a central thread in twentieth-century philosophy of mathematics regards what, philosophically speaking, we should make of metamathematical results, and surely Gödel's are right up there, with 'implications' claimed regarding the very nature of mathematics, of ourselves as its practitioners, etc.

This small group of philosophers created an atmosphere of optimism and purpose that, in turn, had an enormous influence on the development of a few key figures (Quine, Goodman, Putnam) and through them, on generations of philosophers, both here and abroad. At first they joined forces; later, Quine and Goodman, and ultimately Putnam became renegades—splintering is a mark of all revolutionary movements, good and bad. More recently, Putnam has also been rediscovering another lineage, in Dewey, Peirce, and James, one that Quine, anyhow, had claimed all along.

Despite its unified stand against the metaphysics of old, the movement had profound ideological rifts, worth mentioning here only to help identify the prevailing philosophical microclimate, to borrow an oenological neologism. For the 'good guys', the scientistic or scientific philosophers, philosophy made genuine advances, sometimes even established propositions about some subject-matter (the methodology of science? Even the true believers were often reluctant to say what the *subject-matter* might be). The 'other guys', mostly on the other side of the Atlantic, following Wittgenstein, had a seriously different view of philosophy, seeing it perhaps as an unwelcome intrusion on the ordinary life of ordinary people with honest work to do, an intrusion that seemed inevitably to bring confusion and disarray. They were more actively—and pessimistically—preoccupied with *meta*philosophical issues. Philosoph*y* was not a subject—it had no constructive task, no job of its own; philosoph*ers*, on the other hand, did have a task (if not a job): to clean up the mess left by their predecessors. To put a brighter face on it, the task was to show that the Great Philosophical Issues were really only pseudo-questions generated by misuses or extensions of language beyond the bounds of its legitimate purchase. Once this was shown, the 'questions' would go away. A caricature, to be sure; but this was a time of caricature—recall Russell's bitter and caustic 'The Cult of Common Usage',[4] in which he offers five reasons for objecting to the sort of un- and anti-scientific philosophy in vogue around him:

> (1) Because it is insincere;
> (2) Because it is capable of excusing ignorance of mathematics, physics, and neurology in those that had only a classical education;
> (3) Because it is advanced by some in a tone of unctuous rectitude, as if opposition to it were a sin against democracy;
> (4) Because it makes philosophy trivial;
> (5) Because it makes almost inevitable the perpetuation among philosophers of the muddle-headedness they have taken over from common sense. (p. 154)

So much for the villains; the heroes believed

- in science, and, for example, in the study of the concepts of space, time, and causality through their representations in the best science of the time;
- in the arithmetization of analysis, i.e. in the 'reduction' of the reals to the rationals, and the rationals to the natural numbers (and to sets and sequences thereof);

[4] Bertrand Russell, *Portraits from Memory, and Other Essays*, London, Allen & Unwin, 1956. See also his review of Ryle's *The Concept of Mind*: 'What is Mind?' *Journal of Philosophy*, 60/1 (1958).

- in the Frege–Russell analysis of natural number and how it refuted Kant by showing that arithmetic consisted of analytic propositions the knowledge of whose truth was explicable purely as a matter of *linguistic* knowledge, and by extension, that the same was true of the edifice built with the natural numbers (interestingly enough, a view that was being advanced in the absence of a good account of *linguistic* knowledge, something one might have thought should be at the foundation of such a view);
- in Russell's Theory of Descriptions;
- in the semantic conception of truth and in the Tarski hierarchy of metalanguages (of order $< \omega$, of course);
- in the family of c-functions and the development of the concept of Degree of Confirmation as a (partial) solution of the problem of induction;
- in the Deductive-Nomological model of scientific explanation;
- in ever more sophisticated representations of our psychological knowledge, beginning with admittedly crude behaviourist analyses of psychological concepts, refined through their introduction with simple or bilateral 'reduction sentences', ultimately to be regarded as theoretical terms in 'the language of science', in the end as part of a reductive analysis that unified all of science (recall *The Encyclopedia of Unified Science*? How many can name more than two of the volumes that appeared in it?);
- (although this was a later development), in the treatability by 'the methods of empirical science' of fundamental concepts of language, i.e. of linguistics, viz. of meaning, reference, syntax, etc., and hence of all of the issues in philosophy whose resolution depended on answers to questions in which these concepts played an important role.

and in countless other like things. . . . The point is that these had the standing of accomplishments, of philosophical problems laid to rest so that we could move on to others. Naturally, to the extent that the 'solutions' were keyed to the latest and best results of the science of the time—in brief, empirical matters—the stage was set for updates. But the peculiarly *philosophical* mystery was gone.

The last of these, the emergence of the new linguistics, was important in two ways. First, it provided the promise of a reliable tool for addressing linguistic questions, many of which, according to the *credo*, were at the root of philosophical confusion. Secondly, it exemplified once more how philosophy could play a role in the development of knowledge—not, in this case, by being a special branch, or a special sort of knowledge, but by doing the preliminary spadework that eventually spawned new factual and formal sciences, as it had done so many times before.

Here was a case in point: Noam Chomsky's *Syntactic Structures* was published in 1957 and immediately became a bible for philosophers, first on the West Bank of the Atlantic, but then, through John Austin's influence, on the East Bank as well, at least for those who were interested *in the representation of* our linguistic knowledge (as most of us were, since that was the only salvation for a priori knowledge that had any chance of being compatible with current dogma) and, more generally, in the systematic empirical study of language.

As Chomsky himself has frequently admitted, his own work owes much to philosophy and, in particular, to the new philosophy. At the University of Pennsylvania, he had studied not only with Zelig Harris, but with Nelson Goodman as well. Much in his early work on generative syntax is patterned on the work of logicians and thus had a ready appeal for philosophers, who saw it not only as 'precise' and testable, but as precise and testable precisely because the logical structure of the claims represented in a generative grammar seemed to be so transparent, especially when contrasted with the more casual treatment of these same issues by traditional grammarians. The generated strings were either 'acceptable' to speakers, or they were not; strings that were assigned different structures were either ambiguous, i.e. had non-equivalent readings, or they did not. At last there was something more public and general to go on than the grammarian's 'intuition'. The first steps were in syntax, but (for those who had not drunk from the Quinean well and still believed in the possibility of such a science) semantics could not be far behind—someone was bound to join Atlantic syntax with Pacific semantics to create a theory that (like syntax) was also properly empirical, and yet went beyond the 'purely formal' syntactic information that the early Chomsky provided. We were witnessing the launching, with philosophy's help, of yet one more science.

That was philosophy in the late 1950s and early 1960s, or so it now seems to me to have been—full of activity, excitement, lively debate, often to the point of euphoria.[5] Mathematical logic—the temple in which our protagonists paid homage to Gödel, Church, Kleene, Russell-and-Whitehead, Tarski, Turing, Post . . . —was, sociologically speaking, struggling alongside for recognition and beginning to gain a foothold in departments of philosophy and of mathematics across the country (I'm not competent to speak about Europe, although conversations with Kreisel, who had a visiting arrangement with the Sorbonne, and Lacombe, then, if memory serves, relegated to a Lycée in Algeria, painted a bleak picture, at least about France). Intellectually, it was definitely hitting a growth spurt—we were reaching the

[5] This account is meant to depict how it seemed, then, from a certain perspective. To present it here is not to endorse the view from that perspective as an accurate one (even if it may have seemed so then). Of course this is not all the philosophy there was—it is merely all the philosophy that occupied a relatively small, but active and influential, group of philosophers—whom The Sweep of History may well leave under the rug when the Ultimate Synthesis is recorded.

end of the time when logical literacy (I do mean 'literacy', not expertise), 'operationally defined' as the capacity to understand at least the title page of the *Journal of Symbolic Logic*, could amply be had with a logical education consisting of a course in each of model theory, proof theory, recursive functions, and set theory—since

- despite *Meaning and Necessity* and the pioneering work of Lewis (C. I.) and Langford, Fitch, Barcan Marcus, and others, modal logic had not yet achieved respectability; but the ground was already trembling from its distant thundering hoofbeat (was it in 1962 that Hilary Putnam, then at MIT, called me to tell of a young man he had met who was making sense of what, to our eternal shame, we called 'muddle logic'?), and ditto, of course, for the West Coast sport of intensional logic;
- the college of cardinals had barely begun to attract attention;
- hardly anyone was working with weird (e.g. branching, plural, infinite) quantifiers;
- or playing games with them;
- unusual (but interesting) models of ZF were soon to be forced upon us;
- infinitesimals were still obviously fictions, but now that we know how to build everything out of them. . . .

This all added up to a very optimistic picture of philosophy, with pride in a strong record of accomplishment since the launching of the new *credo*, and with promise of much more ahead. We were beginning to attack what we took to be difficult issues in the analysis of science—counterfactual conditionals, nomologicals—the distinction between genuine laws and accidental generalizations, causation. In short, we were busy developing the science of knowledge based on the knowledge of science. The philosophy of science was the queen of the branches, as it had the tools that were to be applied to address all of the problems.

Two 'popular' books heralded the new order: A. J. Ayer's *Language, Truth, and Logic*, and Hans Reichenbach's *The Rise of Scientific Philosophy*—neither a great book, but both excellent exponents of the spirit of the time—they trumpeted the militant affirmation of the new faith in philosophy, in which the fumbling confusions of our forerunners were to be replaced by the emerging science of philosophy. This new enlightenment would replace the old metaphysical views with a new mode of doing philosophy, in which we calculate and collaborate, make genuine advances, build on one another's 'results'—just the way they do it in science. One of the great attractions of the new enlightenment was its simple metaphilosophy—one that let us do philosophy with pride and with a sense of belonging to a common enterprise.

But what had become of the traditional questions of the philosophy of *mathematics*: What does mathematical knowledge consist in? What is it knowledge of? Does mathematics have a subject-matter? Is it concerned with a reality independent of the human mind? Is it a creation of the mind? or convention? Whence does it derive its certainty? They were not much in evidence. Truth is, they had been brushed aside. To the extent that (without provoking accusations of heresy) they could be regarded as questions at all, the party line was that they had been answered, by remarks like Ayer's:

> We see, then, that there is nothing mysterious about the apodeictic certainty of logic and mathematics. Our knowledge that no observation can ever confute the proposition '7 + 5 = 12' depends simply on the fact that the symbolic expression '7 + 5' is synonymous with '12', just as our knowledge that every oculist is an eye-doctor depends on the fact that the symbol 'eye-doctor' is synonymous with 'oculist'. And the same holds good of every a priori truth.[6]

Or Hempel's

> [I]t can be said that the propositions of the system of mathematics as here delimited are true by virtue of the definitions of the mathematical concepts involved, or that they make explicit certain characteristics with which we have endowed our mathematical concepts by definition. The propositions of mathematics have, therefore, the same unquestionable certainty which is typical of such propositions as 'All bachelors are unmarried.'[7]

These views were generally accepted and underlay the generally accepted answers to whatever *epistemic* questions one might raise (except for philosophers of Quinine persuasion, who tended to reject all of the questions). And the 'logical' backing for this float of philosophical currency, accepted by Quineans as well, although with a different slant on its significance, was the well-known and (deservedly) much-admired 'reduction of mathematics to logic'.

In their reaction to this new optimism, my own papers were not just revolting, they took a sceptical turn. The high that progress had induced made the inability of received doctrine to withstand reflective scrutiny all the more disappointing. Of course, everyone knew nothing was ever as simple as Ayer made it out to be—*Language, Truth, and Logic* was the *Manifesto*, not *Das Kapital.* But didn't he have it approximately right? Wasn't the basic point correct, even if he had to fudge the details a bit for popular presentation? Or so almost everyone seemed to think. Hempel's version (quoted above) was quite a bit more sophisticated, and hence all the more seductive (although, as I have argued elsewhere,[8] equally problematic)—

[6] A. J. Ayer, *Language, Truth, and Logic*, 2nd edn., Gollancz, London, 1958, 85.

[7] C. G. Hempel, 'On the Nature of Mathematical Truth', repr. in *Readings in Philosophical Analysis*, H. Feigl and W. Sellars (eds.), Appleton-Century Crofts, New York, 1949, 234.

[8] See 'Frege: the Last Logicist', in *Midwest Studies in Philosophy VI*, 1981, eds. P. A. French, T. E. Uehling, Jr., H. K. Wettstein, University of Minnesota, 1981, 17–35.

no explicit claim there about the synonymy of expressions that were obviously not synonymous, or about synonymy at all, just 'certain characteristics with which we have endowed our mathematical concepts by definition', and who could quarrel with that (pre- 'Truth by Convention' and 'Two Dogmas . . . ,'[9] that is)?

'Two Dogmas . . .' was the most extensive and extensively effective challenge to this entire picture, if effectiveness is to be judged by the number and eminence of the practitioners of the faith whose views changed in response to it. Without undertaking an agonizing reappraisal of that attack and its influence on the evolution of the views under discussion, it is undeniable that, justifiably or not, it branded as tainted the epistemic concepts of meaning, sense, a priori, certainty, necessity, and in so doing, tended to direct attention away from them, leaving the terrain to the more 'secure' syntactic concepts and to truth and reference (or satisfaction).[10] *The effect of this was to undermine the epistemic claims being made for these developments in logic and mathematics* and to focus the evaluation of their philosophic significance on what could be said if one limited one's attention to the concepts that remained 'untouched'—truth, reference, and possibly knowledge, if that could somehow be accounted for in 'extensional' terms. If Quine's attack was misguided, then so was our compliance; and with it, the moratorium that they effected in certain quarters on the discussion of these intensional notions. If it was not, then it is the revival that they have recently undergone that we should lament.

[9] W. V. Quine, 'Two Dogmas of Empiricism', *Philosophical Review*, 60 (1951). Reprinted in many places, the most frequently seen version being the one that appears in W. V. Quine, *From a Logical Point of View*, Harvard University Press, Cambridge, 1953. 'Truth by Convention' first appeared in *Philosophical Essays for A. N. Whitehead*, O. N. Lee (ed.), New York, McKay, 1936, but is repr. in a number of places, including B&P. Why didn't this magnificent paper have the effect in the 1940s and 1950s that it ultimately had in the 1960s and 1970s?

[10] Some (including the present author) have believed that an attack very much to the same effect can also be mounted on these latter concepts, using variants of the arguments that occur in WNCNB, as well as ones that Quine himself uses in his discussions of ontological relativity. Indeed, it may well be Quine's considered view. It takes considerable subtlety to distinguish the views of someone like Putnam in 'Models and Reality' (see n. 2) from those of a philosopher who had jettisoned the concepts of truth and reference in terms of which the debate was carried on (perhaps appropriately so—that is not the question right now). In any event, that is how the recent pragmatist revival strikes me. And the heresy is spreading. The Putnam of 'The John Dewey Lectures', *Journal of Philosophy*, 91/9 (1994), with its partial rejection of some past selves, is a more complex beast and cannot be crammed into a note to prefatory comments.

Even the most constant and long-time friends of truth seem to be defecting: Hempel makes no bones about it in a recent paper, 'The Significance of the Concept of Truth for the Critical Appraisal of Scientific Theories', in which he repudiates the concept of truth as having any analytical value in the philosophy of science: 'The conception of science as a search for truth is strictly untenable' (p. 5), or 'In sum, the question of whether a hypothesis is true or not is not even touched by the most favorable test results; the outcome of empirical tests has no bearing on the truth of a hypothesis' (p. 6).

III. WNCNB

III.A. The problem

As in the television show 'Jeopardy', we had some answers; what was considerably less clear was: 'To what questions were these the answers?' In this era of positive science, everyone agreed, almost regardless of philosophical persuasion (mathematical intuitionists are counter-examples to almost every rule), that the reduction of mathematics to 'logic' was one of the major positive philosophical achievements of the late nineteenth and early twentieth century, so long as one did not look too closely at whether the reduction was indeed to 'logic' or, what comes to the same thing, at what philosophical consequences one would be embracing by agreeing that it *was* logic. *Modulo* that important caveat, there was no dearth of legitimate candidate-bearers of that name, depending on whether the underlying theory was *Principia*-like, or some set theory, or some cleaned-up version of the *Grundgesetze*, or . . . ; and further, within any of these, there was further branching depending on what particular cluster of definitions was chosen to represent the concepts being analysed. In short, one had a collection of largely *mathematical* results, whose adequacy to some *mathematically specifiable* task was largely accepted, but whose philosophical significance remained very much unanalysed, especially for those who agreed with Quine that the concept(s) of meaning underlying Ayer's and Hempel's appraisals could not deliver the epistemic goods needed to uphold their conclusions: namely sanitary accounts of necessity, certainty, and the a priori.

That extensive agreement remained despite these important fault-lines in the overall position was surely due largely to the fact that all of these answers shared, if not the same question, then the 'hard data' in terms of which the purely mathematical portion of the problem was posed—they agreed in broad outline on the relevant aspects of the actual 'use' of the concepts targeted for analysis (or for our purposes, equivalently, 'reduction' or 'explanation'). These were constituted first by the body of mathematical laws that embodied the concepts (e.g. Peano's axioms, presumably 'second-order', the various theorems of analysis, etc.), but of equal importance they also included how their use is embodied in what we tend to call the 'extra-mathematical' employments of the same concepts (as in 'applications' like counting, measurement, etc.), on which Russell had prominently insisted, and which Frege had simply taken as his starting-point.[11] In brief

[11] And that it is more than somewhat misleading to call 'extra-mathematical' ('applied' would be better) since one also counts sides of polygons, integral roots of equations, bridges of Königsberg, etc.).

(intuitionists, strict finitists, and other heretics and revolutionaries aside) there was at least general agreement on the body of mathematical practice, pure and applied, that was being targeted to receive the philosophical benefits being claimed for the reduction—the usual standards of mathematical practice served to define its extent.[12]

But agreement on (a) what to regard as 'the data' and (b) what, formally speaking, would count as a successful 'reduction' or analysis need not bring with it agreement regarding the philosophical significance of such an analysis, or what the existence of the analysis, were one to be found, would show about the body of mathematical and extra-mathematical practice that constituted 'the data'.

To amplify a bit: Given a notion of definition that is sufficiently broad (but still sufficiently narrow to satisfy customary formal constraints, e.g. of eliminability, etc.) arithmetic is evidently reducible to set theory in many ways—via the customary definitions of 'zero', 'number', and 'successor', to be sure, but also in many other ways: Broadly speaking, having identified the target—arithmetic as the theory of $\{0,'\}$, or $\{0,', +, x\}$, successfully axiomatized by Dedekind, Peano *et al.*, and its applications—you need only

(a) identify *the successor function* with some set-theoretic function *f* and *zero* with some element *e* such that the ancestral of *f* with respect to *e* is provably *a progression, i.e.* an ω-sequence,[13]

[12] As in most things, Russell was a little different, and a superficial glance at his work might lead one to think that it pointed in the opposite direction. I have in mind his concern over the axioms of reducibility and infinity in *PM*. I think that by the time the authors of *PM* approached the problem, it was perfectly clear (with only a few exceptions) what would have counted as a successful reduction of mathematics to X (for 'X' read 'logic', or 'set theory')—in the sense of what body of mathematics had to be 'reduced', as well as what would count as reducing it. After all, but for a small glitch (Law V of the *Grundgesetze*), Frege, building on the work of Cauchy, Weierstrass, and others, had *done it*: he had both set the question and answered it. Russell's worries about his follow-up in *PM* were entirely epistemic in nature and concerned the answer; *he took the problem as given*: If the particular attempt embodied in *PM* seemed to Russell to 'fail', it was because what he could confidently call 'logic' just didn't measure up to the task. He therefore had to choose between two modes of failure: a successful 'modelling' with dubious epistemic credentials, and an epistemically sanitary but only partial success—a familiar kind of predicament, that goes with the territory in 'foundational' work. His 'way out'—to express the theorems that required an axiom of infinity for their proof as conditionals of the form

(Ax. Inf. → Theorem)

might strike some as a weasel and give his work the appearance of aiming at a more modest target. The fact is that he just didn't think the target could be hit, given what he took logic to comprise. To his credit, having discovered the inadequacy of his old view to its reductive task, Russell did not simply redefine the true extent of 'logic' and declare victory. That would have struck him as having all of the advantages of theft over honest toil.

[13] Well, 'provably'? It needs to be specified what that would mean. Metamathematically provably? Embedded in some theory in which the notion of an ω-sequence is represented by its ordinary set-theoretic definition and in which the numbers of the reduction are an ω-sequence on that definition? In the former case, the concept of ω-sequence is most naturally

(b) *the numbers* as the ancestral of f with respect to e, and
(c) make explicit provision in the account for the expression of the cardinality relation (e.g. by defining the relation $C(x,n)$ between a number n and a set x of cardinality n in some way such that '$C(x,n)$' holds iff there exists a 1-1 correspondence between x and the numbers less than n).

Everything else is extraneous.[14]

Given only this, however, it remains undetermined whether accounts meeting these conditions

- capture the senses of the expressions defined;
- co-ordinate the (informal, i.e. pre-analytic) singular terms to formal ones *with the same referents*;
- can be used to account for how, in our pre-reduction existences, we knew to be true the propositions of arithmetic that we knew to be true;
- are ways that someone can come to know the content of the mathematics that is represented in the account;
- show that the propositions with which we began (the 'target' propositions) were a priori, or analytic, necessary, or contingent;
- show that the propositions of the reducing theory are the selfsame propositions with which we began, albeit in a different guise, etc.

regarded as (at least) a second-order concept; to demand provability as an ω-sequence may be to confuse levels of analysis. It may suffice that what you pick as your numbers be in fact an ω-sequence. To demand provability is more to place a constraint on the argument that the analysis is adequate than to constrain the analysis itself.

There are further complications as well, which develop if you think of yourself as having developed your *meta*mathematics within first-order ZFC, for example, to which everything (that doesn't severely outstrip it, of course) is reducible—here I have in mind the complications that arise for the analysis from the consideration of 'non-standard' models for ZFC.

The second requirement—that the reduction include an account of cardinality, is deceptively easy to satisfy—or impossible. On the one hand, no resources beyond those needed in the analysis may be required to meet it. However, since anything can be counted, to ask for *a comprehensive* theory of cardinality may be to ask for the moon. What would be the theory *within* which such a general theory of cardinality would be developed? The reductionist's solace here is that whatever infirmities may beset his account because it lacks this degree of generality (because of a slavish adherence to the requirements of consistency) arguably *also* beset the concept being explicated. (If counting is as general as I suggest in the above remarks, then Frege's analysis of number was correct, and he was right to remark, when presented with the Russell paradox, that 'Arithmetic totters'.)

[14] I once thought (and argued in WNCNB) that there was a further requirement, namely that the '<' relation induced by these definitions on the 'numbers' of the reducing theory had to be effectively decidable. This was mistaken, and I hereby recant; the argument, unfortunately longer than would fit in this note, appeared as 'Recantation', in *Philosophia Mathematica*, 3/4 (1996), 184–9.

III.B. The argument

III.B.1. The overall strategy

WNCNB engages these issues by presenting the analysis of number, but constrained as an exercise that aimed, following Frege, at identifying 'which objects the numbers really were', but, following Frege again, in a context in which the obvious answer, namely '0, 1, . . .' wouldn't do.[15] So a constraint was imposed. Although for the sake of the argument, any 'objects', ordered appropriately and presented by means of expressions from whose surface grammar one could not immediately read off that they denoted the numbers would have done equally well, I chose to make them sets. The point of the constraint was to replicate what I took to have been Frege's concern, most forcefully expressed in the opening pages of *Die Grundlagen der Arithmetik*, that in all these centuries of dealing with the numbers, mathematicians had not yet inquired with sufficient care, tenacity, and ingenuity into their nature to have succeeded in discovering it.

This took us from the analysis of number to a question *whose answer* had to take, at least in part, the form

(i) the number n = the object o,

a form we surely owe to Frege. It seemed evident that constraining the answer by requiring that the singular term on the right designate in some prima-facie way a 'logical object', or, for that matter, any object not given to us in canonical numerical notation,[16] created a problem whose solution

[15] Precisely why that answer would not have sufficed for Frege is a contested matter of Frege interpretation whose resolution is not really germane to the present purpose. For my own views on the matter, see my 'Frege: The Last Logicist', *Midwest Studies in Philosophy VI*, Peter A. French *et al.* (eds.), University of Minnesota Press, Minneapolis, 1981, 17–37. In a recent article Hilary Putnam associates himself with a different view of Frege, that he calls 'Kantian' rather than 'Platonist', in which Frege's views on logic (and, by implication, on mathematics as well) are pictured as less 'ontological' than perhaps I would have them, arguing that for Frege logic is 'constitutive' of reason itself.

I am aware that many people (e.g. Michael Dummett) read Frege as a Platonist rather than as a Kantian. However, my colleagues at Harvard, Burton Dreben and Warren Goldfarb, have convinced me long ago that this is a mistake. For a reading of Frege that is influenced by the teachings of Dreben and Goldfarb, see Joan Weiner's *Frege in Perspective* (Cornell). ('Mathematical Necessity Reconsidered', in *Words and Life*, James Conant (ed.), Harvard University Press, 1994, 261)

This is not the place to carry on this debate. The interested reader would do well to consult recent work by Jamie Tappenden on these difficult matters, e.g. 'Fruitful Concepts and Extending Knowledge: Fregean Themes in the Foundations of Mathematics', *Noûs*, 29/4 (1995), 427–67.

[16] That may not be the right way to express this very important condition, for the following reason: *My* favourite number is 17, and has been at least for the last 40 years, ever since Hilary Putnam introduced me to its fascinating properties (maximal randomness is its most important one). Suppose, however (I know it's counter-intuitive), that Kurt is a fast thinker, and, unlike me, fickle. Yesterday he liked 7; this morning it was the Gödel number of the *Principia* numeral for 7 in his epoch-making 1931 paper. This afternoon it will be another; and this evening another still. In fact, as you will have guessed, his patience with numbers declines

was hopelessly underdetermined by any reason we may conceivably have to accept it—by any 'facts' that we might bring to bear on it without begging the very question at issue. It concludes, via a dubious move or two, to be examined below, that numbers couldn't be sets at all—and that the same argument could be extended to any other 'objects' (not themselves obviously numbers) that one might choose to put in the place of sets—a powerful result. The parenthetical reservation is necessary;[17] the argument of sections I and II of WNCNB does not work against the null, 'uninformative' answer: 'the numbers themselves' (in their usual guise, construed *sui generis*), as George Boolos was quick to point out.

The emendation is accepted as a friendly amendment, as it helped qualify, and thus clarify, the import of the first two sections. To be sure, the paper does end with the grandiose claim that 'there are no such things as numbers',[18] but the principal support for that claim was intended to be the arguments of section III; the arguments of sections I and II were meant to bear the much lighter, anti-reductionist, burden of pointing out the consequences of accepting the reductionist premisses in a 'realist' context, i.e. in a context in which all sentences of the form of (i) above are taken to have determinate truth-values. It is regrettable that the author of WNCNB took

exponentially, so that by tomorrow, or the day after, at the latest, he will have favoured each number in turn (although in what order, we do not know). So for each instant t in a three-day interval, it will be true that the expression

(a) 'Kurt's favourite number at t'

denotes a number. If we denote the instant of his i-th defection by 't_i', then for each natural number n (for those who are counting along with us, let his initial fancy count as his 0-th 'defection'), there will be an i such that Kurt begins favouring n at t_i and therefore such that n is the denotation of (a) for every subsequent $t < t_{i+1}$. For each system of notation for instants in the three-day interval, there is therefore a sequence of expressions of the form (a) each of which designates a different number and such that each number is the denotation of at least one such expression, yet no one would argue that any such sequence of expressions would count as a canonical notation for the integers (although, admittedly, its intelligibility may depend on having a canonical notation for substitutions for 't' in (a)). Only those brazen enough to think that Kurt is neither sufficiently fickle nor sufficiently quick to satisfy the conditions of the example will fail to agree that Kurt's favourite numbers escape the 'indeterminacy' arguments in question. Although, admittedly again, we may not know beyond the first two numbers, in some appropriate sense, 'which' number is Kurt's favourite number at t_i.

So some other way of phrasing that condition would be better, perhaps something along the line of '*any "objects" presented by means of expressions from whose surface grammar one could not immediately read off that they denote the numbers.*' That would exclude our fickle Kurt and, if we don't put too much weight on 'read off', it might just do.

[17] Because of a remark by George Boolos. I believe that 'obviously' was the technical term he employed.

[18] WNCNB, p. 73. The sentence ends, somewhat anticlimactically 'which is not to say that there are not at least two prime numbers between 15 and 20.' Again, although it could have been made less preciously, the point is familiar enough: whatever sweeping 'ontological' conclusions one draws in this discussion should not be taken as contradicting established mathematical results. Although revisionism can never be completely excluded, our primary philosophical task is hermeneutic. It follows that, whatever its merits, the reduction does not answer the fundamental questions regarding what makes the true ones true. On the assessment of WNCNB, they wouldn't *be* true.

so few pains to make clear that sections I and II, on the one hand, and section III were really independent of one another (except in so far as the problems of the first suggested that the more radical surgical measures contemplated in the third might be in order), and that the sweeping conclusions of section III were being advanced with considerably less confidence than were those of sections I and II, to which we will return below, after a brief detour through the labyrinth of section III.

III.B.2. 'III. Way out': 'there are no such things as numbers'

WNCNB generated a fair bit of confusion, much of it through the arch title of this section: 'Way out'. There is some textual evidence that the author meant it as a pun on 'the way out' and 'way out' (in the sense of 'far out'), and hoped that the omission of the definite article would suggest that intent, alerting the reader and, as an operator on the whole section, signalling that the theses and arguments advanced in it were being put forth with considerably less conviction than those of the first two—more contemplated than advanced. The apparently almost universal impression has been that the title was meant in the former sense, and that the section contains his proposed solution to the puzzles generated in sections I and II—to the point that commentators tend to supply the 'missing' definite article, referring to the section as 'the way out', while others, because of some similarities between one view contemplated in that section and Dedekind's view, refer to the author as, e.g., 'a neo-Dedekindian'.[19] Hoist by his own petard.

> To *be* the number 3 is no more and no less than to be preceded by 2, 1, and possibly 0, and to be followed by 4, 5, and so forth. And to *be* the number 4 is no more and no less than to be preceded by 3, 2. (WNCNB, sect. III, 70)

Section III does 'argue' that not only could numbers not be sets, they couldn't even be numbers. If the analysis that culminates in the above passage is correct—if *that* is what it is to be the numbers—then no objects could satisfy exactly those conditions, and hence there could be no such things as (even the *sui generis*) numbers. It would be a bit of stretch, even for such a flexible particle as '*neo*', to make that into a neo-Dedekindian conclusion. But the argument for this more far-out conclusion is itself more far out. It does contain a neo-Dedekindian premiss; but it also depends on abstruse metaphysical principles regarding the possible existence of a

[19] Cf. Michael Dummett in the Preface to *Frege: Philosophy of Mathematics*, Duckworth, London, 1991, xii: 'while it is appealing to be a neo-Dedekindian like Paul Benacerraf', a reference to the 'structuralist' remarks in section III(C) 'Arithmetic is therefore the science that elaborates the abstract structure that all progressions have in common merely in virtue of being progressions.' Of course, it is a matter of utter triviality and insignificance what this particular author may have thought or intended; however, it might have helped the ensuing discussion some if the warning signs that had been planted around the trench had succeeded in alerting some readers to take the views with several more grains of salt than they actually employed; hence this confession.

system of objects the essence of each of whose elements consists in the *relations* it bears to the other elements of the system.

The argument takes off from the above characterization of their nature and concludes that numbers couldn't be objects at all, and hence that there could be no such objects as 3, 4, or even 17.[20] It consists of two distinct pieces: (a) a 'structuralist', or 'neo-Dedekindian', analysis of arithmetic as the science of progressions—not of a particular progression, as we are inclined to think, but of progressions in general, and (b) the claim or argument that no system of 'objects' could satisfy the conditions set out in (a). Accordingly, because one person's *modus ponens* is another's *modus tollens*, it is possible to accept the metaphysical principle regarding 'objects' and take the 'evident' falsity of the conclusion to signify that the 'pure' structuralist analysis is somehow flawed. Specifically, it must overlook features of numbers in virtue of which they are possessed of sufficient intrinsic substantiality to ground the relations they also evidently bear to one another[21]—namely that to bear the relations they bear to the others isn't all there is to being one or another of them.[22] In any case, the last two decades have seen structuralism in various forms (including maximally kinky versions, like modal structuralism) receive a good deal of far more expert attention than I could have given it, either then or since, largely at the hands of Hellman, Parsons, Resnik, Shapiro, to mention but a few in alphabetical order. I have to confess to having had a fling of my own with (arithmetical) structuralism; it came early[23] and was relatively brief: If 17-hood is typified by being the seventeenth member of a progression (let's banish 0 for now) then why not say, quite generally, that a number *n* is the

[20] . . . *the* number that must exist, if any do—letting 'E' be a predicate for existence: 17 = $\in_x \Box(\exists y(Ny \& Ey) \rightarrow Ex)$?

[21] That is the path Mark Kalderon explores in his *Structure and the Concept of Number*, Ph.D. diss., Princeton, 1995. I am indebted to him for a number of clarifying discussions of this very murky topic.

[22] Not everyone is discouraged by this kind of argument. Thomas Tymoczko surveyed a number of these issues in a recent article ('Mathematics, Science, and Ontology', *Synthèse*, 88 (1991), 201–28) and advanced the bold thesis that 'all objects are abstract objects' (p. 201), arguing for it by capitalizing on a view (which he attributes to Nicholas Goodman) that 'abstract objects are just very good approximations of concrete physical objects. If you start describing a physical object, and don't complete the description, we can say you've described an abstract object' (pp. 212–13). Tymoczko followed with a parenthetical remark: '(Do we ever really complete descriptions? Probably not, and I'll defend the obvious corollary in the next section of this paper)' (p. 213). To the extent that the abstract/concrete distinction was seen as an obstacle to metaphysical and epistemic purity, the obstacles have now been swept away (unless we wish instead to embrace thoroughgoing scepticism).

Of course, this (deliberately, surely) fudges *incomplete* with *incomplete description*. On the structuralist's conception of 17, the reason it isn't said to be fuchsia isn't that it is really green, with yellow polka dots and we just can't bring ourselves to say it is fuchsia; it is because on that conception, it has no colour at all—indeed, it is the incoherence of the very idea of a 'description' of 17, complete or incomplete, that drives the structuralist to his madness. He won't be so easily rescued.

[23] In *Logicism: Some Considerations*, Ph.D. diss., Princeton, 1960.

relation that an object *a* bears to another *b* iff *a* is the *n*-th member of the progression *b*, or, in set-theoretical lingo, iff it is the set of all ordered pairs <*a*,*b*> such that *a* is an object, *b* is a progression, and *a* is the *n*-th member of *b*, where 'being the *n*-th member' is defined in the usual, non-question-begging way. If arithmetic is the science of progressions, it surely is the science of this, the archetypical progression. But, as Charles Parsons pointed out with his customary gentleness and generosity, there is no such set (which, I suppose, makes it the best possible example of what numbers could not be—if there were some). This banishes the leading candidate along with its cousin, the universal set. But when that is gone, it is hard to know what to invoke in its place. Perhaps one should think about numbers much as one should think about the other 'limiting' concepts whereof one may not speak in too much detail or specificity, or comprehensiveness, like 'set', 'truth', etc., namely that no explicit rendering of them can be a 'complete' rendering. If so, then arithmetic, seen as *the science of number*, does indeed totter.

So much for the argument of section III. Although I continue to be persuaded that there is something right about it, the conceptual baggage it carries so blunts its dialectical edge as to render it useless in this inquiry. Accepting it has to be an *auto-da-fé*.

III.B.3. 'I & II: The education and the dilemma'

But what should we make of the argument of sections I and II? Briefly restated, it set out certain well-known conditions on an account of number (constraining the choice as explained above), and argued

(1) that indefinitely many accounts satisfied all of the conditions;
(2) that, because (i) ('the number n = the object o') has the form of a statement of identity, at most one such account *could* be correct;
(3) (after a detour to examine the credentials of the presumptive leader of the pack, a Frege–Russell style account) that there was no principled way to choose among them, i.e. to decide which sets the numbers really were;

but

(4) that if one of the accounts were the correct one there would be a way to establish which one it was: 'the position that this is an unknowable truth is hardly tenable.' (WNCNB, p. 62)

and therefore

(5) that any feature of an account that identifies a number with a set is a superfluous feature of the account (i.e. not one that is grounded in our concept of number);

consequently

(6) 'numbers . . . could not be sets at all' (WNCNB, p. 62), which also extends, *mutatis mutandis*, to whatever surrogate one might have chosen to put for sets.

Most of the issues that are raised in the above argument have received considerable discussion; very few, if any, have been satisfactorily resolved.

First of all, there is the assessment of whether the front-runner account—the Frege–Russell account, or Frege's account has received its due. This involves several related issues:

(a) the George Boolos–Richard Heck–Crispin Wright resurrection of the account of *Grundgesetze* as a two-step process consisting of
  (i) the derivation or justification of 'Hume's Principle'—that the number of F's = the number of G's (NxFx = NxGx) iff the Fs and the Gs can be 1-1 correlated—in some way that does not make appeal to the infamous Law V of *Grundgesetze*, and
  (ii) the derivation of the basic laws of Arithmetic from Hume's Principle alone, i.e. without recourse to Law V.

  Success on both ends of this attempt would seriously undermine the argument for (3) above, that there was no single, most plausible candidate. But even success here may need bolstering against the arguments of section III—the most plausible candidate may just not be good enough to overcome the metaphysical objections. Here is where

(b) the work by Dummett, Wright, Hale, and others, on Frege's conception of object, on the nature and substantiation of 'metaphysical' claims about them, on 'deflationary' metaphysics, etc. takes on critical importance.[24]

  Where that takes the form of arguing that Hume's Principle is simply 'analytic', I think we must take as provisionally decisive Boolos's recent findings (unpublished, to my knowledge) regarding all the close cousins of Hume's Principle that, like Law V, appear to share Hume's Principle's claim to analyticity, *but that are in fact inconsistent.* Absent a principled way of setting them apart from Hume's Principle, the argument for the analyticity of Hume's principle itself is doomed—at least for the purpose of extracting any epistemic dividends from it.

Secondly, and coming from another direction: no red-blooded realist (of any stripe—metaphysical, internal, external) should accept (4), the bald

[24] See Gideon Rosen's 'The Refutation of Nominalism', *Philosophical Topics*, 21/2 (1993), for a searching analysis of these issues.

statement that if there isn't some a priori proof that some particular candidate reduction is the correct one, there can't be a 'correct' one. So, however congenial that principle may be to any number of people, and it has had its supporters,[25] the need to advert to it is a genuine weakness of the overall argument. Of course, with some hindsight, one could explicitly include among the conditions to be satisfied by any correct account that the demands of reduction be not only *uniquely* but *demonstrably* so, i.e. that no account that doesn't demonstrably outclass *all* competing accounts can be called 'correct' at all. For Frege, such a condition need not be stated; it follows immediately from (his account of) the meaning of identity and the requirement that success in his venture can only come with *proofs* of the laws of arithmetic. But in the present situation, to include it would be a sham and present no dialectical advance. This is related to the next point.

A further weakness, more difficult to assess, but of fundamental importance none the less, is that nothing in the anti-reductionist argument addresses reductionist views based on uses of some principle of parsimony, e.g. Occam's razor.[26] Such arguments are puzzling to say the least. When the principle is used to 'pare down' the assumed population of the universe to the sparest version that will do for some given purpose, the argument in which the principle is invoked is not properly 'reductionist', at least not in the realist sense Frege had in mind, and thus poses no threat to the ontic census-taker, whose task it is to construct an accurate *catalogue* of the 'furniture of the universe'.[27] It is more like the way one should think about

[25] Crispin Wright is among those who has generously come to the defence of this principle: 'However, if numbers *really were* objects of some kind, surely someone who perfectly understood the concept of number should be able, at least in principle, to identify them' ('Skolem and the Skeptic—II', p. 122). Comments by Crispin Wright on 'Skolem and the Skeptic—I', Paul Benacerraf, *Proceedings of the Aristotelian Society*, suppl. vol. 69 (1985), 85–137. Of course, that support wasn't a pure act of generosity; he repaid himself when he found the principle a handy one to use *ad hominem*. I will return to it below, when I turn to the Skolemite arguments.

There is a weak sense of 'to identify them' in which we have been taught that someone who 'perfectly understood' the concept of number *is* able to identify them, e.g. the number n can *always* be identified as the number of its predecessors in the number series, a fact that both Frege and Dedekind put to good use, but that, by itself, does not satisfy the constraints of the problem (even if they have correctly and independently identified 0, which Frege surely thought he had done).

[26] For a telling account of the Occamite's task and trials, see John Burgess's 'Occam's Razor in Physics, Paraphysics, Metaphysics, and Pataphysics', MS, a descendant of which is included in this volume.

[27] There is an evident tension between the phrases 'for some given purpose' and the description of the task of the ontological cataloguer as that of constructing an accurate catalogue of 'the furniture of the universe'. Clearly, for the sake of the present argument, *constructing an accurate catalogue of the furniture of the universe* couldn't count as the 'given purpose'. Less mysteriously, the range of acceptable purposes needs to be specified in some way that leaves 'finding out the truth of the matter', or 'what there really is' and phrases that imply or depend on it, out of the specification. Much of the problem one has making sense of the thorny issues surrounding Occam's razor is rooted in the difficulty of making that distinction in a principled

packing for a trip to the tropics: packing only one of your five cashmere sweaters—you won't need them all—is not to deny the existence of the other four; it is merely to leave them on the shelf. Such restricted applications of parsimony cannot support the conclusion that numbers are sets, or that they don't exist at all (because only sets do). The strongest conclusion they could support is that we can 'do without' including some separate realm of entities in the catalogue if we already list an isomorphic copy of it, presumably because we can 'do' with {17}, our set-theoretical stand-in for 17, whatever we were 'doing' with 17 in the first place—17 won't decay from idleness as {17} takes its place, and {17} will have no trouble slipping out of its basic conceptual set-role to play this new role. It is interesting to note that, for a structuralist, that is as strong as conclusions get. When {17} gets home from the office after a hard day in her set-suit playing her primary conceptual role as set, she will still have enough zip to slip into her number costume and give a flawless performance as 17.

However, there are uses of Occam's razor that do not fit the above characterization so easily (and that are more insidious): The holist, for example, does not admit of subordinate ends such as were used above to blunt the razor's edge, thus leaving the profligate theorist free to collect entities that would otherwise have been shaved off at a swipe. More accurately, whether or not such subsidiary purposes are allowed, the holist makes a further, global, use of the principle whose avowed aim is 'explanatory adequacy', but now of 'everything', counting the theory as one big lump and judging parsimony at this global level. The aim of this manœuvre is to proclaim adequacy to this global task the ultimate test of theory: inclusion in a theory that passes this 'test' becomes tantamount to acceptance into the catalogue, and exclusion, accordingly, banishment from it. Surviving that pass with the razor becomes *constitutive* of existence, in a clever attempt to block

yet 'neutral' manner: Enjoined from multiplying entities beyond necessity, it is natural to ask 'Necessity for what purpose?'

Parsimony sounds like a sensible maxim or rule of thumb, much like 'Don't put too much on your plate!' or 'Don't make your paper longer than it needs to be!'; but absent a context (like one's appetite) that controls how much would be too much, it makes little sense. Because the issue arises most often in the context of providing scientific explanations, the usual context is explanatory. The ground rules become the ones dictated by what an explanatory context demands, and there are as many views about that nowadays as about almost any other philosophical issue.

To complicate matters even further, perhaps the principal fault-line pits the 'realist', for whom an explanation is not adequate if it is not true, against the 'anti-realist', who, to put it somewhat tendentiously, professes more modest aims: Where it is not knowable that the truth has been reached, it is unintelligible to set it as a goal or condition; something akin to 'warranted assertability' must take its place. Of course the final epicycle in this manœuvre will be to declare talk of 'truth', at least in the sense being repudiated, senseless altogether, thus promoting its surrogate into place as the outer limit of our possible aspiration (and thereby undermining the description of the aim as 'more modest'?). This is not the place to enter into this debate, but, if the above even approximates its general outline, it will not be an easy matter to describe the proper role of Occam's razor in a manner that is neutral in the debate between the eliminativist and the old-fashioned realist.

the 'open question' argument: The question 'Yes, Yolanda, it is clear that the assumption of the existence of quiggles is not necessary for the successful performance of this global task, but do they exist anyhow?' is rejected as lacking any clear sense.

The relevance to our problem should be clear enough: Even if the realistically driven reductionist is undermined by *l'embarras du choix*, not so with the holistic Occamite, who is not beholden to any notion of 'getting it right' that transcends the best theory that survives ontic paring. That is what keeps Quine, Field, and their friends in business. But will this process of 'paring' bear scrutiny? For example, do we not need some standpoint from which to survey our 'total' theory in order to judge if it is as spare as it could be? If so, what could that be? Moreover, since there is no guarantee that only one theory will survive these tests and subsequent parings, perhaps a further word of caution is in order, even for a position as well hedged as this: the shoals of WNCNB are very much there and a threat to the incautious holist who doesn't make explicit allowance for the possibility of distinct, but non-equivalent, theories that pass the test equally, both having been pared to the bone; but that word is surely not needed by such seasoned navigators as Field and Quine. ('What could "non-equivalent" come to as applied to theories that pass this (ultimate) test equally well?' we hear them saying.)

III.C. Concluding remarks

In retrospect, I see WNCNB as much less of a 'definitive' statement on the issues it addresses than I imagine I once did, either in its first two sections, or in its more tentative, but also more sweeping section III. The adherent of platonistic reductionism—the brand I have laid at Frege's door (and continue to believe belongs there, despite the rumblings of the 'Harvard School', cf. n. 15, above)—can escape its attack unscathed provided that she either

(a) finds a satisfactory way of defending the Frege–Russell analysis, or some close cousin, perhaps taking her cue from Wright *et al.* and Boolos *et al.*;

or

(b) bites the bullet, sticks to her realist guns, and insists that the impossibility of establishing any answer as the right answer *does not undermine the thought that there is a right answer* (i.e. denies premiss (4) of the replay of the argument as sketched at the beginning of this section).

Or, if the first won't fly (as I increasingly suspect it won't), and the second is repugnant, there is always option (c):

(c) unbridled Occamism in a holist setting, which is Quine's position—why demand more than sets, if sets is all you need (sung to the tune of the Beatles' 'All You Need Is Love')?

Despite admittedly realist sympathies, I find (b) much less appealing than some version of (a), but stripped of pretensions to the epistemic benefits it was supposed to bring, with (c) occupying a kind of middle ground—if not very imaginative, it is probably the safest of the three (if it proves ultimately intelligible).[28] This is in part because an unvarnished, anti-reductionist, insistence on sticking to the *sui generis* model of the numbers runs counter to Frege's demand that we inquire into their nature and prove the laws that govern them, if they are susceptible of proof. Any such proof would have to penetrate the cloak of Peano's primitives and give an analysis of the underlying concepts. Accordingly, what is most appealing about pursuing (a), particularly as compared with (b) and (c), is that it appears to be the only line of inquiry that seems at all sensitive to arithmetical practice—the context of use of the arithmetical expressions, including perhaps the context of their (imagined) 'introduction' into the language. Unless a strong argument can be mounted for its realist underpinnings that at least meshes with a persuasive analysis of mathematical practice, where the roots of mathematical truth must be, (b), by itself, is simply too 'theoretical', too metaphysical; while (c) is too holistic and, even if it can be made to pass the test of intelligibility, runs strongly counter to my own root inclinations in this area, which are fundamentally atomistic. The 'bottom line' for me is that both (b) and (c) feel very remote when it comes to accounting for relatively 'local' features of our experience, which (holist arguments notwithstanding) has to be where meaning and truth are rooted, in arithmetic as elsewhere. It is hard to believe that the fruit can have fallen far from that tree: specifically, that what we read off some abstract formal model of our overall theoretical practice[29] can trump our most immediate intuitions regarding the content of our ordinary, or even our refined, mathematical concepts—at least not without frequent 'reality checks' against the *in*formal practice the model is meant to model.

[28] Except, perhaps, for a dedicated, intemperate, Occamite reductionist who is a Skolemite to boot (see below). That leads to Pythagoreanism—the view that all you need is *numbers*, and hence that numbers is all there is—a dramatic reversal. It has never been clear to me why someone who got as far down the slippery slope as Quine did should not settle on Pythagoreanism as the true doctrine. But perhaps it's just my arithmophilia (thanks to John Cooper for saving me from 'numerophilia') showing again. These are troubled waters and the proper subject of another study.

[29] Like Putnam, in 'Models and Reality', in B&P, pp. 421–44.

IV. METAMATHEMATICS IN PHILOSOPHY

The above remarks and the ones in the present section are addressed to us all as philosophical 'consumers' of metamathematics—philosophy in the grand tradition, the sins of which are amply illustrated by WNCNB. Although I will be challenging conclusions drawn from the use of formal models in the study of the meaning of mathematical expressions and concepts, I do not question the value of that sort of study overall—that has worked too well. But we should not lose sight of their lineage: We construct and interpret them, which is how they derive what meaning *they* have, which in turn gives them their capacity to represent ours. This simple fact should guide us in drawing philosophical consequences from the metamathematical results in which they are implicated.

This should not be controversial. But we sometimes find ourselves inferring constraints on our own ability to mean this or that in our pre-formal talk from (formal) features of the formal representation of our speech or thought—for example when we discover some formal 'incompleteness' result about some formalized theory or language[30] and infer *from that alone* that there is a corresponding incompleteness in our own ability to 'mean' something that the formalized theory has demonstrably not completely captured—as if it were somehow self-evident that our original meaning had been sufficiently captured by relevant features of the formalism, been mirrored in it, if you like, to entitle us to read the 'limitations' we discover in the formalism as tokening corresponding limitations at the point of origin, the mathematical practice it was intended to model. Sometimes such a conclusion is inescapable—as in Tarski's analysis of the semantical paradoxes. But even in such relatively clear cases, the philosophical lesson that the formal work teaches us must be spelled out with considerable care. Surely it is *at most* that our pre-formal concept of truth is either incoherent, or else lacks some of the features that the formal analysis shows jointly to lead to paradox. Further and independent argument is needed to show that it is one or the other, or, even if it is the latter, that a specific one, e.g. semantic self-reference, is at fault. Much of the philosophical work comes in choosing the lesson.

Two further examples, from other areas of the philosophy of mathematics—how we should respond to (a) the Gödel theorems on the incompleteness of arithmetic, and (b) the Löwenheim–Skolem theorems—will illustrate this point more transparently. Both are controversial,

[30] We will use 'theory' and 'language' interchangeably, except when it matters, in which case a *theory* will be a set of sentences (in some language) closed under logical consequence (of the order appropriate to the language), and a language, a set of sentences in some vocabulary—whether 'interpreted' or 'uninterpreted', or whether that is even a relevant parameter (natural languages are neither) will depend on and should be clear from the context.

although perhaps to different degrees; both should help illustrate where the truth about mathematical truth could not lie—what mathematical truth could not be.[31]

IV.A. Some problem cases: Gödel and Löwenheim–Skolem

One case-study of philosophical implications is Gödel's incompleteness theorems. We are sometimes urged to conclude from what we know about our metamathematical accomplishments and abilities, for example that we can prove Gödel's theorems in what seems to be their full generality about all formal systems, machines, mechanical processes, etc., that we are therefore free of the shackles of mechanism, i.e. that we are endowed with certain super-mechanical powers, powers no mere machine could have. I have argued that we might, for all we know, *be* (protoplasmopsychophysical realizations of) machines—arguments based on the incompleteness theorems to the contrary are simply confused and require a supplementary philosophical premiss that does the lion's share of the work but that is usually unstated, or if stated, unsupported.

Another philosophical lesson we are sometimes urged to learn from the Gödelian incompleteness results is perhaps more familiar and certainly more immediate. Because bivalent truth outruns formal provability, for any consistent formal system of arithmetic, the concept of bivalent truth is inherently flawed and must be replaced with a concept of truth-in-arithmetic that is more closely modelled on our abilities as provers. This, of course, then splinters into countless possibilities, depending on how bounded one believes our 'proving' abilities to be.

I will also discuss the mysterious 'Skolem Paradox', and offer some resistance to the Skolemite conclusions that

(a) because our first-order formalizations of set theory have countable models, etc., they cannot express, or 'capture' the concept of 'non-denumerable set';

and that

(b) because we cannot 'capture' or 'express' the concept of 'non-denumerable set' via a first-order axiomatization of set theory, we *have* no such concept.

Here my point has to be that the notion of *'expressible' in a first-order (henceforth FO) theory* is parasitic on the notion of *an interpretation of a*

[31] See 'God, the Devil, and Gödel', *Monist*, 51/1 (1967), 9–33; and 'Skolem and the Skeptic', *Proceedings of the Aristotelian Society*, suppl. vol. 69 (1985), 85–115 for attempts to defuse arguments alleging philosophical implications of metamathematical results, arguing that the alleged 'implications' are not implications at all but require for their substantiation a significant injection of philosophical serum.

FO theory, which in turn relies on the prior notion of a *quantifier domain*—a point that, as I have argued elsewhere, Skolem himself makes (against Zermelo) in his classic 1922 paper that is taken as the *locus classicus* of his own expression of the sceptical view that we have come to associate with his name (wrongly, I think; he held the view, all right, but did not advance it in that paper). Given these dependencies, it is illegitimate to infer constraints on what can be 'expressed' in a particular FO language, the language of ZFC (Zermelo–Fraenkel Set theory with the Axiom of Choice), for example, from what doesn't flow about *all* FO languages (theories) from the very concept of a FO language (theory).

In all three cases we have on the one hand some clear metamathematical results, Gödel's incompleteness theorems or the Löwenheim–Skolem theorems—'Whew! Well, that's *half* the job!'—and on the other, their alleged philosophical consequences: 'Not every closed sentence of first-order Peano Arithmetic has a truth-value'; 'We are not (Turing-) machines'; or 'We have no conception of the uncountable independent of what is invariant across all models of first-order ZFC.' In each case the argument for the alleged consequences depends on a crucial *philosophical* premiss, not always easy to identify, but without which there would not even be the semblance of an argument. I call the task of identifying and justifying that premiss the 'Princess Margaret problem', and the required (although sometimes only covertly stated) premiss 'the Princess Margaret premiss' or 'PMP'. Some stellar examples of PMPs:

- There is something human mathematicians can do that no machine can do—for any (theorem-proving) machine, find its Gödel number and, given its Gödel number, prove its Gödel sentence, something it (the machine) manifestly cannot do.
- 'Truth' for our arithmetic, if it is to characterize *our* mathematico-linguistic abilities, must be bounded by our abilities as *provers*. Hence only a semantics that is grounded in the assertability conditions of sentences can accurately reflect the meaning of arithmetic sentences.
- Our conceptual abilities are bounded by what can be expressed in a First-Order Language simply in virtue of its first-order structure.

In each case, it is impossible to embark on the philosophical voyage without first taking on the appropriate PMP baggage. We will begin with a more detailed discussion of the Skolemite case because it is in certain ways more puzzling—and certainly more far-reaching in its implications, especially if one were, *mutatis mutandis*, to apply in other contexts all of the consequences of the PMP that is wheeled in when needed to do the philosophical dirty work in the case of set theory (the Putnam of 'Models and Reality' is the only philosopher I know of to have bitten that particular bullet). We will then return for a brief look at the Gödel cases, which have received considerably more discussion in the literature and on which we have nothing new to add, except to point out how they fit this pattern, and conclude with some general remarks about this class of problems.

IV.A.1. Skolemite arguments: the general case

Since the question seldom comes up in ordinary conversation, it is unclear how many people take the arguments of the Skolemite sufficiently seriously to conclude about themselves:

(1) that they do not have an 'absolute' concept of uncountability—or, more delicately put, that the concept of uncountability that they might possess or be expressing at any given moment is hopelessly indeterminate, in some sense 'relative' to whatever 'model', say of first-order ZFC, their interlocutors may choose to impute to them for hermeneutic purposes; and
(2) that, moreover, they are not alone: the same holds true of *us*, and for the same reasons—we are no better able than *they* to transcend the expressive and conceptual limitations that the Skolem 'paradox' revealed.

I believe that the arguments that have been advanced for those conclusions are seriously flawed; but also powerful and gripping—and that there are in fact many closet Skolemites among us (where by a 'Skolemite' I mean less someone who subscribes to any particular argument than someone who is persuaded, perhaps in ways unknown to themselves, of its relativistic conclusions). Indeed, recently, overly permissive treatment of such heresies has emboldened a number to come out of their relativistic closets.

IV.A.1.a. The core argument

Every Skolemite argument, however sophisticated, contains the following core:

It is well known that, by purely 'structural' (i.e. uninterpreted) axioms, we can characterize every finite cardinality, in the sense that, for each natural number *n*, we can write first-order axioms that are satisfiable over a domain D iff D has exactly *n* members. Ditto for 'infinite' as a 'cardinality' (ignoring for now the finer Cantorian points about different orders of infinity, which, in any case, are in dispute).

But things get stickier once the Cantorian (call her 'Georgina') tries to make these finer distinctions, even only the crude distinction that divides the infinite sets into the *countable* and the *uncountable*: whereas she can, as we saw, write first-order axioms that are satisfiable, but not satisfiable in any *finite* domain, she cannot write a set of first-order axioms that are satisfiable in an *uncountable* domain but not in any *countable* domain. Therefore, she cannot, by means of uninterpreted first-order axioms, i.e. via the first-order structure of her axioms, characterize the notion of 'uncountable', in this peculiar sense of 'characterize'.

'So,' Georgina might well respond, 'who ever thought one could? At best, this shows that the countable/uncountable distinction outruns the expressive resources of *pure* first-order syntax and semantics. Good to know, but given the origins of the distinction at the heart of Cantorian set theory, no surprise, and surely no cause for alarm.' Not yet, anyway.

IV.A.1.b. The 'sucker play'

Since it is *à la mode* to give 'formal' characterizations of concepts, i.e. characterizations that are presented in a suitably formalized language, it is appropriate to ask Georgina to meet this challenge by doing just that with regard to this concept that she believes she possesses. What does she need to employ in order to give a suitably formal characterization of 'uncountable'? Well, the customary definitions of 'countable' and 'uncountable' make that plain: since a set or domain D is *countable* iff D can be mapped 1-1 onto the integers, and it is *uncountable* if it is infinite (OK so far—it is equinumerous with a proper subset of itself) but *no 1-1 function maps it onto the integers*, what would do nicely would be a set (or predicate or function) 'quantifier' in terms of which to express that condition.[32] She could do it either in a second-order theory (using a predicate or function quantifier), or in some first-order *set* theory, with a set quantifier (or in some hybrid, of course, but these need not detain us here). In each of these cases she will (apparently) be able to use the expressive resources of her language to frame a straightforward definition of 'uncountable'—i.e. to introduce the predicate $U(x)$—and to 'prove', using the customary axiomatic and definitional resources of her theory, of some entity π, picked out by a singular term t of her language, that it is 'uncountable', as that is defined, by giving a formal proof of $U(t)$. No problem.

[32] The scare quotes around 'quantifier' are there to remind us that it is a term of art that applies most clearly and unequivocally to formal objects, i.e. to expressions of a formal language, less obviously either to linguistic expressions of her language or, perhaps even, to concepts of her language. Indeed, if it is not already plain, it will be the burden of this cautionary note to point out that to speak of the resources available to Georgina in terms of 'quantifiers', 'predicates', 'singular terms', and the like, with the properties these would standardly have in some first- or second-order formalization of (her?) language, is already to build the elements of its sceptical answer into the statement of the problem.

Of course, making this plain is by itself no argument against the sceptic's answer. But, it should put us on our guard and in a better position to judge how compelled we must be by his argument, particularly if, as may well turn out, he happened to have smuggled his conclusion into the statement of the problem. In that case, the price we pay for rejecting the sceptical conclusion could be as high as eschewing the use of these devices in representing the logical form of ordinary mathematical reasoning—a stiff price indeed. But it could also lead us to a better grasp of the limits of the applicability of the method as a tool for the analysis of our global practice, even when it is useful for the partial representation of more local pieces of the practice. That remains to be seen; but the possibility must be left open at the start. Surely, our repeated encounters with illegitimate totalities, forbidden predicates, and the like, should have pointed us in that direction and made us sensitive to that possibility: there are good reasons independently of the Holy War Against Scepticism for regarding some of these moves with suspicion.

At this point, if the conversation is to continue, a new constraint must be introduced: Those of Georgina's answers that are framed in a second-order language must somehow be disallowed, or else the game ends with a speechless Skolemite. So she humours the Skolemite and sets aside her second-order answers. (For the sake of brevity, we have omitted from our story a discussion that actually took place and that the sophisticated reader can fill in, wherein the Skolemite claimed that he, and no doubt Georgina as well, *cannot understand* the second-order language, except perhaps by interpreting it in a two-sorted first-order theory, and hence that the 'limitation' to first-order languages is not a *real* limitation, i.e. imposes no genuine constraint. The even-more-sophisticated reader can add an epicycle in which the 'enlightened' Skolemite allows the second-order answers but insists on treating its logic with only Henkin models. A reasonably complete treatment would treat these cases in some detail, but bypassing them here will considerably simplify the exposition and should not alter the moral.)

What had Georgina accomplished? To accommodate the Skolemite, whose arguments cannot even be stated without reference to some formalization, she was asked to 'formalize' her proof—i.e. to present $U(t)$ as a 'formal theorem' of some formalized theory—as the combinatorially derived string of a language whose syntax is explicitly exhibited, whose rules of combinatorial derivation are those of *F*irst-*O*rder *L*ogic, within a theory whose axiomatic resources are also explicitly exhibited. She complied (noting in passing and for the record that there were many ways of complying), but noting also that if her account was also to be taken as a representation of the *meaning* of $U(t)$ as pre-formally employed, she would need in presenting her representation to draw upon resources not obviously among the ones that were supplied in her *FOLK ART* (*F*irst-*O*rder *L*anguage *K*it *A*nd *R*epresentational *T*utorial, where Georgina was first introduced to the basic apparatus of a first-order language and taught how to deploy it)—simply because basic FOLK is insufficient, by itself, to represent important features of the simplest of sentences, as in the following example, borrowed from Dorothy Parker's engaging vignette, 'Horsie',[33]

(p) 'The cat is on the mat.'

Even the most fine-grained and revealing representation of that sentence which uses only FOLK is not up to representing it as being about cats and mats, as opposed to, say, about some prime number between 13 and 19. Appending this explanation, Georgina completed her assignment and turned it in.

But she hadn't bargained on having a Skolemite for a 'TA'. In explaining her grade of 'F', he points out that, despite her apparent success, because she worked in first-order ZFC, she had failed to 'express' any concept of

[33] From 'Horsie' in *Dorothy Parker Stories*, Wings Books, New York, 1992, 199.

uncountability—through her defined predicate $U(x)$, or in any other way: If consistent, her 'theory'(i.e. the set of strings in formalized ZFC that is doing proxy for her theory)[34] admits of *countable* models. What is more, it admits of countable models in which the domain consists of sets and the fundamental relation of the model, $\in$, is a sub-relation of any 'standard' or 'full' model which she might suppose herself to have taken as her starting-point. He grants her (but only for the sake of argument; he believes no such thing) that there is some intuitive model that she may 'have in mind' and, with a few simple conjuring tricks, points out that π, *the very set*[35] that she may have taken as her paradigmatic uncountable set (the referent of t) can be enumerated by some function that *he* can define, but that, coincidentally, happens not to be within reach of her quantifiers (i.e. in the domain of the model of her theory that he produces and imputes to her). Since no such function is in the range he has supplied for her quantifiers, her sentence—that no 1-1 function enumerates t—is, of course, true in his model as well.

Not only can he count π, he can explain why she could not (I should caution the reader here that the rhetorical device of casting this in the form of a dialogue obscures the fact that there is some question whether she is equipped to understand his explanation, given the interpretation of her theory that he attributes to her): Given that $U(t)$ was a theorem, and that t designates π, it follows that π was uncountable *from the perspective of her theory*, since $U(t)$ was a theorem, π had to satisfy the definition of $U(x)$. It did so, but not because (as she thought) it was 'absolutely' non-denumerable—whatever that might mean, but because the Skolemite, in (re-)constructing his model of her theory, cleverly kept the enumerating functions safely out of the reach of her quantifiers. The truth is that π is 'really' countable (the Skolemite's enumeration shows that); her contention that by her predicate $U(x)$ she had expressed an 'absolute' notion of *non*-denumerability in her theory, that π, the referent of her singular term t satisfied, was entirely illusory. What appeared to certify the uncountability of π, namely the provable absence of *any* function that would enumerate π—in the form of a proof that every function that enumerated a subset of π enumerated a *proper* subset of it—did so only because, *using only the resources allotted to her by the Skolemite* (in his model), Georgina had only a very limited pot from which to pick functions to supplement her FOLK for the purposes of representing her argument. As it happens, it

[34] Put aside for now any thought that, because of 'indeterminacies' in the concepts of *finite string, formalized*, and the like, the singular term within the parentheses to which this note is appended may have no determinate referent. Entering into such considerations would launch us into yet another epicycle, possibly an infinite loop.

[35] Perhaps not *the very set*; at best *his* model's value for the very same term t. But *contra* the Skolemite, that may be too easy; there may be something to contend about here: is there a perspective that is neutral between the Skolemite's and Georgina's and from which it can be determined whether they are talking about the very same set?

was a pot from which all the enumerating functions had cleverly been excluded.

Naturally, she complains that in constructing and formalizing the theory she had been careful to specify that the domain was to contain all sets < some given rank (she forgets which), but at the very least, all the coverings of the integers with elements of {0,1}, to use Cantor's suggestive terminology. Not finding them there (where 'there' is in the model the Skolemite attributes to her), but—to add insult to injury—hearing that some of them found their way into the Skolemite's domain instead, Georgina cries 'Foul!' and demands a tuition refund, or at least a better grade, but to no avail. Her lament—that he is able to enumerate what *he* assigns as a referent for *t*, whereas she is not, but only because the model he imputes to her has impoverished quantifiers (it left out a whole bunch of subsets and functions)—is met with a blank stare. He merely reiterates the fact that, given *any* consistent theory that purports to sport an uncountable domain, he can find[36] a model and a function that enumerates it, and if the model is a transitive set model, that also enumerates *every set in it.*

IV.A.1.c. Abject surrender

Conclusion: given these ground rules, try as she may to endow an open sentence of her theory with the meaning 'uncountable', Georgina is bound to fail—in the sense that the Skolemite's rules, at least as he views them, will always permit interpretations that place only demonstrably finite and countable sets in its extension. Despondent, she takes her 'F', conceding that she is unable effectively to express this critical concept and, worse yet, that there is probably really none such to express. If it doesn't fit this mould, it doesn't exist.[37]

As we saw, the 'core' version of the argument had little appeal and Georgina easily shrugged it off, but as the Skolemite gave away more and more to his opponent—call such concessions 'the sucker play'—the argument became more compelling: The more the Skolemite conceded 'for the sake of argument', the more it looked as if the target 'concept', the one that was meant to be captured, *was being* captured; and the closer one came, the more it looked as if what *was* being captured was *all there is to capture* and the more difficult it became to see where any 'more' was going to come from. So when it became clear that, given these ground rules, one would always 'come up short', it was tempting to close the gap by declaring it to

[36] Given the resources needed to prove the version of the Löwenheim–Skolem theorem he needed to turn this trick, 'can find' should be swallowed only after liberal salting.

[37] Were she a tad more contentious, she could probably salvage her grade by insisting that the teacher had set a problem that, on his own view, had no solution, and that, therefore, no student should receive an 'F' for failing to solve.

be imaginary and conceding that there was nothing left to get—and that the Skolemite had carried the day.

But something is surely amiss here. Starting from a perfectly good, established, mathematical theory—Cantorian set theory, or classical analysis (the theory of functions), with established results: 'the paradise which Cantor has created for us', to use Hilbert's glorious phrase, we

(1) invent (devise) structures—formalized languages, for the purpose of representing our logico-mathematical reasoning—what is common to all disciplines and free of extra-logical 'content'.

To that end, we

(2) take these shells—uninterpreted calculi—and confer meaning on some of their vocabulary, for example by stipulating of some that they are the 'logical particles' and obey certain laws. This gives the skeleton its skeletal, preferred, 'interpretation'.

    Since it is unlikely that the laws governing any English, or German, or French expressions correspond exactly with those stipulated to govern the logical particles of first-order logic—that one is therefore not stipulating that the formal symbols be synonymous with the 'ordinary', the details of exactly how this takes place are not pellucid. But we do somehow specify in the 'metalanguage' what the sentences of the constructed language are to mean, at least to some extent—by articulating how the semantic values of sentences compounded with these logical particles depend on the semantic values of simpler parts, i.e. by giving its model theory.

    If even this much can't be made reasonably clear, so much the worse for Skolemite and other 'metalinguistic' incompleteness arguments.

(3) When, as in the present case, we are representing a genuine theory in this formalized idiom, we must say more: we must specify what will be encompassed in the domain of quantification and what subsets of the Cartesian product of that domain are assigned to the various items of 'non-logical' vocabulary, again by making some kind of specification in a language whose expressions already have meaning.

    How Georgina was to do that in the case of set theory is either entirely trivial but unilluminating about the structure of the resulting theory ('∈ designates the membership relation over the domain of sets.' Or '. . . over the cumulative hierarchy of ranks < ?'), or else it is hopelessly complicated and we don't really have a clear idea of how it is to be done in a revealing way.

But however simple or complicated, when only step (2) of the above procedure had been completed, all that had been set in place was the *shell* of the structure at which we were aiming. What could possibly justify circling back and declaring that merely completing step (2) of the procedure *already* constrained the interpretation of the language, including that of the *non-logical* vocabulary, in ways that rendered ineffective any further specifications of meaning (for '$\in$' or '$\forall$', for example)? Or, worse yet, that it uncovered previously unsuspected bounds on the meaningfulness of the expressions *of the semantical metalanguage* in which these further specifications would naturally be made? Yet that is exactly the tenor of the Skolemite's argument. That would be to take what meaning one of our ordinary expressions may have to be limited to whatever may be completely specifiable by the first-order *model-theoretical* structure of the object language in which we were to represent it. Where is the argument that representability by that structure is a proper surrogate for meaning? There plainly is none in the Skolemite argument laid out above.

Absent an independent argument that trimming models by applying the Löwenheim–Skolem argument loses nothing essential to meaning (a Princess Margaret premiss that would suit the purpose), the employment of the Skolemite argument to reach conclusions about what we had meant in the 'informal' theory and 'represented' in the formalized version is grossly illegitimate, particularly if the consequences are counter-intuitive to the point of threatening to undermine central (and otherwise unquestioned and unproblematic) accepted portions of the practice we were meant to represent and thereby to illuminate. In this case, the damage is done by attributing a severely truncated scope to Georgina's universal quantifier—for no independent and principled reason. It is to Putnam's credit, as well as to that of Crispin Wright and Hartry Field,[38] all of whom have recently helped themselves to this 'model-theoretic' argument, that they see the problem and see it *as* a problem: the Putnam of 'Models and Reality' is no ordinary Skolemite, as much of that paper is occupied with the details of how what he considers to be the crucial 'epistemic' facts are preserved under all of the various options considered (to no ultimate avail, of course). And the same is true of Field and Wright. Each advances some Skolem-

[38] 'Indeterminacy in Logical and Mathematical Concepts', unpublished MS draft, dated 6/9/93. The final version is included in this volume. As with Field's other writings on this and related subjects, I find this rich in subtle and persuasive arguments that deserve full and detailed treatment, rather than the light brush that is all that time and space will permit in the present context. I would like to emphasize that the thrust of the present paper is to disagree far more with the tone and outlook of Field's view than with the detail of his arguments, which, given a particular set towards the philosophical significance of metamathematical work, I find both intriguing and illuminating. What we face here is the classic *modus ponens–modus tollens* problem, in the following form: A style of analysis of the content of our logico-mathematical discourse leads to the conclusion that the discourse is far more indeterminate than we might antecedently have expected. On this everyone agrees. Some (Field, Putnam, Wright) accept the indeterminacy; what I am suggesting is that one should prefer to wonder wherein the analysis is deficient that it should lead to such a bizarre conclusion. I address Wright's remarks (cf. n. 25) in some detail below.

independent reasons for thinking that the indeterminacy concluded through the L–S argument is an appropriate take on the way things are. So if their overall argument is to be successful, it has to be on the following basis: How well do their accounts capture what is essential to the faithful representation of mathematical practice? That's the 'Princess Margaret' portion of the task.

IV.A.2. Skolemite arguments: The Wright approach

Crispin Wright joined the issue in a probing response to 'Skolem and the Skeptic' that I mentioned above (cf. n. 25). His remarks illustrate the point I would like to urge in a particularly illuminating way, so I will take them as a point of focus to explore the matter a bit further.

As we saw above with Georgina, the Cantorian laments that the Skolemite (he is 'the Skeptic' in that paper) simply leaves something out each time he applies some version of the Löwenheim–Skolem theorem (henceforth 'LST') in a sceptical argument. I contend that his lament is justified because part-and-parcel of representing the 'full-blooded' concept of set in some formalized language involves supplying the quantifiers of the formal theory[39] with 'all' the sets, or at least with all those whose existence is implied or presupposed by the *in*formal theory.[40] For Georgina, that means a supply sufficiently ample to make the interpreted theory into as good a duplicate of the informal (but perfectly legitimate) theory of sets that we are trying to represent as we can make it.[41] In the argument, the Skeptic (Wright, who defends him) 'gives away' a lot of structure: He grants the classical theorist that she can imagine transitive sub-models of the standard model. It's what I have called 'the sucker play'—the closer to the real thing you grant your opponent, the more likely she is to think she is getting the genuine article; in Putnam, this takes the form of building in all of the theoretical and observational constraints we may wish to add—provided they are expressed in the pure first-order language of the theory. Then,

[39] Of course this is a place where we should be saying 'language' instead of 'theory' since we are talking of quantifiers and interpretations and the like, and since in our jargon it is a language that receives an interpretation.

[40] This is a misleadingly 'neutral' way of stating the issue, since one might easily frame the dispute as one regarding just what *is* implied by the informal theory.

[41] Note that nothing is being said about the *dynamics* of the practice, although that would be clearly relevant to modelling the practice. Our focus here is necessarily static and synchronic: we aim at a snapshot of the theory as it is accepted/practised today. But the matter of dynamics is an interesting one, particularly for set theory, for two reasons:

(1) it is in the nature of things for any representation of the type we are considering (any snapshot), if consistent, to cry out for extension, and
(2) it should also play an important role in any successful attempt to meet Wright's very legitimate challenge that a successful account flesh out the features of our 'understanding' of the concept of set—and of mathematical concepts generally—that may *not* have been captured in a 'Skolemite' model or in a first-order formalization of the theory.

when you compare the stripped-down model you are peddling to the fancy one she had thought all along she was getting, you will be better situated to persuade her that the stripped-down model you offer is the best that money can buy; any 'feature' of any model offered by any other dealer that you can't match is one that dealer can't deliver either. Here is how Wright goes about it:

> SMT [the Sub-Model Theorem] entails that a commitment to these axioms [the axioms of ZFC], plus possession of the intended grasp of '∈'—the sole non-logical primitive—need involve no commitment to the uncountable; someone who accepts it need not say or do anything at variance with acceptance of all the standard set-theoretic axioms interpreted *as* set-theoretic axioms. . . . just as—according to Benacerraf's argument—a full and complete explanation of the concept of number is neutral with respect to the identification of the integers with any particular objects, so—the Skeptic will urge—a full and complete explanation of the concept of set is neutral with respect to the existence of uncountable sets. But if there really were uncountable sets, their existence would surely have to flow from the concept of set, as intuitively satisfactorily explained. ('Skolem and the Skeptic'—II, 124)

In brief, any model satisfying the Sub-Model theorem is the genuine article; but SMT shows that those too are saddled with the sceptical conclusion, therefore they all are. (Although it is not usually his way to be arch, surely Wright has tongue-in-cheek here; he couldn't be aiming at showing what a bad argument the argument of WNCNB on which he relies really is.) His real arguments are these.

First, he correctly anticipates that the reply would be that something was missing from the above account and challenges me as follows:

> [T]he residue is contained, presumably, in the *informal* explanations to which, Benacerraf reminds us, Zermelo intended his formalization to answer. . . . Yet it is notable that Benacerraf nowhere ventures to supply the missing informal explanation—the story which will *pack enough into the extension of 'all sets'* to yield Cantor's theorem, under its intended interpretation. (ibid. 124–5; emphasis added)

For Wright, the combined effect of the sceptical argument and the Cantorian's inadequate reply shifts the burden of proof to her. If it is to be argued that the interpretation of the quantifiers is inadequate, and 'more . . . *needs to be packed into the interpretation of the quantifiers*, then an appropriate informal explanation is owing *of the residue*' (p. 125; emphasis added).

Wright then notes that we

> have absolutely no right to the ideas that a countable set model . . . has to be 'non-standard', a misrepresentation of our intuitive intentions, *unless* we buy Cantor's reasoning as a piece of informal mathematics, something that defeats *any* candidate for a 1-1 correlation of integers and reals. (p. 133)

Which is, of course, precisely what the Cantorian does buy. Wright, of course, will soon baulk at that purchase, for very curious but instructive reasons. Wright rehearses Cantor's diagonal argument for the uncountability of the reals and then says: 'However, in order to imbue Cantor's observation with its intended significance, we need to make a substantial assumption' (p. 133).

Although he doesn't tell us what that assumption is, I think it is safe to assume that it has much to do with the previous quotation. Instead, he (apparently) changes the subject and points out that if we initially restricted our attention to a computable array of *effectively computable* decimals, we could, also by a diagonal argument, conclude instead that there was a computable decimal not in the array. 'Cantor's argument takes on a quite different significance: rather than showing something about uncountability, it shows that there is no recursive enumeration of all recursively enumerable decimals' (p. 134).

Certainly; but how is that relevant? The issue was never whether the diagonal form of argument alone invoked uncountability. But there is a method to his cunning. After pointing out that, *mutatis mutandis* the same holds of subsets of the natural numbers (and *effectively enumerable* subsets of the natural numbers) he continues:

> So, *under the restrictions*, the result is, again, that there is no effective . . . enumeration of all effectively . . . enumerable subsets of natural numbers. And again, we know that the finite and effectively enumerable infinite subsets of natural numbers form a countable set. So the moral is simple: before the Diagonal Argument, or the informal proof of the power set theorem, can lead us to a conception of the intended range of the individual variables in set theory to regard any countable set model as a non-standard truncation, *we need to waive the restrictions. And in order to understand the waiver, we need to grasp the notion of a non-effectively enumerable denumerably infinite subset of natural numbers* (p. 134; emphasis added).

Wright identifies the Princess Margaret premiss for us: to understand the diagonal argument as establishing that the power set of the integers is nondenumerable one must

(a) first run the argument in terms of an effective enumeration of effectively decidable sets of integers; that would show that there is some subset not in the enumeration;
(b) and secondly, eliminate the restriction on the array to an effectively enumerable array of effectively enumerable sets—*by explicitly adjoining the non-effectively enumerable sets* of integers to the array and allowing ourselves to look unrestrictedly for just *any* set of integers.
(c) Success in these twin ventures would yield our desired conclusion.

But to bring that off, he continues, the Cantorian must display a grasp of

[T]he notion of a *non-effectively-enumerable* denumerably infinite subset of natural numbers.

That is what, if [s]he is in the business of giving explanations, the Cantorian needs to explain. (ibid. 134–5, emphasis added)

Understandably (particularly for someone with constructivist leanings) Wright baulks at the thought that she can explain that notion, at least not by any route that he would be inclined to recognize as legitimate, i.e. as given by a *law*, or *rule*, or as the extension of a *property* that is itself capable of 'positive' characterization. We are therefore likely to be left without the needed explanation. This is not an unfamiliar argument: complementation —even complementation relative to the set of subsets of N—is suspect; to be branded as legitimate, every set must receive a 'positive' definition on some 'acceptable' basis.

However, what is surprising, and difficult to justify even for someone with a 'constructivist' bent, is the implicit thought (unless I have misunderstood him, which is more than possible) that the only route to a legitimate conception of 'subset of N' is *as the union of* the notions 'effectively enumerable subset of N' and '*non*-effectively enumerable subset of N', each of which needs to receive an independent, 'positive' construal. It is a lot to ask of Georgina. Moreover, as an account of the historical roots of any notion of subset of integers that we may have, it is surely anachronistic, since its component notions of 'effectively enumerable subset of N' and '*non*-effectively enumerable subset of N' were not yet around when Cantor was doing his thing.[42] That Georgina must proceed in this way if she is to claim an understanding of subset sufficient to substantiate Cantor's theorem would be a Princess Margaret premiss *par excellence*.

Let us review the passages I chose to emphasize. The small point I have been trying to make through their consideration—and it is a small point—is that typically we begin with a mathematical practice—as we did here (a) that was fairly clear (not pellucid—just fairly clear); (b) that did not stand in need of *special* explanation or justification; (c) whose quantifiers were not in need of 'packing' (with an extra complement of sets, just for the trip to platonic heaven?). If we must do any quantifier stocking at all, it has to be those of the formal theory that we stock, since they are the ones that

[42] Since there are uncountably many degrees of unsolvability, is it also true that to understand 'subset of N' we must first understand 'non-effectively enumerable to degree d' and its complement, since the power set of N is the union of the two? The higher the degree of unsolvability, the better the argument that we don't have the *general* notion that is the union of the two?

It is imaginable that Wright means just that—since it is precisely such sets that would trouble the 'constructivist' more than any others, and the more arcane, the more troublesome. But this at least places the view where it can be discussed and defended—as a view that is troubled about the possible existence of sets of integers membership in which cannot be 'assessed' by any epistemically sanitary means. It is a large and respectable philosophical tradition, but one that gets no help from Skolemite arguments, since it supplies the Princess Margaret premiss that is needed to make them go.

started out empty. If we short-change them in the process, that's our fault, not Cantor's.

The Diagonal Argument, applied to an unconstrained array of infinite decimals, proves that no countable array contains them all—*precisely because it was unconstrained*:

Constrain it as in Wright's example and you get a different theorem, not surprisingly related to the nature of the constraint. To get an unconstrained version from the constrained version, you need simply return to the *status quo ante*, which you do simply by omitting the constraint; you don't need to accept a *further* constraint, namely to demonstrate effective control of some non-effective notion.

If this is right, the Skolemite has no case until he successfully constrains the class of models that may be considered (by supplying a suitable PMP and arguing for it) which he would need to do by some argument independent of LST.

Burden-of-proof arguments are unsatisfying and this is unquestionably a burden-of-proof argument (albeit in self-defence, against the challenge to the Cantorian to explain herself 'If not this (pointing to the doctored up diagonal argument), then what?' which is the ultimate burden-of-proof argument). What is the Skolemite's demand for an explanation, e.g. of the nature of the 'residue' (what residue?), if not a burden-of-proof argument? In each case—those of the old-time Skolemites, as well as the new and ingenious crop that Wright has been launching—the constraint that leads to the restricted result came from 'outside', i.e. from some a priori view of what we could conceivably 'understand'.

Wright's more positive views, as expressed in his speculative section, strike me as compatible with the push that I am giving him—to find a plausible independent argument to support the PMPs he deploys. After rightly taking me to task for making the Cantorian's task too easy by attributing an overly restrictive motivation to the Skeptic, he seems to strike the right balance when he looks for ways of understanding what understanding we may have had to begin with. To look elsewhere for the roots of mathematical truth is to look where they could not conceivably be, since all of these formal modelling techniques are ways of representing what had already to be there to be represented. When they seem to show, by *reductio*, that something that we formerly thought to make perfect sense does not, we need to be sure that we have produced a *genuine reductio*, and not an *ersatz*, made-up one, as the Skolem 'paradox' surely is. The best comparison I can suggest is with trying to persuade someone who understands ordinary counting that she really has no concept of 17 and 18 as distinct and intelligible cardinalities because both get translated as 'more' into a language that has only '1', '2', '3', . . . '13', 'more' (and no more) as the available choices.

IV.B. Gödel's theorems and concluding remarks

I will close with a word about Gödel's theorems. Although defections are coming in fast (Putnam, Field) there is still, or at least used to be, pretty general agreement that one's knee-jerk response to the first incompleteness theorem is also the correct response. The theorem heralds what its name suggests: an incompleteness in *formalized* arithmetic. However hard one may squeeze, one cannot extract from it the thought that, although we once believed we had a concept of arithmetic truth, now that we see that the sentences true in arithmetic cannot be exactly those corralled by any plausible formal 'proof' procedure (if bivalence is also to be preserved), we must concede that we did not. The incompleteness *that Gödel demonstrated*, the incompleteness of his first incompleteness theorem, was shown to exist in the calculus, not in our conception. Not that one couldn't be lurking in our conception as well—of course one could—but that just hasn't been shown; as with the Skolemite, to go that extra mile requires a Princess Margaret premiss and a separate argument to support it.

The same holds for libertarian arguments to the effect that our abilities transcend those of any machine, outstrip in proof-power any formal system, i.e. do not constitute or cannot be adequately represented as a recursively enumerable set of sentences etc. Some have been offered by John Lucas; perhaps by Gödel himself—that is unclear; most recently, Roger Penrose[43] has attracted a fair bit of attention by renewing 'anti-mechanistic' arguments based on the Gödel theorems. In his review of Penrose's latest work, Putnam takes him to task for what he (Putnam) calls 'a straightforward case of a mathematical fallacy' that, he says, Penrose mistakenly believes to be merely a 'philosophical disagreement with the logical community' over what we may conclude from Gödel's theorems. Without trying to adjudicate this particular dispute, it should be easy to see how such a disagreement could arise: If Penrose claimed that his 'mentalistic' conclusions *followed from* Gödel's theorems, he would indeed be guilty of a mathematical fallacy (or a logical one, at least). But if he were willing to acknowledge the PMP to which he was appealing, and with the aid of which he was (validly) deriving his philosophical conclusion, however unpalatable, he would be guilty of no such thing—bad philosophy maybe, which is really worse—but not a mathematical blunder. Let M be some metamathematical result in which we may be interested, C the philosophical conclusion we 'derive' from it, and P the PMP to which we must invariably have recourse: If we did our work well, M, P $\vdash$ C, i.e. M and P together entail C. But if so, then it is also true that the metamathematical premiss *alone* has a 'purely

[43] In *The Emperor's New Mind: Concerning Computers, Minds, and the Laws of Physics*, Oxford University Press, 1989, as well as in his more recent *A Search for the Missing Science of Consciousness*, reviewed by Hilary Putnam in *New York Times Book Review*, 20 Nov. 1994, p. 7.

philosophical' consequence, albeit a weaker one, viz. (P → C), since in that case, (P → C) would be entailed by M alone. Of course, authors who brandish metamathematics as the authority for their philosophical conclusions are not always 'up front' with the particular PMPs which, if pressed, they would have to supply in order to achieve logical purity; some may even doggedly insist that M ⊢ C, *tout court*. So, how we should represent the error of omitting mention of P in our assessment of the implicational situation can vary considerably and, sometimes, be very much a matter of hermeneutic nuance. I did not bring up Penrose (and his predecessor John Lucas) in order to re-examine their arguments, but merely to remind us of the general point they illustrate, namely that you need a Princess Margaret premiss to obtain either the libertarian or the conceptual limitation conclusion from Gödel's theorem.

This is a genuine fork in the hermeneutic road. There is no knock-down argument there—only a choice and a suggestion regarding which is the right way to go, at least until we are given compelling reasons to do otherwise. But note the parallel with the Skolemite's reaction when confronted with the possibility of producing countable models in perpetuity: he insists that the incompleteness is in our conception, that, contrary to what we may think, we have no conception of the (absolutely) uncountable. But to be entitled to that conclusion, he must supplement his argument with another that suggests that we do not have a sufficiently rich conception of 'all sets', or 'all functions from the power set of ω onto ω', or, perhaps and most radically of all, of 'all'. In the case of arithmetic truth, the premiss to be supplied is that a notion of arithmetic meaning *not* grounded in formal proof is too impoverished to deliver a conception of truth worthy of the name, at least not for arithmetic.

It should therefore not surprise us to find Putnam, at least the most recent Putnam that I have found on this issue, locating the (Gödel) incompleteness *in our conception*. The following passage, taken from 'Mathematical Necessity Reconsidered',[44] expresses that very view. He has been advancing a heavily 'epistemic' notion of truth, one grounded in a 'thick' notion of 'understanding' in which, from a page chosen at random, 'logical truths do not have negations that we (presently) understand' (p. 256, but 17 in the original typescript). The case with mathematical truth is similar—established mathematical theorems *have* no negations that we really understand. But now he considers an objection:

> It might be argued, however, that there must be more to the truth of the theorems of mathematics than my story allows, for the following reason: even if the theorems of mathematics are consequences of principles whose negations could not, as far as

[44] 'Rethinking Mathematical Necessity', in *Words and Life*, James Conant (ed.), Harvard University Press, 1994.

we now know, intelligibly be true—principles such that the idea of *finding out that they are false* has not been given a sense—still there are statements of mathematics whose truth value no human being may ever be able to divine, even if more axioms become accepted by us on grounds of 'intuitive evidence' or whatever. Certainly there are (by Gödel's theorem) sentences whose truth value cannot be decided on the basis of the axioms we presently accept. Yet, given that we accept the principle of bivalence, many of these sentences are true (as many as are false, in fact). And nothing *epistemic* can explain the truth value of such undecidable statements, precisely because they *are* undecidable. To this objection I can only answer that I am no longer able to attach metaphysical weight to the principle of bivalence; but a discussion of *that* issue would take another paper at least as long as this one. (p. 259, his emphasis)

It is generally accepted that 'epistemic' considerations tend to point towards more restrictive interpretations of what we are capable of 'understanding' (that is surely what Putnam thinks here). I have been arguing that in both the Skolemite case and in the philosophical interpretation of Gödel's theorems, they point in the opposite direction. Of course, that may simply be a confusion on my part, since I take 'epistemic' considerations to include ones of meaning, but without accepting as axiomatic that these are ineluctably entwined with a semantics of 'assertability conditions', as opposed to a 'truth-conditional' semantics, or even some other, as yet undreamt-of, kind.

2

Is Platonism Epistemologically Bankrupt?

BOB HALE

Platonism about mathematics, as I shall understand it here, is an ontological thesis. The Platonist holds that we may take the surface syntax of mathematical statements—with its apparent involvement of singular reference to and quantification over numbers of various sorts, sets, and other mathematical entities—as a reliable guide to their logical form, and hence to their truth-conditions. Given that many such statements are true—as the Platonist, in company with most other philosophers of mathematics and working mathematicians, also believes—it follows that there exist such objects as numbers and sets. Since it appears to make no sense to ask where numbers or sets are located, or when they came into existence and how long they will last, the Platonist concludes that they are abstract objects, lying outside space and time. It does not strictly follow that no event or state of affairs whose proper description involves reference to numbers can be capable of standing in causal relations with other events or states of affairs. But it is at least very plausible to think that the states of affairs to which, on the Platonist's view, true *purely* mathematical statements owe their truth must be incapable of so standing. This apparent causal isolation of mathematical facts, as the Platonist conceives them, lies at the heart of various objections to Platonism. In this paper, I shall be concerned with one of them. For near enough two decades, one principal source of the conviction that Platonism is untenable has been epistemological in character. Broadly, the thought has been that the Platonist conception of mathematics effectively precludes any credible, non-mystifying, account of our knowledge of mathematical truths. Most commonly—and originally—this charge has been developed against the background assumption of a broadly causal conception of knowledge. It was on precisely this conflict—between Platonist truth-conditions for mathematical statements on the one hand, and a causalist conception of knowledge on the other—that Paul

Thanks are due, as usual, to Crispin Wright for his constructive criticism and encouragement; I have benefited, too, from discussion of an early version of this material at a seminar held in the University of Michigan, and of a more recent version presented at the conference *Philosophy of Mathematics Today*, held in Munich, July 1993. I am also very grateful to anonymous referees for the *Philosophical Review* for perceptive criticism and constructive suggestions.

Benacerraf focused in 'Mathematical Truth'.[1] And much of the ensuing discussion[2] has followed suit, i.e. has seen the issue as essentially about whether some causal constraint upon knowledge that is sufficiently exacting to rule out knowledge of mathematical propositions, Platonistically construed, can be both well motivated and defensible.

The thesis of this paper is a strictly limited one: baldly stated, it is that there is no *special* epistemological problem confronting a Platonist construal of pure mathematical statements. I do not claim that there is no problem at all, because I think we do not yet have an adequate general account of how we can have knowledge of, or justifiable beliefs in, necessary truths. I do claim that, given a satisfactory epistemology of necessary truth in general, there is no additional problem for Platonism. It is an assumption of this paper that there are necessary truths and that such truths can be known. Whilst I shall try to respond to the most important reason that may be given for denying that, even if there are necessary truths, mathematical truths cannot be among them, I shall offer no defence here of the general assumption. So none of the arguments that follow should move the outright sceptic about necessary truth; and by the same token, none of them will be at the service of those philosophers—Quine being the most notable among them—who combine endorsement of Platonism with scepticism about necessity.

When the epistemological challenge is pressed along the lines sketched above—as presupposing a specifically causal *analysis* of knowledge—there is a fairly obvious line of response apparently open to the Platonist. For it is reasonably clear that not just any causal constraint upon knowledge, or justified belief, is bound to be in tension with the presumed acausality of abstract objects or states of affairs involving them. Thus it is, we may suppose, a plausible requirement for a belief to rank as knowledge, or as properly justified, that some appropriate causal connection should obtain between the belief and the subject's reasons, or—in the case of perceptual beliefs—between the belief and relevant features of the situation in which the subject acquires it. This kind of causal condition might be needed to exclude cases where, although able to supply perfectly adequate reasons to believe that *p*, or suitably placed to observe that it is so, I would still believe that *p* even if I did not have those reasons, or was not so placed—because I have been brainwashed, say, or have unwittingly ingested hallucinogenic substances. There is no clear ground for supposing this kind of causal condition to be unsatisfiable in the case where *p* deals—at least as construed by the Platonist—in matters abstract. Constructing or following proofs and

[1] *Journal of Philosophy*, 70 (1973), 661–79.

[2] e.g. Mark Steiner, *Mathematical Knowledge*, Cornell University Press, Ithaca, 1975; Crispin Wright, *Frege's Conception of Numbers as Objects*, Aberdeen University Press, 1983, esp. Sect. xi; Bob Hale, *Abstract Objects*, Blackwell, Oxford, 1987, esp. ch. 4.

doing calculations are dateable and locatable processes which can perfectly well have effects—such as inducing headaches, fatigue or excitement—and there is no reason why these should not include mathematical beliefs. On the face of it, the kind of causal constraint on knowledge[3] that *p* which *would* give trouble here would be one requiring an appropriate causal linkage between the subject's belief that *p* and the truth-conferring state of affairs that *p* itself. But it is far from clear that any such strong constraint can be enforced without jeopardizing claims to knowledge, or reasonable belief, of kinds we should normally regard as defensible.[4]

1. FIELD'S GENERALIZATION OF BENACERRAF'S PROBLEM

In view of the possibility—and, in my view, considerable plausibility—of the kind of response just sketched, it is important to enquire whether an effective epistemological challenge to Platonism can be mounted without reliance upon any proposed causal constraint on knowledge that may itself be vulnerable to objection as unduly restrictive, independently of the present issue. In the title essay of his collection *Realism, Mathematics and Modality*,[5] Hartry Field seeks to motivate mathematical anti-realism by developing just such a challenge. Acknowledging that the objection he is presenting is a reformulation of the problem posed by Benacerraf, he remarks:

> Benacerraf formulated the problem in such a way that it depended on a causal theory of knowledge. The present formulation does not depend on *any* theory of knowledge in the sense in which the causal theory is a theory of knowledge: that is, it does not depend on any assumption about necessary and sufficient conditions for knowledge. Instead, it depends on the idea that we should view with suspicion any

[3] It is a nice question how a causal constraint of this more exacting type should be formulated, in the case of justified belief. Assuming that justified beliefs need not be true, it cannot be required that a justified believer's belief that *p* is suitably causally related to the fact that *p*, since there may be no such fact. A causal theory of justification might require that the causal process involved be reliable. It is a further question—too large for adequate discussion in a note—whether this is best understood as giving rise to a strong causal theory in my sense; if so, then the resulting theory would—or so I should argue—be open to the objection to strong causal theories of knowledge suggested in the text. If not, the theory could still underpin a version of the epistemological objection to be discussed in the remainder of this paper. It is also open to question whether even this stronger type of constraint is strong enough, by itself, to give trouble—further argument would seem to be needed to establish that if a statement involves reference to abstract objects, the state of affairs (if any) which renders it true is thereby debarred from figuring in any sort of causal explanation. For a clear appreciation of this point, see Wright, op. cit. 85 ff.

[4] For detailed presentations of this line of thought, see the works cited in n. 2.

[5] Hartry Field, *Realism, Mathematics and Modality*, Blackwell, Oxford, 1989, see essay 7, Sect. 2.

claim to know facts about a certain domain if we believe it impossible to explain the reliability of our beliefs about that domain.[6]

In brief, Field's generalized formulation of Benacerraf's challenge has it that on a realist[7] view of mathematics, our mathematical beliefs should be pretty reliable, but that, in view of the realist's conception of their subject-matter, the high degree of reliability they enjoy must be quite inexplicable.

In more detail, the problem, according to Field, can be put—without invoking the notion of knowledge or causal conditions supposedly constraining it—as follows:

The mathematical realist believes that his or her own states of mathematical belief, and those of most members of the mathematical community, are to a large extent disquotationally true. This means that those belief states are highly correlated with the mathematical facts: more precisely (and put without talk of truth or facts), [the realist must accept] that for most mathematical sentences that you substitute for '*p*', the following holds:

(1) If mathematicians accept '*p*', then *p*

. . . the fact that [this holds] for the most part is surely a fact that requires explanation: we need an explanation of how it comes about that mathematicians' belief states and utterances so well reflect the mathematical facts. But there seems prima facie to be a difficulty in principle in explaining the regularity. The problem arises partly from the fact that mathematical entities, as the Platonist conceives them, do not causally interact with mathematicians, or indeed with anything else. This means that we can't explain the mathematicians' beliefs . . . on the basis of the mathematical facts being causally involved in the production of those beliefs . . . or on the basis of some common cause producing both.[8]

Field adverts briefly to the idea of a non-causal explanation, but complains that:

[6] Field, *Realism, Mathematics and Modality*, 232–3. Penelope Maddy, *Realism in Mathematics* (Clarendon Press, Oxford, 1990) likewise stresses the need to meet this more general form of concern, as does W. D. Hart, in his review of Steiner's *Mathematical Knowledge, Journal of Philosophy*, 74 (1977), 118–29.

[7] Here I use the term 'realist' as Field does, to refer to any position which takes mathematical statements to be (at least) disquotationally true, when taken at face value. So used, it is pretty well interchangeable with 'Platonist', in my usage. Realism in this sense is to be distinguished from—or at least does not obviously coincide with—realism in the sense made current by Michael Dummett, in which a realist about a given class of statements holds that statements belonging to that class may be true, or false, in a way that essentially transcends our capacity, even in principle, to recognize them as such. The question of the relations between Platonism and realism, in this sense, about mathematics is to a large extent unresolved, and cannot be discussed here.

[8] Field, *Realism, Mathematics and Modality*, 230–1. There is, as John Burgess observed in discussion, some question how far Field succeeds in posing the problem without reliance on talk of truth or of facts; but I shall refrain from pressing this worry here.

[It] is very hard to see what this supposed non-causal explanation could be. Recall that on the usual Platonist picture, mathematical objects are supposed to be mind- and language- independent; they are supposed to bear no spatio-temporal relations to anything, etc. The problem is that the claims that the Platonist makes about mathematical objects appear to rule out any sensible strategy for explaining the systematic correlation in question.[9]

The trouble for the Platonist arises—so the thought here appears to be—not just from the acausality of mathematical objects, but from their apparently total disconnection from us (their lack of spatio-temporal relations, etc.) which, Field thinks, obstructs any sort of explanation of the correlation, whether causal or non-causal in character.[10]

2. BELIEFS ABOUT CONSISTENCY

Field emphasizes that he is not claiming that the Platonist can do *nothing* towards explaining the general regularity he schematically formulates as:

(1) If mathematicians accept '*p*' then *p*.

For 'as mathematics has become more and more deductively systematized, the truth of mathematics has become reduced to the truth of a smaller and smaller set of basic axioms; so'—Field allows—'we could explain the fact that the mathematicians' beliefs tend to be true by the fact that they have been logically deduced from axioms, if we could just explain the fact that what mathematicians take as axioms tend to be true'.[11] That is, the explanatory problem reduces to that of explaining how it comes about that, for most mathematical sentences we can substitute for '*p*':

[9] Field, *Realism, Mathematics and Modality*, 231.

[10] In 'Realism and anti-realism about mathematics' (*Realism, Mathematics and Modality*, 53–78; see esp. 69), Field emphasizes a parallel problem over reference, stressing that what makes the problem an apparently insoluble one for the Platonist is 'not merely that [mathematical entities] exert no causal influence on us, but also the fact that they don't stand in other physical relations to us which could serve to explain the referential link'. It is not clear just what sort of relaxation of a narrowly causal account of reference Field might have in mind here, or what cases he thinks there may be in which we can refer to objects which stand in non-causal, but none the less physical, relations to us. There is clearly a question arising here whether there could be developed a generalized version of the challenge to Platonism based on reference, paralleling Field's generalization of the epistemological challenge. But I have no space to pursue this question here.

[11] Field, *Realism, Mathematics and Modality*, 231. The suggestion that the problem can be reduced to one about a small set of axiomatic beliefs is, as John Burgess has remarked, in some tension with the idea that what is called for is an explanation of a *tendency* to get things right, or of a good correlation between mathematicians' beliefs and the facts. But there surely remains an explanatory question confronting the Platonist, even if Field's formulation of it is somewhat unhappy—if mathematicians' axiomatic beliefs are true, that ought not to be just a fluke.

(2) If most mathematicians accept '*p*' as an axiom, then *p*

holds true.

Indeed, the Platonist can even get a little bit further, according to Field. For he can go *some* way towards explaining the regularity which consists in the fact that most instances of (2) hold. For if most instances of (2) hold, then it follows that most instances of

(3) If mathematicians accept '*p*' as an axiom, then '*p*' is logically consistent with everything else they take as axioms

also hold. And this latter fact, Field believes, does not resist explanation. Indeed, though Field does not explicitly say so, it is pretty clear that it had better *not* resist explanation. For the belief that most instances of (3) hold—the belief, that is, that we are good at picking consistent sets of mathematical axioms—is very nearly equivalent to, and is certainly required by, the belief that we are good at picking out mathematical theories that are *conservative*, in Field's sense.[12] And if there could be no explanation of that, his own position would be vulnerable to the general epistemological challenge which he takes to undermine Platonism. So it is vital to Field's own anti-

[12] In Field's own formulation: A mathematical theory *S* is conservative if, for any nominalistic assertion *A* and any body of such assertions *N*, *A* is not a consequence of *N* + *S* unless *A* is a consequence of *N* alone (cf. Field's 'On conservativeness and incompleteness', *Journal of Philosophy*, 81 (1985), 239–60, repr. in *Realism, Mathematics and Modality*, 125–46; see 125). On a thumbnail, Field's aim is to show how a nominalist may take mathematical statements at face value—and thus endorse the Platonist account of their truth-conditions—but avoid commitment to mathematical entities by denying that such statements are ever (non-vacuously) true. He sees his main task as that of undermining the Quine–Putnam indispensability argument—that we should accept (at least a large body of) mathematical statements as true because they play an indispensable role in successful scientific theories, especially physical theories. This task is to be discharged by arguing that mathematical theories need not be true to be good, as Field succinctly puts it. The programme has two parts, in both of which the idea that standard mathematics is conservative plays a key part: Field argues first, that conservativeness is enough to justify the use of mathematics as a means of deduction within, for example, physical theories; and second, that Platonistically formulated physical theories (i.e. theories whose formulation involves reference to numbers, functions, etc.) can be replaced by nominalized versions—here the appeal to conservativeness is required to show the adequacy of the nominalistic replacement theories. For a clear explanation of Field's programme, see his *Science without Numbers*, Blackwell, Oxford, 1980, Introduction and ch. 1, and his 'Realism and anti-realism about mathematics', *Philosophical Topics*, 13 (1982), 45–69, repr., with a new postscript, in his *Realism, Mathematics and Modality*, 53–78. Although I am fundamentally opposed to this positive programme, I ought to stress that it is no part of my aim in this paper to develop arguments against it—this I have done elsewhere, in *Abstract Objects*, ch. 5, 'Nominalism' in A. D. Irvine (ed.), *Physicalism in Mathematics*, Kluwer, 1990, 121–43, and most recently in a joint paper with Crispin Wright 'Nominalism and the contingency of abstract objects', *Journal of Philosophy*, 89 (1992), 111–35. See also Wright's 'Why numbers can believably be', *Revue Internationale de Philosophie*, 42 (1988), 425–73. Field responds in his *Realism, Mathematics and Modality*, 43–5, and more recently in 'The conceptual contingency of mathematical objects', *Mind*, 102 (1993), 169–84, to which Wright and I reply in 'A reductio ad surdum?': Field on the contingency of mathematical objects', *Mind*, 103 (1994), 169–84.

realist programme that the regularity enshrined in (3) should be acceptably explicable. And Field thinks it is:

For instance, part of the explanation of (3) is surely that the mathematical community has over time managed to weed out the inconsistencies in earlier mathematics—in Newton's beliefs on calculus, in Fourier's on functions, in Cantor's on sets, and so forth. (I think in fact that this is a large part of the explanation of (3); this is connected to the fact that a large part of the reason most of us believe that modern set theory is consistent is the thought that if it weren't consistent someone would probably have discovered an inconsistency in it by now.)[13]

(3), then—in contrast with (1) and (2), in Field's view—does represent a genuine regularity. And if it can indeed be explained, either in Field's or in some other way, this could provide an acceptable starting-point for the required explanation of the other regularities, corresponding to (1) and (2). But Field is confident that we cannot move from this starting-point, because there is, as he observes, a large—in his view, unbridgeable—gap between the consistency of an axiomatic theory and its truth, and so a correspondingly large gap between explaining the regularity (3) and explaining the regularity (2).[14]

It might be thought, however, that there is a way to bridge Field's gap. Elsewhere,[15] in the course of discussing how, as a nominalist, he ought to understand the key notion of conservativeness, Field concedes that an explanation along familiar model- or proof-theoretic lines is ruled out—as involving an unacceptable reference to sets of one kind or another—and opts for a primitively modal construal: roughly, and focusing on the notion of consistency—which is required for conservativeness, and in terms of which it may, of course, be defined—we are to take a theory to be consistent just in case its axioms are possibly collectively true. But if this is how the notion of consistency is to be understood, then—given that (3) records a genuine and explicable regularity—Field ought, it seems, to hold that

(4) For most '*p*', if most mathematicians accept '*p*' as an axiom, it is possible that *p*

does so too. But if every mathematical truth is a necessary truth, then it holds, for all mathematical '*p*', that:

(5) It is necessary that *p* or it is impossible that *p*.

And from (4), together with the premiss that (5) holds for *all* mathematical '*p*', it follows that

(6) For most mathematical '*p*', if most mathematicians accept '*p*' as an axiom, then it is necessary that *p*.

Whence, by the Law of Necessity, we have (2)—*quod erat explicandum*.

[13] Field, *Realism, Mathematics and Modality*, 232. [14] Ibid. 232. [15] See, ibid. 127.

Tempted as he may be by this simple-minded response, the Platonist would do well to resist. For one thing, there is room for doubt about the principle upon which the explanatory argument just sketched proceeds—to the effect that given an explanation why some statement *A* holds true, together with a deduction of *B* from *A* (along, perhaps, with supplementary premisses), we have (or at least can find) an explanation why *B*.[16] And doubts might be raised, too, about the sort of explanation Field thinks is available for (3).[17] A third—and potentially lethal—objection is as follows. Whether or not Field himself would be disposed to accept (4) as true and explicable, the fact is that we cannot safely assume that *every* philosopher who doubts the tenability of a Platonist construal of mathematical statements must be prepared to grant as much. The most that has to be conceded—or so it may be claimed—is that mathematicians are pretty reliable at picking axioms that are consistent in a purely *syntactic* sense. And someone could grant that much and yet quite consistently refuse to allow that our best mathematical theories comprise statements that are possibly true. There thus remains, it seems, a crucial gap—without the intermediate step from (3) to (4), the availability of an explanation for the regularity (3) formulates affords no assurance—even granted the additional premiss (5)—that (6) (and thence (2)) record genuine and explicable regularities. The obvious strategy of seeking to meet the epistemological challenge by building upon the explanation (i.e. of (3)) Field grants to obtain the required explanation of (2) (and hence of (1)) thus appears to break down. Can the Platonist do better? In the next section, I shall try to show that he can.

3. BELIEFS ABOUT LOGICAL CONSEQUENCE

I shall argue that the Platonist can make much more capital than Field anticipates from his concession that the problem of explaining how it is that mathematicians tend to be reliable in their mathematical beliefs in general can be reduced to that of explaining how they come to be reliable in their axiomatic beliefs.

As a first step we should observe that the concession is a special case of

[16] The quickest way to see this, perhaps, is to note that any given truth about the physical world for which we can give an explanation will entail very general existential statements—such as that there exist physical objects—which have a strong claim to be viewed as expressing brute, i.e. inexplicable, contingencies.

[17] Field's view appears to be that the best explanation of our belief in the consistency of ZF, say, is in terms of its actually being consistent—that the most likely explanation of our failure, thus far, to locate an inconsistency is that there is none to be found. But it is quite unclear how this probability claim might be established. I am indebted to Crispin Wright for helping me to appreciate this difficulty.

a quite general point which ought to be conceded by anyone disposed to insist upon naturalistic explanations of our belief-forming reliability in regard to any given subject-matter: that whenever we have, or can find, a suitable explanation of our reliability in forming beliefs belonging to a certain base class—concerning any subject-matter—there can be no insurmountable obstacle to our extending that explanation to cover our reliability in forming beliefs reached by deductive inference from beliefs belonging to that base class. In very many cases, the actual process by which we arrive at our beliefs will involve the making of deductive inferences, so that if the presumed good correlation between our beliefs and the facts is not to be a miraculous coincidence, it will have to be the case that we tend to carry out this part of the process properly—we tend, that is, by and large, to make valid inferences rather than fallacious ones—and it is hard to see how a satisfying explanation of the good correlation could avoid appealing to this fact. To be sure, the general concession is, strictly, a merely *conditional* one: the conclusion that deductive reasoning produces a genuine expansion of the range of beliefs over which we tend to be reliable *could* be resisted, but the price of resistance is high. It would involve either denying that there is any suitable base class of reliable non-inferential beliefs which might be expanded by deductive reasoning, or denying that such reasoning can ever be reliable. It is hard to see how the first course could avoid endorsement of global irrealism—that is, the thesis that none of what we are pleased to call beliefs are even minimally truth-apt;[18] and the second entails a very radical form of solipsism, on which we are at best reliable in our beliefs about the here and now. Granted that neither of these extreme positions is sustainable, it follows, not only that we do in fact tend to be reliable in making deductive inferences, but that this tendency admits of a suitable explanation.

Amongst the matters on which we hold beliefs are matters of logical consequence. The second step is to observe that—whilst there is certainly room for some slippage—it would be quite remarkable, to say the least, if, although we tend to make valid inferences rather than invalid ones, our beliefs about what follows from what were wildly out of line with the facts of the matter. If that is right, it has to be allowed that there is a fairly good correlation between our beliefs about logical consequence (briefly, our logical beliefs) and the relevant facts. And there ought, accordingly, to be available a suitable explanation of our tendency to hold true logical beliefs.

Although it is arguably no more than a contingent fact that our logical

[18] Serious doubts about the very coherence of global irrealism are raised by Crispin Wright; see his 'Kripke's account of the argument against private language', *Journal of Philosophy*, 81 (1984), 759–78 and his further discussion of the issues in *Truth and Objectivity*, Harvard University Press, 1992, ch. 6.

beliefs generally[19] correlate fairly well with the facts, it can (step 3) perfectly well be an ingredient of the Platonist's position—though it is scarcely to be reckoned distinctive of it—that the facts about logical consequence themselves are non-contingent or necessary; and that that *A* is a logical consequence of *X* is something which we know (when we do know this) a priori—or at least it is something that we have (normally, anyway) a priori grounds to believe. Furthermore, there is apparent no good reason why the Platonist should not draw upon this feature of his overall position in responding to the epistemological challenge. Of course, should he do so, he must be prepared to face other objections (from sceptics about logical necessity). The present point is simply that the Platonist cannot be required to confront the epistemological challenge in the form: explain the (putatively) good correlation between mathematical beliefs and the facts, *on the assumption that logical beliefs are at best contingently true.*

On the contrary assumption about logical consequence, among the statements which we can know, or justifiably believe, there will be some that are necessarily true, and knowable (or justifiably believable) a priori; and whatever constraints are to be imposed on what is to count as an acceptable explanation of our tendency to gets things right in regard to some given subject-matter should leave space for an explanation of our tendency to arrive, by a priori means, at some true beliefs whose truth is a matter of *necessity.*

It will not be necessary to pursue here the large and difficult question of how such an explanation might run. It is enough, for present dialectical purposes, to observe that—on the assumptions currently in play—a suitable explanation of our tendency to form true logical beliefs is in principle available, and to make one purely negative point about its character. This is that the explanation will not be straightforwardly causal, if by that we understand an explanation that consists either in displaying facts about logical consequence as causal antecedents of the corresponding logical beliefs or in tracing both to a common cause. Explanations of this sort, appealing to a (very probably, further explicable) counterfactual dependence between our beliefs and the relevant facts are just what we might, plausibly, expect to be able to give in the case of our perceptual beliefs, for example, and perhaps—more generally—whenever we are concerned with empirical beliefs about matters of contingent fact. But on present assumptions, the truth of our logical beliefs is a matter of necessity; and where beliefs of a certain sort are, when true, necessarily so, there is simply no purchase for the idea that our tendency to get things right is, at bottom, a tendency for

[19] Generally, not invariably—because it can be argued that assent to certain logical truths is a necessary condition of grasp of the logical concepts they feature, so that there can be no possible world in which we are routinely wrong about them. I thank an anonymous reader for a reminder—which should not have been needed, but was—of this point.

our beliefs to vary (counterfactually) in step with the facts. For in such a case, the relevant facts could not have been otherwise, so that it is merely vacuously true that, had things been otherwise, we would not have held the beliefs we do. There is no useful explanation that runs: 'Had it not been the case that *A* follows from *X*, we would not have believed that *A* follows from *X* . . .'.[20] This is not to say that an appropriate explanation will be wholly non-causal; the crucial point is that, on present assumptions, there must be space for an explanation of the good correlation between our logical beliefs and the facts which does not involve regarding the latter as bearing causal relations to other things.

The bearing of these reflections upon the generalized epistemological challenge to Platonism should be evident enough. The Platonist—or at least, the kind of Platonist whose position I am anxious to defend—will hold that true mathematical statements are, like true statements of logical consequence, necessarily so. Why, then, may the Platonist not respond that an explanation of our tendency to form true mathematical beliefs can run along essentially the same lines as an explanation of our tendency to arrive at true beliefs about other matters of necessity, such as matters of logical consequence?

This proposal is, of course, quite schematic, and can hardly be expected to command final assent in the absence of a satisfying account of the epistemology of necessary truth in general—and this, I agreed at the outset, we do not yet have. The task of providing one, though certainly pressing, is too large an undertaking for the remainder of this paper. There are, however, some quite general objections which may be urged against the proposal, the force of which is entirely independent of the detail of any specific account of the epistemology of necessity. In my closing section, I shall try to neutralize what I take to be the most important of them.

4. CONTINGENCY AND NECESSITY

The proposal of the preceding section will, of course, cut no ice with philosophers who reject the notion of necessity outright—but as I remarked at the

[20] I take this to be the point David Lewis is making in *On the Plurality of Worlds*, Blackwell, Oxford, 1986 when he writes: 'the department of knowledge that requires causal acquaintance is not demarcated by its concrete subject matter . . . [but] by its contingency. . . . nothing can depend counterfactually on non-contingent matters' (111). Field interprets the passage quite differently, as attempting to block the demand for explanation altogether. I agree with Field—but not with his reasons—that Lewis has produced no argument which succeeds in doing that. But it seems quite clear from the context that Field has misinterpreted Lewis, whose point is just that when matters of necessity are in question, *causal* explanations are inappropriate—Lewis explicitly acknowledges (113) the legitimacy of the demand for an explanation, and frankly admits that he has none to offer.

outset, it was no part of my plan to produce an answer to the epistemological problem that would convince them, since they will reject any form of Platonism that involves taking mathematical truths to be necessary, regardless of any other objections that might be made to it, and, in particular, quite independently of the specifically epistemological objection with which we are concerned. Setting aside—here, anyway—such blanket scepticism about necessity, I shall address a more sharply focused objection with which many will, I suspect, feel considerable sympathy. For it may be urged that, even if house room should be accorded to the notion of necessary truth itself, and a satisfactory account of our knowledge of such truths can be given, the present proposal is doomed to failure because the parallel between logical beliefs and mathematical ones on which it relies breaks down at the crucial point, and does so because mathematical statements, even if true (when taken at face value), are never *necessarily* so. In consequence, even if we can expect to be able to construct an explanation of our tendency to get things right about certain non-contingent matters—such as matters of logical consequence—such an explanation cannot be extended to cover our mathematical beliefs, precisely because even if such beliefs were true, they could not possibly be necessarily true.

That Field, in particular, would press such an objection is clear enough. The denial that mathematical truths (if such there be) are necessary—in any sense germane to the defence of Platonism, anyway—is, in effect, the real substance of his central objection to what he takes to be[21] an attempt by David Lewis to deflect the epistemological challenge by arguing that a demand for an explanation of our reliability is misplaced, when the beliefs whose reliability is in question are, when true, true by necessity. In what sense, he asks,[22] are mathematical facts supposed to be necessary?

> They are not logically necessary, nor do they (at least on the face of it) reduce to logically necessary truths by definition. . . . They are of course *mathematically* necessary in the sense that they follow from the basic laws of mathematics. Similarly, the existence of electrons is presumably *physically* necessary, i.e. follows from basic physical laws. But Lewis does not think that the epistemological problem of explaining how our 'electron' beliefs can reliably indicate the existence of electrons is really a pseudo-problem, just because the existence of electrons is physically necessary; so why should the fact that the existence of numbers is mathematically necessary show that the corresponding epistemological problem about numbers is a pseudo-problem?

There is no question but that Field is right to suppose that the Platonist can make no headway if the only sense in which mathematical statements

[21] See preceding note. Obviously the force of the objection is unaffected by the misreading of Lewis, and equally obviously, if the objection is good at all, it is fatal to the present proposal.

[22] Field, *Realism, Mathematics and Modality*, 235. This is the source of the next quotation.

can be necessary is a merely *relative* one (that is, they are necessary in the sense that they are logical consequences of some set of propositions—mathematical axioms, say—which are not themselves necessary in their own right). Acceptance of mathematical statements as necessary in that sense is perfectly consistent with holding them to be false (as Field does), or even, for that matter, with denying that they are genuine statements, apt for truth-value, at all.[23] If appeal to the necessity of accepted mathematical statements is to subserve the defence of Platonism, the sense of necessity in question must be an *absolute* one, in which necessity entails truth. Field is confident that there is no such sense in which mathematical statements can qualify as necessary. The only truths which, if any are, are absolutely necessary are, he contends,[24] logical truths, or truths reducible to such by definition. And he is confident that mathematical truths do not fall into this category. But why?

What backs Field's confident pronouncement that mathematical truths are not logically necessary is, in part at least, his adoption of an austere conception of logical necessity, as outlined in the introduction to his book.[25] According to this conception, only those true statements are necessary which are true solely in virtue of their logical form. Logical form here does not necessarily mean just first-order form; Field seems prepared—in contrast with notable sceptics such as Quine—to allow that what is usually termed 'higher-order logic' is indeed part of logic, and that certain statements essentially involving higher-order quantifiers are logically or necessarily true ('$(\forall x)(\exists F)Fx$' is a plausible example)—provided at least that they carry no implications about how many individuals there are. The fact remains that whilst Field does not restrict logical necessity to first-order logical truth, he will admit as logically necessary only those statements that are true solely in virtue of their mode of composition out of sentential operators[26] and quantifiers of various levels. They must, for example, contain no essential occurrences of predicates, or other non-logical vocabulary. If this conception is strictly applied, even statements—e.g. 'All vixens are female'—which admit of transformation into truths of logic by definitional substitution will not rank as logically necessary; though as we have seen, Field seems prepared—again in contrast with Quine—to allow an extension of the absolute notion of logical necessity to cover such statements. But beyond that he will not go.

If the category of (absolutely) necessary truth is to be demarcated in this way, then Field is probably right to deny that mathematical truths are

[23] Though, of course, on such a view, it would have to be held that the consequence relation, among mathematical sentences anyway, is a purely syntactic one.

[24] Cf. Field, *Realism, Mathematics and Modality*, 237 n. 8.

[25] Ibid. 30–8.

[26] This includes modal operators, subject to a restriction the intended effect of which is to prevent any distinctively essentialist theses from qualifying as logically true.

absolutely necessary. Logicism—in the quite strong form in which it would, if correct, allow us to count mathematical truths as logically necessary in this sense—no longer seems tenable.[27] But pretty clearly, that is not the end of the issue. Granted that mathematical truths are not logically necessary in the austere sense Field prefers to employ, what good ground is there to believe that this is the only viable or respectable sense that is both relevant and available?

It is worth noting, to begin with, just how oddly restrictive the profferred demarcation is. It allows us, as we have seen, to reckon a true statement as logically necessary, even though it may not be, as formulated, true solely in virtue of its logical form alone—but only provided that it admits of reduction, by definitional substitution, to some other statement which is so true. So it lets in things like 'All vixens are female.' But it excludes statements like 'If one event precedes another and that in turn precedes a third, then the first precedes the third.' So it dignifies, as logically necessary, any statement we might properly take to be true solely in virtue of features of the concepts deployed in it, but only when the relevant features can be so encapsulated in an explicit definition of some constituent non-logical expression that it is possible to transform the original statement into an instance of some formally valid schema. But where relevant conceptual features cannot be thus encoded in an explicit definition, as is plausibly the case with statements like the one about temporal precedence above, it rules against counting them as logically necessary, for all that our inclination to regard them as conceptual truths—as true solely in virtue of features of their ingredient concepts—may be every bit as secure as in cases where explicit reduction is possible. But—at least once it is recognized that there can be statements that are non-contingently true, but only in virtue of the senses of non-logical expressions used in their formulation—cases of the latter sort, though common enough, can hardly be viewed as constituting anything other than a special subclass. To draw the line in the way proposed is, on the face of it, to assign a significance to the possibility of explicit definition which there is no clear reason to think it possesses.[28]

[27] The qualification is not merely decorative. It is far from clear that nothing properly describable as a version of logicism is defensible, at least as far as elementary number theory goes. For a forceful defence of number-theoretic logicism, see Wright, *Frege's Conception*, ch. 4.

[28] To forestall possible misunderstanding, let me say straightaway that there is no question but that we can effect a demarcation of logical truths along the general lines Field proposes. Nor is there any doubt that there are good technical and theoretical reasons for effecting a distinction between logical truths and others in just that way. My point is that it is another question entirely whether the distinction thus effected can properly be assigned the *philosophical* significance with which Field wishes to invest it, whether it can bear the philosophical load he must ask it to carry, if he deploys it in defence of his denial that mathematical truths can be absolutely necessary.

To the extent, then, that Field's confidence that mathematical statements cannot be absolutely necessarily true—or, therefore, absolutely necessarily false—relies upon assigning to his preferred way of demarcating the class of logically true statements the philosophical significance he appears to wish to assign to it, that confidence is, I contend, inadequately grounded. A division of logical truths from the rest can doubtless be effected along the lines proposed, but no compelling reason is yet apparent for supposing that among the rest there are no statements possessing the kind of absolute necessity which Field is—some of the time, anyway—prepared to recognize as belonging to logical truths austerely circumscribed. But there is another, and deeper, reason underlying—and, Field would claim, justifying—his confidence on the point. This emerges in a footnote to which reference has already been made. The relevant portion runs thus:

> In this discussion I have avoided taking a stand on whether even logical necessity should be viewed as 'absolute' necessity. One view, to which I am attracted, is to reject the whole notion of 'absolute' necessity as unintelligible. Another view, also with some attractions, regards the notion as intelligible but regards the only things that are absolutely necessary as logic and matters of definition; in particular (as Hume and Kant held) there can be no ' absolutely necessary' entities, and so mathematics (taken at face value) cannot be absolutely necessary since it implies the existence of mathematical entities of various sorts.
>
> Whichever of these two views one prefers, the fact that mathematics involves existential commitment in a way that logic doesn't makes for a sharp distinction between mathematical necessity and logical necessity.[29]

[29] Field, *Realism, Mathematics and Modality*, 237 n. 8. Although Field here floats the idea of rejecting the notion of absolute necessity altogether, he gives no indication whatsoever of the grounds upon which he would base such a rejection. Indeed, it is not even clear whether what he envisages is rejection of the notion of absolute necessity coupled with retention of that of relative necessity, or rather, more radically, rejection of the notion of necessity altogether. Neither course—it seems to me—is free from difficulties. If he takes the former, some account is clearly owing of what the necessity of logic and matters of definition is relative to, but it is anything but clear what plausible answer he can give. A further—closely related but distinct—difficulty is that there must be serious doubt about the stability of this option: in so far as Field explains what relative necessity is, its being relatively necessary that A is a matter of As being a *logical consequence* of some specifiable set of truths K which are not themselves necessary. The corresponding conditional (asserting that if K then A) is presumably itself necessary, but how could this necessity be other than absolute. In short, it appears that the very notion of relative necessity is bound to involve the absolute notion. In view of this, the latter course may seem the more attractive. And there have, of course, been philosophers—most notably, perhaps, Quine—who believe we should take it. But (a) there are, I think, strong reasons to doubt the coherence of the resulting position see esp. Wright, *Wittgenstein on the Foundations of Mathematics*, Duckworth, London, 1980, 319–23, 415–20, and, for an improved formulation of the central argument, his 'Inventing logical necessity' in Jeremy Butterfield (ed.), *Language, Mind and Logic*, Cambridge, 1986, 189–94, and (b) it is anyway quite unclear how Field can opt for the more radical course, given the crucial, if limited, employment he wishes to make of modal notions in his preferred account of mathematics. The issues here are—I need hardly say—both profound and very difficult. But in view of the considerations just mentioned (especially the last, together with the fact that Field clearly holds that even if his second view is

The notion that logic is distinguished from other disciplines by its complete lack of existential commitment is so familiar and so widely accepted that it is, perhaps, hard to see how it could—relevantly and soberly—be challenged. And if we elect to construe logical truth in Field's austere sense, there can, perhaps, be no question of doing so, since it is, in effect, built into that characterization of logical truth that no truth of that kind can carry existential commitment. But that is not the question which properly concerns us here. The crucial question is, rather, whether *any* absolutely necessary truths can carry such a commitment, bearing in mind that this class need not comprise only logical truths in Field's sense. And since, evidently, no absolutely necessary truth can entail any statement—existential or otherwise—which is, at best, merely contingently true, the issue concerns whether any existential statements can be necessarily true. Field is sure that none can be, because—following Kant—he holds that no legitimate definition of any concept can by itself ensure that the concept defined has application. If we don't stand firm on that much, Field contends, then we throw open the floodgates to all sorts of monsters. In particular, we shall be powerless to resist the notorious ontological proof of God's existence.[30]

I lack space for a full discussion of the issues this raises here. But let me conclude with a brief indication why I do not think we should be bustled by this not unfamiliar set of considerations into thinking that defence of the view that the existence of numbers is necessary must collapse into transparent and gross implausibility at this point. I shall confine myself to three points.

First, Field's attempt to discredit that view by association with the ontological argument can perfectly well misfire, since the latter can, and should, be faulted on grounds which do nothing to call into question the general idea that there may be necessary existential truths. To mention just one, the argument might be rejected on the grounds that it fails to escape the following dilemma, which does not rule out necessary existential truths in general.[31] Either the proposed explication of the concept of God as a being

adopted, mathematics can't be absolutely necessary), I think it is not unfair to proceed on the assumption that a notion of absolute necessity is available. Argument to the effect that mathematical statements can't—in view of their existential commitments—be logically or absolutely necessary can be found in other papers by Field e.g. 'Is mathematical knowledge just logical knowledge?' *Philosophical Review*, 93 (1984), 509–52 (repr., with additions, in *Realism, Mathematics and Modality*, 79–124; the relevant passage is 79–81).

[30] Field's predilection for this gambit—attempting to discredit any species of Platonism which sees the existence of numbers as necessary by contriving the appearance that it must keep (bad) company with the ontological argument—is exemplified in several places. See e.g. his critical notice of Wright (1983), repr. as 'Platonism for cheap?' in *Realism, Mathematics and Modality*, 147–70, and the Introduction to the same collection, 43 n. 23, as well as the passage from Field's *Philosophical Review* paper cited in n. 29.

[31] I owe this way of formulating the point to Wright. See his 'Why numbers can believably be', 455–6. Wright draws attention to several other important disanalogies between the broadly

no greater than which can be conceived is to be construed as a universally quantified biconditional:

$$(\forall x)(x = \text{God} \leftrightarrow \text{no being greater than } x \text{ can be conceived})$$

in which case, we can accept it and still have room to deny that anything is identical with God, or it is to be understood as a singular identity:

$$\text{God} = (\iota x)(\text{no being greater than } x \text{ can be conceived})$$

in which case we can consistently reject it as false, on the ground that nothing satisfies the description on the right-hand side.

Second, it is important to stress that we have so far seen no compelling grounds for thinking that all absolutely necessary truths must be, as Field claims, either logical truths in his austere sense or definitional transformations of such. It follows that—unless such grounds can be provided—defence of the view that certain existential statements are necessarily true need not involve taking issue with the claim which Field thinks settles the issue in his favour—the claim, that is, that no proper definition of a concept can by itself ensure that the concept defined has application. To think otherwise is to overlook the possibility that certain existential statements may be consequences of—or may themselves be—truths that are absolutely necessary, without being (definitional transformations of) logical truths in Field's narrow sense. The kind of argument by which (Fregean) Platonism is best supported makes no illicit attempt simply to define numbers into existence. That there are numbers is to be accepted as a consequence of the fact that various statements are true whose semantic character is as their surface syntax suggests—in particular, in point of incorporating bona fide singular terms which, if they have reference at all, refer to numbers. This suggests that the crucial issue should now focus on the status, as necessary or contingent, of these statements.

This brings me, effectively, to my third point. Statements deploying the concept of (cardinal) number can be divided into two sorts: purely arithmetic statements, most simply exemplified by finite equalities and inequalities, such as '$15 + 8 = 23$' and '$7^2 < 51$', and applied statements of number, such as 'The number of male philosophers greatly exceeds that of female philosophers' and 'The number of electrons in the universe is very large indeed.' Statements of the first kind are, if Platonism of the kind here in question is correct, true by necessity; statements of the latter sort are at best

Fregean argument for number-theoretic Platonism and the ontological argument which, in my view, collectively scotch the whole idea that anyone running the former must be committed to endorsing the latter. Field makes a fresh attempt to saddle the Fregean Platonist with a version of the ontological argument—quite unlike anything Anselm may have had in mind—in his 'The conceptual contingency of mathematical objects', *Mind*, 102 (1993). For an explanation why it fares no better than its predecessors, see Hale and Wright, 'A reductio ad surdum?: Field on the contingency of mathematical objects'.

contingently true. And there, so it may seem, is the rub. For to appeal to the necessity of statements of the first sort, in the present context of argument, would be obviously unsatisfactory, since that is precisely what is in dispute. But if the ultimate premisses of an argument to the existence of numbers must include statements of the latter sort, then the conclusion that numbers exist can be no less contingent than they are. It thus appears that any attempt to argue for the necessary existence of numbers—at least along anything resembling these lines—must get stuck on one or other horn of a lethal dilemma.

There are, however, at least two reasons to think that the dilemma is a false one.

First: even were it true, as our statement of the dilemma assumes, that there are no necessarily true applied statements of number—roughly, statements concerning how many things fall under some specific concept(s)—it is arguable that no matter what the exact character of any possible universe, there would be some true statements of number to be made about it. It is, no doubt, a highly contingent matter that the universe contains, for example, male and female philosophers. It is, presumably, contingent that it contains males and females of any species at all. It is even no more than contingently true that the universe contains electrons, protons, etc. Quite generally—or so it may be contended—whatever specific sortal concepts we pick, it will be contingent that there are so-and-so many instances of the one, and that the number of instances of the one stands in such-and-such an arithmetical relation to the number of instances of the other. Granting all that, it may still be necessary that, whatever the universe was like, there would be some true statements of number to be made about it—some truths to the effect that there are so-and-so many instances of such-and-such a concept. Thus one possible response to the envisaged dilemma would be that it is not required, for a non-question-begging defence of the claim that numbers necessarily exist, that there be *specific* statements of number that are necessarily true—it is enough that as a matter of necessity, *some* statements of number are true. If this weaker claim could be sustained, it would provide the basis for a defence of the claim that the existence of numbers is no contingent matter that steers between the horns of the dilemma.

In the widely employed jargon of possible worlds, the weaker of the two claims just contrasted might be expressed as that in any possible world, some statements of number are true, the stronger claim being that there are some statements of number which are true in all possible worlds. But this way of expressing the matter risks confusion, since it may obscure a distinction—crucial here—between what is *true of* a possible world and what can be *truly asserted in* a possible world. And it may then appear that even the weaker claim is open to an obvious objection. For to say that in some possible world, a certain statement (or some statement of a certain type) can be truly asserted might—though it need not be—understood as implying that there is, in that world, some creature endowed with thought and language capable of asserting it. And since it is plainly possible that there should exist no such creatures at all, it cannot be the case that in any possible world some statements of number can be truly asserted. But of course, that is not how the weaker claim is intended to be understood, and this is why it is better expressed—in possible-worlds jargon—as the claim that of any possible world, some statements of number (i.e. some statements of the type '$NxFx = n$') are true. Nor is it needful, in order to avoid the unwanted implication, to suppose that possible worlds—the actual world and other, merely possible, worlds—come, so to speak, ready carved into kinds or types of object independently of the classificatory endeavours of creatures such as ourselves, or that concepts somehow float free of conceptual schemers. The second-order quantifier implicit in the claim that in any possible world, some statements of the type '$NxFx = n$' are true can be understood as ranging over concepts we actually have, or at least could form.

There are, however, other and better reasons for dissatisfaction with the line of argument sketched above, at least as it stands. For one thing, even if the argument succeeds, its conclusion is a very weak one, in two respects: first, it establishes at best that in every possible world, there are *some* numbers—which numbers there are, in any given world, will depend upon which statements of number are true of that world, and this is evidently something which can vary across different possible worlds; and second, it is hard to see how the argument can possibly yield enough numbers—enough for serious arithmetic purposes, at a minimum—given that the statements of number to which it appeals are to concern how many concrete objects there are or could be of this or that sort. For another, it may be argued that even this assessment of how much the argument establishes is unduly generous. For—or so it may be claimed—it is perfectly possible to construe true statements of number about other possible worlds in a way that does not demand the existence of numbers in any but the actual world: just as we can (and, indeed, must be able to) make sense of the evident possibility of a world in which there are no creatures capable of thought and language without supposing that anything in such a world is equipped with the concept of such creatures, so—it may be argued—we can appreciate the possibility of a world in which the number of aardvarks is, say, 10^{13} without supposing that that or any other number exists in that world. To see the key idea, consider the statement that although Bill is in fact the tallest man there is, there might have been a man taller than Bill. The fact is that, whilst the envisaged possibility would be realized in a world in which Bill exists alongside someone taller than he is, we can make perfectly good sense of it by supposing a world in which Bill does not exist, but there is a man whose height exceeds the height Bill actually has. That is, the supposition that the statement that someone is taller than Bill is true of a certain possible world does not require—though it does not preclude—Bill's existence in that world; his existence in the actual world is enough. And in much the same way, it may be contended, we can see how it can be true of another possible world that the number of aardvarks it contains is 10^{13}, without having to suppose that that number exists in that world; it is enough that it exists in the actual world. Generalized, the counter-thought is that we can always construe statements of number true of other possible worlds with their numerical terms having reference to numbers existing in the actual world alone. To say that there might have been 10^{13} aardvarks does not have to be understood as claiming that in some possible world *w*, the number of aardvarks = 10^{13}—so as to require the existence *in w* of the number of aardvarks; it can be taken instead as asserting that for some *w*, the number of aardvarks-in-*w* = 10^{13}, where the numerical terms make reference to actual world numbers. Thus contrary to initial appearances, the argument given fails to establish the existence of any numbers in any other world at all.

Potent as this counter-argument may at first seem, it is fundamentally flawed. According to the position it advocates, the truth of statements of number about other possible worlds affords no compelling ground for taking numbers to exist in those other worlds, because the ingredient numerical terms can always be construed as having reference to numbers existing in the actual world alone. This claim makes a very substantial assumption about what numbers actually exist. Since for (at least) every finite cardinal *n*, and more or less any choice of sortal concept *F*, it is true of some possible world that $\mathrm{N}xFx = n$, the actual world will have to contain (at a minimum) *all* the finite cardinal numbers. What assurance can a proponent of the counter-argument provide that it does so? We may take it that she will hold that the actual world contains a finite cardinal *n* for each actually true statement '$\mathrm{N}xFx = n$', where *F* is any concrete sortal concept. Clearly, however, this will yield arbitrarily large finite numbers only if some such concept actually has (at least) countably many instances. And of this, it seems, we can have no assurance whatever. To sustain the counter-argument what is needed is an a priori ground for thinking that there are, in the actual world, all the numbers demanded by true statements about how many things of any given sort there might have been—that is, an argument which establishes, independently of any assumption about which statements of number happen to be true of the actual world, that the actual world contains all the finite numbers. But precisely because such an argument must be free of any assumption which is at best contingently true of the actual world, it will be at the service of those of us who take the existence of numbers to be a matter of necessity. The counter-argument is thus radically unstable.

Evidently an argument of the type just envisaged would provide the core of a much stronger response to the original dilemma, quite free of the admitted limitations of the first response just essayed. Can such an argument be given? I believe so. For the assumption that there are no necessarily true statements of number overlooks an obvious train of thought—in effect, that which underlies Frege's own recipe for establishing the existence of the whole infinite sequence of finite cardinal numbers. If, following Frege, we take each cardinal number to be essentially the number belonging to a suitable concept—that is, if we take canonical terms for numbers to be terms of the form 'the number of *F*s'—then what is required, and suffices, to establish the existence of a given cardinal number is to exhibit a concept which is guaranteed to have appropriately many instances. In particular, and more precisely, if we take 0 to be the number belonging to the concept $x \neq x$, then what must be shown is that the number of x such that $x \neq x$ (briefly, Nx: $x \neq x$) exists in every possible world (i.e. $\Box \exists y\ y = \mathrm{N}x{:}\ x \neq x$). But $\exists y\ y = \mathrm{N}x{:}\ x \neq x$ straightforwardly follows from Frege's criterion of identity for cardinal numbers, i.e. that for any F and G, the number of *F*s is identical with the number of *G*s iff the *F*s and the *G*s are in one-one correspondence, together with the supplementary premiss that no object is self-distinct. Since both premisses are necessarily true, so is the conclusion.[32] The number 0 is thus guaranteed existence in each possible world, irrespective of which—if any—concrete sortal concepts are there instantiated. And it follows, by the laws of identity, that concept $x = 0$ is uniquely instantiated, so that it is further necessarily true that the number of objects falling under this concept is 1. So the number 1 is likewise guaranteed existence in every possible world. By familiar moves, we can—for each finite cardinal n—identify a concept which necessarily has exactly n instances, whence the existence of n itself is guaranteed.

I conclude that the existential commitments of mathematical statements, Platonistically construed, pose no insuperable obstacle to their being necessarily true, and hence that no reason is yet apparent for thinking that an explanation of our tendency to have true mathematical beliefs cannot run along essentially the same lines as an explanation of our tendency to have true beliefs in other cases—such as beliefs about logical consequence—where the propositions believed are, if true at all, necessarily so. Just how

[32] More fully, the suggested derivation runs as follows:

(1) $\Box\forall F\forall G\,(\mathrm{N}xFx = \mathrm{N}xGx \leftrightarrow Fx \approx Gx)$	
(2) $\Box \sim\exists x\ x \neq x$	
(3) $\mathrm{N}x\ x \neq x = \mathrm{N}x\ x \neq x \leftrightarrow x \neq x \approx x \neq x$	1 $\Box$E, $\forall$E
(4) $\sim\exists x\ x \neq x$	2 $\Box$E
(5) $x \neq x \approx x \neq x$	4 (any 1-1 relation will do)
(6) $\mathrm{N}x\ x \neq x = \mathrm{N}x\ x \neq x$	3,5 detachment
(7) $\exists y\ y = \mathrm{N}x\ x \neq x$	6 $\exists$I
(8) $\Box\exists y\ y = \mathrm{N}x\ x \neq x$	7 $\Box$I

such an explanation should run is a large and difficult question on which—to stress—I have ventured no positive suggestion here. Pressing as it is, it must be reserved for another occasion,[33] as must the equally pressing question how far the kind of argument rehearsed in these last few paragraphs may be extended to secure the necessary existence of numbers of other kinds, and (pure) sets in general.

[33] Some preliminary discussion, incorporating suggestions about the direction a responsible epistemology of necessity might take, may be found in the contributions by Wright and myself to the symposium 'Necessity, caution and scepticism', *Aristotelian Society*, Suppl. vol. 63 (1989), 175–238.

3

Do We Have a Determinate Conception of Finiteness and Natural Number?

HARTRY FIELD

I'm going to begin with a rather familiar worry, about whether we have a determinate understanding of second-order quantification. Then I will consider an analogous worry about whether we have a determinate understanding of the notion of finiteness. At first blush the case for the second worry seems quite analogous to the case for the first; but for various reasons doubts about the determinacy of the notion of finiteness may seem harder to take seriously than doubts about the determinacy of second-order quantification. I'll be considering a number of different reactions to this situation.

1. INTRODUCTION

1A. First-order set theory

I'll start off the discussion not with second-order quantification, but with the notion of set; for the moment I'll tacitly take our logic to be first-order. How determinate is our understanding of the notion of set?

I think that familiar considerations against the determinacy of our notion of 'set' are fairly strong. I'll remind you of two of them. The first concerns very central set-theoretic questions such as the size of the continuum, that are unsettled not only by the standard axioms of set theory but by any axioms that anyone has been able to think of that are intuitively compelling. The continuum example seems to me an especially striking one because it isn't simply that there is an undecidable 'yes'/ 'no' question, of whether the

This paper is a shortened version of 'Are Our Logical and Mathematical Concepts Highly Indeterminate?', which appeared in *Midwest Studies in Philosophy*, 19. I distributed a first draft of that paper at the Munich conference, and gave a talk based on it that roughly corresponds to the shortened version here. I received extremely helpful comments on the early draft from many of the conference participants, especially Penelope Maddy, Stewart Shapiro, and Bill Tait; outside of the conference, comments by Shaughan Lavine and Vann McGee also led to improvements. I am grateful to *Midwest Studies* for allowing me to reprint the paper in shorter form.

continuum hypothesis is true; it is rather that virtually every answer to the question of the size of the continuum is known to be consistent with the standard axioms of set theory, and with all known intuitively compelling extensions. (More exactly, it is known that one can consistently take the size of the continuum to be any infinite cardinal not cofinal with ω.) And even after arbitrarily choosing an answer to the size of the continuum, the size of its power-set is still almost totally undetermined; and so on.

One sometimes hears it said that the continuum hypothesis should be regarded as false, on the ground that it makes the set-theoretic hierarchy too small. Certainly plenitude has its attractions: if you're going to have sets at all, you might as well have lots and lots of them. But of course increasing the number of sets doesn't obviously increase the size of the continuum; it can decrease it, by increasing the number of 1-1 correspondences. And even granting the assumption that a higher answer to the size of the continuum means a more plenitudinous set theory, one can hardly argue against the continuum hypothesis on the grounds that there is a more plenitudinous answer that is consistent, since this argument would tell equally against any other conceivable answer.

I guess I think that if one had to plump for a single answer, the generalized continuum hypothesis might be best simply because it seems the least arbitrary. But I think it is more natural to say that we shouldn't regard there as being a unique answer: for each infinite cardinal κ not cofinal with ω, we could refine our conceptions of set and membership into conceptions set_κ and $\in_\kappa$ for which the size of the continuum is κ; and none of these conceptions of set and membership is privileged over any other. This point of view is of course a natural one for someone with fictionalist sympathies: each of these answers to the size of the continuum is an equally good fiction. But it seems to me that it is also natural from a Platonist point of view, if Platonism means simply that there are lots of abstract entities: indeed, the principle of plenitude seems to lead to precisely this viewpoint. What could be more plenitudinous than regarding all the consistent concepts of set and membership as instantiated side by side?[1] (By 'instantiated side by side', I mean that we refuse to single out one instantiation as privileged and to regard all others as merely 'unintended models' generated by the completeness theorem. For a clearer elaboration and a discussion, see the Appendix to the longer version of this paper (cited in the unnumbered footnote).) Of course, refining the concepts of set and membership to settle the size of the continuum would leave many other questions radically unset-

[1] I'm talking about consistent conceptions of *pure* sets. For impure sets the situation is slightly more complicated, since theories postulating them can be internally consistent yet not consistent with the actual physical world: see Field (1989, 56–7), for a clearer statement of this and an example. This situation can't arise however if the impure set theories are *conservative* in the sense of Field (1989); so the view in the text can be extended to impure concepts of set and membership by substituting 'conservative' for 'consistent'.

tled, including the size of its power-set; there is no way to refine our concepts so that all questions will be settled. The picture then is that there are multiple mathematical universes, in which the answers to many set-theoretical questions differ; and that nothing about our current concepts of set and membership serve to pick out a unique one, or even to narrow down the field enough so that questions like the size of the continuum get unique answers.[2]

The second consideration against the determinacy of 'set' is the one pressed by Putnam in the first half of Putnam (1983). It is based (in part) on the fact that even if one took the truths of set theory to be settled, there are alternative interpretations of 'set' and '∈' that would generate these truths, by Löwenheim–Skolem and compactness arguments. That of course is not in itself a problem: it is natural to say that these interpretations are unintended. But Putnam's question is, *in virtue of what is 'the intended one' uniquely intended?* Unless an answer to this question can be given, it would seem that the whole idea of a unique intended interpretation must be given up; if we put aside the fictionalist view that there is simply nothing in reality answering to these notions, the conclusion should be that 'set' and '∈' are referentially indeterminate. And I think Putnam argued pretty convincingly in the first half of Putnam (1983) that in the case of 'set' there is a very serious problem in giving any account of 'intendedness' according to which one interpretation is intended and all the others aren't. (I'll need to review this argument eventually, to distinguish it from an argument that Putnam gives later in his paper that attempts to generalize the conclusion to all concepts; that other argument seems to me clearly fallacious, but has nothing to do with the much more persuasive argument in the first half that is confined to the concept of set and related concepts.[3] But for now let me simply say that Putnam has raised a challenge that seems prima facie very difficult

[2] Note that I am not identifying the view that there are multiple 'disjoint' universes of sets, different ones for different answers to the continuum question, with the view that our use of 'set' leaves the size of the continuum indeterminate. The first does not really entail the second: conceivably even if there are such multiple universes, there is something in our usage that determines which universe we are talking about, though it is hard to see what facts about our usage could do this. The second doesn't entail the first either: even if there is a privileged universe, there might be nothing in our practice to determine that 'set' applies to everything in the universe or that '∈' applies to the membership relation of the universe. But elaboration of this requires a different sort of argument from the one based on undecidable sentences; I turn to it next.

[3] Putnam's attempt to extend his conclusions to non-mathematical concepts rests on confusing the view that reference is determined by e.g. causal considerations with the view that reference is determined by a description theory in which descriptions containing the word 'cause' play an especially prominent role. The fallacy has been often pointed out: see for instance Lewis (1984). The first half of Putnam (1983) doesn't rely on this fallacy. It relies instead on the fact that in the case of 'set' nothing like a causal theory of reference seems as if it could play much role; so that our theory of sets has to be the main determinant of reference, and that theory has non-standard models.

in principle to meet: explain in virtue of what we 'intend' a standard as opposed to a non-standard interpretation of 'set'.) If the challenge is impossible to meet, the natural conclusion would seem to be that *even taking the truths of set theory to be settled*, 'set' and '∈' are referentially indeterminate. And of course once one grants the referential indeterminacy of these notions, it becomes all the more natural to question whether claims like the continuum hypothesis have a definite truth-value on our current understanding of 'set' and '∈'.[4]

It is often assumed that if we regard concepts like 'set' and '∈' as indeterminate, we can't go on using classical logic in connection with them, but must use intuitionistic logic instead. But if there is a good argument for this, I don't know it. Indeterminateness seems formally a lot like vagueness, and there seems to be nothing at all incoherent about the idea that vague and indeterminate concepts obey classical logic. Of course, 'definitely true' is not a bivalent concept—a sentence A and its negation can both fail to be definitely true—but one doesn't have to give up the law of excluded middle, one can say that the disjunction $A \vee \neg A$ is definitely true even though neither disjunct is. An account of definite truth along these lines—supervaluation semantics—is obvious, intuitive, and well known. (See for instance Fine (1975).)[5] It is thus not at all obvious that the practice of standard mathematics presupposes the determinacy of the concept of 'set'.

1B. Second-order set theory

It is sometimes said that the worries about the determinacy of 'set' can be obviated by appealing to *second-order* set theory. This seems to me a serious mistake; I argue this at some length in the longer version of the paper. I do not deny of course that we can make a distinction between full interpretations and general (Henkin) interpretations of the second-order language, or that the Löwenheim–Skolem theorem and compactness theorems used in producing the non-standard interpretations in Putnam's argument fail if interpretations must be full. But I argue that none of this is of any help in undermining Putnam's argument (and that related considerations about full models are of no help in undermining the continuum hypothesis argument).

[4] The second consideration favouring indeterminacy is really a generalization of the first: if one could find a way to argue that we somehow manage to intend a unique interpretation of 'set' and '∈', then of course there would be a uniquely correct answer to the size of the continuum in this intended interpretation, whatever the difficulty in finding that answer. But the first consideration is worth separate statement to make clear the importance of worries about undecidable sentences.

[5] Of course, the understanding of a notion like 'set' is prior to the understanding of a formal truth theory for sentences containing this notion. The claim is only that the supervaluational semantics gives as good a representation of the semantics of an indeterminate notion of set as a standard model-theoretic semantics for set theory would give of the notion of set were that notion determinate.

Rather, the situation is that if the arguments have force in the first-order context, they have equal force in the second-order context and serve to show there that there is an indeterminacy in our second-order quantifiers and our concept of full model. (The main point was made some years back in Weston (1976).)

Of course, it might be that Putnam's argument (and the continuum hypothesis argument) are wrong: I will have more to say about this later. Here my claim is only that if Putnam's argument is correct in a first-order context, then it is also valid in a second-order context, and establishes that second-order quantification is also highly indeterminate.

1C. Finiteness

The basic Putnamian argument for the indeterminacy of our notion of set and of second-order quantification is that nothing in our practice with these notions seems as if it could possibly determine their extensions (or more generally, their contributions to truth-conditions) at all uniquely (even up to isomorphism). To the extent that this is a problem for these notions, it would seem to be equally a problem for our notion of finiteness (and our notion of natural number).[6] We can define the quantifier 'There are only finitely many' set-theoretically: $\exists_{fin}xA(x)$ iff $\{x|A(x)\}$ is in 1-1 correspondence with the predecessors of some natural number n. Or we can take it as a primitive. In the second case, our understanding of 'finite' seems to be exhausted by the inferential rules we lay down for it, where we allow as before both positive rules as to what follows from what and formalizable negative rules about what doesn't follow from what. This is ultimately so in the first case as well, because the definition is in terms of notions like 'set' and 'natural number' the understanding of which is exhausted by inferential practice. Whether 'finite' is defined or primitive, there will be non-standard interpretations of our practice, according to which many formulas of form $\exists_{fin}xA(x,y)$ will be satisfied by objects c for which there are really infinitely many b such that $<b,c>$ satisfies A(x,y). In particular, 'natural number', though defined by a condition entailing 'Each natural number has only finitely many predecessors', will be satisfied in some interpretations of our overall theory of numbers by non-standard 'natural numbers' with infinitely many predecessors; and 'finite sequence of expressions' ('sequence with only finitely many initial segments') will be satisfied in some interpretations of our overall theory of finite sequences by non-standard 'finite sequences' with infinitely many initial segments. And so if the basic Putnamian argument is correct, it looks as if we can't have a

[6] I talk mostly about the notion of finiteness rather than the notion of natural number, to emphasize that the problem I am concerned with does not depend on the ontology of mathematics: it arises for mathematical fictionalists as well as for everyone else.

determinate conception of finiteness, or a conception of the natural numbers that is determinate even up to isomorphism.

This conclusion seems prima facie much harder to swallow than the analogous conclusion about the notion of set or second-order quantification. It seems harder to swallow both because the notion of finiteness (or natural number) seems much more elementary, and because it seems presupposed even by our notion of inferential procedure. After all, an inference, or proof from given assumptions, is a finite sequence of expressions in which each is either a premiss or a logical axiom or follows from predecessors in the sequence by certain rules; if we have no determinate notion of finite, then there is going to be an indeterminacy even in our notion of proof from given assumptions. And thus it seems we can't even talk coherently about what our inferential procedure that supposedly fixes our indeterminate truth-conditions is. Surely, it would seem, we have talked ourselves into an unacceptable situation.

There are three possible reactions to this situation. The first is that despite Putnam's argument (and the argument from the continuum hypothesis),[7] we do have determinate conceptions of sets and/or second-order quantification. If so, then presumably we have a determinate conception of finiteness too.[8] The second possible position is that the radical conclusion about finiteness is to be swallowed somehow. And the third, intermediate, position is that there is a relevant difference between the cases of 'set' and second-order quantification on the one hand and the finiteness case on the other. Of course, each of these three positions subdivides, e.g. the first subdivides over the issue of *just where* the Putnam argument (which I still haven't stated very explicitly) goes wrong.

In the remainder of the paper I will discuss at least one version of each of these three positions, without, I'm afraid, reaching any totally definite choice between the second and third.

2. REJECTING THE PUTNAMIAN ARGUMENT: DISQUOTATIONAL TRUTH CONDITIONS

One way to argue that the argument in the first half of Putnam (1983) is mistaken even in the case of sets (as is the continuum hypothesis argument)

[7] A reply to Putnam's argument would give rise to a reply to the continuum hypothesis argument as well: see n. 4. The direct analogue of the continuum hypothesis argument in the finiteness case concerns Gödel's Theorem, which I will discuss in Sect. 3A.

[8] The latter is definable in set theory, and even without set theory, in second-order logic with relation variables or function variables. And though not definable in second-order logic with set variables only and no device for forming ordered pairs, it seems to me that there is no serious possibility of holding such second-order quantification determinate and finiteness not.

would be to argue that it is *incoherent* to suppose that our notions of set and of second-order quantification are indeterminate. Putnam himself suggests a version of this strategy at the very end of Putnam (1983), albeit cryptically; his version seems to rely on his 'internal realism without metaphysical realism', in particular, on his doctrine that neither sets nor anything else exist independently of our descriptions of them. But I think that there is a clearer version of the strategy that appeals to the disquotational properties of truth, and that is the version I will discuss.

Of course, to say that the conclusion of the arguments for indeterminacy is incoherent is not in itself to tell us just where these arguments go wrong; so for the 'incoherence' strategy to be ultimately satisfactory, it must be supplemented to include such a diagnosis. But I will concentrate only on whether it is incoherent to suppose that the notion of 'set' is indeterminate.

The first step in the argument for the incoherence of the indeterminacy claim is to assume either 'strongly disquotational' notions of truth and satisfaction, on which the disquotation schemas

(T) 'p' is true if and only if p

and

(S) For all x, x satisfies 'F' if and only if F(x)

are analytic; or at least 'weakly disquotational' notions of truth and satisfaction that give these schemas a pretty much indefeasible status.[9] The next step is to infer from this that

(T*) 'p' has the truth conditions that p

and

(S*) 'F(x)' has the satisfaction conditions that x is F

have the same analytic or indefeasible status. And the final step is to infer that 'x is a set' has perfectly determinate satisfaction conditions, namely the condition of being a set, and that 'There is a set of cardinality between that of the rationals and that of the reals' has perfectly determinate truth-conditions, namely that there is a set of such cardinality. To suppose that sentences containing 'set' don't have definite truth-conditions would be incoherent; and similarly for sentences containing second-order quantifiers.

I do not deny the appeal of this line of argument (though I suspect that only the strongly disquotational version of it could be used to make even an initially plausible diagnosis of exactly how the arguments for indeterminacy go wrong). Nonetheless, I shall question the third (final) step. In my

[9] I'm going to ignore considerations about the semantic paradoxes. (Perhaps the claim about (T*) below could be made equally plausible by the weaker claim that the inference from ' "p" is true' to 'p', and the converse inference, are always legitimate.)

view it is highly problematic even on the strongly disquotational notions of truth and satisfaction according to which (T*) and (S*) are analytic.

That there may be a problem in the third step becomes evident when one reflects on vagueness. Surely any position on truth must be compatible with certain words such as 'bald' having borderline cases; but for someone (say Jones) to be a borderline case of 'bald' is for there to be no definite fact of the matter as to whether Jones satisfies the predicate. My general claim is that indeterminacy in the notion of 'set', or in the second-order quantifier, is rather analogous: there are certain objects which neither definitely are in the extension of 'set' or the range of the second-order quantifier nor definitely outside it. But it is best to look at the vagueness case in slightly more detail.

There are two main approaches to vagueness. On one, the law of excluded middle fails when it comes to vague sentences: if Jones is a borderline case of vagueness we cannot accept the sentence

> Jones is bald or Jones is not bald.

This view, combined with the disquotation schemas, implies that we also cannot accept either

> 'Jones is bald' is true or 'Jones is not bald' is true.

And this gives fairly clear sense to the idea that 'bald' doesn't have determinate satisfaction conditions.

This approach to vagueness could easily be carried over to the kind of indeterminacy that has been argued to infect the notion of set or second-order quantification. If we adopted this approach, then the indeterminacy in the notion of set or of second-order quantification consists in the fact that the law of excluded middle doesn't apply to it. The failure of excluded middle is fully compatible with the disquotation schemas (T) and (S) and (T*) and (S*).

I have said though that my own preference is to adhere to classical logic despite the indeterminacy. To see how this works, it is necessary to see how to make sense of vagueness without abandoning classical logic and without modifying the disquotation schemas (at least for one's primary notion of truth). How, that is, is the advocate of these schemas and of classical logic to say what differentiates a vague predicate like 'bald' from a fairly definite predicate like '(absolutely) square', given that the instances of (S) involving each of them are analogous? The only way to do this, I think, is with an object-language notion of 'definitely' or 'determinately' that extends first-order logic. 'Definitely p' implies 'p', but not conversely; so although either Jones is bald or Jones is not bald, it doesn't follow that either Jones is definitely bald or Jones is definitely not bald. (In classical logic the disquotation schemas do imply that either 'Jones is bald' is true or 'Jones is not bald' is true; but this is compatible with neither 'Jones is definitely bald' nor 'Jones is definitely not bald' being true, or equivalently, with 'Jones is bald' being neither definitely true nor definitely not true.) Analogously, it doesn't follow from the disquotation schemas that for any x either x is definitely a set or x is definitely not a set; or that for any x and y, either x is definitely a member of y or x is definitely a non-member of y. And so even adhering to excluded middle, it is compatible with the disquotation schemas that 'set' and '∈' don't have definite satisfaction conditions.

(There is an ambiguity that is a possible source of confusion. To say that '∈' has definite satisfaction conditions could be interpreted to be true simply on the grounds that

(2) For any x and y, it is definitely the case that [<x,y> satisfies '∈' if and only if $x \in y$].

That of course does follow from the disquotation schemas plus classical logic plus reasonable assumptions about the definitely operator. But (2) implies only that '∈' *definitely has satisfaction conditions*; what is relevant to the claim that it *has definite satisfaction conditions* is not (2), but

(3) For any x and y, [<x,y> definitely satisfies '∈' if and only if $x \in y$].

The claim of indefiniteness is that this fails (as does any other sentence of form

For any x and y, [<x,y> definitely satisfies '∈' if and only if . . .]

in which a formula not involving 'definitely' goes in for the dots). If there is any confusion about this, consider the analogue for 'bald' in place of membership.)

I have been pressing an analogy between an uncontroversial phenomenon, vagueness, and a highly controversial one, referential indeterminacy in such notions as set membership and second-order quantification. The analogy is subject to a weak interpretation and a strong interpretation. The weak interpretation is simply that the two formal treatments I have considered for vagueness can be used for the referential indeterminacy of set membership too, showing how the claim of referential indeterminacy is compatible with acceptance of the disquotation schemas. The stronger reading is that the analogy is not simply a formal one: that the reasons that make 'bald' vague are indeed rather similar to the reasons that make 'set' indeterminate. The weak reading of the analogy is enough to undermine the view that it is incoherent to suppose our concept of 'set' indeterminate, but I am inclined to accept even the stronger reading of the analogy. What makes 'bald' vague—what makes it the case that there can be people who aren't definitely bald or definitely not bald—is that there is nothing in our training with the word 'bald' that could plausibly be thought to determine its extension, plausibly be thought to determine whether or not people like Jones are bald. What makes 'set' and 'membership' indeterminate is that there is nothing in our training in the use of these terms that could plausibly be thought to determine their extension, e.g. that could plausibly be thought to determine the size of the continuum. If this is right, then the analogy to vagueness is not purely formal.

3. COULD OUR CONCEPTION OF FINITUDE BE INDETERMINATE?

I now turn to a view on the other extreme, one which says that even the notion of finitude is referentially indeterminate. Actually I will begin by considering an especially radical version of this position: I'll call it radical indeterminism. (The longer version of the paper discusses a number of variants of it.) Then in Section 3B I will return to a more general consideration of the view that the notion of finiteness is referentially indeterminate.

3A. Radical indeterminism and Gödel's Theorem

Radical indeterminism has it that at any time, a person's concept of finitude is exhausted by the maximal mathematical theory of finitude that he or she implicitly accepts at that time: more specifically, any mathematical claim about finitude that has different truth-values in different models of someone's maximal theory has no determinate truth-value on that person's current conception of finitude.[10] (Since claims about natural numbers are in effect claims about linear orderings in which every object has only finitely many predecessors, the same holds for mathematical claims about natural numbers.) If there is indeterminacy as to what exactly is the maximal theory a person implicitly accepts, then there is indeterminacy as to what indeterminate concept the person employs; and so there will be no determinate answer as to which questions have determinate answers. For part of my discussion I will ignore such 'second-level' indeterminacy in discussing the radical indeterminist view, but I will argue in the end that nothing of importance turns on my doing so. (Also, for stylistic ease I will sometimes ignore

[10] There will be *non*-mathematical claims about finitude that we *will* want to regard as factual even though they aren't decided by any theory we accept: it may be that no theory we accept implies that there are only finitely many stars; but if in fact there are less than $10^{10^{10}}$ of them, and our mathematical theory implies that if there are less than $10^{10^{10}}$ Fs then there are only finitely many Fs, then of course we want to say that it is determinate that there are only finitely many stars. Here the failure of the mathematical theory to imply that there are only finitely many stars is due to its not having access to all the facts about stars, so we can grant that there is a definite fact here without granting that there are mathematical facts or facts specifically about finitude that transcend our theory.

interpersonal variation, and speak of the maximal theory 'we' implicitly accept—or if second-level indeterminacy is taken into account, the class of reasonable candidates for such a theory.)

In stating the radical indeterminist view I have employed the notion of 'implicitly accepting' a theory, but of course, this is quite vague. Roughly, we could say that I *implicitly accept* a sentence if I could fairly easily be brought to explicitly accept it without feeling I was learning anything essentially new; though this would require qualifications to rule out things I was brought to accept by fallacious arguments that I could be brought to see were fallacious. The extent to which what I implicitly accept goes beyond what I explicitly accept depends on how broadly we interpret 'fairly easily' and 'essentially new', but it seems clear that on any reasonable interpretations, what we implicitly accept is not closed under consequence.[11] We could roughly explain what it is to implicitly accept a mechanical inference rule in a similar (and similarly vague) manner. On any way of making precise the notion of implicitly accepting, take a person's *maximal theory* to be the closure of the set of sentences that he or she implicitly accepts under the set of mechanical rules that he or she implicitly accepts. So the class of reasonable candidates for a person's maximal theory is the class of theories that we get on different ways of making 'implicitly accepts' precise. This makes it clear that for any person at any time, there is indeed a non-empty class of reasonable candidates for the person's maximal theory at that time.[12] And a consequence of the radical indeterminist view is that any sentence not decided by any such theory has no definite truth-value on that person's current conception of finitude.

It seems to me that on any reasonable way of making 'implicitly accepts' precise, the set of sentences and inference rules that a person implicitly accepts at a given time is almost certainly recursive (and might well be finite). Also, for most of us, what we implicitly accept (on any reasonable way of making this precise) entails the axioms of arithmetic. Consequently, each reasonable candidate for our maximal theory that is consistent has a Gödel sentence that is undecidable in it. Let us now make the simplifying pretence that there is a uniquely best candidate for the maximal theory we accept at a given time. (I don't believe anything of substance turns on this simplifying pretence, though this will eventually require a bit of discussion.) And let us suppose this maximal theory M to be consistent. Then the radical indeterminist view has it that if G is the (any standard) Gödel sentence of

[11] The set of what we implicitly accept, as I've just defined it, might be inconsistent. It might be better to redefine 'implicitly accept' so that this couldn't happen, but it is simplest to stick to the definition given and to take the radical indeterminist as holding that if this set is inconsistent, then none of our mathematical sentences has a determinate truth-value.

[12] There is no obvious reason to think that this class of reasonable candidates has a largest member; for instance, it is not obviously closed under union. In other words, there may be no maximal reasonable candidate for being a person's maximal theory.

M, G has no determinate truth-value (on our current understanding of the terms in M). This seems prima facie much more surprising than the claim that the continuum hypothesis has no determinate truth-value. Among other things, few claim to have much certainty as to what the alleged determinate truth-value of the continuum hypothesis is; but (especially if we know what our maximal mathematical theory is—and I don't see any principled reason why we can't know this), don't we have available a good informal argument for believing the (any standard) Gödel sentence of this theory to be true? And if there is such a good argument for its truth, there is certainly a good argument for thinking that it has a truth-value, hence a good argument against radical indeterminism.

Against this, I will argue that the informal arguments for the truth of the Gödel sentence of our maximal mathematical theory are unsound; but I need to explain away the illusion to the contrary. (The discussion will assume a Platonist perspective; I say something about the fictionalist perspective in the longer version.)

The illusion to the contrary arises because we do have a very compelling argument for the truth of the Gödel sentence of Peano Arithmetic, and even for the truth of the Gödel sentence of Zermelo–Fraenkel set theory. This argument depends on an informal inductive argument that since the logical and non-logical premisses of the theory in question are true and the rule of inference (*modus ponens*) preserves truth then the theorems are true.[13] (Given this, we can argue that if the Gödel sentence is provable it is true; but by construction of the Gödel sentence together with a Tarski biconditional, we also have that if it is unprovable it is true; so it is true.) That argument that the theorems of the theory are true is what isn't formalizable in the theory in question. The most fundamental reason why it isn't formalizable within the theory in question is Tarski's Theorem: there is no adequate definition of truth formalizable in the theory. But we do I think have a concept of arithmetic truth (definable in set theory), and of truth in ZF (definable only in more powerful set theories); and the informal inductive argument involving truth seems quite compelling, so I think that we do have grounds for regarding the Gödel sentences of these theories as true.

One way to block this informal argument for the Gödel sentence of our maximal theory would be to argue that Tarski's Theorem shows that we have no truth predicate for the language of this theory (since if we had one, it would be part of that theory, given that that theory is maximal). But this has two problems. First, the claim that we employ a language for which we have no truth-concept is hard to swallow. (Doesn't the word 'true' express such a concept?) And second, this claim is *not* entailed by Tarski's Theorem;

[13] For simplicity I assume a formalization of logic with *modus ponens* as the only inference rule. On formalizations with a rule of generalization, the induction isn't on truth but on having a true universal closure.

rather, Tarski's Theorem entails only that any such predicate must be non-classical in that it can't obey the Tarski biconditionals without restriction. That applications of 'true' *to sentences not containing 'true' and related terms* should obey the Tarski biconditionals is hard to deny; to avoid denying this while allowing the predicate to have unrestricted application, we must suppose that the predicate 'true' is not definable in substantially more basic terms in our maximal theory, but instead is primitive and governed by axioms that are part of the maximal theory.[14]

But if we do have a truth-predicate for the language of our maximal theory M in that theory, and if as I shall assume we can use that truth-predicate in inductions, then can't we inductively argue that all theorems of M are true? And can't we use this to argue for the Gödel sentence of M in the way just reviewed?

I think it is clear that such an argument can't be correct: for if it were correct, there would now be nothing to stop its being given in M itself, which is of course impossible if M is consistent. But to make this more convincing, we need to look at the details of why the argument can't be carried out (either informally or in M).

To see why the reasoning for the truth of the Gödel sentence is unsound, you don't need to preclude use of the truth-predicate in inductions: indeed, we shouldn't preclude it, for its use in inductively arguing that the theorems of Peano Arithmetic and ZF are true seems perfectly fine. The problem rather is that to prove inductively that the theorems of M are true, you need to establish

(a) that all axioms of M are true

and

(b) that all rules of inference of M are truth-preserving.

Now, even if M didn't contain a truth-predicate, there might be difficulty in arguing for (a), if M isn't finitely axiomatized. Certainly there could be a problem with a formal argument for (a): the assumption that a given axiom is true seems to amount to little more than the assumption of the axiom

[14] The idea that we should avoid postulating an increasing hierarchy of classical truth-predicates by using instead a single non-classical truth-predicate that avoids Tarski's Theorem has become pretty standard at least since Kripke (1984). Kripke proposed that we define the predicate set-theoretically: however, to do so he had to take the predicate as applicable only to sentences whose quantifiers are confined to a specific set. (If we had a definition of 'true' in terms of other vocabulary, then certain applications of 'true' *to sentences containing only that vocabulary* would have to violate the Tarski biconditionals (which is implausible if that vocabulary is substantially more basic): for the Tarski biconditionals fail for certain sentences containing 'true', and the definition would then give rise to failures for sentences in the defining vocabulary.) The idea of regarding a non-classical truth-predicate as undefined and unrestricted in its application and governed by explicit axioms has now become common: see Feferman (1984), McGee (1991), and for an especially useful survey, Friedman and Sheard (1987).

itself, but the generalization that all of them are true is a more powerful claim, one that might well go beyond the power of the theory M. A case could be made that we would still have an informal argument for (a), at least if we assume that the person whose maximal theory is M knows that he believes each axiom of M and does so for good reason; this is a rather delicate matter to decide. Fortunately, there is no need to decide this (or to decide on the finite axiomatizability of M either); for *the problem with adhering to (a) and (b) becomes very much more serious when one remembers that M contains a (non-classical) truth-predicate.* For now the axioms of M must be taken to include axioms governing this truth-predicate, and the rules of inference must include special rules governing the truth-predicate; and it turns out that on all reasonable axiomatizations of the non-classical truth-predicate, either the assertion of the truth of these truth axioms or the assertion of the truth-preservingness of these special rules of inference (or both) is inconsistent with the axioms governing 'true'. In other words, even if we assume that there is no special problem with knowing that all of the axioms of M *not involving 'true'* are true, and no special problem in knowing that rules like modus ponens *that don't specially involve 'true'* are truth-preserving, this isn't enough for the inductive argument: the inductive argument requires that we know that even the axioms of M that employ 'true' are true and that even the rules that employ 'true' are truth-preserving. And the fact is that *on all the reasonable axiomatizations of the non-classical truth-predicate, the assertion of the truth of THESE axioms of M and/or the truth-preservingness of THESE rules of inference of M would actually be inconsistent with M.* I leave the demonstration of this claim to a technical note.[15]

[15] In illustrating where the inconsistency will occur, let's confine our attention to theories that employ full classical logic. Then the assertion that all logical and non-logical axioms of M are true and that all rules of M are truth-preserving implies that T(x) holds whenever x is a theorem of classical logic; indeed, whenever x is a theorem of Peano Arithmetic (which I take to be part of M). So in what follows we can assume as a background theory this plus the truth-preservingness of *modus ponens*.

Most theories of truth employ both the rules of T-Introduction (A/T<A>) and T-Elimination (T<A>/A) (as well as perhaps the rules ¬A/¬T<A> and ¬T<A>/¬A, which I won't consider here). The claim that T-Introduction is truth-preserving implies the schema

T-REP T<A> ⊃ T<T<A>>;

and the claim that T-Elimination is truth-preserving implies the schema

T-DEL T<T<A>> ⊃ T<A>.

But the two rules of inference together with these two schemas is inconsistent in the background theory just mentioned. (T-REP, T-DEL, and T-Introduction entail that every sentence is true; adding T-Elimination yields inconsistency. For a proof, see Friedman and Sheard (1987, 16), whose notation I am following.) So you can't accept both rules together with the claim that they are truth-preserving.

To give up T-Elimination seems thoroughly unattractive. Besides, the situation with T-Introduction is unacceptable on its own: in combination with the further axiom that there is

This shows why we can't prove G, or True<G>, in M.[16] Does it also show that we can't informally argue for it? I think it does. The informal argument would still have to be based on an inductive argument that the theorems of M are true; and the induction would have to employ the very same concept of truth employed in M, since it is the most general truth-concept we have. (If we had a richer truth-concept on which we could make inductions, then

at least one sentence that is not true, T-Introduction and the claim that it is truth-preserving (T-REP) are inconsistent. (Proof: Let L be the liar sentence, for which we can prove L ≡ ¬T<L>. So T<L> ⊃ ¬L; using T-Introduction and the law T<A⊃B> ⊃ (T<A> ⊃ T<B>), we get T<T<L>> ⊃ T<¬L>; with T-REP, this gives T<L> ⊃ T<¬L>. But it is easy to see (using the background theory) that if at least one sentence is not true then not both T<L> and T<¬L>; so ¬T<L>. So by the unused half of the liar sentence biconditional, L. So by T-Introduction, T<L>; contradiction.)

Keeping T-Elimination while restricting T-Introduction is a possibility; but the restriction of T-Introduction means that it is not obvious that all axioms can be declared true. Indeed, I think that the only attractive view that restricts T-Introduction does so to keep the axiom T<A> ⊃ A; but this is inconsistent with its own truth, as Montague's theorem (or indeed, more elementary reasoning involving only a liar sentence) shows. I imagine there are sharper results about the inconsistency of adding assumptions (a) and (b) in the text to one's truth theory than I have given here; but these should give the general flavour.

The proof of the Gödel sentence G requires not only the inductive proof I have been discussing: it also requires a restricted version of the conditional T<A> ⊃ A, or at least a restricted version of the principle ¬A/¬T<A>. For the induction would give you

If Provable<G> then True<G>,

from which we can infer

If ¬G then True<G>;

but unless we have True<G> ⊃ G, we can't get to ¬G ⊃ G and thence to G. (And unless we have G ⊃ True<G>, we can't get to True<G>.) Alternatively, by formalizing the proof of the first incompleteness theorem we can get to G from CON, the formalization of the consistency of M; but then even if we had the claim that all theorems of M are true, we'd need to infer ¬T<0=1> from ¬<0=1> in order to infer CON. I think though that reasonable axiomatizations of truth will allow at least restricated versions of these principles.

[16] Some may think that in my account of why the Gödel sentence of M is unprovable I am overestimating the importance of the fact that our maximal theory employs a truth-predicate for which some of our usual assumptions about truth must break down, and underestimating the importance of the gap between asserting of each of a class of sentences that it is true and asserting that all of them are true. Vann McGee has suggested to me an interesting argument for this view: his argument is based on 'Craig's trick' for obtaining bizarre recursive axiomatizations of recursively enumerable sets of sentences. Let M* be the Craig reaxiomatization of the set of consequences of M not containing the truth-predicate. M* has a Gödel sentence G that is undecidable in M*, and hence must be undecidable in M (since G doesn't contain the truth-predicate). But for each axiom A of M*, T(A) follows from M (since A doesn't contain the truth-predicate); and M also implies the truth-preservingness of the rules of inference of M*. So here the obstacle to proving G can only be the gap between proving T(A) for each axiom A of M* and proving the generalization that all the axioms of M* are true; the role of the non-classical nature of truth-predicates seems to have disappeared.

But I don't think that the role of the non-classical nature of the truth-predicate really has disappeared. Suppose that N is a typical theory that does not employ a truth-predicate, and that N* is the Craig reaxiomatization of the class of theorems of N that meet a certain (decidable) condition C. If we believe N, and N is finitely axiomatized, then we have available a formal argument (in a metatheory employing a truth-predicate for the language of N) for the truth of all the axioms of N*. (There is also a corresponding though somewhat more

we would have a more powerful mathematical theory than M, contrary to the assumption that M is maximal.) But the considerations just mentioned (and elaborated in n. 15) show that *the principles that would be required in the induction are not merely unavailable to us, they are incorrect in that they are inconsistent with M*. So the problem isn't just that the informal argument employing those principles goes beyond M; rather, the problem is that the informal argument is unsound.[17,18]

I have been arguing that we have no good informal argument for the Gödel sentence of our maximal mathematical theory; if this is right, it removes one reason for thinking that there are mathematical claims unsettled by this theory that have determinate truth-value on our current mathematical concepts. But it may be thought that my argument crucially presupposes the assumption that there is a uniquely best candidate for a person's maximal theory at a given time.

I don't think it does crucially presuppose this. Suppose that there is a class

questionable informal argument when N isn't finitely axiomatized but when we know that we believe each axiom of N and do so for good reason.) For in these circumstances we can argue that all theorems of N are true, hence that all the theorems of N meeting condition C are true, and the axioms of N* are simply cleverly coded analogues of these theorems of N. In the case of the maximal theory M on the other hand, this argument for believing that all the axioms of M* are true breaks down, since it depends on the lemma that all the theorems of M are true, which we don't get because M contains the non-classical truth-predicate. So even though M* doesn't contain the non-classical truth-predicate, the analysis of why we have no reason to think all its axioms are true does depend on the non-classical truth-predicate in the theory M from which it was obtained.

[17] This is so even for the Craig theorem version of the argument discussed in the previous note: there, the argument is unsound because the premiss that all the axioms of M* are true is not determinately true. (It is not formally decidable in M. Equivalently, it can fail in non-standard models of M, even though for each 'genuinely finite' sentence A in M* the claim that A is true is determinately true, because in non-standard models the range of quantification includes non-standard sentences.)

[18] There is a second strategy for trying to informally argue from M to its Gödel sentence G. The argument would begin by noting that $M \cup \{\neg G\}$ is a highly unattractive theory: it is ω-inconsistent, and its ω-inconsistency is provable in M. (The ω-inconsistency might arise from an inconsistency in M, in which case ¬G would be true.) The next step would be to argue from this unattractiveness in $M \cup \{\neg G\}$ to the claim

(1) that if we found ourselves committed to $M \cup \{\neg G\}$ we would have strong motivation to revise our commitments: either by dropping ¬G or by dropping part of M.

(Any reason to believe ¬G would be a reason to take the second alternative in (1): this is not surprising, since ¬G is equivalent (given a small fragment of M) to the standard formalization of the claim that M is inconsistent.) The third step would be to somehow get from (1) to

(2) We have grounds based on the axioms and rules of M for disbelieving ¬G, and hence (by excluded middle) for believing G.

But the gap between (1) and (2) is a big one on any view: (1) may suggest that the axioms and rules of M give grounds for thinking that *if ¬G holds we should revise M*, but that isn't the same as giving grounds for denying ¬G. (Maybe certain axioms and rules of M themselves give grounds for thinking we should revise M, even without the assumption that ¬G holds!) And the gap between (1) and (2) is especially big for the radical indeterminist, who denies that G has determinate truth-value prior to a commitment to either G or ¬G.

of candidates for my maximal theory; to be slightly crude about it, let's give each a numerical degree from 0 to 1 measuring how good a candidate it is for being my maximal theory, where the numerical degrees add to 1. Then for each theory M_i that gets a non-zero measure d_i for being my maximal theory, its Gödel sentence G_i gets a non-zero measure of at least d_i (and typically much more)[19] for being epistemically unwarranted. And so we can consistently suppose that each such G_i gets no fully definite truth-value.

I suppose that somebody might think that if G_i also has a non-zero degree of *warrantedness* then it must have a truth-value, for anything having no truth-value would be absolutely unwarranted. But this seems to me a mistake: the right conclusion is that if it has a non-zero degree of warrantedness then *to that degree* it has a truth-value. If we think that if there were a best candidate for a maximal theory then anything unsettled by that theory would have no truth-value, then the proper conclusion of there being no best candidate for the maximal theory is that there is also no best candidate for the set of sentences that have truth-values. There is no reason to assume that any sentence that is a possible candidate for having a truth-value definitely has one.

There is also another way to see that the indeterminacy as to what my maximal theory is shouldn't affect the conclusion I have drawn. For surely (if my mathematical beliefs are consistent) there is a recursively axiomatized consistent theory M so powerful that it includes any reasonable candidate for my maximal theory (as I have defined that—see n. 12). But then I have no good informal argument for its Gödel sentence, if what I have said is correct; so again, considerations about informal arguments for Gödel sentences can't undermine the view that sentences not formally decidable in M don't have definite truth-values on my current mathematical concepts.

It may be thought that the argument of this section gives insufficient weight to the fact that the mathematical theories we accept involve schemas: in particular, the schema of mathematical induction.[20] It is natural to interpret such schemas as not merely involving the assumption of each instance expressible in our language, but to involve a commitment to new instances as new terms are added to our language. But then, it seems, the commitment to the Gödel sentence of our maximal theory follows: we get the Gödel sentence from the induction schema, once we add a truth-predicate for our language.

I accept the view that accepting a schema involves not merely accepting the actual instances in the language at a time, but also taking on a commitment to new such instances as terms are added to the language. Even

[19] Since G_i will typically fail to follow from many of the other M_js, as well as M_i.

[20] If the induction principle is stated as a single sentence quantifying over sets or properties, then comprehension schemas about set or property existence are relevant instead.

so, I think that the argument sketched is faulty. For on my view, we don't get the Gödel sentence from the induction schema once we add a truth-predicate for the language: indeed, we already have a truth-predicate for our language, so there is nothing to add. Of course, the truth-predicate we have is non-classical, i.e. doesn't obey the Tarski biconditionals; couldn't we add a classical truth-predicate that does obey the Tarski biconditionals restricted to sentences that don't contain that classical predicate (but that may contain the non-classical one)? Yes, but then to deduce the Gödel sentence of the old theory we need to use not just instances of the induction schema involving this truth-predicate, but those Tarski biconditionals. Indeed, we need to add more: we need to add the generalization that a universal claim is true if and only if all of its instances are. These added principles are doing a large part of the work, not just the extension of the induction scheme to the new predicate; so the argument that the eventual commitment to the Gödel sentence of M was already built into our practice seems to me faulty.

3B. More basic worries about the indeterminacy of finitude

The view that our concepts of finitude and natural number are referentially indeterminate does not require the radical indeterminism outlined in Section 3A: one could posit a more limited referential indeterminacy that allowed the truth of the Gödel sentences to be settled while not allowing the truth of all sentences to be settled. (One could even in principle posit a referential indeterminacy so limited that every sentence of arithmetic got a determinate truth-value; but I doubt that indeterminacy not affecting truth-values can be made plausible.) But even such a limited referential indeterminism about finitude raises some deep worries: indeed, it raises worries that may seem to many more compelling than the worries based on Gödel's Theorem.

Before trying to get clear what these deeper worries might be, I need to make clearer what the indeterminacy claim involves. One often hears it characterized—including by some of its advocates (Skolem (1967); Putnam (1975, 22) and (1983, 2))—as the view that the notion of finiteness makes sense only *relative to a model*. The motivation for this characterization is that set theory implies that there can be a model M of set theory in which a certain set A satisfies the condition 'has only finitely many members', but where an expansion of the domain of the model to include new sets of natural numbers (but no new members of A) can lead to the same set satisfying the condition 'has infinitely many members'. (The set 'looks finite' in M because M contains a function mapping it 1-1 onto an object b satisfying the predicate 'natural number' in M; but in the larger model, b does not satisfy the predicate 'natural number'.) One way of arguing for relativity would be to conclude from this that set theory implies that the set A isn't really finite or infinite, but only finite or infinite relative to the model. But that conclusion is simply wrong: indeed, in the simplest examples of this situation, set theory implies that the set is infinite. The proper conclusion is just that set theory implies that there are models in which satisfying 'has finitely many members' is compatible with infinitude. (That this is the proper conclusion has been frequently pointed out by those who take the finite/infinite distinction to be determinate. I'm saying that we can and should recognize that this is the proper conclusion even if we regard the finite/infinite distinction as indeterminate.)

A slightly better variant of the above argument for relativity would be that the same set would be finite if M were the whole universe of sets and infinite if the larger M′ were; so the set is finite relative to M and infinite relative to M′. But among its other problems, this assumes that we can intelligibly talk of the same set occurring in a possible universe 'given by M' and a possible universe 'given by M′' (whatever exactly it might mean to say that a universe is 'given by' a model); and the identification of sets across universes is dubious at best. (There is no difficulty in making cross-world identifications of sets that are uncontroversially hereditarily finite: such sets can be cross-world identified recursively, by a 'cross-world extensionality principle', that is, by the principle that sets in different worlds that can be independently shown to have the same members should be identified. But that is no help in the case of a set A that is declared finite in M and a set B that is declared infinite in M′, where M and M′ are conceived as separate universes rather than embedded in a common universe of sets; for there A and B cannot be independently shown to have the same members by the cross-world extensionality criterion.)

To characterize the view that our notion of finiteness is indeterminate in terms of a relativity in the notion strikes me as thoroughly misleading. It suggests that the normal (unrelativized) notion of finiteness needs to be abandoned, and that in its place we put a new relativized notion of finite, not subject to the same difficulties. But I think this is doubly wrong. First, I don't think the normal notion does need to be abandoned, even if it is referentially indeterminate, any more than the referential indeterminacy of 'bald' forces us to give up the notion of baldness; indeed, the notion of finiteness is surely far too indispensable to be abandoned. Second, I don't think that there is any hope of introducing a relativized notion of finiteness that avoids the indeterminacy. The assertion 'There are finitely many As relative to model M' (where 'M' is a singular term for a model for some language L, and L contains a formula to which the model assigns the set of As as its extension) does not seem to me any more referentially determinate than the unrelativized assertion that there are finitely many As: models after all are sets of a certain type, so any indeterminacy in the notion of set is going to give rise to indeterminacy as to which model M we are referring to. And the problem doesn't arise only when we try to refer to a particular model: to say 'There is a model (of such and such a sort perhaps) such that there are finitely many As relative to that model' is equally indeterminate (if the basic Putnamian indeterminacy argument is correct), because of the indeterminacy as to what sets there are.

So let us dismiss the claim of the *relativity* of finitude as simply a mistake; the claim of interest is that the notion of finitude is referentially indeterminate, but not to be either relativized or abandoned.

I think that what is primarily worrying about the idea that our notion of finiteness is referentially indeterminate is that our notion of inferential procedure relies on this notion: any indeterminacy in our notion of finiteness will induce an indeterminacy in our notion of proving a claim from given assumptions according to a given proof procedure P, for it will induce an indeterminacy as to which discretely ordered strings of expressions (with first and last elements) that meet the other conditions for being a P-proof are genuinely finite and hence are genuine P-proofs. And the idea that it is indeterminate which strings of expressions are P-proofs (where P is a definite proof procedure) seems extremely hard to swallow.

I am not completely convinced, though, that the idea is one that it is impossible to live with. After all, any purported proof we come across will have less than (say) $10^{10^{10}}$ component formulas (each with less than say $10^{10^{10}}$ symbols). The indeterminacy we are contemplating, motivated as it is by the model-theoretic argument in Putnam (1983), will not be so radical as to question that this is definitely finite, so it will not question that there is a definite matter of fact as to whether it is a proof. Slightly more generally, the model-theoretic argument does not question that there is a perfectly definite fact as to whether any given sentence (of less than $10^{10^{10}}$ symbols) is *$10^{10^{10}}$-provable*, that is, provable (in the given proof procedure) with a proof of less than $10^{10^{10}}$ component formulas each with less than $10^{10^{10}}$ symbols. And isn't this all the determinacy we have a right to be confident of?

Of course, we can set-theoretically describe infinite sequences of formulas in a discrete ordering with first and last elements; and when someone describes one, we all declare unequivocally that it is not a genuine proof. But that is also compatible with the indeterminacy of 'proof': in all cases where what we've described is clearly infinite, it clearly fails to meet the conditions we impose on proofs, since we impose finiteness as one of the conditions. That we reject discrete orderings that are clearly infinite as not genuine proofs is obviously not enough to show the determinacy of our concept of genuine proof.

Let us be clear on the consequences that we arrive at if we take the 'Models and Reality' argument as showing that even the notion of finiteness is indeterminate. First, the argument only implies that there is indeterminacy about what is unprovable by any proof *however long*, not that there is indeterminacy about what is $10^{10^{10}}$-unprovable. (So though it may be indeterminate whether the Gödel sentence of our maximal mathematical theory is true, as *radical* indeterminism claims, we should not expect any indeterminacy as to the truth of a sentence asserting its own $10^{10^{10}}$-unprovability.) Second, even the unbounded notion of unprovability isn't illegitimate, and it will have perfectly determinate applications in certain cases, including those where we can *prove* unprovability in our maximal theory (at least, where we can $10^{10^{10}}$-prove it!). And even where it may have no determinate application, nothing prevents full classical reasoning using the notion.

Many will find this position radical enough; but actually it seems to lead naturally to a still more radical consequence so far suppressed. The consequence is that even the operators of first-order logic make no determinate contribution to truth-conditions. The argument for this is straightforward in outline: our understanding of these operators is determined in part by the proof conditions of sentences containing them; to the extent that the notion of proof becomes indeterminate, the operators become indeterminate too.

Some may feel that this line of argument overestimates the role that proof theory plays in an understanding of logical operators. I disagree, and think that those who think so attribute to the theory of truth a role in understanding that it could not conceivably have. (Which isn't to say that it has no role at all.) But I will not pursue this matter here. Let me simply say that the considerations earlier in this section suggest that the extent of indeterminacy in the logical operators needn't be great: it isn't obvious that there need be any indeterminacy at all in their contributions to the truth-conditions of sentences of humanly accessible length; indeed, maybe the only indeterminacy concerns strings of symbols for which it is indeterminate whether they are of finite length.

But there may be reasons for suspecting more indeterminacy in the contributions that our first-order logical notions make to truth-conditions. One worry is this: to the extent that arguments for the indeterminacy of more powerful operators are based on model-theoretic arguments like Löwenheim–Skolem and compactness theorems, they beg the question in favour of the determinacy of the first-order operators; for the notion of model builds in the contribution that the first-order operators make to truth-conditions. This is certainly a fair worry, and I am inclined to think that taking it seriously would raise substantial questions about the determinacy of the first-order quantifiers (though not about the determinacy of the truth-functional connectives, over and above the worries of the previous paragraph). But this is not a subject I will pursue here.

4. DOES PUTNAM'S ARGUMENT REALLY EXTEND TO THE NOTION OF FINITENESS?

Despite what you may believe, I'm really a commonsensical guy, and would like to avoid the radical conclusions of Section 3 if I can. Here is my best attempt to do so.

I begin by noting that the argument in the first half of Putnam (1983) isn't based solely on the fact that there are non-standard models of set theory and non-standard models of the second-order quantifier. If it were, it would surely extend to the notion of finiteness; but if you look at how the argument actually does work, the extension is very much more problematic. Putnam's argument is based rather on the fact that for set theory and second-order quantification there are non-standard models *of a particular sort*; and in the case of finiteness, there are no non-standard models *of that sort*. In the case of finiteness, you get the non-standard models by the compactness theorem, rather than the downward Löwenheim-Skolem theorem that Putnam uses in his argument. And that, I claim, makes a very important difference.

To explain this, I need to review the details of Putnam's argument. Putnam begins by recognizing that there is no need for an advocate of the view that 'set' and '∈' have determinate extensions[21] to think that their extensions are made determinate solely by our uses of these terms in pure mathematics. For anyone who has learned the notions of set and membership can apply them to the world around us; indeed, these notions (and related notions like that of a function, that are usually defined in terms of set and membership) are used heavily in the usual formalizations of physical theory. It is the entire range of uses of 'set' and '∈' that are available to determine their extensions.

But what is it about our usage of the words 'set' and '∈' that determines their extension, to the extent that their extension is determinate? Let us concede (as Putnam implicitly does) that the contribution that our first- and second-order logical operators make to truth-conditions is determinate *up to the ranges of the quantifiers.*[22] (I think this concession is in fact pretty reasonable: it seems to me that our inferential practice does largely determine the contribution that our logical operators make to truth-conditions (range

[21] I'm speaking loosely: the naïve interpretation of 'set' and '∈' is such that they have no extension, since too many things satisfy them to be incorporated into a set. So a more accurate formulation of Putnam's conclusion would be that there is no determinate answer to the question of what satisfies the words 'set' and '∈', or that it is indeterminate how they contribute to the truth-conditions of sentences containing them.

[22] I'm assuming here that there is no non-first-order logical operator in the language except possibly a second-order quantifier. If other kinds of non-first-order operators are in the language, such as $\exists_{\text{fin}}$, the concession would have to be stated more carefully, so as not to rule them determinate by fiat.

aside, in the case of the quantifiers).) Let us concede further (as Putnam quite explicitly does in the first half of his paper, before taking it back in the second half) that the interpretation of our 'observational vocabulary' is fairly fixed: it is fairly determinate what properties the observational predicates stand for. Or to put it in extensional terms, let us concede that it is determinate whether or not a given object (or n-tuple of objects) satisfies a given observational predicate, given that it is in the range of our first-order quantifiers. (Again I think this is quite plausible, ordinary vagueness aside: our practice of accepting and rejecting sentences containing observation predicates does largely determine whether any object or n-tuple of objects that we quantify over satisfies the predicates.) In fact, let's extend this a bit, from the 'observational vocabulary' to the entire non-mathematical vocabulary of the language. And let us take that non-mathematical vocabulary to include the predicate 'non-mathematical', so that it is determinate of any object whether or not it satisfies 'non-mathematical' (or 'mathematical'). Even conceding all this determinacy in the non-mathematical and logical vocabulary, Putnam argues that nothing determines close to a unique answer to the question of which things satisfy the predicate 'set'.

The argument is that the only reasonable candidate for what determines the extension of the concepts of set and membership is our entire body of beliefs and deductive practice involving these concepts; and that the way this body of beliefs and deductive practice fixes their extension is that their extension is to be such that to the extent possible, the beliefs are to be true and the deductive practices are to preserve truth. Even allowing the beliefs in question to include our non-mathematical beliefs, and assuming the interpretation of the logical and non-mathematical concepts in that body of beliefs to be fixed in the way just described, there are three possible sources of indeterminacy. The first—which I don't want to emphasize, but must mention in order to formulate the others properly—is that not all of our beliefs can possibly come out true, *however* 'set' and '$\in$' are interpreted: any person's beliefs are inconsistent, and even if you agree on a way to strip off the inconsistencies, any person's beliefs imply false observational claims; so only a proper subset of their beliefs can be made to come out true on some allowable interpretation, that is, on some interpretation that accords with the constraints imposed on the interpretations of the logical and non-mathematical vocabulary. And this means that there will be more than one choice of a maximal set of beliefs that can be made to come out true on some allowable interpretation or other. Presumably one such maximal set is more relevant (to the determination of the extension of 'set') than another to the extent that it is larger and contains more central beliefs, and to the extent that the observational beliefs in it are ones that the observational procedures we accept would license. But it is highly unlikely that any one such maximal set of beliefs will uniquely come out most relevant (on any reasonable explication of 'most relevant'); that is one source of indeterminacy. But let us follow Putnam and put this source of indeterminacy completely aside: let's assume that a given such maximal set S of our beliefs has been identified.

A second source of indeterminacy, also not terribly interesting, is that for any allowable interpretation of the given maximal set S of beliefs and any permutation f of its universe *which leaves the non-mathematical objects fixed*, the new interpretation that results by replacing each object o by f(o) in the extension of every predicate is also allowable (that is, it also satisfies the constraints on the physical vocabulary). But this isn't terribly exciting, since in the case of mathematics few would expect more than uniqueness of interpretation up to isomorphism.

The most important source of indeterminacy, and the one that Putnam emphasizes early in the paper, is the downward Löwenheim–Skolem theorem: any interpretation of the given maximal set of beliefs S will have a countable sub-interpretation that also satisfies S. *The fact that it is a sub-interpretation is important, for it guarantees that if the original interpretation satisfies the constraint on the interpretation of the non-mathematical vocabulary, the sub-interpretation will too.*[23] Now, any natural interpretation of the given set of beliefs S will contain objects y that are 'genuinely uncountable sets in the interpretation': that is, objects y such that there are uncountably many x such that <x,y> satisfies 'Set(y) & x∈y'. The alternative interpretations constructed by the downward Skolem–Löwenheim theorem violate this and so seem highly 'unintended', and yet they satisfy all of the constraints, *including the constraints on the physical vocabulary*.[24]

From this presentation, it should be clear why the Putnam argument as it stands does not go over to show that there is any indeterminacy in the notion of finiteness. For if we use a compactness argument instead of a downward Löwenheim–Skolem theorem argument to get the non-standard interpretation, there is not the slightest reason to think that the constraints on the physical vocabulary will be preserved. (And a downward Löwenheim–Skolem argument won't give you a non-standard interpretation of 'finite': finite sets satisfy 'finite' in every interpretation in which the axioms

[23] In other words, if the original interpretation is such that for any two objects x and y in its domain, <x,y> satisfies 'heavier than' if and only if x is heavier than y, then this must be so in the new interpretation as well; and analogously for each other physical predicate. (And for non-physical predicates too; though by combining the downward Löwenheim–Skolem argument with the trivial permutation argument of the previous paragraph, we would get models for which the extension to the non-physical predicates fails.)

[24] It is worth noting that the transitive collapse of the alternative interpretation, which is isomorphic to the alternative interpretation, also satisfies the constraints on the physical vocabulary. For transitive ∈-interpretations the 'naturalness' condition is more snappily formulated as the condition that the interpretation contain uncountable sets, so the transitive collapse violates naturalness in this snappier formulation.

of set theory come out true, since 1-1 correspondences between finite sets are themselves finite and hence exist in every such interpretation.)[25]

But even if Putnam's argument as it stands doesn't carry over to the finiteness case, it could still be that the standards he has imposed on an answer to the question of what determines the extension of 'set' and '$\in$' and 'finite' are enough to prevent 'finite' from having a determinate extension. To argue that this is not so, one needs a positive account of how the extension of 'finite' (or, the contribution to truth-conditions of '$\exists_{\text{fin}}$') might be determined. I will close with a modest suggestion about this; it is not entirely attractive, but I know none that is better.

The suggestion is that *if the physical world is as we typically think it is*, our physical beliefs are enough to determine the extension of 'finite'. Consider the one-place predicate B(Z), defined as

> Z is a set of events which (i) has an earliest member and a latest member; and (ii) is such that any two of its members occur at least one second apart.

We typically think that any set Z of events that satisfies this will contain only finitely many members; moreover, that there is no finite bound on the size of the sets that satisfy it. Call these 'our cosmological assumptions'. (The first says in effect that time is Archimedean, the second that it is infinite.) From our cosmological assumptions (plus some set theory), we can infer the following:

> (*) $\forall$Y{Y is a set $\supset$ [Y is finite $\equiv$ $\exists$Z$\exists$f(B(Z) & f is a function that maps Y 1-1 into Z)] }.

(Y is any set, not necessarily a set of events.) Now of course the theory S consisting of set theory plus our cosmological assumptions (which implies (*)) has non-standard models, in which certain infinite sets satisfy B(Z) and hence in which certain infinite sets satisfy the predicate 'finite'. However, if our first cosmological assumption is true, any such model must assign a non-standard extension to the formula B(Z); and *in particular, it must either contain things that satisfy 'event' which aren't events, or it must contain pairs of events which satisfy 'earlier than' or 'at least one second apart' even though*

[25] This formulation is really correct only for transitive $\in$-interpretations. (For interpretations generally, the actual finitude of a set is completely irrelevant to anything, since any mathematical object can play any role whatever in the interpretation.) The correct formulation is that in any interpretation in which the axioms of set theory come out true, *objects that are 'genuinely finite sets in the interpretation'* satisfy 'finite'; where to say that an object y is a 'genuinely finite set in the interpretation' means that it satisfies 'set' in the interpretation and is such that for only finitely many x does <x,y> satisfy '$\in$' in the interpretation.

Throughout the rest of the paper I will adopt the policy of talking about general interpretations in terms that are strictly appropriate only to transitive $\in$-interpretations. This saves considerable verbiage; the risk of misunderstanding is small; and it is routine to supply the correct reformulations.

the first isn't earlier than the second or the two aren't one second apart. Either way, the model will violate the constraint on the interpretation of the physical vocabulary, namely that the extension of such a predicate in the model can only contain things that actually have the corresponding property. In sum: if the first cosmological assumption holds then *no model of S in which 'event' and 'earlier than' and 'at least one second apart' satisfy the constraints on the interpretation of the physical vocabulary can be one where any infinite sets satisfy B(Z).* (Here 'infinite set' really means 'object y such that for infinitely many x, <x,y> satisfies '$\in$' in the interpretation'—see n. 25.) But it's easy to see that no pair of an infinite set and a finite set can satisfy '$\exists$f(f is a function that maps Y 1-1 into Z)' in any model; *so no allowable model of S can be one where any infinite sets satisfy 'finite'.* And of course there are no models of S where any finite sets fail to satisfy 'finite', since S includes set theory and genuinely finite sets satisfy 'finite' in every model of set theory; so (if the first cosmological assumption holds) *in every allowable model of S. a set satisfies 'finite' if and only if it is genuinely finite.*

This would be cold comfort if our second cosmological assumption weren't also true: if it were false, no model meeting the constraints on the interpretation of our physical vocabulary could satisfy the theory S (since finite sets bigger than any sets that actually satisfy B would be constrained to be both finite and infinite); in that case, the interpretation of 'finite' couldn't be fixed by S. And dropping the second cosmological assumption from S wouldn't restore the situation: for then (*) would have to be replaced by (*′) $\forall Y\{Y \text{ is a set} \supset [\exists Z\exists f(B(Z) \;\&\; f \text{ maps } Y \text{ 1-1 into } Z) \supset Y \text{ is finite}]\}$; and while we could still argue that any allowable interpretation of (the weakened) S is such that only finite sets of physical events satisfy B in the interpretation, it wouldn't follow that only finite sets satisfy 'finite' in the interpretation (since nothing would now prevent a set from satisfying 'finite' even if the interpretation contains no 1-1 mapping from it to a set of events satisfying B). In particular, there would be nothing to rule out non-standard derivations in a given formal system, derivations that contain infinitely many formulas even though the set of formulas in them satisfies the predicate 'finite'. So if we want to argue that notions like 'provable in such and such a system' have a determinate extension on grounds like those here considered, you need the second cosmological assumption as well as the first.[26] But with the two assumptions together, the constraints on the

[26] To a large extent the argument of the last two paragraphs extends to the case where we don't have set theory but have a primitive quantifier $\exists_{fin}$. More fully, in that case, the truth of our cosmological assumptions would rule out at least the most noxious sort of indeterminacy in our concept of finitude e.g. it would rule out indeterminacy in the notion of proof.

To discuss this, we first need a sufficiently comprehensive notion of model for a language containing $\exists_{fin}$ as primitive: one that allows non-standard models. So: take a model for this language to be a model in the usual sense together with a distinguished collection of subsets of the domain (perhaps required to be closed under the subset relation), and regard a formula of form $\exists_{fin}$xB as true in the model (relative to an assignment function) if the set of objects in

physical vocabulary determine a privileged class of interpretations in which 'finite' determinately stands for what it should, thus providing a way around all the nasty consequences contemplated in Section 3.

I will close with four comments to help clarify this proposed solution. First, the proposal is *not* that (*) is an adequate definition of 'finite', in a sense that makes (*) anything close to analytic. Nothing prevents us from discovering that time is finite in extent; what's true is only that if it is finite in extent, then temporal considerations of the sort just reviewed don't determine a unique extension for 'finite'.[27] If time is finite but our

the domain that satisfy B (relative to the rest of the assignment function) is in the distinguished collection. (A standard model will be one where the collection contains all and only the finite sets.)

Now suppose there is a 1-place predicate W(x) such that the following are part of our overall theory:

(i) Any x such that W(x) is an event;
(ii) There is an earliest x such that W(x), but no latest such x;
(iii) For any distinct x and y, if W(x) and W(y) then x and y occur at least one second apart.

And make the cosmological assumption

(iv) $\forall x[W(x) \supset \exists_{fin} y(W(y) \ \&\ y \text{ is earlier than } x)]$.

(The other cosmological assumption, that time is infinite, is now built into (ii) and (iii).) Then (i)–(iv) will of course have non-standard models; however, if our cosmological assumption is true, no model of (i)–(iv) in which 'event' and 'earlier than' and 'at least one second apart' have extensions in accord with the constraint on the non-mathematical vocabulary can be one in which anything satisfying W(x) can have infinitely many 'predecessors' satisfying W(x). This imposes a strong constraint on the 'distinguished collection of subsets' that serve as the 'finite subsets' of the model.

In particular, consider any formula Y(x,z) for which we believe $\forall z \exists_{fin} x Y(x,z)$. (Example: 'z is an expression of English and x is a subexpression of z.') Often—and in this example—we will be able to define a formula A(x,y) such that the following are part of our overall theory:

(v) $\forall x \forall x' \forall w \forall w'[\text{if } A(x,w) \text{ and } A(x',w') \text{ then } w = w' \text{ iff } x = x']$;
(vi) $\forall z \forall x[Y(x,z) \supset \exists w(W(w) \ \&\ A(x,w))]$

(In effect A(x,y) defines a 1-1 mapping of $\cup_z\{x|Y(x,z)\}$ into $\{x|W(x)\}$; in this example, of the expressions of English into the W events.) Then any reasonably comprehensive axioms for $\exists_{fin}$ will surely imply

(*) $\forall z\{\exists_{fin} x Y(x,z) \equiv \exists u \forall x \forall w(W(u) \ \&\ W(w) \ \&\ Y(x,z) \ \&\ A(x,w) \supset w \text{ is earlier than } u)\}$

But in that case, the models that our cosmological hypothesis restricts us to are all ones in which for any object o satisfying $\exists_{fin} x Y(x,z)$, there are only finitely many o′ such that <o′,o> satisfies Y(x,z). Thus at least the most noxious non-standardness has been eliminated.

[27] Modified temporal considerations might. For we could reduce the strength of the second cosmological assumption, by weakening B(Z) as follows:

> Z is a set of events, no two of which are simultaneous, which (i) has an earliest member and a latest member, and (ii) is such that any two members with no member between them are the same distance apart.

The asumption that this weaker version of B(Z) is satisfiable by arbitrary large finite sets is the assumption that time is infinite *in the large or small*, rather than in the large as before. But you don't get something for nothing: the assumption that only finite sets satisfy B(Z) is correspondingly strengthened to the assumption of Archimedeanness *in the large and small* rather than just in the large: that is, events infinitessimally far apart as well as infinitely far apart must now be ruled out.

cosmological assumptions about *space* are true, then conceivably an analogue of (*) for space could determine a unique extension for 'finite'. But if both are finite in extent—as is surely a serious empirical possibility—the view being suggested here is that the notion of finiteness would have genuine indeterminacy of the sort discussed in Section 3. I'm sure that some will feel that making the determinateness of the notion of finite depend on cosmology is unsatisfactory; perhaps,[28] but I don't see how anything *other* than cosmology has a *chance* of making it determinate.

The second clarificatory comment is that it is essential to the solution that the cosmological assumptions be *actually* true: you can't solve the problem modally, using the mere *possibility* that they be true (even if the concern is simply to fix the extension of 'finite' as applied to pure sets, so that worries about sets changing their members in counterfactual situations don't arise). Why not? For instance, if the world were finite, couldn't we fix the extension of 'finite' for pure sets by the condition

(**) $\forall Y\{Y$ is a pure set $\supset [Y$ is finite $\equiv \Diamond\exists Z\exists f(B(Z)$ & f is a function that maps Y 1-1 into Z)] }?

No: this move to the mere possibility of a set Z of events satisfying the condition does indeed let all genuinely finite pure sets into the extension of 'finite', even if they have more members than the number of seconds in the history of the universe; but the same flexibility that allows them in also allows in many infinite pure sets, because of the possibility of an infinite set Z of events satisfying B.[29] The basic problem is that the move to possibility means giving up the use of the actual world as in some sense a standard of finiteness, so there is no longer a way to keep the non-standard finite sets out.

My third clarificatory comment is that even if we had overwhelming evidence for our cosmological assumptions, it is not clear that scepticism about the determinacy of the notion of finiteness would be foreclosed: the sceptic could point out that since the physical theory that includes the cosmological assumptions has non-standard models, all our evidence is compatible with the physical world itself constituting a non-standard model of our

[28] It may be worth noting though that the two cosmological assumptions are also presupposed by certain intuitive explanations of finiteness: for instance, that a finite set is one that a being who lived forever and couldn't count more than one object per second could finish counting. (The proposal in the text is a considerable improvement over that intuitive explanation. First, there may be obstacles to counting a set other than infinitude: consider for instance a finite set of real numbers that aren't nameable in English. More important, the intuitive explanation gives incorrect results about finiteness when the cosmological assumptions fail, rather than making 'finite' indeterminate.)

[29] It would of course do no good to strengthen the condition B(Z) to require that Z be finite: the game is to look at interpretations of set theory plus (*) that satisfy the physical constraints, and if an extra occurrence of 'finite' is added into (*) it has to be subjected to the possibility of reinterpretation also.

theory, that is, a model in which time is non-Archimedean but satisfies all the sentences that we use to express the claim that it is Archimedean. If the physical world itself is such a non-standard model of our theory, then the above argument for the determinacy of finiteness doesn't go through: that argument worked by using comparison to the physical world to rule out the non-standard models. Actually there is some question in my mind as to whether it is even coherent to suppose that the physical universe might constitute a non-standard model of our theory.[30] But even if this sceptical possibility is coherent, it is not particularly believable: so the sceptical possibility does not undermine the claim that if we have evidence for the cosmological assumptions we should believe the notion of finiteness to be determinate.

My final clarificatory comment is about how far the argument here can be pushed: if cosmological considerations can make 'finite' quite determinate in extension (thus somewhat constraining the determinations of 'set' and second-order quantification as well), can they make less elementary notions determinate (thus providing further constraints on 'set' and second-order quantification)? It might be thought that if the argument works for 'finite' it works equally for 'has cardinality $2^{\aleph_0}$'. For define B*(Z) as 'There are points of space a and b such that Z is the set of points on the line segment between them', where the quantifier is restricted to non-mathematical entities, and take as cosmological assumptions that there are sets Z such the B*(Z) and that any Z such that B*(Z) has cardinality $2^{\aleph_0}$. (The first of these cosmological assumptions would be false if lines in space weren't made up of physical points (but points were only mathematical abstractions). The second assumption could err in either direction: there might be more than $2^{\aleph_0}$ points in any line segment in physical space if line segments contain points that are infinitesimally close; there might be fewer if line segments in physical space are ordered like intervals in some subfield of the real numbers.) Let S be set theory plus these cosmological assumptions. S implies that for any pure set Y, Y is of cardinality $2^{\aleph_0}$ iff there is a 1-1 mapping from a Z such that B*(Z) onto Y; can't we use this to argue as with 'finite' that in any allowable interpretation of S, a set satisfies 'has cardinality $2^{\aleph_0}$' if and only if it genuinely has cardinality $2^{\aleph_0}$?

No, for two reasons. First, even if we could assume that in any allowable interpretation the sets that satisfy B*(Z) have cardinality $2^{\aleph_0}$, we still would have no guarantee that the interpretation won't have sets of cardinality $2^{\aleph_0}$ that fail to satisfy 'has cardinality $2^{\aleph_0}$' in the interpretation; for the

[30] In discussing such questions it may be important to distinguish two versions of the sceptical hypothesis: one in which our lives are of genuinely finite duration even though there are pairs of events whose temporal separation is merely pseudo-finite, the other (suggested to me by both Hilary Putnam and Shaughan Lavine) in which our lives are themselves merely pseudo-finite.

interpretation might not contain the 1-1 correspondence to a Z such that B*(Z). No such thing can happen in the finitude case, since any 1-1 correspondence between finite sets exists in every interpretation of set theory. Second, we have no right to assume that in any allowable interpretation the sets that satisfy B*(Z) have cardinality $2^{\aleph_0}$. Indeed a downward Löwenheim–Skolem argument shows that there are allowable interpretations where only countable sets satisfy B* (and also allowable interpretations where only sets of any given uncountable size less than $2^{\aleph_0}$ do, if we assume the denial of the continuum hypothesis). Such sets will of course also satisfy 'has cardinality $2^{\aleph_0}$' in the interpretation; so being of cardinality $2^{\aleph_0}$ is neither necessary nor sufficient for satisfying 'has cardinality $2^{\aleph_0}$' in interpretations that satisfy the constraints on the physical vocabulary.

One might be tempted to strengthen the constraints on the physical vocabulary a bit: one might be tempted to assume that the domain of first-order quantification over the physical universe was determinate. (The constraints above guaranteed that in any allowable interpretation, quantifiers over the physical ranged *only* over physical things; the further constraint now being contemplated is that they include *every* physical thing.) In that case, 'has cardinality $2^{\aleph_0}$'would become slightly more determinate: it could no longer be satisfied in allowable interpretations by sets of cardinality less than $2^{\aleph_0}$. But it could still fail to be satisfied in allowable interpretations by sets of cardinality $2^{\aleph_0}$. The gain in determinacy would thus not be all that great, and would certainly fall far short of establishing the determinacy of 'set', or of second-order quantification generally. (Indeed, for cardinals greater than the cardinality of the physical universe we couldn't get even this much determinacy, as a modified downward Löwenheim–Skolem argument shows; and it is hard to see how to get this much determinacy for uncountable cardinals less than $2^{\aleph_0}$.) Moreover, it is not completely clear to me that the strengthening of the constraints on interpretation is reasonable: I don't think it is obviously determinate that our quantifiers range over all points of space rather than over some well-behaved sub-collection of them (and it is worth stressing that I did not assume this to be determinate in arguing that suitable cosmological assumptions could make determinate the notion of finiteness). In short, the prospects of getting useful determinacy results for notions less elementary than 'finite' are minimal.[31]

[31] You get a bit more determinacy if you take arbitrary *regions* of physical space to count as non-mathematical, and construe arbitrary quantification over them to be determinate. Indeed, if you assume that every region contains a minimal region (a point) which is itself taken as non-mathematical, and make a completeness assumption about regions in addition to assuming that the regions form a Boolean algebra with zero element removed, then quantification over sets of physical points would be determinate on these assumptions, because the assumptions guarantee that it is closely associated with quantification over physical regions. Note though that because this view requires that regions are non-mathematical entities, we still wouldn't have determinate quantification even over all sets of non-mathematical entities: that would require that quantification over sets of regions be determinate, which even our strong assumptions give us no reason to suppose.

REFERENCES

Feferman, Solomon (1984), 'Toward Useful Type-Free Theories, I', in Robert L. Martin (ed.), *Recent Essays on Truth and the Liar Paradox*, Oxford: Oxford University Press, 237–87.

Field, Hartry (1989), *Realism, Mathematics and Modality*, Oxford: Blackwell (revised edn. 1991).

Fine, Kit (1975), 'Vagueness, Truth and Logic', *Synthese* 30: 265–300.

Friedman, Harvey, and Sheard, Michael (1987), 'An Axiomatic Approach to Self-Referential Truth', *Annals of Pure and Applied Logic* 33: 1–21.

Kripke, Saul (1984), 'Outline of a Theory of Truth', in Robert L. Martin, *Recent Essays on Truth and the Liar Paradox*, Oxford: Oxford University Press, 53–81.

Lewis, David (1984), 'Putnam's Paradox', *Australasian Journal of Philosophy* 62: 221–36.

McGee, Vann (1991), *Truth, Vagueness and Paradox*, Indianapolis: Hackett.

Putnam, Hilary (1975), 'The Thesis that Mathematics is Logic', in Putnam, *Mathematics Matter and Method: Philosophical Papers, i*, Cambridge: Cambridge University Press, 12–42.

——(1983), 'Models and Reality', in Putnam, *Realism and Reason: Philosophical Papers, iii*, Cambridge: Cambridge University Press, 1–25.

Skolem, Thoralf (1967), 'Some Remarks on Axiomatized Set Theory', in van Heijenoort (ed.), *From Frege to Gödel*, Cambridge, Mass.: Harvard University Press, 290–301.

Weston, Thomas (1976), 'Kreisel, The Continuum Hypothesis and Second Order Set Theory', *Journal of Philosophical Logic* 5: 281–98.

4

Logical Consequence: Models and Modality

STEWART SHAPIRO

My long-standing interest in the notion of logical consequence became urgent when I was working on my recent book on higher-order logic (1991). The most common complaint is that second-order logic is out of bounds, as *logic*, because its consequence relation is not effective. Even though higher-order logic is squarely within the prevailing model-theoretic tradition, its set of logical truths is not recursively enumerable—it is not even in the Kleene hierarchy. Advocates of second-order logic have been accused of such absurdities as attributing occult powers to the mind and rejecting Church's thesis (see, for example, Burgess (1993)). The force of this argument depends crucially on what logical study is supposed to accomplish, our present topic. A common slogan is that logic is to codify the pre-theoretic norms of correct reasoning, the notion of logical consequence. This is surely correct, as far as it goes, but it does not go very far. What is this pre-theoretic notion of logical consequence, and what is it to codify something? Detailed answers to these questions are rare, even in works on philosophical logic. For example, in two widely read monographs both entitled *Philosophy of Logic* (Quine (1986) and Putnam (1971)), the question of what logical theory is all about is barely asked.

Several intriguing and complex questions lie in the vicinity: questions about the relationship between formal languages and natural languages, questions about logical form, questions about logical and nonlogical terminology, and general matters of epistemology and metaphysics. The purpose of this paper is to motivate the problems and provide the main outlines of solutions. The upshot is that when properly adjusted, the model-theoretic formulation of consequence provides a reasonable mathematical model of

This paper benefited considerably from the discussion at the conference, 'Philosophy of Mathematics Today'. The participants, who are the authors of the other articles in this volume, were most generous. I would also like to thank Nuel Belnap, Jill Dieterle, John Etchemendy, Ivor Grattan-Guinness, Robert Kraut, David McCarty, Penelope Maddy, Gregory Moore, Michael Scanlan, Barbara Scholz, Gila Sher, Mark Steiner, Neil Tennant, and Mark Wilson for reading an earlier version of this project and providing useful criticism. I apologize for any omissions from the list. I also thank the students who participated in two extremely productive seminars on this and related topics: John Chaplin, Karen Cordrick, Morgan Jackson, Jill Dieterle, Ty Lightner, and Pierluigi Miraglia.

logical consequence, one that answers to some of the modal, semantic, and epistemic intuitions underlying the pre-theoretic notion of consequence. As such, model theory generates insight into correct reasoning, and can be used to clear the vagueness from that notion.

1. HISTORICAL AND PHILOSOPHICAL BACKGROUND

Since Aristotle, the notion of consequence has been on the philosophical agenda. In the last few centuries, a number of influential mathematicians turned their attention to this concept. The list includes Leibniz, Bolzano, Boole, Schröder, Frege, Hilbert, Russell, and, more recently, Gödel, Church, and Tarski. Now, of course, logic is a respected branch of both philosophy and mathematics. And we all know where we are, with our model theory, proof theory, etc. There have been, and still are, a variety of characterizations of the intuitive idea that a sentence (or proposition) Φ is a logical consequence of a set Γ of sentences (or propositions). The following list is not exhaustive. The parenthetical names are meant to suggest directions, not historical accuracy.

First of all, there are modal, or metaphysical definitions:

(1) It is not possible for every member of Γ to be true and yet Φ false (Aristotle).
(2) Φ holds in every possible world in which every member of Γ holds.

There are also epistemic characterizations:

(3) There is a deduction of Φ from Γ by a chain of legitimate, gap-free (self-evident) rules of inference (Leibniz, Frege).
(4) It is irrational to maintain that every member of Γ is true and Φ is false.

And there are linguistic, or semantic, characterizations:

(5) Φ holds in every interpretation of the language in which every member of Γ holds (Hilbert, Tarski).
(6) The truth of the members of Γ guarantees the truth of Φ in virtue of the meanings of the terms.
(7) The truth of the members of Γ guarantees the truth of Φ in virtue of the meanings of a special collection of the terms, the 'logical terminology' (Tarski).
(8) There is no uniform substitution of the nonlogical terminology that would render every member of Γ true and Φ false (Bolzano, Quine).
(9) The truth of the members of Γ guarantees the truth of Φ in virtue of the forms of the sentences (or propositions).

Other slogans are that logical consequence is truth-preserving, logic is absolutely general, and logic is topic neutral. It might be added that there is a tradition, with roots back to antiquity, that argues that Φ is not a logical consequence of Γ unless Γ is *relevant* to Φ.

No doubt, some of the pre-theoretic notions are related to others. For example, (2) has been proposed as an explication of (1), in terms of one or another of the now common 'possible worlds semantics' of modality; and (5) may be offered as an explication of (2) and even (1), depending on how 'interpretations' are related to 'possibilities', a question that will occupy us shortly. One can also think of (7) or (8) as related to (9), if the 'logical terminology' of a sentence or proposition is what marks, or determines, its form. To say the least, however, some of these links are controversial; and at least prima facie, some of the characterizations are at odds with others.

In recent decades, harmony was achieved, partly through the development of the Hilbert programme and the Gödel completeness theorem and partly through Tarski's landmark (1935). Most philosophers and mathematicians have settled on a *model-theoretic* formulation of consequence, often called 'Tarskian'. In a background set theory, one characterizes a class of *models* for the language in question, and defines a relation of *satisfaction* between models and sentences. Then Φ is said to be a logical consequence of Γ if Φ is satisfied by every model that also satisfies every member of Γ. The subsequent development of model theory, via axiomatic set theory, has led to the rigorous codification of this notion, well studied in the field of mathematical logic. Logical consequence is sometimes held up as one of the few success stories of philosophy, as an issue that has been settled, once and for all.[1]

In a typical book or class on symbolic logic, one or two of the preformal characterizations of logical consequence are given at the outset, with or without a brief discussion. Very quickly, however, the reader or student finds himself proving theorems about the relation of satisfaction between certain strings on a finite alphabet and certain sets of *n*-tuples. The question before us here is the relationship between this model-theoretic notion of consequence and the ones given at the start of logic texts and courses. If we have truly succeeded, what is it that we have accomplished?

It is curious that even historically, there were few attempts to relate the model-theoretic notion of consequence to the more traditional metaphysical, modal, epistemic, and semantic characterizations, at least not by advocates of model theory. One may ask, for example, just what the various 'models' are supposed to be, and what they have to do with logic. Do they have something to do with modality, as in characterizations (1) and (2), or perhaps something semantic, as in (5)–(9)? And how do the epistemic matters of deduction and what one can rationally maintain ((3)–(4)) fit in?

[1] Intuitionism and the aforementioned school of relevance logic are two notable challenges to this harmony. We will consider another one in due course, Etchemendy (1990).

It might be added that the historical emergence of model theory is a story that has yet to be told.

The problem of interpreting logical consequence via model theory is complicated by the fact that many of the concepts involved in the preformal characterizations are now under serious attack. One of the most influential contemporary philosophers, Quine (e.g. 1960), has argued against the use of semantic notions like 'meaning' in philosophical programmes. Such notions, he claims, are not precise enough for respectable scientific analysis. Characterizations (6), (7), and, probably, (5) invoke such items directly. In any case, there is no consensus today on the notion of 'meaning', even among those who resist Quine's scepticism. If one is to rely on a semantic characterization of consequence, the relationship between 'meaning' and model theory must be addressed.

On the contemporary philosophical scene, there is also some suspicion of modal notions. Again, Quine is a major influence. In (1986) he writes, 'We should be within our rights in holding that no formulation of any part of science is definitive so long as it remains couched in idioms of . . . modality.' Even so, the notions of *logical* possibility and *logical* necessity enjoy wide acceptance. This is probably due to the fact that the logical notions have supposedly been explicated by model-theoretic ones. From the present perspective, of course, this puts the cart before the horse.

The scepticism towards modality does not dominate contemporary philosophical logic. Many influential writers, including Kripke, Lewis, Field, Hellman, Chihara, and Hintikka, make extensive use of modal notions in work on philosophy of mathematics and science, propositional attitudes, tense logic, modality, etc. However, there is no consensus on the extensions, let alone the proper explications, of the various modal notions, with the possible exception of the logical ones. Moreover, most of the work makes essential *use* of the model-theoretic notion of consequence. Many treatments attempt to explicate modal notions with a variation of model theory, relying on a theory of 'possible worlds'. In effect, the bulk of contemporary work on modality uncritically *presupposes* the correctness of the model-theoretic conception of logical consequence. The idea seems to be to *extend* the success of logic to other areas.

2. WHAT IS LOGICAL THEORY ABOUT, ANYWAY?

Clearly, we need to know what sort of question 'model theory' answers before we can assess how good this answer is. There are similar questions about another branch of logic, proof theory. What is the purpose of a deductive system? What is it attempting to capture? It will prove instructive to keep this case in mind.

We start with the slogan that logic is the study of correct reasoning. I take it that our focus is on reasoning that is carried out in a *language*, or something like a language, such as a 'language of thought' or a realm of propositions.[2] As such, reasoning involves the use or apprehension of linguistic items by the reasoning subject. So we must consider the role of a *natural language*, like English, or perhaps a natural language augmented with the special terminology and symbols of a branch of mathematics.

In contrast, model-theoretic consequence and, for the most part, proof-theoretic consequence are rigorously defined only for *formal* languages. Strictly speaking, a formal language is a mathematical object, a recursively defined set of strings on a fixed collection of characters. So, in order to see how model theory and proof theory relate to correct reasoning, we must first probe the relevant relationships between formal languages and natural languages, or whatever the medium of reasoning is. Why even use the word 'language' for both? One surely cannot claim, without further ado, that English just *is* a recursively defined set of strings on a fixed alphabet with the type of syntax typically found in logic books.

What follows are a few positions on this matter. I don't claim that the items below exhaust the terrain, nor are they necessarily exclusive.

One view is that a formal language accurately exhibits actual features of a natural language, or a certain fragment thereof. For example, some philosophers, like Montague (e.g. 1974), Davidson (e.g. 1984), and Lycan (1984) claim that declarative sentences of natural language have underlying *logical forms* and that these forms are displayed by formulas of a formal language (see also the papers in Davidson and Harman (1972)). Both of these claims are crucial.[3] Other writers hold that declarative sentences

[2] If, as the traditional intuitionists contend, there is such a thing as reasoning that cannot be represented in anything like a linguistic structure, it is beyond the purview of anything recognizable as logic. From the other perspective, one can wonder if linguistic resources are enough. In some branches of mathematics, reasoning involves pictures and diagrams. One could argue, perhaps, that these are mere heuristic aids, and are not part of the real reasoning. Or else one can try to show that these pictures have something like a syntax and semantics, and thus bring pictures into the linguistic fold. In conversation, John Etchemendy, Kenneth Manders, and Sun Joo Shin have pressed the importance of this.

[3] I am indebted to Robert Kraut here. There is no need to quibble over whether a given author holds that each unambiguous declarative sentence (in context) has a unique logical form, or whether each sentence exhibits a number of forms at once. There is an important difference in emphasis between the present question (concerning natural and formal languages) and the agendas of the aforementioned authors. Their goal is to provide semantic accounts of a wide range of natural language sentences, such as those dealing with modality, propositional attitudes, action, and perception. The programme is to systematically render such sentences into a formal language, be it first-order, higher-order, modal, etc. Although the 'translations' typically do not preserve anything like the grammatical form of the original natural language sentences, it is claimed that they do preserve *logical form*. For example, in various possible worlds accounts, modal operators are rendered as quantifiers. A programme is confirmed if 'intuitive consequences' among the natural language sentences are translated as model-theoretic consequences in the formal languages. On the other hand, present concern is with

express *propositions*; and formulas of formal languages somehow display the forms of these propositions. Notice that these are *realist* positions in that on such views, a given formal language is either correct or incorrect, depending on whether it displays the relevant mathematical structure underlying whatever is expressed by declarative sentences of natural languages.

Another view, held in part by Frege and, perhaps, Leibniz, is that because natural languages—even natural languages of mathematics—are hopelessly vague and ambiguous, they should be *replaced* by formal languages.[4] A similar view, held by Quine (e.g. 1960, 1986), is that a natural language should be *regimented*, cleaned up for serious scientific and metaphysical work. One desideratum of the enterprise is that the logical structures in the regimented language should be transparent. It should be easy to 'read off' the logical properties of each sentence. Quine sometimes uses the expression 'canonical form' for the relationship between declarative sentences and their regimented counterparts. A regimented language is similar to a formal language regarding, for example, the explicitly presented rigour of its syntax and its truth conditions.

In truth, however, no one has actually used a formal or regimented language for extensive science or philosophy, at least not all by itself. For the most part, such languages are poor instruments of communication, because they are so unwieldy, hard to scan, etc. To state the obvious, Quine still writes in English and Frege stuck to German, for the most part. The view at hand is better characterized as a proposal that we *envision*, or, to be honest, *pretend* that we are speaking and writing in a formal or regimented language. But when viewed this way, the connection between correct reasoning and model theory or proof theory is even harder to make out.

Virtually all mathematicians and many philosophers occasionally mention and even use formulas in a formal language to help disambiguate, or otherwise clarify what is meant. This suggests that one might think of a formal language as a proposed *addendum* to a natural language. Then our

the relationship between a formal language and the fragment of natural language it comes closest to, extensional or mathematical discourse. Our question concerns the relationship between model theory and the pre-theoretic notion of consequence *for that fragment*. In effect, present focus is on what seems to be a *presupposition* of the foregoing programmes—that model theory somehow gets it right for part of natural language.

[4] Frege also held that there is a fully objective realm of 'thoughts', and that sentences of a respectable, scientific language should reflect the composition of the thoughts they express. Fregean thoughts resemble propositions, as traditionally conceived. On our present topic, Frege (1879) wrote: 'I can best make the relation of my ideography to ordinary language clear if I compare it to that which the microscope has to the ordinary eye . . . Considered as an optical instrument, [the eye] exhibits many imperfections, which ordinarily remain unnoticed . . . [A]s soon as scientific goals demand great sharpness of resolution, the eye proves to be insufficient.' This metaphor suggests a realist interpretation, like that of the previous paragraph. It is not easy to delineate the descriptive and the prescriptive elements in Frege's philosophy of logic.

present question concerns the relationship between this addendum and the original language. What does the consequence relation of the addendum tell us about correct reasoning in general?

The Frege-Quine views begin with the observation that natural languages are vague and ambiguous, suggesting that the situation is hopeless. In reply, one might point out that the context of expression ought to resolve any ambiguities, and there are many rigorous theories of the semantics of vague terms. However, once again, virtually all of these theories presuppose the model-theoretic notion of consequence. In effect, model theory is an unanalysed starting-point of this research. So even if one or another of the various formal semantics of vagueness were correct, the problem at hand would remain.

On all the views so far, it would seem that in studying the model-theoretic or proof-theoretic consequence relation on a formal (or regimented) language we are studying what the author claims is either the real logical consequence relation of natural language (whatever that is) or what ought to be the consequence relation. Without rejecting either of these orientations outright, the view to be developed here hesitates at this point. We do not assume (or deny) that there is such a thing as *the* underlying logical form or forms, of most, if not all, of the declarative sentences of a natural language fragment (even in context), nor of the propositions supposedly expressed by them. On the other hand, for all its faults, natural languages do serve us well, at least when tools to increase precision and clarity are appended. Thus, we do not envision that we have replaced natural languages, much less help lobby for this replacement, earnestly or otherwise. A formal language is a *mathematical model* of a natural language (of mathematics), in roughly the same sense as, say, a Turing-machine is a model of calculation, a collection of point masses is a model of a system of physical objects, and the Bohr construction is a model of an atom. In other words, a formal language displays certain features of natural languages, or idealizations thereof, while ignoring or simplifying other features.[5] Advocates of the aforementioned views might claim that formulas in a formal language are mathematical models of logical forms, propositions, or sentences from the language of thought. The present orientation is consistent with this.

With mathematical models, which features one focuses on, which are idealized, and which are ignored depends on the purposes at hand, on why one is developing a model in the first place. Here, of course, our goal is to shed light on correct reasoning, and perhaps to improve reasoning. So, presumably, we should focus on and idealize those features of natural language that bear on correct reasoning, whatever those may be.

[5] Sánchez-Miguel (1993) also speaks of a formal language as a mathematical model, or what he calls a 'mock-up', of natural language.

From the present perspective, a mathematical model is a tool that is perspicuous, easy to work with, and easy to study. Of course, a model should also be 'realistic' in that some of its features do correspond, more or less, to features of what it is a model of—natural languages (of mathematics) in the case at hand. There should be a balance between simplicity and closeness of fit. Notice that, typically, there is no question of 'getting it exactly right'. For a given purpose, there may be bad models—models that are clearly *incorrect*—and there may be good models, but it is unlikely that one can speak of *the* 'correct model'. There is almost always a gap between a model and what it is a model of. Typically, one can make a model more 'realistic' (i.e. more correct) at the cost of making it more cumbersome and difficult to study and use—as happens, for example, when volumes are added and friction is considered in models of physical systems. But we should not be too hasty here. Sometimes, we do end up with compelling reasons to believe that we have 'got it right', that there is a unique mathematical model for a given purpose. For example, one might argue that this is the case with computability and recursiveness, via Church's thesis (see my (1993*b*), which includes some more mundane examples). Nevertheless, we need not assume in advance that we will find uniqueness, and model theory should not be faulted for failing to accomplish this grand dream.

An intriguing, but philosophically complicating factor is that we are out to model a *normative* notion, namely *correct* reasoning. The nature of normativity is one of the most subtle and baffling items on the agenda of contemporary philosophy. Unlike the case with what may be called 'descriptive models', it seems plausible here that a good mathematical model, if generally accepted, or generally accepted by 'experts', can come to affect, or even to constitute what counts as 'correct reasoning'.[6]

Notice that there are two uses of the word 'model' in play here, one in 'mathematical model', which refers to the relationship between a mathematical structure and what it represents, and the other in 'model-theoretic consequence', which refers to the various items *within* model theory, the

[6] This may distinguish models in the natural sciences from models in the social sciences. Burgess (1992) makes a distinction between a 'descriptive' theory which attempts to describe a given phenomenon, such as the practice of mathematical proof, and a 'prescriptive' theory which attempts to say what the phenomenon should be. The aim of some intuitionist writers is prescriptive, while the aim of traditional logicians is descriptive. The latter take the practice of providing proof, and try to give a formal account of *it*. As indicated, the situation is more complex than this, since here, the target of the descriptive enterprise is itself a normative concept. There is some philosophical literature on the use of mathematical models in the natural and social sciences, some of which relates this notion to the models of model theory (e.g. van Fraassen (1980), and the papers from a conference on the subject, published in *Synthese*, 12 (1960)). Once again, however, most of this work presupposes the correctness of the model-theoretic account of consequence. In effect, the models of model theory are a *starting-point* of this research.

familiar n-tuples. The extent to which the use of the same term 'model' is appropriate remains to be seen.

Typically, some features of a given mathematical model correspond to features of the reality that is being modelled and some don't. The latter are called *artefacts* of the model, and let us call the former 'representors'. In a point-mass model of a physical system, the co-ordinate system, the units of measure, and the notation for numbers are artefacts of the model. They don't correspond to anything in real physical systems. The metric and various relationships between forces and distances, like the inverse-square principle, are representors, not artefacts. Notice that, in a given case, it may not be clear what is representor and what is artefact, and perhaps the boundary is not sharp. For example, it took some doing to see the fact that the parts of the Bohr model are spheres with sharp boundaries is, at best, an artefact. One can also view the controversies over absolute space as debates over the extent to which parts of the co-ordinate system are artefacts. It seems that different theorists can agree that a model is itself a good representor of something while disagreeing about which parts are representors and which are artefacts. In the case at hand, it must be determined which features of formal languages, deductive systems, and model-theoretic semantics correspond to features of correct reasoning and which features don't.

Typically, a formal language consists of variables, connectives, quantifiers, parentheses, and schematic letters of various types, together with formation rules. Clearly, parentheses are artefacts, serving to make the formulas unambiguous. Of course, natural languages also have devices for disambiguating, but there is no reason to think that these correspond to parentheses in formal languages, nor is ambiguity under study here. It is 'idealized out'. On the other hand, variables, connectives, and quantifiers do have counterparts in natural languages, more or less, and so they are representors. Their deductive and semantic behaviour is at the core of the various formal concepts of consequence.[7]

Deductive systems are mathematical models of actual or possible chains of reasoning in whatever is modelled by the formal language. Ideally, a chain of reasoning should be correct if and only if it is represented by a deduction in the given deductive system (or if there is a deduction that represents an expansion of the chain of reasoning, filling any gaps). From this perspective, the purpose of a deductive system is to capture an epistemic notion of logical consequence, like item (3) or (4) in our list. Since Aristotle, some

[7] The role for variables, connectives, and quantifiers is part of the issue of the boundary between logical and nonlogical terminology, a matter that will occupy us shortly. It is not clear whether this boundary is an artefact or not. Notice, incidentally, that schematic letters represent both adjectives and common nouns. The difference between these is 'idealized out'.

philosophers have claimed that various components of a deductive system have counterparts concerning reasoning, usually on the epistemic side. Rules of inference represent 'primitive' or 'minimal' steps in a chain of reasoning, steps whose correctness is obvious, or have no gaps, or cannot profitably be broken down into parts. Then one might claim that every such step is represented in the rules of a deductive system, a 'completeness' of sorts. Some authors, like Hacking (1979) and Tennant (1987), claim that 'logical terminology' is characterized by conservative introduction and elimination rules of natural deduction systems, and suggest that such rules somehow represent the way that the 'logical terminology' of natural language is (or could be) learned. This would make introduction and elimination rules of the natural deduction format more than artefacts. At present, I presume a more relaxed view that most of the components of deductive systems are artefacts, and that we are only interested in the end-product, the notion of 'deduction' or maybe just 'deductive consequence'. Only this need correspond to anything concerning real reasoning in natural language.

Finally, what are we to make of the various models within model theory? What do they represent, or what, if anything, are the models models of? First, one might claim that models aren't models of anything—they are just artefacts of the logical system. Model theory would be only an elaborate technical device, a tool to study the deductive system. On this view, a model-theoretic semantics only has value to the extent that its consequence relation matches the deductive consequence relation. If there are soundness and/or completeness results, the theorist can determine what is and what is not deducible by studying the models.

This orientation towards model theory is consonant with philosophical views that take the deductive, epistemic formulation of logical consequence to be primary, item (3) in our list. Some logicists may have had a similar view, in holding that there is no perspective outside the envisioned formal language in which one can verify its axioms and rules of inference. The perspective has been called 'logic as language' (see Goldfarb (1979) and van Heijenoort (1967)). Ironically, a similar view towards deductive systems fits the 'social-practice' schools of philosophy, as articulated by the later Wittgenstein, Sellars, Brandom, and Robert Kraut. A common slogan is 'meaning is use'. Once one has identified the 'allowed moves' in the practice of reasoning, which is what a deductive system is out to model, there is nothing else to accomplish. There is no sensible question of going on to 'justify' the axioms and rules of inference in terms of a 'pre-theoretic' standard that correct rules must adhere to.[8]

[8] This perspective highlights the so-called 'paradoxes of implication' that were prominent in the writings of earlier logicians, and are still emphasized by relevance logicians. For example, in ordinary, everyday reasoning, one is probably not allowed to start drawing arbitrary conclusions once a contradiction is discovered.

On the other hand, even from the social-practice orientation, there is more to deductive reasoning than giving deductions. There is also a practice of challenging or refuting purported inferences. This is done by giving a counter-argument, an argument in the same form as the one put forward, but with true premisses and false conclusion. The realm of models may be thought of as representing the source of counter-models in natural language reasoning. So model-theoretic consequence would represent those natural language arguments that can't be refuted. There is no a priori reason to think that this class of arguments corresponds to those that can be deduced.[9] Clearly, it would be internally incoherent if there were a natural language argument that was both deducible and refutable by counter-argument, but there is no compelling reason to think that every natural language argument that can't be refuted can, in fact, be deduced.

Putting the 'logic as language' and the social-practice traditions aside, it is common among logicians and philosophers to think of the model-theoretic notion of consequence as primary. A deductive system must answer to the model theory, not the other way around. It is the *deductive system* that is said to be 'sound' or 'complete', *vis-à-vis* the model theory. If there is a mismatch between them, as happens with second-order languages, it is the deductive system that is at 'fault'. The conventional wisdom is that effective, sound second-order deductive systems don't have enough axioms or rules (to deduce every semantically valid argument), not that the semantics doesn't have enough models (to refute every deductively invalid argument).[10]

One desideratum of our account is to preserve at least part of the intuition behind this conventional wisdom. Presumably, the models correspond to something concerning correct reasoning, and this something is related to whatever it is that natural language deductions must answer to. Without adjudicating the question of priority, I do think there are different informal notions of consequence in play, and that model theory captures some of those. They are related to the semantic and modal items in our list. There is a rich interplay between such notions and the epistemic ones.

As a first approximation, I propose that the satisfaction relation between models and formulas of formal languages corresponds, more or less, to whatever it is that makes declarative sentences of natural languages true or false, i.e. to truth-conditions. To borrow an elegant phrase from Hodes

[9] This is an informal counterpart of soundness and completeness. Notice, incidentally, that there is no a priori reason to think that the class of arguments that can't be refuted is 'effective'.

[10] Someone who takes a line opposite to the conventional wisdom might claim that standard semantics doesn't have enough models, and go on to advocate Henkin semantics, which has more. Incidentally, an approach close to the conventional wisdom is not precluded by the social practice orientation to logic. Model theory is the result of the practice correcting itself, from within. I thank Robert Kraut for this point.

(1984), truth in a model is a model of truth. This recapitulates the intuition that valid arguments are 'truth-preserving'.

Although this is no more than a sketch, I don't have much to add by way of clarification. On this approach, the relation of *satisfaction* is a representor—it represents truth or truth-conditions. It would seem, then, that the models of model theory also represent something. But what? Etchemendy (1990) indicates two orientations towards semantics, which correspond to two answers to this question. One is that a model represents a *possibility* or a *possible world*, in which case model-theoretic consequence is a model of a *modal* notion of consequence, items (1) and (2) in our list. In the spirit of possible worlds semantics, the universe of models represents all possibilities. This programme is called 'representational semantics'. The other orientation is that a model is an *interpretation* of the *language*, giving alternate *meanings* or denotations or extensions to the nonlogical terminology. The resulting notion of consequence would correspond to some of the semantic notions, such as items (5), (7), and (9). The programme is called 'interpretational' semantics.

Consider a natural language sentence, 'Snow is black and Chicago is in Idaho.' In a sentential formal language, this would be represented by a formula $p\&q$. In sentential logic, the 'models' are lines in a truth-table. Consider the 'top' line in which p and q are both true. On the representational view, this would stand for a possibility that snow is black and Chicago is in Idaho, a possibility concerning the colour of snow and the location of a city. On an interpretational view, this line would stand for an interpretation of the language in which the sentences 'Snow is black' and 'Chicago is in Idaho' both mean something which is true. In the interpretation, 'black' might denote white and 'Idaho' might denote Illinois.

On the interpretational orientation, the distinction between logical and nonlogical terminology is central. The nonlogical terminology is what gets reinterpreted in the different models, while the logical terminology stays fixed. On the other hand, the logical–nonlogical boundary does not seem to play a central role in representational semantics. There we are out to characterize, or model, all of the possibilities—and the linguistic structures play a minimal role. So the extent to which the notion of logical terminology is an artefact seems to turn on the difference between interpretational and representational semantics.

3. THE CRITIQUE

Etchemendy (1990) launches a sustained, multi-faceted attack on the model-theoretic characterization of logical consequence. He claims that it systematically misses the relevant pre-theoretic notion of consequence.

Etchemendy does not say much about what the pre-theoretic notion is, relying instead on a number of intuitions.

It seems that model theory is not a good model of representational semantics nor is it a good model of interpretational semantics. First, the representational approach: Consider the sentences 'Bill is taller than Al and Al is taller than Bill' and '6 > 4 and 4 > 6'. Both are symbolized as *Rab&Rba*. This formula is satisfied by a model in which the domain is the set of natural numbers, the denotation of *a* is 2, the denotation of *b* is 7 and the extension of *R* is 'relatively prime'. Clearly, we can't think of this model as a *possibility* concerning the relative heights of Bill and Al or the relative sizes of 6 and 4. The model has nothing to do with either of these. Moreover, it simply isn't possible that Bill is taller than Al and Al taller than Bill, nor is it possible that 6 > 4 and 4 > 6. So there can't be a model that represents this. On this approach, models are supposed to represent only what is possible, not what is impossible. For reasons like this, Etchemendy argues that model theory be thought of as an interpretational semantics. And it does seem that the interpretational approach does fit much of what Tarski does in the classic (1935). In assigning denotations and extensions to the nonlogical terminology, we are interpreting, or reinterpreting it.

Nevertheless, from the representational perspective, one thing that model theory has going for it is the central idea of a 'domain-of-discourse'. This would stand for what the contents of the universe would be in the indicated representation, possibility, or possible world. However, it seems that the notion of a domain is foreign to the interpretational project. If quantifiers are logical terms, then in assigning denotations and extensions to the nonlogical terminology, the range of the quantifiers stays as it is—the real universe.[11] If quantifiers are nonlogical, anything goes. 'All' might be interpreted as 'none', 'few', or 'hardly any'.

Etchemendy proposes a way around this hurdle. We are to think of the quantifier, 'something', for example, as having two components, 'some' and 'thing'. The 'some' part is logical, and not open to reinterpretation, while 'thing' is nonlogical. So, for example, if the domain-of-discourse is the set of natural numbers, then 'thing' would be interpreted as 'natural number', and 'something' would come to 'some natural number'.[12]

[11] The notion of consequence developed in Tarski (1935) does not allow for varying domains-of-discourse, and so Tarski's notion of consequence is not the current model-theoretic one. But this exegetical matter need not be settled here.

[12] On this plan, all of the (other) nonlogical items would have to be interpreted over the set of natural numbers. Etchemendy calls this a 'cross-term restriction', suggesting that it is based on interlocking meanings of some of the terms. The proposal is analogous to insisting that the interpretation of 'bachelor' be a subset of the interpretation of 'unmarried'. Incidentally, the idea that the quantifiers are nonlogical, since they refer to varying domains-of-discourse, is suggested by the Enderton text (1972), which lists the universal quantifier '∀' among the 'parameters' of a formal language.

Etchemendy's major thesis is that no matter what, the model-theoretic notion fails to capture the modal nature of logical consequence. Even when the extension comes out correct, there is no analysis of consequence—the coincidence is an accident of the framework of the formalism.

Consider an inference from 'Andrew Jackson was President' to 'Andrew Jackson was male'. If the only nonlogical item in these sentences is 'Andrew Jackson', then the inference comes out correct on an interpretational semantics. On any other interpretation for 'Andrew Jackson' the conclusion would be true if the premiss is. But there is nothing 'logical' about this inference. The underlying generalization, all Presidents were male, is true, but not on logical grounds. It is an accident of history. Etchemendy refers to this phenomena as a substantive generalization. The fact that there are such generalizations tells against the interpretational orientation towards consequence.

From the opposite perspective, if we take 'or' to be nonlogical, then an inference to 'snow is white or Chicago is in Idaho' from 'snow is white' comes out invalid. Just reinterpret 'or' as 'and'. Here, it seems, we have lost all contact with modality.

The straightforward response to the first example would be to declare that 'President' and 'male' aren't really logical terms, and shouldn't be considered as such. If we could interpret 'President' as, say, 'Professor', then we would see that the first argument is not valid. Moreover, 'or' really is a logical term, and we can't undermine the second argument, about snow and Chicago, by varying it. But where do we draw the logical–nonlogical boundary? Etchemendy argues that there is no sensible, objective way to draw this line, referring to the very notion of a logical constant as a 'myth'. Clearly, it won't do to just say 'I know a logical term when I see one.' At least in (1935), Tarski also despaired of finding an objective criterion:[13]

> [N]o objective grounds are known to me which permit us to draw a sharp boundary between the two groups of terms. It seems to be possible to include among logical terms some which are usually regarded by logicians as extra-logical without running into consequences which stand in sharp contrast to ordinary usage . . . I also consider it quite possible that . . . we shall be compelled to regard such concepts as 'logical consequence' . . . as relative . . . which must, on each occasion, be related to a definite, although in greater or less degree arbitrary, division of terms into logical and extra-logical.

[13] In a letter of (1944), Tarski reiterated this point: 'sometimes it seems to me convenient to include mathematical terms, like the $\in$-relation, in the class of logical ones, and sometimes I prefer to restrict myself to terms of "elementary logic". Is any problem involved here?' Before Tarski, Bolzano (1837) also proposed an interpretational account of consequence, and he acknowledged the 'relativity' of his consequence relation. Bolzano gives 'Caius is human, therefore Caius has an immortal soul', with only 'Caius' as 'variable', as an example. Bolzano distinguishes between inferences whose validity can be known by logic alone and those that require other knowledge, such as that of theology. But, like Tarski (1935), Bolzano was sceptical of there being a clearly demarcated boundary between logical and nonlogical items.

There is, of course, a large literature on the logical–nonlogical boundary, which I won't go into here. Suffice it to note that most of it invokes notions like rationality, apriority, modality, and formality, items that are prima facie foreign to the interpretational approach. I think it must be conceded that if one sticks to an interpretational orientation exclusively, there is no objective way to mark the logical–nonlogical boundary. I take this as a reason not to adopt an exclusively interpretational approach. Etchemendy says that the alternative is a representational semantics, which model theory does not deliver. Against this, we need to combine the two approaches, of which more later.

When Tarski (1935) expressed scepticism of drawing the boundary objectively, he did not claim that his account captures the pre-theoretic notion no matter what boundary is chosen. Quite the opposite:

> This division is certainly not quite arbitrary. If, for example, we were to include among the extra-logical signs the implication sign, or the universal quantifier, then our definition of the concept of consequence would lead to results which obviously contradict common usage.

In sum, it doesn't follow from the lack of a sharp, objective border that any old border will do.

Etchemendy's relentless attack has other threads. He argues that even if we somehow settle on a logical–nonlogical boundary, interpretational semantics via model theory is still a bad account of logical consequence. With some boundaries, the theory gets the extension wrong and, even where the extension is correct, the account provides no compelling *justification* for this correctness. In short, model theory does not provide what a good philosophical analysis should deliver.

There is a problem with the 'size' of the universe. Etchemendy's ploy to take the particle 'thing' in 'something' as nonlogical, does allow us to consider domains-of-discourse that are *smaller* than the real universe, or to be precise, domains that are subclasses of the range of unrestricted quantifiers of ordinary language. But there is no way to invoke domains that are (or may be) *larger*. Suppose, for example, that the universe were finite, as a strict finitist might maintain. Let the universe have, say, exactly 12 objects. Let Φ be a (first-order) sentence that is satisfied in all and only models whose domains have at least 13 objects. Under our supposition about the size of the universe, $\neg\Phi$ is an interpretational logical truth, a consequence of the empty set. For a different example, let Ψ be a first-order sentence that is satisfiable only in models whose domains are infinite, e.g. $\forall x(sx = sy \rightarrow x = y) \,\&\, \neg\exists x(sx = 0)$. Then, unless the actual universe is infinite, $\neg\Psi$ is an interpretational logical truth. It is counter-intuitive for the class of logical truths to depend on the size of the actual universe, but that is exactly what happens from an interpretational point of view. Moreover, even the finitist should agree that $\neg\Phi$ and $\neg\Psi$ are not logically true, even though the interpretational perspective does not yield this.

The straightforward reply, of course, is to point out that the universe is not finite. However, even though Etchemendy himself is not a finitist, he holds that examples like these undermine the model-theoretic account, as an *analysis* of logical consequence. Presumably, the infinitude of the universe cannot be *demonstrated* on logical grounds alone (contra Frege). It follows that the correctness of the model-theoretic account of consequence also cannot be demonstrated on logical grounds alone, since this correctness depends on the infinitude of the universe. In other words, even if the model-theoretic account can be made to deliver the correct extension of logical consequence, this depends on 'accidental' features of the universe, features that (for all we know) might fail to hold.

Etchemendy notes that when we consider higher-order languages, the situation is worse. For example, there is a sentence that is logically true if and only if the axiom of choice holds, and there is a sentence logically true if and only if the continuum hypothesis fails (see Shapiro (1991), chs. 3, 5, and 6). As above, Etchemendy's complaint is not that choice fails or that the continuum hypothesis is true, but that the structure of the set-theoretic hierarchy is a matter of *mathematics*, not logic. The rejoinder to this is one of the main themes of my (1991). There is no need to draw a boundary between mathematics and logic. It is a Quinean point, despite the fact that Quine remains a staunch opponent of second-order logic. One should expect the study of correct reasoning, especially correct reasoning in mathematics, to have substantial mathematical theses.

Notice, incidentally, that all of the aforementioned examples come out as logically true on a purely *representational* orientation to semantics, if they are true at all. Indeed, if the axiom of choice is true, then it is *necessarily* true, and so there is no representation of the way the world might be in which the axiom of choice is false. Therefore, it is a representational logical truth, a consequence of the empty set. The same goes for either the continuum hypothesis or its negation. Similarly, if the natural numbers exist, they exist necessarily, and so there is no representation of the way the world might be in which the universe is finite. Thus $\neg\Phi$ and $\neg\Psi$ are both representational logical truths. Since Etchemendy uses such sentences as *counter-examples*, his view seems to be that the real pre-theoretic notion of logical consequence is not purely representational. It has an epistemic component as well, perhaps along the lines of items (3) and (4) in our list.[14]

[14] Etchemendy confirmed this in correspondence, suggesting that there must be a relation of *justification* between premisses and conclusion. The epistemic component of logical consequence is briefly discussed in his (1990, 93).

4. PUTTING THE PIECES TOGETHER: WHAT HAVE WE GOT AND WHY IS IT WORTH HAVING?

The pending question is to determine what, if anything, model theory is a model of. To do this, we should settle the dichotomy between representational and interpretational semantics. The demand that we go one way or the other is a false dilemma. It's not that I have a third option in mind (as for example, Sher (1996)). Rather, the interpretational and representational perspectives are not exclusive.

Coffa (1991) argues that since Kant, a central item on the agenda of philosophy has been to account for the necessity of logic and mathematics without invoking Kantian intuition. According to Coffa, the most successful line was what he calls the 'semantic tradition', traced from Bolzano through Frege to the Vienna Circle. The idea is that the necessity we are after lies in the use of language. Instead of speaking of essences or a priori intuition, we speak of meaning. This suggests that interpretational semantics may be our path to modality.

Notice that the representational–interpretational dichotomy is on a par with the analytic–synthetic distinction. At least by the end of the semantic tradition, an analytic truth is one that is true in virtue of meaning, while a synthetic statement is true or false in virtue of the way the world is. With interpretational semantics, we take a sentence or argument and vary the meanings of the terms. That is, we vary the *language*. With representational semantics, we vary the world, considering the sentence or argument with the same meaning, but in different possible worlds.

Due to Quine's influence, it is widely (but not universally) held today that the analytic–synthetic distinction is philosophically suspect. Most philosophers do not doubt that there is a difference between the role of language and the role of the world in determining the truth-values of sentences, but many hold that this distinction is not all that sharp and that the distinction cannot play roles assigned to it by the members of the semantic tradition.

Whatever the fate of this Quinean doctrine, I think that one should be sceptical of the interpretational–representational dichotomy. The semantic tradition suggests that much. There are, to be sure, clear cases of interpretational semantics. Suppose, for example, that someone took the conjunction 'and' in 'Snow is white and snow is not white' as nonlogical and considered 'Snow is white or snow is not white', as a 'model'. The last sentence cannot really be thought of as depicting a *possibility vis-à-vis* the original statement. Nevertheless, there are a large number of cases where we can go either way, as the above examples indicate. Moreover, it is very common to speak of the models of model theory as somehow representing possibilities, although it is added that it is *logical* possibility that is invoked and not some metaphysical possibility. I take it that the modal intuitions underlying logical consequence are worth preserving, if we can.

I propose to add another item to the above list of 'intuitive' notions of logical consequence, perhaps in a section by itself entitled 'conglomeration':

(10) Φ is a logical consequence of Γ if Φ holds in all possibilities under every interpretation of the nonlogical terminology in which Γ holds.

In the jargon of possible worlds, one models a given sentence or argument by first choosing a possible world and then reinterpreting the nonlogical terminology in that world. Thus, our item (10) blends the modal ((1)–(2)) and semantic ((5)–(9)) characterizations of logical consequence. The epistemic characterizations are brought into the picture indirectly.

My claim is that model-theoretic consequence can be made into a good mathematical model of this notion (10) and that both the intuitive notion so characterized and its mathematical model are useful tools for shedding light on the normative/modal/semantic notion of correct reasoning in natural language, the target of logic. I will use the remaining space to further sketch this claim and defend it.

There are two 'parameters' available to our logical theorist. One is the boundary between logical and nonlogical terminology, and the other is the strength of the set-theoretic metatheory—assumptions about the size and breadth of the universe of models. To get a good mathematical model of logical consequence, and sustain the relevant intuitions, both parameters must be adjusted properly.

To illustrate part of this claim, consider some extreme 'settings' for our parameters. Suppose that we insist that there is but one 'possible world', namely the real universe, and that every lexical item is 'logical'. That is, we don't vary anything on either the 'world' or the 'language' side. Then, as Tarski noted, 'logical consequence' coincides with what is sometimes called *material consequence*: Φ is a consequence of Γ if and only if either Φ is true or at least one member of Γ is false. Second, if every item is nonlogical, then 'logical consequence' is trivial: Φ is a consequence of Γ only if Φ is in Γ. Third, if every lexical item is 'logical' and we postulate a realm of 'possible worlds' representing every metaphysical possibility, then logical consequence coincides with what Etchemendy might call 'representational consequence', and model theory would be 'representational semantics'. In this case, we vary the world, but not the language. Fourth, if we say that there is but one possible world, the real universe again, but declare some terms nonlogical in order to vary their meanings, we end up with what may be called 'interpretational consequence'. Here we vary the language, but not the world.

Clearly, standard model theory does not coincide with any of these extremes, and none of them is a good articulation of logical consequence. As I see it, both 'world' and 'language' get varied, in an interrelated, controlled manner.

Let's start with the logical–nonlogical boundary. I noted above that from an *exclusively* interpretational perspective, there is no principled way to draw even a fuzzy boundary. So much the worse for that approach. Here we are trying to get at least some modality into the picture.

One desideratum of the present framework is that the logical terminology has a straightforward interpretation in every model. That is, given any model M and any logical term t, it should be clear how t is to be interpreted in M. This requirement tells against taking singular terms, like '6' or 'Bill Clinton', as logical. It would not be clear what such terms would denote in models in which 6 or Bill Clinton are not in the domain-of-discourse. In general, even though we are trying to maintain some modality as part of logical consequence, it is wise to avoid such metaphysical nightmares as 'transworld identity' in developing logical theory. The idea is to keep modality, but make it tractable. After all, logic has been more successful than metaphysics, and we do well to not hold logic hostage to metaphysical disputes if we do not have to. Again, this suggests that singular terms not be logical. We vary their denotata at will.[15]

Notice that such metaphysical problems do not occur with standard logical terminology. Connectives, truth-functional connectives, and quantifiers make sense in any model, with their usual meanings. We don't have to strain intuitions in order to understand such items or, to use the metaphor, to 'find' them in the models. Notice, incidentally, that the same goes for the membership or predication relation of higher-order logic. This is of a piece with Frege's motivation for including such items in logic. The requirement that logical terminology make sense in every model, without change of meaning, is related to the intuition that logic be 'topic-neutral'. Logic should apply anywhere, no matter what we are talking about.

Similarly, consider an inference like '17 > 2 and 17 is prime, so 17 is odd'. On our account (10), this can be blocked if either there is a possible world with large even primes, or if 'prime', '17', etc. are nonlogical. The former route invokes metaphysical issues about various types of modality—metaphysical, mathematical, etc. The latter does not.

To use Etchemendy's terminology, the main desideratum here is that there should be no 'substantive generalizations' involving only logical

[15] For similar reasons, predicates like 'President' and 'courageous' should not be logical. If the domain is the set of natural numbers, it is not clear what the extension of 'courageous' should be. One could declare that these predicates have the empty extension, but this seems counter-intuitive. If someone insists that such non-topic-neutral terms be logical, the best approach would be to develop a free logic.

terminology essentially. The sorts of examples that Etchemendy brings are handled by the fact that our notion of consequence (10) invokes other possibilities, varying the world as well as the language. Even if we could construe 'President' and 'male' as logical (see n. 15), the inference from 'Andrew Jackson was President' to 'Andrew Jackson was male' is undermined by considering a possibility in which there are non-male Presidents. However, we have no assurance, yet, that *every* substantive generalization is ruled out. To be precise, let Φ be any formula. If it is true that Φ holds in every possibility for every interpretation of the nonlogical schematic letters, then Φ should be necessarily true. To mimic Tarski's (1935) procedure, let Φ' be the result of replacing all nonlogical terms in Φ with appropriate (perhaps higher-order) variables, and then binding the result with prenex universal quantifiers over all free variables. For example, $(Pa \vee \neg Pa)'$ is $\forall X \forall x(Xx \vee \neg Xx)$. Notice that Φ' has no nonlogical terms, so there is nothing to reinterpret. In general, Φ is a logical truth if Φ' is true in all models, $\forall M(M \models \Phi')$. Thus, a modal notion, logical truth, is understood in terms of the plain truth of a generalization. Etchemendy calls this 'Tarski's fallacy'. The point here is that there is no error if, whenever the generalization $\forall M(M \models \Phi')$ is true, then Φ has the requisite modal status.

To handle this requirement, we turn to our other parameter, the strength of the metatheory. The underlying idea is that the modal nature of logical consequence is represented in the universe of models. It is, in effect, an exchange of modality for ontology, or in the present framework, a *modelling* of modality with ontology. As Putnam (1975) puts it, in another context:

> [M]athematics has . . . got rid of *possibility* by simply assuming that, up to isomorphism anyway, all possibilities are simultaneously *actual*—actual, that is, in the universe of 'sets'.

To put it negatively, the model-theoretic programme will fail if there are not enough models to represent every 'possibility', in the appropriate way. The metatheory is designed to make sure that this does not happen.

It turns out that what is 'the appropriate way' depends on the logical–nonlogical boundary. To illustrate this, suppose we construe a predicate N for 'natural number' as logical and it happened that every model contained the natural numbers. Then $\exists$xNx would be a *logical* truth. Frege might have believed this, but it seems counter-intuitive today. We can block it either by making N nonlogical or by including models that don't have any natural numbers. Of course, standard logics do both.

Next, to follow up another of Etchemendy's complaints, suppose that there is no model whose domain is infinite, as a finitist would contend (i.e. suppose that the axiom of infinity were false). Then, as noted, a first-order sentence like $\neg[\forall x(\mathrm{s}x = \mathrm{s}y \rightarrow x = y) \,\&\, \neg\exists x(\mathrm{s}x = 0)]$ would be logically true, something even a finitist should deny (unless she held that the finitude of the universe is necessary on logical grounds). The proper response is that if there aren't any infinite models, then model theory is a poor mathematical model of logical consequence. The only conclusion to draw, however, is that model theory is not available to a finitist.[16] He should not use it. But this does not prevent the rest of us from invoking an axiom of infinity in the metalanguage, in order to bring our model theory in line with the intuitive notion. We don't need to claim that the axiom of infinity is true on logical grounds, since we are out to *model* the intuitive notion of consequence, not to give some sort of philosophical analysis of it. We don't have to prove *on logical grounds alone* that our model is a good one. In defending our account, we use whatever resources are available.

This point can be illustrated with another example. If our finitist were correct, then the standard analysis/explication of computability would also fail. There would not be enough Turing-machines to represent some (possible) algorithms. But this does not prevent the rest of us from accepting Church's thesis.

As far as cardinality goes, an axiom of infinity is enough for first-order languages. The downward Löwenheim–Skolem theorem entails that if a (countable) argument has a counter-model at all, it has one in a model whose domain is at most countable. The situation with second-order logic is more complex. There are sentences, with no nonlogical terminology, that are satisfiable only in domains of Mahlo cardinality, domains greater than measurable cardinality, etc. So, if these sentences are not to be declared logically false, the metatheory must have some rather strong axioms of infinity (see Shapiro 1991, ch. 6)—unless the possibility of such large domains can be ruled out on logical grounds. This illustrates the fact that the proper settings of our two parameters are not independent. Second-order logic, which amounts to taking certain quantifiers and other terminology as logical, requires richer assumptions about the set-theoretic hierarchy than first-order logic does.

So far, the only constraints on the metatheory we have considered concern the cardinalities of the models. Is this the only type of constraint we need to consider in order to avoid substantive generalizations? With standard first-order or higher-order languages, the answer is 'yes'. This is a consequence of what may be called the *isomorphism property* of the formalism: If two models M, M' are isomorphic *vis-à-vis* the nonlogical items in a formula Φ, then M satisfies Φ if and only if M' satisfies Φ. The isomorphism property is a manifestation of the intuition that logical truth and logical consequence should be a matter of 'form', to the extent that

[16] The same example also shows that a finitist should not believe that first-order logic is complete, since Gödel's completeness theorem depends on an axiom of infinity (in the metatheory). The separate question of whether model theory is available to a nominalist is interesting, but won't be pursued here. See Shapiro (1993*a*).

isomorphism preserves 'formal' features of various models (whatever those are). I would submit that the isomorphism property is a necessary condition of any proposed model theory worthy of the name. It is an explicit requirement in abstract studies of model-theoretic logics (e.g. Barwise and Feferman 1985).

Suppose that a given sentence Φ is false in a model M whose domain is d. Given any set d' with the same size as d, there is a model M' isomorphic to M whose domain is d'. M' is obtained by interpreting the nonlogical items of Φ in d'. By the isomorphism property, Φ is false in M'. Also, let N and N' be models whose domains have the same cardinality. Then Φ is satisfied by N under every interpretation of the nonlogical terminology if and only if Φ is satisfied by N' under every interpretation of its nonlogical terminology. In short, the only thing that can prevent a given formula from being a logical truth, or the only thing that can prevent a given formula from being a consequence of a set of formulas, is a model of a certain *size*.

These considerations might provide comfort to those who are sceptical of modality. If the isomorphism property holds, then in evaluating sentences and arguments, the only 'possibility' we need to 'vary' is the size of the universe. If enough sizes are represented in the universe of models, then the modal nature of logical consequence will be registered. It is a pleasing situation. We avoid most of the sticky metaphysical questions about modality, moving to interpretational semantics instead. The only 'modality' we keep is 'possible size', which is relegated to the set-theoretic metatheory.

In sum, we can be sure that there are no substantive generalizations in formal languages because the logical–nonlogical boundary is chosen to satisfy the 'isomorphism' property and because the set-theoretic hierarchy is big enough. The isomorphism property is closely related to Tarski's (1986) and Sher's (1991) proposal for the logical–nonlogical boundary. Their idea is that a term is logical if its extension is invariant under any permutation of the domain. Tarski sees this criterion as an extension of Klein's *Erlanger* programme for characterizing various geometries in terms of what is invariant under different sorts of permutations. I submit that the Tarski–Sher criterion is a necessary condition on the notion of a logical term. Sher takes it as registering the intuition that logic is 'formal'.[17]

Notice that these considerations do not settle the question of whether the intuitive, preformal notion of logical consequence is 'formal', whatever that might mean, nor does it say whether there is even a fuzzy logical–nonlogical boundary for natural language, the language of thought, etc. These matters seem to depend in part on the extent to which the logical items of a given formal language and model theory correspond to items in

[17] I won't pass judgement on their further claim that this criterion is also sufficient, because, as noted above, I don't think there is a strong need to fix the logical–nonlogical boundary once and for all (with Tarski 1935, 1944, see n. 13 above).

natural language. Strictly speaking, the connectives and quantifiers of first-order formal languages do not occur in natural languages, and the natural language counterparts of these items are not exact equivalents. Beginning logic teachers and texts usually note some of the differences at the outset, but don't make much of them after that. Again, there is a gap between model and modelled. All we need to claim here is that with first-order or higher-order languages, the match here is close enough to yield a good model of correct reasoning.

I close with a few remarks on the connection with the epistemic aspects of logical consequence, items (3) and (4) in our list. The blended interpretational–representational approach (10) proposes a standard that real chains of inference must adhere to, but there is no reason to demand that every logical consequence can be verified by a chain of legitimate, gap-free reasoning. This is analogous to the idea that a deductive system should be sound, but not necessarily complete.

More can be said, however. It is often claimed that ordinary reasoning should be 'truth-preserving', but it is not clear what this means. The slogan could reflect the modal component of logical consequence, in the sense that if the premisses of a good argument were true, then the conclusion would *have to be.* Or it could reflect an intuition that the argument have a valid *form*, assuming that one can delimit the form–content distinction in natural language, or whatever the vehicle of reasoning is. At the very least, however, ordinary consequence should adhere to *material* consequence. Using the rules, we can't reach a false conclusion from true premisses. Tarski (1935) notes that with his notion of consequence, we get at least this much: 'it can be proved, on the basis of this definition, that every consequence of true sentences must be true.' This applies to the blended interpretational–representational notion (10), provided only that the real world is possible. It applies to model theory to the extent that there is a model that represents the world (i.e. has enough objects).

Etchemendy points out that the simple 'theorem' that Tarski describes does not depend on where the logical–nonlogical boundary is drawn. It holds if 'President' is logical and if 'or' is nonlogical. Indeed, the theorem is a simple universal instantiation. If every interpretation of the nonlogical terminology is a material consequence, then so is the original interpretation. Etchemendy then claims that, because of this, the theorem fails to highlight the modal nature of consequence, accusing Tarski of an elementary modal fallacy.[18]

[18] Let $IC(\Gamma,\Phi)$ be a formula asserting that Φ is an interpretational (or Tarski) consequence of Γ, and let T be a truth-predicate for both formulas and sets of formulas. Then the theorem is $IC(\Gamma,\Phi) \rightarrow (T(\Gamma) \rightarrow T(\Phi))$, which entails $\Box[IC(\Gamma,\Phi) \rightarrow (T(\Gamma) \rightarrow T(\Phi))]$. Etchemendy accuses Tarski of inferring $IC(\Gamma,\Phi) \rightarrow \Box(T(\Gamma) \rightarrow T(\Phi))$, which would, of course, be a modal fallacy. In any case, Etchemendy is right that the above theorem, by itself, fails to establish that

More to the point here, Etchemendy also argues that Tarski's notion of consequence is useless on the epistemic front. In general, we can't know that a conclusion follows (in Tarski's sense) from the premisses without first knowing that the conclusion is true if the premisses are. That is, we can't know that an argument is a consequence in Tarski's sense without first knowing that it is a material consequence. For example, we can't know that 'Andrew Jackson was male' follows from 'Andrew Jackson was President' unless we already knew that there were no non-male Presidents.

Etchemendy's remark is correct in the peculiar sense that philosophers and mathematicians sometimes use the phrase 'in general'. That is, with the *most unfavourable* logical–nonlogical divisions, we can't know that a given argument is an interpretational consequence without first knowing it is a material consequence. But this situation is not universal, and, as above, it is not the one Tarski intended.

In the same sentence as the one just quoted, Tarski wrote, 'the consequence relation which holds between given sentences is completely independent of the sense of the extra-logical constants which occur in these sentences.' This is clearly correct, and does reflect an important intuition. However, as Etchemendy's analysis makes clear, it is not enough. In a way, the consequence relation should depend *only* on the 'senses' of the *logical* terminology, and not on any accidental features of the world. But this last does not hold with all logical–nonlogical boundaries. In particular, it fails if there are substantive generalizations with only logical terminology. With more standard boundaries, such as those of first-order and higher-order logic, one can establish in the metatheory that some argument *forms* are model-theoretic consequences, using only the truth-conditions for the logical terminology (and other facts about the sizes of models). Consider, for example, typical soundness theorems, where one shows that a given deductive system is 'truth-preserving'. In such cases, model-theoretic consequence does connect with the epistemic intuitions underlying logical consequence. We *know* an argument in natural language is 'truth-preserving' to the extent that some of the terminology in it corresponds to the logical terminology of a model-theoretic consequence in the formal language, and we can know that the formal counterpart is valid from the truth-conditions of the logical terminology in the model-theoretic semantics. In other words, if we know that model theory is a good mathematical model of logical consequence (10), then we know that we won't go wrong using a sound

interpretational semantics, by itself, reflects anything modal. Whether Tarski thought otherwise, and whether such a giant of logical theory committed this elementary fallacy (wittingly or otherwise), depends on how we are to read his word 'must'. Sometimes this word is used to emphasize a conclusion, without asserting any modal status for it: 'It rained last night, so my clothes must be ruined', 'the lights are out, so a fuse must have blown'. On the exegetical front, we have a choice between attributing an elementary logical error to Tarski or finding a more charitable reading of the text.

deductive system. Also, we can know that an argument is a logical consequence (and thus a material consequence) via a set-theoretic proof in the metatheory. In either case, true premisses yield true conclusions.

With standard first-order and higher-order languages, the match between model and modelled is so close that, at least for mathematical language and reasoning, model-theoretic consequence has come to *constitute* the standard of correct inference, in roughly the same sense that a good dictionary constitutes the standards of correct spelling and word usage. One is likely to reject a proposed inference just because it does not correspond to a model-theoretic consequence in an established formal language. The purpose of this paper is to show how this is as it should be.

REFERENCES

Barwise, J., and S. Feferman (1985), *Model-theoretic Logics*, New York, Springer-Verlag.

Bolzano, B. (1837), *Theory of Science*, tr. by R. George, Berkeley, University of California Press, 1972.

Burgess, J. (1992), 'Proofs about proofs: A defense of classical logic', in *Proof, Logic and Formalization*, ed. by Michael Detlefsen, London, Routledge, 8–23.

——(1993), review of Shapiro (1991), *Journal of Symbolic Logic*, 58: 363–5.

Coffa, A. (1991), *The Semantic Tradition from Kant to Carnap*, Cambridge, Cambridge University Press.

Davidson, D. (1984), *Inquiries into Truth and Interpretation*, Oxford, Clarendon Press.

——and G. Harman (eds.) (1972), *Semantics of Natural Language*, Dordrecht, Reidel.

Enderton, H. (1972), *A Mathematical Introduction to Logic*, New York, Academic Press.

Etchemendy, J. (1990), *The Concept of Logical Consequence*, Cambridge, Mass., Harvard University Press.

Frege, G. (1879), *Begriffsschrift, eine der arithmetischen nachgebildete Formelsprache des reinen Denkens*, Halle, Louis Nebert, tr. in J. van Heijenoort (ed.), *From Frege to Gödel*, Cambridge, Mass., Harvard University Press, 1967, 1–82.

Goldfarb, W. (1979), 'Logic in the twenties: The nature of the quantifier', *Journal of Symbolic Logic*, 44: 351–68.

Hacking, I. (1979), 'What is logic?', *Journal of Philosophy*, 76: 285–319.

Hodes, H. (1984), 'Logicism and the ontological commitments of arithmetic', *Journal of Philosophy*, 81: 123–49.

Lycan, W. (1984), *Logical Form in Natural Language*, Cambridge, Mass., MIT Press.

Montague, R. (1974), *Formal Philosophy*, ed. by R. Thomason, New Haven, Yale University Press.

Putnam, H. (1971), *Philosophy of Logic*, New York, Harper Torchbooks.

Putnam, H. (1975), 'What is mathematical truth?' in H. Putnam, *Mathematics, Matter and Method: Philosophical Papers*, i, Cambridge, Cambridge University Press, 60–78.

Quine, W. V. O. (1960), *Word and Object*, Cambridge, Mass., MIT Press.

——(1986), *Philosophy of Logic*, 2nd edn., Englewood Cliffs, NJ, Prentice-Hall.

Sánchez-Miguel, M. (1993), 'The grounds of the model-theoretic account of the logical properties', *Notre Dame Journal of Formal Logic*, 34: 107–31.

Shapiro, S. (1991), *Foundations without Foundationalism: A Case for Second-Order Logic*, Oxford, Oxford University Press.

——(1993*a*), 'Modality and ontology', *Mind*, 102: 455–81.

——(1993*b*), 'Understanding Church's thesis, again', *Acta Analytica*, 11: 59–77.

Sher, G. (1991), *The Bounds of Logic*, Cambridge, Mass., MIT Press.

——(1996), 'Did Tarski commit "Tarski's fallacy"', *Journal of Symbolic Logic*, 61: 653–86.

Tarski, A. (1935), 'On the concept of logical consequence', in A. Tarski, *Logic, Semantics and Metamathematics*, Oxford, Clarendon Press (1956), 417–29.

——(1944), 'A philosophical letter of Alfred Tarski', *Journal of Philosophy*, 84: 28–32.

——(1986), 'What are logical notions' (ed. by John Corcoran), *History and Philosophy of Logic*, 7: 143–54.

Tennant, N. (1987), *Anti-Realism and Logic*, Oxford, Oxford University Press.

Van Fraassen, B. (1980), *The Scientific Image*, Oxford, Oxford University Press.

Van Heijenoort, J. (1967), 'Logic as calculus and logic as language', *Synthese*, 17: 324–30.

5

Tarski's Thesis and the Ontology of Mathematics

CHARLES CHIHARA

1. WHAT IS TARSKI'S THESIS?

According to some philosophers, there is a widespread belief that model theory provides us with an accurate conceptual analysis of the intuitive, pre-analytic notions of *logical consequence* and *logical truth*, at least when the sentences to which these notions are to be applied are those of certain formalized languages.[1] To undermine this belief is the principal goal of John Etchemendy's book. Vann McGee has continued this attack in his article (McGee 1992). In this paper, I shall assess some of the objections of these two philosophers, focusing especially on Etchemendy's 'contingency argument', which raises questions about the existence and nature of mathematical objects.

The thesis that *a sentence of a formalized language is logically valid* (or *logically true*) *iff it is true in every model* is called by McGee 'Tarski's thesis' (McGee 1992, 273). Actually, McGee also applies the term to the thesis that a sentence of a formalized language is logically consistent iff it is true in at least one model. The term makes reference to the great Berkeley logician, even though it is admitted that Tarski's notion of model differs from the notion we use today, because 'Alfred Tarski is credited with the analysis of logical validity as truth in every model' (McGee 1992, 273).

Etchemendy also attacks a third version of 'Tarski's thesis', according to which, if Γ is a set of sentences of a formalized language, and ϕ is a sentence of this language, then ϕ is an *intuitive* logical consequence of Γ iff ϕ is a model-theoretic consequence of Γ (Etchemendy 1990, 4–5). I shall refer to this last statement of Tarski's thesis as 'the third version', 'the first version' being restricted to the one involving logical truth and validity, and 'the second version' referring to the one involving logical consistency and satisfiability. All three forms of Tarski's thesis assert an extensional identity between an intuitive notion and a model-theoretic notion in a way that

[1] Such a claim is made many times in (Etchemendy 1990).

brings to mind Church's thesis. However, as Etchemendy points out, Tarski's thesis has not been subjected to the kind of scrutiny and controversy that Church's thesis has received (Etchemendy 1990, 6).

I shall begin by taking up one of McGee's objections, even though Etchemendy raised his objections before McGee did. One reason for proceeding in this way is that McGee's objections are generally clearer, to some extent because McGee does not concern himself much with historical questions: he does not discuss the writings and arguments of Tarski, whereas it is sometimes difficult to distinguish Etchemendy's criticisms of Tarski's thesis from his criticisms of Tarski.[2] I wish to emphasize that my investigation of Tarski's thesis is not a historical one: I am not concerned, as is Etchemendy, with what the great logician may or may not have claimed or argued.[3]

Clearly, an investigation of Tarski's thesis requires that model-theoretic validity (or truth in all models) be sharply distinguished from logical validity or logical truth. Model-theoretic validity is a property of both uninterpreted and interpreted sentences of logical languages; whereas logical truth is a property of only *interpreted sentences*, that is, sentences whose terms and symbols have been given (or have) meaning. The concept of logical truth has been explained as follows: 'A true sentence of a given language is said to be a *logical truth* of the language if the sentence's being true follows solely from the meaning of the logical words it contains' (Massey 1970, 29).

To what extent is Tarski's thesis accepted by logicians? It depends upon what is meant by the phrase 'formalized languages' occurring in the statement of the thesis. Unless we specify, or at least delimit, the class of languages in question, it would be doubtful that anyone believes Tarski's thesis. Thus, there undoubtedly are some logicians who would be willing to accept the truth of Tarski's thesis when the languages in question are restricted to those of, say, just first-order logic. For example, in *The Encyclopedia of Philosophy*, the author of the entry 'Logic, Modern', after sketching the semantics of first-order logic and describing how 'truth in a model' and 'validity' are defined, states that the 'concluding steps are (a) to identify the intuitive concept of *logical truth*, with the precise concept of *valid sentence*

[2] Indeed, the casual reader of Etchemendy's book may get the impression that Tarski himself is the central target of the many objections raised in the book. However, Etchemendy emphasizes at the very beginning that his concern is 'not historical' (p. 7). He states: 'This book consists of a single, extended argument. The conclusion of the argument is that the standard, semantic account of logical consequence is mistaken' (p. 8). The best analysis of Etchemendy's criticisms of Tarski's account of logical consequence that I know of is to be found in Craig Bach's Ph. D. diss. 'Tarski and the Concept of Logical Consequence', University of California, Berkeley, 1996.

[3] Gila Sher is also very much concerned with what Tarski wrote and thought about the concept of logical consequence. But she arrives at a very different evaluation of Tarski's views than does Etchemendy. See Sher (1991, esp. ch. 3).

and (b) to certify those argument forms as valid whose conclusions are *consequences* of their premises.' But many of these logicians would not believe the thesis if we included among the formalized languages higher-order languages or modal languages. Many, for example, would not consider the modal quantifier 'It is necessary that' to be a purely logical notion, especially if it could mean 'It is metaphysically necessary that' or 'It is physically necessary that.' Since McGee never makes clear just what languages are to be included under the term 'formalized languages', it is not at all obvious that many logicians or philosophers in fact subscribe to Tarski's thesis.

Suppose, for now, that we regard the thesis to be restricted to the sentences of just first-order logic. Is Tarski's thesis generally accepted? I believe that there are many logicians—mathematical logicians in particular—who simply take the model-theoretic definitions of 'consequence' and 'validity' as mathematical *definitions*, which may or may not provide us with correct conceptual analyses of certain pre-analytic notions; just as many mathematicians and philosophers take the usual epsilon-delta definition of continuity as just a definition, which need not conform to any pre-existing notion of continuity that Leibniz, Newton, and other seventeenth- and eighteenth-century mathematicians may have had.[4] Such a position, after all, conforms to Quine's famous doctrine that *explication is elimination*.[5]

This is not to say that these scholars believe that the pre-analytic notion of validity and consequence can be *completely eliminated* in favour of the model-theoretic notions: there is no evidence that I know of suggesting that they wish to abandon these intuitive notions, in favour of the model-theoretic ones, when reasoning in a natural language. But *within the framework of first-order logic*, 'consequence' and 'validity' are taken to be defined model-theoretically; and they simply do not concern themselves about the intuitive notions when working within this framework.

There are logicians who think that there is some sort of correspondence between the pre-analytic notions and the model-theoretically defined ones, but this correspondence is not regarded as so precise and accurate as is described by Tarski's thesis. For example, as Herbert Enderton presents the situation in his textbook, (Enderton 1972), we have 'real life' deductions which we try to model by means of mathematics. How well do the systems of mathematical logic we construct model the 'real life' deductions we make? All that Enderton claims is that first-order logic is 'admirably suited to deductions encountered in mathematics' (Enderton 1972, 2).

[4] At one point, Etchemendy asks: 'How do we know that our semantic definition of consequence is extensionally correct?' He goes on to say: 'That [this question] now strikes us odd just indicates how deeply ingrained is our assumption that the standard semantic definition captures, or comes close to capturing, the genuine notion of consequence' (Etchemendy 1990, 4–5). I do not think that this diagnosis is correct for some people: for some logicians, the question is similar to: How do we know that our epsilon-delta definition of continuity is correct?

[5] Quine (1960, 260).

There are also logicians who question Tarski's thesis for certain sentences of classical first-order logic. Take the sentence

$$(\exists x)x = x$$

which is a model-theoretically valid sentence of first-order logic with identity. Is this a logical truth? Gerald Massey does not think so, writing:

[W]hether there is something rather than nothing is an empirical or factual question, not a problem of meaning analysis. We hold therefore, that (interpreted) wffs which are valid in every non-empty domain but which are not valid in the empty domain . . . are not, strictly speaking, logical truths. (Massey 1970, 260)

Although it is doubtful that Tarski's thesis, as McGee formulates it, is very widely accepted, the following argument of McGee's for rejecting the thesis is of interest and suggests further ways of clarifying the thesis.

2. THE AI OBJECTION

McGee considers a language which includes, among its logical constants, the usual constants of first-order logic, in addition to a new quantifier '$\exists^{AI}$' which means: 'There are absolutely infinitely many'—a notion that McGee attributes to Cantor.[6] More specifically,

$$\exists^{AI}x(\phi x)$$

is defined to be true just in case *there are as many things x that satisfy ϕ as there are things in a proper class*. One can argue that this new quantifier is a logical constant on the grounds that it is a cardinality quantifier and all cardinality quantifiers are logical constants.[7]

Let Γ be the sentence

$$\exists^{AI}x(x = x).$$

Suppose that Γ is added to the axioms of ZFC to yield a set theory that is interpreted in the standard way, that is, the domain is taken to be the collection of all sets and the only non-logical constant, a binary predicate, is taken to be the membership relation. Under this interpretation, Γ is true, since the totality of things that exist constitutes a proper class. Yet Γ is false in all models, since all the models of our language have sets as domains. In other words, Γ is false in all models by the model-theoretic definition of the new quantifier; even though, under the intended interpretation, it is true.

[6] McGee (1992, 279, n. 16). For a detailed discussion of Cantor's notion of 'absolutely infinite collections', see Hallett (1986, esp. ch. 4).

[7] Gila Sher (1991) gives several arguments for classifying cardinality quantifiers as logical constants.

Therefore, it is argued, model-theoretic inconsistency does not correspond to logical inconsistency. McGee concludes that 'if we set out with the purpose of constructing a formal language for which Tarski's thesis fails, we can do so' (McGee 1992, 279).

The above objection can serve to illustrate how both McGee and Etchemendy regard the 'formal languages' of Tarski's thesis: they are 'interpreted' languages whose sentences are regarded as having meanings that are independent of the model-theoretic semantics of the language—it is in virtue of these meanings that the sentences can be judged to be logically true, logically consistent, and the like. As McGee understands the situation, the model-theoretic semantics may or may not accurately reflect these independent meanings.

The question is: Does McGee's 'absolutely infinite' example constitute a genuine counter-example to Tarski's thesis? Obviously, it depends—to return to an earlier theme—upon whether the new logical language described above is included among the 'formalized languages' to which Tarski's thesis applies. I wonder just how many logicians would subscribe to Tarski's thesis if it were made explicit that the extensional identity of model-theoretic validity and logical truth is supposed to cover even sentences from the above language. Why might one not expect Tarski's thesis to hold for formalized languages of the above sort? The question as to whether a sentence is logically true or not depends upon what the logical terms or symbols occurring in the sentence mean (i.e. upon what intuitive sense is given to the logical terms or symbols in the sentence). The question of whether the sentence is model-theoretically valid or not, on the other hand, depends upon the model-theoretic semantics of the language. But unless the semantical properties assigned to the logical terms and symbols in the sentence, by the model-theoretic semantics, accurately reflect the intuitive meanings given to these terms and symbols, there is no reason to think that 'valid' should match up with 'logically true'.

Take the sentence '$F^1a \vee -F^1a$' of first-order logic. Suppose that the connective '$\vee$' is interpreted to mean 'whenever', even though the model theory treats all the logical constants in the standard way. Then, when the predicate 'F^1' is understood to mean 'is in Berkeley' and the individual constant 'a' is taken to stand for John Searle, the above first-order sentence would mean 'John Searle is in Berkeley whenever it is not the case that John Searle is in Berkeley.' Thus, we would have an example of a sentence that is model-theoretically valid (since it is true in all models) even though it is false when the language is interpreted in the way described above. Of course, *the intuitive meaning being assigned to the logical constant '$\vee$' occurring in the first-order sentence does not correspond at all to the semantic features attributed to this connective by the model theory.*

If Tarski's thesis is to be at all plausible, then, it must be restricted to formal languages that are so interpreted that *the meanings of all the logical terms in the language conform to the model-theoretic semantical features of the language*. Thus, although the $\exists^{AI}$ example is an example of a model-theoretically valid sentence that is not logically valid, it is not clear that it constitutes a genuine counter-example to Tarski's thesis, *when the thesis is reasonably construed*. Obviously, if one wishes to introduce a special quantifier of the above sort, then one needs to tailor the model theory to represent accurately the semantic features of such a logical term. If one wants to have a language with a quantifier **Q** that requires, for the truth of a simple assertion that **Q**x(ϕx), domains the size of proper classes, then the model theory for that language should not restrict the domains of all the models to sets.

What this example does suggest, however, is that there are theoretical risks in the practice, which the majority of mathematical logicians have adopted, of expressing first-order versions of set theory in logical languages whose model theory requires domains that are sets, even though the intuitive meaning given to the quantifiers requires a domain that is a proper class.[8]

3. ETCHEMENDY'S UNDERGENERATIONAL OBJECTION

What grounds, then, does Etchemendy give for his claim that *Tarski's thesis is false* when the thesis is restricted to just the sentences of first-order logic? Let us focus on the first version of Tarski's thesis according to which a sentence is logically true iff it is model-theoretically valid. Now Etchemendy allows that one direction of the biconditional holds: he allows that if a sentence of first-order logic is model-theoretically valid, it is a logical truth.[9] That leaves only the other direction to reject. Thus, Etchemendy's claim that Tarski's thesis is false comes down to this: Tarski's thesis 'undergenerates', that is, the model-theoretic notion does not generate enough logical truths. Specifically, it is claimed that *there are logical truths that are not model-theoretic validities*.

But what logical truths of first-order logic, according to Etchemendy, are not model-theoretically valid? They turn out to be what most philosophers would classify as *analytic truths*. For example, it is claimed that there are logical truths of the language of *first-order arithmetic* that are not model-

[8] Few philosophers have written about this practice. See Lear (1977) for some worries about it.

[9] '[F]or first-order languages the model-theoretic account does not overgenerate: no argument declared valid by the model theory will be invalid' (Etchemendy 1990, 154).

theoretically valid.[10] From a Fregean perspective, such truths are analytic; and if such analytic truths are to be classified as logical truths, then it is not at all surprising that some 'logical truths' are not model-theoretically valid.

Those of you who are familiar with *Tarski's World*, the Etchemendy and Barwise computer program for the MacIntosh, will not be surprised at the above claim that some sentences of first-order arithmetic are logical truths, even though they are not model-theoretically valid. The language of *Tarski's World* is an 'interpreted' language: it is interpreted in the sense that the meanings of the connectives, quantifiers, and predicates of the language are fixed in advance of any model-theoretic definitions. The quantifiers are restricted to ranging over the objects in a 'world', whose objects are geometric solids located on a grid. There are a great many different worlds that one can bring up on the computer screen. (Clearly, the semantical approach here is representational.) There is a binary predicate that means 'is to the left of' and there is one that means 'is to the right of'. If 'L^2' is the former and 'R^2' is the latter, then it can be seen that the sentence

$$(x)(y)(L^2xy \rightarrow R^2yx)$$

is true—indeed necessarily true in virtue of the meanings given to the symbols in the sentence. This is given as an example of a sentence that is *logically valid*, even though it is not so regarded in standard logic texts.[11]

Let us distinguish two senses (or kinds) of logical truths: there are those first-order sentences that are true just in virtue of the meanings of the *logical constants* occurring in them[12]—these are logical truths in the strict sense (they correspond to what Massey classified as 'logical truths'); and there are those sentences that are true in virtue of the meanings of any symbol or term occurring in them—these are 'logical truths' in the loose sense. The latter sort of logical truths are what most philosophers call 'analytic truths'. Similarly, we can distinguish two senses (or kinds) of logical-consequence relations: a strict and loose sense. In the strict sense, some sentence ϕ is a logical consequence of a set of sentences Γ iff ϕs being true follows solely from the truth of the members of Γ and the meanings of the logical constants occurring in these sentences. I shall call the latter loose sort of logical consequence 'analytic consequence'.

We can now see clearly that there is another ambiguity infecting Tarski's thesis: in the first version, the term 'logical truth' can have either the strict or the loose meaning; in the third version, the term 'logical consequence' can stand for strict logical consequence or analytic consequence.[13]

[10] Etchemendy (1990, 151). Cf. also p. 148. [11] Barwise (1991, 102).

[12] I shall not attempt to characterize the meaning of 'logical constant'. However, see Sher (1991, ch. 3), for a promising analysis of 'logical terms'.

[13] That there is such an ambiguity infecting Etchemendy's discussion of Tarski's thesis became evident to me in the discussion following Etchemendy's presentation of his paper 'On

Etchemendy's objection that the model-theoretic notion of logical truth 'undergenerates' (since there are analytic truths that are not model-theoretically valid) requires taking the term 'logical truth' in Tarski's thesis to have the loose meaning. In short, he is taking Tarski's thesis to be the thesis that a sentence of an interpreted first-order language is an analytic truth iff it is model-theoretically valid. *But no sensible philosopher or logician thinks that this version of Tarksi's thesis is true.*

I said in the beginning of this paper that I would not be concerned, in my investigations, with historical questions, in particular with questions of what Tarski may have claimed or argued. But here, I feel compelled to make one comment about Tarski's views on logical consequence. In the important paper to which, according to Etchemendy (Etchemendy 1990, 1), our contemporary methods of defining logical truth and consequence can be traced, Tarski wrote:

> Since we are concerned here with logical, i.e. *formal* consequence, and thus with a relation which is to be uniquely determined by the form of the sentences between which it holds, this relation cannot be influenced in any way by empirical knowledge, and in particular by knowledge of the objects to which the sentence [ϕ] or the sentence of the class [Γ] refer. The consequence relation cannot be affected by replacing the designation of the objects referred to in these sentences by the designation of any other objects. (Tarski 1956, 414–15)

I understand Tarski to be claiming the following: Suppose that ϕ and all the members of Γ are sentences of an interpreted logical language, and suppose that ϕ is a (formal) consequence of Γ. Then, even if one changed the denotations of all the non-logical constants of the language (and thereby altered the designations of the objects referred to in the above sentences), ϕ would still be a consequence of Γ. If I am right, then one can infer, from this passage, that Tarski was not attempting to capture the intuitive notion of analytic consequence with his model-theoretic definition of consequence, since the relation of analytic consequence holding between sentences can obviously be affected by replacing the designation of objects referred to in these sentences by the designation of other objects.

If we consider the more sensible version of Tarski's thesis, according to which a sentence is a logical truth (in the strict sense) iff it is model-theoretically valid, what examples of sentences that are logical truths (in the strict sense) but that are not model-theoretically valid does Etchemendy provide? None at all. Thus, we can see that either Etchemendy has a sound objection against a thesis that no sensible person espouses, or he has pro-

the (Mathematical) Analysis of Truth and Logical Consequence' at the meeting of the Logic Club of the Group in Logic and the Methodology of Science held on the campus of the University of California, Berkeley, on 23 February 1993.

duced no examples to back up his claim that the model-theoretic analysis of logical truth undergenerates.[14]

4. THE CONTINGENCY PROBLEM

There is another kind of objection to Tarski's thesis that Etchemendy raises, which is more subtle than the above. He argues that even when the model-theoretic criterion gets things right, as it frequently does for the first-order case, it does so only by a kind of fluke. When it is seen how the model-theoretic analysis sometimes gets things right, it is argued, it will also be seen that the analysis 'has not captured the ordinary concept of logical truth' (Etchemendy 1990, 115).

In first-order logic with identity, one can express the statement that there are at least n objects, for any cardinal number n. For example, there are at least two objects can be written:

$$(\exists x)(\exists y)(x \neq y).$$

Let σ_n be a first-order sentence that expresses the statement that there are at least n objects. Then consider the sequence of *cardinality sentences*

$$-\sigma 2, -\sigma 3, -\sigma 4, \ldots$$

These sentences express the statements that there are fewer than n objects, for n = 2, 3, 4, . . . Now Etchemendy argues that

> [I]f the universe is finite, the present account will mistakenly pronounce an infinite number of the [above cardinality sentences] logically true; in that case, the account will be extensionally incorrect. Of course, if the universe is infinite, none of these sentences will be declared logically true. But is that because the account has captured our ordinary notion of logical truth? After all, these sentences are not *in fact* logically true, but neither would they *be* logically true if the universe were finite. (Etchemendy 1990, 113–14)

Etchemendy allows that the set theory presupposed by our model-theoretic analysis guarantees that none of the above sentences comes out logically true: the Axiom of Infinity asserts that there are infinitely many things in the universe. According to Etchemendy, the mere fact that we need to appeal to such a powerful non-logical assumption about the world in order

[14] Etchemendy claims that 'once we focus on any interesting first-order language, such as the language of elementary arithmetic, it seems clear that the standard model theory does undergenerate. It is only our uncritical adoption of the model-theoretic analysis that has obscured this simple point' (Etchemendy 1990, 151). I doubt that this point in question is obscure: it is just that hardly any one would regard that sort of 'undergeneration' as counting against any sensible version of Tarski's thesis.

to obtain the needed result shows that the model-theoretic analysis is defective.

To 'dramatize' his objection, Etchemendy asks us to suppose that the model-theoretic definitions really did capture our ordinary notion of logical truth. In that case, he asserts, 'it would be inconsistent—not just wrong, but *inconsistent*—for a finitist to hold that none of the [above cardinality sentences] is logically true' (Etchemendy 1990, 115). But this just shows, thinks Etchemendy, that the model-theoretic account is wrong. The finitist can reasonably claim that, although the universe is in fact finite, it could have been infinite. And from this possibility, 'the finitist should obviously be allowed to conclude that none of these sentences is logically true' (Etchemendy 1990, 115).

McGee has expressed a similar sort of problem with Tarski's thesis, calling it 'the contingency problem'. 'What sentences are valid ought not to be a matter of contingent fact,' he argues, 'and Tarski's thesis would appear to make it so' (McGee 1992, 274). McGee's response to this problem is to rely upon the metaphysical assumption that 'the objects of pure mathematics are metaphysically necessary existents and that the truths of pure mathematics are metaphysically necessary truths' (McGee 1992, 275). This allows him to argue that the Axiom of Infinity is not a contingent truth about the actual world, as Etchemendy has suggested, but is a necessary truth that holds in every possible world.

Etchemendy has anticipated this response. He asserts that, even if the Axiom of Infinity is a necessary truth, it is not a logical truth. Thus, he argues, the 'output of Tarski's account is still influenced by extralogical facts' (Etchemendy 1990, 116). Evidently, Etchemendy believes that no satisfactory account of logical truth can be influenced by any extra-logical facts.

I, too, am not satisfied with McGee's suggested way of saving Tarski's thesis from Etchemendy's contingency objection—but not because I accept Etchemendy's response. I do not believe that any sets exist. So I do not believe that it is a necessary truth that sets exist. However, I believe that Tarski's thesis can be defended against the contingency objection, without resorting to the doctrine that the axioms of set theory are necessary truths. In the following, I shall describe one way this can be done.

The propositional calculus

Let $\mathscr{L}$ be an *interpreted* propositional language with the logical connectives

$$- \vee \ \& \rightarrow \leftrightarrow$$

and the sentential letters

$$P_1, P_2, P_3, \ldots$$

A valuation of $\mathscr{L}$ is a function

$$f: N \rightarrow \{0,1\}.$$

Letting 0 play the role of true and 1 play the role of false, truth under a valuation is defined in the standard truth-functional way. Then, we can state a propositional form of the third version of Tarski's thesis as follows:

> For any sentence ϕ of $\mathscr{L}$ and any set Γ of sentences of $\mathscr{L}$, ϕ is a logical consequence of Γ iff, for every valuation f, if every member of Γ is true under f, then ϕ is true under f

(where 'logical consequence' has the strict meaning described earlier).

Let us now restate Etchemendy's contingency objection with respect to this version of Tarski's thesis. Suppose that only finite sets exist. Then, since valuations are infinite sets, whose members are ordered pairs, there would in fact be no valuations at all.[15] In that case, in a trivial way, for every valuation f, if every sentence of Γ is true under f, then ϕ is also true under f. By the above version of Tarski's thesis, every sentence would be a consequence of every set of sentences. This, supposedly, shows that Tarski's thesis is defective.

Let us suppose that we became convinced that only finitely many things exist and hence that only finite sets exist. Would we then conclude that there are no valuations? Would we then become convinced that every sentence of $\mathscr{L}$ is a logical consequence of every set of sentences of $\mathscr{L}$? Surely not. The valuations to which we refer, in the above version of Tarksi's thesis, are supposed to describe or, in some way, represent the various possibilities regarding the truth-values that the totality of statements expressed by the sentential letters of $\mathscr{L}$ can have; so the valuations describe the different ways the world could have been, in so far as the truth-values of the sentential letters are of concern. If we become convinced that there are no infinite sets and hence no such things as valuations, then we would look for some other way of describing these possibilities. But there would be no need to change the fundamental conceptual analysis of the logical-consequence relation for the language $\mathscr{L}$.

How else could we describe the possibilities in question? We would like to be able to describe infinite sequences of zeros and ones—without the familiar framework of set theory or any other theory that presupposes the *existence* of infinitely many things. Also, we need to be able to speak about all such sequences—a non-denumerable totality—in a way that does not assert the actual existence of any entities. How can this be done? My Constructibility Theory is designed to deal with just this sort of task.[16] It is, like

[15] I am taking functions to be defined within set theory as sets of ordered pairs.

[16] As developed in Chihara (1990).

Principia Mathematica, a 'no-class' logical theory within which mathematics can be developed. Formalized within a type theoretical framework, its theorems are assertions about what open-sentences it is possible to construct. Since, essentially, the set theory of Simple Type Theory can be developed within the Constructibility Theory, it is clear that we have a sufficiently powerful instrument for both describing the infinite sequences of zeros and ones needed to define the models of the language $\mathscr{L}$ and also for theorizing about such models. This time, whether or not infinitely many things exist will not affect what sentences of $\mathscr{L}$ are or are not logical consequences of some set of sentences. Furthermore, all the true statements of the Constructibility Theory are necessary truths, and hence there is no problem of letting the question of whether some sentence is or is not a logical consequence of some set of sentences depend upon 'substantive claims about the size of the universe'.

It might be objected, recalling a move of Etchemendy's, that even if the truths of the Constructibility Theory are necessary truths, they are not *logical truths*. Thus, Etchemendy could still argue, the output of this account of the logical consequence relation for $\mathscr{L}$, 'remains dependent on completely non-logical facts'.

But why is this a damaging objection? It is true that, in order to represent all the possible ways the world could have been, in so far as the truth-values of the sentential letters are of concern, there is a reliance upon certain necessary, but non-logical, truths. But no reason has been given to think that such an account of logical consequence must be defective. Is an analysis of some geometrical phenomenon defective just because it relies upon some non-geometrical facts, say facts of arithmetic. Suppose, for example, that I use the Constructibility Theory to represent the number of possible ways that a plane convex polygon of n sides can be divided into triangles by diagonals. Will it be objected that the output of this account remains dependent on completely *non-geometrical* truths? If so, why is that an objection that should cause any concern?

It is our logical analysis of the language $\mathscr{L}$ that provides the insights needed to describe the different possible ways that the world could have been, in so far as the truth-values of the sentential letters are of concern. The Constructibility Theory is used to represent all these possibilities. The fact that we need to rely upon certain necessary truths that are not strictly logical in nature in order to obtain an appropriate representation of all the possibilities does not imply that our logical analysis of the language is defective.

First-order logic

Thus far, I have been responding to Etchemendy's contingency objection within the framework of propositional logic. It is easy to see that the situation is not significantly altered when we view the objection in the context of first-order logic. The Constructibility Theory can be used to describe and represent the models of first-order languages; so there will not be any reliance upon any axiom or assumption about the cardinality of the universe or upon the existence of infinite sets.

5. IS TARSKI'S THESIS TRUE?

I have been evaluating various objections that have been made to Tarski's thesis. What, it might be asked, is my own evaluation of the thesis itself? Before responding to this question, I should like to clarify the thesis in the ways suggested by my earlier discussions of the thesis:

(1) I shall concern myself with Tarski's thesis only in so far as it is restricted to the sentences of first-order logic.
(2) The interpretations or meanings given to the symbols of the language in question must conform to the semantics of first-order logic, as was seen to be required from an investigation into McGee's 'absolutely infinite' example.
(3) The terms 'logical consequence' and 'logically true' are to be understood in the strict way discussed earlier.

With these qualifications, is Tarski's thesis true? Take the first version of the thesis. Is the set of logical truths identical to the set of model-theoretic validities? If a logical truth is a sentence that is true solely by virtue of the meanings of the logical constants occurring in it, then I do not accept the thesis. It seems to me clear that what counts as a model-theoretic validity depends upon more than just what the logical constants mean. Reconsider the example

$$(\exists x)x = x.$$

Does the guaranteed truth of this sentence depend solely on the meanings of the logical constants occurring in it? Clearly, this sentence might not be true if interpretations of the language with empty domains were permitted. Or consider the sentence

$$(x)Fx \vee (\exists x)-Fx.$$

Might we not have an interpreted first-order language, a sentence of which had the above form and yet was not true? We could imagine that the predicate 'F' had such unclear boundaries that whether the referent of 'a' was a member of the referent of 'F' had no definite answer. In that case, it could be argued, there would be some doubt that the sentence '$(x)Fx \vee (\exists x)-Fx$' must be true. Suppose, for example, that the language is interpreted in such a way that there is only one object in the domain and that there is no fact of the matter as to whether or not this one object has the property assigned to 'F'. To this example, it could be responded that we do not have an appropriately interpreted first-order language—the language has not been given, it might be said, a 'legitimate interpretation'. But this just shows that what guarantees the truth of all sentences of the above form must encompass more than just what the logical constants mean: it must include specifications of what is to count as a 'legitimate interpretation' of the language.

Evidently, we need a third sense of 'logically true' that lies somewhere between analytically true and true solely by virtue of what the logical constants mean: a sentence would be logically true, in this third sense, iff its truth were guaranteed solely by virtue of what the logical constants mean and what the requirements are for something to be a legitimate interpretation of the language. We can simplify this definition somewhat since we can specify that one of the requirements for something to be a legitimate interpretation of the language is that the logical constants must be given the meanings determined by the usual set-theoretical semantics. I shall call logical truths of this third kind 'general semantical truths', since they are sentences whose truth is guaranteed by only 'general' semantical features of the language—there is no need to appeal to those 'specific' features of the language that are due to how the parameters of the language are fixed or interpreted—we can give the following definition:

> A sentence of an interpreted first-order language is a *general semantical truth* (or *is general-semantically true*) iff its truth is guaranteed solely by virtue of what the requirements are for something to be a legitimate interpretation of the language.

Regarding the third version of Tarski's thesis, we need an analogous third kind of logical consequence: this would be a kind of consequence based solely upon the form that interpretations of the language must take. The relation of logical consequence of this third kind will be called 'general semantical consequence'.

Is Tarski's thesis true when the intuitive term 'logically true' that occurs in the thesis is specified to mean general-semantically true? It seems to me that the thesis is quite plausible. To support this claim, I shall give, not a rigorous proof, but an argument for its correctness that I find convincing. I shall argue that *a sentence of a first-order language is general-semantically true iff it is model-theoretically valid.*

A justification of Tarski's thesis

Let us suppose that ϕ is a sentence of a first-order language that has been given interpretation $\mathscr{T}$ and suppose that ϕ is model-theoretically valid.

Then if the interpretation $\mathscr{T}$ of the language is to be one within the scope of Tarski's thesis, it must conform to the requirements laid down for an interpretation to be legitimate; and so $\mathscr{T}$ must determine a domain for the quantifiers and referents for the non-logical constants in the way specified. In short, the interpretation $\mathscr{T}$ must yield a set-theoretical interpretation of the language, that is, a structure. Since ϕ must be true under all legitimate interpretations of the language, it must be true under $\mathscr{T}$, and hence true. So ϕ is a general semantical truth.

For the other direction, let us suppose that ϕ is a general semantical truth. Then its truth is guaranteed solely by what is required of interpretations to be legitimate. But if its truth is so guaranteed, then it must be true under every legitimate interpretation of the language. So it must be model-theoretically valid.

Obviously, what I have presented here is not a rigorous justification of the above form of Tarski's thesis. The 'if' direction is especially vague, so an effort should be made to sharpen the reasoning. As a first step in this direction, I shall clarify the notion that *the truth of a sentence is guaranteed solely by what is required of interpretations to be legitimate*. I shall cash out this notion in terms of provability by 'logic'. In other words, it is logic that is to guarantee the truth of the sentence. (Here, I shall not be precise about just what counts as 'logic'.) Thus, the truth of a general semantical truth must be provable by logic from premisses that only provide general semantical information about the logical language: the premisses will give the requirements that an interpretation must satisfy to be legitimate. Thus, if a sentence of the language is a general semantical (logical) truth, then starting with the assumption that the language has been given a legitimate interpretation, if the target sentence is a valid sentence of the (interpreted) language, then we can go on to infer by logic that it is true.

A more precise justification

Let us suppose that ϕ is a sentence of a first-order language that has been given a legitimate interpretation $\mathscr{T}$ and suppose that ϕ is model-theoretically valid. Since ϕ is valid, it can be proved by logic to be valid. (For example, ϕ can be derived from the empty set of premisses, so that, from the soundness of the derivational system, it can be proven to be valid.) Since we are given, by one of the premisses described above, that the interpretation $\mathscr{T}$ of the language is legitimate, it can be inferred that $\mathscr{T}$ must determine a domain for the quantifiers and referents for the non-logical constants in the way required. It can then be inferred that the interpretation $\mathscr{T}$ must determine a structure **S**, which is such that a sentence of the interpreted language is true if it is true in **S**. Since we have a premiss that tells us that ϕ is a sentence of this interpreted language, we can conclude that ϕ is true if it is true in **S**. From the above conclusion that ϕ is valid, it can be inferred that ϕ is true in **S**. Thus, it can be inferred that ϕ is true. So ϕ is a general semantical (logical) truth.

Whether or not this reasoning is totally convincing, I can find nothing in what Etchemendy and McGee have written to refute this form of the thesis.

REFERENCES

Barwise, J., and Etchemendy, J. (1991), *Tarski's World 3.0*, Stanford, CSLI.

Chihara, C. S. (1990), *Constructibility and Mathematical Existence*, Oxford, Oxford University Press.

Enderton, H. (1972), *A Mathematical Introduction to Logic*, New York, Academic Press.

Etchemendy, J. (1990), *The Concept of Logical Consequence*, Cambridge, Mass., Harvard University Press.

Hallett, M. (1986), *Cantorian Set Theory and Limitation of Size* (paperback edn.), Oxford, Oxford University Press.

Lear, J. (1977), 'Sets and Semantics', *Journal of Philosophy*, 74: 86–102.

Massey, G. (1970), *Understanding Symbolic Logic*, New York, Harper & Row.

McGee, V. (1992), 'Two Problems with Tarski's Theory of Consequence', *Proceedings of the Aristotelian Society*, 92: 273–92.

Quine, W. (1960), *Word and Object*, New York, John Wiley & Sons.

Sher, G. (1991), *The Bounds of Logic*, Cambridge, Mass., MIT Press.

Tarski, A. (1956), 'On the Concept of Logical Consequence', *Logic, Semantics, Metamathematics*, Oxford, Oxford University Press.

II. Mathematics, Science, and Method

6

Naturalizing Mathematical Methodology

PENELOPE MADDY

'Naturalism' is a term much bandied-about these days, especially in discussions of scientific methodology. Many have supported the incursion of naturalistic thinking into the philosophy of mathematics as well; I myself have attempted to adopt a naturalistic approach to the study of set-theoretic methodology. However, in recent efforts to spell out more precisely what 'naturalism' comes to in these contexts, I have come to suspect that much of what I once regarded as 'naturalistic' philosophy of mathematics does not deserve that title, that the act of 'naturalizing' has deeper consequences than have heretofore been acknowledged.

My goal here is to trace these consequences. I begin with a discussion of naturalism itself in Section I, describe a case-study and criticize one purportedly naturalistic approach to it in Section II, and conclude, in the final Section III, with an analysis of what's gone wrong and a hint at how to go right.

I. NATURALISM

Like most popular philosophical notions, 'naturalism' means different things to different people. Consider, for example, the recent survey issue of the *Philosophical Review*: Tyler Burge, in his overview of philosophy of language and mind since mid-century, uses 'naturalism' as a synonym for 'physicalism', the view that there are no mental states over and above ordinary physical ones and that mental phenomena should be explained within materialistic physical science;[1] only pages later, Philip Kitcher devotes his entire essay to a history and analysis of 'naturalism', this time understood, crudely, as the claim that scientific results in psychology and elsewhere are central to the pursuit of epistemology.[2] My own usage differs from both of these, so I begin with a brief characterization.

My thanks go to Matthias Schirn and to my fellow participants for a productive and most enjoyable week.

[1] T. Burge, 'Philosophy of language and mind: 1950–1990', *Philosophical Review*, 101 (1992), 3–51, esp. 31–2.

[2] P. Kitcher, 'The naturalists return', *Philosophical Review*, 101 (1992), 53–114.

For methodologists like myself, the object of study is scientific and mathematical practice, so the fundamental meta-philosophical question concerns the proper relations between philosophy and the sciences.[3] I take naturalism to be an answer to precisely this question. The canonical source is Quine, for whom naturalism is the 'abandonment of the goal of a first philosophy'.[4] The Quinean naturalist[5] 'sees natural science as an inquiry into reality, fallible and corrigible but not answerable to any supra-scientific tribunal.' On this view, the philosopher occupies no privileged position from which to critique the practice of natural science; if philosophy conflicts with that practice, it is the philosophy that must give. As a philosopher of mathematics, I extend this compliment to the practice of classical mathematics as well.[6] Perhaps a more descriptive name for this view than 'naturalism' would be 'philosophical modesty'.

By reducing naturalism to this single meta-philosophical claim, we include another influential philosopher under the rubric. I have in mind the philosophical modesty of the late Wittgenstein, for example, in his well-known remark:[7] 'Philosophy may in no way interfere with the actual use of language . . . It also leaves mathematics as it is.' Only mathematics is explicitly mentioned, but it is also unlikely that Wittgenstein would advocate revisions of natural science on philosophical grounds. I say this because a philosophical objection to an established scientific claim or method would certainly be controversial, and Wittgenstein explicitly insists that (proper) philosophy contains no controversial theses at all:[8] 'Philosophy only states what everyone admits. . . . If one tried to advance *theses* in philosophy, it would never be possible to question them, because everyone would agree to them.' So, the late Wittgenstein joins Quine in our naturalist's camp.

This is not to say that Quine and Wittgenstein agree outside the narrow confines of our definition of naturalism: indeed, they differ dramatically. Admitting that philosophy enjoys no special prescientific access to the truth, Quine is content to conduct his inquiry within natural science:[9]

> [I]t is within science itself, and not in some prior philosophy, that reality is to be identified and described. . . . The naturalistic philosopher begins his reasoning

[3] Terminology: I often use 'science' to cover both natural science and mathematics; I resort to 'natural science' when mathematics is to be excluded and the context seems insufficient to determine this by itself.

[4] W. V. Quine, 'Five milestones of empiricism', in *Theories and Things*, Cambridge, Mass., Harvard University Press, 1981, 67–72. The quotation comes from p. 72.

[5] Ibid.

[6] It isn't clear that Quine would approve this extension. I discuss this question in passing in 'Indispensability and practice', *Journal of Philosophy*, 89 (1992), 275–89. For a more detailed discussion and development of this and many other themes of the present paper, see my *Naturalism in Mathematics*, Oxford, Oxford University Press, 1997.

[7] *Philosophical Investigations*, New York, Macmillan, 1953, §124.

[8] *Philosophical Investigations*, §§599, 128.

[9] 'Things and their place in theories', in *Theories and Things*, 1–23, esp. 21, and 'Five milestones of empiricism', 72.

within the inherited world theory as a going concern. He tentatively believes all of it, but believes also that some unidentified portions are wrong. He tries to improve, clarify, and understand the system from within. He is the busy sailor adrift on Neurath's boat.

For Quine, the scientific approach is the best we have, a powerful extension of common sense; indeed, for him, 'our own science . . . is the only point of view I can offer.'[10]

Wittgenstein's attitude towards natural science could hardly be more unlike Quine's. Inspired by Spengler, he took its rise as a sign of cultural decay:[11]

> I was walking about in Cambridge and passed a bookshop, and in the window were portraits of Russell, Freud and Einstein. A little further on, in a music shop, I saw portraits of Beethoven, Schubert and Chopin. Comparing these portraits I felt intensely the terrible degeneration that had come over the human spirit in the course of only a hundred years.

For Wittgenstein, the pursuit of scientific knowledge blinds its practitioners to the ultimately superior insights of art and religion:[12]

> Man has to awaken to wonder—and so perhaps do peoples. Science is a way of sending him to sleep again.
>
> It isn't absurd, e.g., to believe that the age of science and technology is the beginning of the end for humanity; that the idea of great progress is a delusion, along with the idea that the truth will ultimately be known; that there is nothing good or desirable about scientific knowledge and that mankind, in seeking it, is falling into a trap. It is by no means obvious that this is not how things are.

From these and other passages, it is clear that Wittgenstein does not imagine himself, as a philosopher, to be working within the context of natural science.

But despite their differences, both our naturalists face a further pressing meta-philosophical question: stripped of the philosopher's privileged evidential standards, what is left for a methodologist to do? In Quine's case, commentators complain that the naturalized methodologist's investigation can only describe the methods actually used, that such investigation must relinquish any normative goals. For Wittgenstein, the question is even more stark: what can a philosopher accomplish simply by stating 'what everyone admits'? I postpone discussion of Wittgenstein's meta-philosophy until later, resting content for now with a few observations on Quine's.

There is no quarrel that Quine's naturalization of epistemology represents a non-trivial departure from non-naturalistic philosophy:[13] 'I agree

[10] 'Responses', in *Theories and Things*, 173–86, esp. 181.

[11] R. Rhees (ed.), *Recollections of Wittgenstein*, Oxford, Oxford University Press, 1984, 112. The incident is recalled by Wittgenstein's student and friend Maurice Drury. This anti-scientific aspect of Wittgenstein's thought is well set out in Ray Monk's excellent biography, *Ludwig Wittgenstein: The Duty of Genius*, New York, Penguin Books, 1990.

[12] L. Wittgenstein, *Culture and Value*, Chicago, University of Chicago Press, 1980, 5e, 56e.

[13] W. V. Quine, *Pursuit of Truth*, Cambridge Mass., Harvard University Press, 1990, 19.

. . . that repudiation of the Cartesian dream is no minor deviation.' But it is important to realize that to relinquish (non-existent) peculiarly philosophical modes of evidence is not, for the Quinean, to relinquish all modes of evidence; to give up a (wrongly) presumed right to extra-scientific critique is not to give up all critique. For the Quinean, the point is just that if there is criticism to be made, it must be carried out within the scientific practice itself, that is to say, using the ordinary evidential standards of that practice. Current scientific practice need not be taken as gospel, but as a starting-point, as prima facie gospel only, subject to ordinary scientific critique. So there is room for critique of standards, room for the normative, because science itself involves such normative judgements:[14]

> Our speculations about the world remain subject to norms and caveats, but these issue from science itself as we acquire it. Thus one of our scientific findings is the very fact, just now noted, that information about the world reaches us only by forces impinging on our nerve endings; and this finding has normative force, cautioning us as it does against claims of telepathy and clairvoyance. The norms can change somewhat as science progresses. For example, we once were more chary of action at a distance than we have been since Sir Isaac Newton.

How, then, to raise yet another meta-philosophical question, does the philosophical methodologist differ from any another scientist? If she uses the same methods to speak to the same issues, what need is there for philosophers at all? The answer, I think, is that philosophical methodologists differ from ordinary scientists in training and perspective, not in the evidential standards at their disposal. This claim is best clarified and supported by example; I will try to illustrate it here in the treatment of a particular set-theoretic case.

But first one final meta-discussion. Having absorbed the admonition to be modest in their dealings with scientists, would-be naturalistic philosophers might legitimately wonder, 'how modest?' How much of what the scientist says is to assume the described status as prima facie gospel? For example, it may seem reasonable—I hope it does—to side with the scientist against a van Fraassenite when it comes to accepting Perrin's Brownian motion experiments as conclusive evidence for the existence of atoms; it is quite another to give even prima-facie credence when a scientist idly announces that cosmologists are in the business of discerning the thoughts in the mind of God. Now the contrast, in these examples, is easy to draw; without any foreign, pre-philosophical standard, the naturalized methodologist could successfully criticize the mind-of-God remark by scientific standards, surely with the agreement of most other scientists. But suppose the entire scientific community were converted to this belief. Suppose they agreed in citing something or other, say the beauty of cosmological theo-

[14] 'Responses', 181.

ries, as evidence. The naturalized methodologist might then point to inconsistencies between that purported evidential link and the standards of the rest of science, but at this point we are treading into science fiction, an unpromising ground for the naturalist.

What we are up against is the problem of separating the scientist's unscientific claims (about scientific topics) from her properly scientific claims. Wittgenstein touches on the same distinction in the case of mathematics, the distinction between what mathematicians do and what they are inclined to say about it. The former, the actual practice of mathematics, we are to leave as is, but the latter is open to philosophical criticism:[15]

> [W]hat is caused to disappear by such a critique are names and allusions that occur in the calculus, hence what I wish to call *prose*. It is very important to distinguish as strictly as possible between the calculus and this kind of prose.

This corresponds to the aforementioned difference between the scientist's claims about atoms and Brownian motion—part of the scientific 'calculus', if you will—and her claims about discerning the thoughts of God—an accretion of pseudo-scientific 'prose'.

Inevitably, there are cases much harder to classify than the two we've been using, and just as inevitably, these cases are the most important for our purposes. Consider, for example, mechanism, the nineteenth-century view that all physical phenomena can be explained in terms of forces acting on straight lines between bits of matter, with intensities a function of the distance between those bits of matter. Einstein and Infeld, in their popular exposition *The Evolution of Physics*, classify such principles as 'philosophical', as 'exceeding the bounds of science'. But it is obvious that mechanism was not imposed on physics by the exercise of some extra-scientific standard; to quote Einstein and Infeld in full:[16]

> These questions, although exceeding the bounds of physics, are intimately related to it, since science forms the material from which they arise. Philosophical generalizations must be founded on scientific results. Once formed and widely accepted, however, they very often influence the further development of scientific thought by indicating one of the many possible lines of procedure.

Furthermore, such broad methodological maxims can be overturned, as mechanism was overturned by the accumulating weight of anomalies in electrodynamics and optics, and especially, by the development of the vastly successful field concept. So whatever the terminology Einstein and Infeld find apt, we cannot view methodological maxims like mechanism as unscientific prose. Quine gives another example in his passing reference to action at a distance.

[15] *Wittgenstein and the Vienna Circle*, Oxford, Blackwell, 1979, 149.

[16] A. Einstein and L. Infeld, *The Evolution of Physics*, New York, Simon & Schuster, 1966; original edn., 1938, 51.

A more delicate case arises for various issues included under the general heading of 'realism'. It seems clear, for example, that the existence of physical things independently of their being observed is a fundamental tenet of the common sense of which our natural science in an extension. Furthermore, belief in the objective existence of unobservables like atoms served as a methodological guide as far back as the 1820s: Berzelius reasoned that if chemical substances are really composed of molecules, if talk of molecules is more than a convenient fiction, then the atoms in those molecules must have definite spatial arrangements, and hence, in sufficiently complex cases, the same atoms might combine in different spatial arrangements to form molecules with different chemical properties, in other words, isomers.[17] Some such forms of intra-scientific realism might well be overturned by quantum mechanics, but this just confirms that they should be classified as scientific maxims rather than extra-scientific prose.

However, other forms of realism are not so easily defensible. To see this, first consider van Fraassen's version of anti-realism: we have good reason to believe in what we observe, but good reason to refrain from believing in unobservables. This is not to say that unobservables should be banned from science; indeed, if intra-scientific realism about unobservables is scientifically efficacious as a methodological maxim, van Fraassen endorses it:[18]

> [I]n that case even the anti-realist, when asked questions about *methodology* will *ex cathedra* counsel [realism]! We might even suggest a loyalty oath for scientists, if realism is so efficacious. . . . the interpretation of science, and the correct view of its methodology, are two separate topics.

It is only at this level of 'interpretation' that van Fraassen refuses to believe in unobservables. Within our scientific theory, he admits that 'electron' corresponds to something in the actual world while 'flying horse' does not, but he recommends that we 'bracket' this commitment to electrons, step back to an extra-scientific standpoint, and evaluate our true 'epistemic attitude' from there.[19] From this extra-scientific standpoint, he would have us refrain from belief in electrons.

This is not, strictly speaking, unnaturalistic, because it remains philosophically modest: no change in scientific practice is recommended. It grants the correctness of scientific evidential standards within science, while denying their correctness on some higher plane. However, if it cannot be condemned as outright unnaturalistic, it does clearly qualify, in the pejorative sense, as 'prose', as talk about scientific topics (here the existence of unobservable entities) that is unsupported by ordinary scientific standards.

[17] I first read of this example in R. Miller, *Fact and Method*, Princeton, NJ: Princeton University Press, 1987, 456–7.

[18] *The Scientific Image*, Oxford, Oxford University Press, 1980, 93.

[19] Ibid. 81–2.

So far so good. Now suppose a scientist is moved to dispute with van Fraassen. This scientist insists that the existence of atoms is not just supported by all relevant scientific evidence, but that atoms *really* exist. None of the intra-scientific evidence that supports intra-scientific realism is relevant to this claim; van Fraassen has already conceded all that. Thus this scientist's foot-stomping realism is as much an extra-scientific claim as van Fraassen's anti-realism, and as such, also naturalistically condemnable as mere 'prose'.

Wittgenstein muddles this distinction between various intra-scientific realisms and extra-scientific realism when he cautions us against[20] 'what a mathematician is inclined to say about the objectivity and reality of mathematical facts'. In fact, some of what mathematicians say about the reality of mathematical things has methodological force and is supported by good mathematical evidence. I have in mind, for instance, the principle that mathematical things exist independently of our definitions of them, which comes, for example, to the claim that existence proofs should not be required to uniquely define the entity whose existence they establish. The contrary maxim—call it 'definabilism'—has a history stretching back to the mid-eighteenth century; it was overthrown, much as mechanism was, by a series of anomalies and the rise of a more effective rival maxim.[21] In this case, a realist-style principle in mathematics must be classified as an intra-mathematical methodological maxim, as part of responsible mathematical practice. There remain, of course, other versions of realism about mathematical things that would have to count as extra-mathematical, as 'prose', but a blanket rejection of all talk of 'objectivity' or 'realism' in mathematics is too quick.

I will not attempt to provide a general account of the distinction between proper science and prose, trusting instead to our instinctive ability to classify such particular cases as we come across in practice. Notice that Quine and Wittgenstein differ again in their attitudes towards this prose: for Quine, it is beyond the pale, safely ignored; for Wittgenstein, it is subject to philosophical analysis. Once again, I will postpone discussion of how this Wittgensteinian 'philosophy' functions. First, after all this meta-talk, we need a concrete case to focus our thoughts.

II. A CASE FROM THE THEORY OF SETS

The concrete example I have in mind is a question in set theory that is consistent with, but independent of the standard axioms of Zermelo–Fraenkel

[20] *Philosophical Investigations*, §254.
[21] I trace this episode in 'Does V equal L?', *Journal of Symbolic Logic*, 58 (1993), 15–41.

(ZFC). The most familiar example of such a question is the well-known Continuum Hypothesis (CH), but I want to concentrate on a simpler case: the Lebesgue measurability (or not) of Σ^1_2 sets of real numbers.[22] I prefer this question because it is more concrete—unlike CH, it speaks only of certain easily definable sets of reals—and because it is a mathematically natural question—it was originally raised by analysts in the 1920s as part of an ongoing research programme.[23] Only much later was it shown to be beyond the range of ZFC.

At this point, many set theorists operate on another version of intra-mathematical realism, this time the idea that set theory transcends any particular axiom system, that Σ^1_2 sets of reals are either Lebesgue measurable or not independently of the strength or weakness of our current axiom system, which comes to the claim that this question should be treated (tentatively, of course, as in all matters scientific) as a legitimate mathematical question, worthy of pursuit, despite its independence of ZFC. This maxim gains support from the success of the previously mentioned anti-definabilist realism, from the presence of powerful new axiom candidates and established methods for generating more, and from a general scientific preference for giving an interesting question the benefit of the doubt, that is, for considering it a real problem till it proves itself a pseudo-problem. From this perspective, these set theorists note that this question, unlike CH, has the advantage of being decided by fairly simple additional hypotheses: the axiom of constructibility ($V = L$) implies that there is a non-Lebesgue measurable Σ^1_2 set; the axiom of measurable cardinals (MC) implies that all Σ^1_2 sets are Lebesgue measurable. It could be settled, then, by deciding between these two contrary claims: $V = L$ and MC.

To this end, set theorists offer arguments for and against these claims. For example, $V = L$ is powerful—it answers many open questions, including both this one and CH—it is safe—that is, relatively consistent—and it is economical—that is, in a sense, minimal. MC, on the other hand, asserts the existence of a level of the iterative hierarchy beyond those guaranteed by ZFC; it answers to the conviction that the set-theoretic universe is vast, that our axiom system should aim to maximize the range of sets whose existence it implies. So, the central question is how to adjudicate between the competing arguments for $V = L$ and MC. Though MC is preferred by most set theorists, there is no clear evidential standard by which to assess the arguments conclusively. And, finally, the problem of supplying such a stan-

[22] I have discussed this case before e.g. in *Realism in Mathematics*, Oxford, Oxford University Press, 1990, ch. 4, and in 'Indispensability and practice'.

[23] A Σ^1_2 set is obtained from a Borel set by one application of complement and two applications of projection. (Projection reduces, for example, a set of points in the plane to its shadow on a co-ordinate axis.) Equivalently, a Σ^1_2 relation can be defined from a Borel relation by adding two quantifiers ranging over real numbers.

dard is not a standardly mathematical one; a methodologist's perspective is required.

One possible response to this case would be to adopt an austere version of Carnap's principle of tolerance: the choice between ZFC + V = L and ZFC + MC is a purely conventional choice between linguistic frameworks; either is as acceptable as the other (assuming both are consistent); no properly mathematical reason can be given for a choice between them because any proper mathematical reason is internal to a particular mathematical language.[24] This response solves the problem by declaring it a pseudo-problem. However, it can only aspire to naturalism if the relevant version of intra-mathematical realism—the version leading to the set-theorist's conviction that this is a legitimate question to pursue—is reclassified as mere prose. Given that forms of intra-mathematical realism have played important roles in the development of set theory—e.g. in the rejection of definabilism—this seems an unattractive course. If the maxim that sets transcend particular axiom systems is to be overthrown, this should be done on mathematical grounds. So far, given the fruitfulness of the pursuit of ever stronger axiom systems, I see no such grounds.

We see, then, that to be even potentially naturalistic in this case, a proposed methodology must (tentatively) allow the legitimacy of the question of whether or not Σ^1_2 sets are Lebesgue measurable. My own previous effort to outline such a position began from the familiar Quine–Putnam indispensability arguments: we have good reason to believe in mathematical things because they play an indispensable role in our best scientific theory of the world, and as Quinean naturalists, we take our best scientific theory of the world to be the appropriate perspective from which to judge reality. Unfortunately, pure indispensability theory provides an inadequate account of mathematics, again from a naturalized perspective, because it replaces ordinary mathematical evidential standards—e.g. a statement is true if it can be proved from axioms—with others imposed from outside—a statement is true if it is indispensable to our best natural science.

To avoid this consequence, I developed a modified indispensability theory that separates the ontological from the epistemological. On this view, the indispensability of mathematics to natural science counts as evidence for the existence of mathematical things—an ontological

[24] I call this an austere version, because Carnap would allow us to give pragmatic reasons for choosing one linguistic framework over the other. (See e.g. R. Carnap, 'Empiricism, semantics, and ontology', repr. in P. Benacerraf and H. Putnam (eds.), *Philosophy of Mathematics*, 2nd edn., Cambridge, Cambridge University Press, 1983, 241–57. The principle of tolerance goes back to his *Logische Syntax der Sprache*, Vienna, Springer-Verlag, 1934.) Unfortunately, as Quine points out, it is difficult to distinguish these pragmatic reasons from the usual criteria of theory choice in science, and the required separation between internal and external questions tends to collapse. (See Quine, 'Carnap and logical truth', repr. in his *Ways of Paradox*, revd. edn., Cambridge, Mass., Harvard University Press, 1976, 107–32.)

conclusion—but the proper methods for investigating the nature of those mathematical things are those of mathematics—an epistemological conclusion. Following Gödel, I saw the mathematical methods themselves as running parallel to those of natural science: mathematical intuition provides us with knowledge of elementary mathematical truths and axioms—analogous to the elementary scientific knowledge provided by sense perception—and higher level mathematical axioms are supported, like high-level scientific theories, by their ability to organize and explain lower-level facts. Gödel's version of the mathematical intuition, posited at the most basic level of this science–mathematics analogy, was notoriously mystical-sounding; this I replaced with an account of mathematical perception and intuition based in neuroscience and cognitive psychology. The result is what I've called 'set-theoretic realism'.[25]

For the set-theoretic realist, the continuum exists in the same sense that atoms exist; where Berzelius was right to consider the spatial arrangement of atoms in molecules a legitimate question, the set-theorist is right to consider the Lebesgue measurability of Σ^1_2 sets a legitimate question. So the position satisfies the first requirement for a naturalistic theory of the case in question. Furthermore, its account of mathematical-knowledge acquisition is undoubtedly 'naturalistic' in Kitcher's sense of 'epistemology naturalized': it presupposes no cognitive abilities beyond those familiar to psychology; it happily relies on the various results of research in psychology. I have even argued that a version of set-theoretic realism could count as 'naturalistic' in Burge's sense of 'physicalistic'.[26]

None of this is enough to show that it is naturalistic in the sense adopted here, but before I take up that question, let me observe that the science–mathematics analogy inherent in set-theoretic realism goes beyond the various intra-mathematical realisms touched on so far. That sets have location in space-time, that sets of medium-sized physical objects can be objects of perception—these are unlikely candidates for intra-mathematical maxims with any methodological force. In fact, set-theoretic realism is outright extra-mathematical: not only do mathematical things exist by all relevant mathematical standards, but they *really* exist because they are indispensable to natural science. Our system of classification as developed so far suggests that the only alternative is to classify set-theoretic realism as prose, but more careful attention reveals it as enjoying an intermediate status: extra-mathematical, but not extra-scientific. To see this, we need only widen our perspective, include mathematics among the various sciences, and understand the set-theoretic realist as providing an account of mathematical knowledge within psychology.

[25] This view is developed in my *Realism in Mathematics*.
[26] See *Realism in Mathematics*, 150–9.

But even if it escapes classification as mere prose, set-theoretic realism can still be tested by the standards of naturalism, that is, in so far as it is extra-mathematical, indeed, in so far as it goes beyond, say, physics, it must not recommend reform of the actual practice of mathematics or physical science on spurious grounds. Recommendations of reform from outside these well-entrenched practices can only be naturalistically justified if they come from within natural science and if they are based on scientific grounds sufficiently strong to outweigh that entrenchment. In the present case, any call for reform would seem to be based, at worst, in a priori philosophical views about the nature of mathematics or the nature of knowledge, or, at best, on current psychological theories of cognition. The a priori philosophy is 'mere prose', and thus, naturalistically irrelevant, and current psychological theories would have to be classified as less strongly supported than the mathematics or physics against which they are to be balanced. Under these conditions, set-theoretic realism had better not recommend reform of scientific practice, and for all its other so-called 'naturalistic' virtues, I think it must be judged to fail on this most fundamental point.[27] Its first fault, a conflict with the practice of physical science, arises from the original indispensability defence of mathematical ontology.

The indispensability argument depends on a picture of natural science as an overall theory T, confirmation of which confirms all its parts. The trouble is that careful attention to scientific practice reveals a very different structure: in particular, some parts of successful theories are explicitly viewed as useful fictions, as deliberate, convenient falsifications (e.g. the treatment of matter as continuous in fluid dynamics). Confirmation of the theory as a whole is not taken to confirm the literal truth of these parts. This problem is especially acute for those intensely mathematized parts that the indispensability defence would use as cornerstones in its case for the existence of mathematical things; e.g. most of the many extremely successful applications of continuum mathematics treat as continuous physical structures that are confidently believed not to be in fact continuous, and in the few remaining cases, like that of space-time, the jury is still out.

So, the indispensability argument requires a well-confirmed, uniformly literal theory T, while actual scientific practice exhibits a patchwork of 'idealizations', 'models', and 'approximations', in short, useful falsehoods. An unnaturalized methodologist might argue on a priori grounds that actual scientific practice fails to conform to some higher principles—e.g. that there is no legitimate sense in which experiment can confirm one part of a theory and not another—and that that practice should be revised. This avenue is closed to the naturalist. A particularly stubborn naturalist might

[27] I present this argument in greater detail in 'Indispensability and practice' and 'Naturalism and ontology', *Philosophia Mathematica*, 3 (1995), 248–70.

attempt to argue against practices like mathematical idealization on strictly scientific grounds, but it is hard to imagine how science itself would provide good reasons to abandon such undeniably effective methods. Another equally ridiculous approach would be to argue that scientific talk of idealizations, models, approximations, etc., is all mere prose, with no scientific backing or methodological force. Barring all these, the true naturalist must resolve the conflict in favour of the actual scientific practice, leaving the indispensability arguments in serious disarray.

A second fault of set-theoretic realism, the one centring on mathematical practice, arises from the science–mathematics analogy.[28] Though initially helpful, this analogy, like any other, eventually breaks down, and when it does, the realism of set-theoretic realism begins to falter. To see this, consider once again the case of Σ^1_2 sets and their Lebesgue measurability. On examination, the central argument for preferring MC to V = L is probably some principle of plenitude: the more things, or rather, the more different types of things in the set-theoretic universe, the better. The motivation for this vague idea is not hard to find: if one of the goals of set theory is to provide models for all mathematical entities, then the more different types of models it is capable of providing, the better. It isn't obvious (at least to me) how to spell out a precise and detailed argument based on this outline, but my hunch is that this is the way to go in defending our preference for MC.

Compare this case with that of atoms, also mentioned earlier. The modern atomic theory began in the early 1800s with Dalton's work on chemical combination, and by 1860, it played an essential role in both chemistry and physics. Its predictions were well confirmed, it explained a wide range of seemingly unrelated phenomena, it had all the virtues a theory should have. However, scientists were not uniformly convinced of the existence of atoms. The sceptics didn't advocate removing atoms from science, but they did suggest that only the 'directly verified' consequences of atomic theory should be literally believed. The question wasn't settled until around 1913, when theoretical work of Einstein and experimental work of Perrin found 'direct verification' for the existence of atoms in the phenomenon of Brownian motion.

Now consider this general form of argument: such-and-such an hypothesis is justified because it produces a theory with desirable properties x, y, and z. If this is a good style of argument, then there was a good argument for atoms in 1860, but scientists legitimately wanted more. Yet our argument for MC has exactly this form. A form of scepticism in the set-theoretic case comparable to that in the atoms case would go like this: 'Yes,

[28] In 'Indispensability and practice', I develop a more elaborate version of this objection that again casts doubt on the indispensability arguments themselves. Here, for simplicity, I treat only the science–mathematics analogy. See also 'Naturalism and ontology'.

MC is a nice hypothesis in the way you indicate, that is, it does yield more types of things, and it is nice to have as many types as possible, but if V *does* equal L, then the world of V = L *does* contain all the types possible. What's needed is an argument that V = L is *false*.' But if the argument I sketched for MC is (in the main) correct and complete, this style of sceptical rejoinder, though appropriate in physics, is inappropriate in set theory, and here the mathematics–science analogy breaks down.

Worst of all, the *way* the science–mathematics analogy breaks down here counts against the realistic core of set-theoretical realism, that is, against the idea that mathematics, like science, is the study of a realm of objects that exist independently of us. If it is legitimate to argue, crudely, from 'this theory has properties we like' to 'this theory is true', if this is not just a form of wishful thinking, then I doubt we can avoid the conclusion that we are to some extent free to develop the theory in whatever way suits our needs. No objective, independently existing realm of real mathematical entities would permit us that degree of latitude. A loyal set-theoretic realist would have to insist that the suggested sceptical response is appropriate in set theory as well as in science, and thus relinquish any claim to true naturalism.

It should be noted that this challenge to the extra-mathematical realism of set-theoretic realism does not jeopardize the various intra-mathematical realisms that serve as methodological maxims. Thus, for example, it is still reasonable for set theorists to proceed on the principle that sets transcend particular axioms systems, that is, to treat the open questions of set theory as legitimate questions and to continue to seek new hypotheses that might decide them. The legitimate support for this maxim is the previous success of axiomatization and the fruitfulness of the current search for new axioms. This is not to say that the maxim will never be overthrown, or even that it applies with equal force to all open questions. The Lebesgue measurability of Σ^1_2 sets differs from CH in this regard: the former seems not only legitimate, but answerable, while the latter may one day come to be regarded as somehow ill-formed or otherwise unanswerable. The key point is that such a development in the case of CH, if it were to transpire, would arise from actual practice and its frustrations, not from an extra-mathematical, philosophical declaration that CH is meaningless.

To sum up, I have introduced a case-study, a real situation in contemporary set theory, and set it as a challenge to the naturalistic methodologist. I then described how my own set-theoretic realism, originally designed as a naturalist treatment for such cases, manages to satisfy the most immediate naturalistic requirement (our open question is declared legitimate) and to conform to various alternative so-called 'naturalisms', e.g. Kitcher's and Burge's, but ultimately betrays the most basic naturalistic stricture of philosophical modesty. In the end, the science–mathematics analogy is no more naturalistic than the austere positivist's analogy between mathematics and bookkeeping, or the formalist's analogy between mathematics and games.

The picture of set-theoretic methodology sketched here may suggest yet another analogy, this time between mathematics and creative fiction: the set theorist continues the set-theoretic story constrained by various mathematical canons of reasonableness, just as the novelist continues a fictional story constrained by aesthetic canons of appropriateness; mathematical objects, like fictional objects, are incomplete in some metaphysical sense, etc. Given the broad disanalogies between aesthetic standards and mathematical ones, I doubt that this analogy would be even as illuminating as the discredited analogy between mathematics and science. I am left wondering where the true naturalist should look for a metaphysics to faithfully underlie the actual practice of mathematics, and bothered by the observation that the philosophy of mathematics is so often based on limited metaphors.

III. WHAT WENT WRONG

To gain some perspective on the problem of finding a truly naturalistic approach to mathematical methodology, I propose to return at last to our long-postponed discussion of Wittgenstein's meta-philosophy.[29] In our original discussion of naturalism, we noted that for Quine, the naturalized methodologist works within science itself, that is, using the ordinary evidential standards of science itself, but that for Wittgenstein, science is to be avoided:[30] 'The philosopher is not a citizen of any community of ideas. That is what makes him into a philosopher.' Wittgenstein's philosopher leaves language, natural science, and mathematics as they are, and makes no controversial claims:[31]

> Philosophy simply puts everything before us, and neither explains nor deduces anything . . . we may not advance any kind of theory. . . . We must do away with all *explanation*, and description alone must take its place.

So the question is: how does a Wittgensteinian naturalist proceed?

First, as we've seen, we are to separate the actual practice from the

[29] It is, strictly speaking, inaccurate to describe Wittgenstein's views on the nature of philosophy as a meta-philosophy. For him, metaphilosophy is part of philosophy (see *Philosophical Investigations*, §121), so it too is precluded from making controversial claims. What he gives might best be thought of as a psychology of philosophy, where 'psychology' is understood in a Freudian, rather than a scientific sense. See my 'Wittgenstein's anti-philosophy of mathematics', in Johannes Czermak and Klaus Paul (eds.), *Wittgenstein's Philosophy of Mathematics*, Proceedings of the 15th International Wittgenstein Symposium, Vienna, Verlag Hölder-Pichler-Tempsky, 1993, 52–72.

[30] *Zettel*, Berkeley, Calif., University of California Press, 1970, §455.

[31] *Philosophical Investigations*, §§126, 109.

prose—the 'philosophical fog'[32]—that surrounds it. That fog arises, according to Wittgenstein, 'when language goes on holiday',[33] that is, when we inadvertently allow ordinary forms of language to be lifted from their proper context and attempt to apply them in a context-free, 'philosophical' sense. When this happens, we have deliberately cast aside the ordinary evidential standards of the ordinary context, and its mundane details naturally appear drab and useless:[34]

> If I am inclined to suppose that a mouse has come into being by spontaneous generation out of grey rags and dust, I shall do well to examine those rags very closely to see how a mouse may have hidden in them, how it may have got there and so on. But if I am convinced that a mouse cannot come into being from these things, then this investigation will perhaps be superfluous. . . . But first we must learn to understand what it is that opposes such an examination of details in philosophy.

Having cast off the only legitimate standards, convinced that attention to ordinary practice is irrelevant, we (understandably) find ourselves at a loss as to how to settle the questions that inevitably arise. This gives us the illusion that we are up against a deep philosophical problem, and we are off on the misguided business of formulating and debating controversial philosophical theses.

Under the circumstances, to put forward yet another philosophical theory is simply to exacerbate the problem. What's needed instead is treatment:[35] 'The philosopher's treatment of a question is like the treatment of an illness.' The Wittgensteinian philosopher's job is to uncover the linguistic sleight of hand that led us into the fog, that started us philosophizing in the first place:[36] 'What *we* do is to bring words back from their metaphysical to their everyday use.' To succeed is not to present a true philosophical theory, but to put a pseudo-problem to rest, to dispel the fog. Some legitimate questions may remain, but these should be answerable on the basis of the ordinary evidential standards, in terms of the mundane actual practice, the 'grey rags and dust' we once refused to examine. These answers, then, will be commonplace descriptions; they will state only 'what everyone admits'.[37] Ordinary language, language not on holiday, will remain unchanged, as promised.

In the case of mathematics, the prose, the fog, the mistaken philosophical theorizing is Platonism: a widespread variety of extra-mathematical *and* extra-scientific realism that sees mathematics as the study of an otherworldly reality of non-spatiotemporal entities. The linguistic slip here is the appropriation of talk of 'reality' from its home in the language of natural science into our discussion of mathematics:[38]

[32] *Culture and Value*, 57e. [33] *Philosophical Investigations*, §38.
[34] Ibid. §52. [35] Ibid. §255. [36] Ibid. §116. [37] Ibid. §599.
[38] *Lectures on the Foundations of Mathematics*, Ithaca, NY, Cornell University Press, 1976, 240.

[I]f you forget where the expression 'a reality corresponds to' is really at home . . . What is 'reality'? We think of 'reality' as something we can *point* to. It is *this*, *that*. . . . Professor Hardy is comparing mathematical propositions to propositions of physics. This comparison is extremely misleading.

We are led into this slip when we try to account for the interest of mathematics:[39] 'One would like to say of it e.g.: it introduces us to the mysteries of the mathematical world. *This* is the aspect against which I want to give a warning.' Once there, out of context, our 'mathematical reality' talk brings along a raft of deep philosophical questions: what are these other-worldly mathematical things?, how can we ordinary human cognizers acquire knowledge about them?, why are they so useful in the science of this world?, etc.

To avoid all this, to dispel the fog, we return to our point of departure: the question of the interest of mathematics. In ordinary terms, mathematical results are of interest for a wide range of reasons, perhaps most conspicuously, because they make possible the larger part of higher scientific theorizing. This everyday answer doesn't extend to pure mathematics, however, so we are moved to rhapsodize instead about 'the mysteries of the mathematical world'. We import the term 'reality' from natural science, and we are quickly off into deep philosophy. The solution, or treatment, is to resist this impulse, to rest content with the everyday answers.

This all sounds innocent enough: the practice itself and our ordinary mathematical understanding of its interest remain intact; all that's lost is a fanciful interpretation. No naturalistic scruple has been violated; philosophical modesty is sustained. However, at this point, Wittgenstein makes a startling prediction. On his reading, there are parts of mathematics, in particular, parts of pure mathematics like set theory, that seem to have interest *only* because we are deceived by these fanciful interpretations:[40]

[T]his doesn't mean that certain mathematical propositions are *wrong*, but that we think their interest lies in something in which it does not lie. I am *not* saying transfinite propositions are *false*, but that the wrong pictures go with them. And when you see this the result may be that you lose your interest.

The misunderstandings we are going to deal with are misunderstandings without which the calculus would never have been invented, being of no other use, where the interest is centered entirely on the words which accompany the piece of mathematics you make.

Without the 'philosophical fog' of Platonism, he suggests, our interest in set theory will disappear:[41]

[39] *Remarks on the Foundations of Mathematics*, Cambridge, Mass., MIT Press, 1978, II, §40.
[40] *Lectures on the Foundations of Mathematics*, 141, 16–17.
[41] *Philosophical Grammar*, Berkeley, Calif, University of California Press, 1974, 381.

What will distinguish the mathematicians of the future from those of today will really be a greater sensitivity, and *that* will—as it were—prune mathematics... Philosophical clarity will have the same effect on the growth of mathematics as sunlight has on the growth of potato shoots. (In a dark cellar they grow yards long.)

Here the Wittgensteinian methodologist directly criticizes only the surrounding prose, not the existing practice, but Wittgenstein predicts that the future of that practice will be indirectly shaped by elimination of that prose.

I see no violation of naturalism in this eventuality, should it occur. If practice is being improperly influenced by extra-mathematical, indeed extra-scientific talk, then the naturalist would seem duty-bound to eliminate that inappropriate interference; particular scientific projects should be pursued for good scientific reasons, not for illusory philosophical ones. Where Wittgenstein goes wrong, it seems to me, is in his assessment of the interest—I should say, the purportedly illusory interest—of pure mathematics. Mathematicians pursue pure mathematical results for a wide variety of motivations: to develop new techniques, to clarify connections with other parts of mathematics, for the inherent fascination of the practice itself. They (tentatively) act on various intra-mathematical realisms—allow existence proofs that don't explicitly define the object shown to exist, treat sets as less than fully characterized by any particular axiom system—because these maxims have arisen out of the practice of mathematics and proved fruitful. The move to extra-mathematical realism, if it comes at all, comes late in the game, in response to philosophical criticism of these realistic maxims.

To illustrate, consider an underlying theme in Gödel's famous defence of Platonism in 'What is Cantor's continuum problem?'[42] Gödel is convinced that it makes sense to pursue the CH even if it should prove independent of the existing axioms; in other words, he supports a version of intra-mathematical realism. In defence of this maxim, he describes a philosophical Platonism about sets—a 'well-determined reality, in which Cantor's conjecture must be either true or false', consisting of objects 'remote... from sense experience', that 'clearly do not belong to the physical world'.[43]

What's most interesting for our purposes is that he immediately defends this philosophical stance with an intra-mathematical methodological point:[44]

Such a belief is by no means chimerical, since it is possible to point out ways in which the decision of a question, which is undecidable from the usual axioms, might nevertheless be obtained.

Even more striking, in the end, he dismisses his Platonism as inessential to the real problem at hand:[45]

[42] Repr. in P. Benacerraf and H. Putnam (eds.), *Philosophy of Mathematics*, 470–85.
[43] Ibid. 476, 483. [44] Ibid. 476. [45] Ibid. 484–5.

[T]he question of the objective existence of the objects of mathematical intuition . . . is not decisive for the problem under discussion here [i.e. the meaningfulness of CH]. The mere psychological fact of the existence of an intuition which is sufficiently clear to produce the axioms of set theory and an open series of extensions of them suffices to give meaning to the question of the truth or falsity of propositions like Cantor's continuum hypothesis.

These passages suggest that what ultimately supports his intra-mathematical realism is not his philosophical Platonism but an elaborate series of sound mathematical supports: there are mathematically natural ways of extending the system of axioms; new axioms can also be confirmed by their consequences; etc. From this perspective, the intra-mathematical supports are what matter, and they would survive the demise of philosophical Platonism.

Gödel, of course, was a particularly philosophically minded mathematician, but in these passages he hints at a line of thought that I suspect can be traced in the reflections of many mathematicians. Baldly stated, that line of thought goes like this: the practice is sound, here's a batch of prose to protect it from philosophical attack. If that prose is found wanting, the practice remains sound; what's then needed is some new prose. In other words, whatever happens philosophically, extra-mathematically, mathematicians are not likely to stop doing good mathematics, that is, mathematics based on a sound and fruitful practice.[46]

My claim, then, is that Wittgenstein went wrong in supposing that the mathematician's interest in pure mathematics arises solely from a fog of misguided platonistic prose. Here Wittgenstein has missed the many intra-mathematical sources of interest; to use his own image, he has failed to examine the rags and dust with sufficient care. I've described this Wittgensteinian oversight in some detail because I think it also illuminates where I went wrong in my treatment of the set-theoretic case study we've been considering here. Beyond a naturalized epistemology in Kitcher's sense, my goal was a philosophical account of set theory that ratifies the set theorist's conviction that the Σ^1_2 set question is legitimate, that justifies the evidential practices in actual use, and that does all this while remaining naturalistic, that is, without conflicting with actual practice. When set-theoretic realism violated these last naturalistic strictures and was discarded on that account, I set about searching for a new metaphysical position that would succeed where the other had failed. I touched, for example, on the possibility of a fictionalist analogy.

From our strictly naturalistic perspective, this entire project must strike us as wrong-headed. The set theorist's conviction that the Σ^1_2 question is legitimate needs no justification beyond the intra-mathematical ones it

[46] I made this same observation in the previous section, in connection with the demise of intra-scientific, but extra-mathematical set-theoretical realism.

already has. As Gödel's example shows, the intra-mathematical conviction is often used to support the extra-mathematical prose, not the other way round. Similarly, the evidential practices in actual use are justified and can only be justified intra-mathematically, not by some extra-mathematical theorizing. An analogy like that between mathematics and science might call attention to various possibly neglected aspects of mathematical practice, it might serve as a guide to the intra-scientific, but extra-mathematical naturalizing of mathematical epistemology, but in so far as it serves to motivate or purportedly justify the actual methods of mathematics, it is stepping beyond its naturalistic bounds. In short, if, for the naturalist, philosophy cannot criticize actual methods, it cannot justify them, either. Extra-mathematical prose is simply irrelevant to the fundamental methodological story, which resides exclusively in the practice itself.

This suggests that a properly naturalized mathematical methodology must do without the sorts of metaphors we've grown accustomed to. These might well be suggestive, even heuristically useful, but they will eventually come to obscure important aspects of the methods under study, namely those aspects not covered by the metaphor. Our mistake has been of the sort Wittgenstein describes: we look at the practice of mathematics, we think it needs justification, our philosophical prejudices lead us to ignore the ordinary mathematical justifications available right there in the practice, we then project irrelevant philosophical theories instead. Wittgenstein's moral is: attend to the practice itself; everything you need is there before you. We can't quite say that because part of the interest of our case is that the practitioners themselves are uncertain how the practice should be extended. What we can say is this: whatever extensions we might propose must be purely intra-mathematical, must make sense from a purely mathematical perspective, must be more rags and dust, not new extra-mathematical prose.

Let me close with this question: what does the unadorned practice tell us about our test case? Briefly, our review noted that set theorists prefer maximizing theories. We wondered what justifies this preference; we wondered why maximizing theories are more likely to be true, in some extra-mathematical sense of true. But the set theorist's mundane motivation for that preference was there before us: maximizing theories are preferred because one of the goals of set theory is to provide as many types of structures as possible. These humble parts might be used to produce a genuine justification. To pursue this idea, we must examine the reasons behind this particular goal, ask exactly what is meant by maximizing, and determine how effectively MC so maximizes. There are all serious questions, without easy answers, but for our purposes here, it is most important to notice how far they are from the questions we faced earlier, for example, the questions that sought yet another metaphysics. True naturalism sets us on a very different road, and this, I submit, is how it should be.

7

Occam's Razor and Scientific Method

JOHN P. BURGESS

1. OCCAM'S RAZOR AND MODERN NOMINALISM

Entities are not to be multiplied beyond necessity, says the legendary razor of the legendary William of Occam, chief of the legendary medieval nominalists. My topic will be the role of this dictum in the philosophy of the last half-century and the science of the last half-millennium. My topic will thus be history, but it will not be the history of medieval philosophy. I will not be questioning the relation between the modern nominalists and the medieval *nominales*, or between the legendary William of Occam and the historical William Ockham. Before posing the question that *will* concern me, some preliminaries will be necessary, which I will try not to multiply beyond necessity.

The modern nominalism that will concern me is that which consists in disbelief in abstract entities. Nominalism is not a unified school of thought. One distinction that can be made among nominalists is between the *forthright*, who concede their position incompatible with the acceptance of modern mathematics and science, and the *evasive*, who on one ground or another[1]—perhaps the mathematicians don't really mean it when they say that *there are numbers greater than one hundred that are prime*, or perhaps this doesn't really imply that *there are* (*such things as*) *numbers*—deny the incompatibility. The evasive positions require examination; I have partially examined them elsewhere; I do not propose to examine them further here. I will proceed here on the common assumption of the forthright nominalists and all anti-nominalists, that a nominalist must reject current mathematics and science.

Another distinction that can be made among nominalists and anti-nominalists alike is in their attitude towards attempts in the literature to reconstruct current mathematics and/or science nominalistically. (Such

This paper appears here in almost exactly the same form in which it was given at the conference 'Philosophy of Mathematics Today'. The few departures are indicated in the notes.

[1] In discussion at the conference, Paul Benacerraf warned against trying to enumerate all the possible grounds. His own paper dealt, in part, with related issues.

reconstruction often but not always takes the form of a reconstrual: The new theory takes the form of the words of the old theory with new meanings assigned.) Some nominalists propound arguments against abstracta that in no way depend on the outcome of such reconstructive projects. Typically these derive from a certain reading or misreading of one or the other of two papers of Paul Benacerraf (1965, 1973). Some anti-nominalists also propound arguments for abstracta that in no way depend on the outcome of reconstructive projects. Typically these derive from an old paper of Michael Dummett (1956).[2] (I defended a non-Dummettian variety of non-Occamite anti-nominalism in an earlier paper (1983).)

However, many on both sides of the issue hold that nominalism stands or falls conditional on the success or failure of reconstructive projects of the kind alluded to. These philosophers I will call *Occamites*. For they assume a modernized, double-bladed version of Occam's razor. Their assumption is that *abstract entities are to be multiplied as far as but not beyond scientific necessity*. Among Occamites, the distinction between nominalists and anti-nominalists coincides with the distinction between optimists and pessimists about reconstructive projects. Pessimists, anticipating failure, offer an *indispensability* argument against nominalism: Current scientific theories are to be accepted because no adequate nominalistic alternatives can be made available. Optimists, hopeful of success, offer a *dispensability* argument for nominalism: Current scientific theories are to be rejected because no adequate nominalistic alternatives can be made available. Both are typically inspired or provoked by such polysemous publications of Hilary Putnam as his paper (1967) and his book (1971).

My main question about Occamism is suggested by another work of Benacerraf and Putnam, the opening section of their editorial introduction to their standard anthology (1983). There they group the authors anthologized according to their conception of the proper relation between epistemology and mathematics: 'To put it somewhat crudely, if some piece of mathematics doesn't fit [an epistemological] scheme, those in the first group tend to throw out the mathematics, while those in the second tend to throw out the scheme.'

This seems much the same distinction between two conceptions of the proper relation of philosophy to science encountered elsewhere under a different terminology. On the one hand, on a more traditional *externalist* or *alienated* conception, the philosopher remains a foreigner to the scientific community, judging scientific theories by an external philosophical standard. On the other hand, on the more recent *internalist* or *naturalized* conception, the philosopher becomes a citizen of the scientific community,

[2] Even as Dummett himself has retreated from his old position, it has been extensively developed by the St Andrews school, represented at the conference in the papers of Bob Hale and Crispin Wright.

judging scientific theories by internal scientific standards.[3] Typically, the first group aims to provide a philosophical foundation for or critique of science; while the second group regards philosophical 'foundations' as less well founded than the science they seek to found, and extra-scientific philosophical 'critiques' as more open to criticism than the science they seek to criticize—'doubts more dubious than what is being doubted', in a phrase used in a different context by Georg Kreisel.

What generally is the relation between the conditional versus unconditional distinction and the internal versus external distinction? Well, whether a theory ought to be accepted may be dependent on what alternatives are available; but it is a virtual tautology that the theories currently accepted in science are acceptable by the standards of science given the alternatives currently available. It would seem to follow that an unconditional or non-Occamite nominalist, one who is prepared to argue that current theories are still to be rejected even if all attempts to develop novel nominalistic alternatives should fail, must be classed as an externalist, as one not judging acceptability by the standards of science.

The specific question that will most concern me is whether the inverse holds: May a conditional nominalist be classed as an internalist? In other words, is the preference for alternatives avoiding abstracta over theories involving abstracta a standard of science for the acceptability of theories? In yet other words, is Occam's razor a rule of scientific method? I will call this the *tonsorial* question. It has been, I suggest, implicit throughout the modern debate on nominalism from its very beginning.

2. NOMINALISM IN THE 1940S AND 1950S

Contemporary understandings of many of the buzz-words and catch-phrases I have been and will be using derive largely from just three authors and a handful of papers from the late 1940s and early 1950s. The three authors are Nelson Goodman, Rudolf Carnap, and W. V. Quine. Among the handful of papers I cite four. First, there is the joint paper of Goodman and Quine (1947), which launched modern nominalism and its first reconstructive project. Largely technical, it has a philosophical introduction; but the philosophical position there should be associated with Goodman, not Quine. Or at least, when Quine has cited the paper in later work he has dissociated himself from that position, adding to his bibliography a note:

[3] At the conference the paper of Penelope Maddy was explicitly addressed to the defence and illustration of naturalism, and several other participants (Geoffrey Hellman, Michael Resnik, Stewart Shapiro) professed allegiance to it, though they may not all have understood it in quite the same sense I do.

Lest the reader be led to misconstrue passages in my later writings by trying to reconcile them with the appealingly forthright opening sentence of the cited paper, let me say that I should now prefer to treat that sentence as a hypothetical statement of conditions for the construction in hand.

Second, there is the critical paper by Carnap (1950), which launched modern anti-nominalism. Third and fourth, there are the responses of the original nominalists to the original anti-nominalist. Goodman's later position paper (1956), includes a section (3) responding to critics, which in turn includes an item (viii) responding to Carnap. Quine wrote a separate response (1951). By the time he wrote it he had abandoned nominalism, though not for Carnap's reasons but rather for reasons of his own.

Goodman was a nominalist, Carnap an anti-nominalist, Quine first one, then the other. For each it may be asked whether his nominalism or anti-nominalism (as the case may be) was *Occamite* and conditional, or rather *unconditional* and non-Occamite. All four possible combinations were, I suggest, represented by these three writers.

To commence with Goodman, in his response to Carnap he emphasizes his toleration of current scientific practice as useful: 'The nominalist does not presume to restrict the scientist.' He does so, however, in terms that leave no question but that he does not accept current scientific theory as credible: 'The scientist may use platonistic class constructions, complex numbers, divination by inspection of entrails, or any claptrappery that he thinks may help him get the results he wants.'

He also expresses his hopefulness about finding a nominalistic reconstrual: 'But what [the scientist] produces then becomes raw material for the philosopher, whose task is to make sense of all this . . . [to] interpret [it] in understandable terms.' He does so, however, in terms that leave no question but that for him such a reconstrual would involve giving a meaning to words that at present lack one: 'Nominalism is a restraint that a philosopher imposes upon himself, just because he feels he cannot otherwise make real sense of what is put before him.'

Now, whether a theory is credible and ought to be *accepted* may be dependent on what alternatives are available; but whether a theory is intelligible and can be *understood* cannot be thus dependent. A nominalist who maintains that current scientific theories make no sense and lack a meaning—no matter how optimistic that nominalist may be about the prospects for eventually being able to give it a meaning—must be classed as unconditional or non-Occamite. A more direct indication that Goodman's nominalism was not conditional on the success of any project of reconstrual or reconstruction is that he persisted in it even after the failure of his joint project with Quine.

Quine, by contrast, then—as Goodman puts it in his position paper—'defected' from nominalism. This is the best indication that the nominalism of the early Quine was Occamite or conditional. As for the later Quine, he is famous as an advocate of indispensability arguments, or in my terminology, Occamite or conditional anti-nominalism. Thus he was an Occamite early and late, the only change being from optimism to pessimism about the prospects for adequate nominalistic reconstruction.

As for Carnap, the argument of his critical paper was, to put it very crudely, as follows: On the one hand, there are specific existence questions, as when one asks *whether there are numbers that are greater than one hundred and are prime*. Such questions are raised by mathematicians and other scientists, and within the mathematical community there is more or less definite agreement under what conditions an affirmative answer to such a question is confirmed. On the other hand, there are general ontological questions, as when one asks *whether there are really any such things as numbers at all*. Such questions are: 'raised neither by the man in the street nor by scientists, but only by philosophers' (and indeed only by those philosophers who are discontented with the trivial answer implied by current and longstanding mathematical theory, that *of course there are numbers, since there are numbers that are greater than one hundred and are prime*). But within the community of such philosophers there is no very definite agreement under what conditions an affirmative answer to such a question is confirmed: 'the controversy goes on for centuries without ever being solved.' And the existence or non-existence of agreed confirmation conditions is what makes the difference between sense and nonsense, according to the kind of theory of meaning presupposed by Carnap. Insofar as this argument seems directed equally against any position that *so much as takes there be a non-trivial question* about the existence of numbers, Carnap's anti-nominalism is unconditional.

As to the distinction of internalism versus externalism, there is no question that Carnap (by the time of the paper under discussion) professed internalism. Or as Benacerraf and Putnam say: 'There is no question that [he] has abandoned any *critical* function for epistemology.' Carnap's paper is addressed to colleagues who share a commitment to a scientific orientation in philosophy.

The best indication of such a commitment on the part of the early Quine is perhaps the fact that he is almost the only colleague whom Carnap cites by name. As for the later Quine, he is famous as an advocate of naturalized epistemology, or in my terminology, internalism.

To conclude with Goodman, as already indicated, a nominalist classed as unconditional must be classed as externalist. A more direct indication is that in the introduction to the joint paper Goodman invokes not scientific method but: 'a philosophical intuition that cannot be justified by appeal to anything more fundamental.' Internalists do not invoke occult faculties of 'philosophical intuition'!

Thus one of the trio, Goodman, was an externalist, and the other two, Quine and Carnap, professed or were committed to or advocated internalism. Still, they obviously cannot both be classified as *genuine* internalists. For if Occam's razor *is* a tool of scientific method, then the internalism of Carnap, who neglects that tool—who seems willing to accept whatever is customary or convenient for science without lengthy inquiry into whether it is indispensable or necessary, whether one somehow by hook or crook can make do without it—is specious. If Occam's razor is rather a weapon of extra- if not anti-scientific philosophy, then it is the internalism of Quine, who wields that weapon, that is spurious.

It is in this sense that the tonsorial question was implicit already in the debate between Carnap and Quine, though it became explicit only much later. For it did not become explicit until after the period of quiescence in the debate over nominalism that set in after the failure of the joint project of Goodman and Quine.

3. NOMINALISM IN THE 1970S AND 1980S

The earlier period of activity in connection with nominalism ended at the latest by the beginning of the 1960s. By the end of that decade, the influence of the works, already cited, of Benacerraf and Putnam came to be felt, the latter especially inspiring the Occamite variety of nominalism. Renewed optimism about nominalist reconstructive projects was in large part owing to a new conception of what would constitute an acceptable reconstrual or reconstruction, a credible alternative.

Goodman's and Quine's overall position involved many restrictions besides those of nominalism. As Goodman put it in his response to one item (iii) in the list of objections in his position paper: 'Nominalism no more guarantees philosophical soundness than the refusal to eat poison guarantees physical well-being.' For one thing, besides the nominalistic rejection of abstract entities, Quine at least tended also towards the *physicalistic* rejection of *mental* entities, as have most subsequent nominalists. Yet in a volume of critical essays on Quine's philosophy edited by Hahn and Schilpp, containing interesting responses by Quine to many philosophical commentators, Quine emphasizes in response to Goodman that: '[M]y physicalism was additional [to my nominalism]. Nominalism as I understood the term could also countenance—unlike me—an irreducible mental realm.'

For another thing, while nominalism and physicalism are both *ontological* positions (in Quine's sense of the term), positions as to what sorts of things there are, Quine and Goodman alike adopted also certain *ideological* positions (again in Quine's sense of the term), positions as to what kinds of predicates make sense. Above all, they were committed to *extensionalism*, banning *modality*, as Quine emphasizes in response to Charles Parsons:

> [L]ong ago, Goodman and I got what we could in the way of mathematics . . . on the basis of an nominalist ontology and without assuming an infinite universe. We could not get enough to satisfy us. But we would not for a moment have considered enlisting the aid of modalities. The cure would have been far worse than the disease.

It is in just this respect that there has been a shift of opinion as to what would constitute a credible alternative to current mathematics and science. As Quine puts it:

> Putnam [has] argued . . . that modality could be useful in avoiding or minimizing . . . recourse to mathematical objects; and Parsons has been critically scanning this prospect. My own scale of values, however, is the reverse; my extensionalist scruples far outweigh my nominalistic ones. Avoidance of modalities is as strong a reason for an abstract ontology as I can well imagine.

One common objection of those who have critically scanned neo-nominalist projects is anticipated by Quine's remark: The objection has been that the benefit of economy in abstract ontology is outweighed by the cost of extravagance in modal ideology. Quine also remarks: 'Actually I am not even clear on what to count as existence in a modal framework.' This anticipates another objection, that the benefit of parsimony in abstract ontology is only *apparent*; that *really* the abstract ontology is as prodigal as ever. Typically the claim is that *overt* commitment to mathematica has merely been replaced by *covert* commitment to possibilia.

Both of these common objections concede the claim that economy or parsimony of abstract ontology *is* a benefit that (if really achieved and not outweighed) makes a theory preferably acceptable. This claim is one that Quine, as an Occamite, takes for granted. It would moreover be futile to dispute this claim with one prepared to appeal for justification of it to an occult faculty of 'philosophical intuition'. The controversy could go on for centuries and never be solved. However, when the question is raised among those committed to a scientific orientation, appealing only to scientific method, it becomes precisely what I have been calling the tonsorial question. Though not the most common objection to neo-nominalism, the tonsorial question *was* raised in the literature critically scanning it.

I raised it myself in an earlier critical paper (1990). I noted there that the question is essentially historical. I asserted there that the claim that proposed nominalistic alternatives are superior by scientific standards to current theories in physics is one for judgement by the physicists. But I added:

> [W]hat mathematical physicists must judge in the long run, naturalized epistemologists may consider in the short run. Once the distinctive features of the

alternatives have been surveyed, the future decision of physicists whether to accept them will provide new evidence as to how much weight such features have with the scientific community. But the past decisions of physicists already provide some evidence, against which the naturalized epistemologist may in the meantime test claims. . . . [T]he claim that economy of mathematical ontology has weight with the scientific community . . . must be tested . . . against the evidence of past decisions of physicists.

Unfortunately, the professionals who are best equipped to do so have been reluctant to address questions of this kind. One reason is that historians traditionally have given priority to factual reconstruction of the past over interpretive generalization about it, understanding what the past meant to those who lived through it, rather than what it can mean to those of us living today. At the present moment, historians of science tend to concentrate on social influences on decisions of the scientific community, almost to the exclusion of methodological influences. At the same time, historians of philosophy tend to concentrate on the thought of an important individual in its original and integral form, rather than the chopped up or watered down version that is usually what influences the larger community.[4] So the question is left to amateurs like myself by default. What can we hope to accomplish?

The first step would be to try to identify cases for historical study. These would be episodes where a theory was resisted by the scientific community or a segment thereof, and where on one plausible interpretation what motivated such resistance was a preference for economy or parsimony in assuming abstract objects. As I noted:

All, or almost all, proposed changes in the mathematical apparatus of physics that have received ultimate acceptance have increased its power and freedom. But some proposed changes have received only delayed acceptance by the scientific community, after temporary indecision; and it is such cases that provide evidence for the counterweight of some other feature as a scientific standard . . .

In each of these cases, the proposed change involved a decrease in ontological economy . . . At first glance, each might be interpreted as evidence of the weight of ontological economy of mathematical apparatus as a scientific standard.

The (few) cases I identified included, for instance, the ups and downs of *infinitesimals* in calculus and analysis.

In retrospect I recognize it as a limitation that I considered only examples of seeming resistance to *mathematical* abstracta. A comprehensive col-

[4] Thus, if Heyting is exposed as having been marketing a 'lite' version and not the genuine Brouwerian brew, as suggested by Mic Detlefsen at the conference, this makes him less interesting to the historical scholar; whereas, if it is the 'lite' version that was actually consumed by the mathematical masses, it is more important than the genuine stuff when considering something like the tonsorial question.

lection of cases should include examples from both the mathematical and the whole range of empirical sciences. (A comprehensive collection of cases should also include examples both from the more recent and from the earlier history of science, though there is a danger of anachronism if one goes too far back. For natural science emerged from natural philosophy, and until the separation is complete, the distinction between externalism and internalism, and with it the tonsorial question, makes at best limited sense.)

The second step would be to try to identify in each case what would be the main rival interpretation. In my paper I suggested a rival interpretation of the cases there included, noting: 'But in each of these cases, the proposed change also involved a decrease in rigor and/or a danger of inconsistency. ... [and] rigor and consistency are already usually conceded ... to be weighty scientific standards.' In general, the rival interpretation might be that the objects at issue were not really supposed to be *abstract* in the usual sense of causally inert, but rather *concrete* or causally active; or it might be that the issue was not really *ontological* at all but rather *ideological*. Or perhaps several issues were intertwined, in which case the question would become which of these was ultimately of greatest scientific importance. Cases could be classified according to the type of rival interpretation involved.

These two steps of collection and classification would resolve the general tonsorial question into a series of more specific questions about the proper interpretation or analysis of a series of specific episodes or controversies. Doubtless much analysis of specific controversies could then be found in the work of scholars to whom it would never occur to address explicitly the question in its general form. A pattern might emerge from the analyses of these specific controversies, suggesting a conjectural answer to the general question, an interim judgement to the effect that one side has a good prima facie case. Such an interim judgement was indeed suggested in my paper: 'the burden of proof seems to be more on those who would insist that ontological economy of mathematical apparatus is also a weighty scientific standard.'

Trollope writes somewhere in his autobiography that an author should not quote himself. I have badly violated his ordinance. I do not know whether it will be considered a mitigating or and aggravating circumstance when I report that my paper *fell deadborn from the press*. Doubtless its constricted scale and crabbed style is largely to blame. How welcome then is the fact that, independently of my paper, another philosopher has reached similar conclusions about the philosophical importance and historical character of the tonsorial question, and a similar conjecture about its probable answer—and has moreover produced a treatment of the issue as ample in scope as it is admirable in style. I refer to the long chapter (5) in the recent doctoral dissertation (1992) of Gideon Rosen.[5] This work remains unpublished, and will not be freely quoted here. But here is a snippet from the end of the crucial section (§5.4):

> My conjecture is that from the scientist's standpoint—and here I mean the speculative theorist and not just the experimenter or the engineer—*the fact that* [*a*] *new theory manages to eschew commitment to abstract objects and so achieve a striking ontological economy constitutes no reason at all to prefer it.* I would change my mind if the nominalist . . . could point to a single substantial instance in the recent history of science to suggest that *this sort* of economy has ever mattered to scientists. After all, if it were truly important we should expect to find professional physicists engaged in the nominalization project. But of course we don't. And this suggests (at least to me) that . . . purging physics of a commitment to numbers serves no goal recognized either implicitly or explicitly in the practice of science.

An aspect of Rosen's high level of philosophical professionalism is a prudent abstention from amateur historicizing. This leaves an opening for the less prudent, who can perhaps hope to add a discussion of specific cases to his treatment of general issues. This is an opening I will take. As it happens, the most important case from the recent history of the mathematical sciences has already been extensively discussed in another guise in the literature.

4. A CASE FOR STUDY FROM THE MATHEMATICAL SCIENCES

Quine's response to Carnap was, to put it crudely, that his argument rested on a false dichotomy and so failed. Such an objection has considerable validity at least against my crude presentation of the argument. It is only too true that there is no agreement among philosophers as to what it takes to settle an ontological question. It is true enough that there is much agreement among mathematicians as to what it takes to settle an existential question. Still, there is no absolute dichotomy here. For only a couple of decades before the debate between Quine and Carnap it was quite untrue that there was agreement among mathematicians.

In the 1920s a significant segment of the mathematical community adhered to, or at least sympathized with, L. E. J. Brouwer's intuitionistic constructivism; or if not with the genuine Brouwerian version, at least with some version or other of intuitionism; or if not with the intuitionistic variety,

[5] Since the conference, Rosen has joined me in a project for a joint book, *A Subject with No Object: Strategies for Nominalistic Interpretation of Mathematics*, Oxford University Press. Some of the issues discussed in this paper are revisited in the closing chapter of the book, which also contains discussion of what has here been called the 'evasive' kind of nominalism and is there called the 'hermeneutic' kind.

at least some variety or other of constructivism. And a sympathizer with any constructivism is suspicious of a wide class of existence proofs in classical mathematics, namely *non-constructive* or *indirect* proofs, which purport to establish the existence of a mathematical object with a given property without providing any specific example of such a mathematical entity. Here, then, is a comparatively recent example of a controversy in the mathematical sciences. Both ontological and ideological interpretations of the controversy have been given.

One interpretation is ontological. Two-thirds of the way through his position paper (1948), Quine suggests that the recent dispute on the foundations of mathematics was a revival of the legendary medieval dispute on the problem of universals, with intuitionism corresponding to *conceptualism*, where: '*Conceptualism* holds that there are universals but they are mind-made.' This is an idiosyncratic variant on a more common interpretation that associates intuitionism less with medieval conceptualism than with modern idealism. In their crudest form, such interpretations take constructivism to be a form of *mentalism* or *psychologism*, the view that no entities exist transcendently or objectively in the sense of being outside of the minds of human subjects.

On such an interpretation, if constructivists, unlike contemporary nominalists, accepted *some* classical existence proofs, this was not because constructivists, unlike contemporary nominalists, conceded that some abstract entities exist. Rather, it was because constructivists, unlike contemporary nominalists, took mathematical objects to be not abstract entities but mental constructions. And it was because they took mathematical objects to be mental constructions that they rejected those classical existence proofs that purport to establish the existence of a mathematical object without providing a means of mentally constructing any specific example.

Another interpretation is ideological. It is suggested by the dictum of Kreisel that the issue is *not the existence of mathematical objects, but the objectivity of mathematical truth*, and elaborated in an influential paper of Dummett (1973). On such an interpretation, constructivism is a variety of *verificationism*, the doctrine that no truths hold transcendently or objectively in the sense of being independent of verification by human subjects: The very meaning of an assertion consists in its verification conditions. Moreover, the verification conditions for existential generalizations are supposed to be as follows: To verify an existential generalization, such as the assertion that *there are numbers that are greater than one hundred and are prime*, one must specify an instance, such as the assertion that *the number one hundred one is prime*, and then verify that instance. It is for this reason that intuitionists reject indirect proofs.

On such an interpretation, the specification of an instance may be a mental operation, the construction of a mental representation of a mathematical object; and the verification of that instance may also be a mental operation, the execution of a mental computation using that representation of the mathematical object. But nothing in intuitionism requires one to identify a mathematical object like the number one hundred one with the representation of it in the mind of any one person at any one time or place.

In the history of the tonsorial question, the debate between Quine's ontological, mentalist interpretation and Dummett's ideological, verificationist interpretation was the first important case-study. Or rather, it would have been if there had been an explicit debate. Actually, Quine's interpretation was hardly more than a passing suggestion; while Dummett—and in this he may very well be reflecting Kreisel's attitude as well as his own—writes at the beginning of his paper: 'the question is what forms of justification of intuitionistic mathematics will stand up, not what particular writers, however eminent, had in mind.'

I will attempt no definitive resolution of the question here. Still, it may be noted that an interim judgement opposed to any crudely mentalist or ontological interpretation is suggested by passages like the following from (ch. 2, §1 of) Arend Heyting's introductory textbook:

> If a natural number were nothing but the result of a mental construction, it would not subsist after the act of its construction and it would be impossible to compare it with another natural number constructed at another time and place. It is clear that we cannot solve this problem if we cling to the idea that mathematics is purely mental.

Brief as this discussion of intuitionism has been, my discussion in the paper cited earlier of cases from the mathematical sciences in the more remote past was briefer still. Nor will I enter into any lengthier discussion of these mathematical cases here. Rather, I would like to make up for the limitation in my paper cited earlier by discussing—again at no great length—a range of cases from the empirical sciences.

5. SOME HISTORICAL CASES FROM THE EMPIRICAL SCIENCES

Goodman's response to Carnap was intended to allay suspicion that nominalistic intolerance would be an obstacle to the advancement of science. For Carnap's critique had indeed been inspired by worries that nominalistic qualms about such entities as meanings would hamper the development of semantics. Nor is it difficult to understand why such entities would be unpalatable to nominalists. A meaning is an example of an *equivalence type*, what entities that stand in some *equivalence relation* thereby have in common. Their meaning is what utterances that stand in the equivalence relation of synonymy thereby have in common. And quite generally, equivalence types are abstract.[6]

Yet nominalistic qualms about abstracta are not the *only* grounds for objection to meanings. The later Quine, for instance, accepted those paradigmatic abstracta, sets, or classes. Moreover he recognized that these readily provide surrogates for equivalence types, including meanings: The set or class of all entities equivalent to a given one provides a surrogate for its equivalence type. Yet Quine continued to object to meanings, for reasons indicated in his reply to William Alston (who by the way was author of another important early anti-nominalist paper (1958)):

> Hypostasis of meanings is a red herring. I keep urging that we could happily hypostasize meanings if we could admit synonymy. We could simply identify meanings with the classes of synonyms. . . . The point . . . is that the *prior* assumption of an *unexplained* domain of objects called meanings is no way to explain synonymy or anything else. Synonymy, not hypostasis, is the rub. Given synonymy, a domain of meanings is trivially forthcoming for whatever good it would do.

The objection is not ontological, but ideological. The objection is less to the equivalence types, but to the equivalence predicate. Quine objects that the predicate 'synonymous' makes no real sense. Those cases where what at first glance seems to be an ontological issue at second glance seems to be ideological, or be intertwined with an ideological issue of at least equal scientific importance, I will call of *type A*.

Quine seems to have a second objection to the 'reification' or positing of meanings that would arise if his first objection were overcome. Even supposing synonymy were explained in a way that would satisfy him—presumably by a definition in terms of verbal behaviour—and the meaning of an expression were then defined to be the abstract class or set of all expressions synonymous with it, still Quine would raise the question whether such abstracta 'would do any good', would explain anything. Verbal or other human behaviour is presumably to be explained in terms of neurological or other bodily causes. The positing of some concrete neurological or other bodily *correlate* of abstract classes or sets of synonymous expressions would thus be required, and the existence of such hitherto undetected concreta might well be open to doubt even by one who did not question the existence of classes or the meaningfulness of synonymy. Those cases where what at first glance seems to be an issue about abstracta at second glance seems to be about concreta, or be intertwined with an issue about concreta of at least equal scientific importance, I will call of *type B*.

The case of *meanings* thus seems to be of type AB, with both A- and

[6] For reasons explained by Peter Simons in his paper at the conference. A brief account of equivalence types included in the original version of the paper has been omitted here as superfluous, given his fuller and more careful treatment.

B-factors present. The debate among philosophers like Quine and Carnap, who hoped to establish a scientific semantics, obviously is not quite a *scientific* debate. But my own somewhat unsystematic attempts to collect and analyse more scientific examples, which I will now briefly report, have found none of type O, none admitting *only* an Occamite interpretation.

To begin with physics, consider the debate over *substantivalism* and *relationalism*. It began with Newton and Leibniz, which is to say at a time when natural science was by no means fully separated from natural philosophy, but has continued down to the present day. According to traditional lore, the issue among early modern philosopher-scientists was ontological, about the existence of Absolute Space and Absolute Time. And yet even then at least in some part ideological, about the status of such predicates as 'occurring at the same place at different times'. More recent scientific debate has been in large part over the status of such predicates as 'occurring at different places at the same time'. The issue about this predicate of simultaneity, which is applicable to entire concrete, material-energetic particle-events, was whether it makes sense absolutely, or only relatively to a conventional choice of co-ordinates. The mere mention of the name of Einstein should suffice as a reminder of the scientific importance of this purely ideological question. In other words, this debate provides an example of type A.

Turning from physics to biology, a discipline where Aristotelian claptrappery persisted longer than elsewhere, one finds a complex of related issues that at first glance seem to be about abstract ontology, about the existence of *species* as entities over and above organisms. These include issues of classification (synchronically in taxonomy, and more importantly diachronically) in palaeontology. At second glance the issue seems to be more about the status of the equivalence predicate 'conspecific' or 'of-the-same-species-as': Does it make clear sense in application to ancestors and descendants many generations apart? The mere mention of the name of Darwin should suffice as a reminder of the scientific importance of this purely ideological question.

Turning from biology to psychology, consider the debate over the 'reification' of intelligence quotients, or more precisely of certain statistical measures derived from test scores by factor analysis. On inspection, it turns out that the 'reification' objected to does *not* consist in assuming the existence of intelligence quotients. The objection is not that a quotient would be numerical, a number would be abstract, and abstracta are inexistent. Rather, the 'reification' objected to consists in assuming the existence of a concrete correlate of the statistical measure, called 'general intelligence', in the brains of the test-takers, which is supposed to explain their test performance. In other words, this debate provides an example of type B.

Turning from psychology to sociology, the disputed 'reification' of social classes, for instance, is again not an objection to the existence of a class or set in the *mathematical* sense, whose members or elements would be all and only those persons receiving an income of a certain designated range of values or from a certain designated kind of source. Rather, what is often objected to is the attribution of collective *intentionality* and *agency* to social classes, the attribution of beliefs to the bourgeoisie, and preferences to the proletariat, over and above the individual beliefs and preferences of individual bourgeois and proletarians.

Doubtless a great many more examples could be added to my somewhat haphazard collection.[7] Certainly each of the examples collected deserves analysis at greater length than my somewhat breezy remarks. My aim here being to stimulate rather than to settle debate, however, I will not pursue the task of collection and analysis further. Rather, I would like advance a speculation still more general, and I hope still more provocative, than the conjecture that almost every apparent case of application of Occam's razor in science will turn out on analysis to involve ideological A-factors and/or concrete B-factors.

6. ONTOLOGY AND SCIENCE

According to traditional lore, Occam's razor is the general precept about ontology:

(1) *Entities are not to be multiplied beyond necessity.*

I have been considering a more specific precept about abstract ontology:

(2) *Abstract entities are not to be multiplied beyond necessity.*

I have questioned the scientificity of this specific precept, its status as a rule of scientific method. In so doing I have also implicitly questioned the scientificity of the general precept in so far as it makes no distinction of abstract versus concrete.

I have not so far questioned the scientificity of the specific precept about concrete ontology:

(3) *Concrete entities are not to be multiplied beyond necessity.*

In a sense, I do not wish to question its scientificity. I have already suggested that in several cases where scientists exhibited caution about what appeared to be abstracta they were really concreta; and there are many cases where scientists have exhibited caution about entities or substances (phlogiston, caloric, ether) that have never even appeared to be anything but

[7] Several examples, conforming to the pattern of the few I have given, were suggested at the conference, especially by Kit Fine, in discussion following my talk—so many, in fact, that I was unable to take them down.

concreta. I do not question that scientific theorizing *conforms* to the above precept.

I question its scientificity only in the sense that I question whether scientific theorizing is *governed* by the above precept, with its specifically ontological emphasis, or whether it is rather governed by a more general precept, of which the above is merely the ontological aspect. Towards formulating such a more general precept, let me first reformulate the above precept, making explicit what distinguishes abstract from concrete:

(3′) *Causal agents are not to be multiplied beyond necessity.*

An ideological parallel to this ontological precept would be:

(4) *Causal predicates are not to be multiplied beyond necessity.*

Here a new causal predicate would be a predicate connoting a newly posited mode of causal activity on the part of a previously recognized sort of causal agent. The ideological precept, just formulated in what Carnap called the *formal* mode, can thus be reformulated in what Carnap called the *material* mode:

(4′) *Causal activities are not to be multiplied beyond necessity.*

Taking *agencies* in a comprehensive sense, covering both agents and activities, and neutral between ontology and ideology, the precepts about concrete entities and about causal predicates can be merged into a single formulation:

(5′) *Causal agencies are not to be multiplied beyond necessity.*

To make its meaning more explicit, this in turn may be reformulated:

(5*) *Explanations in terms of extraordinary agencies are not to be resorted to until explanations in terms of ordinary agencies have been exhausted.*

What I am suggesting is that what really governs scientific theorizing is a precept something like this last one.

I would make an analogous suggestion in a related case.[8] I have suggested that abstract entities may be multiplied as far as scientific *convenience*, even if this goes well beyond scientific *necessity*. Still I do not question that scientific theorizing conforms to the precept not to multiply entities beyond *convenience*, that scientists avoid introducing entities that are purely *gratuitous*, neither necessary nor convenient. However, do they not also avoid introducing *predicates* that are purely gratuitous? I would suggest that what

[8] The present paragraph has been added to the original version of the paper in response to a point raised by Charles Chihara in discussion at the conference.

really *governs* scientific theorizing is a precept neutral as between ontology and ideology, according to which purely gratuitous assumptions of *whatever* category are to be avoided.

In suggesting that what really governs scientific theorizing is a precept something like the last one above, what I am suggesting is the following: Scientists don't *care* about the difference between positing new sorts of *things* (denoted by new nouns) and positing new modes of *acting* on the part of old sorts of things (connoted by new verbs). They would not, I am suggesting, necessarily consider a theory that dispenses with one currently recognized family of fundamental particles only by introducing a currently unrecognized fifth fundamental force an improvement on current theory. I speculate indeed that the whole distinction of ontology versus ideology is a matter of indifference to scientists, about which they feel no need to be careful. I will not here attempt even to outline or sketch an argument for this speculative suggestion: I will merely indicate what personal experiences led me to formulate it and put it forward. For me, two quite different writers provided crucial inspiration.

The first author who inspired me was Martin Gardner, a well-known popularizer of mathematics through his former column in the magazine *Scientific American*, and a wellknown debunker of pseudo-science through his current column in the magazine *Skeptical Enquirer*. A couple of years ago my wife, herself a mathematician, gave me for my birthday a copy of one of his later books (1956). As it was a soft-cover edition, it had on the back cover as advertising some quotations from reviews of the hardback edition. One of these read: '[Gardner] wields Occam's razor like a switch-blade.'—*Newsweek*.

My head being full of the issue of nominalism at the time, I was momentarily taken aback. 'Has Gardner joined Goodman?' I asked myself. 'Does he now consider the use of complex numbers in physics to be Platonistic claptrappery comparable to divination by inspection of entrails?' On examining the book I was quickly reassured. Nowhere in it does the author compare acceptance of classical existence theorems to belief in augury or spiritism. He is, indeed, very much concerned with presentday descendants of the old haruspices and mediums, the modern psychics and parapsychologists. His criticism of physicists is limited to criticism of *paraphysicists*, those few who, trained in the experimental methods of physics but ignorant of the performance methods of conjuring, have taken it upon themselves to endorse the claims of certain self-described psychics that they are able to bend metal by mind power alone.

Some paraphysicists have indeed, in order to explain the alleged phenomenon of mental metal-bending, postulated novel causal agents, such as *tachyons*. Others have postulated curious quantum activities of more familiar agents. Others have attempted no specific explanation at all, but merely endorsed the claim that there is a *paranormal* phenomenon admitting no *normal* or *ordinary* explanation. Gardner is impartially critical of ontological, ideological, and non-specific paraphysicists without distinction.

I could not locate anywhere among his criticisms of the current psi scene any passage where he explicitly labels some methodological precept 'Occam's razor', but his favourite dictum, often reiterated, is:

($5^{\dagger}$) *Extraordinary claims require extraordinary evidence.*

This is a precept not of philosophical scepticism about science, but of scientific scepticism about the occult.

The second author who inspired me was, ironically, the arch-Occamite Quine. Specifically, there is in his comparatively late book (1981), a first and almost-title essay a retrospective on the development of his views on ontology, and in that essay a section (§2) giving a more or less final formulation to his views on ontological commitment, and in that section an especially forceful statement about the indifference of lay people, of *non*-scientists to ontology:

> We must recognize . . . that a fenced ontology is just not implicit in ordinary language. The idea of a boundary between being and nonbeing is a philosophical idea, an idea of technical science in a broad sense. Scientists and philosophers seek a comprehensive system of the world . . . oriented to reference . . . more squarely and utterly than ordinary language. Ontological concern . . . is foreign to the lay culture.

But what were Quine's *grounds* for this judgement? Mainly he cites the fact that in ordinary language free use is made of abstract nouns, notably nouns formed from verbs and other parts of speech, even though lay speakers and writers generally do not have even an inchoate 'theory' about any corresponding entities. Often common usage provides no answer even to the very first question that an articulate theory would have to address—What are the *identity* conditions for such entities, when do we have one of them and when do we have two?

On reading Quine's grounds, I immediately asked myself, 'Aren't such features every bit as much a part of *scientific* as of lay or non-scientific prose?' Seeking an example to check my suspicion, rummaging through my files for a technical paper by a philosophically unselfconscious scientist, I happened upon a scientific article, Elliott and Mortensen (1975), whose principal author, being my sister, I can confidently say is as innocent of ontological metaphysics as I am ignorant of organometallic chemistry. Among the sixty-six words in its opening paragraph I counted no fewer than a dozen nouns (italicized below) either formed from verbs or otherwise abstract in the grammatical sense:

> The *importance* of [molybdate ions] in biological dinitrogen *fixation* has been demonstrated. Recently *attempts* have been made to characterize the *role* of molyb-

denum in the catalytic *reactions* of the nitrogenase components and its role in the *regulation* of the nitrogenase *system*. Although the antagonistic *properties* of tungsten have been used extensively in these *studies*, there is presently little *information* concerning the *nature* of *inhibition* by tungsten.

This suggested to me certain amendments to the passage quoted from Quine above. In hopes of provoking discussion and stimulating debate, I will close by proposing my amended version of that passage:

We must recognize . . . that a fenced ontology is just not implicit in scientific language. The idea of a boundary between being and non-being is a philosophical idea, not an idea of technical science in any ordinary sense. Certain philosophers seek a comprehensive system of the world . . . oriented to reference . . . more squarely and utterly than scientific language. Ontological concern . . . is foreign to the scientific culture.

REFERENCES

Alston, William (1958), 'Ontological Commitments', *Philosophical Studies*, 9: 8–17.

Benacerraf, Paul (1965), 'What Numbers Could Not Be', *Philosophical Review*, 74: 47–73.

——(1973), 'Mathematical Truth', *Journal of Philosophy*, 70: 661–80.

—— and Putnam, Hilary (1983) (eds.), *Philosophy of Mathematics: Selected Readings*, 2nd edn., Cambridge, Cambridge University Press.

Burgess, John P. (1983), 'Why I Am Not a Nominalist', *Notre Dame Journal of Formal Logic*, 24: 93–105.

——(1990), 'Epistemology and Nominalism', in A. Irvine (ed.), *Physicalism in Mathematics*, Dordrecht, Kluwer, 1–15.

Carnap, Rudolf (1950), 'Empiricism, Semantics, and Ontology', *Revue Internationale de Philosophie*, 4: 20–40.

Dummett, Michael (1956), 'Nominalism', *Philosophical Review*, 65: 491–505.

——(1973), 'The Philosophical Basis of Intuitionistic Logic', in H. E. Rose and J. C. Shepherson (eds.), *Logic Colloquium '73*, Amsterdam, North Holland.

Elliott, Barbara [Burgess, Barbara] and Mortensen, L. E. (1975), 'Transport of Molybdate by *Clostridium pasteurianum*', *Journal of Bacteriology*, 124: 1295–1301.

Gardner, Martin (1989), *Science: Good, Bad and Bogus* (paperback edn.) Buffalo, Prometheus.

Goodman, Nelson (1956), 'A World of Individuals', in I. M. Bochenski, A. Church, and N. Goodman (eds.), *The Problem of Universals*, Notre Dame, University Press, 155–72.

——and Quine, Willand Van Orman (1947), 'Steps Toward a Constructive Nominalism', *Journal of Symbolic Logic*, 12: 105–22.

Heyting, Arend (1931), 'The Intuitionist Foundations of Mathematics', in 'Symposium on the Foundations of Mathematics' (in German), *Erkenntnis*, 2: 91–121; English trans. in Benacerraf and Putnam (1983).

Heyting, Arend (1956), *Intuitionism: An Introduction*, Amsterdam, North-Holland.
Putnam, Hilary (1967), 'Mathematics Without Foundations', *Journal of Philosophy*, 64: 5–22.
——(1971), *Philosophy of Logic*, New York, Harper.
Quine, Willard Van Orman (1948), 'On What There Is', *Review of Metaphysics*, 2: 21–38.
——(1951), 'On Carnap's Views on Ontology', *Philosophical Studies*, 2: 65–72.
——(1981), *Theories and Things*, Cambridge, Belknap Press.
Rosen, Gideon (1992), 'Remarks on Modern Nominalism', Doctoral diss. Dept. of Philosophy, Princeton University.

8

Beyond Definitionism—But Not Too Far Beyond

GEOFFREY HELLMAN

1. OVERVIEW

Our aim here is twofold: first, to identify the principal advantages and disadvantages of the definitionist foundational programme of predicativist mathematics (originating with Poincaré and Weyl, and developed by Feferman *et al.*), and second, to propose a recipe for liberalizing that programme which overcomes the disadvantages identified while preserving most of the advantages, and which synthesizes (arguably the best of) nominalism, predicative flexible type theory, and modal-structuralism. The resulting system is of proof-theoretic strength exceeding that of full classical analysis (indeed, equal to that of a second-order theory of continua); a great deal of modern analysis—quite possibly all that finds scientific application—can be developed within it quite naturally, i.e. with minimal reliance on codings; and yet it presupposes little more than the notions 'nominalistic sum (or fusion) of atoms' and 'finite fusion of atoms', together with the mere logical possibility of a model of a very elementary theory of finite sets and classes (roughly equivalent to postulating the possibility of an ω-sequence of atoms). The system can be claimed to transcend certain critical limitations of predicativist analysis and certain awkwardness of known nominalistic approaches to mathematics at once, and would seem to provide a well-motivated alternative to Zermelo set theory as a framework for scientifically applicable mathematics.

2. THE VIRTUES AND VICES OF DEFINITIONISM

By 'definitionism' we refer to the view that 'all mathematical objects should be explicitly present[able] by finite symbolic expressions' (Feferman 1985,

Work reported on here has been supported by the National Science Foundation, USA, Award No. SBER93-10667. I am grateful to John Burgess and Allen Hazen for helpful correspondence.

44). Paradigmatic are the natural numbers presented by numerals of one form or another, finite sets as presented by brackets enclosing member designators, ordered pairs presented by bracket notation, and classes presented by formulas with a free variable and with bound variables ranging over 'already' presented (presentable) objects. The latter are *predicative* definitions or specifications, and allow the predicativist to speak of classes as a *façon de parler* in place of more long-winded talk of satisfaction of predicates by (given, unproblematic) objects (cf. e.g. Chihara (1973)). In addition to symbolic expressions themselves, the latter have usually included the natural numbers, taken as unconstructed, basic objects. Predicativist analysis then proceeds by introducing definable sets (and functions) of natural numbers, allowing first only arithmetical quantifiers in the defining formulas, and then allowing definitions of new sets of natural numbers with quantifiers restricted to natural numbers or the first level of sets, and so on, iterating the process (ultimately along any countable ordinal suitably codable as a well-ordering within the natural number system itself). The limits of this sort of process are well understood, and are marked by the Feferman–Schütte ordinal Γ_0 (described from 'outside' in terms of provability of instances of transfinite induction up to Γ_0, just exceeding the strength of predicativist systems).[1] In recent years, Feferman especially has developed certain systems of variable or flexible types over the natural numbers which avoid the awkwardness of ramified systems and allow one to carry out predicative analysis naturally, following classical developments quite closely.[2] They are naturally described as occupying a position *between constructive and classical mathematics* (the title Feferman's (1984)). One of the most recent systems is W, in honour of Weyl (set forth by Feferman in his (1988)). Despite its expressive power and flexibility, W is a conservative extension of Peano Arithmetic.

With systems such as W in mind, let us now list the principal advantages and disadvantages of the definitionist framework.

Advantages:

[1] For an informative, informal discussion of Γ_0 see Smorynski (1982).

[2] Such unramified systems raise interesting problems of interpretation from the predicative point of view. The intended, comprehensive interpretation takes the class quantifiers as ranging over arbitrary 'predicatively specifiable' classes (of finite type over a natural number structure), but in order to make this precise, impredicative notions inevitably would be employed (such as reference to the limit ordinal Γ_0 itself). Alternatively, in the manner of Russell's systematic 'typical ambiguity', one may appeal to a multiplicity of restricted but predicatively describable interpretations: for any given limit ordinal $\alpha < \Gamma_0$, the class quantifiers are interpreted to range over classes of order $< \alpha$ in the ramified hierarchy of classes (of finite type) over a structure for the natural numbers. (A comprehension axiom is understood as introducing a class one order above the highest order of the finitely many class parameters occurring in the defining formula.) The axioms of Feferman's W, for example, are certainly satisfied under such interpretations, which are adequate for understanding standard applications of mathematics. For further discussion, see Hellman (1994).

(1) Epistemic security through consistency; one can hardly ask for better than consistency relative to Peano Arithmetic (transfinite induction below ε_0).

(2) Epistemic security through definiteness: with each set or function definable, we 'know just what we're talking about'.

(3) Avoidance of platonist ontology: given the intelligibility of *satisfaction* (of formulas by unproblematic objects), nominalistic interpretations of predicativist systems can be carried out. (Caveat: Strictly, modal operators are probably needed, e.g. to make sense of talk of an infinity of symbolic expressions themselves. And there must be an account of the natural numbers.)

(4) Retention of classical logic. Unlike intuitionism and Bishop constructivism, predicativism allows full classical reasoning (including proofs of existence by reductio ad absurdum), in our view critical to the viability of a foundational programme. (See e.g. Hellman (1992) and (1997).) This is closely related to the next point.

(5) The great bulk of scientifically applicable mathematics is recoverable, with relatively little reliance on complicated codings. (Here systems such as Feferman's W have an advantage over the Friedman–Simpson subsystems of analysis.)

Now for the bad news:

(1) As usually presented, predicativism takes the natural number system as given. This is especially problematic as there is prima-facie appeal to impredicativity in the justification of mathematical induction, or better, of the existence of a structure satisfying it (as has been argued by Parsons (1983)), and in the proof of categoricity (of second-order Peano Arithmetic) as well.

(2) Relativity to languages or language-users: As mathematical language changes, so do the sets of natural numbers which 'exist', etc. All questions about the continuum become relativized.

(3) Restriction to the countable: Although this can appear an advantage, when one considers the significance of Cantor's diagonal argument, it becomes a liability. Indeed, the diagonal argument can be adapted to predicatively intelligible systems quite generally so that even the predicativist can see that there must be more reals than the given system can specify. In the case of unramified systems, the argument is a bit more subtle than in the ramified case, but it still goes through (see Hellman (1994) for details). Thus, despite the move to unramified systems, it seems that, within a predicativistically interpretable system, one cannot speak of a unique continuum, of '*the* real number system' at all. (At least not seriously—one can, armed with a suitable, predicatively intelligible conservativeness proof, speak of anything one likes as a device for proving statements about the restricted domain one takes to be 'real'.) Predicative proofs of *unicity* of

the real number system (referred to by Feferman (1985)) only achieve a relativized form of unicity.

(4) In spite of advantage (5), *some* physically applicable mathematics appears to transcend the bounds of predicativism, especially the use of non-separable Banach or Hilbert spaces, e.g. in quantum field theory (see e.g. Emch (1972)). How serious this and other limitations may be requires further investigation.

(I refrain from listing as an objection, 'But you cannot do justice to that bold and beautiful subject, the theory of the transfinite—ZFC and beyond, the study of large cardinals, of quasi-projective determinacy, and so forth.' The objections are intended to be audible by the predicativist, so 'Hark! The Cantorian angels sing', cannot be one of them!)

3. QUE FAIRE?

We cannot have everything. In particular, we cannot have thoroughgoing security through definitions (advantage (2)) and at the same time avoid language relativity and countabilism. Similarly, we cannot have the advantages of a logicist analysis of the natural number system (à la Dedekind) together with a complete analysis of the concepts 'finite' and 'infinite' and at the same time avoid all impredicative mathematical existence assumptions.[3] Still we may ask for a motivated system somewhere *between constructive and classical mathematics*. As already noted, Feferman's systems occupy such a place, but they face the above difficulties. The system about to be proposed can be located as falling *between between constructive and classical mathematics and classical mathematics*.

The *main ingredients* (to be adjusted and combined in the right way) are the following:

1. The atomic calculus of individuals (ACI), or mereology (with atoms) (as sketched, e.g. in Nelson Goodman (1977));
2. The elementary theory of finite sets and classes (EFSC) of Feferman–Hellman (1995), to be described below;
3. A theory of predicative flexible types (finite types over a ground-level of individuals), say, Feferman's W, but with the number-theoretic axioms replaced by those of EFSC. (Call the resulting system 'W-EFSC'.)
4. A modal-structural analysis of mathematics in W-EFSC, along the lines of Hellman (1989), applied to the main types of structures of interest, e.g. $\mathbb{N}$-structures (i.e. ω-sequences), $\mathbb{R}$-structures, $\mathbb{C}$-structures, etc., as substructures of models of W-EFSC. These are to be described as concrete structures obeying ACI rather than set-theoretically.

[3] This trade-off emerges clearly from the predicative characterization of the natural number structure, together with existence proof, presented in Feferman and Hellman (1995), which is outlined below.

Instructions on the ingredients and how to combine them

Re 1.: Rather than attempt to rewrite W-EFSC nominalistically, it is more convenient to appeal to ACI in our background (meta-)language when we give a modal-structural interpretation of mathematics in W-EFSC. The crucial assumption is that, given any whole *w* of atoms, arbitrary fusions of those atoms are among the parts of *w*. These are in correspondence with the *subsets* of the atoms of *w* (platonistically speaking), *but they are just as concrete (non-abstract) as the atoms themselves. There is no need to restrict attention to the* definable *parts of w, in order to remain nominalistic*. This emerges explicitly in the 'Σ-comprehension-scheme' of ACI:

$$\exists x[\Phi(x)] \rightarrow \exists y \forall z[y \text{ o } z \leftrightarrow \exists u(u \text{ o } z \,\&\, \Phi(u)], \qquad (\mathrm{C}\Sigma)$$

which guarantees the fusion (nominalistic sum) of all Φ's for non-vacuous Φ, where Φ may contain quantifiers ranging over *arbitrary* individuals. ('o' is the *overlaps* relation, interdefinable with '<', *part-whole.*) Once some atoms are given, so are *all* their parts, not just the ones you or I can describe.

Re 2.: The language $\mathscr{L}$(EFSC) is three-sorted, with *x,y,z*, . . . ranging over individuals, *A,B,C*, . . . ranging over finite sets of individuals, and *X,Y, Z*, . . . ranging over classes of individuals; individual terms are generated from individual variables by the binary operation symbol (,), for pairing; the atomic formulas are of the form s = t, s ∈ *A*, s ∈ *X*, . . . , and formulas are generated from these by the usual classical connectives and quantifiers. *WS-formulas—predicative* or *weak second order formulas*—are formulas in which there are no bound class variables. The underlying logic is classical (three-sorted), with equality defined from inclusion in the usual way for set and class variables. The six axioms of EFSC are easily presented:

$$(\text{WS-CA})\ \exists X \forall x[x \in X \leftrightarrow \phi],$$

where φ is a WS-formula lacking free '*X*' ('WS-CA' is for 'weak second-order comprehension axiom');

$$(\text{Sep})\ \forall A \exists B \forall x[x \in B \leftrightarrow x \in A \,\&\, \phi],$$

where φ is a WS-formula lacking free '*B*' ('Sep' is for 'separation');

$$(\text{FS-I})\,(\text{Empty})\ \exists A \forall x[x \notin A]$$

$$(\text{FS-II})\,(\text{Adjunction})\ \forall a \forall A \exists B \forall x[x \in B \leftrightarrow x \in A \vee x = a]$$

('sets obtained from finite sets by adding one element are finite');

$$(\text{P}-\text{I})\ \ (x, y) = (u, v) \leftrightarrow x = u \,\&\, y = v$$

$$(\text{P}-\text{II})\,(\text{urelements under pairing})\ \exists u \forall x, y[(x, y) \neq u].$$

One can readily define the notion of '$\mathbb{N}$-structure' (satisfying the usual conditions on zero and successor—this sufficing for a 'pre-$\mathbb{N}$-structure'—together with mathematical induction (written with '$\forall X$')); one then proves in EFSC,

(I) *Unicity: Any two $\mathbb{N}$-structures are isomorphic.*

(II) *Existence of pre-$\mathbb{N}$-structures, indeed of special pre-$\mathbb{N}$-structures,*

where a pre-$\mathbb{N}$-structure is *special* iff (i) 'any non-zero element has an immediate predecessor', and (ii) 'every initial segment (in the ordering based on the successor relation of the structure) is finite.'

EFSC is then extended to the theory EFSC* by adding the further 'cardinality' axiom,

(*Card*) $\forall A(\text{Ded-fin}(A))$,

where 'Ded-fin' abbreviates the usual definition of 'Dedekind-finite' ('every one-one function on *A* has *A* as its range'). In EFSC*, one proves

(III) *Any special pre-$\mathbb{N}$-structure satisfies mathematical induction, i.e. is an $\mathbb{N}$-structure, hence $\mathbb{N}$-structures exist.*

One also proves

(IV) EFSC* is of the same proof-theoretic strength as Peano Arithmetic.[4]

Thus, despite Poincaré's influential view that mathematical induction plays a primitive role in mathematics, one can begin with the predicatively unproblematic notion of finite set (of describable objects, we may assume), as governed by elementary axioms of EFSC, and recover the core of Dedekind's (1888) analysis of the natural number system in a predicatively acceptable way. (The restriction to domains of describable objects is built into the predicativist's definitionist framework. Finite subsets of such a domain are then explicitly describable by means of finite disjunctions of the form $x = d_1 \vee x = d_2 \vee \ldots \vee x = d_n$, where the d_i are designators of objects in the given domain.)

Thus, despite appearances to the contrary, the predicativist need not take the natural number system as an unanalysed given; indeed, the first 'disadvantage' of predicativism we listed above can be overcome.

Re 3.: If we seek a framework for carrying out predicative mathematics generally, including theorems relating different structures (such as unicity

[4] Shortly after publication of Feferman-Hellman (1995), Peter Aczel pointed out that (III) is already derivable in EFSC (without (*Card*)) and that therefore (IV) can be strengthened to read that EFSC is of the same proof-theoretic strength as Peano Arithmetic. Reflecting this, henceforth reference will be to EFSC. For the proof and for further discussion of the significance of these and related results, see Feferman-Hellman (forthcoming).

of ℕ, ℝ, etc.), it is better to begin with a framework such as EFSC together with the apparatus of types of W rather than with the number-theoretic axioms (included in W as originally presented). As just indicated, in EFSC one has the machinery for expressing naturally such theorems as Dedekind's categoricity theorem, whereas if we begin with number-theoretic primitives, all our 'universes' must be built out of 'numbers' (objects in the range of the number-theoretic primitives).

Re 4.: The key point here is to postulate the possible existence of a countable infinity of atoms to serve as the natural numbers. This can be stated directly by means of axioms on either a successor or a numerical-ordering relation, in turn definable from the part–whole relation, as Hazen has shown, by making use of nested fusions together with plural quantification.[5] Alternatively one may proceed indirectly by postulating atoms to serve as ordered pairs, beginning with a single atom (call it '0') as urelement under pairing, in accordance with the axioms of EFSC; successor can then be introduced via $x' = (x,0)$. All fusions of such atoms are then available as well, without introducing anything abstract at all. Under coding (conveniently, a number-theoretic coding within an ω-sequence of atoms), negative numbers and rational numbers can be introduced in standard ways, and real numbers can be identified as lower Dedekind-sections in the rationals (ℚ), where here these are taken not as sets but as wholes or fusions of the atoms coding the rationals of the Dedekind-section. In this manner, the classical continuum is recovered nominalistically, but based on just a countable domain of pairwise discrete individuals, not on anything so problematic as space-time points. With just a bit more coding (essentially of countably many pairwise non-overlapping ω-sequences within a single given one), it is also possible to carry out the proof of *unicity of the real number system*, and now this means *genuine, absolute unicity*. The crucial observation here is that the classical proof of unicity of ℝ-structures requires only *sequential completeness* ('Every bounded *sequence* of reals has a least upper bound') rather than full completeness ('Every set of reals has a least upper bound'). (See Hellman (1994) for details.)

When it comes, however, to sets of reals, functions from ℝ → ℝ, etc., we treat them predicatively as in W. That is, *definitionism is relaxed only in the*

[5] Plural quantifiers were introduced in Boolos (1985) as a way of understanding monadic second-order quantification without commitment to classes. A paradigm is given by appealing to English plural nouns and pronomial constructions in 'Some critics admire just themselves (i.e. each one of them)', as interpreting its second-order formalization, '$\exists X \forall y[(X(y) \rightarrow \mathrm{Critic}(y)) \;\&\; \forall y \forall z(X(y) \rightarrow (\mathrm{Admires}(y,z) \leftrightarrow X(z)))]$'. This completely reverses the more familiar argument which—appealing to the standard interpretation of the second-order quantifier—takes the equivalence as showing that the ordinary English sentence with its plurals contains a hidden commitment to classes.

For further information on the use of plural quantifiers with mereology in nominalistic foundations of arithmetic, see Hellman (1994).

domain of the concrete; in the domain of the abstract—i.e. where we need to speak of sets, functions, classes, relations, etc.—definitionism remains in effect. We know how in principal to nominalize talk of definable classes, relations, etc. So we save ourselves the headache of recasting mathematics in nominalistic terms directly by availing ourselves of flexible predicative-type theory. But by postponing this invocation by one stage—that is, by allowing direct (modal) quantification over fusions of 'numbers' as well as over objects serving as natural numbers themselves—we achieve mathematical strength far surpassing that of W or any standard approach to predicative mathematics.

Indeed, there is a way of postponing appeal to definability one stage further: we may appeal to the device of (monadic) plural quantification over the reals (constructed nominalistically as above) in order to achieve the effect of quantification over *arbitrary*, not just *definable*, *sets* of reals. In fact, with some further (easily satisfied) assumptions concerning classifications of (possible) atoms, Burgess, Hazen, and Lewis have shown how to encode ordered pairs, not just of atoms, but of individuals generally, so that relations can be treated via plural quantification over ordered pairs! (See Burgess, Hazen, and Lewis (1991); also Lewis (1993).) In the present setting, this would enable us to bring the entire classical theory of functions of a real or complex variable within modal-structural-nominalist W-EFSC.

Still, it should be noted that, if we wish to speak of sets of sets (or functions) of reals, we shall be confined to specifiable sets of specifiable sets (or functions) of reals, as in the system W or W-EFSC.[6] We cannot go on to collect—even specifiably collect—arbitrary sets of reals, as plural quantifiers for the latter do not provide *anything* (i.e. any objects) at that level that might be further 'collected'. (This after all was the point of plural quantifiers.)

To sum up, our system may be identified with W-EFSC together with

$$\Diamond\exists X, f(\omega(X, f)), \qquad (\mathrm{MSN}\omega)$$

[6] This is true if we limit ourselves to postulating the possibility of an infinitude of atoms guaranteeing an ω-sequence. If we add further postulates, e.g. that a *continuum* of atoms is possible, then with plural quantifiers we can recover the level of classical sets of sets of reals, and then, in virtue of BHL pairing, functions and relations of sets (and functions and relations) of reals, i.e. polyadic classical fourth-order number theory (cf. Hellman (1996)). Moreover, there is a natural way of generalizing this latter move from PA^3 to PA^4: one can adopt an axiom of *atomic realizability*, viz.

> If a structure of individuals (characterized by given axioms) is possible, then so is such a structure of atoms.

Repeated application of this, beginning with a countable infinity of atoms, would generate enough individuals to form a natural model of Zermelo set theory. In the present paper, we have refrained from any modal-existence postulate beyond that of a countable infinity of atoms, as we are focusing on comparisons with predicative mathematics. Moreover, we wish to emphasize how rich a system can be developed on this rather modest basis.

abbreviating the statement that X is a whole of atoms satisfying the conditions of an ω-sequence, where f behaves as the successor-relation. (As already indicated above, there are ways in which this can be stated nominalistically and derived from the postulate, 'possibly there are infinitely many atoms'. 'MSN' stands for 'modal-structural nominalistic'.) The whole system may be dubbed if not pronounced W-EFSC-MSNω.

4. IS THE DISH PALATABLE?

(Is this a matter of taste?)

Well, I can certainly sympathize with the nominalistic inclination voiced by David Lewis when he writes of classes that 'they are profoundly mysterious'. Moreover, I join him whole-heartedly when he writes, 'to reject mathematics for philosophical reasons would be absurd', and I laugh as heartily as he 'to think how *presumptuous* it would be to reject mathematics for philosophical reasons' (Lewis (1993), 14–15). And I agree that 'even if we hold on to some mutilated fragments of mathematics that can be reconstructed without classes, if we reject the bulk of mathematics that's still absurd' (ibid.).

But as the work on W and the above shows, *the vast bulk of mathematics can be reconstructed—even very naturally—without classes*. Of course the vast bulk of the official ZF universe cannot be, but what has been shown is that the vast bulk of *mathematics as standardly practised* survives quite well on the mere possibility of an ω-sequence of atoms ('and all that that entails')!

Until now, nothing has been said about how to treat mixed mathematical-material statements crucial for applications of mathematics, such as 'the number of Fs < the number of Gs', 'the mass in grams of this glass = q', etc. Are classes and functions unavoidable here? No, not with the nominalistic resources of the modal-structuralist (e.g. Hellman (1989)) or with those of plural quantification and pairing as already invoked above. The key idea is to apply these devices now *counterfactually*, under the hypothesis of an ℕ-structure of atoms, leaving intact the material system to be enumerated or measured, etc. (For an elaboration of this idea, see Hellman (1989), ch. 3.) We need pairing between 'numbers' and 'non-numbers' to simulate talk of functions or correspondences between the two domains. This can be accomplished along the lines sketched by Burgess *et al.* (1991) and Lewis (1993), even without the assumption of atomicity in the material domain. In this way, then, the preponderance of material applications of mathematics can be represented in the present framework, without ultimate recognition of classes or relations.

It is now time to return briefly to our above list of advantages and disadvantages of predicativism and to assess the situation with respect to the system W-EFSC-MSNω. We have already commented on consistency strength; if we add plural quantification and the resources for ordered pairs of reals, we end up with a system as powerful as classical third-order number theory. Language-relativity is overcome for the continuum, and, by invoking plural quantification, for one stage beyond that. Without further postulation we encounter it finally at the level of sets of sets of reals and beyond, but it seems an inevitable price we pay for advantage (3), nominalist ontology, and for what remains of advantage (2), epistemic security through definiteness.[6] When it comes to scientific applications, the resources for treating even non-separable metric spaces are now available, for we have in effect the full set of reals to work with. Indeed, in light of the power of W itself, it would be hard to find examples of scientifically applicable mathematics not representable in W-EFSC-MSNω. Whether any such examples will ever arise is an interesting speculative question.

5. *Conclusion*: Commenting recently on the force of proof-theoretic reductions and the surprising strength of systems far weaker than the synthesis proposed here, Feferman concluded:

> The uninformed common view—that adopting one of the non-platonistic positions means pretty much giving up mathematics as we know it—needs to be drastically corrected, and that should also no longer serve as the last-ditch stand of set-theoretic realism. (Feferman (1993))

The results of our own investigations summarized above clearly serve to bolster this conclusion.

Not only should the dish following our recipe be palatable; it constitutes a species' lifespan of meals in itself, entirely capable of supporting the science of microwaves and virtually anything else! *Bon appétit*!

REFERENCES

Boolos, G. (1985), 'Nominalist Platonism', *Philosophical Review*, 94: 327–44.

Burgess, J., Hazen, A., and Lewis, D. '*Appendix on Pairing*', in Lewis (1991).

Chihara, C. (1973), *Ontology and the Vicious Circle Principle*, Ithaca, Cornell University Press.

Dedekind, R. (1888), *Was sind und was sollen die Zahlen*? (Brunswick, Vieweg); trans. 'The Nature and Meaning of Numbers', in *Essays on the Theory of Numbers*, New York, Dover, 1963, 31–115.

Emch, G. (1972), *Algebraic Methods in Statistical Mechanics and Quantum Field Theory*, New York, Wiley.

Feferman, S. (1984), 'Between Constructive and Classical Mathematics', in *Computation and Proof Theory, Lecture Notes in Mathematics 1104*, Berlin, Springer, 143–62.

——(1985), 'Intensionality in Mathematics', *Journal of Philosophical Logic*, 14: 41–55.

——(1988), 'Weyl Vindicated: *Das Kontinuum* 70 Years Later', in *Temi e prospettive della logica e della filosofia della scienza contemporanee*, Bologna, CLUEB, 59–93.

——(1993), 'What Rests on What? The Proof-Theoretic Analysis of Mathematics', *Philosophy of Mathematics*, Part I, Proceedings of the 15th International Wittgenstein Symposium, Vienna, Hölder-Pichler-Tempsky, 442–55.

——and Hellman, G. (1995), 'Predicative Foundations of Arithmetic', *Journal of Philosophical Logic*, 24: 1–17.

————(forthcoming), 'Challenges to Predicative Foundations of Arithmetic', in G. Sher and R. Tieszen (eds.), *Between Logic and Intuition*.

Goodman, N. (1977), *The Structure of Appearance*, Dordrecht, Reidel.

Hazen, A. (unpublished), 'Arithmetic in Mereology and Plural Quantification.'

Hellman, G. (1989), *Mathematics without Numbers: Towards a Modal-Structural Interpretation*, Oxford, Oxford University Press.

——(1992), 'The Boxer and his Fists: The Constructivist in the Arena of Quantum Physics', *Aristotelian Society*, Suppl. vol., 66: 61–77.

——(1994), '*Real* Analysis without Classes', *Philosophia Mathematica*, 2: 228–50.

——(1996), 'Structuralism without Structures', *Philosophia Mathematica*, 4: 100–23.

——(1997), 'From Constructive to Predicative Mathematics', in J. Earman and J. Norton (eds.), *The Cosmos of Science, Pittsburgh Series in Philosophy and History of Science*, University of Pittsburgh Press.

Lewis, D. (1991), *Parts of Classes*, Oxford, Blackwell.

——(1993), 'Mathematics is Megethology', *Philosophia Mathematica*, 3/1: 3–23.

Parsons, C. (1983), 'The Impredicativity of Induction', in *How Many Questions? Essays in Honor of S. Morgenbesser*, Indianapolis, Hackett, 132–54.

Smorynski, C. (1982), 'The Varieties of Arboreal Experience', *The Mathematical Intelligencer*, 4: 182–9.

9

Holistic Mathematics

MICHAEL D. RESNIK

By holism I shall mean epistemic or confirmational holism, that is, the thesis that no claim of theoretical science can be confirmed or refuted in isolation but only as part of a system of hypotheses. Duhem's remarks on the law of inertia and similar scientific claims make a strong case for a version of holism restricted to the theoretical branches of natural science;[1] but, as several philosophers of mathematics have noted,[2] Quine's extension of Duhemian holism to include mathematics and logic is much more problematic. Recently Penelope Maddy[3] and Elliott Sober[4] have tied difficulties with Quine's holism to his indispensability argument for the existence of mathematical objects. My purpose here is to defend Quine's thesis, or at least a close variant of it.

I will begin by reviewing the arguments for Duhem's and Quine's versions of holism, and then point out some prima-facie objections to Quine's view that do not apply to Duhem's. This will bring me to the worries Maddy and Sober have raised about Quinean holism and his views on confirming mathematical principles. Although I will not dispute their observations about the practice of science, I will point out that a version of the indispensability argument survives them. However, indispensability is separable from holism: the former purports to support realism about mathematical objects and truths; the latter purports to support the empirical testability of mathematics. So I will still have to defend holism.

I will do this by proposing that scientific and mathematical theories and their attendant methodologies can be partially ordered as more or less

I would like to thank Mark Balaguer, Geoffrey Hellman, Penelope Maddy, Bijan Parsia, Stewart Shapiro, and Paul Teller for helpful correspondence and comments. I am also grateful to participants in the Munich conference for useful discussion following my presentation of an earlier version of this paper.

[1] Pierre Duhem, *The Aim and Structure of Physical Theory*, trans. by Philip P. Wiener (Princeton, Princeton University Press, 1954).

[2] Charles Parsons, 'Quine on the Philosophy of Mathematics' in *The Philosophy of W. V. Quine* (Library of Living Philosophers), ed. by Lewis Edwin Hahn and Paul Arthur Schilpp (La Salle, Illinois, Open Court, 1986). Charles Chihara, *Constructability and Mathematical Existence* (Oxford, Clarendon Press, 1990).

[3] Penelope Maddy, 'Indispensability and Practice', *Journal of Philosophy*, 89 (1992), 275–90.

[4] Elliot Sober, 'Mathematics and Indispensability', *Philosophical Review*, 102 (1993), 35–57.

global (local) so that the theoretical and methodological background which a theory takes as initially fixed is derived from more global ones. Holist reflections may attend the introduction of new entities, new methods, and new fundamental laws into a theory, but usually they do not extend to the more global theories taken as part of the fixed background. Moreover, once in place the new methods provide the beginnings of the local conception of evidence with which the theory's practitioners work. Mathematics is our most global theory.

This picture of science should do much to meet some of the prima-facie objections to Quine's holism, but with mathematics being the most global theory an a priori–empirical distinction threatens to resurface. In the final section of the chapter I will attempt to deal with this worry.

1. DUHEMIAN AND QUINEAN HOLISMS

If an hypothesis H implies an observational claim O, then from the falsity of O we may infer the falsity of H. As Duhem observed, theoretical scientists are rarely in a position to use this mode of reasoning in rejecting hypotheses. Most statements of their sciences of any interest imply no observational claims when taken by themselves, although each does when taken in conjunction with certain other 'auxiliary' hypotheses, A. In most instances then, the falsity of O implies only that H or one or more of the A is false. Strictly speaking, it is systems of hypotheses rather than individual ones to which the usual, deductively characterized notions of empirical content, confirmation, and falsification should be applied. This is familiar stuff, and I suppose that I learned it years ago through Carl Hempel's philosophy of science text.

Sometime within the last two years I learned that Duhem also defended the law of inertia and similar physical hypotheses against the charges that they have no empirical content and are unfalsifiable. One way of putting the law of inertia, you will recall, is to say that a body remains at rest unless an external force is imposed upon it. Since we can only determine whether something is at rest by positing some observable reference system, this law, taken by itself, implies no observational claims. Furthermore, by appropriately changing reference systems we can guarantee that a body at rest relative to our present system is moving relative to the new one, and thereby protect the law against falsifying instances. All this is troubling, because the law of inertia, *qua* physical law, is supposed to have empirical content and be falsifiable. Duhem responded to this worry about the law by observing that it readily produces empirical consequences when conjoined with auxiliary hypotheses fixing an inertial system; and that in needing auxiliaries to produce empirical consequences, it was no different from many other theoretical principles of science, whose empirical content everyone readily acknowledged. Thus the law's critics could not have it both ways: to the extent that their critique challenged the empirical status of the law of inertia it also challenged that of most other theoretical hypotheses.

Extracting empirical content from the law of inertia also depends upon including mathematical principles among the auxiliary hypotheses. Duhem drew no conclusions from this about mathematics. But Quine did. Using the very strategy Duhem used in defending the law of inertia, he argued that even mathematical principles, which by most accounts are just as unfalsifiable and devoid of empirical content as the law of inertia, share in the empirical content of systems of hypotheses containing them.[5] (Of course, if Quine had not independently discredited other epistemically relevant distinctions between mathematics and empirical science, then defenders of the apriority of mathematics could have simply replied that conjoining mathematical statements with auxiliaries can never falsify them because they enjoy a special epistemic status, such as being true by convention.)

I do not know why Duhem did not come to the same conclusion as Quine did. However, Frege, who anticipated Duhem's remarks on the law of inertia,[6] would have had a reason for not doing so. For him, mathematical principles are logical principles, and as such implied by the conjunction of the law and its physical auxiliaries. Thus, to him, it would be a mistake to say that extracting empirical content from the law depends upon including mathematical principles among the auxiliaries, and also a mistake to conclude that a failed prediction counts against the mathematical principles as well.

Quine would not accept this argument, because he restricts logical implication to that afforded by first-order logic with identity. Still he must recognize at least this much of Frege's point: whatever the limits of logic, systems of logical truths are immune from the direct sort of empirical falsification to which systems of mathematical and scientific hypotheses are subject. Quine urges as much in the following passage from a recent book:

> Now some one or more of the sentences in *S* are going to have to be rescinded. We exempt some members of *S* from this threat on determining that the fateful implication still holds without their help. Any purely logical truth is thus exempted, since it adds nothing to what *S* would logically imply anyway.[7]

I will not speculate on whether this passage indicates that Quine has revised his famous doctrine to the effect that we can revise the laws of logic to save

[5] W. V. Quine, *Pursuit of Truth* (Cambridge, Mass., Harvard University Press, 1990), 14–15.

[6] Gottlob Frege, 'Über das Trägheitsgesetz' (1891), in Gottlob Frege, *Kleine Schriften*, ed. by Ignacio Angelelli (Hildesheim, Georg Olms, 1967), 113–24.

[7] Quine, op. cit., 14.

a hypothesis in the face of conflicting experience. It may have been that his earlier reasoning was proof-theoretic, leading him to conclude that we might save a hypothesis by disallowing the deduction of an offending prediction. I do not see why we may not do this at the semantic level too by changing the definition of implication. The effect would be to refuse to acknowledge the 'fateful implication' as such.

At a deeper level one might wonder how revising logic to save a theory in the face of contrary experience could even be an option for us. For without logic the experience would be neither connected to the theory nor contrary to it. However, while it is true that without some logical framework hypothesis-testing could not take place that does not mean that the framework cannot be provisional. Obviously, revisions in the framework must come very gradually, since after changing it we will need to check on whether previously tested hypotheses still pass muster. Thus instead of denying all instances of, say, the law for distributing conjunction over alternation we might reject certain applications of it to quantum phenomena. In this way there would be no danger of lapsing into total incoherence. Nor need we abandon the norms surrounding deduction. While we may change, for example, what counts as an implication or a contrary, we need not abandon norms that commit us to what our theories imply or that prohibit us from simultaneously maintaining two contraries.

Still, even if we allow the revisions in logic, it enjoys a special methodological status by virtue of fixing the very framework through which hypothesis-testing takes place. Because of the prominent place of logic in its methodology, mathematics also shares the methodological halo with logic, making it easier to protect from empirical refutation than the rest of science. These considerations re-enforce an objection to Quine's holism that one frequently encounters, namely that it fails to respect certain intuitions about mathematics and logic that seem to be firmly supported by our mathematical and scientific experience. According to these intuitions, mathematics and logic are fixed points in our investigation of the world, determining the limits of what we can entertain as serious possibilities (to borrow a phrase from Isaac Levi). In a different vein, Charles Chihara and Charles Parsons, have objected to the concept of mathematical evidence that appears implicit in Quine's views about confirmation. On the one hand, Chihara observes that in deciding whether to add a new axiom to set theory no set theorist is going to investigate its benefits to the rest of science.[8] On the other hand, Parsons has pointed out that Quine's holistic picture of science seems to belie the nature of mathematical investigations and to provide no place for the kind of evidence many believe we have in mathematics.[9]

[8] Chihara, op. cit., 15. [9] Parsons, op. cit., 380–3.

2. HOLISM AND THE INDISPENSABILITY ARGUMENTS

Some philosophers continue to hold, contrary to both Quine and Duhem, that observational evidence can be seen to bear upon specific hypotheses instead of whole systems. Clark Glymour, for example, argues that by taking confirmation as a three-termed relation between the evidence E, a hypothesis H, and a theory T, we can see that for a fixed T, E is relevant to some H in T and not to others.[10] Strictly speaking, this is not incompatible with Duhem's or Quine's holism, since they view confirmation as a two-termed relation. One would need to look at the details of Glymour's view to see whether it implies that mathematical principles are never confirmed or confirmable by empirical evidence. Offhand, I would expect that Glymour's approach would support the contrary conclusion, since by holding fixed certain empirical hypotheses concerning a computer, say, we can confirm mathematical conjectures via computer runs.

On the other hand, Elliott Sober[11] explicitly directs his separatist view of confirmation against Quine's philosophy of mathematics. According to Sober, when scientists design tests, they do so to decide between competing hypotheses. Since they intend to be putting specific hypotheses at risk, they take the data they obtain to reflect upon just these hypotheses and not upon the broader systems to which they belong.

Sober also notes that scientific tests very rarely put mathematical claims at the risk of being falsified. From this he concludes that mathematics cannot share in the confirmation afforded to those hypotheses that do pass experimental testing and, consequently, the indispensability argument based upon holism fails.

Sober's remarks about scientific practice are certainly right. Neither scientists nor mathematicians take scientific experiments as providing evidence for the mathematics used in designing them. Despite this, I will argue in subsequent sections that even within a holist framework one can make sense of scientific and mathematical practice.

Before I do so I want to comment on Sober's case against confirmational holism. I take it that he assumes that scientists use rationally defensible principles of methodology and experimental design and that these conflict with holism. I grant their rationality but deny the conflict with holism. It is rational (in the practical sense) for scientists to fix certain hypotheses (as auxiliaries) while testing others, and to act as if the evidence they obtain bears upon the specific hypotheses being tested. Holists don't deny this, but instead they deny that, independently of holding the 'auxiliaries' fixed, a logical relationship obtains between the hypotheses tested and the

[10] Clark Glymour, *Theory and Evidence* (Princeton, Princeton University Press, 1980).
[11] Sober, op. cit.

evidence. From the holist's point of view, Sober is simply confusing logic with practical rationality. I also doubt that Sober's account of confirmation applies to framework principles, such as the equations of special relativity, and conservation laws, since I doubt that we can put these to the specific sorts of tests that Sober has in mind. Yet one would not want to deny them empirical content or to hold that the general theories containing them have not been tested experimentally.

Penelope Maddy has also connected Quine's indispensability argument with his holism. She observes that many predictively useful scientific theories are openly acknowledged as false (e.g. classical physics), and further that scientists sometimes distinguish between parts of a theory that they regard as true and parts that they currently regard as merely instrumentally useful. From this she infers that belonging to scientific theories that are accepted as true may cover too little mathematics to be of comfort to mathematical realists, as she puts it:

> If we remain true to our naturalist principles we must allow a distinction to be drawn between parts of a theory that are true and parts that are merely useful. We must even allow that the merely useful parts might in fact be indispensable, in the sense that no equally good theory of the same phenomena does without them. Granting all this, the indispensability of mathematics in well-confirmed scientific theories no longer serves to establish its truth.[12]

Maddy and Sober seem to think of the indispensability argument as running along the following lines:

(1) Mathematics is an indispensable part of our best scientific theories.
(2) Mathematics shares whatever confirmation accrues to the theories using it (Quinean holism).
(3) So mathematics shares the confirmation accruing to our best scientific theories.
(4) We are committed to the truth of our best scientific theories (naturalism).
(5) So we are also committed to the truth of the mathematics they contain.

Sober clearly disputes the holism expressed in premiss (2). In distinguishing between the true and merely useful parts of a theory Maddy also questions this holism, and, perhaps, the version of naturalism contained in premiss (4).

Let me grant immediately that whether or not Quine has offered this argument, it can be constructed from his principles. Let me also acknowledge with Maddy, Sober, Chihara, and Parsons that few mathematicians would see (3) as their primary evidence for mathematics, if they saw it as

[12] Maddy, op. cit., 281.

evidence at all. Finally, let me admit to Maddy's worry that if our grounds for accepting mathematics are limited to those we have for accepting our best scientific theories, then we might be justified in accepting very little mathematics. For most of our scientific theories have known limitations or difficulties.

There is another way of formulating an indispensability argument, however, and it makes virtues of some of the vices of the one we have just considered. It arises from reflecting on the role that mathematics plays in science: *it is a background framework within which to formulate laws and theories, build models, and carry out inferences.* In light of this we can formulate an indispensability argument for mathematical realism that separates questions of indispensability from questions of confirmation. Such an argument would run:

(6) In stating its laws and conducting its derivations science assumes the existence of many mathematical objects and the truth of much mathematics.

(7) These assumptions are indispensable to the pursuit of science; moreover, many of the important conclusions drawn from and within science could not be drawn without taking mathematical claims to be true.

(8) So we are justified in drawing conclusions from and within science only if we are justified in taking the mathematics used in science to be true.

Notice that, unlike the indispensability argument that concerned Maddy and Sober, this makes no assumptions about the truth or the evidence for our best scientific theories. Even Newtonian physics, which is still used for practical work, and idealized scientific models, such as Maddy's example of waves in an infinitely deep ocean, come under its scope.[13]

The conclusion (8) of this argument invites the worry that we would not be justified in using mathematics in science unless we had prior evidence of its truth. If by science, we mean disciplines such as physics, chemistry, and biology, then I think we do have quite a bit of independent evidence for mathematics. We need not turn to the a priori either. For the evidential relations between experience and the various branches of mathematics are not very unlike that between experience and the various branches of science. Practice with counting, measuring, survey, and carpentry suggested and confirmed the elementary rules of practical arithmetic and geometry long before they were elevated to the status of inviolable laws and codified into mathematical systems. Moreover, many of the basic techniques of calculus can be checked geometrically or arithmetically, which in turn helps support its generalizations to real, complex, and functional analysis.

Of course, we can also couple my indispensability argument to a pragmatic argument to the effect that we are justified in using mathematics in

[13] By the way, this argument is similar to one of Hilary Putnam. See his *Philosophy of Logic* (New York, Harper & Row, 1971), 57.

science, because we know of no other way of obtaining its explanatory, predictive and technological fruits. Here is a formulation of such an argument:

(9) We are justified in doing science.

(10) The only way we know of doing science involves drawing conclusions from and within it.

(11) We are justified in drawing conclusions from and within science only if we are justified in taking the mathematics used in science to be true.

(12) So we are justified in taking that mathematics to be true.[14]

3. THE ROLE OF MATHEMATICS IN SCIENTIFIC IDEALIZATIONS

Premisses (6) and (7) of my indispensability argument presuppose that when scientists develop an ideal model of a physical situation, they treat as unreal only their simplified descriptions of the physical situation and not the mathematics they use. In effect, they establish claims of the form 'in an idealized physical situation of type I, law L holds', and mathematics functions as their 'underlying logic.' Thus truths of the form 'In an infinitely deep ocean, . . .' are mathematical truths—at least they are provided the truth of no scientific theory is presupposed in deriving them. Otherwise, they are conditional truths of the relevant physical theory, which can be converted to mathematical truths by packing the physical theory into their antecedents, viz. 'In an infinitely deep ocean obeying the laws of Newtonian physics'.

I think an examination of scientific practice will confirm this account. Paul Teller liked the idea that mathematics functions as an underlying logic in reasoning about idealized physical situations when I first presented it to him and Maddy in a conversation in October of 1992. However, he subsequently wrote Maddy and me to call our attention to an article by Thomas Weston[15] and to express his worry that the inferences involved in idealization are not meant to be truth-preserving but rather to be *approximate* truth-preserving. As I see it, his uneasiness is irrelevant, since the inferences concerning the approximate truth of a model come after its mathematical development. We use the mathematics before we see how well the model applies. (Even Weston seems to say this in stating that he is trying to make sense of the practice of deriving conclusions we know are not true from premisses we know are not true.) Moreover, using the notion of approxi-

[14] The material in this section overlaps a fuller discussion of indispensability arguments contained in my 'Scientific vs. Mathematical Realism: The Indispensability Argument', *Philosophia Mathematica*, 3/3: 166–74.

[15] Thomas Weston, 'Approximate Truth and Scientific Realism', *Philosophy of Science*, 59 (1992), 53–74.

mate validity introduced by Weston to draw approximately true physical conclusions still requires mathematics, because the theory of approximate validity is itself formulated in mathematical terms. Instead of eliminating the role of mathematics as an underlying logic Weston's approach would only give it new tasks.

In any case, I also doubt that we use anything like approximately valid reasoning in scientific and practical applications of mathematical models. Instead I think we first use mathematical and scientific principles to develop models that allow us to make calculations that yield values or ranges of values for the quantities that interest us. Then we apply statistical techniques to compare such values with the data we have obtained independently (or expect we would obtain) in order to decide whether the models are 'good enough' for our purposes.

Henry Kyburg[16] has worked out a detailed and precise version of a view quite like this, but I think I can better explain what I have in mind with an example. Suppose that I want to cross Long Island Sound in a small sailboat equipped with a compass and navigational charts. If I am near the New York City end of that bay and the day is clear, I will be able to see from one side to the other. In this case I will not worry much about plotting a course at all, much less the best course I can. But if it is foggy or late in the day or I am at the other, wider end of the Sound, I will not have that luxury. The simplest thing I might do is to derive a compass heading by 'walking' a straight line drawn on my chart over to its compass rose.[17] I might be confident that this will bring me 'close enough' to my destination. On the other hand, I may worry about magnetic variation and swing my straight line course a few degrees to compensate for it. Or I might plot a jagged line instead of a straight one in order to compensate for the tides and the winds. Whatever I do, it is likely that I will make certain simplifying assumptions (e.g. I will almost certainly have used an average wind velocity and probably an average current velocity). Thus at most I can conclude, with a certain degree of confidence, that the course I plot will bring me within a certain distance of my destination. If the costs of missing it are high, then I will want this distance to be relatively small and my degree of confidence to be relatively high. The higher the costs, the more steps I will take to build more factors into my calculations and to reduce my simplifying assumptions. I may even decide that I do not know enough about the tides and weather to risk the trip. Or I may know these well, but be unable to figure out how to calculate a sufficiently exact sequence of headings to counter-

[16] Henry Kyburg, *Science and Reason* (Oxford, Oxford University Press, 1990).

[17] At a first pass we might take my 'model' to consist of my chart and parallel rulers and the lines I draw. But at a deeper level my model is a mini-theory incorporating (at least) some geography, assumptions about the reliability of my chart, and geometric principles underwriting the use of the parallel rulers.

act their effects. Moreover, this might prompt me to consult a mathematician rather than an oceanographer.

As this example illustrates, we combine scientific and mathematical theories to build models that we then assess in the light of our previous experience with model-building and the other factors we know. When we do not find that our models suffice for the applications we want to make of them, then we may try to improve upon their questionable assumptions by either making our idealizations more realistic (e.g. in calculating the effects of the current I can use a tide table to tell me how it will vary at different stages of my voyage) or by improving our abilities to deal with complications (e.g. instead of working with straight lines and parallel rules, I might plot my course using a navigational computer that implements a non-linear, time-dependent function of the tides and winds). This may even include starting over with an entirely different approach, e.g. I might decide to borrow a radio direction-finder, use it to determine my current location every twenty minutes, and make rough-and-ready course corrections on the spot.

Discarding a model is not the same as falsifying the assumptions on which it is based. Before I even begin to calculate I know that a model in which the tides exert no force during my trip is based upon a false assumption. Hence finding that such a model is not good enough is not the same as refuting its assumption about the tides. But that does not mean that we cannot use idealizations to test hypotheses. We build models based upon the hypothesis we want to test, and then decide whether they yield values fitting the data well enough and better than models based upon competing hypotheses. Moreover, as Ronald Laymon points out in a nice discussion of the starlight deflections tests of general relativity, idealizations often supply the keys to mathematically tractable applications of physical theories.[18]

Transferring a refutation from a whole model to one of its hypotheses seems especially tricky, because in the face of a failed prediction proponents of a given hypothesis H might fairly ask 'Why should we reject H when we already know that the prediction was based upon false idealization I?' Let me illustrate how one might independently support an idealization by another boating example.

Suppose that on a calm day I motor across the Sound starting at a point where it is ten miles wide using a boat that travels at ten miles per hour. Let us also assume that I compute my compass heading by drawing a straight line on my chart and 'walking' it to the compass rose. If I arrive at the opposite shore seven miles to east of my intended destination, I will know that this cannot be due entirely to the tides, since at their strongest they run at five miles per hour. In other words, I will know that the bias

[18] Ronald Laymon, 'The Path from Data to Theory' in *Scientific Realism*, ed. by Jarrett Leplin (Berkeley, University of California Press, 1984).

introduced by my idealization is not enough to account for my set-back, and that some of my other assumptions must be at fault. Among other things, I assumed that my chart was accurate, that I had correctly calculated my course, that none of the steel fittings on the boat was too close to the compass, and that I steered a steady course. Now, despite the fact that I have already taken the falsity of my assumption about the tides for granted I will also have reason to doubt these other assumptions, and perhaps, ultimately to conclude that one was false.[19]

Where do we stand now? Sober is right that in practice we rarely, if ever, put mathematical laws to the sorts of specific tests that we apply to some scientific hypotheses. Maddy is right that much of the mathematics used in science takes place in the context of models that incorporate idealized assumptions, and thus often prove good enough without being confirmable even as wholes.

But none of this impugns holism. Because entire models, rather than individual hypotheses, bear logical and statistical relationships to experience, in order to conclude that an experiment bears upon a particular hypothesis of a model we must take its other assumptions, including its idealizations, for granted. Typically justifying this will bring in pragmatic considerations, such as theoretical simplicity and mathematical tractability. Furthermore, the success of scientific model-building does support our *practice* of using mathematics to formulate and develop those models. Moreover, in so far as this practice commits us to the truth of the mathematics it deploys, its success also supports our accepting this mathematics as true.

Still holists must acknowledge that Sober and Maddy are right that it is difficult to see the success of such models, whether it be predictive, explanatory, or technological success, as providing evidence, in the ordinary sense, for the individual mathematical principles used in scientific models or as providing anything that mathematicians would recognize as mathematical evidence. From a practical point of view, we may be justified in believing that the mathematics used in science is true, but we still need an account that will reconcile this global pragmaticism with our usual methods for supporting mathematical claims. I will try to provide one in the next section.

[19] What about scientific explanation? When Newton derived Kepler's law that the orbit of a planet around the Sun is elliptical, he did so by ignoring the gravitational effects of the other planets and heavenly bodies. What made this 'good enough' even though it was based upon false assumptions? I am not sure how to answer this. From the perspective of the sailing example, explanations seem to be gravy that comes with a theory that makes good predictions. When we want to avoid or bring about repetitions of an event, such as an airplane crash, it is important for us to have a detailed and accurate explanation of the event. This will favour explanations based upon theories that are good predictors. From this perspective two things favoured Newton's explanation. First, it was based upon a theory that was useful in predicting; second, although he knew that the orbits could not be exact ellipses (Newton's theory told him that) the model fit his data as well as any other contender.

4. GLOBAL AND LOCAL THEORIES

What we take for granted in applying our models varies with the context. In sketching my last example I presumed that it is reasonable to take it for granted that the tidal currents in Long Island Sound never exceed five miles per hour. But what I there assumed never happens probably could happen, and it might even be reasonable to speculate that it did happen in trying to explain some cataclysmic event.

In practice scientists working in a given context assume that they may take large blocks of theory for granted. Specialized scientific theories, for example, molecular biology, are developed within a framework that draws upon principles of more general theories, that is, chemistry, physics, and mathematics. By appealing to this practice we may roughly rank the sciences in terms of their scopes and methodologies as more or less global (or correlatively, as more or less local). Mathematics is our most global theory as it is presupposed by physics, which in turn is presupposed by chemistry, etc. We not only use mathematical truths in physical derivations, we also use mathematical standards to criticize physical arguments and theories (e.g. we complain about the mathematical respectability of quantum field theory).[20]

Corresponding to this rough division of the sciences has developed a division of labour: mathematicians normally do not meddle in physics nor physicists in mathematics, and biologists and chemists are normally not competent to suggest changes in mathematics or physics even when they might want to see one of them changed. As a result when something goes awry in a relatively local science (say, biology), it is not likely that practitioners of more global sciences (say, physics or mathematics) will hear of it, much less be moved to seek a solution through modifying their own more global theories. Nor is it likely that the specialists in a local theory will tinker with global background theories to resolve local anomalies. This is not just a matter of sociology, it is good sense too. Practical rationality counsels specialists to attempt to modify more global theories only as a last resort; for they probably do not and cannot know enough to tackle the task, and modifying a more global theory is likely to send reverberations into currently quiescent areas of science (Quine's point about minimizing mutilation).

Specialization has also fostered local methodologies and standards of evidence. These can provisionally override more global and holistic perspectives and declare data, obtained via local methods, to bear on this or

[20] I realize that this ranking is not entirely realistic. For example, mathematicians' complaints about physical definitions or arguments frequently fall upon deaf ears, so physics does not always abide by more global standards. On the other hand, physics frequently uses instruments e.g. cloud chambers, whose evidential status presupposes more local theories e.g. the chemistry of gases.

that local hypothesis. I should emphasize, however, that, on the account I am proposing, local conceptions of evidence, in particular those that lead us to see data as confirming specific hypotheses, are ultimately justified in terms of their effect on the health of science as a whole rather than on some a priori basis. Hence the divisions we find in the practice and scope of the various sciences should not be taken as refuting holism or as indicating hard-and-fast epistemic divisions between the sciences or between mathematics and the so-called empirical sciences. Nor do they show the irrationality of modifying some global principle to fix a more local problem.

The methodological picture I have been sketching also allows for a kind of localized holism, that is, one in which scientists look at their speciality as a whole without taking into consideration how accepting an hypothesis or positing new objects or processes might influence other sciences, whether they be more global or more local. Furthermore, when scientists posit new systems of objects and laws governing them they also provide for connections (bridge principles) with statements previously accepted as evidential. Doing so enlarges their local conception of evidence, and permits them to raise and answer questions that previously may have made no sense.

This might remind you of Carnap's distinction between internal and external questions, so let me explain how my conception differs from his. According to Carnap, positing a new system of objects is really to introduce a new language along with axioms and rules of inference that enable us to decide questions formulated in that language. The language's axioms may include both physical and mathematical laws, and its rules of inference may include procedures for making empirical inferences as well as those of logical and mathematical deduction. On this picture, enlarging a local conception of evidence is tantamount to introducing a new linguistic framework. Furthermore, rejecting a framework axiom or rule, whether it be logical, mathematical, or 'empirical', amounts to rejecting the framework as a whole.[21] He also believed that questions about the acceptability of a framework (external questions) are fundamentally different from ones raised within a framework (internal questions). The rules of a framework fix the considerations relevant to deciding its internal questions, and nothing else counts, whereas only pragmatic considerations are pertinent to deciding whether to accept a framework itself.

Carnapian frameworks are built from scratch, so that we do not just tack, say quark theory, onto a pre-existing quantum mechanics, but start out by specifying the logical and mathematical principles of quark theory, and then its fundamental physical laws, and so on, until we introduce the specific

[21] Actually, Carnap replaces the distinction between the logico-mathematical and the empirical with a framework relative analytic–synthetic distinction. See his 'Empiricism, Semantics and Ontology' repr. in *Philosophy of Mathematics* ed. by Paul Benacerraf and Hilary Putnam (Cambridge, Cambridge University Press, 1983).

quark hypotheses. Of course, 'hypotheses' is a misnomer, since in the quark framework the quark hypotheses are just as a priori (and analytic) as its logical laws. Since Carnap recognized no facts prior to or transcending frameworks, framework principles cannot be true in virtue of independently obtaining facts. Instead, they are true by convention.

Unlike Carnap I see global theories as transcending the local theories that are built upon them, and as provisionally fixing a background of facts upon which local theorists may draw. These facts may also furnish them with reasons for preferring one local theory to another. (They might show, for example, that a proposed theory is untestable.)

Carnap did not seem to consider the possibility of appealing to pragmatic grounds to justify adding a new hypothesis to an extant framework, as we might do today in positing, for example, a new type of quark, or a larger infinite cardinality. This is just where Quine would argue that there is no real epistemic difference between 'externally' changing a framework by adding a new hypothesis (expressed in the language of the framework) and 'internally' deciding that the evidence supports the hypothesis. I side with Quine on this.

Quine sometimes gives the impression that the decision to accept a new hypothesis is always to be referred to the system of science as a whole. I can agree to this too but only if we make the additional methodological layers explicit. The decision to accept an hypothesis is often dictated by local methodological principles that in turn have developed from a more global perspective, and that perspective in turn may involve an even more global one, and so on, until we ultimately reach considerations affecting the system of science as a whole.

By the way, in allowing that scientists may take a perspective that is locally holistic, I do not want to suggest that this is invariably the best thing for them to do. The early debates about the axiom of choice and non-constructive mathematics, for example, focused on philosophical and methodological issues restricted to mathematics. But we now know that rejecting the axiom of choice or non-constructive mathematics would have forced very significant changes in mathematical physics too.

Let me now review some of our earlier methodological observations using the global–local distinction. Holism should be uncontroversial if construed as a comment on the logical relationship between theoretical hypotheses and their empirical consequences. To derive a prediction P from such a hypothesis H we must not only appeal to H itself but also to auxiliary assumptions A. If P fails, all that follows is that H or A is false. To conclude *via* disjunctive syllogism that H is false, we must already know that A is true.

Yet it is certainly not practical for scientists to treat every hypothesis as equally vulnerable each time they run an experiment. The path to success lies in taking a more local perspective and tinkering with those hypotheses one knows best. From such a perspective it makes sense to treat the global auxiliaries as (provisional) fixed points. Thus, for example, when a biological experiment shows that either a biological H or a mathematical A is false, there will be (and should be) no question in the mind of the biologist that it is H which is false. This is no refutation of holism, but merely an illustration of the global–local approach to methodology. To see it otherwise is to presuppose that we have infallible (or at least separable) knowledge of mathematics.

I can now say something about the relationship between holism and our ordinary conception of mathematical evidence. Empirical success no more confirms individual mathematical claims than it does individual theoretical hypotheses. However, it does provide a pragmatic justification for positing mathematical objects, truths about them, and principles for applying mathematical laws to experience. From this perspective we may encourage mathematicians to develop their own standards of evidence, so long as the result does not harm science as a whole. Because mathematics is our most global science we should expect that many mathematical methods and principles will be justified by means of considerations neutral between the special sciences, and thus often pertaining to mathematics alone. In this way we can reconcile holism with the features of mathematical practice that Chihara, Maddy, and Parsons have emphasized.

Considering the place of proof in mathematics will illustrate how this works. Early mathematicians took experience with counting, bookkeeping, carpentry, and surveying as evidence for the rules and principles of arithmetic and geometry that were eventually taken as unquestionably true. They began to put more emphasis on deduction after they became aware of the difficulties in deciding certain mathematical questions by appealing to concrete models. Even today we could (and sometimes do) use concrete models to decide what is true in certain structures, e.g. we might simulate a Turing machine on a computer to determine whether it gets into a certain state for a given input.[22] But the advantages of proof to the practice of mathematics are so obvious that it serves mathematics better to frown on experimental approaches. Moreover, proof wins out from the perspective of science as a whole. For requiring mathematics to prove its results increases its reliability, and decreases its susceptibility to experimental refutation. Thus we have a local conception of mathematical evidence in which proof plays a more important evidential role than empirical or quasi-empirical considerations.

[22] Some people think that this is a form of a priori proof, but I would argue that it is a way of using empirical and mathematical premisses to draw a mathematical conclusion in much the same way we use mathematical and empirical premisses to draw empirical conclusions. See my 'Computation and Mathematical Empiricism', *Philosophical Topics*, 17 (1989), 129–44.

The development of non-Euclidean geometries and abstract algebra further promoted the purely deductive methodology of the axiomatic method, through showing mathematicians how to make sense of structures that might not be realized physically. It also promoted a shift from viewing mathematical claims as unqualifiedly true or false to regarding them as true or false of structures of various types. In turn this has further insulated mathematics against empirical refutation. For when the mathematics of a scientific model proves inadequate, it is far simpler to save the mathematics from refutation and conclude that it merely failed to match the physical structure to which it was being applied. (I will subsequently refer to this way of saving a branch of mathematics as *a Euclidean rescue.*) Ironically, the ease with which this rescue may be applied tends to encourage separatists who regard mathematics as true in all possible worlds and immune from empirical refutation.

5. REVISING LOGIC AND MATHEMATICS

Given the way the local methodology of mathematics has developed, we are unlikely to find much in the attitudes of most holists towards the practice of contemporary science and mathematics to distinguish them from separatists. Both are likely to recommend a Euclidean rescue when a scientific model fails. By introducing some new characters, the strict holists, we can bring out a difference in the attitudes of certain holists and separatists. Strict holists maintain that we have no evidence for a claim unless it is part of a system of hypotheses that has been confirmed. While allowing for local conceptions of evidence, they only permit them to apportion confirmation to specific hypotheses within a confirmed system, and prohibit the introduction of independent, e.g. non-empirical, evidence for scientific or mathematical claims. On this view, mathematics with no current application, such as results employing large infinite cardinals, would have little title to truth.

One problem with strict holism is that it is not clear that it could countenance certain metamathematical investigations of even the mathematics that does get used directly in science. It is hard to see how investigations of the consistency of various systems of analysis could be directly useful in *constructing* scientific models or in *deriving* empirical consequences from physical hypotheses formulated in the vocabulary of analysis. Despite this, the information these proofs provide about the reliability of those models is relevant to *evaluating* the practice of science. Similarly, it is often useful for scientists to know of mathematically equivalent ways of formulating scientific theories or of probabilistic approximation methods that might substitute for rigorous calculations. Yet I could imagine cases in which the metamathematical justifications of these equivalences or methods would not figure directly in deriving empirical results. This is one reason why I would advocate a liberal holism, one which would allow the development of a local conception of mathematical evidence that could countenance mathematical truths that have no foreseeable empirical use.

Some of Quine's remarks on higher set theory might suggest that he is a strict holist. But he is not. Consider this passage:

> What now of those parts of mathematics that share no empirical meaning, because of never getting applied in natural science? What of the higher reaches of set theory? We see them as meaningful because they are couched in the same grammar and vocabulary that generate the applied parts of mathematics. . . . On our two-valued approach they then qualify as true or false, albeit inscrutably.
>
> They are not wholly inscrutable. The main axioms of set theory are generalities operative already in the applicable part of the domain. Further sentences . . . *can still be submitted to considerations of simplicity, economy, and naturalness* that contribute to the molding of scientific theories generally. Such considerations support Gödel's axiom of constructibility, 'V = L'. It inactivates the more gratuitous flights of set theory, and incidentally it implies the axiom of choice and the continuum hypothesis. (my italics)[23]

Although the first paragraph clearly shows that Quine believes that securing evidence for unapplied mathematics is problematic, in the second paragraph he goes beyond the strict holist in allowing pragmatic considerations to justify accepting axioms that are not needed in deriving empirical consequences from scientific hypotheses. In particular, if ZFC alone suffices for the mathematical needs of natural science, then from the strict holist's perspective there is no further need to decide the continuum hypothesis or to inactivate 'the more gratuitous flights of set theory'. We can get along with ZFC alone without adding any further axioms to address these further issues.

What other differences might there be between holists and separatists? Strict holists can conceive that a mathematical statement we previously regarded as true might degrade to one whose truth-value cannot be settled.[24] Thus, according to them, it may not always be possible to perform a Euclidean rescue. Quine, though no strict holist, envisions something like this in remarking that the mathematics we recognize as true might be cut back to predicative set theory if it can meet the mathematical needs of science.[25] Separatists, of course, would not permit such developments to curtail otherwise acceptable branches of mathematics.

I am merely speculating, to be sure. I know of no strict holists nor of any mathematics that science clearly no longer needs. But it is to speculations

[23] op. cit., 94–5.
[24] Here I follow Quine in allowing these mathematical statements to be truth-valued.
[25] Ibid.

such as these that we must turn to discern other differences between holists and separatists.

Holists can allow an experiment to refute the mathematics used in designing it by waiving local conceptions of evidence. How it might refute the mathematics in question would depend upon the case at hand. Let us consider some of these cases.

First, let us note that it can happen that scientists use a mathematics so rudimentary and tentative that both its concepts and proof methods fail to be well defined. (This is a reasonable view of the calculus used in late seventeenth- and early eighteenth-century physics. Some conversations I have had with philosophers of physics indicate that this applies to contemporary quantum field theory too.) Now if we developed a new branch of mathematics for a specific type of application and using it proved unreliable, then rather than using a Euclidean rescue it might be more rational to reject the mathematics as ill-formed. Since this option is open to both holists and separatists, I will restrict my attention to rigorously specified mathematical theories.

Suppose that while designing an experiment some physicists need to know whether a certain function has a maximum. They hear about a conjectured consequence S of a new extension of set theory, ZFC + A, by an axiom A, and they subsequently determine that S implies that their function does have a maximum. Let us also suppose that they assume this maximum exists in testing one of their pet physical hypotheses and that their experiment produces a result conflicting with their expectations. We can imagine them deciding that the conjecture must have been wrong—not simply that they had misunderstood or misheard the conjecture.

But what does it mean to say that the conjecture is wrong? If one can be reasonably certain that the set theory in question (ZFC + A) applies to the physics in question, then it might be reasonable to take the experiment as evidence that the conjecture is not one of its theorems. Now a separatist might hold that certain empirical data, such as computer outputs or the results of hand computations, could be sought and used to refute or establish a conjecture.[26] But holists must allow that even experiments, like our physicists', that were not originally designed to test a mathematical conjecture, might count against it. For they countenance no sharp epistemic distinction between experiments that are designed to test a specific claim and those that merely presuppose the claim without testing it.

Perhaps, the physicists will conclude that the conjecture was wrong in a weaker sense of being inappropriate to their physics. Then they might

[26] These separatists would not be committed to the thesis that mathematics is a priori, however. See my 'Computation and Mathematical Empiricism' for a discussion of the type of reasoning I have in mind.

conclude that the set theory in question does not apply to the physics; that is, they will execute a Euclidean rescue.

Let us now suppose that instead of a conjecture the problematic mathematical claim S is a known theorem of the extended set theory, ZFC + A. Then the physicists would almost certainly perform a Euclidean rescue. For what other options would they have? They might examine the theorem's proof on the offchance that it contains an error; but, due to their relative lack of mathematical expertise, they would have little reason to expect this tactic to succeed.

Finally, let us imagine that after repeated unsuccessful efforts to account for the anomalous experiment our physicists call in the mathematicians. Even then it would be rational for the mathematicians to try a Euclidean rescue first, and to attempt to apply a different, perhaps, newly developed, branch of mathematics to the physics.

Would it be rational for the mathematicians to see the experiment as refuting ZFC + A? According to our story, it implies a claim which when combined with physical hypotheses, is contrary to the experimental results. Stretching our imaginations to near the breaking-point, we can think of the mathematicians as arguing as follows: some statement in the model must be false, and the physical ones are beyond question, so the mathematical statement must be false; but since it is implied by ZFC + A this too must be false. But notice how strong a claim this would be. Because we can always save a consistent branch of mathematics via a Euclidean rescue, for our mathematicians to reject the axioms of ZFC + A would be to take them to be inconsistent![27] Of course, they could avoid this step by taking an even more radical one. They could take the physical model as showing that the axioms do not imply the theorem, and then proceed to modify the rules of deduction used in its proof.

The main difference between holists and separatists is that when a well-established piece of axiomatic mathematics is part of a model conflicting with experience, separatists recognize only the Euclidean rescue, while holists also admit the options of rejecting the mathematics as inconsistent or altering its underlying logic. They see nothing in the epistemology of mathematics that excludes using these options in the rare instances when doing so will benefit science more than the Euclidean option.

Of course, the foregoing does not refute separatism, it only helps make holism more palatable to those with separatists' leanings. Moreover, even

[27] In drawing this conclusion our mathematicians had assumed that ZFC + A applied to the physics. Instead of concluding that it is inconsistent, shouldn't they just conclude that it does not apply after all? Yes, this would be a more reasonable choice if they had independent grounds for thinking that ZFC + A was consistent. But holists could find the stronger conclusion warranted if the mathematicians' independent grounds for believing in the consistency of ZFC + A were weak.

this tempered and conservative holism would be unacceptable if, independently of holism, one could establish an epistemically principled division between the empirical and formal sciences. But I do not see much hope of success here. I will conclude by canvassing some of the difficulties confronting separatists.

First, they would need a way of dividing mathematics from the rest of science that explains why we should not use Euclidean rescues to save sufficiently precise scientific theories. Otherwise, it would be reasonable to hold, for example, that the Michelson–Morley experiment did not refute Newtonian physics, but only showed it to be unfit for modelling bodies travelling close to the speed of light. I like this move for the same reason I like using Euclidean rescues in mathematics: it preserves more truths and more theories.

Second, they would also require a similar division between logic and mathematics, or if they lump mathematics with logic, a division between logic and the rest of science. A stronger logic sees more distinctions between structures, decides more mathematical propositions, and attributes more empirical consequences to a theory. A stronger logic thus narrows our options for revising theories. Since a weaker logic requires us to supplement a theory with more auxiliary hypotheses to test it, it leaves us with more options when our predictions fail.

Third, the division between logic and the rest of science must be more than just a matter of methodological convenience for separatists, otherwise they will be indistinguishable from holists. Separatists are committed to a distinct epistemology for logic. Now in practice logicians begin with logical intuitions concerning the correctness of various inferences, which they systematize through the method of wide reflective equilibrium. In this way logic is much like ethics. Many of the reasons that lead philosophers to adopt non-cognitivism in ethics seem equally compelling to me when applied to logic. So to end on a radical note, I do not think that there are logical facts to be known; the truths of logic are just ordinary truths generated via our most global methodology,[28] and in calling something a logical truth we do not ascribe a metaphysical property to it but rather mark it for special treatment.

[28] I have not ranked logic as our most global science, because I do not see it as having a subject-matter. Metalogic does have one, of course, but metalogic is applied mathematics. I develop this view further in 'Ought There to Be but One Logic?' in Jack Copeland (ed.), *Logic and Reality* (Oxford University Press, 1996).

III. Finitism and Intuitionism

10

Finitism and Intuitive Knowledge

CHARLES PARSONS

1. INTRODUCTION

In the introduction to his paper of 1958 on an extension of the finitary method in proof theory, Kurt Gödel remarks that 'finitary mathematics is defined as that of intuitive evidence',[1] by which he evidently means:

(1.1) If a proposition has been proved by the finitary method, then it is intuitively evident.

(1.2) If a proposition is intuitively evident, it can be given a finitary proof.

(1.1) and (1.2) formulate in a succinct way a thesis about the significance of the finitary method that did, in my view, belong to the outlook of the Hilbert school although it is not stated quite so directly in their writings. I have stated it in the way I have because I think it useful to consider (1.1) and (1.2) separately. It is also useful to consider them in connection with two theses that together constitute a mathematical characterization of finitism:

(1.3) Proofs in primitive recursive arithmetic (PRA) are finitist; hence any theorem of PRA is finitistically provable.

(1.4) If a proposition in the language of primitive recursive arithmetic is finitistically provable, then it is a theorem of PRA.

(1.3) is clearly expressed in writings of Hilbert and Bernays; an analysis of finitism that did not yield it would be hard put to it to show that it was

I am grateful to the participants in the Munich conference for their comments, especially to Geoffrey Hellman for subsequent correspondence. Since then the paper has been presented to other audiences, which I also wish to thank; in some cases I am conscious of not having done justice to the comments. I am grateful to Jaakko Hintikka for the invitation to speak at a symposium on Hilbert's Philosophy of Mathematics at Boston University, which led me to focus the paper on finitism.

[1] 'Über eine bisher noch nicht benützte Erweiterung des finiten Standpunktes', *Dialectica*, 12 (1958), 280–7, at 280. The paper is reprinted with an English trans. in *Collected Works, ii: Publications 1938–1974*, Solomon Feferman *et al.* (eds.), Oxford University Press, 1990. However, translations of quotations are my own. In the 1972 English version of the paper, what corresponds to the quotation in the text is the remark that 'finitary mathematics is defined as the mathematics of *concrete intuition*' (ibid. 272, emphasis Gödel's).

faithful to Hilbert's intentions. In discussions of the subject it has been noncontroversial. (1.4) is another matter. It is likely that it was implicit in the Hilbert school's understanding of finitism before 1931.[2] Under the impact of Gödel's incompleteness theorem, Bernays fairly quickly became convinced that an extension of the finitary method was needed to prove the consistency of classical first-order arithmetic (PA).[3] It was not claimed that the proofs obtained in the 1930s by Gentzen and then Ackermann were finitist, and there has been general agreement that they are not. That still leaves a substantial intermediate area, where analyses of finitism have differed. Kreisel in 1958 proposed an analysis according to which finitism goes substantially beyond PRA and it is possible to codify the totality of finitist methods in a system such that one can give (in PRA) a proof of the consistency of PA relative to it.[4] On the other hand, two different analyses

[2] This would, of course, be refuted if one could show that Hilbert and his collaborators understood Ackermann's function to belong to finitist mathematics. This would be implied by the claim of G. Kreisel that in the discussion of the continuum hypothesis in 'Über das Unendliche', (*Mathematische Annalen*, 95 (1926), 161–90, at 181–90) the functions of higher types introduced there are 'accepted as finitist without discussion'. ('Proof theory: some personal recollections', in Gaisi Takeuti, *Proof Theory*, 2nd edn., Amsterdam, North-Holland, 1987, 395–405, at 396.) Hilbert makes no such assertion, and precisely the higher-order character of these functions puts the claim that Hilbert thought them finitist in doubt. The fact that Ackermann's function is mentioned in this connection and defined by a recursion with a type 1 argument (pp. 185–6) would suggest that that too was not considered finitist. But more definite documentary evidence seems to be lacking. (I am indebted here to Wilfried Sieg.)

[3] I have not been able to determine exactly when Bernays came to this view. It is not expressed in Hilbert and Bernays, *Grundlagen der Mathematik I*, Berlin, Springer, 1934; 2nd edn., 1968, although that could well have been in deference to Hilbert, who says in his brief preface that Gödel's theorem only shows that for further consistency proofs 'one must exploit the finitary method in a sharper way than is necessary in considering the more elementary formalisms' (p. vii). Page references to Hilbert and Bernays are to the second edition of this volume, but passages cited occur in the first. Translations are my own.

The view that finitary methods are insufficient to prove the consistency to PA is expressed, with a very slight hesitation, in Bernays' well-known 'Sur le platonisme dans les mathématiques', *L'enseignement mathématique*, 34 (1935), 52–69, at 68. (This paper was presented as a lecture in 1934.)

Bernays' slight hesitation in 1934 arises from the fact that he claims only empirical certainty for the claim that 'elementary combinatorial' proofs can be formalized in a theory like PA. But he is somewhat more emphatic on this point in the immediately following lecture 'Quelques points essentiels de la métamathématique', *ibid.* 70–95, at 88. Bernays, however, goes on to discuss the characterization of the 'finitary point of view'. He mentions the fact that some had not distinguished this point of view from intuitionism, and evidently does not think the explanations given by Hilbert precise enough to exclude this. However,

> Nonetheless, in metamathematical proofs one has always confined oneself to a narrower framework, in view of the tendency natural to elementary evidence. One remained in the domain of reasonings that can be formalized without using bound variables. It is by this limitation that we fell into the difficulties mentioned. In fact, our thesis that one can formalize in $\mathfrak{N}$ [i.e. PA] every proof of an arithmetical theorem that conforms to the finitary point of view is only valid if this point of view is interpreted in the restricted sense. (p. 90, my translation)

[4] 'Ordinal logics and the characterization of informal concepts of proof', *Proceedings of the International Congress of Mathematicians, Edinburgh 1958*, Cambridge University Press, 1960, 289–99.

offered by W. W. Tait in 1967 and 1981 yielded defences of (1.4).[5] Because Tait has offered the most developed defence of it and it has come to be identified with him, I will refer to (1.4) as *Tait's thesis*.

In his paper of 1958, Gödel proceeds to argue that abstract notions, propositions involving which cannot be made intuitively evident in the required sense, are necessary in order to prove the consistency of PA. Then from (1.1) it follows that this consistency cannot be proved finitistically. Gödel does not mention Tait's thesis, although he seemed to adhere to it in two lectures of the 1930s.[6] I will not comment on this argument in a more systematic way, although it is worth noting that, in its surface structure, it avoids appeal to a thesis like (1.4) at the cost of using the presumably controversial epistemic notion of intuitive knowledge.

Like Gödel, I will use this notion, but my main concern will be the relation of (1.1) and (1.3). If (1.3) is assumed, then (1.1) implies that proofs in PRA convey intuitive knowledge. This is rather explicitly argued in writings of Hilbert and Bernays, and my main concern will be to present and assess these arguments.

2. FINITISM AND INTUITION IN THE HILBERT SCHOOL

I have relied so far on understandings of finitism by writers external to the Hilbert school, particularly Gödel. But although Hilbert was no systematic philosopher, I don't think there can be much doubt that finitary mathematics as he understood it was to be based on intuition in such a way that it would be reasonable to characterize a finitist proof as yielding intuitive knowledge in some sense. Hilbert, however, stressed the role of intuition as providing the *objects* of finitary mathematics. Thus consider the following famous passage:

[5] 'Constructive reasoning', in B. van Rootselaar and J. F. Staal (eds.), *Logic, Methodology, and Philosophy of Science III*, Amsterdam, North-Holland, 1968, 185–98; 'Finitism', *Journal of Philosophy*, 78 (1981), 524–46. I should point out that the aims of the present paper are very different from those of Tait's. I do not purport to give the kind of analysis of finitism that would yield (1.3) and (1.4) (or some different mathematical characterization). I simply assume (1.3); of course one of Tait's tasks is to show that his analysis (in particular that of 1981) yields it. Since that analysis does not rely on a notion of intuitive knowledge, the doubts about Hilbert's thesis expressed in Sect. 6 below do not affect Tait's case for (1.3).

[6] 'The present situation in the foundations of mathematics', lecture given to the Mathematical Association of America in December 1933, and a lecture given in January 1938 to an informal circle organized by Edgar Zilsel in Vienna. See *Collected Works, iii: Unpublished Essays and Lectures*, Solomon Feferman *et al.* (eds.), Oxford University Press, 1995, 51–2, 92.

In a note of the 1958 paper commenting on Kreisel's analysis, Gödel treats it as capturing an extended sense of finitism. That would indicate that he still held Tait's thesis for the proper sense of finitism.

As a condition for the use of logical inferences and the performance of logical operations, something must be already given to our faculty of representation, certain extralogical concrete objects that are intuitively present as immediate experience prior to all thought. If logical inference is to be reliable, it must be possible to survey these objects completely in all their parts, and the fact that they occur, that they differ from one another, and that they follow one another, or are concatenated, is immediately given intuitively, together with the objects, as something that neither can be reduced to anything else nor requires reduction.[7]

Hilbert's term 'concrete' (*konkret*) gives rise to confusion. Since he makes clear that he is primarily thinking of symbols of a formalized language and goes on to talk of them as types, his objects are, as this distinction is made in Anglo-American discussions of nominalism, abstract rather than concrete, unless perhaps he intends the talk of types as a *façon de parler* to be reduced to talk of tokens.[8] Otherwise the objects he calls concrete are what I call quasi-concrete, that is, it is intrinsic to them to have instantiations in the concrete in the narrower sense.[9] They might similarly be called quasi-spatio-temporal, but it would not be misleading to call them simply spatio-temporal, as one understands geometric figures to be spatial. In any case, Hilbert's term evidently has an additional meaning of something like 'perceptual', as is shown by its association with intuitive givenness, as in the following brief characterization of the finitary method: 'What is characteristic of this methodological standpoint is that considerations are put forth in the form of *thought experiments* on objects that are assumed to be *concretely present* (*konkret vorliegend*).'[10]

[7] 'Über das Unendliche', 171, trans. from Jean van Heijenoort (ed.), *From Frege to Gödel: A Source Book in Mathematical Logic, 1879–1931*, Cambridge, Mass., Harvard University Press, 1967, 376. These remarks are repeated almost verbatim in several papers of Hilbert, going back to 'Neubegründung der Mathematik (Erste Mitteilung)', *Abhandlungen aus dem mathematischen Seminar der Hamburgischen Universität*, 1 (1922), 157–77, at 163. (This observation is made in Ulrich Majer, 'Geometrie und Erfahrung: Ein Vergleich der Auffassungen von Hilbert und Kant', Report no. 17/93 of the Research Group on Semantical Aspects of Space-Time Theories, Zentrum für interdisziplinäre Forschung, Bielefeld, 1993, 4.)

[8] This is suggested in one place, 'Neubegründung der Mathematik', 163 n. 1, where Hilbert says that he calls 'signs of the same shape (*Gestalt*) "the same sign" for short', thus intimating that being the same sign is thought of as an equivalence relation of tokens rather than strict identity. There is a question whether Hilbert could still maintain this in 'Über das Unendliche', where he insists that nothing infinite is to be found in the physical world, in particular not a potentially infinite sequence of tokens. He could, however, hold some form of modal nominalism. Neither he nor the often more explicit Bernays addresses this issue.

[9] In the corresponding place in 'Neubegründung der Mathematik', Hilbert uses *diskret* instead of *konkret*. Clearly the objects he has in mind are discrete. But nothing obviously follows about what he means by *konkret*. Hilbert certainly had a conception of geometric intuition, of which the objects would presumably be continuous. Cf. Majer, 'Geometrie und Erfahrung' and its revised English version 'Geometry, intuition, and experience: From Kant to Husserl', *Erkenntnis*, 42 (1995), 261–85.

[10] *Grundlagen der Mathematik* I, 20, emphases in the text. Bernays, in a paper roughly contemporaneous with 'Neubegründung', speaks of the domain of the 'concrete-intuitive', within which foundational considerations are to be developed. See 'Über Hilberts Gedanken zur Grundlegung der Arithmetik', *Jahresbericht der Deutschen Mathematiker-Vereinigung*, 31 (1922), 10–19, at 18.

In finitary arithmetic, Hilbert understands the objects to be numerals, strings of occurrences of '1'. Thus according to him the objects of finitary arithmetic are objects of intuition. Exactly how Hilbert understood the term *Anschauung* is a large question, since he often uses it with a quite broad meaning, in keeping with the unphilosophical use of mathematicians.[11] Nonetheless the concept of intuition underlying his account of finitism can be described more precisely, in particular so as to give a reasonably clear sense to the claim that finitary proofs yield intuitive knowledge. I have argued in the past that we can describe in a convincing way an intuition among whose objects are formal expressions of the kind Hilbert is concerned with, so that for example when we have before us a token of the sort of string Hilbert is concerned with, we typically have intuition of the string.[12] Although this brings in an element of reconstruction, I will assume that the appropriate conception of intuition is that worked out in my own writings on the subject. That conception certainly has certain essentials in common with Hilbert's: its objects are quasi-concrete and spatio-temporal; its scope is a minimal generalization of that of ordinary perception so as to take in abstract objects such as sign configurations and possibly geometric figures.

Before we proceed something needs to be said about intuitive knowledge. Intuitive knowledge, as I understand it, is knowledge that is based on intuition in an appropriate way. In the case of singular propositional knowledge, it will typically be based on intuition of the objects the proposition is about. This is clearly in accord with the usage of the Hilbert school. Although I have not found the terms *anschauliche Erkenntnis* and *anschauliche Evidenz* in Hilbert's papers, they do occur in Bernays' writings of the time of his close collaboration with Hilbert.[13] The appeal to thought-experiments in the account of the finitary method in Hilbert and Bernays can be seen as an attempt to convey intuitive knowledge.

That would make plausible that in some cases of singular propositions about strings, our knowledge is intuitive knowledge. This is most

[11] Consider e.g. Hilbert and S. Cohn-Vossen, *Anschauliche Geometrie*, Berlin, Springer, 1932. Here intuition, being geometrical, is still not too far removed from Kant's conception, but the authors are not attempting to base an epistemological account of geometry on it.

[12] 'Mathematical intuition', *Proceedings of the Aristotelian Society*, NS 80 (1979–80), 145–68; 'Intuition in constructive mathematics', in Jeremy Butterfield (ed.), *Language, Mind, and Logic*, Cambridge University Press, 1986, 211–29; 'On some difficulties concerning intuition and intuitive knowledge', *Mind*, NS 102 (1993), 233–45.

[13] In 'Über Hilberts Gedanken', p. 15, Bernays describes his inquiry as whether it is possible to give a foundation for the 'transcendent assumptions' of mathematics by means of 'primitive anschauliche Erkenntnisse'. Bernays' most important philosophical statement before the impact of the incompleteness theorem is 'Die Philosophie der Mathematik und die Hilbertsche Beweistheorie', *Blätter für deutsche Philosophie*, 4 (1930), 326–67, repr. in *Abhandlungen zur Philosophie der Mathematik*, Darmstadt, Wissenschaftliche Buchgesellschaft, 1976. Certainly an idea of intuitive knowledge underlies the discussion of the finitary standpoint, which at one point is described as the 'Standpunkt der anschaulichen Evidenz' (p. 40 of reprint, by which this paper is hereafter cited).

convincing in the mathematically most trivial cases, namely instances of the successor relation and the order relation. But it is still not trivial from a philosophical point of view, because singular propositions about types have general implications concerning tokens.[14]

That is still so far a very weak conclusion, for two reasons. First, it says nothing about general propositions about strings. Second, it says nothing about how far into the sequence of strings our intuition extends. If one maintains that every string can be intuited, then clearly 'can' has to have some abstract sense that doesn't imply practical possibility; the claim would be that it is in this sense (some writers say 'in principle') possible to intuit any string, even though life is too short (and maybe, at some point, the universe too small) for us actually to do it. I am not convinced that Hilbert himself held this, but others have defended a similar view.[15] In Section 7 I will comment on its relevance to the concerns of this paper.

Even if that view were accepted, it would still leave the problem of general propositions. For this concept of intuition to be of interest, it must be the case that significant general propositions can be intuitively known. This, I have said, is the natural way to read Hilbert's view. I think one can attribute to Hilbert the following thesis, a consequence of the theses (1.1) and (1.3):

Hilbert's thesis. A proof of a proposition according to the finitary method yields intuitive knowledge of that proposition. In particular, this is true of proofs in primitive recursive arithmetic.

By primitive recursive arithmetic (PRA) I mean, as usual, the quantifier-free theory in which one has classical propositional logic with free variable reasoning, the elementary axioms concerning successor and identity, introduction of function symbols by primitive recursion, and the rule of mathematical induction.

An argument for Hilbert's thesis would have to have a number of components. I think it would follow from the following subtheses (with respect to the Hilbertian intuitive model):

(2.1) Successor can be seen intuitively to be well-defined.
(2.2) The elementary successor axioms can be known intuitively.
(2.3) In each case of introduction of a function symbol by primitive recursion, if the assumed functions have been intuitively seen to be well-defined, then this is so of the new function introduced, in such a way that the recursion equations are known intuitively.

[14] Cf. 'Mathematical intuition', 153–4.

[15] Myself included, in 'Mathematical intuition'. But cf. 'Intuition in constructive mathematics'.

(2.4) Logical inference preserves intuitive evidence.
(2.5) Inference by induction preserves intuitive evidence.

One finds in writings of Hilbert and Bernays arguments for some of these subtheses. I have addressed some of the issues in past treatments of intuition, particularly the cases of (2.1),[16] (2.2),[17] and (2.5).[18] In what follows I will try to address the remaining issues, particularly with reference to arguments offered by Hilbert and Bernays. We will then see where Hilbert's thesis stands.

3. LOGIC AND INTUITION

Let us consider first subthesis (2.4), the claim that logical inference should preserve intuitive evidence. But note that the logic allowed in PRA is limited to elementary reasoning about identity and classical propositional logic.

Now one could stipulate that intuitive evidence should not depend on reasoning or inference at all. That seems to be the way Descartes uses the term *intuitio*. Clearly the intent of Hilbert and other theorists of mathematics was not to be so strict as that, and such a stipulation would make the notion of intuitive evidence inapplicable to any whole mathematical theory that includes the results of proof. But then some logic, that is some basic forms of reasoning, will have to be allowed as preserving intuitive evidence. The general principle should be that what is allowed is to be absolutely basic to reasoning in general, or to reasoning about objects or about the particular kind of objects with which we are concerned.

A principle that should be considered is that if an inference from, say, premisses *A* and *B* to a conclusion *C* is to preserve intuitive evidence, then the inference should itself be a matter of intuitive insight. We might call this principle 'Descartes' principle', since in the *Rules for the Direction of the Mind* he demands of deduction that the inferences should have this character. One might put this by saying that 'if *A* and *B*, then *C*' must be intuitively known, but that is not right because if one has that knowledge, and the knowledge that *A* and that *B*, the step to *C* is still an inference, which one would naturally describe as a conjunction-introduction to attain '*A* and *B*' and then *modus ponens* to attain *C*. So, in order to avoid the famous Lewis Carroll regress, one can't escape the problem of the status of inferences, distinguished from that of propositions, with respect to intuitive knowledge.

[16] 'Mathematical intuition', Sect. V; 'On some difficulties', Sect. III.
[17] 'On some difficulties', Sect. II.
[18] 'Intuition in constructive mathematics', 226–7.

Descartes' principle is probably counter-intuitive in general, but it is particularly so when interpreted in terms of the present conception of intuition and applied to logical inference. One reason is surely the generality of logic, which extends to propositions that involve reference to objects that are not objects of intuition and concepts that are not in any reasonable sense intuitive.[19]

Logical inference has the character that generally no new intuition of objects is needed to attain the conclusion that was not already needed to attain the premisses. There may be exceptions, such as disjunction-introductions where the new disjunct contains reference to objects not referred to in the premiss. But in these cases, the intuitions would be at most presuppositions of the use of certain parts of the conclusion. One might want to say that intuitive knowledge is not preserved if a new disjunct is introduced that is in some way not intuitive in its content.

No formulation of PRA can do without simple rules concerning identity: instances of $a = a$ and the substitutivity of identity. These would be reasonably understood as minimal inference rules for reasoning about objects, at least mathematical objects. In particular, the most elementary computations proceed by these rules. Given that the intuitive knowledge in question rests on intuition of objects, if we allow any extension at all of intuitive knowledge by reasoning, such basic reasoning concerning identity would have to be allowed. In fact, however, their consequences are quite trivial until they are combined with induction and recursion.

Already this admission has the consequence that there will be degenerate cases of intuitive knowledge, for example statements of the form $t = t$, where t designates an intuited object, and the general statement $a = a$. Logic does not rest on intuition in the way that perceptual judgements rest on perception. In both these cases, intuition plays a certain role in seeing them to be true: in the former because it is intuition that gives the object that is said to be self-identical,[20] in the latter because intuition is involved in understanding the range of the variable and thus in its coming to have a definite sense. (As a pure law of logic, it is schematic.) Still, it would not be right to say that it is by intuition that we know these statements, since they are applications of more general rules that are not restricted in their application to objects of intuition.

[19] I have not tried to introduce or explain a notion of intuitive concept, but if one is clear about what is an abstract concept in the context of Gödel's discussion of finitism, then an intuitive concept could simply be one that is not abstract. It is still not clear what would be the place of logical notions in such a distinction. Evidently it does not follow that identity, the conditional, conjunction, and disjunction are abstract, so long as they are applied to atomic sentences that do not contain abstract concepts. On the other hand, quantification does appear to take us outside the realm of the intuitive.

[20] I leave aside the question whether $t = t$ is to be allowed as true in cases where t does not designate anything. No such case arises in the language of PRA.

Why should the use of propositional logic be held to preserve intuitive knowledge? The question now arises how we understand the connectives. Hilbert and Bernays thought of them as truth-functional. The facts they relied on are the following: Any atomic formula of PRA has the property that for any assignment of numerals to its free variables, it can be decided by computation. (This clearly depends on the computability of primitive recursive terms.) But then for a compound formula, again a truth-value can be computed for any assignment of numerals to the variables. Hence they described true generalizations in this language as 'verifiable'.[21]

This computability of the logical connectives when used with the language of PRA is certainly helpful in convincing oneself that PRA accords with general demands of constructivism. In fact there is a stronger result than this brief exposition suggests: PRA can be formulated as an equation calculus, in which the only logic is reasoning about identity of the sort discussed above. (Such formalisms are often called 'logic-free', but for our present purposes identity counts as a logical notion.) Then propositional connectives obeying the classical rules can be introduced by contextual definition. The use of primitive recursion to this end is minimal; it is in my view not affected by the doubts expressed below about the intuitive character of primitive recursion in general.[22]

Nonetheless one has a more natural formulation of PRA if one takes the connectives, or at least some of them, as primitive. Then the reliance of Hilbert and Bernays on their computability introduces the mathematical 'can' at a point where we wish to avoid it. It may seem that it is unavoidable, for how can we justify the use of classical logic other than by the observation that the truth-functions are computable? We can, however, formulate PRA with positive propositional logic (i.e. the negation-free fragment of intuitionistic), derive intuitionistic logic using the definition of $\neg A$ as $A \to 0 = 1$, and then prove any instance of the law of the excluded middle (see Appendix). The use of primitive recursion in this derivation is again minimal: addition, subtraction, and recursions that amount to a definition by cases between 0 and successor arguments. One might still not be entirely satisfied with it because of the use of the definition of $\neg A$; for example, it leaves the statement that 0 is not equal to 1 trivial. We can, however, think of $\neg A \leftrightarrow (A \to 0 = 1)$ as an axiom schema.[23]

[21] The concept is introduced in *Grundlagen der Mathematik* I, 237. Note that in the case of formulae with free variables but without quantifiers, which is the case that concerns us, a formula is said to be verifiable if 'it can be shown' that every result of replacing the variables with numerals is a true formula. It seems likely that the authors had in mind that this could be shown by a finitistic proof.

[22] Bounded quantifiers can also be contextually defined using primitive recursion. It is not so clear, however, that these recursions preserve intuitive knowledge; see Sects. 5–6.

[23] Note, however, that if instead $\neg(0 = 1)$ is an axiom, the left-right direction of this biconditional is an instance of the characteristically intuitionistic schema $\neg A \to (A \to B)$, although the right-left direction is a simple *modus tollens*.

There would still be questions about the understanding of the connectives in this logic. But it should be clear that what is assumed, the basic logic of identity and minimal propositional logic, is minimal for reasoning about objects. It doesn't have the implications that quantificational logic, particularly classical, was found to have in application to an infinite domain of objects. The case that this weak logic takes us beyond the intuitive would have to be made, and I do not know of anyone who has made it.[24]

4. INDUCTION

Induction and recursion will naturally be seen as the crux of the issue surrounding Hilbert's thesis, since it is they that cash in our conception of generalization about the natural numbers, and we do not have non-trivial arithmetic without that. Earlier,[25] I made the suggestion that induction should be regarded as conservative of intuitive evidence because of its analogy with a logical principle. Where N is the natural number predicate, we naturally think of the principles

$$N0$$

$$\frac{Nx}{N(Sx)}$$

as introduction rules for the natural number predicate. Induction is then the corresponding elimination rule[26]

$$\frac{A(0) \quad \begin{matrix}[A(x)] \\ \vdots \\ A(Sx)\end{matrix} \quad Nt}{A(t)}.$$

The formal analogy of induction to a logical principle is very strong, and it is exploited in proof-theoretic studies of systems containing arithmetic. Is it clear, however, that it is more than a formal analogy?

We can't make for induction the claim we made for logical reasoning in

[24] There is a difficulty of a different kind, which would apply to any claim to the effect that certain rules of inference preserve intuitive knowledge. Kenneth Manders points out that one might hold that only a *perspicuous* proof yields intuitive knowledge. A proof involving a very large number of applications of rules (even within the limits of feasibility) may very well not be perspicuous. I am not sure whether the requirement Manders suggests is a reasonable one.

[25] 'Intuition in constructive mathematics.'

[26] The notation is usual in writing on natural deduction; the bracketed formula is an assumption that is discharged by the inference. x is not allowed to occur free in any other open assumption in the derivation of $A(Sx)$, nor in $A(y)$ unless y is x.

the previous section, that it is basic for reasoning in general, or for reasoning about objects. It has rather the character of being basic to reasoning about objects in a particular domain, the objects obtained from an initial element by arbitrary finite iteration of a given operation. In the case whose intuitive character is at issue, Hilbert's model of strings, the initial element is an object of intuition, and it is intuitively known that the operation (adding a stroke to a string) is well-defined. If we admit some conception of the domain of strings as intuitive, it seems we ought to admit induction as preserving intuitive knowledge.

I say 'some conception', because we can certainly characterize the domain in a way that would have no claim to be intuitive, for example by a Fregean definition using full second-order logic. Even the idea that induction should hold for *any* predicate that we can precisely understand introduces what the participants in the discussion of finitism and intuitive evidence would have regarded as abstract concepts. What is claimed by the finitist is that instances of induction, provided that their content does not involve any non-finitist notions, become evident one at a time. This is in keeping with (2.5): applications of induction *preserve* intuitive evidence; there is no claim to the effect that a principle of induction is intuitively evident.

It follows that it would not be right to say that we understand clearly what it is to iterate a sucessor operation a finite number of times and *then* see that induction is true of the objects resulting from such iteration beginning with 0.

5 PRIMITIVE RECURSION

The discussion so far has made the case for Hilbert's thesis rest to a large extent on recursion, that is on subthesis (2.3). For even if we grant that inference by induction preserves intuitive evidence, this will not lead us very far in arithmetic unless we have some of the function symbols that in PRA are introduced by primitive recursion. What we might call the rule of primitive recursion allows us, given functors ψ, χ of n and $n + 2$ arguments respectively ($n \geq 0$),[27] to introduce a new functor φ of $n + 1$ arguments and to assume the equations

$$\varphi(0, a_1 \ldots a_n) = \psi(a_1 \ldots a_n),$$

[27] If we do not also have a rule of explicit definition, we must allow terms composed from given functors. (I use the term 'functor' simply for singular terms with empty argument places; I do not wish to assume that they designate *functions*; compare the use of the word 'predicate', where it is not assumed that predicates designate properties or relations.)

$$\varphi(Sb, a_1 \ldots a_n) = \chi[b, \varphi(b, a_1 \ldots a_n), a_1 \ldots a_n].$$[28]

Let us consider first a degenerate case, where χ does not depend on its second argument, and the second equation thus takes the form

$$\varphi(Sb, a_1 \ldots a_n) = \chi[b, a_1 \ldots a_n].$$[29]

To convince ourselves that φ is well defined, we only need to convince ourselves that every number is either 0 or of the form Sb. In the case of the intuitive model that concerns us, this is immediate from the way in which it has been explained. Every 'number' is either '1' or has been obtained from it by a succession of adding a '1' on the right, and that is the successor operation. We don't have to understand 'a succession of' such operations to see that every number is either the initial one or a successor.

This trivial case, however, leaves out precisely the feature of primitive recursion that gives it its power, that it enables us to iterate a given function and cashes in the idea that the numbers are obtained by arbitrary iteration of the successor operation. The first non-trivial case is that of addition, which is equivalent to iteration of the successor operation beginning with an arbitrary a.

At this point let us return to Hilbert and Bernays. They do give what would be intuitive arguments for the well-defined character of primitive recursive functions, both in general and in particular cases such as addition. Let us first consider what they say about addition and multiplication. I will shift ground in one respect, and following them treat the sequence of positive integers beginning with 1 rather than the non-negative integers beginning with 0; that is more natural for an interpretation in their intuitive model, since then '1' designates itself.

Then addition reduces simply to concatenation of strings, and one can hardly doubt that it is intuitive in the required sense. The basis recursion equation is replaced by the identity $a + 1 = a + 1$, so that the only recursion axiom that is needed is

[28] The term 'rule of primitive recursion' has its point, because in a formalism with function variables, we could formulate a single pair of axioms

$$R(0, a, f) = a$$
$$R(Sb, a, f) = f[b, R(b, a, f)],$$

which will do the work of the rule and might be called the axiom of primitive recursion. Proof-theoretic work shows that the relation of the rule and axiom of primitive recursion is closely connected with that of the rule and axiom of induction.

[29] This 'degenerate' recursion schema is reducible to a single case: if we set $\xi(0, a, b) = a$ and $\xi(Sc, a, b) = b$, then we can define the functor φ of the text explicitly as

$$\xi[b, \psi(a_1 \ldots a_n), \chi(b, a_1 \ldots a_n)].$$

$$a+(b+1)=(a+b)+1,$$

and that is certainly intuitively obvious if we think of addition as concatenation of strings considered as spatial configurations.

Concatenation does not take place by adding on to the string *a* one term of *b* at a time; rather one can see that $a + b$ is defined without relying on the build-up of b by a sequence of steps. I think it is for this reason that Hilbert and Bernays claim that the general associative law is 'immediate from the definition of addition' (23); in particular, we do not need induction to see its truth.[30] That reinforces the interpretation of them as thinking of addition in the way I am proposing, although their own explanation (22) is a little obscure. It will be less clear that addition is intuitive if we explain $a + b$ as what is got by iterating the successor operation beginning with *a* instead of 1. That we understand an operation that effectively does that is a conclusion.

On this view, addition is a special case, in that iteration is not directly involved in understanding it. Therefore its intuitive character does not show that the same will obtain for primitive recursion in general. But let us go one step further and consider the case of multiplication. Hilbert and Bernays again give a special explanation, which appeals to a kind of replacement: 'Multiplication can be defined in the following way: $a \cdot b$ designates the numeral that one obtains from the numeral *b* by, in its construction, always replacing 1 by the numeral *a*' (p. 24). This is a persuasive picture, but it amounts to saying that the same procedure of generating strings can be carried out with the role of '1' replaced by that of *a*. It may be most persuasive, however, if we suppose that every string can be intuited, an assumption that I have wanted to avoid.

This assumption does not seem to me essential. What is really involved is that we imagine an arbitrary string replacing '1' in the construction of *b*; since it is a bounded spatial configuration, the case is just a more complex version of what is involved in seeing that successor itself is well defined (see n. 12).

Even if you are not persuaded by this, I ask you to grant for the sake of argument that multiplication is intuitively well defined. Unlike the explanation we gave of addition, this is an entering wedge for the idea that iterating an intuitively well-defined operation will yield an intuitively well-defined one. But the situation is rather special, since what happens at each stage is the same, namely the insertion (in effect construction) of *a*. In particular, there is a bound given in advance on the complexity of what is added to the construction at each stage.

The considerations don't easily generalize, and it is particularly with

[30] In this respect, they contrast the associative with the commutative law.

respect to exponentiation that questions have been raised. Already in 1934 Bernays questioned whether the evidence that $67^{257^{729}}$ exists is intuitive.[31] Subsequent writers have expressed doubts about exponentiation on strict finitist or other grounds.[32] These doubts are reinforced by the acceptance of polynomial-time computability as a criterion for feasible computability, so that in general computation of the exponential function is not feasible.

Although I will return to the case of exponentiation, it is not clear that we can deal with it by considerations specific to the case at hand, as we did with addition and multiplication. Let us return to the general case. To make Hilbert's thesis plausible, we need to convince ourselves that primitive recursion introducing a new functor yields intuitively evident equations, provided that the defining equations associated with the functors on the right-hand side of the equations are intuitively evident. That is to say that the thesis of the conservativeness of intuitive evidence extends from induction to recursion. But this is a larger step than one might at first think, just because primitive recursion introduces a new function symbol, one might say a new concept. The analogy between induction and a logical principle does not seem to be of any help at this point.

Hilbert and Bernays do claim that primitive recursion is finitistically acceptable and offer an argument to this effect (25–7). The core of it goes as follows: Suppose (taking the case without parameters) a functor φ has been introduced by the equations

$$\varphi(1) = a$$
$$\varphi(n+1) = \psi(\varphi(n), n).$$

Then, for a given string m, one can by a sequence of m steps obtain an outcome which is either a or a term of the form

$$\psi[\psi(\ldots\psi(a, 1),\ldots, m-2), m-1]$$

which no longer contains φ. 'With this we have reached a computable expression, for ψ is to be an already known function. One has to carry out this computation from inside out, and the numeral obtained as its output is to be assigned to the numeral m' (26).

But this seems to be an argument by Σ_1 induction: A computing procedure is described, and it is shown that it terminates. The key step is to show that in computing $\varphi(m)$ for a given numeral m, one can by a succes-

[31] 'Sur le platonisme dans les mathématiques', 61; p. 265 of the trans. in Paul Benacerraf and Hilary Putnam (eds.), *Philosophy of Mathematics: Selected Readings*, 2nd edn., Cambridge University Press, 1983.

[32] Bernays was doubting not exponentiation but the intuitive character of the evidence that it is well-defined. But cf. D. van Dantzig, 'Is $10^{10^{10}}$ a finite number?' *Dialectica*, 9 (1955–6), 272–7. For more recent instances, see Edward Nelson, *Predicative Arithmetic*, Princeton University Press, 1986, and David Isles, 'What evidence is there that 2^65536 is a natural number?' *Notre Dame Journal of Formal Logic*, 33 (1992), 465–80.

sion of steps replace it by a term that does not contain φ. As a proof, it can be structured in different ways; the one that seems closest to the surface of the text would argue that if $m = 1$, this replacement is trivial, and if it can be carried out for m, then it can be carried out for $m + 1$. Once one has the φ-free term, moreover, there is still the task of computing it, and the depth of nesting of ψ in it is precisely m: another induction seems involved in seeing that its computation terminates.

One might also see the matter the following way:[33] The recursion equations describe a procedure for computing φ. For $m = 1$, the procedure simply outputs a. For $m + 1$, the procedure is to compute $\varphi(m)$, with result say k, and then compute $\psi(k)$. We can easily prove that the computation terminates, by induction using the predicate 'the computation of $\varphi(m)$ terminates', or, if we allow ourselves the function symbol in a free-logical language, '$\varphi(m)$ is defined' or $\exists k[\varphi(m) = k]$. These arguments are clearly not finitist by Hilbert's lights, since they involve unbounded existential quantification; moreover, a Σ_1 induction is only made explicit by a recursion. One can see the considerations as a description of how, parallel to the construction of a string or other numeral m, one constructs the value of $\varphi(m)$. But we haven't been given a non-circular way of showing that the construction must work, unless we give ourselves Σ_1 induction. So our problem has not been advanced.

It is worth pointing out, however, that in 'Die Philosophie der Mathematik und die Hilbertsche Beweistheorie' (see n. 13), Bernays gives the same 'replacement' account of multiplication and attempts to generalize it to exponentiation, in a context in which he points out that exponentiation quickly gives rise to numbers whose representation by numerals of the Hilbertian canonical form is not practically possible:

> Here also there are bounds for the executability of repetitions both in the sense of actual representability and in the sense of physical realization. Consider the example of the number $10^{10^{10}}$. We can reach this number in a finitary way as follows: We start from the number 10, which, in accordance with one of the normalizations given earlier we represent by the expression
>
> 1111111111.
>
> Now let z be any number which is represented by a corresponding expression. If we replace in the above expression each 1 by the expression z, then again a number expression arises, as we can make clear intuitively, which for communication we designate with '$10 \cdot z$'. Thus we obtain the process of multiplying a number by 10. From this we obtain the process of passing from a number a to the number 10^a, in that we let the number 10 correspond to the first 1 in a and, to each attached 1, *the process of multiplying by 10*, and continue until we are at the end of the expression a. The number obtained by the final process of multiplying by 10 is designated by 10^a.

[33] Suggested to me by Tait, 'Finitism', 532.

This procedure offers no difficulty from the intuitive point of view.[34]

In the case of exponentiation, each 1 in the construction of *a* is replaced by a 'process'. It is hard to see how this description can do what Bernays apparently wants, since processes are not clearly objects of intuition. Possibly he means that what the 1 is to be replaced by is *the result of multiplying by 10* what one obtained at the previous stage. But then the argument is again either just a description of a computing procedure or an argument involving Σ_1 induction, since without an inductive hypothesis one does not have such a 'result'.

There is really nothing specific to exponentiation in Bernays' argument; one could replace multiplication by any function that has been seen to be well defined and argue that this will obtain for its iteration. It seems likely that Bernays came to think his 1930 argument unclear; possibly he saw the general argument for primitive recursion in Hilbert and Bernays as replacing it.

I conclude that neither Hilbert nor Bernays offers a non-circular argument for (2.3) and therefore that they do not offer a convincing argument for Hilbert's thesis. It seems to me very doubtful that this obstacle can be overcome. If it could by an argument of the kind given by Hilbert and Bernays, one would expect the argument to be convertible into a proof that, say, exponentiation is well-defined, using hypotheses essentially weaker: unproblematic primitive recursive functions and induction. Unless one allows unbounded Σ_1 induction, one will not get the result: If the functions that one allows are all polynomially bounded, and only Δ_0 induction is allowed, then one will not be able to prove that exponentiation is everywhere defined.

6. WHERE DOES THIS LEAVE US?

It seems to me that the discussion of primitive recursion produces a sort of stand-off, familiar to those of us who try to understand the justification of mathematical induction. If one accepts the introduction of new function symbols by primitive recursion as a procedure that preserves intuitive knowledge, it is not clear that what one asserts to be intuitively known will include items that can be seen on other grounds not to be, unless one makes the demand for practical intuitability of the objects involved. The iterability of procedures that we can see to lead to a result is, plausibly, part of our notion of numbers as the result of iterating the successor operation beginning with 0, or of Hilbert's strings as the result of iterating the putting on of an additional '1' beginning with '1'.

[34] 'Die Philosophie der Mathematik und die Hilbertsche Beweistheorie', 38–9, my trans., emphasis added.

On the other hand the conclusion that even exponentiation must be regarded as everywhere defined in a genuinely intuitive arithmetic does not seem to be forced on us. It seems open to us to hold that what is intuitively known in arithmetic is limited to a fragment of primitive recursive arithmetic that does not include recursions for any function that grows as fast as exponentiation.

Nonetheless I am inclined to the negative conclusion, that intuitive knowledge in arithmetic does not go beyond the fragment of PRA that includes addition and multiplication and allows the introduction of functions by bounded primitive recursion only. We can add to the difficulty we have found in seeing that exponentiation can be intuitively seen to be well defined some other respects in which exponentiation proves to be a barrier. The most important is predicativity. Starting from the observation (made earlier by me and still earlier by Michael Dummett)[35] that induction even as an informal explanation of the natural numbers is impredicative, Edward Nelson in *Predicative Arithmetic* undertook to develop arithmetic on a basis that would avoid this impredicativity (see n. 32). His method for getting some induction would be too complex to explain at this point. It yields Δ_0 induction but does not allow the proof that exponentiation is well defined. Attempts to develop arithmetic in a predicative way using ideas from the ramified theory of types without reducibility have also been made. The most direct again do not yield a proof that exponentiation is well defined.[36] If we add to this the considerations leading to the conclusion that exponentiation is not feasibly computable, it is difficult to maintain that our insight that exponentiation is well defined is intuitive.[37]

This conclusion would still leave open the question whether the functions that can be seen intuitively to be well defined have weaker closure

[35] Michael Dummett, 'The philosophical significance of Gödel's theorem' (1963), in *Truth and Other Enigmas*, London, Duckworth, 1978, 199; and my 'The impredicativity of induction', in Michael Detlefsen (ed.), *Proof, Logic, and Formalization*, London, Routledge, 1992, 139–61 (expanded version of a paper published in 1983).

[36] Daniel Leivant develops a notion of 'ramified provability' that does yield exponentiation. In terms of the ramified theory of types, he shows that if n is a natural number in the sense of having all properties of a certain level possessed by 0 and closed under successor, then 2^n is a natural number in the sense of having all properties of lowest level possessed by 0 and closed under successor. This is unsatisfying because he does not give a single sense in which if n is a number, then so is 2^n. However, this appears to be true in his scheme already for addition and multiplication. Moreover, his constraint of ramification is a powerful one. For the advocate of Hilbert's thesis it only puts off the day of reckoning, since only elementary functions are provably recursive in Leivant's ramified sense, and so iterated exponentiation is not. See 'Intrinsic theories and computational complexity', unpublished. A more satisfying result along these lines is obtained in recent work of John Burgess and Allen Hazen. This puts in doubt the implication of the text that exponentiation will prove a barrier if one understands strict predicativity in terms of ramified theories. That it is such if one understands it as Nelson does is confirmed by the fact that $I\Delta_0$ with exponentiation is not interpretable in Q. The correct conclusion may be that Nelson's criterion is too narrow.

[37] That we even have this insight is questioned by Nelson and others.

properties, such as being closed under *bounded* primitive recursion. The latter would imply that bounded quantification can be introduced into intuitive arithmetic, a plausible result. I am far from sure about questions of this kind. It is possible that the notion of intuitive knowledge is not precise enough to decide them. They could also depend on the answers to difficult mathematical questions. There are quite straightforward questions about restricted recursions whose answers are not known; for example, it is not known whether the polynomial-time computable functions are closed under bounded recursion.

7. THE THESIS THAT EVERY STRING CAN BE INTUITED

If we are not revisionist about mathematics, then it is evident that if 'can' implies some form, however weak, of practical possibility, then this thesis is false; probably already by exponentiation we can offer counter-examples. We can't regard it as literally true that human beings have the ability to intuit any string. It is also not likely that we can construct a real, physical machine that would have that ability.

Now let us suppose that we do allow ourselves to understand the 'can' in terms of abstract mathematical possibility.[38] The considerations that show successor to be intuitively well-defined show that if a string can be intuited, so can its successor. (This is plausibly true also for practical possibility.) One can then argue by induction that every string can be intuited. This might be questioned on the ground that similar reasoning leads to Sorites paradoxes.[39] Given that the context is supposed to be mathematical possibility, I don't think that is a convincing objection. But it does suggest that there is something trivial about the conclusion. It is the concept of number, by way of an application of induction, that tells us that every number can be intuited, rather than some independent insight into what is intuitable yielding a domain of intuitable objects. But just for this reason, I don't think it likely that a convincing argument for the falsity of the thesis will be forthcoming. Objections in the literature to the claim that every string can be intuited seem either to understand the modality in terms of human ability

[38] I largely avoid the term 'in principle'. In actual usage, it is ambiguous; possibility in principle may be abstract mathematical possibility in the sense in which I am understanding it, or it may be constrained by other considerations, for example basic physical considerations. It is with the latter understanding that R. O. Gandy argues that there are finite bounds to what inscriptions can in principle exist and limits to what can in principle be known. If his concern were with intuition, he would very probably say the same about what can in principle be intuited. See 'Limitations to mathematical knowledge', in D. van Dalen, D. Lascar, and T. J. Smiley (eds.), *Logic Colloquium '80*, Amsterdam, North-Holland, 1982, 129–46.

[39] There are of course different views about what part of the reasoning leading to sorites paradoxes should be rejected. I am claiming that in the present situation the appropriate response is to embrace the conclusion that every string can be intuited.

or to amount to a rejection of the application of mathematical modality in combination with epistemic notions. The former objection is based on a misunderstanding.

The second objection, that combination of mathematical modality with epistemic notions (or perhaps other non-mathematical notions) is questionable, is one that I think should be regarded seriously. Full consideration of it will have to be reserved for another occasion.[40] In this section I will assume that we can reason about the possibility of intuition in a mathematical way and try to assess some consequences of this assumption.

Acceptance of the conclusion that every string can be intuited might seem to trivialize every question about what strings can be intuited. Consider, for example, the question, suggested by Bernays' discussion, whether a string of length $67^{257^{729}}$ can be intuited. Surely, if every string can be intuited, then every string of length $67^{257^{729}}$ can be intuited. However, '*x* can be intuited' is on the face of it intensional. The conclusion follows only if it is taken in a *de re* way. Indeed, such a reading underlies the formulation of the thesis that every string can be intuited. The point will be clearer if we consider a slightly different formulation, say: For every string *x*, I can intuit *x*.[41]

It is well known that universal instantiation is not always valid when the variable occurs in an intensional context. Since strings are types, there is just one string of length $67^{257^{729}}$, so that we can let s_0 abbreviate 'the string of length $67^{257^{729}}$'. From 'For every string *x*, I can intuit *x*', it doesn't follow right away that I can intuit s_0.

One might question this *de re* use of 'intuit', but let us see what follows if we pursue it. Let $CI(x)$ abbreviate '$x = 0$ or I can intuit the string of length *x*'. The intensionality of *CI* has two sources, the notion of possibility and that of intuition. If we regard it as coming only from the first source, a rather uncomfortable result will follow. Consider the following reasoning:

Let *R* be a primitive recursive predicate such that $\exists yRxy$ is not recursive. Suppose we define φ by

$$\varphi(x) = 0 \text{ if } \exists yRxy,$$
$$\varphi(x) = 1 \text{ otherwise,}$$

so that φ is also not recursive. Now one might argue: Clearly $CI(0)$ and $CI(1)$, but $\varphi(x) = 0$ or $\varphi(x) = 1$; hence $CI[\varphi(x)]$ for any number *x*. This reasoning can be questioned on the ground that the inference from $CI(0)$ and $\varphi(x) = 0$ to $CI[\varphi(x)]$ is an application of substitutivity of identity in an inten-

[40] For some relevant discussion see my 'What can we do "in principle"?' in Maria Luisa Dalla Chiara, Kees Doets, Daniele Mundici, and Johan van Benthem (eds.), *Logic and Scientific Methods*, Dordrecht, Kluwer, 1997, 335–54.

[41] Nothing will be changed if we say 'Someone can intuit *x*'; indeed, in the case of mathematical possibility, it will make no difference to the outcome whether this is read as 'Someone is such that he can intuit *x*' or 'It can be that someone intuits *x*.'

sional context. (Likewise with 1 instead of 0.) But if the intensionality is merely modal, it can be replied that $\varphi(x) = 0$, if true, is mathematically necessary, and we do have the substitutivity of necessary identity.[42] Thus the proof that for every x, $CI[\varphi(x)]$ is not blocked. It might be objected to on the ground that it is not intuitionistic, but is it clear that the possibility of intuition does not obey the laws of classical modal logic?

Let us, however, explore *CI* a little further. In what follows, I will show that rather plausible assumptions that don't obviously conflict with the intensionality of *CI* imply that $CI(t)$ holds for any term t of PRA and that stronger reasoning whose exclusion is not obviously arbitrary would lead to a larger class of terms for which this holds.

Let φ be an n-place functor; we will say that it *preserves intuitability* if the inference

$$\frac{CI(t_1)\ldots CI(t_n)}{CI[\varphi(t_1\ldots t_n)]}$$

is a correct inference for any terms designating numbers.

The intensional character of *CI* implies that we should not accept universal instantiations whose premisses contain *CI*. But it does seem a reasonable assumption that such instantiation should be accepted for terms t of which $CI(t)$ holds; in other words we should accept the inference

$$\text{UII}\quad \frac{\forall x A(x) \quad CI(t)}{A(t)}.$$

Now let us assume what I have claimed to be evident, that from intuition of a string one can pass to intuition of its successor in the string sense (i.e. the result of adding one more symbol). This implies $CI(n) \rightarrow CI(Sn)$ for every number n. But then, by UII, we have that S preserves intuitability.

I will make one further assumption, namely that in *CI* contexts we can replace one side by the other of one of the defining equations for a recursive function, with arbitrary terms instantiated for the variables. From this it evidently follows that if φ is explicitly defined from functors that preserve intuitability, then φ itself preserves intuitability.

Now let us consider a case of primitive recursion (where for simplicity we assume there is only one parameter)

$$\varphi(0, a) = \psi(a)$$
$$\varphi(n+1, a) = \chi[n, \varphi(n, a), a].$$

[42] It would be assumed that mathematical necessity obeys at least S4, but that is plausible enough.

We assume that ψ and χ preserve intuitability. Then we have for any numerical term s the inference from $CI(s)$ to $CI[\psi(s)]$ and by replacement to $CI[\varphi(0, s)]$. Suppose we can infer $CI[\varphi(b, s)]$ from $CI(s)$. Then since χ preserves intuitability, we have $CI[\chi(\mathrm{b}, \varphi(b, s))]$,[43] and hence by replacement $CI[\varphi(Sb, s)]$. Hence we have by induction $\forall x\{Nx \rightarrow CI[\varphi(x, s)]\}$. By UII, for any term t such that Nt and $CI(t)$, $CI[\varphi(\mathrm{t}, \mathrm{s})]$. That is, φ preserves intuitability.

But this is enough to show that $CI(t)$ holds for any term of PRA.

If we allow more complicated inductions involving the predicate *CI*, we can prove that functors defined by more complex recursions preserve intuitability, for example Ackermann-type functors introduced by nested double recursion. Particularly since these proofs will still be intuitionistic, it is not obvious that they should be ruled out. But that they are possible does indicate that the mathematical possibility of intuition is being conceived rather generously.

Does this give rise to an argument for Hilbert's thesis? Suppose s and t are such that I can intuit a string of length s, and I can intuit a string of length t. Then, surely, by comparing them I can determine whether $s = t$ is true. It thus seems that any closed formula of PRA can be decided in an intuitive way. But what follows about intuitive knowledge of generalizations? The most this argument shows is that if a formula of PRA is true for all values of its variables, then in each particular case this can be known intuitively. But this does not yield intuitive knowledge of the generalization. Even if this obstacle can be circumvented, I see no way to get around the fact that what these considerations yield is the *possibility* of intuitive knowledge, according to a rather liberal kind of possibility. But Hilbert's thesis concerns actual intuitive knowledge, at least given a proof in PRA. Even if the questions about the ideas involved can be resolved, the case for Hilbert's thesis does not seem to be materially advanced. More generally, the relevance to questions about what we know intuitively of questions about what intuitions are possible 'in principle' is not clear.

APPENDIX

For the conventional formulation of PRA, we can assume that the logic allowed is minimal propositional logic, with $\neg A$ defined as $A \rightarrow 0 = 1$. For to obtain intuitionistic logic, it suffices to derive $0 = 1 \rightarrow A$ for any A. If A is atomic it is an equation $s = t$. We introduce by primitive recursion a function symbol φ satisfying $\varphi(0) = s$

[43] Note that $CI(b)$ follows from the assumption Nb, since we have accepted the induction that yields $\forall x[Nx \rightarrow CI(x)]$.

and $\varphi(Sn) = t$. Clearly $0 = 1$ implies $\varphi(0) = \varphi(1)$, whence $s = t$.[44] The other cases are shown in the usual way by induction on the construction of A. First, since $0 = 0$, $0 = 0 \vee \neg(0 = 0)$. From the axiom $\neg(Sx = 0)$, we have $Sx = 0 \vee \neg(Sx = 0)$. So by induction, $x = 0 \vee \neg(x = 0)$. Now using the lemma

$$x = y \leftrightarrow (x - y) + (y - x) = 0.$$[45]

we can infer $x = y \vee \neg(x = y)$, and thus $A \vee \neg A$ for atomic A. By induction on the construction of A, we can derive excluded middle for any A.[46] And since we have intuitionistic logic, we can also derive $\neg\neg A \rightarrow A$.

[44] I owe this device to A. S. Troelstra. φ has as parameters the free variables of s and t.

[45] Subtraction is introduced by recursion. First $\mathrm{pr}(0) = 0$, $\mathrm{pr}(Sn) = n$. Now let $x - 0 = x$, $x - Sn = \mathrm{pr}(x - n)$.

[46] Suppose $A \vee \neg A$ and $B \vee \neg B$. To see $(A \rightarrow B) \vee \neg(A \rightarrow B)$, assume first A. If we assume B, $A \rightarrow B$ and hence $(A \rightarrow B) \vee \neg(A \rightarrow B)$ follow. If we assume $\neg B$, assume $A \rightarrow B$. Then B follows, which is a contradiction, whence $\neg(A \rightarrow B)$, whence $(A \rightarrow B) \vee \neg(A \rightarrow B)$. By $\vee$E we discharge the assumptions of B and $\neg B$. Assume now $\neg A$. By intuitionistic logic $A \rightarrow B$ follows, whence again $(A \rightarrow B) \vee \neg(A \rightarrow B)$. By $\vee$E, we can discharge the assumptions of A and $\neg A$. The case of disjunction is similar, and the others are simpler.

11

Hilbert's Finitism and the Notion of Infinity

KARL-GEORG NIEBERGALL AND MATTHIAS SCHIRN

An ihren Früchten sollt ihr sie erkennen,
heißt es auch für die Theorien.
(By their fruits ye shall know them—
that applies also to theories.)
Hilbert, 'Über das Unendliche', 98 [384]

When Hilbert set about developing his foundational programme in the 1920s, proof theory was still enjoying a state of pre-Gödelian innocence. The principal objective of the programme was to establish once and for all the entire heritage of classical mathematics by means of a finitist metamathematical consistency proof. Hilbert thus aimed at justifying, by using only finitist methods, what he took to be the disputable non-finitist modes

Versions of this material (some in Portuguese or Spanish or German) were presented at the 3rd International Congress of Analytical Philosophy held in September 1995 at the Federal University of Santa Catarina (UFSC) in Florianópolis, the Center of Logic, Epistemology and History of Science of the State University of Campinas (CLE, UNICAMP), the Catholic University of Rio de Janeiro (PUC-RJ), the Federal University of Pernambuco (UFPE) in Recife, the Catholic University of São Paulo (PUC-SP), the Institute of Philosophical Investigations of the National Autonomous University of Mexico (IIF, UNAM), the National Autonomous University of Costa Rica, the Argentinean Society of Philosophical Analysis (SADAF) in Buenos Aires, the International Congress 'Truth: Logic, Representation and World' held in January 1996 at the University of Santiago de Compostela, the University of Sevilla (Dept. of Mathematical Analysis), the History of Logic Conference held in April 1996 at the University of Texas at Austin, the University of Paris, the University of Regensburg, and at the University of Munich. Some topics of Hilbert's programme were also discussed at the University of São Paulo (USP). Thanks to the audiences for interesting discussion. We are grateful to Ulrike Ritter, Robert Bublak, Manfred Harth, Stefan Iwan, and Roberto Torretti for reading an earlier version of this paper. Special thanks are due to Nicholas Jacob-Flynn and Uwe Lück.

The idea of writing a paper on Hilbert's finitism grew out of our joint seminar on his proof-theoretic programme at the University of Munich in the winter term 1993–4. Hilbert's classical papers on his proof theory, stimulating as they are and transparent as they may appear at first glance, are in fact not free from ambiguity and vagueness. We do not blame him for that, of course, nor do we mean to apologize for any remaining shortcomings in our account. Nevertheless, we feel inclined to spell this out here. With the exception of Hilbert (1926) and (1928*a*) the translations of Hilbert's writings are our own. For the most part, we have slightly modified the existing translations. As to the translation of the world 'inhaltlich', we have adopted the neologism 'contentual', coined by the translator of Hilbert (1926), Stefan Bauer-Mengelberg. The references to the English translations of Hilbert (1926), (1928*a*) and Tarski (1935) appear in square brackets.

of inference of a formalized arithmetical system, for instance, the application of *tertium non datur* to infinite totalities or complete induction. Finitist means were considered by him to be unassailable and absolutely reliable, because he believed that they satisfy the methodological principles of representability in intuition and surveyability. It is widely held that Hilbert failed to characterize in a precise manner what he called the 'finitist point of view' (*finiter Standpunkt*) or 'finitist attitude' (*finite Einstellung*). We, for our part, share this opinion. A general characterization of the finitist point of view, which may still serve as a guiding principle, can be found in the first volume of Hilbert and Bernays's monumental work *Grundlagen der Mathematik* (1934, 32).

Our considerations on the beginnings of number theory and algebra have served the purpose of demonstrating in both its application and use the direct contentual reasoning, effected in the form of thought-experiments on intuitively given objects and free from axiomatic assumptions. In order to have a concise expression for it, we shall call this kind of reasoning *finitist* reasoning and shall likewise call the methodological attitude underlying this reasoning the *finitist* attitude or *finitist* point of view. In the same sense, we shall speak of finitist concept formations and assertions by expressing with the word 'finitist' in each case the fact that the reflection, assertion or definition in question proceeds within the limits of both the conceivability in principle of objects and executability in principle of processes, and is thus carried out within the scope of concrete considerations.

There can hardly be any doubt that Hilbert's original metamathematics was designed to show the consistency of formalized mathematical theories T such as Peano Arithmetic (PA) or second-order arithmetic (Z_2) by 'weaker', though more reliable, means than those that could be formalized in T. It was in the light of Gödel's Second Incompleteness Theorem (cf. Gödel (1931)) that proof theorists felt obliged to reconsider the conditions under which a metamathematical consistency proof for PA, for instance, could be accepted. The pivotal question was: Can we prove the consistency of PA, for example, with methods which, even though they extend those of PA, are nonetheless indisputable or at least more reliable than the disputable non-finitary means available in PA? In fact, Gentzen's and Ackermann's consistency proofs for PA rested on the idea that a mode of inference not formalizable in PA, but applied in the proof, such as transfinite induction up to ε_0, could be accepted as a secure metamathematical means by virtue of its (allegedly) constructive character (cf. Gentzen (1936), (1938) and Ackermann (1940)). Nevertheless, these proofs offended against a fundamental insight which Hilbert could have enshrined in the maxim: *Never abandon the base of intuition in your proof-theoretic practice*. They could thus not claim to be in the spirit of his finitist point of view. Remarks such as those of Gaisi Takeuti (1975, 96): 'We believe that our standpoint is a natural extension of Hilbert's finitist standpoint, similar to that introduced by Gentzen, and so we call it the Hilbert–Gentzen finitist standpoint' are therefore to be treated with reserve. Like Solomon Feferman (1988, 366), we do think that the notion of constructive methods is by no means clearer than the notion of finitist methods. On the contrary, the former is in general less clear than the latter.

This paper is devoted to what we take to be the most important aspects of the doctrine of finitism which Hilbert developed in his classical articles on proof theory between 1922 and 1928. Where it seems useful or enlightening, Hilbert and Bernays's *Grundlagen der Mathematik* (1934) will be taken into account. We believe that a critical examination of key issues of Hilbert's finitist metamathematics in the 1920s is not only of considerable interest in its own right, but will also provide the appropriate setting for a discussion of, say, the boundaries of post-Gödelian varieties of finitism. For reasons of space, we must leave this for other articles.[1] We argue that Hilbert's tacit assumptions of infinity concerning both the domain of objects talked about and the domain of expressions of a given formal language are the major weakness of his finitist approach. Special emphasis is placed on the scope of the language of finitist metamathematics. In Hilbert's view, finitist metamathematics is an informal 'theory' couched in an informal language. For reasons to be spelled out later, we shall, for the most part, construe it as a theory stated in a formal language. One aim is to show that Π^0_1-sentences cannot be formulated in that language; another is to demonstrate that there is no need to do this anyway.

1. HILBERT'S CONCEPTION OF METAMATHEMATICS

In the 1930s, proof theorists had come to construe metamathematics right from the start as forming a part of mathematics, or to replace an informal version of metamathematics with a formal one. In the latter case, the formal version of metamathematics is considered to belong intrinsically to mathematics proper. According to the method which has come to be known as 'arithmetization of metamathematics', mathematical expressions are coded by mathematical objects, preferably by natural numbers. Some logicians have even gone so far as to *define* expressions as certain numbers. This

[1] For a detailed discussion of the notion of finitism in Hilbert and Bernays's *Grundlagen der Mathematik* (1934, 1939) see Niebergall and Schirn, 'Extensions of the finitist point of view' (forthcoming). In 'Finitism = PRA? On a thesis of W. W. Tait' (forthcoming), we attempt to show that Tait's influential interpretation of finitism (cf. Tait (1981)), which identifies finitism with Primitive Recursive Arithmetic (PRA), is disputable. Finally, in our paper 'Hilbert's programme and Gödel's theorems' (forthcoming) we try to show that a weak version of Hilbert's programme is compatible with Gödel's Incompleteness Theorems by employing only what are clearly natural provability predicates.

approach involves a primacy of mathematics *vis-à-vis* metamathematics in two respects: first, in a linguistic or syntactic respect (the language of metamathematics belongs to the language of mathematics), and second, in an ontological or a semantic respect (expressions as the intended objects of metamathematical considerations are reduced to or identified with genuinely mathematical objects).

Hilbert's conception of metamathematics differs strikingly in a number of ways from the view we have just sketched. *Firstly*, for Hilbert, metamathematics is a collection of statements about figures or strings of figures, drawn from a finite stock. Unlike the mathematical objects, which are habitually conceived of as abstract, the figures considered in metamathematical operations are concrete, spatio-temporal objects and as such intuitable (or perceptually recognizable) and surveyable. The axioms, formulas, and proofs which constitute the formal edifice of the arithmetical theory are on a par with the numerical figures or numerals of the fragments of number theory which Hilbert constructs by means of contentual, intuitive methods.[2] They are the subject-matter of the contentual, proof-theoretic considerations. A formalized proof, like a numeral, is a concrete and surveyable object which must be given as such to our perceptual intuition. 'It can be communicated from beginning to end' (Hilbert (1926), 97 [383]). The term '(natural) number' is reinterpreted; it is explained by means of the term 'sign', at least in Hilbert (1922). The procedure suggested by Hilbert could thus be seen as reversing the post-Gödelian method of coding signs by numbers.

Secondly: While nowadays metamathematics is conceived of as being formulated and axiomatized in a formal language, Hilbert's finitist metamathematics *qua* contentual theory of formalized proofs (cf. 1926, 99 [385]) is neither formalized nor axiomatized. It is supposed to allow only intuitive and contentual reasoning, and, as Hilbert (1922, 164; 1934, 32) stresses, this kind of reasoning is, by its very nature, free from axiomatic assumptions. Due to his sharp distinction between formalized mathematics and contentual metamathematics, the meanings attached to the word 'to prove' in the two 'disciplines' are fundamentally different. In formalized mathematics, it means *to infer according to the formal rules of the calculus*; in contentual metamathematics, it means *to show by means of contentual, intuitively evident inference*.[3] It is precisely the intuition-based character of metamathematical reasoning that is supposed to guarantee its security and reliability.

[2] Cf. Hilbert's contentual construction of a fragment of number theory in his 1922 paper.

[3] In his paper 'Zur Hilbertschen Beweistheorie' (1927, 2), J. von Neumann characterizes the proofs of formalized arithmetic as an arbitrarily defined logical game. He adds that this game must be largely analogous to classical mathematics. By contrast, metamathematical 'proofs' are specified as concatenations of immediately evident contentual insights.

Thirdly: Although Hilbert's metamathematical vocabulary may embrace natural language analogues of logical expressions which correspond to truth-functional connectives, it does not include specifically mathematical expressions such as '—' and ':'. Hence, the language of metamathematics, as construed by Hilbert, is not a genuinely mathematical language. If we were to construct it in a precise manner, we would presumably arrive at a 'concatenation-theoretic' language, similar to the ones designed by Tarski (1935), Quine (1946) (cf. also Goodman and Quine (1947)), R. M. Martin (1958), or Montague (1961). In contrast to these approaches, in a metamathematical language à la Hilbert, it might happen that the operation of concatenation cannot be carried out unrestrictedly, since there may be only a finite number of figures. In a nutshell: Hilbert's finitist metamathematics, as a collection of sentences held to be true on intuitive grounds, is a non-formalized, non-axiomatized, and surveyable 'theory' with a non-mathematical vocabulary whose intended domain consists of concrete and surveyable figures, in particular, of formulas (e.g. axioms) and proofs.

In what follows, we state a number of theses, partly by appeal to the foregoing outline of the nature of Hilbert's metamathematics, and introduce some useful vocabulary.

(1) Hilbert's finitist metamathematics is not a mathematical or formal theory.

It will henceforth be referred to as M and its set of sentences as L_M. Note that Hilbert says in several places that metamathematics is a body of mathematics. However, for the sake of clarity we do not refer to M as a mathematical theory.

(2) Even if M were couched in a formal language and a mathematical theory 'equivalent' to M (say MM, formulated in L_{MM}) could be found, neither M nor MM need be recursively enumerable, i.e. axiomatizable.[4]

The non-axiomatized character of M derives from the fact that M is not stated in a formal language. It is important to distinguish this clearly from the question whether M is axiomatizable. The latter concerns only the complexity of the set of sentences of M or the set of Gödel numbers of sentences of M within the arithmetical hierarchy.[5] It asks whether we *could* give a recursive set of axioms for M. In any case, the crucial point of thesis (2) is that the set of Gödel numbers both of sentences belonging to MM and of sentences belonging to M need not be recursively enumerable.

[4] 'Equivalent' ought to be understood here in the sense of a paraphrase or an explication of an informal theory through a formal one.

[5] We assume familiarity with the arithmetical hierarchy; cf. Smorynski (1977) and Kaye (1991).

(3) Both M and MM contain only true sentences.
(4) MM may be considerably weaker than (QF-IA).[6]

(QF-IA) is obtained from PA by restricting the induction schema to formulas without unbounded quantifiers (cf. Sieg 1985).

To facilitate our line of argument concerning M, we construe M as a mathematical object or as a theory stated in a formal language.[7] It is only in such a context that we are able to deal in a precise manner with the relations in which M may stand to formalized theories. We attach particular importance to the fact that the formalized version of M enables us to consider the complexity of metamathematical expressions within the arithmetical hierarchy. It is possibly misleading, for instance, to use the term 'Π^0_1-sentence' with respect to M. Strictly speaking, we could only ask whether for sentences in L_M there are 'formal equivalents' in L_{MM} which are Π^0_1-sentences. Nonetheless, for the sake of simplicity we shall occasionally take the liberty to speak of Π^0_1-sentences regarding M. It goes without saying that Π^0_1-sentences play a key role in our reflections on Hilbert's proof-theoretic programme. Consistency assertions for recursively enumerable theories T are naturally formalized as Π^0_1-sentences. After having identified M with MM, the obvious way of establishing the consistency of T is to prove in MM 'the' consistency assertion for T, i.e. the Π^0_1-sentence 'Con_τ'.[8]

Finally, we argue for the following claim:

(5) Unbounded universal quantifications cannot be formulated in L_M, nor can genuine Π^0_1-sentences or Π^0_0-formulas be formulated in L_{MM}.[9]

2. FROM CONTENTUAL NUMBER THEORY TO FORMALIZED ARITHMETIC

In his essay 'Neubegründung der Mathematik' (1922), Hilbert claims, by appealing to Frege's and Dedekind's attempts at laying the logical foundations of number theory, that abstract operations with extensions of concepts

[6] Since we assume in (2) that MM need not be recursively enumerable, the use of the word 'weaker' is in need of explication here. In 'Hilbert's programme and Gödel's theorems', we show how little our habitual conception of the strength of a theory, stemming from our consideration of recursively enumerable theories, works when we compare these with theories which are not recursively enumerable.

[7] In contemporary discussions of Hilbert's programme, this way of construing M is usually accepted without further ado.

[8] The consistency assertion for a theory T is by no means to be considered unequivocal; cf. Rosser (1936) and Feferman (1960).

[9] By *genuine* Π^0_1-sentences (or *genuine* Π^0_0-formulas) we understand those Π^0_1-sentences (or Π^0_0-formulas) which are not equivalent to Π^0_0- or Σ^0_1-sentences over PA.

and contents turn out to be inadequate and insecure. Hilbert considers it a cardinal prerequisite for the use of logical inferences that certain extra-logical discrete objects be intuitively present prior to all thought. It is not quite clear what the phrase 'prior to all thought' is precisely supposed to mean here. It seems that Hilbert has in mind objects to which we have direct cognitive access through intuition, without having to rely on conceptual thinking. In his view, then, logical inferences are operations for which it is essential that they can be carried out with concrete, spatio-temporal objects.[10]

Hilbert (1922) introduces the following methodological point of view. In number theory, we proceed from signs *qua* concrete, intuitable, and irreducible objects. With these we operate, but we also make contentual statements about them. Let us consider the relevant details.

The sign '1' is declared to be a number, further a sign that begins and ends with '1', such that '1' (except the last '1') is always followed by '+' and '+' always by '1'.[11] According to this stipulation, the signs '1 + 1' and '1 + 1 + 1', for instance, are numbers and as such they form the subject-matter of the contentual considerations. Both in 'Über das Unendliche' and in *Die Grundlagen der Mathematik* (1934, §2), Hilbert chooses the process of adding the sign '1' for producing the extra-logical objects of number theory. The signs '1', '11', '111', etc. are now more appropriately called *figures*.[12] These are generated by way of building up longer sequences (or more complex configurations) out of shorter (or simpler) ones. This process in turn can be erased by progressive decomposition. Hilbert and Bernays

[10] Cf. Hilbert (1926, 89 [376]). It is perhaps illuminating to compare this with related remarks in some of Hilbert's later texts. In 'Über das Unendliche' (108 [392]), he expresses the conviction that the very possibility of scientific knowledge depends on certain intuitive ideas and insights, and he adds that logic alone does not suffice. In *Grundlagen der Mathematik* (1934, 2), Hilbert and Bernays point out that formal axiomatics requires, for the purpose of deduction and consistency proofs, certain evident truths. This kind of evidence does not rest on a special epistemic relation to a given subject-matter. Rather, it is one and the same for all axiomatics whatsoever, namely this primitive form of knowledge (Erkenntnisweise) which is the precondition for every exact theoretical investigation. Finally, in his paper 'Die Grundlegung der elementaren Zahlenlehre' (1931), Hilbert stresses that the most fundamental idea of Kant's epistemology still retains its importance: to establish that intuitive attitude a priori and thus to investigate the condition of the possibility of every knowledge. The apriori is said to be nothing more and nothing less than a primitive attitude, the so-called finitist attitude. And Hilbert goes on to characterize this attitude in much the same way as in 'Neubegründung der Mathematik'. His distinction between three sources of knowledge—experience, logic, and that primordial source of knowledge which governs the finitist attitude, that is, intuition—is reminiscent of Frege's distinction between sense perception, the logical source of knowledge, and the geometrical source of knowledge (that is, spatial intuition); cf. Frege (1969, 286 ff., 298 ff.). For a critical discussion of Hilbert's epistemology see Kitcher (1976); see also Parsons's and Detlefsen's articles in this volume.

[11] Hilbert does not use quotation marks when he mentions signs. For simplicity's sake, we shall occasionally subscribe to his practice.

[12] In what follows, we shall refer exclusively to the process of adding 1 when we talk about e.g. the generation or existence of (stroke-) figures or numerals.

(1934) suggest calling these figures *numerals*, adding that their use of this word diverges slightly from the usual one. The numerals '1', '11', '111', etc. are devoid of meaning, comparable to chess pieces on a chessboard. Their pedestrian role is that they are simply there, can be manipulated finitistically according to certain rules, and, as will become evident in a moment, be both referred to and talked about in contentual statements.

As to the precise figurative condition of the numerals, we allow for them, as usual, a certain scope, i.e. small differences concerning both the shape of 1 and its size and the distance when adding a 1 ought not to be taken into account. What we need essentially is only that in the 1 as well as in the addition we have an intuitable object that can be recognized in a unique way, and that we can always survey the discrete parts out of which a given numeral is built up. (Hilbert and Bernays (1934), 21)

It seems clear that we lack precise criteria to decide whether in a given case there are only small differences of the kind mentioned. Michael Resnik (1980) draws attention to further difficulties inherent in Hilbert's approach. How, for example, do we recognize by appeal to intuition that a numeral of length, say 1,000,003, is shorter than a numeral of length 1,000,004? Resnik (1980, 99 ff.) suggests that Hilbert is able to resolve these difficulties only by making a broader use of his criterion of surveyability.

In addition to the signs *qua* figures or numerals, Hilbert uses signs that serve to convey information, for instance, the sign '3' as a name for the numeral '111'. Note that the symbol '1', unlike '2', '3', etc., plays a twofold role in intuitive number theory; it occurs as a meaningless figure as well as a name for this figure.[13] Hilbert employs further the signs '+', '=', '>', and others, conveying assertions. For example, '3 > 2' conveys the fact that the sign 3 (that is '111') extends beyond the sign 2 (that is, '11'), or that the latter sign is a proper segment of the former (cf. Hilbert (1922), 163 f.; 1926, 90 [377]). The signs which serve to convey information are to be distinguished from the numerals or figures which form the subject-matter of the contentual considerations in intuitive number theory. A sign conveying information, taken by itself, is a figure for which Hilbert likewise assumes that it can be recognized in a unique way. Within the theory, however, such a sign is not the object of any contentual consideration, but serves only as a means to formulate facts, assertions, and assumptions (cf. Hilbert and Bernays (1934), 21).

In his 1922 as well as in his 1926 paper, Hilbert uses also Gothic letters $\mathfrak{a}$, $\mathfrak{b}$, $\mathfrak{c}$, etc. as names for numerals. '$\mathfrak{b} > \mathfrak{a}$' is then just as '3 > 2' a sentence of

[13] It is not quite correct when Hilbert says that the sign 3 is an *abbreviation* of the numeral 111; rather, it *denotes* that numeral. In his 1922 paper (166), Hilbert suggests that Gothic letters serve occasionally and provisionally as '*Kurzzeichen*'. The latter are designed to stand for or denote other signs. Hilbert uses the word 'bedeuten' in this context. In his 1926 paper (93 [379]), he likewise says that the letters $\mathfrak{a}$, $\mathfrak{b}$, . . . , stand for numerals.

contentual number theory; it communicates the fact that the numeral $\mathfrak{b}$ extends beyond the numeral $\mathfrak{a}$. Concerning the signs used in intuitive number theory we must, therefore, distinguish between two levels, as it were: between the 'object-level', to which belong the concrete, intuitable numerals as the objects of contentual considerations, and which have no task of presentation or communication, and the 'meta-level', to which belong the signs that serve to impart information, and which, unlike the figures belonging to the object-level, denote something. '2 = 11' or '3 > 11', for instance, would be illicit combinations of signs, if '11' is here understood not as a sign of communication referring to the number eleven, but as a numeral. A numeral is not to be combined with signs of communication such as 2, =, and >.[14]

According to Hilbert, contentual, intuitive number theory, as outlined so far, contains neither axioms nor rules of inference, and consequently no contradiction can arise (1922, 164). The proof of the commutative law of addition '$\mathfrak{a} + \mathfrak{b} = \mathfrak{b} + \mathfrak{a}$' is merely a procedure resting on the construction and decomposition of numerical figures and, by its very nature, distinct from the principle of complete induction. The latter is said to be a more far-reaching formal principle which in turn needs proof.[15]

Turning to related remarks in Hilbert and Bernays (1934, 23 f.), we learn that in contentual number theory the commutative law of addition, unlike the associative law, is obtained by complete induction. From the point of view of intuitive number theory, this mode of inference is to be construed as follows: We are given a statement S which states something about a numeral and has an intuitive content. It is assumed that S holds for 1 and further that we know that S, if it holds for a numeral $\mathfrak{n}$, also holds for the numeral $\mathfrak{n} + 1$. We conclude from this that S holds for each (concretely) given numeral $\mathfrak{a}$. Hilbert and Bernays make it clear that here we are concerned not with a principle in its own right, but rather with a conclusion resulting from the concrete construction of the numerals. Thus, despite first appearances, their account essentially fits together with Hilbert's observation in his 1922 essay to which we referred above. Whereas the induction

[14] In his vigorous attack on the alleged legitimacy of game formalism of his contemporaries Heine and Thomae, Frege (1903) had already insisted on the need to distinguish sharply between the language of the calculating game and the language of the theory of the game. Frege made it clear that it is illegitimate to assign a twofold role to one and the same numerical sign; it cannot function as a meaningless figure of the calculating game and as a meaningful sign of the theory of the game at the same time.

[15] Cf. 1922, 164. In 'Die Grundlagen der Mathematik' (1928*a*, 301 f. [472f.]), Hilbert rejects Poincaré's bold claim that the consistency of mathematical induction could never be proved except through induction itself. In particular, he takes Poincaré to task for not distinguishing between two methods of induction, namely between induction resting on the construction and decomposition of numerical figures and induction proper based on the induction axiom. We think that Hilbert's criticism is unjustified, however.

schema is a number-theoretic truth, Hilbert's construction and decomposition of numerals is carried out with concretely given objects. It is worth emphasizing that 'concatenation-theoretic induction schemata' (cf. Martin (1958); Montague (1961)) are still to be distinguished from Hilbert's 'proofs' by construction and decomposition of figures. Plainly, the procedure described above requires only $\mathfrak{a} - 1$ applications of *modus ponens* to establish for some concretely given numeral $\mathfrak{a}$ that $S(\mathfrak{a})$ and, hence, no induction at all. It is thus fair to say that Hilbert is right in claiming that he does not use metatheoretic induction as a principle in its own right. Yet, in displaying the composition and decomposition of figures as an analogue of metatheoretic induction, he obtains merely a trivial approximation to the ordinary induction principle. This is so, because Hilbert has no right to conclude from the premisses of the 'induction schema' to the universally quantified inductive assertion $\forall x S(x)$. His view seems to license only that we conclude to its metatheoretic counterpart.[16]

Hilbert is, of course, aware that the whole of number theory and analysis cannot be constructed by means of contentual, intuitive methods. These methods break down when we are to deal with statements about functions or 'infinitely many numbers'. For infinitely many numbers we simply cannot write down numerals. It follows that, from the finitist point of view, genuine universally quantified sentences, for example, cannot be analysed as infinite conjunctions, but must be introduced axiomatically.[17]

To prove the consistency of the axioms of second-order arithmetic, Hilbert suggests the following strategy. All statements of classical analysis are converted into formulas that can be concretely exhibited. In other words, the entire mathematical theory, including its axioms, is formalized; as such it replaces or mirrors the contentual reasoning of customary, informal mathematics (cf. Hilbert (1922), 165; 1923, 180; 1928*a*, 291 [465]). Mathematics proper *qua* formalized theory is joined by contentual metamathematics which serves the sole purpose of carrying out finitary consistency proofs (and soundness proofs) for formalized mathematics. The consistency assertion for a formalized arithmetical theory T is said to amount to the claim that there is no proof in T of '1 ≠ 1', i.e. that '1 ≠ 1' cannot be obtained as an end formula from the axioms of T (1926, 383 [97]). And Hilbert believes that to show that '1 ≠ 1' is not among the provable formulas of T is a task which can be performed within the scope of intuition (1926, 383 [97]; cf. Hilbert and Bernays (1934), 44).

[16] Unless he uses an ω-rule. Concerning Hilbert's introduction of a restricted ω-rule in 'Die Grundlegung der elementaren Zahlentheorie' (1931), see Niebergall and Schirn, 'Extensions of the finitist point of view' (forthcoming).

[17] Hilbert and Bernays (1934, 34 f.) stress that in the modes of inference applied in number theory we already encounter a transgression of the finitist standpoint when we allow existential statements on the 'integers' instead of on numerals and without taking into account the possibility of really determining the number in question.

3. OBJECT-LANGUAGE AND META-LANGUAGE

So much for our brief outline of Hilbert's transition from contentual number theory to formalized arithmetic and its attendant metatheory. We shall now deal with the relationships between the language of intuitive number theory (henceforth referred to as L_S), the language of formalized arithmetic (referred to as L_{FA}) and the meta-language of L_{FA}, i.e. the language of finitist metamathematics M which we earlier called L_M. Examining the relationships between L_S, L_{FA}, and L_M will help us to gain a clear idea mainly about (a) certain problematic features of L_{FA}, arising especially from Hilbert's distinction between real and ideal statements, (b) the nature and the scope of L_M, and (c) the context in which we encounter Hilbert's tacit assumptions of infinity. Hilbert's conception of real and ideal statements, the specific limitation of the expressive power of L_M as well as his implicit assumptions of infinity will be subjected to further scrutiny in subsequent sections.

We take it that the finitist construction of contentual number theory in 'Neubegründung der Mathematik' serves Hilbert, though not exclusively, as an exemplary preparation for the ensuing characterization of typically metamathematical considerations. However, despite the kinship between L_S and L_M, which we shall spell out in a moment, it would certainly be a mistake to consider intuitive, finitary number theory itself to be a part of metamathematics proper. The 'proofs' that can be carried out in finitary number theory are neither formalized nor do they (expressly) form the subject-matter of contentual considerations. While intuitive number theory, as outlined by Hilbert, still embraces 'object-language' and 'meta-language', his distinction between formalized mathematics and attendant contentual metamathematics goes hand in hand with a twofold, albeit implicit stipulation: the object-language is the language of formalized mathematics, and the meta-language is the language of metamathematics.

Whereas L_S is an informal language consisting only of contentually interpreted sentences, L_{FA} is a formalized mathematical language, embracing '=', '+', '>', '<', etc., truth-functional connectives, quantifiers (or the ε-operator), and variables.[18] L_{FA} presumably does not contain stroke-figures or strings of stroke-figures.[19] Although Hilbert does not construct L_{FA} according to strict syntactic rules, we may gain a reasonably clear idea about its well-formed formulas.

[18] Due to the fact that '$\mathfrak{b} > \mathfrak{a}$' and '3 > 2', for example, state something about objects *qua* 'signs', L_s is a language of concatenation theory rather than one of contentual mathematics. It is only the identification of numbers with stroke-symbols that eventually turns L_s into a mathematical language. Notice that the use of '+' or '>' in L_s is misleading in so far as it suggests an arithmetical context. In L_s, '+' or '>' do not, however, stand for number-theoretic addition or the greater-than relation; rather, these signs stand for the composition of stroke-figures or, respectively, for the relation *extends beyond*.

[19] In this connection see, however, Hilbert (1922, 165).

In view of the fact that L_S and L_{FA} share a number of symbols, it is natural to ask whether there are sentences belonging to both languages, and in case there are, whether the roles they perform are the same or different. '$\forall xy(x > y)$', for example, is a sentence belonging to L_{FA}, and we can conclude from it to '3 > 2', for instance, through instantiation. On the grounds that we can infer a sentence B from a sentence A only within one and the same language, we should assume that '3 > 2' belongs to L_{FA} as well. And we already know that '3 > 2' is a sentence of L_S (cf. 1926, 93 [380]). Hence, L_{FA} would contain contentually interpreted sentences. This conclusion conflicts with the view, suggested by certain remarks of Hilbert, that L_{FA}, conceived of as a full-fledged formalized language, is devoid of meaning; the terms of L_{FA} seem to lack a denotation and its sentences seem to have no meaning and no truth-value at all. Thus, in 'Neubegründung der Mathematik' (165), Hilbert underscores the fact that once the distinction between formalized mathematics and informal metamathematics has been drawn, we can strictly and systematically separate the formulas and formal proofs from the contentual reasoning. The latter belongs, as it were, to a higher level than the manipulation of the formal apparatus of formalized arithmetic; metamathematics is its proper place (cf. Hilbert (1922), 165). In 'Über das Unendliche' (94 f. [381]), Hilbert points out that it will be consistent

> to divest the logical signs, too, of all meaning, just as we did the mathematical ones, and declare that the formulas of the logical calculus do not mean anything in themselves either, but are ideal statements. . . . In a way that exactly corresponds to the transition from contentual number theory to formal algebra we regard the signs and operation symbols of the logical calculus as detached from their contentual meaning. In this way, we now finally obtain, in place of the contentual mathematical science that is communicated by means of ordinary language, an inventory of formulas that are formed from mathematical and logical signs and follow each other according to definite rules.[20]

Now, if both the mathematical and the logical signs of L_{FA} are denied all meaning, it is hard to see how there could be any meaningful sentence compounded out of them, unless Hilbert were to provide an interpretation for certain L_{FA}-formulas at a later stage.[21] In the light of these considera-

[20] In 'Die Grundlagen der Mathematik' (1928*a*), Hilbert points out that contentual number theory, when it is embedded in algebra, appears to be formalized.

[21] Hilbert (1923, 180; cf. 1928*a*, 291 [465]) observes that the axioms and provable sentences, which originate in the interplay between the derivation of new provable formulas from the arithmetical axioms by means of *formal* inference on the one hand and the addition of new arithmetical axioms as well as the consistency proof by means of *contentual* inference on the other, are the copies of the thoughts constituting the ordinary procedure of classical mathematics. In doing so, he presumably wants to convey that under a given interpretation the formulas of L_{FA} are converted into sentences of classical mathematics. It is clear, however, that formalized mathematics is entirely independent of such an interpretation. Cf. Heyting (1934, 38).

tions, it seems that '3 > 2' *qua* sentence of L_{FA} means nothing, and this is at odds with its status as a contentual sentence that serves to convey information.

It is only fair, though, to observe that Hilbert does not explictly say that all well-formed formulas of L_{FA} are devoid of meaning, and he thus escapes what could perhaps be a head-on conflict when he characterizes the finitary sentences of formalized arithmetic expressly as meaningful (cf. 1926, 93 [380]; 1928*a*, 297 f. [470]). At any rate, it is only the non-finitary sentences of L_{FA} that are free from ambiguity: they are definitely meaningless. To avoid confusion, we take it from now on that concerning its well-formed formulas L_{FA} divides into a meaningless and a meaningful part. It remains to inquire into the relationship between non-finitary sentences on the one hand and finitary sentences on the other, as well as into Hilbert's possible motive for admitting meaningful sentences in L_{FA} at all. His use of the symbols '=', '+', '<' and '>' in L_{FA} seems to give rise to a special problem and will likewise be discussed in due course.

In algebra we have, according to Hilbert, an increase in the number of finitary objects. A sentence, when it is combined 'with the contentual indications', is still regarded as admissible from the finitist point of view, such as (1) '$\mathfrak{a} + \mathfrak{b} = \mathfrak{b} + \mathfrak{a}$', for example, where $\mathfrak{a}$ and $\mathfrak{b}$ are arbitrarily given numerals. Nevertheless, Hilbert does not 'select this form of communication'; rather, he takes the formula (2) '$a + b = b + a$', which is no longer 'an immediate communication of something contentual at all, but a certain formal object' (1926, 93 [380]). Unlike (1), the non-finitary formula '$a + b = b + a$' does not mean anything, any more than stroke-figures do. However, by substituting names of numerals for 'a' and 'b' in (2), we can derive finitary, that is, meaningful, sentences such as '2 + 3 = 3 + 2', for instance. But why should it matter for Hilbert that in formalized arithmetic meaningful sentences be derivable at all? Why should he admit that contentual statements occur not only in the meta-language L_M of L_{FA}, but also in L_{FA} itself?[22] Surely, the derivation of the 'old finitary statement' '2 + 3 = 3 + 2' from '$a + b = b + a$' by way of substitution is a purely syntactic operation and, hence, independent from questions of meaning. If in L_M the meanings of the finitary sentences of L_{FA} were to play a crucial role, L_M would emerge as a semantic meta-language. The right way to look at L_M is, however, to regard it as a purely syntactic meta-language.

In spite of this assessment, one might wish to argue that for Hilbert it is still of interest to be able to rely on contentually interpreted sentences in L_{FA}. Consider the following example. Let S and T be two theories formulated in L_{FA} and suppose that their consistency has been established by

[22] It is plain that L_{FA}, in contrast to L_M, does not contain statements about L_{FA}-formulas and L_{FA}-proofs.

metamathematical means. Suppose further both that $S \vdash a + b > b + a$ and $T \vdash a + b = b + a$, and that some of the customary properties of '>' and '=' are provable in S and T. For obvious reasons, one would prefer T to S. By means of substitution, we obtain $S \vdash 2 + 3 > 3 + 2$ and $T \vdash 2 + 3 = 3 + 2$. Now, if the formulas '2 + 3 > 3 + 2' and '2 + 3 = 3 + 2' are conceived of as purely formal objects without any consideration of meaning, an argument that would favour T and rule out S is not in sight; on assumption, both theories are consistent. If '2 + 3 > 3 + 2' and '2 + 3 = 3 + 2' are interpreted (in the ordinary way), however, one can argue that T is preferred to S, on the grounds that it is more fruitful or more useful not only for arithmetic taken by itself, but also for science as a whole. Whether Hilbert has something like this in mind when characterizing the finitary sentences of L_{FA} as meaningful is open to question. Still, certain remarks made by him both on the success and fruitfulness of theories (1926, 98 [384]) suggest that he may have endorsed the 'holistic' idea just developed. The following passage from 'Über das Unendliche' is revealing in this context: '[I]f justifying a procedure means anything more than proving its consistency, it can only mean determining whether the procedure is successful in fulfilling its purpose. Indeed, success is necessary; here, too, it is the highest tribunal, to which everyone submits' (1926, 81 [370]). Consistency is already a good thing in itself. Yet what really counts for a theory is, in the end, consistency plus success or fruitfulness. And the latter can hardly be said to be independent from science as a whole.

As to the symbols '=', '+', '<', and '>' in L_{FA}, they have a Janus-faced nature. When these occur in infinitistic, that is, ideal, sentences like '$\forall xy(x > y)$' or '$a + b = b + a$', they are meaningless formal objects. As constituents of 'statements of communication' like '3 > 2' or '2 + 3 = 3 + 2', however, they are meaningful signs of communication. If that were not so, the statements of communication themselves would lack a meaning, at least on acceptance of the principle (call it K) that the meaning of a complex expression, in particular of a sentence, is a function of the meanings of its parts (and the way these are combined to form the complex expression). Suppose that we are given the derivation of '2 + 3 = 3 + 2' from '$a + b = b + a$' by means of substitution. Unlike their occurrence in the former formula, '+' and '=' mean nothing in the latter. This does not matter, however, since the derivation in question is, as we have pointed out, a purely syntactic operation. It is perfectly reasonable to assume that in the two formulas the signs '+' and '=' are used in a uniform manner. If Hilbert wished to account for the twofold semantic aspect of, e.g. '+', by using '+' as a meaningless symbol and by choosing, e.g. '⊕' for the contentually interpreted sign for addition, he would be unable to derive, by means of substitution, finitary or real formulas from non-finitary or ideal formulas.

We face a special difficulty when we consider a sentence like (i) '$a + 2 =$

2 + *a*', for example. Let us assume that it is a well-formed formula of L_{FA}. How are we to interpret (i) and its components? Unlike the signs of communication, the variables '*a*', '*b*', etc. of L_{FA} are definitely meaningless; and they cannot obtain a meaning as constituents of sentences even when they are combined with signs of communication, such as in (i), that normally mean something. On the strength of principle K, we must conclude that (i) means nothing. Of course, if we subscribed to a version of Frege's context principle saying that the meaning (or sense) of a subsentential expression consists exclusively in its contribution to determining the meaning (or sense) of any sentence in which it may occur, '+' and '2', as components of (i), could hardly be said to have a meaning at all. In short, if (i) is well formed in L_{FA}, it is not to be conceived of as a formula with a semi-formal and a semi-contentual face. This holds especially when we endorse the above-mentioned version of Frege's context principle. It is, rather, a meaningless formal object which, we think, Hilbert would have classified as an ideal statement.[23]

Let us say a little more about the meta-language L_M and its relation to L_S. The set of well-formed terms of L_M should at least comprise names of expressions of L_{FA}, e.g. names of the L_{FA}-universal quantifier, of L_{FA}-function signs, of the sign '>' of L_{FA}, of L_{FA}-terms like '2', '3' and '$\mathfrak{a}$', '$\mathfrak{b}$', and of variables of L_{FA}. By adopting, somewhat anachronistically, a device from Tarski (1935, 287 [172]), Hilbert could have introduced specific terms of the meta-language of a structural-descriptive character, that is, special names of concrete L_{FA}-expressions, of strings of such expressions and of structural relations existing between them. Clearly, L_M and L_S have a couple of features in common. Both languages contain only finitary sentences, the so-called statements of communication. And in both languages we talk about concrete, spatio-temporal objects or about figures intuitively represented by the mind's eye. As regards L_S, these objects are stroke-figures, while concerning L_M they are formulas, axioms, and proofs. We take it that Hilbert was aware that names of stroke-figures need not form a part of L_M and that he would not have included them, had he set about constructing L_M in a systematic way. Thus, despite their affinity, L_S and L_M are different languages.[24]

[23] Cf. Hilbert (1928*a*). Take the truth-functional compound (ii) '$\forall xy(x + y = y + x) \rightarrow 2 + 3 = 3 + 2$'. It consists of an infinitistic and a finitary component and is unquestionably a well-formed formula of L_{FA}. If we accept principle K, (ii) is just as meaningless as the antecedent. Now, in the light of Hilbert's division of all well-formed formulas of L_{FA} into real and ideal statements, (ii), too, ought then to be characterized as an ideal statement.

[24] It seems that an arbitrary sentence of L_s (or an arbitrary finitary sentence of L_{FA}) has no complete semantic analogue in L_M, even we grant that a sentence like '$\mathfrak{a} > \mathfrak{b}$', for example, actually occurs in L_M. As symbols of L_M, '$\mathfrak{a}$' and '$\mathfrak{b}$' could only be terms denoting expressions of L_{FA}. In Hilbert (1922), capital Gothic letters are introduced as signs of communication standing for formulas, and small Gothic letters as signs of communication referring to functionals. It is plausible to assume that, e.g. '$\mathfrak{A} > \mathfrak{C}$', is a well-formed sentence of L_M as imagined by

4. ASSUMPTIONS OF INFINITY

Let U be a sentence or a set of sentences. By proving U or in recognizing the truth of U, we make an assumption of infinity if and only if

(a) U says under a contentual interpretation that there is an infinite object; or (b) U, when interpreted contentually, is satisfiable only in an infinite domain.

For (a), we may take '$\exists x(x = \omega)$' as an example for U (say in ZF); for (b), we may take the theory PA; and for both (a) and (b), we may take the theory ZF.

In order to make (a) and (b) more precise, we ought to explain the expression 'infinite'. It seems to us that there are three possible ways of doing this.

(i) 'Infinite' is defined in a generally accepted and precise context, for instance, in axiomatic set theory.

(ii) An axiomatization for 'infinite' is provided. Although we do not know of any such axiomatization, generalized quantifiers might be considered an example.

(iii) It is presupposed that the expression 'infinite' is already understood, and a formal explication is regarded as unnecessary.

(i) and (ii) fail to provide us with any information about the meaning of the defined or axiomatized expression 'infinite', apart from what is captured in the definitions or axioms. If we want to establish that the definition or axiomatization is adequate, that is, that the defined or axiomatized expression 'infinite' may be accepted as an explication of 'infinite' as a constituent of ordinary language, we must already understand more or less what 'infinite' means. In other words, every attempt to fix the meaning of 'infinite' in a precise manner in terms of (i) or (ii) is ultimately justified by appeal to (iii). The question of how we come to understand 'infinite' as an ordinary language expression will be set aside here.

As far as Hilbert's relevant texts on proof theory are concerned, he does not offer any explicit explication of the expression 'infinite'. This applies especially to the axiom 'If a predicate A applies to an object τA, then it applies to all objects *a*', which he marks out as the 'primary source of all transfinite concepts, principles, and axioms' (1923, 183).[25] This distinguished

Hilbert, where the sign '>' is used with the same meaning as in L_S. On this assumption, the symbols '>' and '<' belong to the vocabulary of L_M, where they mainly serve to convey assertions about L_{FA}-formulas and L_{FA}-proofs.

[25] This axiom embodies the core of Zermelo's axiom of choice and is designed to provide a justification for the transfinite modes of inference in mathematics. Cf. also Hilbert (1926, 96 [382]; 1928*a*, 292 [466]; 1928*b*, 316 f.). Soon after 1923, Hilbert reformulated the axiom by

transfinite axiom as well as the other 'ordinary' transfinite axioms derivable from it (Hilbert (1926), 96 [382]) do not deserve their name; they fail to give us any formal explication of the term 'infinite'. Unlike the sentences of metamathematics and those belonging to the finitary part of formalized mathematics, Hilbert's so-called transfinite axioms are uninterpreted; and they would be satisfiable in finite domains even if we were to provide an interpretation for them. So, it seems that Hilbert belongs to those advocating (iii).[26]

Hilbert faces a problem here. On the one hand, the finitist point of view owes both its attraction and strength to the exclusive use of finitary (intuitively evident and absolutely reliable) methods.[27] On the other hand, the finitist (meta-) mathematician seems to be at a loss to say what 'finite' or 'infinite' is to mean. Now, the fact that a theory T is constructed in such a way that its correctness is probable, or assumed, or guaranteed, although the correctness cannot be proved in T itself nor even stated in the language of T, is nothing new and scarcely considered a stumbling-block by mathematicians, even when they focus on important mathematical theories such as ZF, for instance. One essential reason for accepting ZF as a foundation of mathematics is its assumed soundness, that is, its truth in the set-theoretic universe V. Yet in the language of ZF we cannot even express the truth of ZF in V. Seen from this angle, Hilbert's situation is comparable to that of the present-day mathematician. The conception which makes Hilbert's metamathematical approach attractive and important, namely that of finiteness, can be formulated within the framework of his finitism just as little as the conception favouring ZF, namely that of set-theoretic truth, can be formulated in the language of ZF.

The fact that Hilbert omits to subject the notion of infinity to a mathematical treatment does not, of course, imply that he dispenses with assumptions of infinity. In the first half of 'Über das Unendliche', he concludes, by appeal to his previously established results, that reality is finite. In what follows, he examines the question whether the infinite is at least admissible

means of the transfinite choice function $\varepsilon_x A(x)$. '$A(a) \rightarrow A(\varepsilon_x A(x))$' is referred to as the logical ε-axiom. From a contentual point of view, the ε-symbol can be interpreted as follows: '$\varepsilon_x A(x)$' denotes an object a of which the statement '$A(a)$' certainly holds, if it holds of any object at all. As is well known, Hilbert defines both the existential and the universal quantifier by means of the ε-operator; cf. Hilbert (1928*a*, 1928*b*).

[26] In his 1926 essay, Hilbert mentions initial segments of Cantor's second number class, but presumably does so only for the sake of illustrating (the nature and the scope of) assumptions of infinity made in the mathematics of his day. We suppose that *qua* finitist metamathematician, Hilbert refrains from claiming understanding of set-theoretic definitions of infinity, because these are definitely non-finitary.

[27] Bernays (1935, 210) questions the alleged security and reliability of finitist metamathematical reasoning. The chance of an oversight or a mistake in a long finitist argument is said to be especially great. Kreisel (1958, 211) believes, however, that a satisfactory characterization of 'finitist proof' would be based on Hilbert's notion of reliability.

in our rational thinking. The upshot is this: '[N]owhere is the infinite realized; it is neither present in nature nor admissible as a foundation in our rational thinking. . . . The right to operate with the infinite can be secured only by means of the finite' (1926, 108 [392]).

However, if infinitely many or infinite objects have no 'well-justified place in our thinking', not even in our mathematical reasoning, then we are forced to ask how matters stand with those assumptions of infinity that are made in analysis or, even more extensively, in set theory. Hilbert might have responded as follows: If we renounce interpreting these mathematical theories, we know that their acceptance does not involve any existential assumptions at all, let alone assumptions of infinity. Clearly, Hilbert does not modify ZF *qua formal* mathematical theory; he can still prove '$\exists x(x = \omega)$' in ZF. However, construed as a sentence of formalized mathematics, '$\exists x(x = \omega)$' cannot be used to assert anything at all, because as such it is devoid of meaning. By contrast, when we assert '$\exists x(x = \omega)$' in classical, informal mathematics, we are perhaps disposed to acknowledge the existence of an infinite object.

Granted that formalized mathematics does not comprise assumptions of infinity, it remains to find out whether the advocate of finitist *metamathematics* makes such assumptions. Admittedly, the idea that he relies on assumptions of infinity seems prima facie preposterous. What he does in his proof-theoretic practice is no more and no less than this: he encounters or generates finitely many finite and surveyable figures, operates with them and makes statements about them.[28] A glance at Hilbert's characterization of numerals (cf. e.g. 1926, 89 [377]; Hilbert and Bernays (1934), 20 f.) suggests, though, that he stipulates precisely this: '1' is a numeral (i.e. a stroke-figure) and, given that $\mathfrak{b}$ is a numeral, the 'concatenation' of $\mathfrak{b}$ with '1' is a numeral. According to the customary, naïve conception of the infinite, Hilbert's stipulation conveys that there are infinitely many numerals, just as the existence of a successor for every natural number n distinct from all its predecessors guarantees the existence of infinitely many natural numbers. Furthermore, since an adequate explication or reconstruction of the notion of natural number rests on the assumption that there are infinitely many natural numbers, Hilbert would be entitled to identify the natural numbers with numerals only if he had good reasons to assume the existence of infinitely many numerals. In what follows, we want to scrutinize further assumptions of infinity which Hilbert makes about the domain of expressions. We consider three examples: quantification, formalized proofs, and what, following Hilbert's jargon (1926, 91 [379]), may be called 'capability of being negated'.

[28] Remember that Hilbert and Bernays (1934) speak also of thought-experiments which the finitist proof theorist effects on intuitively given objects.

First example: Quantification

In 'Über das Unendliche' (Hilbert (1926), 90 f. [377 f.]), we find a passage which is worth quoting in full:

> The largest prime number known up to now (39 digits) is $\mathfrak{p}$ = 170 141 183 460 469 231 731 687 303 715 884 105 727. By means of Euclid's well-known procedure we can, completely within the framework of the [finitist] attitude . . . , prove the theorem that between $\mathfrak{p}$ +1 and $\mathfrak{p}! + 1$ there certainly exists a new prime number. This statement itself, moreover, accords completely with our finitist attitude. For '*there exists*' here serves merely to abbreviate the statement:
>
> Certainly $\mathfrak{p} + 1$ or $\mathfrak{p} + 2$ or $\mathfrak{p} + 3$ or . . . or $\mathfrak{p}! + 1$ is a prime number.
>
> But let us go on. Obviously, to say
>
> There exists a prime number that (1) is $> \mathfrak{p}$ and (2) is at the same time $\leq \mathfrak{p}! + 1$
>
> would amount to the same thing, and this leads us to formulate a statement which expresses only a part of Euclid's assertion, namely: there exists a prime number which is $> \mathfrak{p}$. So far as content is concerned, this is a much weaker assertion, stating only a part of Euclid's statement; nevertheless, no matter how harmless the transition appears to be, there is a leap into the transfinite when this partial statement, taken out of the context above, is enunciated as an independent assertion.

Note that '$\mathfrak{p}$' is a constant. The statement 'There exists a prime number between $\mathfrak{p} + 1$ and $\mathfrak{p}! + 1$' is an existentially quantified statement claiming the existence of finitely many objects, and thus, according to Hilbert, an abbreviation of a finite disjunction. So construed, it is in accordance with the finitist attitude. By contrast, to regard an existential quantification over an infinite domain as an abbreviation of an infinite disjunction would be illicit for Hilbert. For clearly, an infinite disjunction would be an infinite object, and the assumption that such an object exists, is, from the finitist standpoint, completely unjustified. As a matter of fact, Hilbert takes this caveat to heart:

> The statement '$\mathfrak{p} + 1$ or $\mathfrak{p} + 2$ or $\mathfrak{p} + 3$ or . . . *ad infinitum*' is, as it were, an infinite logical product [sum], and such a transition to the infinite is no more permitted without special investigation and perhaps precautionary measures than the transition from a finite to an infinite product [sum] in analysis, and initially has no meaning at all. (1926, 91 [378])

It is, of course, not mandatory to regard an existentially quantified sentence as an abbreviation of a (possibly infinite) disjunction. Present-day logicians usually do not construe it in this way.[29] In 'Die logischen Grundlagen der Mathematik' (1923, 181), Hilbert speaks only of the logical equivalence of finite conjunctions and universal quantifications over finite domains. Now, whatever interpretation of quantification Hilbert may have

[29] If in a language of predicate logic infinite disjunctions can be formulated at all, they are not necessarily equivalent to existentially quantified statements.

preferred at the end of the day—be it the 'abbreviation' conception, the 'equivalence' conception or some other conception—he seems to consider quantification to be something that we do not properly understand, unless it is definable in truth-functional terms. While a modern logician, when confronted with such a situation, would quite naturally appeal to the method of axiomatizing quantification, to Hilbert, by contrast, this device would perhaps have appeared as a pioneering step. Still, due to the fact that Hilbert leaves the non-finitary part of formal mathematics and, hence, unbounded quantifications uninterpreted, axiomatization of quantification does not bring him closer to an understanding of quantification.

Returning to the example concerning prime numbers, we face the question: How can Hilbert claim to know that 'There exists a prime number which is $> \mathfrak{p}$' is a shorthand expression for or is equivalent to '$\mathfrak{p} + 1$ is a prime number or $\mathfrak{p} + 2$ is a prime number or $\mathfrak{p} + 3$ is a prime number or . . . *ad infinitum*'? It is conceivable that the natural numbers end with $\mathfrak{p}! + 20$, i.e. that there is no natural number greater than $\mathfrak{p}! + 20$, or alternatively, that there is no numeral extending the numeral $\mathfrak{p}! + 20$. If that were the case, 'There exists a prime number which is $> \mathfrak{p}$' could be construed as a mere abbreviation of '$\mathfrak{p} + 1$ is a prime number or $\mathfrak{p} + 2$ is a prime number or $\mathfrak{p} + 3$ is a prime number or . . . $\mathfrak{p}! + 20$ is a prime number' and would consequently pose no obstacle at all to the finitist. It is only on the grounds that Hilbert (a) presupposes tacitly the existence of infinitely many numbers or numerals, and (b) confines himself, at first glance, to considering a truth-functional analysis of quantification, that he gets involved in problems from his finitist point of view. If we appeal to number-theoretic statements, (a) and (b) imply that he would have to acknowledge the existence of infinite conjunctions or infinite disjunctions. Since these are finitistically inadmissible, Hilbert eventually abandons (b). He might instead have relinquished (a) and should, indeed, have done so to keep finitism intact.

Second example: Formalized proofs

According to Hilbert, the characteristic marks of proofs in a formalized mathematical theory are concreteness, intuitability, surveyability, and complete communicability.[30] Intuitively speaking, concretely given proofs are finite objects. When Hilbert sets about establishing the consistency of, e.g. PA, by finitist (proof-theoretic) means, he has to show for every proof figure $\mathfrak{a}$ in PA that '$1 \neq 1$' cannot be the end formula of $\mathfrak{a}$. Suppose now that the domain of concretely given proofs embraces only finitely many elements; for the sake of illustration, let us say, 100 PA-proofs. It will then be a straightforward matter to show metamathematically the consistency of PA.

[30] In *Grundlagen der Mathematik* (1934, 44), Hilbert characterizes a proof in formalized arithmetic as 'a succession of processes each of which belongs to a supply of relevant actions specified from the beginning'.

Just inspect each of the PA-proofs. If you convince yourself that none of these ends with '$1 \neq 1$', the consistency of PA is finitistically guaranteed. Ideally, in the hypothetical situation we are considering, we ought to encounter only proofs ending with some formula distinct from '$1 \neq 1$'. We may even grant that some proof ends with '$3 \neq 3$', provided that none of the 100 PA-proofs ends with '$1 \neq 1$'. If no single PA-proof proceeds from the proofline, say, '$3 \neq 3$', to a proofline '$1 \neq 1$', the consistency of PA is warranted. In this way, we could give finitist metamathematical consistency proofs also for theories of type $T + \{\psi, \neg\psi\}$, as long as we make sure that each proof generated in such a theory is not a proof of '$1 \neq 1$'. We could, for instance, continuously prove '$1 = 1$', with distinct proof lengths and redundant prooflines.

Needless to say, this caricature of a consistency proof was not intended by Hilbert. What counts is rather the *(un-)provability* of '$1 \neq 1$' in a theory T. It is only in this context that we are able to understand why in his sketchy remarks on the methodology of consistency proofs Hilbert suggests modifying concretely given proofs in a certain way: they are turned into proofs displaying overtly that they do not contain '$1 \neq 1$' as a proofline. We urge the reader to cast a glance at the consistency proof for a weak subtheory of PA in Hilbert and Bernays (1934). In particular, the authors describe operations on proofs that can be carried out only on condition that in the initial proof not all objects of the universe have already been exhausted (cf. 224 f.). Of course, it is conceivable that all objects of the universe have in fact been exhausted. If they have, and if Hilbert intends to speak only of actually given proofs, then he is debarred right from the start from producing proof variants which he needs for further proof-theoretic considerations. His treatment of the idea that the operations generating the proof variants can be carried out suggests, then, clearly that he is concerned with *provability*. But what are the syllables 'bility' in 'provability' to mean? If they are to mean that a proof possibly exists, then Hilbert would merely have sought refuge in the use of an unexplained 'modalism'. At any rate, the standard method for the elimination of possible objects consists, especially in mathematics, in the introduction of objects of a higher type (e.g. functions) which in their turn are infinite.

Third example: Capability of being negated

Intuitively appealing as the modal treatment of the existence of proofs—there *could* be infinitely many of them!—may appear, there is little textual evidence that Hilbert subscribed to it. As far as we know, he almost never speaks of figures, say numerals or proofs, which *could* be given, but only of figures that *are* (concretely) given (cf. Hilbert (1928*b*), 320). But let us now turn to our third example which is taken from 'Über das Unendliche' (1926,

91 [378]). The relevant passage is meant to explain the mode in which we encounter the transfinite when we deal with universally quantified statements. The statement

(+) If $\mathfrak{a}$ is a numeral, we must always have $\mathfrak{a} + 1 = 1 + \mathfrak{a}$

is said to be incapable of being negated from the finitist viewpoint. Hilbert hastens to add that this is due to the fact that, from this viewpoint, (+) can be construed not as a conjunction of infinitely many numerical equations, 'but only as a hypothetical judgement that comes to assert something when a numeral is given' (91 [378]). If there are only finitely many numerals, (+) can be transformed, without alteration of meaning, into a finite conjunction containing no quantifiers. This would be a clear case in which (+) is capable of being negated. We conclude, therefore, that the claim that (+) is incapable of being negated can be sustained only if infinitely many numerals do exist.

In our discussion of three paradigm cases involving assumptions of infinity, we left room for the possibility that the universal quantifier, governing implicitly (+) or the talk about arbitrary proofs, ranges over possible objects (*possibilia*). If that were indeed Hilbert's response to the conflict between his advocacy of finitism and his tacit appeal to assumptions of infinity, he would not have been moving one single step towards a solution. Plainly, to presuppose that there is an infinite domain of possible objects *qua* domain of interpretation does not avoid such assumptions. Rather, it entangles Hilbert in further difficulties.

From an exegetical point of view, we should like to emphasize that Hilbert's relevant texts on his proof-theoretic programme do not provide an unequivocal clue as to whether he makes or intends to make assumptions of infinity and, if he does, what their precise nature is. It seems that in this respect he shies away from putting his cards on the table. Moreover, the question whether Hilbert *qua* metamathematician actually makes assumptions of infinity could be given a definite answer only if he were to state explicitly both the vocabulary and the formation rules of the formal object-language he is considering.[31] Nonetheless, in the light of the available 'evidence', we believe that he proceeds from the assumption that there are infinitely many numerals (and infinitely many other expressions) of the formal object-language, and this belief has, of course, shaped our analysis. The fact that in vindicating finitist metamathematics Hilbert failed to dispense once and for all with assumptions of infinity regarding

[31] As a matter of fact, Hilbert does so, to some extent, in his 1922 paper (cf. 165 ff.). In particular, he explains 'A $\rightarrow$ B' as a formula, if A and B are formulas. Thus, we encounter here the standard metalogical assumption of the existence of infinitely many well-formed formulas. The characterization of the formal language of arithmetic in Hilbert (1922) is, however, too sketchy to serve as evidence for answering definitely the question above.

not only the domain of objects talked about, but also the domain of expressions of a given formal language, we regard as the vulnerable spot of his approach.[32]

5. REAL AND IDEAL STATEMENTS

In the last two sections of this paper, we shall discuss two intimately related topics. The first concerns Hilbert's introduction of ideal statements as opposed to real statements in 'Über das Unendliche' and 'Die Grundlagen der Mathematik' which we have already touched upon in Section 3. The second concerns the search for viable alternatives to the standard formalization of consistency assertions.

Hilbert's distinction between real and ideal statements lies at the very heart of his metamathematics.[33] He calls the finitary, contentually interpreted sentences *real statements* and distinguishes between elementary real statements and those of (allegedly) problematic character. The latter are bounded quantifications.[34] Hilbert stresses that in formalized arithmetic ideal statements are adjoined to the real statements 'to ensure that the customary laws of logic again hold one and all' (1926, 94 [380 f.]). However,

[32] Unless a satisfactory explication of the potential infinite is forthcoming we are sceptical as to the use of this notion. This is, of course, not to rule out that there may be cases of infinity where the distinction between *potential* and *actual* appears appropriate. Hilbert seems to align himself with Cantor not only as regards the nature of the potential infinite, but also concerning the nature of the actual infinite (cf. Hilbert (1926), 85 [373]). The actual infinite is said to be given when we consider the totality of the natural numbers as a completed entity (that is, as the set of the natural numbers). This conception deviates from our own view. Thus, we maintained that the assumption of the unique existence of a successor for every natural number goes hand in hand with an assumption of infinity. To make such an assumption does not require, in addition, that the existence of the set of the natural numbers be postulated as an object *sui generis*. In particular, we spoke only of assumptions of infinity *simpliciter*, because we do think that the characterization of the potential infinite as something that is becoming or coming to be is hard to understand (cf. Hilbert (1926), 85 [373]). In the present context, it is perhaps worth recalling that according to Cantor, the potential or syncategorematic infinite is an improper infinite (an *indefinitum*); it has only a reality borrowed from the actual or categorematic or proper infinite (cf. Cantor (1932), 165, 373, 404, 411). As far as the actual infinite is concerned (cf. 401, 405), Cantor distinguishes between the increasable actual infinite or *transfinitum* and the non-increasable actual infinite or *absolutum* (405).

[33] Hilbert's talk of formulas to which contentual communications of finitary statements correspond we find unclear and possibly misleading. In particular, we fail to see in what sense there should be a difference between the old finitary statements familiar from L_s and now forming a part of L_{FA} and communications of finitary statements.

[34] Hilbert (1928*a*, 297 [470]) emphasizes that in his proof theory only the real statements are directly capable of verification. Due to the fact that ideal statements are considered to be devoid of any meaning, it is hard to see how they could be capable of verification at all. At any rate, they could not be verified by contentual considerations. Another point worth noting is that ideal statements could and, indeed, should be regarded as ontologically harmless by the finitist. Since Hilbert leaves them uninterpreted, they do not embody any finitistically illicit assumptions of infinity whatsoever.

the extension of the original, narrow domain of finitary statements through ideal statements is considered to be legitimate only if no contradiction creeps into the original domain. And in order to be sure about this, a finitist metamathematical consistency proof is required.

We shall now argue for the claim that genuine Π^0_1-sentences, i.e. sentences in which unbounded universal quantifiers occur essentially, are ideal statements and that such statements cannot occur in L_M or L_{MM}. The argument is very simple.

(1) All sentences of L_M (or L_{MM}) are meaningful.

(2) Ideal sentences or formulas are meaningless.

(3) If ψ is a formula built up out of the variables '*a*', '*b*', etc. and truth-functional connectives, and containing at least one free variable, then ψ is an ideal formula.

From (3) we conclude to

(4) If the language under consideration is an arithmetical one, every genuine Π^0_0-formula is ideal.

From (1), (2), and (4) follows the claim: No genuine Π^0_0-formula belongs to L_M (or L_{MM}).

When Hilbert (1926) comes to introduce ideal statements, he considers only quantifier-free languages. For this reason, we confined ourselves to speaking of Π^0_0-formulas. It is clear that, once it has been established that genuine Π^0_0-formulas cannot belong to L_{MM}, L_{MM} cannot contain Π^0_1-sentences either. For if a Π^0_1-sentence were to belong to L_{MM}, the Π^0_0-formula derivable from it by way of eliminating the quantifiers would likewise form a part of L_{MM}.

So far, so good. Nevertheless, someone familiar with Hilbert's essays on proof theory might object that this is not yet the end of the matter. He might wish to draw attention to remarks in 'Über das Unendliche' which seem to suggest that universally quantified statements about numerals or expressions can be formulated without offending against the canons of finitism. To see whether the objection has any force, let us take a closer look at two sample sentences (see Hilbert 1926, 91 [378], 93 [380])[35]:

(+) If $\mathfrak{a}$ is a numeral, we must always have $\mathfrak{a} + 1 = 1 + \mathfrak{a}$.

(*) If $\mathfrak{a}$, $\mathfrak{b}$ denote numerals, then always $\mathfrak{a} + \mathfrak{b} = \mathfrak{b} + \mathfrak{a}$.

We extracted (*) from Hilbert's remark '[…] is still admissible from our finitist point of view, as, for example, the sentence that always $\mathfrak{a} + \mathfrak{b} = \mathfrak{b} + \mathfrak{a}$, where $\mathfrak{a}$, $\mathfrak{b}$ denote certain numerals […]'.

[35] Cf. Sect. 4.

Are (+) and (*) finitistically admissible? To begin with, (i) (+) and (*) seemingly contain '**𝔞**', respectively '**𝔞**' and '**𝔟**', as 'natural language equivalents' of variables.[36] (ii) Since (+) and (*) convey an assertion about figures, they seem to belong to L_s or L_M. Taken together, (i) and (ii) seem to imply that in L_s or L_M we may formulate 'natural language equivalents' of Π^0_1-sentences.

In Sections 2 and 3 we have seen that Hilbert enlarges the vocabulary of the language of contentual number theory to include small Gothic letters which function as signs of communication. L_M may, of course, also contain small Gothic letters, but these designate expressions of L_{FA}. Whatever of these languages '**𝔞**' and '**𝔟**' may belong to, (i) is at variance with Hilbert's use of Gothic letters as closed terms.

Two further points call for attention. It is unclear first, what the phrase 'must always' in (+) is intended to convey and, second, whether the embedded statement '$\mathfrak{a} + 1 = 1 + \mathfrak{a}$' ought to appear in quotation marks. If we interpret 'is always' as 'is true', 'is universally valid' or 'is provable', for example, (+) is turned into

(++) If **𝔞** is a numeral, '$\mathfrak{a} + 1 = 1 + \mathfrak{a}$' is true (universally valid, provable),

while (*) is transformed into

(**) If **𝔞**, **𝔟** denote numerals, '$\mathfrak{a} + \mathfrak{b} = \mathfrak{b} + \mathfrak{a}$' is true (universally valid, provable).

The difference between the sentences (++) and (**), which are supposed to be explications of (+) and (*) respectively, is an important one. It is incorrect to say that '**𝔞**' *is* a numeral, but correct that it *denotes* a numeral.

(+++) 'If **𝔞** denotes a numeral, "$\mathfrak{a} + 1 = 1 + \mathfrak{a}$" is true'

is therefore well-formed. By contrast, if we were to replace (++) with

(++′) If **𝔞** is a numeral, then $\mathfrak{a} + 1 = 1 + \mathfrak{a}$,

and (**) with

(**′) If **𝔞**, **𝔟** denote numerals, then $\mathfrak{a} + \mathfrak{b} = \mathfrak{b} + \mathfrak{a}$,

then we would accept (++′) as well-formed and reject (**′) as ill-formed.[37]

[36] It might be considered inappropriate to ask whether the Gothic letters *qua* signs of communication are constants or whether they might be variables whose (intended) domain of interpretation contains as elements stroke-figures or expressions of L_{FA}. Strictly speaking, the distinction between variables and constants makes sense only in the context of familiar formalized languages. However, since (+) and (*) are stated in an informal language, it seems that we cannot reasonably ask whether they contain variables or not. 'Informal language' need not mean 'ordinary language'; it can also be interpreted as 'language of informal mathematics', for instance. In fact, we believe that this interpretation fits the setting of Hilbert's metamathematical considerations.

[37] If we say: 'If **𝔞** is a numeral, "$\mathfrak{a} + 1 = 1 + \mathfrak{a}$" is true', we obtain: since 3 is a numeral and '**𝔞**' takes on numerals as values, the sequence of signs X which begins with three strokes,

Our initial question whether (+) and (*) are finitistically admissible can now be given two answers. The first is at hand if we construe (**) as an explication of (*): ‘If $\mathfrak{a}$, $\mathfrak{b}$ denote numerals, then always $\mathfrak{a} + \mathfrak{b} = \mathfrak{b} + \mathfrak{a}$’ is finitistically admissible. The quotation taken from Hilbert (1926, 93 [380]), from which we extracted (*), suggests another explication of (*). What Hilbert wanted to convey is that for every given name (of a numeral) $\mathfrak{a}$, $\mathfrak{b}$, the statement ‘$\mathfrak{a} + \mathfrak{b} = \mathfrak{b} + \mathfrak{a}$’ is true. This explication motivates the second answer:

Let $\mathfrak{a}$ and $\mathfrak{b}$ be given; ‘$\mathfrak{a} + \mathfrak{b} = \mathfrak{b} + \mathfrak{a}$’ is then finitistically admissible.

If we transfer this line of thought to (+), we see that (+) may be only a rather misleading formulation. What Hilbert had in mind when formulating (+) was presumably this: given an arbitrary expression ‘$\mathfrak{a}$’ (of the right sort), ‘$\mathfrak{a} + 1 = 1 + \mathfrak{a}$’ is a well-formed formula and true. If his finitist point of view were developed in a systematic way, it would rule out that (+) be formulated at all. It would allow the formulation only of sentences like ‘$\mathfrak{a} + 1 = 1 + \mathfrak{a}$’ or ‘$\mathfrak{a} + \mathfrak{b} = \mathfrak{b} + \mathfrak{a}$, for given expressions ‘$\mathfrak{a}$’ and ‘$\mathfrak{b}$’. In short, if we are given ‘$\mathfrak{a}$’, then ‘$\mathfrak{a}$’ is a sign of communication, in particular, a closed term, and the sequence of expressions ‘$\mathfrak{a} + 1 = 1 + \mathfrak{a}$’ is a well-formed sentence of L_s, L_{FA} and possibly also of L_M. The problem arising from (i) and (ii) has disappeared, because (+) and (*) are not formulated at all in L_S or L_M.

It springs to mind that (+) is not a well-formed sentence of either L_s or L_M. The simple reason is that if ‘$\mathfrak{a}$’ is considered a variable, (+) is an ideal statement. Ideal statements are inadmissible both in L_s and L_M, however. (+) cannot be contained in L_{FA} either. There are two reasons, we think. First, (+) contains natural language expressions; it is not stated in a formal language. Second, Hilbert (1926, 93 [380]) expressly introduces variables ‘*a*’, ‘*b*’, etc. In L_{FA}, in addition to the Gothic letters, distinguishing the role of the former sharply from that of the latter. Obviously, in L_{FA} he wishes to exclude the formulation of a sentence like ‘If $\mathfrak{a}$, $\mathfrak{b}$ are numerals, we must always have $\mathfrak{a} + \mathfrak{b} = \mathfrak{b} + \mathfrak{a}$’ and instead replace it with ‘$a + b = b + a$’.[38]

The upshot of our analysis is, then, that (+) does not belong to any of the Hilbertian languages considered so far. This by no means clashes with the claim that for every given expression ‘$\mathfrak{a}$’ the statement ‘$\mathfrak{a} + 1 = 1 + \mathfrak{a}$’ belongs to L_s, L_{FA}, or L_M. Plainly, in (+++) ‘$\mathfrak{a}$’ is a variable ranging over communi-

followed by a sign for plus, followed by a sign of communication referring to the numeral ‘1’, etc., is true, and therefore a well-formed formula. To put it in another way: X is equal to 3 $\frown$ ‘+’ $\frown$ ‘1’ $\frown$ ‘=’ $\frown$ ‘1’ $\frown$ ‘+’ $\frown$ 3, where ‘$\frown$’ is the concatenation symbol, ‘ “+” ’ a name of the sign for plus, etc. Or yet in other words: X is identical with ‘111 + 1 = 1 + 111’. Clearly, this identification makes no sense, because X is ill-formed. We would rather expect that ‘3 + 1 = 1 + 3’ be explained as well-formed. As has just become evident, this is not the case, however. Replacing (++) with ‘For every sign of communication $\mathfrak{a}$ “$\mathfrak{a} + 1 = 1 + \mathfrak{a}$” is true’, we obtain, by means of instantiation, ‘ “3 + 1 = 1 + 3” is true’ and thus the desired result. ‘3’ is, of course, a sign of communication.

[38] Cf. Sect. 3.

cation signs. Thus (+), construed as (+++), belongs to a meta-language of L_s or of L_{FA} or of L_M, as the case may be. It seems most likely that Hilbert did not even imagine such meta-languages. Be this as it may, due to its infinitistic nature, (+) ought in no way to be considered a sentence belonging to L_s or L_M.[39] The question as to whether L_M, despite the limitation of its expressive power, may still allow the formulation of familiar metamathematical claims will now be given close attention. We shall take *consistency* as a paradigm of a metamathematical property.

6. ALTERNATIVES TO THE STANDARD FORMALIZATION OF CONSISTENCY ASSERTIONS

If we take Hilbert's conception of finitism at face value, genuine Π^0_1-sentences are inadmissible in L_M (or more exactly: in L_{MM}). This leaves Hilbert with the fundamental difficulty to formulate a metamathematical consistency assertion. Since the standard formalizations of consistency assertions for recursively enumerable theories are Π^0_1-sentences, it seems that it must be possible at least to formulate them in the language of metamathematics. Surely if there is no way of stating the consistency of a theory T in a finitary manner, then Hilbert's project of carrying out finitary consistency proofs for formalized mathematical theories is bound to fail right from the start, independent of Gödel's Second Incompleteness Theorem.[40]

The obvious way of establishing metamathematically the consistency of, say, PA is as follows: 'PA is consistent' is appropriately formulated in L_M (or better: in L_{MM}) as 'Con(PA)', and 'Con(PA)' is shown with metamathematically admissible methods. Here we do not commit ourselves to the standard formalization of the consistency assertion for PA—which we fix as 'Con_{pa}'—but endorse any *adequate* formalization.

To gain a sufficiently clear idea about the nature of an adequate formalization of 'PA is consistent', we consider briefly the standard metatheoretic definition of consistency. The complexity of 'Con(PA)' within the arithmetical hierarchy is of special interest here. 'PA is consistent' can be defined within natural language by appeal to informal mathematical terminology, e.g. as 'Not every statement in L_{PA} is PA-provable' or 'Falsum is not PA-provable'. The latter version amounts to 'There exists no proof for Falsum in PA' or 'For all PA-proofs *b* and statements A, if *b* is a proof for A, then

[39] Notice that even if (+) (or (*)) were amenable to a finitist interpretation, it would still resist formulation within the bounds of Hilbert's finitist standpoint; and this is true even if (+) is explicated as (+++) or (++'). Another point worth emphasizing is that if '$\mathfrak{a} + 1 = 1 + \mathfrak{a}$' is considered a sentence for every sign of communication '$\tilde{a}$', the finitist is committed to acknowledging the existence of infinitely many signs and thus faces a predicament (cf. Sect. 4).

[40] It is further plain enough that even if it were possible to formulate Con_τ finitistically, there would be no guarantee that Con_τ can be proved by finitary means alone.

A is distinct from Falsum.' '*b* is a PA-proof for A' clearly abbreviates '*b* is a sequence of prooflines such that every line is an axiom of PA or follows from earlier lines by the rules of PA (e.g. by *modus ponens*)'. Now, when we paraphrase the above-mentioned sequence of informal definitions in the formal language of first-order logic, we arrive at the standard formalized consistency assertion '$\mathrm{Con}_{\mathrm{pa}'}$' which is Π^0_1 in a formula pa′ defining PA.[41] Apart from the possibly unusual representation pa′, this is in fact the standard formalization of 'PA is consistent'. Following customary usage, we shall employ the 'natural' representation for PA (called 'pa'); we may choose it to be a formula of complexity Π^0_0.[42] Plainly, in order to be able to show the consistency of PA in MM in the sense of this version, i.e. in order to prove $\mathrm{Con}_{\mathrm{pa}}$ in MM, we must be able to formulate in L_{MM} at least genuine Π^0_1-sentences (or Π^0_0-formulas with free variables).

Michael Detlefsen (1979) and Aleksandar Ignjatović (1994), and in a similar vein W. W. Tait (1981), seem to feel at ease to state a formalized version of finitist metamathematics MM in a language which contains formulas of at least complexity Π^0_1. They consequently regard a formulation of the consistency assertion for PA as '$\mathrm{Con}_{\mathrm{pa}}$' as unproblematic. In contrast to their approaches, we deny that Π^0_1-sentences can be formulated in L_{MM}. However, since Hilbert's chief aim is to *show*, by finitary means, the consistency of, e.g. PA, it remains to devise alternatives to the standard formalization of consistency assertions. Here is what we refer to as version (A) for axiomatizable theories S and T with representation τ, which we consider to be inspired by Hilbert himself.

Version (A): S proves the approximative consistency of T :⇔
$\forall n\ \mathrm{S} \vdash \neg\mathrm{Proof}_\tau(n, \bot)$.

The formalized proof predicate is the standard one. In this context, it is important to rely on the assumption that the representation τ of T is in Π^0_0. On this assumption, then, '$\neg\mathrm{Proof}_\tau(n, \bot)$' is a Π^0_0-sentence and can therefore be formulated in the language of finitist metamathematics.

For PA, version (A) differs from the provability of '$\mathrm{Con}_{\mathrm{pa}}$' in two important respects.

(1) '$\neg\mathrm{Proof}_{\mathrm{pa}}(n, \bot)$' is Π^0_0.

(2) $\forall n\ \mathrm{PA} \vdash \neg\mathrm{Proof}_{\mathrm{pa}}(n, \bot)$.

So, for the restricted case in which the consistency of PA ought to be provable in PA, Hilbert's programme can be carried out, provided that a consistency proof along the lines of contemporary proof theory has been

[41] The formalized proof predicate will be written as '$\mathrm{Proof}_{\mathrm{pa}'}(x, y)$' and the provability predicate as '$\mathrm{Pr}_{\mathrm{pa}'}(y)$'.

[42] The formula of complexity Π^0_0 exists by virtue of a theorem of William Craig (1953).

replaced with an approximative consistency proof. The view that approximative consistency proofs tally with Hilbert's intentions and ought not to be disparaged or rejected out of hand will now gain additional support.

In his 1923 paper, Hilbert emphasizes that the fundamental idea of how a proof of the consistency of a set of axioms is carried out is always this: 'We assume that we are concretely given a proof *qua* figure with the end formula $0 \neq 0$; [. . .] We then show by means of a contentual consideration that this cannot be a proof which meets our requirements' (184; cf. Hilbert (1931), 491). Thus, if we carry out the process of formalization which led us from M to MM, in particular if we replace proofs *qua* figures with proofs *qua* numbers and contentual considerations with proofs in MM, we see that there are at least three ways of explicating, in a precise manner, what Hilbert's conception of finitist metamathematical consistency proofs may have been.[43]

(a) $\forall n$ (If n is a proof for $\perp$ in S $\Rightarrow$ MM $\vdash$ 'n is no proof in σ')[44]

(b) $\forall n \forall \psi$ (If n is a proof for ψ in S $\Rightarrow$ MM $\vdash \overline{\psi} \neq \overline{\perp}$)

(c) $\forall n$ MM $\vdash \neg\text{Proof}_\sigma(n, \perp)$.

Fortunately, we need not choose between (a), (b), and (c), since we can prove: If S is recursively enumerable with Π^0_0-representation σ, and if MM decides correctly every Π^0_0-sentence, then (a), (b), and (c) are equivalent to one another. Thus, in order to establish finitistically the consistency of a theory T, it seems that Hilbert sought to provide an 'approximative consistency proof' for T in MM.[45]

In stating Hilbert's principal objective, we deliberately used the phrase 'to *show* by finitary means'. Unlike the standard method of carrying out consistency proofs, method (A) is unsuitable for proving 'PA is consistent' in MM. Still, the primary goal is to establish whether PA is consistent. It is of secondary importance whether this goal is achieved by proving in MM 'PA is consistent' or by pursuing a method which only allows to show the

[43] We cannot definitely rule out that Hilbert conceived of consistency proofs in the sense of the modern conception, which is MM $\vdash \forall x\, \neg\text{Proof}_\sigma(x, \perp)$.

[44] ' "n is no proof in σ" ' is to be construed here as a formalization of 'n is no proof in S', which is Π^0_0 if σ is Π^0_0.

[45] So far we rendered phrases like 'We assume that we are concretely given a figure $\mathfrak{a}$; then . . .' formally always by '$\forall a(\text{figure}(a) \rightarrow \ldots)$', i.e. by a (meta-)quantification over expressions. Perhaps we did so at the cost of abandoning the 'constructive' character of the original phrase. For various reasons, we shall nevertheless adhere to our mode of paraphrasing. If all *possible* proofs cannot be transformed into inconsistency proofs, then the same applies to all concretely exhibited proofs. Furthermore, we are not sure whether Hilbert intended formulations like the one quoted above to differ in meaning from a universal quantification over expressions, nor do we see how such a difference could be accounted for in a precise manner. Finally, our preferred paraphrase of Hilbert's talk of concretely given proof figures is widely accepted in the pertinent literature on his proof theory.

consistency of PA, provided that, from the finitist point of view, this method is sound. Suppose that

(*) $\forall n$ MM $\vdash \neg\text{Proof}_\tau(n, \bot)$

holds. Since M can be conceived of as a collection of intuitively evident truths, and, hence, MM can be regarded as sound, we obtain

(**) $\forall n$ '$\neg\text{Proof}_\tau(n, \bot)$' is true.

In other words,

(***) T is consistent.

So, in order to show, by finitary means, that a theory T is consistent, it suffices to show: for all n, MM $\vdash \neg\text{Proof}_\tau(n, \bot)$.

We have already observed that PA proves its own approximative consistency. Yet Hilbert aims at showing the consistency of the whole of classical mathematics in a theory which is presumably quite weak. In this context, the question may arise: Can we prove in weak theories like (QF-IA) or Q (i.e. Robinson Arithmetic)[46] the approximative consistency of strong theories such as ZF? It seems that we cannot do this. We can merely prove: for all n, Q $\vdash \neg\text{Proof}_{pa}(n, \bot)$, if PA is consistent or correspondingly: for all n, Q $\vdash \neg\text{Proof}_{zf}(n, \bot)$, if ZF is consistent. Clearly, in proving this we do not advance one single step; if we carry out an approximative consistency proof for a theory T in such a weak theory as MM, we are bound to assume that T is consistent.

Version (B): As regards method (A), one might object that, although the transition from (*) to (**) is correct by virtue of the specific nature of M, the finitist is unable to show this with his restricted means. One could accordingly demand that the following principle be provable in MM:

(+) If for every n MM $\vdash \neg\text{Proof}_\tau(n, \bot)$, then T is consistent (or: then for every n '$\neg\text{Proof}_\tau(n, \bot)$' is true).

An obvious extension of (+) would be

(++) If for every n MM $\vdash \psi(n)$, then for every n $\psi(n)$ (or: then '$\psi(n)$' is true; for arbitrary formulas ψ).

This requirement calls for various comments.

(i) Neither (+) nor (++) belongs to L_{MM}, but presumably to a meta-language of L_{MM}. It seems, though, that Hilbert did not consider any hierarchy of meta-languages.[47] It is conceivable that M or MM are coded again

[46] Cf. Tarski, Mostowski, and Robinson (1953).

[47] Notice in this context that the finitist metamathematician as such cannot make any statement about his proof-theoretic practice. In particular, he is incapable of making any statement

in the formal object-language, and hence that the meta-language of L_{MM} is converted into a meta-language of the formal object-language. We believe, however, that this manœuvre is not in the spirit of Hilbert's view.

(ii) Nonetheless, by formalizing (++) in a natural way, we obtain the well-known uniform reflection principle RFN[mm] for MM, i.e. $\forall x \mathrm{Pr}_{mm}(\ulcorner \overline{\psi(x)} \urcorner) \rightarrow \forall x \psi(x)$. Similarly, by formalizing (+) we arrive at $\forall x \mathrm{Pr}_{mm}(\ulcorner \neg \mathrm{Proof}_\tau(x, \perp) \urcorner) \rightarrow \forall x \neg \mathrm{Proof}_\tau(x, \perp)$. For convenience, we refer to this sentence as (F+). Both RFN[mm] and (F+) have at least complexity Π^0_1.[48] Supplementing method (A) with the requisite of MM-provability of (F+) or RFN[mm] therefore undermines the motivation for (A).

(iii) Both the uniform reflection principle RFN[τ] and restrictions of it along the lines of (F+) are, for a wide range of theories T with representation τ, not T-provable (cf. Feferman (1960); Smorynski (1977)). So, it seems that the requirement of MM-provability of (+) and (++) cannot be met.

In his essay 'Hilbert's program and the omega-rule' (1994, 331 f.), Ignjatović presents what he calls 'almost finitistic' consistency proofs. In order to show 'almost finitistically' the consistency of a theory T (with representation τ), '$\forall x \mathrm{Pr}_{qf\text{-}ia}(\ulcorner \overline{\neg \mathrm{Proof}_\tau(x, \perp)} \urcorner)$' must be provable in (QF-IA). Thus, an 'almost finitistic' consistency proof is a natural formalization of an approximative consistency proof in (QF-IA).[49] It seems to us that this approach is only of technical interest. Even in the light of a reformulation of sentences of the form '$\forall x \mathrm{Pr}_\sigma(\ulcorner \overline{\neg \mathrm{Proof}_\tau(x, \perp)} \urcorner)$', designed to reduce their complexity, the resulting sentences are no less complex than Π^0_1 and, hence, cannot be formulated in L_{MM}. It is, moreover, only for such theories T which are approximately as 'strong' as (QF-IA) itself that their consistency can be proved 'almost finitistically'.[50] It follows that within the setting suggested by Ignjatović a finitary consistency proof for a significant part of classical mathematics is beyond reach. For those who wish to carry out (internal) consistency proofs, the presumed virtue of the method suggested by Ignjatović is that (QF-IA), for instance, proves ('almost finitistically') its own consistency. However, that a theory like (QF-IA), for example, proves its own consistency in a certain sense is likewise guaranteed by version (A). And to be sure, the notion of an approximative consistency proof, as defined by version (A), after all fares better than Ignjatović's notion of an 'almost

about his metamathematical consistency proofs; he just carries them out and this is the end of the matter. He could make statements about his consistency proofs only by moving one step higher as it were, i.e. to the meta-language of L_M.

[48] RFN[mm] is equivalent to $\forall x(\mathrm{Pr}_{mm}(\ulcorner \overline{\psi(x)} \urcorner) \rightarrow \psi(x))$, which is Π^0_1 if mm is in Π^0_0 and ψ in Π^0_1 (cf. Smorynski 1977).

[49] Hao Wang (1952) was probably the first to have drawn attention to such a formalization as a way of giving 'consistency proofs' for a theory T within T.

[50] In (QF-IA) we can prove e.g. $\forall x \mathrm{Pr}_{qf\text{-}ia}(\ulcorner \overline{\neg \mathrm{Proof}_{qf\text{-}ia}(x, \perp)} \urcorner)$, i.e. we can show almost finitistically the consistency of (QF-IA) (cf. Feferman (1962); Smorynski (1977)).

finitistic' consistency proof, since the latter gives rise to a problem which the former avoids.

Why should we believe that T is consistent, when in (QF-IA) we prove '$\forall x \mathrm{Pr}_\sigma(\ulcorner\neg\mathrm{Proof}_\tau(x, \bot)\urcorner)$'? And why should we believe that in proving this we conduct a finitary consistency proof for T? One possible way of answering these questions is this: '(QF-IA) is sound and "$\forall x \mathrm{Pr}_\sigma(\ulcorner\neg\mathrm{Proof}_\tau(x, \bot)\urcorner)$" is therefore true. Let us assume now that S is sound; then "$\forall x \neg\mathrm{Proof}_\tau(x, \bot)$" is true and T is therefore consistent.' The choice of (QF-IA) for S then suggests itself for the simple reason that we need concern ourselves only with the soundness of (QF-IA). Yet the soundness of (QF-IA) is secured by virtue of the assumption that (QF-IA) = MM; for, clearly, the soundness of MM is beyond doubt. This argument is akin to the one adduced in relation to (non-formalized) approximative consistency proofs. We have now only saddled upon ourselves the additional complication that we need to argue for the soundness of S, if S is not identical with MM.

(iv) Even if with respect to method (B) RFN[mm] or (F+) were MM-provable, a consistency proof for T in MM would not gain persuasive power. For here again it holds: in order to be able to show that T is consistent, we must assume not only that RFN[mm] or (F+) is MM-provable, but also that RFN[mm] or (F+) is true. By assuming the truth of RFN[mm] or (F+), we make, however, a further assumption of soundness.

So, why do modern proof theorists prefer the standard method for carrying out consistency proofs and do not rest content with a consistency proof for a theory T in, e.g. PA, according to method (A)? A possible answer is close at hand. If we wish to guarantee the consistency of T by conducting an 'ordinary' consistency proof for T in PA (or MM), we are bound to presuppose the soundness of PA (or MM). By contrast, if we are concerned to warrant the consistency of T by carrying out an approximative consistency proof for T in PA according to method (A), we must, in addition, either have ample reason for the consistency of T or simply presuppose it. In short, even if we succeed in adducing a strong argument for the soundness of PA, what we achieve with this in the case of an approximative consistency proof is merely that we establish the consistency of T relative to the assumption that T is consistent. Needless to say, this is far from giving a consistency proof for T in a convincing way.[51]

To end our discussion, let us summarize. Whether we prove in a theory S the consistency of a theory T, the approximative consistency of T or almost finitistically the consistency of T, we must always *presuppose* that S is sound[52] in order to guarantee that T *is* consistent. Here is a case in point. Suppose that $S \vdash \mathrm{Con}_\tau$. It does not, of course, follow from this assumption

[51] Cf. Resnik (1974, 138) who argues in a similar vein.

[52] If T is axiomatizable with Π_0-representation τ, it suffices to assume Π^0_1-soundness.

that T is consistent; and this is true independent of the question whether S is equal to T or not: S might be inconsistent.[53] By contrast, if we assume that S is sound, then T is consistent; yet the soundness of S had simply to be presupposed. Moreover, even if S were to coincide with T and thus Gödel's theorems were false, i.e. even if S were to prove its own consistency, we would lack any guarantee whatsoever that S is consistent. We would instead end up with a truism: If S is consistent, then S is consistent. It has further been observed that, owing to the incontestable soundness of MM, an ordinary or approximative consistency proof for a theory T in MM is, indeed, fairly persuasive: it warrants the consistency of T. However, MM will serve to prove the approximative consistency of theories T only if they approximately have the same 'strength' as MM itself. Due to this circumstance and the supposed weakness of MM, we are far from giving a consistency proof for, say, analysis in MM. Put succinctly, Hilbert's programme, when it is construed in the standard way, cannot be carried out by conducting approximative instead of ordinary consistency proofs. We pursued a modest aim, however. We only tried to show that the finitist metamathematician who feels obliged to dispense with genuine Π^0_1-sentences can still, in some reasonable sense, state that a formalized theory T is consistent. He is therefore not deprived of the possibility of establishing the consistency of T.

Setting the notion of an approximative consistency proof aside, we may ask whether other 'non-standard' ways of proving in PA the consistency of PA can be contrived. The answer is 'yes', provided that we are prepared to use unnatural provability predicates.[54] The use of such predicates is disputable, however. If we had good reasons to believe that Π^0_1-sentences can be formulated in L_{MM}, we could suggest that the soundness of a theory T be proved in an axiomatizable theory S and 'Con_σ' be proved in T. And we might regard this as a way of establishing the consistency of S in S by employing only what are clearly natural provability predicates. At first glance, the proposal seems to founder on Gödel's Incompleteness Theorems. On closer examination, however, this is less obvious, as we shall try to show in a sequel to this paper.[55]

REFERENCES

Ackermann, W. (1940), 'Zur Widerspruchsfreiheit der Zahlentheorie', *Mathematische Annalen*, 117: 162–94.

[53] If T is axiomatizable with Π^0_0-representation τ, the consistency of S already implies its Π^0_1-soundness. In this case, then, the consistency of S is not only necessary, but also sufficient to guarantee the consistency of T.

[54] Cf. Feferman (1960).

[55] See Niebergall and Schirn, 'Hilbert's programme and Gödel's theorems' (forthcoming).

Bernays, P. (1935), 'Hilberts Untersuchungen über die Grundlagen der Arithmetik', in Hilbert (1935, 196–216).

Cantor, G. (1932), *Gesammelte Abhandlungen mathematischen und philosophischen Inhalts*, E. Zermelo (ed.), Berlin, Springer-Verlag; repr. Hildesheim, 1966.

Craig, W. (1953), 'On axiomatizability within a system', *Journal of Symbolic Logic*, 18: 30–2.

Detlefsen, M. (1979), 'On interpreting Gödel's second theorem', *Journal of Philosophical Logic*, 8: 297–313.

Feferman, S. (1960), 'Arithmetization of metamathematics in a general setting', *Fundamenta Mathematicae*, 59: 35–92.

—— (1962), 'Transfinite recursive progressions of axiomatic theories', *Journal of Symbolic Logic*, 27: 259–316.

—— (1988), 'Hilbert's program relativized: proof-theoretical and foundational reductions', *Journal of Symbolic Logic*, 53: 364–84.

Frege, G. (1903), *Grundgesetze der Arithmetik. Begriffsschriftlich abgeleitet*, II, Jena, H. Pohle; repr. Darmstadt and Hildesheim, 1962.

—— (1969), *Nachgelassene Schriften*, H. Hermes, F. Kambartel, and F. Kaulbach (eds.), Hamburg, Felix Meiner.

Gentzen, G. (1936), 'Die Widerspruchsfreiheit der reinen Zahlentheorie', *Mathematische Annalen*, 112: 493–565.

—— (1938), 'Neue Fassung des Widerspruchsfreiheitsbeweises für die reine Zahlentheorie', *Forschungen zur Logik und zur Grundlegung der exakten Wissenschaften*, 4: 19–44.

Gödel, K. (1931), 'Über formal unentscheidbare Sätze der Principia Mathematica und verwandter Systeme I', *Monatshefte für Mathematik und Physik*, 38: 173–98.

Goodman, N., and Quine, W. V. (1947), 'Steps toward a constructive nominalism', *Journal of Symbolic Logic*, 12: 97–122.

Heijenoort, J. van (1967), *From Frege to Gödel: A Source Book in Mathematical Logic, 1879–1931*, Cambridge, Mass., Harvard University Press.

Heyting, A. (1934), *Mathematische Grundlagenforschung, Intuitionismus, Beweistheorie*, Berlin, Springer-Verlag; repr. Berlin, Heidelberg, New York, 1974.

Hilbert, D. (1899), *Grundlagen der Geometrie*, Leipzig, Teubner; seventh revised and enlarged edn., 1930.

—— (1922), 'Neubegründung der Mathematik', *Erste Mitteilung, Abhandlungen aus dem Mathematischen Seminar der Hamburger Universität*, 1: 157–77; repr. in Hilbert (1935, 157–77).

—— (1923), 'Die logischen Grundlagen der Mathematik', *Mathematische Annalen*, 88: 151–65; repr. in Hilbert (1935, 178–91).

—— (1926), 'Über das Unendliche', *Mathematische Annalen*, 95: 161–90; repr. in D. Hilbert, *Hilbertiana*, Darmstadt, Wissenschaftliche Buchgesellschaft, 1964, 79–108; English trans. 'On the infinite', in van Heijenoort (1967, 367–92).

—— (1928*a*), 'Die Grundlagen der Mathematik', *Abhandlungen aus dem Mathematischen Seminar der Hamburger Universität*, 6: 65–85; abbreviated version repr. in Hilbert (1899, 289–312); seventh revised and enlarged edn., 1930; English trans. 'The foundations of mathematics', in van Heijenoort (1967, 464–79).

—— (1928*b*), 'Probleme der Grundlegung der Mathematik', *Atti del Congreso nazionale dei matematici, Bologna 3–10 settembre 1928*, 1: 135–41; with supplementations repr. in *Mathematische Annalen*, 102 (1929), 1–9, and as appendix X in Hilbert (1899, 313–23); seventh revised and enlarged edn., 1930.

—— (1931), 'Die Grundlegung der elementaren Zahlenlehre', *Mathematische Annalen*, 104: 485–94.

—— (1935), *Gesammelte Abhandlungen*, III, Berlin, Springer-Verlag; 2nd edn., 1970.

—— and Bernays, P. (1934), *Grundlagen der Mathematik*, I, Berlin, Springer-Verlag; 2nd edn., with modifications and supplementations, 1968.

—— (1939), *Grundlagen der Mathematik*, II, Berlin, Springer-Verlag; 2nd edn., with modifications and supplementations, 1970.

Ignjatović, A. (1994), 'Hilbert's program and the omega-rule', *Journal of Symbolic Logic*, 59: 322–43.

Kaye, R. (1991), *Models of Peano Arithmetic*, Oxford, Oxford University Press.

Kitcher, P. (1976), 'Hilbert's epistemology', *Philosophy of Science*, 43: 99–115.

Kreisel, G. (1958), 'Hilbert's programme', *Dialectica*, 12: 346–72; revised version with a postscript in P. Benacerraf and H. Putnam (eds.), *Philosophy of Mathematics: Selected Readings*, Cambridge, Cambridge University Press, 1983, 207–38.

Martin, R. M. (1958), *Truth and Denotation*, Chicago, Kitcher.

Montague, R. (1961), 'Semantical closure and non-finite axiomatizability I', in *Infinitistic Methods*, Warschau, Pergamon Press.

Neumann, J. von (1927), 'Zur Hilbertschen Beweistheorie', *Mathematische Zeitschrift*, 26: 1–46.

Quine, W. V. O. (1946), 'Concatenation as a basis for arithmetic', *Journal of Symbolic Logic*, 11: 105–14.

Resnik, M. (1974), 'On the philosophical significance of consistency proofs', *Journal of Philosophical Logic*, 3: 133–47.

—— (1980), *Frege and the Philosophy of Mathematics*, Ithaca, NY, Cornell University Press.

Rosser, J. B. (1936), 'Extensions of some theorems of Gödel and Church', *Journal of Symbolic Logic*, 1: 87–91.

Sieg, W. (1985), 'Fragments of arithmetic', *Annals of Pure and Applied Logic*, 28: 33–72.

Smorynski, C. (1977), 'The incompleteness theorems', in J. Barwise (ed.), *Handbook of Mathematical Logic*, Amsterdam, North-Holland, 821–65.

Tait, W. W. (1981), 'Finitism', *Journal of Philosophy*, 78: 524–46.

Takeuti, G. (1975), *Proof Theory*, Amsterdam, North-Holland; 2nd enlarged edn., 1987.

Tarski, A. (1935), 'Der Wahrheitsbegriff in den formalisierten Sprachen', *Studia Philosophica*, 1: 261–405; English trans. 'The concept of truth in formalized languages', in Tarski, *Logic, Semantics, Metamathematics*, 2nd edn., ed. J. Corcoran, Indianapolis, Hackett Publishing Co., 1990, 152–278.

—— Mostowski, A., and Robinson, R. M. (1953), *Undecidable Theories*, Amsterdam, North-Holland.

Wang, H. (1952), 'Truth definitions and consistency proofs', *Transactions of the American Mathematical Society*, 73: 243–75.

12

Constructive Existence Claims

MICHAEL DETLEFSEN

1. INTRODUCTION

It is a commonplace of constructivist thought that a claim that an object of a certain kind exists is to be backed by an explicit display or exhibition of an object that is manifestly of that kind. Let us refer to this requirement as *the exhibition condition.* The main objective of this essay is to examine this requirement and thereby to arrive at a better understanding of its epistemic character and of the role that it plays in the two main constructivist philosophies of this century—the intuitionist programme of Brouwer and Weyl, and the finitist programme of Hilbert. Gödel gave what is the currently prevailing conception of the relationship between the intuitionist and finitist conceptions of the exhibition condition in his 1958 essay 'Über eine bisher noch nicht benützte Erweiterung des finiten Standpunktes':

> Bernays . . . teaches us to distinguish two components of the finitary viewpoint; namely, first, the constructive element which consists in the demand that we be permitted to speak of mathematical objects (mathematischen Objekten) only to the extent that we can exhibit or through construction actually produce them; secondly, the specifically finitist element, which demands in addition that the objects about which one makes statements (Aussagen), with which the constructions are executed, and which one obtains by their means, are 'intuitive'. In the final analysis, that means that they are spatio-temporal configurations (Anordnungen) of elements whose constitution (Beschaffenheit) is irrelevant aside from whether they are identical or distinct. (In contrast to this, the objects of intuitionist logic are meaningful statements and proofs (sinnvolle Aussagen und Beweise).) (p. 282, my translation)[1]

So, Gödel conceived of the exhibition condition itself—that is, the demand that claims of existence be backed either by an explicit exhibition of an object of the kind claimed to exist, or by a specification of a means of effecting such an exhibition—as the common feature which qualifies both finitism and intuitionism as forms of constructivism. He believed them to be distinguished not by different conceptions of the nature of exhibition or

[1] See Kreisel (1958–9, 148) and Parsons (1967, 205) for statements of similar views (indeed, views which may well be simply affirmations of Gödel's).

the exhibitory process itself, so much as by the nature of the objects exhibited—finitary exhibition being restricted to spatio-temporal concrete objects, intuitionism allowing as well such abstract objects as the mental constructions which constitute their proofs.[2]

We believe this characterization of Gödel's to be deficient, and shall argue that a more illuminating treatment of the differences between finitism and intuitionism reveals that it is *the epistemological conception of the nature of exhibition itself* which most fundamentally distinguishes the two. There is not a single understanding of the nature of the exhibition condition shared by the two, but rather two different conceptions based on two radically different epistemological ideals. We shall argue, therefore, that the fundamental divergence of finitism and intuitionism is not over the nature of the possible *objects* of exhibition, but rather, and far more basically, over the question of the nature and epistemic significance of exhibition itself.[3]

Our main interest, however, is not to criticize Gödel, but rather to increase our understanding of the exhibition requirement and the similarities and differences between finitism and intuitionism. The bulk of our efforts will therefore be directed at these aims rather than at developing a detailed description of the deficiencies of Gödel's view.

[2] As Gödel put it:

> P. Bernays has repeatedly called our attention to the fact that since the consistency of a system cannot be proved by means weaker than those of the system itself, it is necessary to exceed the bounds of finitary mathematics, in the sense of Hilbert, in order to prove the consistency of classical mathematics or even just of classical number theory. Since finitary mathematics, in this sense, is defined as that which is *intuitive evident*, this means (as Bernays also explicitly stated in his 1935, 62, 69) that in order to prove the consistency of number theory, one must use certain *abstract* concepts. By abstract (or non-intuitive) concepts is here understood essentially those that are of the second or higher order; that is, those that do not consist of properties of or relations between concrete objects (e.g. combinations of signs), but rather depend upon mental constructions (*Denkgebilde*) (e.g. proofs, meaningful propositions and the like). In the proofs, use must be made of insights concerning these latter which derive not from the combinatorial (space-time) properties of the representing sign-combinations, but rather only from their sense (*Sinn*). (Gödel 1958, 280; my translation)

[3] There may be connections between the two matters, of course. But they are not well brought out simply by emphasizing the abstract–concrete distinction. Rather, as we shall see, the fundamental distinction is that between a conception of exhibition (the intuitionist) which sees it as giving rise to an essential epistemic asymmetry between the exhibitor and the non-exhibitor, on the one hand, versus a conception of exhibition (the finitist) which sees it as giving rise to an essential epistemic symmetry between exhibitor and non-exhibitor. Forcing exhibition to produce concrete objects may contribute towards creation of an epistemic symmetry between exhibitor and non-exhibitor because of the public inspectability of concrete objects. Allowing it to issue in abstact objects as well, however, does not seem to have any similar tendency to produce an asymmetry between exhibitor and non-exhibitor. The reason why is that abstract objects are not in general asymmetrically accessible by human epistemic agents. Moreover, it is not the abstractness or concreteness that is of the essence, but rather the symmetry or asymmetry of access.

2. A LITTLE BACKGROUND

Like the other philosophies of mathematics of the late nineteenth and early twentieth centuries, finitism and intuitionism are, albeit in different ways, offspring of Kant's general critical epistemology and the response of early nineteenth-century geometry to it. Chief among the influences of the critical epistemology were (1) Kant's conception of what is required for knowledge of existence, and (2) his demarcation between the so-called judgements of the understanding,[4] on the one hand, and the principles of reason, on the other.

Of these two, the former seems to have played an important positive role in shaping both intuitionism and finitism and, in particular, in giving rise to their common belief in the need for an exhibition requirement. Kant's conception of what is required for knowledge of existence is essentially given in the following remark (CPR: A: 639; B: 667):

> In whatever manner the understanding may have arrived at a concept, the existence of its object is never, by any process of analysis, discoverable within it; for the knowledge of the existence of the object consists in the fact that the object is posited in itself, outside the mere thought of it.

Thus, for Kant, knowledge of existence requires more than mere consistent conception of a thing. It requires, in addition, that the object be 'given' in some special way, a way that impresses it upon us in a manner independent of its conceptual representation, and, so, in a way that the mere consistent thinking of it never could. This Kantian belief in the need for an extra-conceptual 'givenness' was what, in the end, motivated both Brouwer and Hilbert to adopt their respective forms of the exhibition requirement as conditions on the acceptability of existence claims. They differed greatly, however, in their ideas of what this extra-conceptual 'givenness' was to consist in, as we shall see below.

The other main Kantian influence—the demarcation between the judgements arising from intuition and the understanding, on the one hand, and the principles of reason, on the other—played both interestingly similar and interestingly different roles in shaping the philosophies of Brouwer and Hilbert. Hilbert consciously patterned the pivotal structural feature of his entire mathematical epistemology—namely the distinction between the so-called *real* and *ideal* parts of mathematics—after Kant's distinction between

[4] The phrase 'judgements of the understanding' is, however, a little misleading. What Kant opposed the principles of reasoning to were not judgements arising wholly from the faculty of the understanding, but rather judgements arising from the 'agreement' between representations of the understanding (concepts), on the one hand, and representations of sense, on the other. Thus, what are commonly referred to as judgements of the understanding actually involved both sensibility and the understanding.

the genuine judgements and inferences arising from intuition and the understanding, on the one hand, and the instrumental or regulative pseudo-propositions and pseudo-inferences of reason, on the other. Brouwer, on the other hand, conceived of the Kantian apparatus rather differently. He saw it as applying to the use of logic in the development of our mathematical knowledge. Genuine mathematical reasoning, according to him, does not admit the use of truly logical reasoning.[5] Still, he acknowledged, logical reasoning can play a kind of regulative, *meta*mathematical role in guiding the development of our genuinely mathematical knowledge. That is, he admitted that a logic shorn of the unrestricted use of the law of excluded middle found in classical logic—in other words, what since Heyting's formulation of it in 1934 has been referred to as 'intuitionist logic'—might enjoy a certain limited utility as a means of locating the linguistic images of propositions for which there exists a genuine mathematical proof. Thus, Brouwer allowed that logic, properly constrained, might attain a certain epistemic value as an instrumental device for (i) reminding the mathematical knower of genuine constructions of bygone times and for (ii) communicating to others those 'propositional sites' as it were where they might best invest their epistemic efforts in trying to develop a genuine proof and thus attain some genuine mathematical knowledge.[6] In the end, then, Brouwer, too, like Hilbert, admitted that there is some at least restricted epistemic benefit in the use of a certain instrumental apparatus like Kant's apparatus of reason to help us in developing our genuine mathematical knowledge.

Of the two Kantian influences on Brouwer and Hilbert just described, we shall pay particular attention to the first in this essay. We mention the second for two reasons only. Firstly, we want to warn the reader away from

[5] One finds, I believe, no similar need for freedom from logic in Hilbert's finitary or real reasoning, though the kind of logical reasoning of which he saw it making use was not full classical logic. Indeed, according to Hilbert, it is precisely because the logic of finitary reasoning is not classical that we need to introduce the ideal propositions and proofs. Cf. Hilbert (1925, 379–80), Detlefsen (1993, 285–304) for more on this point.

[6] In Brouwer's own words

> [O]n account of the highly logical character of usual mathematical language the following question naturally represents itself: Suppose that an intuitionist mathematical construction has been carefully described by means of words, and then, the introspective character of the mathematical construction being ignored for a moment, its linguistic description is considered by itself and submitted to a linguistic application of a principle of classical logic. Is it then always possible to perform a languageless mathematical construction finding its expression in the logico linguistic figure in question? After careful examination one answers this question in the affirmative (if one allows for the inevitable inadequacy of language as a mode of description) as far as the principles of contradiction and syllogism are concerned; but in the negative (except in special cases) with regard to the principle of excluded third. (Brouwer 1952, 14)

Cf. Brouwer (1908, 108–9; 1933, 443; and 1954, 523–4) for similar remarks. Cf. also Detlefsen (1990) for a fuller discussion of this matter.

what is apparently an easy-to-form misimpression of Brouwer's views on logical reasoning—namely that he thought of it as being of utterly no epistemic value whatever. Perhaps this will help to forestall hasty dismissals of Brouwer's views on the grounds that logical reasoning *is* of some apparent use in procuring knowledge of mathematical truths, and in teaching and communicating mathematics to others. Secondly, we want to warn the reader away from a similarly troublesome misunderstanding of Hilbert's views—namely one which confuses his views on so-called existence claims in ideal mathematics, which he sometimes expressed by saying that consistency suffices for existence, with those of his conception of existence claims in real mathematics, which is where he saw the need for an exhibition requirement. One who fails to make this distinction will never be able to make proper sense of Hilbert's real–ideal distinction and, in particular, his distinction between those existential sentences which are literally meaningful and those which are not (cf. Hilbert 1925, 380). Without this, she will not be able to make any significant progress towards understanding Hilbert's programme generally.

Let us now consider one final piece of background information before proceeding to the more substantive discussion of the exhibition condition itself. This concerns the nineteenth-century reaction to Kant and, especially, the reaction to his idea that arithmetic and geometry, though strictly separate and distinct, are none the less ultimately of the same basic epistemic character. For convenience sake, let us refer to this as *Kant's symmetry thesis*.

For the most part, the nineteenth century rejected this thesis, and the dissenters included both mathematicians and philosophers. On the mathematical side, the leader of the opposition seems to have been Gauss who, well before the discoveries of Bolyai and Lobachevsky, announced his disagreement with the symmetry thesis. Indeed, if the 54-year-old memories of his thoughts as a 15-year-old are to be trusted, Gauss's doubts began to develop within a decade of the first publication of the *Critique of Pure Reason* (1781).[7] Moreover, these doubts apparently only increased with

[7] It is not, by the way, clear that Gauss's memories *are* to be trusted. For though in one place (viz. a letter to Schumacher dated 28 November 1846, cf. *Werke*, VIII, 238–9) he offers a fairly specific dating (viz. the year 1792), which is in rough agreement with the dating cited above, in another place (viz. a letter of 6 March 1832 to the elder Bolyai, commenting on the work of the younger, cf. *Werke*, VIII. 220–1) he suggests a date in the late 1790s ('30–5 years' before the writing of the letter). In the passage in question, Gauss seems basically to be concerned to make a claim to priority, which he 'derives' from a profession of humility. He thus wrote his old acquaintance (who had overlapped with Gauss during the latter's student days in Göttingen) that 'humility' forbade him praising the work of his son since to do so would be to praise his own work, the 'entire content' of the son's work, including both 'the results to which he was led and the route by which he was led to them', having been 'almost entirely' those that had occupied Gauss's own 'Meditations' off and on for past '30–5 years'. (The German: 'Jetzt Einiges über die Arbeit Deines Sohnes. Wenn ich damit anfange, "daß ich solche

time, for, in an 1829 letter to Bessel (cf. *Werke*, VIII. 200), Gauss spoke of an ever increasing (and, by then, nearly 40-year-old) conviction 'that we cannot found geometry in a completely *a priori* manner'.[8]

Nor need we trust only in Gauss's memory. For an 1817 letter to Olbers gives independent corroboration that he had indeed become convinced of the non-*apriority* of geometry well before the announcement of the celebrated results of Bolyai and Lobachevsky.[9]

> I have become ever more convinced that the necessity of our geometry cannot be proven, at least not from the vantage of human understanding and in a way that is accessible to human understanding. Perhaps in another life we'll gain insights into the nature of space that are currently unattainable. Until then, however, we must not classify geometry with arithmetic, which is purely *a priori*, but rather assign it the same status as mechanics.[10]

nicht loben darf": so wirst Du wohl einen Augenblick stutzen: aber ich kann nicht anders; sie loben hieße mich selbst loben: denn der ganze Inhalt der Schrift, der Weg, den Dein Sohn eingeschlagen hat, und die Resultate, zu denen er geführt ist, kommen fast durchgehends mit meinen eigenen, zum Theile schon seit 30–35 Jahren angestellten Meditationen überein.')

It is perhaps worth noting that this claim of priority by Gauss so offended the younger Bolyai that he is said never to have spoken to Gauss again. Considering the fate of many another at whom Bolyai took offence (viz. the thirteen men he once shot in duels in a single afternoon), though, it seems that Gauss got off lightly.

[8] The German: 'wir die Geometrie nicht vollständig a priori begründen können.'

[9] Janos Bolyai announced his results in 1823. The letter containing the announcement, which was written to his father and dated 3 November is translated in Bonola 1955. Lobachevsky presented his findings in a lecture at Kazan delivered 12 February 1826.

[10] Cf. Gauss 1976, 651–2. The German is:

> Ich komme immer mehr zu der Überzeugung, daß die Notwendigkeit unserer Geometrie nicht bewiesen werden kann, wennigstens nicht vom menschlichen Verstande, noch *für* den menschlichen Verstand. Vielleicht kommen wir in einem andern Leben zu andern Einsichten in das Wesen des Raumes, die uns jetzt unerreichbar sind. Bis dahin müßte man die Geometrie nicht mit der Arithmetik, die rein *a priori* steht, sondern etwa mit der Mechanik in gleichen Rang setzen.

Gauss expressed essentially the same view in an 1830 letter to Bessel (cf. *Werke*, VIII. 201), which Kronecker would later quote with approbation in his (i.e. Kronecker's) only philosophical writing (cf. Kronecker 1887, 265, n. 2).

> My innermost conviction is that geometry has a completely different position in our *a priori* knowledge from arithmetic. . . . we must humbly admit that, though number is purely a product of our intellect, space also has a reality external to our intellect which prohibits us from being able to give a complete specification of its laws *a priori*.

[The German is: 'Nach meiner innigsten Überzeugung hat die Raumlehre in unserm Wissen a priori eine ganz andere Stellung, wie die reine Größenlehre . . . wir müssen in Demuth zugeben, dass, wenn die Zahl bloss unsers Geistes Product ist, der Raum auch ausser unserm Geiste eine Realität hat, der wir a priori ihre Gesetze nicht vollständig vorschreiben können.']

It may seem curious that Kronecker, who is best known for his 1886 remark that God created the natural numbers and that everything else is 'Menschenwerk', should adopt Gauss's characterization of the 'in principle' difference between geometry (and mechanics) and arithmetic. One might be able to make some sense of saying that that which is 'purely a product of our intellect' is also God's creation by positing God as the creator of our intellects. But how is one to understand a claim to the effect that that which 'has a reality external to our intellect' is,

After the announcement of these results, Gauss described them, both in the letter to Bolyai mentioned above, and also in his 1832 treatise, *Theoria residuorum biquadraticorum*, as simply adding to the evidence for rejecting Kant's conception of geometrical knowledge.[11] Thus, in the former, we find him saying that 'it is precisely in the impossibility of deciding *a priori* between Σ [Euclidean geometry] and S [Bolyai's non-Euclidean geometry] that we have the clearest proof that Kant was wrong to claim that space is only the form of our intuition'.[12] Similarly, in the latter, he made criticism of Kant's argument for the ideality of space (cf. §13 of the *Prolegomena*) a major theme, and explicitly claimed that, contrary to Kant, it is possible to show that space has 'a real significance' (eine reelle Bedeutung) of its own independent of the forms of our intuition.[13]

Gauss thus denied Kant's symmetry thesis, and, partly through his influence, partly through the influence of others (e.g. Bolyai, Bessel, and Kronecker), this eventually became the prevailing viewpoint amongst nineteenth-century philosophers of mathematics. As a consequence, arithmetical thinking was moved closer to and geometrical reasoning further away from the core of the human intellect. But even though there was a general consensus concerning the need for asymmetrical epistemologies for arithmetic and geometry, there remained the potentially divisive issue of how, precisely, these differences were to be conceived. Here, three main alternatives emerged. The first, favoured by Gauss and Bolyai and described above, was to 'demote' geometrical knowledge to a posteriori (or at least inferior a priori) status and to preserve full a priori status for arithmetical knowledge only. The second, favoured by the logicists Frege and Dedekind, employed what might be regarded as a subtler reading of Kant. It allowed geometry to retain its full a priori status while seeking a more fundamental basis still for arithmetical thinking—a basis that would have its fundaments in rational thought *per se*, and not just in that part of rational thought

for all that, still 'Menschenwerk'? The answer, I believe, may lie in a distinction which Kronecker made between arithmetic in a 'narrower' sense (by which he meant the arithmetic of the natural numbers) and arithmetic in a 'wider' sense (by which he meant an agglomeration of areas including algebra and analysis as well as the arithmetic of the natural numbers (cf. Kronecker 1887, 265)). If one takes arithmetic in the narrower sense to be the work of God, and understands the 'alles andere' of 'Die ganzen Zahlen hat der liebe Gott gemacht, alles andere ist Menschenwerk' as signifying not the likes of geometry and mechanics (disciplines lying wholly outside arithmetic understood in even the 'wider' of Kronecker's two senses), but rather those parts of arithmetic in the wider sense (e.g. algebra and analysis) that had not as yet been reduced to arithmetic in the narrower sense, one obtains a view that is consistent with both of Kronecker's remarks.

[11] The letter to Bolyai refers to the argument in Gauss 1832 as providing a different, but equally strong, refutation of Kant's conception of geometry.

[12] 'Σ' and 'S' are the respective designations used in Bolyai 1832 (cf. §15) for 'the System of Geometry resting on the hypothesis of the truth of Euclid's Axiom' and 'the system founded on the contrary hypothesis'.

[13] Cf. Gauss 1899, 194.

having to do with mathematics. The third, that of the intuitionists, divided geometry into a mathematical part and a spatial-intuitive part. The mathematical part was then absorbed into mathematics through its 'arithmetization' (after the manner of Descartes). The spatial-intuitive part, on the other hand, was simply excluded from mathematics. Thus, what had traditionally been known as geometry was partly absorbed into arithmetic, partly diverted into non-mathematics. The net result, however, was a centralization of arithmetic (and its underlying intuition of time) which moved it to the very heart of human mathematical knowledge and the human epistemic enterprise in general.[14]

Brouwer was explicit in his denial of any appeal to a sensuous intuition of space in mathematics, and he believed this to be the unmistakable message of the work of Bolyai and Lobachevsky.

> [T]he most serious blow for the Kantian theory was the discovery of non-Euclidean geometry, a consistent theory developed from a set of axioms differing from that of elementary geometry in this respect that the parallel axiom was replaced by its negation. . . . [But] however weak the position of intuitionism seemed to be after this period of mathematical development, it has recovered by abandoning Kant's apriority of space but adhering the more resolutely to the apriority of time. This neo-intuitionism considers the falling apart of moments of life into qualitatively different parts, to be reunited only while remaining separated by time as the fundamental phenomenon of the human intellect, passing by abstracting from its emotional content into the fundamental phenomenon of mathematical thinking, the intuition of the bare two-oneness. This intuition of two-oneness, the basal intuition of mathematics, creates not only the numbers one and two, but also all finite ordinal numbers, inasmuch as one of the elements of the two-oneness may be thought of as a new two-oneness, which process may be repeated indefinitely; this will give rise still further to the smallest infinite ordinal number ω. Finally this basal intuition of mathematics . . . gives rise immediately to the intuition of the linear continuum, i.e. of the 'between', which is not exhaustible by the interposition of new units and which therefore can never be thought of as a mere collection of units.
>
> In this way the apriority of time does not only qualify the properties of arithmetic as synthetic a priori judgements, but it does the same for those of geometry, and not only for elementary two- and three-dimensional geometry, but for non-Euclidean and n-dimensional geometry as well. For since Descartes we have learned to reduce all these geometries to arithmetic by means of the calculus of co-ordinates. (Brouwer 1912, 127–8)[15]

[14] Hilbert's response to Kant was different still from all the three just given. It does not, however, in my view, constitute a denial of the symmetry thesis and so it is not included on the current list.

[15] The pagination here and throughout is that of Brouwer 1975. Also it should be noted that, as the last paragraph of the quoted passage makes clear, Brouwer did not believe that geometry would vanish with the abandonment of spatial intuition. Rather, it would be saved, even purified, through its arithmetization.

For Brouwer, then, all of pure mathematics was to be reduced to arithmetic, and Kant's original pair of intuitions, one of space, the other of time, was to be replaced by a single, fundamental intuition of time—an intuition in which 'we become conscious of time as change per se' (Brouwer 1907, 61).[16]

This, of course, does not in and of itself explain how arithmetic knowledge is to be seen as especially basic to human cognition. To do that, Brouwer had to fashion an arithmetic epistemology which revealed its basicness to human thought in general. What he arrived at, we believe, was an epistemology having certain Kantian elements, but also one which draws heavily upon the nineteenth-century opposition to Kant and, in particular, upon the epistemological ideas of the nineteenth-century romantic idealists.

3. 'IN THE BEGINNING WAS THE ACT':[17] EXHIBITION AS CREATION

Superficially, it might seem that Brouwer's main modification of the Kantian viewpoint was his alteration of the role to be played by temporal intuition in the production of mathematical knowledge. For Kant, it was the basis for only a part of our mathematical knowledge—namely that embodied in arithmetic. For Brouwer, on the other hand, it was seen as the basis for the whole of our mathematical knowledge—geometry as well as arithmetic.

But though this surely describes a large-scale difference between Brouwer and Kant, it does not get us to the heart of the philosophical difference separating the two. To do that, I think, we have to look 'inside' Brouwer's conception of temporal intuition and consider its ultimate nature. We get a glimpse of this in Weyl's summary of the intuitionist's reaction to Kant: 'the numbers are to a far greater measure than the objects and relations of space a free product of the human mind and therefore transparent to the mind.'[18] Somehow, arithmetic retains its status as

[16] For a more detailed discussion of Brouwer's rejection of spatial intuition, the reader is referred to the Appendix of this paper.

[17] 'Im Anfang war die Tat'. This is from Goethe, cf. *Faust* I (line 1237, in the Studierzimmer) (p. 48 in Goethe 1948). See also, lines 1224, 1229, and 1233. Cf. also Fichte, *Wissenschaftslehre*, 468.

[18] Cf. Weyl 1949, 22. Brouwer said similar things, some of which were even stronger (cf. van Stigt 1990, 122–3, 129, 135, 174, 177, 188 for further discussion and references). Basically, he seems to have seen one's knowledge of her creations as a form of self-knowledge, and he seems to have thought of this kind of self-knowledge as the only genuine knowledge possible. Gödel, too, in his Gibbs lecture (cf. 15–16 of the manuscript version), refers to such a principle, basing it on some presumed relationship of omniscience and omnipotence that a creator has with respect to her creations. He did not, however, endorse it.

mathematics and (non-arithmetizable) geometry loses its status as mathematics because the former but not the latter is made 'transparent' to the mind by being a creation of it.

Looking further inside Brouwer's conception of temporal intuition, we see that he, unlike Kant, took temporal intuition ultimately to consist in some kind of basic self-awareness or self-knowledge not unlike the so-called *intellektuelle Anschauung* which the post-Kantian romantic idealists (e.g. Jacobi, Fichte, Schelling, Schopenhauer, and Goethe) emphasized in their writings. According to this conception, time is merely a part of what we existentially experience or *live* (as opposed to intellectually comprehend) in *being aware of* our own being, where the pivotal feature of this existential self-knowledge or self-awareness is its *immediacy*, conceived as a kind of epistemic connection with ourselves which is not mediated by representation. In immediate self-knowledge or self-awareness, we are related to ourselves in an epistemically significant (indeed, basic) way that does not involve our representing ourselves to ourselves and thence comprehending that representation. We simply sense or are aware of ourselves directly, without comprehension under a concept.

Part of the motive behind the search for such a *representation-less* form of knowledge seems to have been a particular conception of the nature of error; namely that it is essentially a case of *mis*representation. Representation-less knowledge would therefore have the crucial trait needed for a Cartesian foundation of knowledge—namely incorrigibility or infallibility. Another part of the motive, however, and one that is of considerably greater importance for our purposes here, is more Kantian in origin. It concerns the connection between representation or conception and existence; namely that since, as Kant emphasized, no concept or representation contains or *gives* the being or existence of that which it represents, it follows that knowledge of existence comes by way of something other than our grasp of concepts or representations.

The romantic idealists therefore sought a form of knowledge which provides a kind of contact with existence that grasp of concepts or representations never could. Their reasoning went something like this: We have knowledge of our own existence through a representation-less existential awareness of ourselves. Such representation-less grasp of existence is essentially the only avenue to knowledge of existence that is at our disposal. All knowledge of existence must therefore ultimately be rooted in our existential grasp of our own being. Hence, all knowledge of existence must somehow incorporate or graft the thing-to-be-known into one's own being so that its being can come to be known through the existential grasp that a person has of herself. In other words, in order to know that a thing exists, we must ultimately *live* or *be* that thing as we live ourselves. Schelling summarized the overall position as follows:

[N]othing is immediately certain except the proposition *I exist*, and this is the most particular of all propositions (since it has no meaning outside the confines of immediate consciousness) and also the *absolute presupposition* which must be accepted if anything else is to attain to certainty. The proposition *there are things external to us* will thus only be certain . . . to the extent that it is identical to the proposition *I exist*, and its certainty can only match that of the proposition from which it derives its certainty.[19] (Schelling 1800, 344)

Such ideas seem clearly to be present in Brouwer's writings. A particularly striking and important example of this is his description of the pivotal act of 'turning into the Self'.

[Y]ou know that very meaningful phrase 'turn into yourself'. There seems to be a kind of attention which centres round yourself and which to some extent is within your power. What this Self is we cannot further say; we cannot even reason about it, since—as we know—all speaking and reasoning is an attention at a great distance from the Self; we cannot even get near it by reasoning or by words, but only by 'turning into the Self' as it is given to us. . . . Now you will recognize your Free Will, in so far [as] it is free to withdraw from the world of causality and then to remain free, only then obtaining a definite Direction which it will follow freely, reversibly. (Brouwer 1905, 2)[20]

[19] In the same vein, Schopenhauer (cf. The *World as Will and Idea*, III, 59) remarked that

[E]verything *objective*, everything external, since it is always only something apprehended, something known, remains also always indirect and secondary, therefore absolutely never can become the ultimate ground of explanation of things or the starting-point of philosophy. . . . only that which is given in self-consciousness fulfills this condition, that which is within, the *subjective*.

Cf. also Schopenhauer, *The World as Will and Idea*, I, 145, 163.

Other romantic idealists offered different variations on this theme. Perhaps particularly worth mentioning here is that of Friedrich Schlegel's, where he explicitly distinguishes between intuitive knowledge and proof or explanation by saying that 'in the end, the main thing is that man should *know* something and should assert it. To prove or even just to explain is in most cases entirely superfluous' (cf. Schlegel 1960, I, 2, pp. 197–8; my translation). We cannot enter further into the questions raised by Schlegel's remarks here, though we think they are of considerable potential importance to the working out of a constructivist epistemology along romanticist lines.

[20] In the section following that in which the above-quoted remark occurs (a section entitled 'Man's Fall Caused by the Intellect'), Brouwer went on to describe how the emergence of Intellect led to the abandonment of the Self (loc. cit., 3–4). He there characterized science as 'an infatuation of desire restricted to the human mind' (ibid.). The attempt to find the foundations of science were said eventually to cause one to 'knock up' a 'theory of knowledge' (ibid.), and when all such attempts to fill the gaps in our understanding at last failed, the more prudent would revert to the ultimate philosophical 'plug'—viz. an 'Ego' that comprehends but which is itself incomprehensible (ibid.). The less prudent, on the other hand, would 'keep on and on until they grow mad; they grow bald, short-sighted and fat, their stomachs stop working properly, and moaning with asthma and gastric trouble, they fancy that in this way equilibrium is within reach and almost reached' (loc. cit., 5). 'So much', he concluded, 'for science, the last flower and ossification of culture' (ibid.). There are many more passages along the same lines in this essay. Cf. those on truth as manifestaton of the Self in the sections entitled 'Immanent Truth' and 'Transcendent Truth' (in which latter, the Intellect is also described as 'this gift of the Devil'). Talk of the free and reversible activity of the creative self also figures again prominently in Brouwer 1948 (cf. 480).

It is important because it is in the so-called act of 'turning into the Self' that the seminal idea of Brouwer's entire intuitionist epistemology is found—namely that which he articulated in the so-called *First Act of Intuitionism* which 'completely separates' mathematics 'from mathematical language and hence from the phenomenon of language described by theoretical logic' (cf. Brouwer 1981, 4–5). Brouwer's 'turning into the Self' is motivated by the need for a knowledge of the self that cannot be had by grasping a concept or representation of the self because, as he insists, 'all speaking and reasoning is an attention at a great distance from the Self; we cannot even get near it by reasoning or by words, but only by "turning into the Self" as it is given to us.' It is therefore precisely that which is called for by the First Act of Intuitionism with its demand that we 'completely separate' mathematics from representation and reasoning.

It is striking because of the strong conceptual and phraseological resemblance it bears to Fichte's description of the road to knowledge: 'Look into yourself. Turn away from everything that surrounds you and towards your inner life. This is the first demand that philosophy makes on its followers. What matters is not what is outside you, but only what comes from within yourself.' (*Wissenschaftslehre*, 422)[21]

All in all, then, and despite differences of detail that separate Brouwer's ideas from those of his romantic idealist predecessors, we believe that he adhered to a romantic idealist conception of knowledge in his mathematical epistemology and that he saw mathematical knowledge—and particularly, knowledge of existence—in fundamentally the same way that they viewed knowledge generally. Mathematics is 'created by a free action'[22] of the human intellect—an action of which we have a kind of immediate awareness through our awareness of our Selves (cf. Brouwer 1907, 96–7; 1948, 488, 494; 1951, 4; 1955, 551). To exist in mathematics means 'to be constructed by intuition' (Brouwer 1907, 96),[23] so that our knowledge of the

[21] The German: 'Merke auf dich selbst: kehre deinen Blick von allem, was dich umgibt, ab, und in dein Inneres: ist die erste Forderung, welche die Philosophie an ihren Lehrling tut. Es ist von nichts, was außer dir ist, die Rede, sondern lediglich von dir selbst.' Cf. also *Wissenschaftslehre*, 456–63, 466–8.

[22] The fuller remark is: 'Mathematics is created by a free action independent of experience; it develops from a single aprioristic basic intuition, which may be called invariance in change as well as unity in multitude' (Brouwer 1907, 97). He goes on to say, by way of distinguishing his view from that of F. Meyer, that this 'first act of construction has *two* discrete things thought together . . . ; F. Meyer . . . says that *one* thing is sufficient, because the circumstance that I think of it can be added as a second thing; this is false, for exactly this *adding* (i.e. this setting it aside while the former is retained) *presupposes the intuition of two-ity*; only afterwards this simplest mathematical system is projected on the first thing and the *ego which thinks the thing*.' It is interesting to compare this with the reasoning employed by Dedekind in his famous proof of the infinite.

[23] In fact, Brouwer began this remark with the claim that 'mathematics is quite independent of the material world'. He therefore seems to have been in disagreement with Schelling who said that (cf. Schelling 1800, 350): 'The whole existence of mathematics depends upon intui-

objects of mathematics is basically a product of our existential awareness of ourselves.

The exhibition requirement in Brouwer is therefore, I believe, to be understood in terms of the supposedly special knowledge that a freely creative agent has of her free creations. The idea behind it is roughly that *our epistemic position with respect to that which we create is better than our epistemic position with respect to that which we do not create*. The reasoning behind this idea would seem to be as follows: (i) in a genuinely free act of creation, a creator gets exactly what she intends (such, at any rate, is how we propose to understand the importance of an act of creation's being a *free* act);[24] (ii) if a creator gets in a free creative act exactly what she intends, then a creator's access to her creations is essentially that which she has to her own creative intentions.[25] Thus, (iii) a creator's knowledge of a freely created object is essentially that of her knowledge of her own creative intentions. Assuming, then, that one's knowledge of her own creative intentions is somehow especially clear and direct, it follows that one's knowledge of her creations should be likewise clear and direct.

Special attention must, however, be paid to the chief element of the above line of reasoning; namely to that premiss which says that a creator gets exactly what she intends in her free acts of creation.[26] For convenience sake,

tion and so exists only in intuition, but this intuition is external. Moreover, the mathematician is not concerned directly with intuition (as an act of construction) itself, but only with the construct which can clearly be given an external representation. Philosophy, however, looks solely to *the act of construction itself*, which is a purely internal thing.' Brouwer therefore appears to have wanted to make mathematics into the kind of thing that Schelling described philosophy, and not mathematics, as being, and not what he (i.e. Schelling) described mathematics as being.

[24] Foreshadowings of this idea are found in the writings of the Christian neoplatonists (e.g. Augustine, Boethius, Anselm) and Nicolas of Cusa. The latter even coined a term for the notion—*visio intellectualis*—of a power by whose means one is somehow able to resolve a plurality of opposites into a unity. Such a power (variously referred to as *perfect intellection* or *nonsensible intuition*) was also considered by Kant, though, for him, only God and not humans could possess it (cf. CPR: B 307, 311–12). He conceived of it as an ability to grasp objects freely (i.e. without their being forced on us) and immediately through concepts alone (GS, VIII. 389).

[25] Presumably, the reasoning would not proceed in quite this way. For the creator would not only have to get what she intends, but also *know* that this is so in order for the inference to go through. This in turn would seem to require that she know her creative act to be free. Such knowledge, however, might also qualify as some kind of self-knowledge, and so sustain the basic epistemic benefits sought. In order to obtain a fully worked-out account, clearly more would have to be said. In the interests of space, however, we leave such discussion for another occasion. It is the basic idea that is important, and the basic idea is that stated above.

[26] A problem that we shall only mention here is what might be called the 'latency problem'. This is the problem of the extent to which creative acts are to be permitted to *contain* other creative acts as latent in them. To the extent that such containment is permitted, creative acts would seem to be less than perfectly transparent, and their immediacy therefore less than complete. There would therefore appear to be a tension between permitting latent containment of creative intentions in one another and the basic creativist ideal of the immediacy of the creator's access to his creations. Hence, the creativist would seem to be committed to minimizing latent creation, and thus to minimizing the role of anything like logical inference or

let's refer to this as the *thesis of creatorly infallibility*. What I should now like to do is to consider two different ways of thinking of this thesis: one which, for lack of a better term, I will call *metaphysical*, the other *logical*. I begin with the logical conception.

On this way of understanding the thesis—a way which bears recognizable affinities to certain ideas commonly associated with Wittgenstein—there is no *logical* distance between existence in mathematics and creatorly intention. Mathematics, so the view goes, is an activity under our sovereign control. We *decide* what is to be included in it and what not. As principal parts of that activity, mathematical proofs, including existence proofs, are therefore to be seen as results of our legislative intentions. They do not serve evidentially to compel us to accept their conclusions. They do not, that is, function to remove doubt from their conclusions through *penetration of* the truth. Rather they function to *logically exclude* error by establishing a proof as a *norm* governing or determining what is to count as mathematical activity.

Creation (i.e. that which underlies claims to existence) on this view is thus ultimately an exercise of a kind of *legislative power*; and the special epistemic status of our mathematical creations is accordingly the result of our having a *freedom* or *power* to *decide* what is and what is not to be included in that activity which constitutes mathematics.

We do not believe that Brouwer and Weyl had such a logical conception of the creatorly omnipotence thesis in mind. Rather, they seem to have favoured a view in which creatorly infallibility is principally understood not in legislative or prescriptive but rather in descriptive terms—terms ultimately having to do with the special epistemic *penetration* that a creator is supposed to have of her creations. This penetration is ultimately a product of the descriptive knowledge that the creator has of herself and, in particular, of her creatorly intentions. It is not legislative in character because the self of the creator's self-awareness (which is ultimately the object of her knowledge) is not (at least not on the present conception) a product of her legislative decision. The freedom of the free creative self consists not in her ability legislatively to determine her basic nature, but rather in her being free to act in accordance with that nature. Acting freely in this way, the creator's free creations stem or emanate or derive from or are a part of herself. Hence, her epistemic access to them is a function of her epistemic access to herself, and that, according to the romantic idealists, is supposed to be particularly good. More specifically, on the romantic idealists' con-

logical development in mathematics. Brouwer, of course, adhered to this. He also, however, maintained that all of mathematics is based on 'a single aprioristic intuition', and this would seem to require latent containment of a vast body of creative acts in the space of a single intuition (viz. the ur-intuition of time).

ception, one's knowledge of one's own existence sets the standards for knowledge of existence generally. Therefore, to the extent that one's awareness of self carries with it awareness of a construction, one's knowledge of the existence of the object constituted by that construction is likewise assured.

Arguments of a related sort go back at least to the Schoolmen. They (at least some of them) conceived of Divine creation as creation *ex nihilo*, or creation from sheer creative intention. The reasoning was as follows: The objects of God's creative intentions emanate or flow from His being and are therefore a part of it. But God's being sets the standards for existence generally. Therefore, to be an object of creative intention in the mind of God is to exist.[27]

I think the argument of the romantic idealists works in something like the same way—except, of course, that it is embedded in an idealist rather than a realist setting. The key idea is the same however: namely that our existence sets the standard for existence generally or, if one prefers a more epistemological formulation, that our knowledge of our own existence sets the standard for knowledge of existence generally. We are therefore in the position of 'gods' with respect to those existents whose construction flows from our creative selves.[28] They exist because we exist and they are a part of us.

We have here then, I believe, what are at least the epistemological rudiments of the intuitionist's conception of the exhibition requirement. Through conferring on mathematical entities the same epistemic force or immediacy as the existence of our own willing, acting Selves, this conception of exhibition presents a solution to Kant's fundamental requirement that objects be somehow 'posited in themselves, outside the mere thought of them'. It provides that extra epistemic force exceeding what can be pro-

[27] Note that this is neither an argument from the power or omnipotence of God nor an argument from His legislative authority. The reasoning is not that to exist as an object of God's creative intension is to exist because God has the power or legislative authority to do all that He intends. Rather, it is that *in being* an object of God's creative intention that object of intention *shares in His being* and so exists because God's being sets the standards for existence generally. Hence, far from making use of God's power as a premiss, this reasoning actually serves as a model for explaining God's creative omnipotence (conceived of as His ability to make good on all His creative intentions). God is creatively omnipotent because in being an object of His creative intention, an object has all the being it ever could have—namely the being of His being.

[28] The idea that we are God-like with respect to our intellectual lives was also an idea emphasized by Dedekind. Like Brouwer and Weyl, he held that the numbers were 'free creations of the human mind' (cf. Dedekind 1887, §73). Moreover, he went on to say that we, their creators, are 'a god-like race, possessing, without any doubt, creative power not only with respect to material things (e.g. railroads and the telegraph), but also, indeed especially, with respect to intellectual things' (cf. *Werke*, III. 489). This is my translation of the German: 'Wir sind göttlichen Geschlechtes und besitzen ohne jeden Zweifel schöpferische Kraft nicht blos in materiellen Dingen (Eisenbahnen, Telegraphen), sondern ganz besonders in geistigen Dingen.'

vided by 'mere thought'—even thought of a logically perfect sort—that Kant viewed as necessary for knowledge of existence and which he saw as being the special office of intuition. The Brouwer–Weyl conception of exhibition would therefore appear to be one which fully merits the title of 'Intuitionism'.

As we shall now see, this intuitionist conception of the exhibition requirement, with its basic reliance upon the primacy of self-knowledge, differs radically from the conception of the exhibition requirement found in Hilbert's finitary epistemology.

4. 'IN THE BEGINNING... IS THE SIGN': EXHIBITION AS PUBLIC DISPLAY

At the heart of intuitionist epistemology, therefore, was a pronounced *individualist* theme—an identification of the self as not only the ultimate *source* of all knowledge but also the ultimate *object* of all knowledge. Hilbert, on the other hand, in effect denounced the subjectivity of such a viewpoint and opted instead for a constructivist epistemology having an inter-subjective understanding of the exhibition requirement. He found the intuitionists' focus on the inner life and self-knowledge too subjective a basis upon which to found mathematical knowledge. Indeed, he saw the romantic idealism of the intuitionists as emblematic of the very type of subjectivism from which he believed scientific inquiry is supposed to liberate us. As he put it, 'it is part of the task of science to liberate us from arbitrariness, sentiment and habit and to protect us from the subjectivism that already made itself felt in Kronecker's views and, it seems to me, finds its culmination in intuitionism.'[29] He therefore consciously adopted an ideal of knowledge which emphasized objectivity or inter-subjectivity rather than subjective certainty.[30]

In Hilbert's finitary epistemology we therefore find a more *communitarian* conception of knowledge according to which the domain of mathematical existents is to be open to all members of the human (or, at least,

[29] Hilbert 1927, 475.

[30] The intuitionists, though they would surely have admitted the dominance of objectivity as an ideal of Western epistemology, would none the less have decried it as deeply mistaken. Their view was well expressed by Jacobi who wrote (cf. *Werke* II. 11; my translation)

> Ever since the time of Aristotle, there has been an increasing tendency in the philosophical schools to subject immediate knowledge to non-immediate, the original and fundamental faculty of intuiting or perceiving to the faculty of reflection, which operates by abstraction—to subject the original to its copy, the essence to the word, Reason to Understanding.

To them, then, Hilbert's finitist campaign slogan—'Am Anfang . . . ist das Zeichen' ('In the beginning . . . is the sign', cf. Hilbert 1922, 163) would have been anathema. For it represented only the latest expression of the wearying and regrettable tendency of Western epistemology to subject 'essence to word' and the immediacy of being to the abstraction of representation.

the human-scientific) community equally. This does not, of course, imply access to a realm of things as they are in themselves, unconditioned by (possibly pervasive) traits of human cognition. Rather, it suggests only the possibility of attaining a kind of exploitable trans-communitive convergence with respect to judgements concerning the basic characteristics of mathematical existents. The corresponding conception of the exhibition requirement is therefore essentially this: *though the exhibition of an exhibited object may actually be 'staged' or produced by a single individual or group of individuals belonging to an epistemic community, it is none the less equally accessible to all members of that community*.

This trans-communitive symmetry of epistemic access to the objects of exhibition is the core feature of Hilbert's finitary conception of exhibition. Because of it, there is no significant feature of an exhibited object that can either be known only or known better by an exhibitor than a non-exhibitor. The finitary conception of exhibition is therefore starkly at odds with the intuitionist conception which conceives of exhibition as inducing an irreducible asymmetry of epistemic access between the exhibitor (i.e. the creator) and the non-exhibitor (non-creator).

The primary motive for this objectivist or communitarian ideal of exhibition seems to have been a desire to facilitate communication. This in turn would facilitate such communitarian goods as convergence of judgement concerning the properties of the basic objects of mathematical thought (i.e. what, in Hilbert's view, could be described as *finitary* objects), which in turn might lead to such communitarian epistemic benefits as, say, a productive, fair, and efficient division of epistemic labour.

The resulting view is, we believe, both deeply Kantian and deeply Leibnizian in character. Before proceeding to a more detailed explanation of this odd remark, however, let me first provide a little general background for the sake of orientation.

Like the intuitionists, Hilbert, too, regarded the discovery of non-Euclidean geometries as cause for rejecting Kant's conception of geometry as based on an a priori intuition of space (cf. Hilbert 1930, 383).[31] Unlike the intuitionists, however, he equally opposed the idea of basing arithmetic on a Kantian a priori intuition of time, believing this to have been discredited by Einstein's relativistic theory of time (cf. Hilbert 1930, 384). He therefore rejected both Kant's belief in an a priori intuition of space and an a priori intuition of time. In this respect, then, he was more strongly opposed to Kant's mathematical epistemology than were the intuitionists.

[31] Indeed, he went so far as to say that 'Geometry is nothing other than that part of the collected framework of physical concepts which deals with the possible chains of connections amongst fixed bodies in the world of actual things' (Hilbert 1930, 383), and that 'in recent times, the view, introduced by Gauss and Helmholtz, concerning the empirical nature of geometry has become a secure result of science' (ibid. 384).

Yet despite his rejection of both Kant's a priori intuition of space and a priori intuition of time, Hilbert did not reject a priori intuition altogether. He did not, that is, become a mathematical empiricist.[32] Instead, he maintained what was a basically Kantian epistemology, at least in its structural elements. In particular, he argued explicitly for a 'third' source of knowledge in addition to experience (Erfahrung) and logic (Deduktion) (cf. ibid. 382–3), and he described his own conception of a priori knowledge as that which remains when the Kantian theory of the a priori is stripped of its 'anthropomorphic garbage' (ibid. 385). Moreover, he remained committed to Kant's general distinction between a part of thought (what Hilbert termed *real* thought) whose business is the making of literal, genuine judgements and inferences, and a part of thought (*ideal* thought) which serves as an instrument for regulating and developing the growth of the real part. Finally, like the intuitionists, he relied heavily on Kant's observation that genuine knowledge of existence requires more than merely a clear, consistent conception of the would-be existent.[33] In combination, these ideas allow us to resolve what would otherwise be a serious conflict in Hilbert's thought; namely that created by the fact that he sometimes claimed that consistency suffices for existence claims in mathematics and at other times insisted existence claims require a supporting exhibition. The conflict is resolved by realizing that existence claims in *ideal* mathematics must be distinguished from existence claims in *real* mathematics. Consistency (or at least real-soundness) may suffice for the former, but the latter require supporting exhibitions.

We noted a moment ago that Hilbert described his view of the a priori as that part of Kant's theory that remains after it is stripped of its anthropomorphic garbage. But what part is this? For our purposes, it can be seen as consisting in the establishment of the so-called 'finitary setting' (die finite Einstellung)[34]—a kind of framework or structure or, to use Hilbert's term, 'base setting' (Grundeinstellung) of our mathematical and scientific knowledge generally (cf. Hilbert 1930, 383). This base setting, according to Hilbert, consists of those judgements or cognitions which represent the minimum necessary for all objective (i.e. inter-subjective) human thought and communication. It is founded on the innate capacity humans have to grasp the shapes of concrete signs, and to reckon with them in a variety of elementary ways (cf. Hilbert 1925, 376 and 1927, 464–5).[35] Hilbert saw the basic

[32] See Nelson 1971, 153 for an assessment of the causes of the 'degeneration' of post-Kantian philosophy into a reenactment of the pre-Kantian struggle between dogmatic empiricism and dogmatic rationalism.

[33] We say this knowing full well, of course, that Hilbert said such things as that existence in mathematics is simply consistency. One must distinguish between existence in the ideal part of mathematics and existence in the real part. More on this in a moment.

[34] Hilbert here refers the reader to his 1925 essay for a description of this 'setting'.

[35] For the most part, Hilbert claimed no precise knowledge of the basic laws of this capacity, though in his later years he apparently came to think that they were in some sense well described by the laws of Primitive Recursive Arithmetic.

structure of his mathematical epistemology (his 'programme', if you will) as intended to provide a kind of Kantian 'deduction' of this basic knowledge as the irreducible epistemological prerequisite for all human mathematical knowledge and, indeed, for all human scientific knowledge generally. In his own words:

> There have in fact been philosophers—and Kant is the classical representative of this viewpoint—who claimed that aside from logic and experience we have yet a certain a priori knowledge of reality. Now I admit that even construction of a theoretical framework requires certain a priori insights, and such insights always underlie the acquisition of knowledge. I believe that in the final analysis mathematical knowledge rests on intuitive insight (anschauliche Einsicht) of just this type, and that we must take a certain such a priori approach even to construct number theory. Therein is contained the meaning of the most general fundamental thought (der allgemeinste Grundgedanke) of Kantian epistemology: namely the philosophical problem of determining the basic a priori intuitional setting of all conceptual knowledge and experience and therewith to investigate the condition of its possibility. I believe that this is essentially what is done in my researches concerning the principles of mathematics. The a priori is thus nothing more and nothing less than a base setting (Grundeinstellung) or expression (Ausdruck) of certain irremissible preconditions (unerläßliche Vorbedingungen) of thought and experience. However, the border between what we reserve for the a priori, on the one hand, and what we regard as requiring experience, on the other, must be drawn differently from the way in which Kant drew it. Kant greatly overestimated the role and compass of the a priori.[36]

So, structurally, Hilbert's epistemology bears a strong similarity to Kant's. It preserves a literal–regulative division and also preserves a place for a kind of a priori cognition, albeit one which diverges in various ways from Kant's. This a priori cognition, which Hilbert believed to consist in our ability to identify and reckon with spatial shapes or forms in a variety of elementary ways, formed the basis of his conception of exhibition. He assigned it three principal traits. The first is that it is a form of cognition in which the forms or shapes (Gestalten) of possible concrete signs are 'given'

[36] Hilbert (1930, 383); my translation. Compare this with the description given in Bernays 1933 (my translation).

> Investigation in the foundations of mathematics has shown . . . that a certain kind of pure-intuitive knowledge must be used as the starting-point for mathematics and that not even logic, as a theory of judgements and inferences, can be developed without making use of this kind of intuitive knowledge. We have in mind the intuitive representation of discreta from which we derive our most basic combinatorial representations and, in particular, that of succession. Constructive arithmetic develops in accordance with this elementary intuitive knowledge.

Compare it also with the description in Hilbert (1922, 163), where the finitary standpoint is described as that 'firm philosophical setting . . . demanded both for the founding of pure mathematics and for scientific thought, understanding and communication generally'. A similar description is given in Hilbert (1927, 465).

to us in an 'immediate' way ('in der Vorstellung gegeben' and 'unmittelbar anschaulich gegeben', cf. Hilbert 1925G, 171). The second is that it is a form of cognition in which the contents are immediately clear (unmittelbar deutlich).[37] The third is that it is a form of cognition whose judgements are open to repeatable confirmation ('wiedererkennbar', ibid.), where, we suggest, repeatability is to be understood not only in an intra-subjective but also in an inter-subjective sense.[38]

Finitary intuition was thus fitted with three features: immediate givenness, immediate clarity/distinctness, and inter-subjective confirmability. We should now like briefly to comment upon these features. Perhaps the first thing to note in this connection is that the immediacy of the 'givenness' of finitary intuitions is not the same as that of the intuitionists' intuitions. For the intuitionists, immediacy served the end of certainty.[39] The naked self, disclosed to itself through a kind of representation-less, existential awareness, was supposed to make misrepresentation, and hence error, impossible. Thus, as noted above, immediacy for the intuitionists stood for something like the difference between the *merely apprehended* and the *lived.*

The immediate 'givenness' of finitary intuition, on the other hand, had little if anything to do with representationlessness—however that might be understood. Rather, it had to do with a certain phenomenological aspect of intuition that Kant emphasized in his distinction between intuition and the understanding: namely the *involuntariness* of intuition. Kant thought it important to distinguish between those representations—his so-called concepts—which are the result of our own cognitive initiative, so to speak ('selbst ausgedachten', to use Kant's terminology, cf. CPR: A: 639; B: 667;

[37] It may be that 'immediately distinguishable' would be a better translation than 'immediately clear'. For there is a tradition in German philosophy (including Leibniz and Kant) of distinguishing 'Klarheit' from 'Deutlichkeit'. Roughly, the former signifies a kind of internal transparency of an idea and the latter a kind of distinguishability of an idea from all other (simple) ideas. Cf. G. Gabriel, 'Klar und Deutlich', in *Historisches Wörterbuch der Philosophie* (J. Ritter and K. Gründer (eds.), Schwabe & Co., Basel/Stuttgart, 1976), vol. 4, 846–7. Whether Hilbert was mindful of this tradition in his use of the term is, however, unclear. I would like to thank Gereon Wolters for calling this to my attention, and Christiane Schildknecht for a useful discussion of the history of the distinction.

[38] Cf. also Hilbert (1922, 162–3) for a similar description. It is also worth noting that Hilbert insists on the 'repeatability' of finitary intuition despite the fact that he uses the term 'Erlebnis' in describing it as 'immediate experience prior to all thought' (unmittelbares Erlebnis vor allem Denken) (cf. Hilbert 1922, 162; 1925G, 171). I mention this because 'Erlebnis' is the word that is often used for personal knowledge or experience, and was widely employed by the post-Kantian idealists in exactly this way. The term generally employed in philosophical discussions of knowledge by experience is 'Erfahrung', and this is the term used in Hilbert (1930) when he talks about experience as one of the fundamental sources of knowledge. This is perhaps some small further confirmation that Hilbert did not see finitary intuition as reducible to sensory intuition, and that he was sincere in treating it as a separate 'third' source of knowledge.

[39] This was not 'certainty' in the sense defined by Kant in the CPR: A: 820–2; B: 848–50.

cf. also A: 51, 68; B: 75, 93), and those that are forced or imposed upon us regardless of our own cognitive initiatives—his so-called intuitions. Hilbert apparently followed Kant in marking such a distinction, and regarded finitary intuitions as involuntary (i.e. gegeben) and possessed of what might be called *original* assertoric force. That is, they were believed to command our cognitive assent and to do so *immediately* (i.e. 'unmittelbar', or as part of the very substance of the intuition itself) rather than in a derivative way by means of some sort of *reflection on* the intuition.[40]

This involuntariness or 'givenness' of finitary intuitions is, I believe, a crucial feature of Hilbert's conception of existence in real mathematics and hence of his conception of the exhibition condition. For, in real mathematics, Hilbert seems to have accepted the Kantian idea that knowledge of existence cannot be secured through mere conception, even conception satisfying the strictest conditions of coherency or consistency.[41] That is, he seems to have believed that, in addition to clear conception, there is a need for the object to be impressed upon us with some extra-conceptual epistemic force—a force which underwrites its objectivity (or inter-subjectivity) and assures us that it is not our own private invention or imagination. This, presumably, is what is supposed to happen in finitary exhibition.

Neither Kant nor Hilbert, however, believed that involuntariness was the only condition necessary for the establishment of a given type of intuitive cognition as objective in nature. Kant's argument for this was essentially as follows: there can be idiosyncrasies of an individual mind that are phenomenologically involuntary; consequently, phenomenological involuntariness is not, in and of itself, a sufficient reason for regarding something as inter-subjective (i.e. 'not idiosyncratic to one's individual mind'). He therefore introduced a distinction (cf. CPR: A: 820–2; B: 848–50) between two types of cognitive or epistemic compulsion: one a *subjective* compulsion, which he called 'Überredung' (and which might be translated as 'mere persuasion'), and the other an *inter-subjective* or *objective* compulsion, which he called 'Überzeugung' (and which might be translated as 'rational conviction').

The former was characterized as being 'privately valid' and having its ground 'only in the special character of the subject' (CPR: A: 820; B: 848). Thus though, subjectively speaking, it might have the same compulsive force

[40] Cf. Nelson (1912, 190–2) for a useful and interesting discussion of the immediacy of intuition that proceeds in a somewhat different direction from the one we take here.

[41] Cf. CPR: A: 150–3; B: 189–92 for Kant's discussion of the place of consistency in knowledge. Consistency, he maintained, is a sufficient condition for analytic knowledge, but it is never sufficient for synthetic knowledge. As he said: 'even if our judgement contains no contradiction, it may connect concepts in a manner not borne out by an object, or in a manner for which no ground is given . . . and so may still, in spite of being free from all inner contradiction, be either false or groundless' (CPR: A: 150; B: 190).

as Überzeugung, the 'givenness' of Überredung derives from certain individual characteristics of the judging subject and so is private or subjective in a way that decreases its epistemic value. Überzeugung, on the other hand, is a form of epistemic compulsion which affects everyone (or at least everyone 'in possession of reason' (ibid.)) in the same way. The criterion for deciding whether a given intuitive judgement is a case of 'Überzeugung' or merely a case of 'Überredung' is therefore whether the grounds producing compulsion in the judging subject are such as would compel assent in *every* being possessing human reason. For such inter-subjectively binding cognitions, Kant believed, there is 'at least the presumption that the ground of the agreement . . . rests upon the common ground, namely upon the object' (ibid.).

We believe that Hilbert had something similar in mind in his insistence on the 'Wiedererkennbarkeit' or repeatable confirmation of finitary intuition (cf. Hilbert 1925G, 171). This would establish a kind of inter-subjectivity of intuition which, in combination with its 'givenness' would give the object of a finitary exhibition an existence which goes beyond the mere thinking of it and, at the same time, insure that this 'givenness', is inter-subjective and not the mere idiosyncracy of the individual exhibitor.

The epistemic asymmetries between creator and non-creator which, as seen earlier, were so important to the intuitionist exhibition, are thus eliminated from finitary exhibition.[42] It yields an object which both exhibitor and non-exhibitor alike can, and, indeed, must judge by its *inter-subjectively* confirmable effects.

Because of these differences between finitary and intuitionistic exhibition, we believe it to be a serious mistake to think of them, as Gödel did, as descendants of a single generic conception of exhibition which merely place different restrictions on the abstractness/concreteness of the objects to be exhibited. Rather, they differ most radically and fundamentally over the very conception of exhibition itself.

5. CONCLUSION

What, ultimately, the differences between the finitist and intuitionist conceptions of exhibition come to, however, is a question that is both too deep and too complex to be adequately dealt with here. Nonetheless, I would

[42] We are assuming here, of course, that there are no laws relating an agent's creative intentions to the effects of her actions that are so certain that, knowing them, a creator could deduce the effects of her creative acts with greater confidence than could be achieved through observation of those effects. Such an assumption seems plausible enough, however, once one has given up on any logical or metaphysical connection between creators' intentions and the effects of their creative acts.

like to roughly indicate some of my ideas on these more basic matters.

Communitarian conceptions of knowledge, such as Hilbert's, are, I believe, motivated by two fundamental kinds of concerns: on the one hand, concerns of epistemic justice or equity, and, on the other, concerns of epistemic co-operation or efficiency. Individualist conceptions of knowledge like those of Brouwer, on the other hand, are driven by quite different concerns: namely concerns of epistemic appropriation or agency or authenticity, and concerns of autonomy.

As an excellent example of epistemic communitarianism motivated by concerns of justice and equity, let us consider Leibniz. His *characteristica universalis* and *calculus ratiocinator* were, it seems to me, inspired by two basic observations: firstly, that native intellectual ability—and, in particular, what we might think of as native *intuitional* ability—is distributed across humanity in a relatively uneven way; secondly, that the ability to will simple, elementary acts such as those involved in a computation is distributed across humanity in a relatively even way. He therefore, in a generally reformationist spirit, sought a means of effecting a *redistribution* of the inequitably distributed epistemic shares determined by birth and upbringing by substituting the relatively equitably distributed capacity to will elementary acts of syntactical or calculary manipulation for the inequitably distributed native intellectual or intuitional ability. In this way he hoped to alleviate at least some of the grosser and more morally repugnant inequities (both of an intellectual, but also, derivatively, of a political and economic nature) of birth. He believed that the *calculus ratiocinator*, with its extremely elementary operating procedure, could be operated by one having only the most rudimenatary intellectual ability and that, diligently applied, it could bring the full range of the humanly knowable within the epistemic reach of such a calculator. Will-power was therefore to take the place of intuitional ability with the consequence that the only reason for ignorance would be laziness or lack of industry. The lazy, however, have only themselves to blame for their lack of epistemic (and other) goods.[43]

There is little to indicate that Hilbert himself was motivated by such lofty moral ideals in the design of his programme. He seemed, none the less, to be moved by considerations of co-operation and division of labour. Structured in the right way, this would make the total knowledge available to a community greater than that attainable by any one or any subgroup of its individual members (an ideal that was also emphasized by Leibniz). Attempts to divide epistemic labour, of course, produce a corresponding

[43] This is not quite right since Leibniz also acknowledged that will or industry is not entirely evenly distributed across humanity (despite the fact that its maldistribution is nothing like that concerning native intuitional ability). Nonetheless, he took it as a duty for individuals and governments to do what they could to encourage the discouraged and generally to assist those suffering from a deficit of industry or will.

need for an epistemically 'inexpensive' means of controlling the quality of the divided contributions so as to keep inferior constituent products from contaminating the corporate product and to guard against free-riders and saboteurs. It may be that formalization of the kind created by Frege and Russell, and advocated by Hilbert in his foundational writings, provides such a means. For such formalization both (i) forces public disclosure of the 'credentials' of any candidate vying for admission to the community's corpus of approved methods or beliefs, and, because of the elementariness of the operations involved in determining conformity with a formalism, (ii) shows some signs of being able to do so without causing a serious drain on epistemic resources that might otherwise be diverted to more direct and efficient production of epistemic goods.[44] Formalization therefore provides the makings of a public standard for individual epistemic contribution that a whole community might embrace in its attempts to safeguard the quality of the common epistemic product and monitor the evenness of participation in its production.

But though Hilbert may have been moved by considerations of epistemic co-operation if not considerations of epistemic justice in framing his conception of exhibition, neither desideratum seems much to have moved Brouwer. This is no doubt in part due to his mistrust of all things corporate. He viewed what ordinarily passes for communication and co-operation as nothing more than the inventions of crafty and determined imperialists. Both, he believed, are at bottom simply covert attempts at 'will-transmission'. Communication and, especially, communication of 'reasoning' is an attempt, by means of an artfully concocted myth of 'reason', to compel the communicatee to do what the communicator wants him/her to do. Co-operation is likewise a kind of sleight of hand whereby one agent attempts to bring the negative power of the label of 'uncooperativeness' to bear in bringing about a certain piece or trait of desired behaviour. Both, then, are attempts to hide what is really the imposition of the will or preference-system of one individual or group on another under a kinder, gentler name—'communication' or 'co-operation'—a name that would put refusal to capitulate in a bad light. Language, formalization, communication, and co-operation are therefore all tools of the imperialists of the will in their attempts to conform the will of the autonomous individual to their own.

Partly, however, it was also due to the fact that Brouwer, like the romantic idealists before him, had very high standards of what I will call *epistemic appropriation* or *epistemic agency*. Brouwer was concerned to make math-

[44] This latter feature gives the reason why the operation of a quality-control calculus must be elementary. The total epistemic effort required to operate the calculus must remain small in comparison to the total amount of effort available to the community for the pursuit of epistemic goods in general. Otherwise, the epistemic effort diverted from activities having the potential to be more directly productive of epistemic goods cannot be justified.

ematics matter. He therefore sought its origins in the deepest recesses of the human cognitive soul. The view, I believe, can be roughly described as follows. It is one thing for a proposition to be plausible or what we might call merely believable, and quite another thing for it to be incorporatable into a set of epistemic cares and concerns around which a human cognitive life might be centred. This is related to the question of epistemic agency; that is, the question of when a belief becomes the belief *of* an agent. The answer, I think, is that a belief becomes a belief of a given agent when but only when it comes to occupy a place of centrality in her human epistemic life—when, that is, it is either linked to the deepest parts of her individuality, or when it is seen as belonging to that common set of cognitive matters that are determinative of human epistemic life generally and which thus make one's epistemic life a distinctively *human* epistemic life.

I believe that Brouwer thought of things in this way. He was a kind of mathematical 'existentialist' in that he sought a conception of mathematical knowledge that makes it basic to our human existence. For this to be so, he believed, mathematical knowledge must be intimately related to our most basic knowledge of ourselves—that is, to that 'existential' awareness that we have of ourselves as willing, acting beings. Such views are, to be sure, extraordinary when viewed in the light cast by the bland and vapid views that have dominated the philosophy of mathematics of the past few decades. They are not, however, for all that, either silly or out of place.

These remarks, of course, only skip across the surface of what are very deep waters. Still, I hope, they may serve to indicate in general terms where I see the deepest differences between finitist and intuitionist epistemology and, in particular, between their respective conceptions of the exhibition condition as lying. They connect, in the end, I believe, with matters as basic as those which concern the relation of the individual to society and our private to our public selves.[45]

APPENDIX: BROUWER'S REJECTION OF SPATIAL INTUITION

Brouwer presented some of his arguments in favour of rejecting spatial intuition (loc. cit., 63–4) as part of a critique of Russell's views of space. Contrary to Russell, he claimed that space is but the 'form of externality' in our sense experience and need not be admitted as a necessary condition of that experience: 'empirical space is an arbitrary creation to enable us . . . with the aid of mathematical induction to

[45] A more detailed discussion of some of the issues touched on in the concluding section may be found in the author's 'Peace, Justice and Computability: Leibniz' Program and Church's Theorem'.

bring *different* causal sequences . . . together *under one point of view*.' A world of external objects is therefore a created 'idealizing summary of real experiences as parts of a fictive totality of possible experiences' (ibid. 64). It does not entail that 'invariants which merit the name of objects will enter into real experience' (ibid.).

Later in the same work (cf. 68–71), Brouwer addressed himself directly to Kant, and specifically to Kant's argument for the thesis that 'the perception of an external world by means of a three-dimensional Euclidean space is an invariable attribute of the human intellect; another perception of an external world for the same human beings [being] a contradictory assumption'. Brouwer took Kant to have partially based his defence of this thesis on a two-part premiss to the following effect: (i) 'we can obtain no external experiences barring those placed in empirical space' (ibid.), and (ii) we 'cannot imagine those experiences apart from empirical space' (ibid.). He objected to the second part, and therefore to the overall conclusion of the ensuing argument, as follows:

> [W]e *obtain* our experiences apart from all mathematics, hence apart from any space conception; mathematical classifications of groups of experiences, hence also the creation of the space conception, are free actions of the human intellect, and we can arbitrarily refer our experiences to this catalogization, or undergo them unmathematically.
>
> Consequently the addition to the first premiss, that we cannot imagine our known external experiences apart from our space conception, is definitely false. With it the conclusion basing the *apriority* of three-dimensional Euclidean geometry mainly on this addition, must also be rejected. (Brouwer 1907, 68)[46]

What we see here, then, is a rejection of the Kantian framework that goes well beyond the rejection of his symmetry thesis and its presupposition of separate intuitions of space and time. It is a rejection of the entire Kantian conception of the a priori as a whole. For Kant, apriority consisted in inseparability from external experience. For Brouwer, it did not. In his view, 'not only does . . . mathematics exist independently of all experience, but all experience is also independent of all mathematics. Human experience is not *passively subjected* to any single mathematical system; not even to the co-ordinate of time; nor even to the time-continuum devoid of measure. . . . Mathematics develops out of its basic intuition in a self-multiplication guided by *an entirely free choice*' (loc. cit., 70, emphases mine).

[46] What Brouwer refers to here as 'the addition to the first premiss' is what we have identified above as its second part. Brouwer goes on in this same passage to attack even the validity of Kant's argument. Briefly, his argument (cf. 68–9) is that we can both construct and relate our experience to geometrical schemas other than Euclid's. Because of this, says Brouwer, 'the human intellect . . . can . . . disengage itself very well from three-dimensional Euclidean geometry.' One might also wonder at the 'hence' that links Brouwer's claim that we obtain our experiences apart from all mathematics to the further claim that we obtain our experiences apart from any space conception. This would seem to require that the 'space conception' be included in mathematics. But didn't the preceding quote from Brouwer suggest that he rejected the idea that spatial intuition is genuinely mathematical? One can resolve this seeming contradiction by reading 'space conception' as pertaining to arithmetized geometry and bearing in mind that Brouwer viewed even the primordial intuition of time as something that is imposed on experience only through an act of will.

On Brouwer's view, then, whatever necessity or apriority belongs to mathematics is not, as it was for Kant, due to any putative inseparability of mathematics from experience that it possesses. Rather, it is the result of certain organizational features of the human intellect; in particular, the fact that the most fundamental of all human intellectual phenomena is the primordial intuition of the two-ity of time (ibid. 70–1) and that mathematics is based on a 'concentration' of the sequences arising from this intuition within the human intellect (ibid. 53). The organizational tools available to the human mind (i.e. the tools available to it for organizing its thoughts and experiences) are those which derive from its ability to arrange things into temporal sequences. The mind is able to extract the order-types of these sequences without immediate thought of applicability, though it none the less has these order-types available to it as possible orderings which might be 'projected into reality' (ibid.) should the opportunity arise.[47] A priori judgements are therefore those which reflect the human mind's most basic organizational abilities, and these are intellectual, as opposed to purely sensible, in character. As Brouwer put it (1907, 53–4):

Proper to man is a faculty which accompanies all his interactions with nature, namely the faculty of taking a *mathematial view* of life, of observing in the world repetitions of sequences of events, i.e. of causal systems of time. The basic phenomenon therein is the simple intuition of time, in which repetition is possible in the form: 'thing in time and thing again', as a consequence of which moments of life break up into sequences of things which differ qualitatively. These sequences thereupon concentrate in the intellect into the mathematical sequences, not *sensed* but *observed*. . . . after only a little development of the method, the observed sequences will no longer consist solely of phenomena observed independently of man's will, but will be complemented by phenomena evoked by man (deeds without any direct instinctive aim, but executed only to complete the causal system so that it will become more easily managcable).

This in brief, then, is Brouwer's reaction to the developments in nineteenth-century geometry. It constitutes a very considerable departure from Kant's views—greater, we would say, than that represented by the logicism of either Frege or Dedekind.

REFERENCES

Bernays, P. (1933), 'Die Grundgedanken der Fries'schen Philosophie in ihrem Verhältnis zum heutigen Stand der Wissenschaft', *Abhandlungen der Fries'schen*

[47] Of this departure from instinctive development, Brouwer makes the following interesting remark which connects his views with certain of those developed later by Wittgenstein: 'the non-instinctive nature of this intellectual action renders the certainty that the parts of the sequence really belong together, anything but perfect. Consequently, it can always be falsified, which is observed in the discovery that "the rule no longer applies"' (Brouwer 1907, 53). This suggests that Brouwer did not regard a priori judgements as invariably possessed of high certainty.

Schule, NS, 5/2: 97–113.

Bolyai, J. (1832), *The Science of Absolute Space*, English trans. of the Latin original by G. B. Halsted in Bonola 1955.

Bonola, R. (1955), *Non-Euclidean Geometry*, English trans. of 1912 Italian edition by H. S. Carslaw, New York, Dover.

Brouwer, L. (1905), 'Life, Art and Mysticism', in Brouwer 1975, 1–10.

——(1907), 'On the Foundations of Mathematics', in Brouwer 1975, 11–101.

——(1908), 'The Unreliability of the Logical Principles', in Brouwer 1975, 107–11.

——(1912), 'Intuitionism and Formalism', in Brouwer 1975, 123–38.

——(1933), 'Volition, Knowledge and Language', in Brouwer 1975, 443–6.

——(1948), 'Consciousness, Philosophy and Mathematics', in Brouwer 1975, 480–94.

——(1951), *Brouwer's Cambridge Lectures on Intuitionism*, D. van Dalen (ed.), Cambridge, Cambridge University Press, 1981.

——(1952), 'Historical Background, Principles, and Methods of Intuitionism', in Brouwer 1975, 508–15.

——(1954), 'Points and Spaces', in Brouwer 1975, 522–48.

——(1955), 'The Effect of Intuitionism on Classical Algebra of Logic', in Brouwer (1975), 551–4.

——(1975), *Brouwer: Collected Works*, I, A. Heyting (ed.), Amsterdam, North-Holland.

——(1981), *Brouwer's Cambridge Lectures on Intuitionism*, D. van Dalen (ed.), Cambridge, Cambridge University Press.

Dedekind, R. (1887), *The Nature and Meaning of Number*, English trans. by W. Beman, New York, Dover.

——1932 *Werke*, III, Braunschweig, F. Vieweg & Sohn.

Detlefsen, M. (1990), 'Brouwerian Intuitionism', *Mind*, 99 (369): 501–34.

——(1993), 'Hilbert's Formalism', *Revue internationale de philosophie*, 47(186): 285–304.

Fichte, J. G. (1797), *Werke*, III (*Wissenschaftslehre*), Leipzig, Felix Meiner.

Gauss, C. F.. (1832), *Theoria residuorum biquadraticorum*, II; repr. (with a German summary) in Gauss 1863–1903, II.

——(1863–1903), *Werke*, I–XI, Leipzig, Teubner.

——(1899), *Briefwechsel zwischen Carl Friedrich Gauss und Wolfgang Bolyai*, F. Schmidt and P. Stäckel (eds.), Leipzig, Teubner.

——(1976), *Briefwechsel mit H. W. M. Olbers*, Hildesheim, Georg Olms.

Gödel, K. (1958), 'Über eine bisher noch nicht benützte Erweiterung des finiten Standpunktes', *Dialectica*, 12: 280–7. Cf. the English trans. in Gödel (1990).

——(1990), *Kurt Gödel: Collected Works*, II, S. Feferman *et al.* (eds.), Oxford, Oxford University Press.

Goethe, J. W. (1948), *Faust und Urfaust*, Wiesbaden, Dieterich'sche Verlagsbuchhandlung.

Heijenoort, J. van (1967), *From Frege to Gödel: A Source Book in Mathematical Logic, 1879–1931*, Cambridge, Mass., Harvard University Press.

Heyting, A. (1958), 'Intuitionism in Mathematics', in R. Klibansky (ed.), *Philosophy in the Mid-Century*, Florence.

Hilbert, H. (1902), 'Mathematical Problems', *Bulletin of the American Mathematical Society*, 8: 437–79.

——(1922), 'Neubegründung der Mathematik: Erste Mitteilung', in Hilbert 1935, 157–77.

——(1925), 'On the Infinite', in van Heijenoort 1967, 367–92.

——(1925G), 'Über das Unendliche', *Mathematische Annalen*, 95: 161–91.

——(1927), 'The Foundations of Mathematics', in van Heijenoort 1967, 464–79.

——(1930), 'Naturerkennen und Logik', in Hilbert 1935, 378–87.

Jacobi, F. (1815), *Werke*, II (Leipzig, Gerhardt Fleischer); reissued by Wissenschaftliche Buchgesellschaft, Darmstadt, 1980.

Kant, I. (1900), *Gesammelte Schriften*, Berlin, Reimer.

Kreisel, G. (1958–9), 'Review of Wittgestein's *Remarks on the Foundations of Mathematics*', *British Journal for the Philosophy of Science*, 9: 135–58.

——(1971) Review of M. E. Szabo (ed.), *The Collected Works of Gerhard Gentzen*, *Journal of Philosophy*, 68: 238–65.

Kronecker, L. (1887), 'Über den Zahlbegriff', in *Werke*, iii, pt. 1, 249–74.

Nelson, L. (1912), 'Die Unmöglichkeit der Erkenntnistheorie', *Abhandlungen der Fries'schen Schule*, NS, 3/4: 583–617.

——(1949), *Socratic Method and Critical Philosophy: Selected Essays* (ed. and trans. by T. K. Brown), New York, Dover.

——(1971), *Progress and Regress in Philosophy*, II, J. Kraft (ed.), Oxford, Blackwell.

Parsons, C. (1967), 'Mathematics, Foundations of', in P. Edwards (ed.), *The Encyclopedia of Philosophy*, V, New York, Macmillan.

Schelling, F. (1800), *System des transcendentalen Idealismus*, Hamburg, Felix Meiner.

Schlegel, A. and F. (1960), *Athenaeum*: 1798–1800, Stuttgart, J. G. Cotta'sche Buchhandlung

Schopenhauer, A. (1883–6), *The World as Will and Idea*, I, III, R. Haldane and J. Kemp (trans.), London, Trübner & Co.

Stigt, W. van (1990), *Brouwer's Intuitionism*, Amsterdam, North-Holland.

Weyl, H. (1949), *Philosophy of Mathematics and Natural Science*, Princeton, Princeton University Press; page references are to the 1963 repr. by Atheneum Press.

IV. Frege and the Foundations of Arithmetic

13

On the Harmless Impredicativity of $N^=$ ('Hume's Principle')

CRISPIN WRIGHT

I. INTRODUCTION: DUMMETT'S OBJECTIONS TO 'NEO-FREGEANISM'

Michael Dummett's *Frege: Philosophy of Mathematics* and its famous precursor, *Frege: Philosophy of Language*, contrast sharply in their estimates of Frege's achievements in those respective subjects.[1] The guiding conviction of *Frege: Philosophy of Language* was that even admirers of Frege's writings have tended to a merely superficial appreciation of his contribution, that he is justly viewed as the father of analytical philosophy itself, the first philosopher to perceive the centrality of the philosophy of language, and especially the theory of meaning, in all philosophy, and the inventor of the tradition of systematic thought about meaning within which so much of important recent and contemporary philosophy belongs. By contrast, *Frege: Philosophy of Mathematics*—notwithstanding the kindest possible concluding accolade[2]—returns a predominantly negative verdict on Frege's accomplishment in relation to the specific questions which were most important to him. In particular, to the questions at which *Grundlagen* and *Grundgesetze* are primarily directed—the character of the subject-matter of number theory and real analysis and the nature of our knowledge about it—Frege's distinctive Platonist-cum-logicist answer was, according to Dummett, demonstrably 'catastrophically wrong'.[3]

My thanks to all who participated in the discussion at the Munich conference and especially to Bob Hale and Stewart Shapiro.

[1] Michael Dummett, *Frege: Philosophy of Language*, London, Duckworth, 1973; *Frege: Philosophy of Mathematics*, London, Duckworth, 1991.

[2] '[T]he greatest philosopher of mathematics yet to have written', *Frege: Philosophy of Mathematics*, 321.

[3] *Frege: Philosophy of Mathematics*, 292. The emphatic tone of this assessment is sustained in numerous passages in Dummett's text. Compare 'the central flaw in [Frege's] entire philosophy of arithmetic', p. 232; 'the big error . . . in [Frege's] philosophy of mathematics' and '[Frege's] philosophy of arithmetic was, indeed, fatally flawed', p. 321; 'The inconsistency of Frege's . . . system was not a mere accident [but] . . . an error affecting his entire philosophy', p. 223; 'the task at which Frege so lamentably failed', p. 317; and so on.

This of course has long been the usual view of the matter. Indeed, Frege himself was among the earliest to subscribe to it. The dominant opinion throughout this century has been that Russell's Paradox totally exploded Frege's approach, showing that the nature of numbers, and our knowledge of them, cannot be explained and justified in anything like Frege's way.

Thirteen years ago I published a short book[4] (henceforward *Frege's Conception*) which challenged this view, arguing that Frege's Platonism—the conception of numbers as specific objects providing the proper subject-matter of arithmetic, but given to us independently of any special 'intuition'—admitted of an epistemologically responsible, powerful defence along the very lines suggested by Frege's writing, and that—the paradox notwithstanding—(something to the purpose of) the logicism too was salvageable as a programme, at least as far as number theory was concerned, and maybe further.

It merits emphasis that Dummett's negative assessment is qualified by his conviction that, at some general level and in a way which I do not think he succeeds in making very clear, the logicist conception of mathematics is an insight,[5] and by his acceptance that the defence of Platonism, based on the Context Principle, which is offered in *Frege's Conception* is both true to the intent of *Grundlagen* and effective against the philosophical mind-set which has typically driven nominalists and reductionists of various stripes.[6] Given these broad sympathies, the question arises where specifically Dummett would take issue with the pro-Fregean lines of argument offered in *Frege's Conception*.

We are spared the need to speculate, for the latter book comes in for quite extensive criticism in *Frege: Philosophy of Mathematics*, with four principal points of abrasion emerging:

(i) In the chapter on 'Abstract Objects' in *Frege: Philosophy of Language*, and others of his earlier writings,[7] Dummett contended that Frege's use of the Context Principle to justify the ascription of reference to numerical singular terms, while correct as a counter to the nominalist *denial* of the reality of abstract objects, serves to vindicate only a 'thin' (contrast: *realistic*) conception of reference for such terms—a conception therefore at odds with Frege's full-blooded realism about the items for which they stand. The coherence of this idea of Dummett's was questioned in some detail in *Frege's Conception*.[8] In his new book Dummett repeats the claim—or some-

[4] Crispin Wright, *Frege's Conception of Numbers as Objects*, Aberdeen University Press/Humanities Press, 1983.

[5] See e.g. *Frege: Philosophy of Mathematics*, 308–9.

[6] See e.g. pp. 155–6, 181–3, 207–8, 231, 236, and 309–10.

[7] See also for instance Michael Dummett, *The Interpretation of Frege's Philosophy*, London, Duckworth, 1981, 424–7 and 452–7.

[8] See *Frege's Conception* ch. 2, sect. x.

thing very close to it[9]—insisting that the required distinction between two conceptions of reference is valid, and that the Fregean has no right to any but a 'non-realistic' notion in application to contextually defined singular terms.[10]

(ii) The core of the Fregean defence of a Platonist account of a region of discourse is that a referential construal of the relevant range of abstract singular terms may be explained and justified by *abstraction*—by laying down equivalences like $N^=$, that (where 'Nx:Fx' stands for the number of things which are F):

> For all F, G: Nx:Fx = Nx:Gx ↔ F and G are equinumerous (i.e. may be put into one–one correspondence)

or the corresponding principle for the identity of directions,

> For all lines, a, b: the direction of a = the direction of b ↔ a and b are parallel.

In the view of the Fregean, such a principle, stipulating that the truth-conditions of its two sides are to coincide, allows us to pass from a prior understanding of the type of context featured on its right-hand side to a grasp of the type of context which features on the left. Naturally, no one would deny that we can do that if the left-hand side is taken merely as a notational variant, an *unstructured* equivalent of the right. What is distinctive about the Fregean perspective is that we are to take it, in our passage across the biconditional, that the syntax of the left-hand side is *just what it seems*, and that the familiar expressions it contains—in the case of the direction principle, for example, the symbol for identity and the embedded names of lines—are to have their customary meaning. The epistemology of the abstract objects spoken of in contexts of the kind typified on the left-hand side is thus taken to be just that of the presumed (relatively) unproblematic states of affairs depicted by the kind of context featuring on the right.

[9] It is moot how close. Dummett writes (p. 191):

> 'Ultimately, Wright fails to find this intermediate view'—that attributed to Dummett in *Frege's Conception*—'coherent: he doubts if there is any tenable position between the austere'—eliminative reductionist—'and robust'—realist—'interpretations. As concerns contextual definitions, properly so called, I shall here maintain an intermediate view, perhaps one more austere than that which Wright had in mind. I shall however spend no time in discussing either how faithfully Wright represents the views I expressed in *Frege: Philosophy of Language*, or how far those I advance here diverge from them.'

This is regrettable. Clarity would have been best served if Dummett had taken the trouble to respond to the *Frege's Conception* criticisms, and to define any respects in which the new 'intermediate' view represents a change of mind. For an attempt to do the work for him, and argument that the new view is no stronger, see Bob Hale's 'Dummett's Critique of Wright's Attempt to Resuscitate Frege', in *Philosophia Mathematica*, (3) vol. 2 (1994), 122–47; sect. 4: 133–8.

[10] See *Frege: Philosophy of Mathematics*, 189–99, 234–5, and 239 for development of this line of criticism.

Is this line so much as coherent? Dummett apparently argues that it is not—or at least that anything recognizable as a Fregean notion of sense must be a casualty of it. For Frege, sense determines reference. So if the senses of the two sides of, say, the direction principle are the same, they must purport the same references. The Fregean Platonist must therefore acknowledge the presence of reference to directions on the right-hand sides of instances of the direction principle, even though it has to be possible—in his view—fully to understand the sense of such clauses while wholly unaware of the concept of direction. And that is to violate an important element in the *transparency* of Fregean sense: that it must be transparent to one who grasps the sense of an expression what references to what kinds of objects its featuring in (or being itself) a true sentence would involve.[11]

(iii) The third area of disagreement concerns the 'Caesar problem'—the indeterminacy which Frege thought afflicted $N^{=}$ as an account of numerical identity, as manifested in its incapacity to decide statements of the form,

Nx:Fx = Julius Caesar,

or more generally, any statement of the form,

Nx:Fx = q,

where 'q' is a term not given in the form, 'Nx:Fx', nor introduced as an abbreviation for such a term. As is familiar, it was in effect the Caesar problem that moved Frege to identify the cardinal numbers with extensions—a futile response, even prescinding from the inconsistency of his theory of extensions, since the problem must arise for extensions too. I absolutely agree that, for a variety of reasons, one of which will be very salient later, it is essential to find some constructive response to the difficulty. But in Dummett's view, the defence of a Fregean philosophy of arithmetic offered in *Frege's Conception* depends on an inadequate solution to the Caesar problem, indeed one actually canvassed and rejected by Frege himself.[12]

[11] For elaboration, see *Frege: Philosophy of Mathematics*, 168–79 and 194–5.

[12] Specifically, Dummett interprets the *Frege's Conception* proposal as a licence to treat 'the way in which the object q is introduced as a property of q, which it is not' (*Grundlagen* §67; *Frege: Philosophy of Mathematics*, 160). This is a bad misreading of what is actually suggested at pp. 113–17 of *Frege's Conception*. In particular, the proposal, so interpreted would, as Dummett emphatically observes, have the effect that numbers cannot be classes—since they are not introduced as such—whereas I explicitly remarked (*Frege's Conception*, 115) that, pending further refinement, the effect of the proposal would seem to be precisely the opposite. Maybe Dummett attributed that to mere confusion. In any case, what *is* presupposed by the *Frege's Conception* proposal is not what Dummett—and Frege—object to but rather that the way in which it is (correctly) introduced may teach one something of the properties of a *concept*, and *thereby* something about objects falling under it. I will return to this briefly in Sect. V below. For further discussion of Dummett's misreading, see Hale's 'Dummett's Critique of Wright's Attempt to Resuscitate Frege', 130–3.

(iv) Finally, Dummett contends that

The inconsistency of [*Grundgesetze*] was not a mere accident (though a disastrous one) due to carelessness of formulation. [Frege] discovered, by August 1906, that it could not be put right within the framework of the theory, that is, with the abstraction operator as primitive and an axiom governing the condition for the identity of value-ranges: but the underlying error lay much deeper than a misconception concerning the foundations of set theory. It was an error affecting his entire philosophy.[13]

And, if so, one inherited by *Frege's Conception*. For Dummett, the inconsistency of Basic Law V points to a fundamental and irremediable flaw in the whole conception of how abstract objects may be given to us which permeates Frege's work, and my book.

Each of these four claims involves, if cogent, a fundamental line of criticism, and deserves treatment in detail. Considerations of space demand some restriction of focus, however, and—while strenuously disagreeing with each of them—I therefore elect here to concentrate on that to which Dummett himself appears to attach the greatest importance, the fourth.[14]

II. BAD COMPANY?

A strict definition should provide for the elimination from any context in which it occurs of the expression, or type of expression it defines. $N^=$, and Basic Law V, are not, by this criterion, strict definitions. Numerical singular terms, as introduced via $N^=$, and singular terms standing for extensions or, more generally, for courses of values, as introduced by Basic Law V, will not always be eliminable via the paraphrases licensed by those two principles. The reason is that predicates (open sentences) containing singleton occurrences of singular terms of those kinds will be wanted to feature among those to which the cardinality operator, and its analogue for courses of values (the abstraction operator), may permissibly be applied. It is, of course, vital for Frege's purposes that this be so, since his proof of the infinity of the number series depends upon its being allowable to number numbers which precede a given number, and hence to bind with the cardinality operator predicates, like

$$y = Nx\colon x \neq x,$$

[13] *Frege: Philosophy of Mathematics*, 223

[14] Each of the four lines of objection is usefully addressed in Bob Hale's paper cited above. The second objection was anticipated and treated in my 'Field and Fregean Platonism', in A. D. Irvine (ed.), *Physicalism in Mathematics*, Dordrecht, Kluwer, 1990, 73–93, sect. 4.

which feature occurrences of that operator which $N^=$ provides no means to eliminate.[15]

One effect of this consideration is to the Fregean's advantage. Even were Dummett to succeed in making out a contrast between 'thin' and robust, or 'realistic' singular reference, and in showing that contextually definable abstract singular terms qualify only for the former, it would still remain to be demonstrated that numerical singular terms, introduced via $N^=$, are in the same boat—and only if they are would Dummett's claim be justified that Frege's Platonism about number, to the extent that it is sustained by the Context Principle, cannot provide a sufficient foundation for his arithmetical realism.[16]

But of course there is a concomitant disadvantage: since $N^=$ does not provide a strict *definition* of every necessary and legitimate kind of use of terms formed using the cardinality operator, the question becomes urgent whether and how $N^=$ is really satisfactory as an *explanation* of that range of numerical singular terms. No doubt it would be too simple to assume that whatever is not a strict definition of a symbol is not a satisfactory explanation of it either. Still, definition, whether explicit or contextual, is one readily intelligible form of explanation. If $N^=$ succeeds as an explanation of uses of the cardinality operator for which it provides no strict definition, we need to hear a story about how.

Here is the point of entry of Dummett's fourth criticism—what he presents as the most serious and fundamental criticism of Frege's entire approach. In Dummett's view, $N^=$ does *not* provide the necessary explanation, and the general model it incorporates, of how the use of a range of abstract singular terms, and with it our conception of an associated population of mathematical objects, may be explained and justified, is altogether misconceived. Why does he think this?

Dummett writes:

> [I]f the context principle, as expounded by Wright, is enough to validate the 'contextual' method of introducing the cardinality operator, it must be enough to validate a similar means of introducing the [class] abstraction operator.

[15] Just to elaborate a little. The statement

(a) $Nx{:}x{\neq}x = Nx{:}x{\neq}x$

—the statement that the number of things which are not self-identical is identical to itself—can be transformed via $N^=$ into a theorem of pure second-order logic with identity, to the effect that there is a one–one correspondence between the instances of the concept, *not self-identical*, and themselves. But now consider:

(b) $(\exists y)(y = Nx{:}x{\neq}x)$.

This is, of course, an immediate consequence of (a) in any logic suitable for the Fregean's purpose. But its singleton occurrence of the numerical operator is one which the resources afforded by $N^=$ provide no means to eliminate.

[16] Dummett is under no illusion about this, of course (*Frege: Philosophy of Mathematics*, 205).

This is why the mere fact that, on his view, it is unnecessary to define the cardinality operator in terms of classes or of value-ranges does not entitle Wright to ignore the problem of the abstraction operator. For Frege's method of introducing [the latter] was, notoriously, *not* in order. It rendered his system inconsistent; and that inconsistency forced him eventually to acknowledge that his entire enterprise had failed. If the context principle, as stated by Wright, was sound, there could have been no inconsistency . . . we may take [Wright] as concerned to vindicate, by appeal to the context principle, a method of introducing the cardinality operator which Frege did not in fact adopt: namely, by laying down the criterion of the identity for numbers [$N^=$], and supplementing it by some solution to the Julius Caesar problem . . . To all appearances, this would exactly resemble Frege's method of introducing the abstraction operator in *Grundgesetze*. In this case, we therefore have three options: to reject the context principle altogether; to maintain it, but declare that it does not vindicate the procedure Wright has in mind; and to formulate a restriction upon it that distinguishes the cardinality operator from the abstraction operator. Wright does none of these things: he maintains the context principle in full generality, understood as he interprets it, and defends the appeal to it to justify ascribing a reference to numerical terms, considered as introduced in the foregoing manner, without stopping to explain why an apparently similar method of introducing value-range terms should have led to contradiction. He owes us such and explanation.[17]

Two things about this. First, it should not seem at all obvious that I *do* owe any such explanation. Dummett here writes as though nothing at all were known about the likelihood of similar inconsistency afflicting $N^=$. But in fact, as Boolos and Burgess have independently observed,[18] the system which results from superimposing $N^=$ onto a suitable foundation of higher-order logic is consistent if classical analysis is. Dummett is right that there is a structural similarity between Frege's own procedure with Basic Law V and the way I proposed that a Fregean might deploy $N^=$; but the question is why—when there is relative security about the consistency of $N^=$—the fate of Basic Law V should nevertheless be thought to cast a pall of suspicion over its status. So far from my needing to explain why, in the teeth of the contradiction, it might still be justifiable to deploy $N^=$ in the fashion proposed, the onus is rather upon the critic to explain why, in the teeth of the presumable consistency of $N^=$, it is not.

Second, it anyway isn't true that I didn't 'stop to explain' why an apparently similar manner of introducing value-range terms should have

[17] *Frege: Philosophy of Mathematics*, 188–9.

[18] See the discussion at pp. 6–10 of George Boolos's 'The Consistency of Frege's *Foundations of Arithmetic*', in J. Thomson (ed.), *On Being and Saying: Essays in Honor of Richard Cartwright*, Cambridge, Mass., MIT Press, 1987, 3–20, and John Burgess's review of *Frege's Conception* in *Philosophical Review*, 93 (1984), 638–40; also Harold Hodes's 'Logicism and the Ontological Commitments of Arithmetic', *Journal of Philosophy*, 81 (1984), 123–49—(see remarks at 138 concerning an equivalent of $N^=$ involving branching quantifiers). For a detailed proof, see the first appendix to Boolos and Heck, '*Die Grundlagen der Arithmetik*, §§82–3', in this volume.

led to contradiction. On the contrary, a perfectly good explanation of that is provided by the comparative discussion[19] of the derivation of Russell's Paradox from Basic Law V and the outcome of the corresponding reasoning from $N^{=}$. As is familiar, in order to obtain Russell's Paradox from Basic Law V:

$$(\forall F)(\forall G)(\{x\colon Fx\} = \{x\colon Gx\} \leftrightarrow (\forall x)(Fx \leftrightarrow Gx)),$$

we first derive:

$(\forall F)(\exists y)\ y = \{x\colon Fx\}$—'naïve' abstraction for extensions—

and then take F to be the concept of non-self-membership, which we may express as:

$$(\forall G)(z = \{x\colon Gx\} \rightarrow -Gz),^{20}$$

thus arriving at Russell's class,

$$\{z\colon (\forall G)(z = \{x\colon Gx\} \rightarrow -Gz)\}.$$

Call this class **r** and suppose it satisfies the condition on its own members, i.e.:

$$(\forall G)(\mathbf{r} = \{x\colon Gx\} \rightarrow -G\mathbf{r})$$

and take for G that very condition. Then, since $\mathbf{r} = \{x\colon Gx\}$, we have $-G\mathbf{r}$; so **r** does not satisfy that condition. Hence: $(\exists G)(\mathbf{r} = \{x\colon Gx\}\ \&\ G\mathbf{r})$. But the condition which makes this existential statement true must, by Basic Law V, be coextensive with that by reference to which **r** was defined. Hence **r** must, after all, satisfy the latter. And so on, back and forth.

When the paradox is presented in this rather cumbersome fashion, rather than, as is normal, in terms of a primitive symbol of class membership, it is evident both why there might seem to be the prospect of a similar antinomy with $N^{=}$, and why that prospect is not fulfilled. Again, we may first use $N^{=}$ to obtain

$(\forall F)(\exists y)(y = Nx{:}Fx)$—'naïve' abstraction for numbers—

and then take as F a precisely analogous condition to Russell's, viz.

$$(\forall F)(z = Nx{:}Fx \rightarrow -Fz),$$

thus arriving at the potentially rogue number:

$$Nz\colon\ (\forall F)(z = Nx{:}Fx \rightarrow -Fz)$$

[19] *Frege's Conception*, 155–6

[20] No particular importance attaches to this way of characterizing the notion.

—the number of numbers which fall under none of the concepts of whose instances they are the number. Call this number **s.** We now ask whether **s** falls under its own defining condition, i.e. whether

$$(\forall F)(\mathbf{s} = Nx{:}Fx \rightarrow -F\mathbf{s})$$

If so, then, taking F as that very condition, we immediately have that it fails to do so. So far, so bad, as it were. But can we complete the other half? We have:

$$(\exists F)(\mathbf{s} = Nx{:}Fx \ \& \ F\mathbf{s}),$$

but now the next step fails us. We cannot infer, from the fact that **s** falls under *some* condition of whose instances it is the number, that it falls under its originally defining condition above. We cannot do so precisely because, in contrast with the situation of classes, where we have that

$$(k = \{x{:}\ Fx\} \ \& \ k = \{y{:}\ Gy\}) \rightarrow (Fk \leftrightarrow Gk),$$

we clearly cannot take it generally that:

$$(k = Nx{:}Fx \ \& \ k = Ny{:}Gy) \rightarrow (Fk \leftrightarrow Gk).$$

All that results, then, is a proof that **s** falls under some concept of whose instances it is the number. That, so far from being paradoxical, is surely true for all cardinal numbers generally, finite as well as infinite.[21]

A simple, head-on response to Dummett's demand that a restriction be formulated to distinguish the case of the cardinality operator from that of the class-abstraction operator is merely to exploit this difference in the obvious way. If:

$$(\forall F)(\forall G)\ \Sigma x{:}Fx = \Sigma x{:}Gx \leftrightarrow Fx \approx Gx,$$

is a proposed Fregean abstraction whose right-hand side expresses an equivalence relation on *concepts*, the question may be put whether that relation is or is not such that a Σ-object's falling under *some* concept, F, of which it is Σx: Fx entails that it falls under *every* F of which it is Σx:Fx. Say that the relation is *Russellian* if so. Then if, but only if, such a relation is Russellian will it be possible to confound the mooted principle by an application of the reasoning which leads to Russell's Paradox. So the demanded restriction is merely that the relevant equivalence relation should not be Russellian. Coextensiveness violates this restriction; one–one correspondence does not.[22]

[21] Actually, not quite all. Zero is evidently an exception. But it would seem to be the only exception, all other numbers falling under *some* concept to which they belong. Thus, as J. L. Bell has observed to me in discussion, in a domain consisting just of cardinal numbers, **s** is in fact Frege's number 1—the number belonging to the concept, $x = 0$. (Elsewhere, where κ is the number of non-numbers, **s** will be $\kappa + 1$.)

[22] Someone who thinks such a restriction would be 'ad hoc' (though I am not very clear

III. DUMMETT'S REAL OBJECTION

These reflections address the letter of Dummett's quoted complaint. But he would undoubtedly complain they are mere skirmishing in the context of the point he primarily intends. They would be *apropos* if all that could be amiss with an explanatory schema of this general kind was inconsistency. But explanations can have less dramatic shortcomings.[23]

Dummett's thinking about the issues here is strikingly of the kind which Wittgenstein set himself against in his notorious discussions of inconsistency and the paradoxes in the *Remarks on the Foundations of Mathematics*—thinking which would have it that Russell's Paradox is merely the symptom of an *underlying* disorder, whose insidious workings can be present even in cases where that especially distressing symptom is not

what the force of the complaint would be) may be interested to think through how any second-order abstraction on a Russellian relation is at the service of a development in some ways akin to *Cantor's Paradox*; that is, will call for objects which, on pain of contradiction, may not be reckoned to occur within an original (putatively all-inclusive) domain. Suppose to the contrary that it is possible to abstract on a Russellian relation, R, in such a way that, for each concept F defined over the domain, an object, z, can be found within the domain whose properties are those of the relevant abstract, ΣF. Then there should be a one–one map, M, containing all and only pairs such that, for each concept F and the corresponding equivalence class of concepts under R, {F, G, ...}, one term is a z, lying within the domain, whose properties are exactly those of ΣF and the other is {F, G, ...}. Consider a particular such pair, ⟨z; {F, G, ...}⟩. Since each of F, G, etc., have the same Σ-abstract, z must have the properties of each of ΣF, ΣG, etc. So it follows from the Russellian character of R that z instantiates all or none of the concepts, F, G, etc. That immediately suggests the Cantorian thought. Form the concept C instantiated just by each z which instantiates none of the concepts in its M-correlate. C will be an element of an equivalence class under R, {C, . . , . .}, which, by hypothesis, will have an M-correlate, k, in the original domain. Does k instantiate C? If it does, then, by the definition of C, it instantiates none of the concepts in its M-correlate. C is one such concept; so k does *not* instantiate C. But in that case the Russellian character of the abstractive relation ensures that k instantiates no other concept in {C, . . , . .}. So k instantiates none of the concepts in its M-correlate, and *is* therefore an instance of C . . . . . . Thus no element of the domain can play the role of ΣC.

Clearly it will not do to attempt a Fregean abstraction on an equivalence relation which is *inflationary* (to borrow Kit Fine's useful term) in this way: one which, on any domain of objects, generates more equivalence classes of concepts than there can be corresponding objects in the domain. That is a restriction whose point could have been appreciated before we had any inkling of Russell's Paradox.

[23] An important line of objection to the neo-Fregean perspective on $N^=$, about which I shall say nothing in this paper, branches off at this point, based on the observation that there are principles of the same general type as $N^=$ and Basic Law V which although, unlike the latter, consistent, are inconsistent with the former. An example would be the abstraction on the second-order equivalence relation: 'F and G are such that finitely many things are one or the other but not both.' The system generated by adding that principle to second-order logic has finite but no infinite models. And a problem is therefore raised along the lines: what allows $N^=$ to rank as (something like an) analytic principle at the expense of such competitors which seem, formally, perfectly analogous and no less legitimate?

This objection is original to and was developed by George Boolos in his 'The Standard of Equality of Numbers', in G. Boolos (ed.), *Meaning and Method: Essays in Honor of Hilary Putnam*, Cambridge, Cambridge University Press, 1990, 261–77. It is discussed in detail in my 'The Philosophical Significance of Frege's Theorem', in Richard Heck, Jr. (ed.), *Language, Thought, and Logic,* Oxford, Oxford University Press, 1997, pp. 201–44.

present.[24] For Dummett, Frege's whole conception, as embodied in the Context Principle, of what it takes to justify and explain the use of a range of singular terms is disfigured by a flaw, at work in the generation of the contradiction, but also present in cases, like $N^=$, where a contradiction is not forthcoming.

What is this flaw? It appears to be Dummett's view that explanatory principles like $N^=$ and Basic Law V, taken in the way intended, fail to be explanatory because *circular*—because they presuppose the very conceptual resources they are supposed to explain. It is very familiar that there is no deriving Russell's Paradox unless the rogue set is taken to lie within the range of the universal quantifier by reference to which, in a step of naïve abstraction, it is defined—we have to be able to pass from

$$(\forall x)x \in \mathbf{r} \leftrightarrow -(x \in x)$$

to

$$\mathbf{r} \in \mathbf{r} \leftrightarrow -(\mathbf{r} \in \mathbf{r}).$$

In the context of the derivation of the paradox licensed by Basic Law V, the point emerges as the reflection that the courses of values, or more specifically extensions, for which Basic Law V enables us to introduce singular terms, are taken to lie within the range of the universal quantifier which occurs on its right-hand side. And this feature is, of course, also present in $N^=$: the fully explicit expansion of the statement that there is a one–one correspondence between F and G naturally involves both first-order and higher-order quantification, and the referents of the numerical singular terms which $N^=$ enables us to introduce are themselves taken to lie within the relevant range of first-order quantification.

In Dummett's view, it seems, this impredicativity imports a vicious circularity: it is viciously circular to attempt to define a term-forming operator by reference to a domain of quantification which is intended to include the referents of the very expressions which that operator enables us to form. An analogue of Basic Law V which was purified of this feature would not allow the derivation of the paradox—so to that extent, the feature is responsible for the paradox. But, so apparently thinks Dummett, the feature is objectionable even in the consistent case.[25]

[24] See e.g. *Remarks on the Foundations of Mathematics*, pt. III, §80. I am not, of course, suggesting that one should take Dummett's concern lightly purely on this account!

[25] Here are some illustrative quotations (all from *Frege: Philosophy of Mathematics*):

> Suppose it [N=] had been presented as an axiom governing the cardinality operator, taken as primitive, as Axiom V governs the abstraction operator in *Grundgesetze*. . . . So considered, it would have been as objectionably impredicative as the analogous specification for value-ranges: for the truth of any statement of identity between numbers would depend on the extensions of two predicates defined on a domain which included the cardinal numbers, and whose composition the axiom was supposed to be playing an essential part in determining. (p. 226)

For Fregean Platonism, it suffices, in order to justify the claim that a certain class of expressions possess objectual reference, to observe first that their syntactic function is that of ordinary, uncontroversial singular terms;[26] and second, that they so function in statements which, in the light of some canonical explanation of their truth-conditions, we are justified in claiming to be true. Now Dummett, as I read him, does not intend to question this broad conception of the matter—on the contrary, he regards it as incorporating a decisive insight against what he himself regards as the misbegotten forms of nominalism that have benighted discussions of mathematical existence in this century. What Dummett *does* hold is that the force of this defence of Platonism stands in need of qualification in response to the following dilemma. Of any higher-order Fregean abstraction we may ask: Is the field of the relation cited on its right-hand side to include concepts under which may fall the very items to which the terms introduced on the left-hand side purportedly refer? If not, then the equivalence may constitute or contribute towards what is genuinely a contextual *definition* of the relevant terms; but the notion of reference which we will thereby be entitled to apply to them will be at most the etiolated or 'thin' notion with which Dummett wants to contrast a full-blooded, 'realistic' counterpart. But if, on the other hand, the relation does include such concepts within its field, then, according to Dummett, we have not yet met the conditions for the Fregean Platonist defence to be activated: specifically,

> The context principle allows us to ascribe a reference to mathematical terms provided that we have fixed the truth-conditions of sentences in which they occur;

> Frege leapt to the conclusion that the basis for introducing any new range of abstract terms must consist in the determination of the truth-conditions of identity statements involving them. In a certain sense, this was not far from the truth. It led, however, to the root confusion that allowed him to believe that he could simultaneously fix the truth-conditions of such statements and the domain over which the individual variables were to range. This belief was a total illusion. . . . we have *first* to attain a grasp of the intended domain of the individual variables: it is only after that that we can so much as ask after the meanings of the primitive non-logical symbols. (p. 232)

> [Frege] had, however, fatally overlooked the circularity of the whole procedure: that of specifying the criterion of identity in terms of the truth of sentences of the theory, and, more generally, that of attempting simultaneously to specify the domain and the application of the primitive predicates to its elements. (p. 233)

Also see esp. the passage from p. 236 quoted in the text below.

[26] 'Proof-theoretic' might be a little happier than 'syntactic', since it is the behaviour of the terms in inferential contexts which is important. Dummett's pioneering discussion of how to characterize this notion of a singular term is in *Frege: Philosophy of Language*, ch. 4. Bob Hale points up certain difficulties for Dummett's proposals in 'Strawson, Geach and Dummett on Singular Terms and Predicates', in *Synthese*, 42 (1979). The issue is the focus of §ix of *Frege's Conception* and of ch. 2 of Hale's *Abstract Objects*, Oxford, Blackwell, 1987. Linda Wetzel responds to Hale in 'Dummett's Criteria for Singular Terms', *Mind*, 99 (1990), 239–54. Hale responds to Wetzel and to other difficulties in two recent papers: 'Singular Terms (1)' in Matthias Schirn (ed.), *Frege: Importance and Legacy*, Berlin, Walter de Gruyter, 1996, 438–57; and 'Singular Terms' in B. McGuinness and G. Oliveri, eds., *The Philosoply of Michael Dummett*, Dordrectt, Kluwer, 1997, pp. 17–44.

—and, Dummett should have added, we are justified in supposing that some of those truth-conditions are met—

> but Frege was *completely mistaken* [my emphasis] about how we can go about fixing such truth-conditions.[27]

And again

> What must be disputed . . . is Frege's—and Wright's—idea of what is sufficient for determining the truth-conditions of sentences containing terms for newly introduced kinds. Impredicative specification of the conditions for the truth of identity statements involving one or two such terms is *not* sufficient, contrary to Frege's belief and to that of his disciple [*sic*!] Crispin Wright. It fails to fix truth-conditions for all sentences containing the new terms, when these terms are formed by attaching an operator to a predicate or functional expression; and it fails to do so because of the lack of an independent specification of the domain, which it attempts, but fails, to circumscribe simultaneously with its determination of the truth-conditions of sentences containing the new terms.[28]

The crux of Dummett's fourth objection, then, is the question: Does the necessary impredicativity of $N^=$—necessary if, as Frege's strategy for proving the infinity of the number series requires, it is to be possible to apply the cardinality operator to concepts under which numbers fall—does this impredicativity prevent $N^=$ from counting as an adequate explanation of the numerical terms whose use it is supposed to explain and legitimate?

IV. MOTIVATIONS FOR THE WORRY

Ever since the concern first surfaced in the wake of the paradoxes, discussion of the issues surrounding impredicativity—when, and under what assumptions, are what specific forms of impredicative characterizations and explanations acceptable—has been signally tangled and inconclusive. It seems to me that, despite his repeated attempts at formulation, it remains extremely difficult to fill out Dummett's specific concern here in convincing detail. The vicious circularity of a purported explanation ought to consist in its unavoidable presupposition, on the part of someone who is to follow it, of the very conceptual resources which following it is supposed to impart. By that standard, so I contend, $N^=$, though indeed ineliminably impredicative as the Fregean must intend it, is not viciously circular. I shall canvass two, as it seems to me, defective lines of thought which might suggest otherwise; the final section will then offer considerations which, if sound, point the way to a settlement in the Fregean's favour.

[27] *Frege: Philosophy of Mathematics*, 234. [28] Ibid. 236.

(i)

Some of Dummett's formulations—that just quoted, for example[29]—suggest a train of thought along the following lines:

A statement involving quantifiers is determinate in content only if the range of quantification is fixed in advance. Hence it is only if we have fixed in advance what the range of the objectual quantifiers in fully explicit expansions of the right-hand sides of instances of $N^=$ is to be, that the left-hand-side statements, whose use that principle purportedly explains, can have any clear meaning. Yet for someone so far innocent of the concept of natural number, that domain can be satisfactorily characterized only if no natural numbers are included in it. It follows that $N^=$ can work as an explanation only if constrained to be predicative—and hence useless for Frege's purposes.

One response would be as follows. Suppose we grant that the determinacy in content of a quantified statement demands some prior specification of the range of its quantifiers. Even so, it is by no means evident that such a specification cannot be given in the kind of case which interests us without introducing a circle of presupposition. It depends on what *kind* of specification of the domain is supposed to be necessary. Someone innocent of the concept, *tiger*, for instance, may nevertheless possess the concept, *feline quadruped*; and for such a person there need, presumably, be no indeterminacy in the content of quantified statements whose range of quantification is understood to be precisely the class of feline quadrupeds. Analogously, the line of thought outlined could therefore presumably be met if we could disclose some more *inclusive* sortal concept, which someone might have grasped who did not yet possess the notion of cardinal number, and which could be used to specify the range of the quantifiers in $N^=$.

However, candidates do not thrust themselves forward. Frege himself—at the time of his innocence before Russell's Paradox—would presumably have thought that we do indeed possess various such concepts, in the notion of *extension* (class), for instance, or indeed in the general notion of an *object*. But if that is to be the response to the objection, then a defence is now owing of the idea that there is a determinate totality of *all classes*, or of *all the objects that there are*, which may intelligibly be stipulated as the range for certain uses of quantification.

I don't think that is the way for the Fregean to go. It cannot be a satisfying response to the outlined train of thought that it may always be answered by the citation of a determinate domain of sufficient generality—if only, at the limit, the domain of 'all objects'—if one has any feeling at all for the Russellian thought[30] that it is a deep misconception of certain very

[29] For additional relevant references to earlier writings of Dummett's, see Richard Cartwright's paper cited in n. 32 below.

[30] Russell's thought is taken up by Dummett in his notion of an 'indefinitely extensible'

general ranges of objects to think of them as constituting totalities—*domains* in the intended sense—at all.

What should immediately be said is, rather, that it is not and *cannot* in general be a prerequisite for a quantified statement to be determinate—or determinate enough—in content that the range of its quantifiers should have been specified in advance. The target of first-order predicate logic, for instance, is the codification of statement forms whose instances will hold true of *every* determinate non-empty domain, independently of its further detail. Such statements are clearly determinate in content if we are able to recognize that they have this feature; but the feature is precisely one that they possess *independently* of the detail of any specified domain of quantification—and their content accordingly cannot depend on the prior specification of an intended domain.

The point is not, however, or not merely, that the contrary assumption is wrong in particular cases. Rather, it is obscure how it is even as much as coherent. If it is only in a context in which the domain of quantification has been independently determined that quantified statements can have determinate content, the obvious question to ask is: What are the *means* whereby that domain may or may not have been determined? It's hardly credible that a range of items might be specified without the use of generality of any kind. But then the demanded prior specification will *itself* involve further quantifications.[31] So a vicious regress is generated—vicious because the determinacy of any quantified statement will now depend upon an infinity of prior specifications.

Two points, then, are salient. First, the contention that an understanding of quantified statements depends quite generally upon a prior specification of a domain of quantification is neither true nor coherent. Second, as the case of first-order logical laws brings out, some counter-examples to this incoherent contention are of *unrestricted generality*. There simply are intelligible, unrestrictedly general thoughts—thoughts which are therefore available to thinkers who are capable of simultaneously grasping the idea that some of the objects within the range of the former may, if only temporarily, lie outwith their classificatory repertoire. Doubtless we could do with further explanation of such generality: of what it is to grasp such an unrestrictedly general thought, of whether classical forms of quantificational reasoning continue to be appropriate in such cases, and so on. But the phenomenon is unquestionable. And we ought therefore to be open to the idea

totality. See 'The Philosophical Significance of Gödel's theorem', *Ratio*, 5 (1963), 140–55 (repr. in M. Dummett, *Truth and Other Enigmas*, London, Duckworth, 1978, 186–201). The notion looms large in the concluding chapter of *Frege: Philosophy of Mathematics*, and in Dummett's 'What is Mathematics About?' in his anthology, *The Seas of Language*, Oxford, Clarendon Press, 1993, 429–45.

[31] Cf. Hale, 'Dummett's Critique of Wright's Attempt to Resuscitate Frege', at 143.

that the unrestricted, hence impredicative generality that features on the right-hand sides of (instances of) $N^{=}$ is an example of it.[32]

(ii)

A sympathizer with Dummett may want to persevere:

> Let it be that it is incoherent to think that it is only after we have first specified a determinate domain that a quantified statement has any definite meaning. Nevertheless, it is undeniable that the truth-conditions of a quantified statement depend upon the range of objects to which its author intends it to answer. What has to be the case in order for an utterance of the sentence, 'All householders were worse off under the Poll Tax' to be true, depends on the intended range of 'All householders': conditions sufficing for the truth of the sentence when that range is restricted to householders in Britain cease to do so when it is widened to include Continental Europeans. It should be recognized, accordingly, that the truth-conditions of statements on the right-hand sides of instances of $N^{=}$, and *hence the meanings thereby assigned to their left-hand-side equivalents*, vary as a function of the range of the individual quantifiers on the right-hand sides. That sets up a simple dilemma for the Fregean: if that range is taken to embrace the referents—cardinal numbers—of the type of expression supposedly being explained, then no one innocent of the concept of number is yet in position to understand what that range is, nor therefore to latch on to the meanings intended for the left-hand-side statements. If, on the other hand, the range is restricted to entities of which someone who knows nothing of numbers can have some concept, then the truth-conditions of statements on the right-hand sides, and hence the meanings assigned to their left-hand-side equivalents, will not be the ones that the Fregean has in mind; and $N^{=}$ will fail to convey the intended information. In short: either $N^{=}$ fails to be intelligible to the intended audience, or it fails to assign the intended meanings to the statements it is supposed to explain.

In one respect this line of thought gets us no further. For at the first horn of the dilemma posed, it too makes the suspect assumption that quantification whose range includes objects of a certain kind can be intelligible only to someone who is *au fait* with the concept of that kind of object. But there is an additional weakness which is worth highlighting: the casual movement between the notions of meaning and truth-conditions. It is harmless to think of the truth-conditions of a quantified statement as varying in tandem with variation in the range of quantification only provided the point is precisely not confused with one about *meaning*. The contribution of its range of

[32] A very useful recent discussion of unrestrictedly general quantification, which came to my attention only after drafting this paper is Richard Cartwright's 'Speaking of Everything', *Noûs*, 28 (1994), 1–20. Cartwright's discussion is focused specifically on Dummett's and makes a cogent case for disentangling the question of whether and what sense is made by unrestrictedly general quantification from the question whether there can be any single all-inclusive objectual *domain* (construed in the usual model-theoretic sense of that term).

quantification to the purport of a quantified statement is best compared to the contribution effected by context—the location and time of its utterance, identity of the speaker, and so on—to the purport of a statement involving *indexicals*. There is correspondingly no harm in describing the truth-conditions of a sentence like 'I was here yesterday' as varying in the mouths of different speakers; but there is nevertheless a clear sense in which its *meaning* does not so vary but makes a uniform contribution, in conjunction with the context of utterance, towards determining whether an utterer speaks truly. So too with quantified statements: 'All householders were worse off under the Poll Tax' should be seen not as varying in meaning from context to context, according as the range of quantification varies, but as imposing a *uniform condition* whose being satisfied or not will depend on what the range of quantification is taken to be.

Now, the germane point is that, while the truth-conditions of the left-hand sides of instances of N⁼ will indeed vary in response to variation in the domain of the objectual quantifiers involved on the right-hand sides, the most basic explanatory brief which N⁼ has to discharge is one of which it is actually no part to convey such truth-conditions. That brief—akin to that of explaining the meaning of a sentence like 'I was here yesterday' in such a way as to convey an understanding of what would be said by it on any particular occasion of its use for variable speaker, place, and time—is to explain what condition a given population of objects has to meet if a particular 'Nx:Fx = Nx:Gx' is to be true of that population—to impart a piece of information which in conjunction—but only in conjunction—with an understanding of some intended range of quantification, will enable the recipient to grasp the truth-conditions of a particular sentence of that kind. There is thus no merit in the suggestion that the only kind of understanding of the right-hand sides available to a novice is one which is unsuitable to facilitate the understanding of the left-hand sides which the Fregean will intend N⁼ to bestow. What N⁼ is supposed to bestow, in the first instance, is only, for each particular F and G, a knowledge of how things have to be in a given population if 'Nx:Fx = Nx:Gx' is to be true of it. Why may not that condition be grasped in full readiness, as it were, for populations which include numbers by someone who doesn't yet know what numbers are?

Call this the *range-unspecific* grasp of (an instance of) N⁼. I have suggested that there is no evident obstacle to the capacity of N⁼ to impart at least this. And now the view of the Fregean will be, in addition, that there is no obstacle to the intelligent fusion, as it were, of this range-unspecific understanding of N⁼ with the kind of unrestrictedly general understanding of the objectual quantifiers on (instances of) its right-hand side emphasized above. It is all the same to the range-unspecific understanding of N⁼ what the intended range of its objectual quantifiers is supposed to be. And it is all the same to the unrestrictedly general understanding of a quantified statement what actual, perhaps undreamed of types of object may turn out to fall within its scope. How is a problem of circularity generated by the union of these understandings? Why can they not be yoked together to produce exactly the state of information from which a recognition of the infinity of the series of natural numbers may follow, just as Frege believed?

A thought to consolidate that question. Quite early in *Grundlagen*, directly after the famous passage about Euclidean geometry, Frege writes:

> [W]e have only to try denying any one of [the fundamental propositions of the science of number] and complete confusion ensues. Even to think at all seems no longer possible. The basis of arithmetic lies deeper, it seems, than that of any of the empirical sciences, and even than of geometry. The truths of arithmetic govern all that is numerable. This is the widest domain of all; for to it belongs not only the actual, not only the intuitable, but everything thinkable. Should not the laws of number, then, be connected very intimately with the laws of thought?[33]

Frege's idea is that this absolute generality of number theory—its applicability wherever logic is applicable[34]—ought to predispose us to the idea that arithmetic is analytic in just the way that—he takes it—logic is. And Dummett for his part emphasizes the point in several places and seems disposed to regard it as one of Frege's abiding insights.[35] But whether or not the point suffices to establish a sense in which arithmetical thought is already logical, even before it comes to any logicist constructions, there seems no denying the *datum*: number theory does have this generality—*and must, therefore, be applicable in particular to its own proper objects*: the finite cardinal numbers.

The corollary is obvious. Since numerical concepts and operations are to be applicable to objects of any kind at all, any satisfactory explanation of them must encompass this generality and hence—if it has recourse to objectual quantification in the statements that make it up—must include the numbers themselves within its range. But that is to say that any stated explanation of the notion of cardinal number in general—assuming, as is surely inevitable, that it involves quantification at all—must either involve *impredicative* quantification or fail to encompass its target notion in full generality. The feature of $N^=$ which Dummett views as hopelessly compromising Fregean aspirations is thus a feature which any account of identity for cardinal numbers in the full generality which Frege, with Dummett's approval, attributes to that notion, has to have.

33 *Grundlagen*, §14.

34 The generality is not quite absolute, of course, but belongs with that of predicate logic. Schemes of thought are conceivable—for instance, as articulated in a tensed feature-placing language of the kind envisaged in chs. 6 and 7 of P. F. Strawson's *Individuals*, London, Methuen, 1959—which would naturally be subject to ordinary propositional logic but in which the idea of cardinal number would have no application.

35 See *Frege: Philosophy of Mathematics*, 43–6, 224, and 308.

In sum: if—as Dummett thinks—Frege is right to ascribe this generality to our arithmetical notions, then it is a fact of our intellectual lives that the concept of (sameness of) cardinal number has a reflexivity which it will take recourse to impredicative quantification to capture. That an attempted account of the notion uses such quantifiers is therefore, not cause for complaint, but a necessary condition of its completeness. And the impression that vicious circularity is thereby imported must accordingly depend on some misconception or other.

V. UNDERSTANDING N⁼

We have reviewed two unpersuasive lines of argument for the claim that the impredicativity of N⁼ bars it from playing the stipulative-explanatory role that the neo-Fregean would assign to it, and suggested that the very generality of the concept of cardinal number must be in tension with the existence of any fundamental difficulty in this area of the kind Dummett supposes. Even if all this is correct, however, there remains on the neo-Fregean a clear obligation to offer something more positive: an account of how N⁼ actually can work as an explanation—how someone innocent of the concept of cardinal number can use the principle to advance not merely to a general grasp of the condition any range of objects and pair of concepts, F and G, must meet if 'Nx:Fx = Nx:Gx' is to be true of them, but to an understanding of the specific uses of the cardinality operator which the neo-Fregean reconstruction of arithmetic demands.

As stressed earlier, this obligation actually arises independently of a concern about impredicativity. For as noted, N⁼ is not a strict definition in the way the corresponding principle for identity of directions is (or can be). The latter can participate in a regimen of stipulations whose collective effect is to provide for the contextual eliminability of all singular reference to and quantification over directions.[36] N⁼, by contrast, provides no general means for the contextual definition of occurrences of the cardinality operator: contexts formed by the deletion of one term from a statement of numerical identity and its replacement by a bound variable remain recalcitrant. For all that has so far been said, therefore, it is open to an awkward customer to claim that such occurrences of the operator are wholly unaccounted for. What is true as far as impredicativity is concerned is that *if* N⁼ can indeed provide all that is necessary for an understanding of such uses, and in general for a construal of a Fregean reconstruction of sentences of number theory along the lines, for example, outlined in *Frege's Conception*, then—prescinding from the Caesar problem, which is another matter—

[36] See *Frege's Conception*, 29 f.

there can be no well-motivated doubt about its effectiveness as an explanation. And if its impredicativity puts in jeopardy neither its consistency nor its explanatory power, what is there left for a worry about impredicativity to be about?

In order to deliver what is needed—to show that $N^=$ provides the basis for an understanding of the kinds of use of the cardinality operator which Frege's reconstruction of arithmetic demands but which that principle does not eliminatively define—one would naturally think in terms of constructing a comprehensive ranking of such uses by degree of complexity and of trying to show how a grasp of the constructions prior to any particular stage in the ranking provides for the construal of those at that stage. I shall turn to that thought shortly. First, though, it will prove expedient to focus on one central, though as we shall see atypical component in the general task: that of explaining how $N^=$ provides the resources to construe Frege's definitions of zero and its successors.[37]

How should one start? The question is under what conditions someone can understand the right-hand side of (an instance of) $N^=$ who doesn't yet know what numbers are. No problem, to be sure, if numbers are excluded from the range of the individual quantifiers. But what we need, rather, is a base class of substitutions for its predicate variables for which, even when its individual quantifiers are construed impredicatively, $N^=$ can function as an explanation of the associated contexts of numerical identity without any presupposition of a prior understanding of arithmetical concepts. The vital initial reflection is therefore that there is indeed such a class of cases. In a large range of examples, an ability to verify whether F and G are equinumerous will not depend on knowing what numbers are or whether any of them are F or G (even if some of them are). Examples will include concepts like 'man born in Chipping Norton between the wars', 'Julio-Claudian emperor', 'apple on the table', or 'book on the shelf'. Even if the number 3 is in fact an apple (the one on top of the pile), someone who doesn't know that is in no way disadvantaged when it comes to determining whether there are exactly as many apples on the table as books on the shelf. Even if Julius Caesar is a number (the number zero), someone who doesn't know that is in no way disadvantaged when it comes to determining whether there were as many Julio-Claudian emperors as Tudor monarchs.

Say that F and G are *unproblematic* just in case the knowledge whether or not either or both are instantiated by some number is not needed by one who is to determine whether they can be one–one correlated. Then, when F and G are unproblematic, $N^=$ gives a perfectly clear, in no way circular,

[37] The argument to follow generalizes and completes the considerations advanced in *Frege's Conception*, §xvii, 132–45.

account of the truth-conditions of 'Nx:Fx = Nx:Gx', even when the objectual variables in the *explanans* are intended to include numbers in their range.

Now introduce a faultlessly rational subject—I'll follow Gareth Evans's example and call him Hero[38]—who has mastered an appropriate higher-order logic, in which the *definiens* of $N^=$ can be formulated, together with a range of such unproblematic concrete sortal concepts and a battery of singular terms standing for instances of them but who so far has no specifically arithmetical vocabulary. Then our task is to show how, on receiving $N^=$, and thereby coming to understand 'Nx:Fx = Nx:Gx' when F and G are drawn from his unproblematic stock, Hero somehow has all he needs in order to advance to an understanding of Frege's definitions of zero and its successors.

There are two aspects to this accomplishment. Hero has to understand the numerical predicates, like 'y = Nx:x ≠ x', which Frege's definitions exploit, and to understand the singular terms formed from them, like 'Ny:[y = Nx:x ≠ x]', given as *definientia* of the standard numerals. Both, however, will arguably be encompassed by an understanding of an appropriate range of identity statements. For reflect: since it ought to be reckoned sufficient for an understanding of any predicate that someone should understand any statement which results from completing it with any term that he understands (or comes to understand), it should suffice for an understanding of, for instance, the above numerical predicate that, for each term, 't', that he understands or comes to understand, Hero understands the corresponding 't = Nx:x ≠ x'. Likewise it ought to suffice for an understanding of the associated term, 'Ny:[y = Nx:x ≠ x]' that, where 't' is any term that he understands (or comes to understand), Hero should understand the corresponding identity statement, 't = Ny:[y = Nx:x ≠ x]'.

Let us take it that an understanding of any such identity-statement will be ensured by an appropriately generated knowledge of its truth-conditions. Then what needs to be shown is how Hero can determine the truth-conditions of each of the series of increasingly complex identity-statements which are associated as indicated with an understanding of the numerical predicates which Frege's definitions of the numerals successively utilize and of the numerical terms which successively result. More specifically, designate each predicate, 'Ax', for which 'Nx:Ax' is Frege's *definiens* for the number, n, *Frege's predicate for n*; and designate 'Nx:Ax', likewise, as *Frege's term for n*. Then for each 't' which he understands or comes to understand, Hero needs to be able to determine the truth-conditions of each of the infinite series of statements,

[38] Evans so names his imaginary subject of experience in 'Things Without the Mind', in Zak van Straaten (ed.), *Philosophical Subjects: Essays Presented to P. F. Strawson*, Oxford, Clarendon Press, 1980, 76–116.

$$t = Nx{:}x \neq x,$$

whereby he understands Frege's term for 0 and predicate for 1;

$$t = Ny{:}\,[y = Nx{:}x \neq x],$$

whereby he understands Frege's term for 1;

$$t = Nx{:}x \neq x \ \vee\ t = Ny{:}\,[y = Nx{:}x \neq x],$$

whereby he understands Frege's predicate for 2;

$$t = Nz{:}\,[z = Nx{:}x \neq x \ \vee\ z = Ny{:}\,[y = Nx{:}x \neq x]],$$

whereby he understands Frege's term for 2;

$$t = Nx{:}x \neq x \ \vee\ t = Ny{:}\,[y = Nx{:}x \neq x] \ \vee$$
$$t = Nz{:}\,[z = Nx{:}x \neq x \ \vee\ z = Ny{:}\,[y = Nx{:}x \neq x]],$$

whereby he understands Frege's predicate for 3;

$$t = Nw{:}\,[w = Nx{:}x \neq x \ \vee\ w = Ny{:}\,[y = Nx{:}x \neq x] \ \vee$$
$$w = Nz{:}\,[z = Nx{:}x \neq x \ \vee\ z = Ny{:}\,[y = Nx{:}x \neq x]]],$$

whereby he understands Frege's term for 3;
. . . and so on.

Note now that every oddth member of that series from the third onwards is a disjunction which Hero will understand if he understands its constituent disjuncts—since he has (higher-order) logic. And each constituent disjunct of such a disjunction occurs in its own right earlier in the series. So the task reduces to that of showing how Hero is in position to know the truth-conditions of each identity-statement of the form,

$$t = n_f,$$

where 'n_f' is the Fregean *definiens* of the numeral for n, and 't' is either an expression of the form 'Nx:Fx' which he already understands (or comes to understand) or else a concrete singular term from his initial repertoire.

Take the latter case first; suppose that 't' is one of the initial concrete singular terms, or indeed any other term which Hero comes to understand as denoting, if anything, then a concrete entity of some kind. Then he will be able to know the truth-conditions—and indeed the truth-value—of '$t = n_f$' provided, and so far as I can see, only provided he has a general solution to the Caesar problem: some principled account of the distinction between numbers and things of other kinds which allows him to determine that no concrete object can be a number.

The Caesar problem is often viewed as the ineluctable nemesis of any form of mathematical Platonism. Certainly it needs an extended treatment, not to be ventured here. But I should at least explain my own continuing feelings of optimism about it. Is it really plausible that Hero is in no position to make *anything* of the question whether any particular object of a kind previously familiar to him coincides with an object denoted by one of the numerical terms which $N^=$ introduces? There is one thought which surely should be salient to him. Admitted that nothing directly follows from the actual statement of $N^=$ about the external identity, so to speak, of the objects of which it speaks, a constraint *is* exerted by the fact that the principle is offered as the explanation of a *kind*: a determination of the sort of thing that numbers, *qua* numbers, are. For one who receives $N^=$ in this way, it will be understood to be of the *essence* of cardinal numbers that facts about identity and distinctness among them are constituted in facts about one–one correspondence among concepts.[39] So nothing can be a number whose every distinction from other numbers is not constituted in such facts. Since this is not true of persons, or stones, or concrete objects generally—since identity and distinctness among such things is grounded in facts of quite different kinds—Hero has good grounds for the thought—with which of course, as Frege noted, we sympathize[40] (and what source of information do we have that Hero lacks?)—that there is no co-reference between any of his initial stock of terms and any of the numerical singular terms introducible via $N^=$.

This is, to stress, merely to make a suggestion about how the general distinction which, seemingly, it comes quite naturally to us to recognize, between the abstracts associated with principles like $N^=$ and many ordinary kinds of object, is grounded in the concepts concerned. It is not a suggestion which will cover all cases; in particular, it will not cover the relations between Fregean abstracts and *classes* generally since it *is*, plausibly, of the essence of equivalence classes under a given relation that identities and distinctions among them are grounded in facts about the obtaining of that relation. Here, though, it seems to me, it is plausible to regard the indeterminacy as real: Frege's identification of cardinal numbers with equivalence classes of concepts under one–one correspondence, for instance, strikes one merely

[39] This aspect of abstraction principles, conceived as forms of conceptual innovation, goes unrecognized in Dummett's discussion in *Frege: Philosophy of Mathematics*; see e.g. the criticism offered at 161–2 of the *Frege's Conception* suggestion about the Caesar problem (which more elaborate formulation was essentially an attempt to develop the idea presently entertained) and Dummett's suggestion at 126 of the identifiability of directions and lines-through-a-fixed-point. For discussion of Dummett's misinterpretation of the point of the *Frege's Conception* proposal, and Hale's own related proposals, see Hale's 'Dummett's Critique of Wright's Attempt to Resuscitate Frege', 122–47.

[40] In *Grundlagen* §66 when the Caesar problem is first brought up in connection with the abstraction for directions, Frege makes a remark which, transposed to the case of $N^=$, would run as follows: 'Naturally no one is going to confuse [Caesar] with [the number 0]; but that is no thanks to our definition of [number].' A sceptic about the solubility of the Caesar problem had better feel that, to the contrary, such a confusion is precisely what is invited; otherwise, there has to be something which discourages it and which has somehow been implanted in our understanding, so that the Caesar problem is less that of making a distinction than that of formulating the means whereby it is already drawn.

as an available move—one which nothing intuitive either mandates or forbids. If that is right, then its neutrality on such examples is actually to the credit of the mooted proposal: if there *is* indeterminacy in some instances, a correct account of what determines the determinate ones will not say anything about those.

The other class of cases to consider are identity-statements of the form, '$t = n_f$', where 'n_f' is the Fregean *definiens* of the numeral for n, and 't' is an expression of the form 'Nx:Fx' which Hero already understands. We may take it that it will suffice for a knowledge of the truth-conditions of these statements that Hero should know, for each particular 'n_f', *how many* F's there have to be if '$Nx{:}Fx = n_f$' is to be true; and that this knowledge will certainly be captured by knowledge, for each 'n_f', of a truth of the form:

$$Nx{:}Fx = n_f \leftrightarrow (\exists_n x)\,Fx,$$

where '$(\exists_n x) \ldots x$' expresses the numerical quantifier, 'There are exactly n . . .' as standardly recursively defined.[41] That definition can, of course, be given in fully explicit form in higher-order logic. It therefore suffices for the immediate purpose to reflect that, with the numerical quantifiers so defined, each instance of the above schema may be derived from $N^=$ in higher-order logic and is therefore available to Hero.[42]

Our result, then, is that $N^=$ provides the resources for a construal of any identity-statement linking one of Frege's numerals with an antecedently understood term, 't', in two cases: where 't' is understood to stand for a type of object which cannot be a number if the essentialist proposal about the Caesar problem outlined above is right; and where 't' is (an abbreviation of something) of the form, 'Nx:Fx'. These cases are not exhaustive; for instance, they exclude terms for classes. Nevertheless the result is enough, I argued, for the conclusion that the principle explains the meaning of Frege's numerals.

There are two respects in which this finding falls short of the generality that must ultimately be accomplished. First, nothing explicit has so far been said about how Hero is to construe the results of binding the free variable in a numerical predicate not by the cardinality operator but by the ordinary first-order quantifiers. Rectifying this omission raises no difficulty. If a predicate is agreed to be determinate in content, then the contents of its quantifications may presumably be conceived as determined in line with whatever semantics for the quantifiers, classical or constructivist, is found agreeable. But a predicate ought to be regarded as determinate in content whose every completion by a term whose meaning is understood is likewise determinate in content. So it must suffice to provide meanings for the quan-

41 For details, see the Appendix. (To the best of my knowledge, the relevant clauses were first formulated in *Grundlagen* §55.)

42 A proof is sketched in the Appendix.

tifications of numerical predicates that each of their instantiations to a term which is previously understood, or which comes to be understood, is determinate in content. Ultimately, therefore, provided we have under semantic control all the terms which Hero's basic vocabulary allows us to form, quantification will look after itself.

The second omission is another matter. Hero's initial vocabulary comprises higher-order logical vocabulary, an indefinite stock of concrete singular terms, a range of unproblematic predicates, and the cardinality operator as governed by $N^=$. Many other numerical predicates, both pure and mixed,[43] besides Frege's predicates for the natural numbers will, of course, be formulable in terms of these materials. And each such predicate will be attachable to the cardinality operator to form a term whose contribution to the meaning of identity-statements containing it $N^=$ will somehow have to be harnessed to explain. But obviously it cannot in general be expected that, for each such predicate, Bx, it will be possible for Hero to prove a result of the form,

$$Nx{:}Fx = Nx{:}Bx \leftrightarrow (\exists_n x)Fx.$$

For instance, since Hero's base language is adequate for the formulation of number theory, we know we can construct a predicate, θx, within it meaning: is a member of a pair of twin primes. Since there is no surety that anyone can know how many twin primes there are, we are in no position to expect that Hero will be able to recover the truth-conditions of contexts of the form,

$$Nx{:}Fx = Nx{:}\theta x,$$

by the kind of method which works for contexts involving Frege's terms for the natural numbers.

The *general* form, then, of the means whereby $N^=$ fixes the meaning of numerical singular terms cannot be that outlined for the Fregean numerals, which are something of a special case. And the general account the Fregean needs, of how Hero is empowered to construe *any* numerical predicate, cannot be based on the principle that it will suffice to be able to associate a numerical quantifier with it in the fashion illustrated for Frege's predicates for the natural numbers. So how to proceed?

Well, let us review the prospects for an induction on the degree of complexity of expressions in Hero's language, as earlier anticipated. Let a *numerical term* be any term of the form, 'Nx:Ax', where 'Ax' is a predicate expressible in Hero's language. If 'Nx:Ax' and 'Nx:Bx' are a pair of such terms, let 'Nx:Ax = Nx:Bx' be their *corresponding identity-statement*. Let S

[43] These notions to be understood in the obvious way: pure numerical predicates are any formulable using just the N-operator and the resources of Hero's logic, while mixed numerical predicates deploy additional non-logical constituents.

be any such identity-statement formulable in Hero's language. Define the *rank* of S as follows:

(i) S is of rank 0 if 'Ax' and 'Bx' are each unproblematic;
(ii) S is of rank n + 1 if there are numerical terms occurring within 'Ax' and/or 'Bx' whose corresponding identity-statement is of rank n, and for no such terms is the corresponding identity-statement of rank higher than n.

Now extend the idea of rank first to numerical terms, and then to statements in general, in the obvious way:

(i) A term 'Nx:Ax' is of rank n just in case 'Nx:Ax = Nx:Ax' is of rank n.
(ii) An arbitrary sentence, S, of Hero's language is of rank n just in case some numerical term occurring within S is of rank n and no numerical term occurring within S is of rank higher than n.

Hero understands all numerical identity-statements of rank 0. Suppose he also understands all numerical identity-statements of rank n: Does it follow that he understands all of rank n+1?

Clearly that will follow if an understanding of all numerical identity-statements of rank n ensures an understanding of all statements (in Hero's language) *of any kind* of rank n. For an understanding of all statements of any kind of rank n will ensure the comprehensibility of all *right-hand sides* of $N^=$ of rank n; and that in turn will suffice for an understanding of the truth-conditions of every numerical identity-statement of rank n+1. But given that the syntax and non-numerical constituents of any statement of rank n are familiar to Hero, the point ought to carry on grounds of mere compositionality that he will understand any statement of rank n provided he understands all *numerical terms* of rank n. So our question—the crux of the whole issue—is precisely whether an understanding of all *numerical identity-statements* of rank n ensures an understanding of all *numerical terms* of rank n. If so, Hero will be able to construe all numerical identity-statements of whatever rank and indeed all arithmetical contexts, pure and applied, whose formulation his language allows.

It might seem that the point does not require very much argument. Surely, if someone understands an identity-statement, he is bound to understand the ingredient terms. But that of course is too quick in the present context. To know the truth-conditions of a numerical identity-statement, as determined via $N^=$, need not be—it might seem—to be empowered to conceive it full-bloodedly *as* an identity-statement. Such a full-blooded conception will demand that one's understanding of the ingredient terms comply with certain platitudes about what it is to understand any singular term—or more specifically, in view of the semantic complexity of numerical terms, any *descriptive* singular term. It cannot be assumed without further ado that an understanding of the right-hand side of an instance of $N^=$, coupled with the instruction that one is to treat the left-hand side in all respects as if it were a statement of identity, somehow ensures that the latter is genuinely so conceived.

So, what are the germane platitudes? They are two, it seems to me. To understand a descriptive term, it is arguably necessary—and will certainly suffice, which is what matters for our purpose—to know, first, what *kind* of thing, if anything, it stands for; and second, what would qualify a thing of that kind to *be* the thing it stands for. Let 'Nx:Ax', then, be any numerical term of rank n. Does Hero have an adequate conception of the kind of thing it may stand for? And does he, if so, have a grasp of the specific condition which a thing of that kind must meet in order to qualify as its referent?

'Nx:Ax' is apt to refer, if to anything, then to a cardinal number. So Hero has the requisite general conception if he knows what cardinal numbers are—which he does if he understands that cardinal numbers are individuated by facts about one–one correspondence among concepts and is apprised of a solution of the Caesar problem to whatever extent is necessary in order to have a conception of the difference between numbers and other kinds of thing. So this issue breaks no new ground: Hero's understanding complies with the first platitude provided the Caesar problem does indeed admit of solution.

What of the second: Does Hero have a grasp of the specific condition which a number should meet in order to qualify as the referent of 'Nx:Ax'? By hypothesis, he understands all numerical identity-statements of rank n, and so understands 'Ax' (since it occurs in the *definientia*, via $N^=$, of a range of numerical identity-statements of that rank which contain 'Nx:Ax' as one related term) But—as an upshot of our lemma concerning the Fregean numerals—Hero also knows, for each finite number, n_f, that for that number to be the referent of 'Nx:Ax', it is necessary and sufficient that there be exactly n A's—a condition, to stress, which he can formulate and appreciate. So he does indeed grasp, for each finite number, what condition it must meet in order to be the referent of 'Nx:Ax'.

True, this condition varies as a function of which number we consider; but that merely betokens that the descriptive content of 'Nx:Ax' incorporates a certain, quite common type of reflexivity—and is easily given a uniform statement once explicit provision is made for that. The condition for a number to be the referent of 'Nx:Ax' is that 'Ax' should instantiate the numerical quantifier associated with *that number*. (Compare: 'the MP with the largest absolute majority in the House of Commons': in order to qualify as the referent, a member must be he/she for whom the number of votes cast in *his/her* constituency minus the total number cast in *that* constituency for any other candidate is greater than the corresponding number for any other MP.) True, also: the understanding described amounts at best to a grasp of what it is for a *finite* number to be the referent of 'Nx:Ax'. But—assuming it clear that more is required—corresponding connections will in any case presumably hold between whatever infinite cardinals Hero is in position to conceive and appropriately specialized higher-order definable numerical quantifiers ('There are countably infinitely many . . .', 'There are uncountably infinitely many . . .', and so on).

Once the lemma concerning the special case of the Fregean numerals is in place, then—and assuming, as always, the means to bury Caesar—really quite simple considerations suffice to supply the generalization which was owed. The *proof-theoretic* unification which $N^=$ supplies for the fundamental laws of arithmetic is accordingly matched by an *explanatory* unification: to have understood the principle as an explanation of the concept of cardinal number is to be apprised of a complete explanation of all possible uses of the N-operator within a higher-order language of Hero's kind[44]—which is to have a basis on which all the distinctive concepts of number theory can be defined.

It follows that nothing can be essentially involved in the epistemology of number theory that is not involved in an understanding, and knowledge of the truth of 'Hume's Principle'. The neo-Fregean view is that such knowledge may be had, for relative cheap, after the fashion of knowledge of *any* bona fide Fregean abstraction—that such principles may properly be viewed as keeping company with definitions and other freely undertaken forms of concept-determination. That comparison should continue to be the focus of the debate about the neo-Fregean project. My point here has been that, whatever else may or may not prove to be wrong with it, it is certainly not spoiled at the outset by the impredicativity in which Dummett sees the doom of Frege's enterprise.

APPENDIX: ON THE PRINCIPLE, N^q

The schematic principle,

$$Nx{:}Fx = n \leftrightarrow (\exists_n x)Fx$$

is christened N^q in Bob Hale's *Abstract Objects*,[45] and is there invoked to scotch the charge of Harold Hodes[46] that we are quite at liberty to interpret the N-operator as denoting *any* one–one function to natural numbers from equivalence classes of concepts under one–one correspondence. Apart from the use to which I have put it

[44] Michael Dummett quite correctly observes (this volume, p. 383) that certain cases are not covered by the foregoing considerations. The remaining details are outlined in my response to him (this volume, pp. 399–402).

[45] Hale, *Abstract Objects*, 223.

[46] Hodes, 'Logicism and the Ontological Commitments of Arithmetic', 134–5.

above, in grounding the interpretation of the Fregean numerals, the principle is important both for its application to Hodes's argument, and because of the role it plays in explaining how Frege's account of cardinal number illuminates the applications of arithmetic. Hale, however, presents the principle as something that we might use to *supplement* $N^=$ in a response to Hodes's challenge. That it is no supplement may be seen as follows:

Stage-setting

Assume the standard recursive definitions of the numerically definite quantifiers:

$$(\exists_0 x)Fx \leftrightarrow (\forall x)-Fx$$

$$(\exists_{n+1} x)Fx \leftrightarrow (\exists x)(Fx \ \& \ (\exists_n y)(Fy \ \& \ y \neq x))$$

and let 'n_f', as before, abbreviate Frege's *definiens* for n. Define 'Pxy'—immediate predecession—as:

$$(\exists F)(\exists w)(Fw \ \& \ y = Nv{:}Fv \ \& \ x = Nz{:}[Fz \ \& \ z \neq w]).$$

Define 'Nat(x)'—x is a natural number—as:

$$x = 0 \vee P^*0x,$$

where 'P*xy' expresses ancestral predecession.
Let '$(\exists R)(F\ 1\text{–}1_R\ G)$' express that there is a one–one correspondence between F and G.

We shall use three main lemmas from the proof of the Peano axioms from $N^=$ outlined in the concluding section of *Frege's Conception* (numbering as there assigned):

Lemma 51: $(\forall x)(Nat(x) \rightarrow x = Ny{:}[Nat(y) \ \& \ P^*yx]$,
—every natural number is the number of its ancestral predecessors.
Lemma 52: $(\forall x)(Nat(x) \rightarrow -P^*xx)$,
—no natural number ancestrally precedes itself.
Lemma 5121: $(\forall x)(\forall y)(Nat(x) \ \& \ Nat(y) \rightarrow (Pxy \rightarrow (\forall z)(Nat(z) \ \& \ (P^*zx \vee z{=}x) \leftrightarrow (Nat(z) \ \& \ P^*zy)))$
—if one natural number immediately precedes another, then the natural numbers which ancestrally precede the second are precisely the first and those which ancestrally precede the first.

Finally, recall that Frege's 0 is $Nx{:}x \neq x$ and that each successive $(n{+}1)_f$ is $Nx{:}[x{=}0 \vee \ldots \vee x{=}n_f]$. Each of these objects qualifies as a natural number in the light of the above definition of 'Nat(x)'. *Proof*: 0_f qualifies by stipulation; $(n{+}1)_f$ qualifies if n_f does—take 'F' in the definition of 'Pxy' as '$[x{=}0 \vee \ldots \vee x{=}n_f]$' and 'w' as '$n_f$' to show that $P(n_f, n{+}1_f)$; then reflect that $Pxy \rightarrow P^*xy$ and that P*xy is transitive. (*Frege's Conception*, Lemmas 3 and 4 respectively.)

Proof of N^q *for Frege's natural numbers*

Base: To show

$$Nx{:}Fx = 0_f \leftrightarrow (\exists_0 x)Fx,$$

it suffices to reflect that the lhs holds just if $(\exists R)(Fx\ 1\text{–}1_R\ x \neq x)$, which in turn holds just if $-(\exists x)Fx$.[47]

Hypothesis:
Suppose $Nx{:}\ Fx = n_f \leftrightarrow (\exists_n x)Fx$.
We need to show that it follows that

$$Nx{:}Fx = (n+1)_f \leftrightarrow (\exists_{n+1}x)Fx.$$

PROOF

Left-to-right:
Consider any F such that $Nx{:}\ Fx = (n+1)_f$. By lemma 51 and the reflection that $Nat(n_f)$, $n_f = Nx{:}\ [Nat(x)\ \&\ P^*xn_f]$. So by the Hypothesis, $(\exists_n x)(Nat(x)\ \&\ P^*xn_f)$. But by lemma 52, $-P^*n_f, n_f$. So $(\exists_n x)(Nat(x)\ \&\ (P^*xn_f \vee x{=}n_f)\ \&\ x{\neq}n_f)$. So $(\exists y)(Nat(y)\ \&\ (P^*yn_f \vee y{=}n_f)\ \&\ (\exists_n x)(Nat(x)\ \&\ (P^*xn_f \vee x{=}n_f)\ \&\ x{\neq}y)\,)$. So by the recursion for the quantifiers, $(\exists_{n+1}x)(Nat(x)\ \&\ (P^*xn_f \vee x{=}n_f)$. But by lemma 5121 and since $P(n_f, n{+}1_f)$, we have that $(\forall x)(Nat(x)\ \&\ (P^*xn_f\ \vee\ x{=}n_f) \leftrightarrow Nat(x)\ \&\ P^*(x,\ n{+}1_f)\,)$. So $(\exists_{n+1}x)(Nat(x)\ \&\ P^*(x, n{+}1_f)\,)$.

That establishes the desired result for one concept of which $(n+1)_f$ is the number. But by $N^=$, any G such that $(n + 1)_f = Nx{:}Gx$ will admit a one–one correspondence with that concept. So a lemma to the following effect will now suffice:

$$(\forall F)(\forall G)(\exists R)(F\ 1\text{–}1_R G) \rightarrow ((\exists_{n+1}x)Fx \leftrightarrow (\exists_{n+1}x)Gx)$$

A proof by induction—strictly, at third order—suggests itself:

Base:
It suffices to show $(\forall F)(\forall G)(\,(\exists R)(F\ 1\text{–}1_R\ G) \rightarrow (\,(\forall x) - Fx \leftrightarrow (\forall x) - Gx)$.

Hypothesis:
Suppose $(\forall F)(\forall G)(\,(\exists R)(F\ 1\text{–}1_R\ G) \rightarrow (\,(\exists_n x)Fx \leftrightarrow (\exists_n x)Gx)$.
Consider any H such that $(\exists_{n+1}x)Hx$. Then $(\exists x)(Hx\ \&\ (\exists_n y)(Hy\ \&\ y{\neq}x)\,)$. Let a be such that $Ha\ \&\ (\exists_n y)(Hy\ \&\ y{\neq}a)$. Let J be one–one correlated with H by R. Let b be such that $Jb\ \&\ Rab$. Then R one–one correlates $Hx\ \&\ x{\neq}a$ with $Jx\ \&\ x{\neq}b$. So, by the Hypothesis, $(\exists_n x)(Jx\ \&\ x{\neq}b)$. So $(\exists x)(Jx\ \&\ (\exists_n x)(Jx\ \&\ x{\neq}b)\,)$. So $(\exists_{n+1}x)Jx$.

Right-to-left:
Consider any F such that $(\exists_{n+1}x)(Fx)$. Then there is some a such that $Fa\ \&\ (\exists_n y)(Fy\ \&\ y{\neq}a)$. So by the Hypothesis $Ny(Fy\ \&\ y{\neq}a) = n_f$. So, by $N^=$, there is an R such that $(Fy\ \&\ y{\neq}a)\ 1\text{–}1_R\ (Nat(x)\ \&\ P^*xn_f)$. Let $R^\#$ correlate $(Fy\ \&\ y{\neq}a)$ with $(Nat(x)\ \&\ P^*xn_f)$ in just the fashion of R, and let it also correlate a with n_f. Then $(Fy)\ 1\text{–}1_{R^\#}\ (Nat(x)\ \&\ (P^*xn_f \vee x{=}n_f)\,)$. But, as established above, $(\forall x)(Nat(x)\ \&\ (P^*xn_f \vee x{=}n_f) \leftrightarrow Nat(x)\ \&\ P^*(x, n{+}1_f)\,)$. So $Nx{:}Fx = (n+1)_f$.

[47] As George Boolos has reminded me, Frege himself observes, at *Grundlagen* §75 and §78, that he is in a position to obtain proofs of N^q for 0_f and 1_f respectively.

14

Neo-Fregeans: in Bad Company?

MICHAEL DUMMETT

For a time, Frege believed that he had solved the central problem of the foundations of mathematics. He believed so firmly until 1902, when Russell wrote to announce the contradiction, and hopefully until 1906, when he recognized his inability to resolve it. He then concluded that he had failed in his life's ambition, to provide sure foundations for arithmetic. Crispin Wright, however, believes that he ought not to have despaired so soon. He should have recast his whole exposition: for, if only he had realized it, he *had* solved the central problem.

What was the central problem? It was, for each fundamental mathematical theory, to secure, without begging the question and without appealing to facts of experience or (save in the case of geometry) to intuition, the existence of the objects of the theory: by a fundamental theory is here meant one whose objects cannot be explained by constructing them from those of any other theory taken as already known.

The context principle, enunciated in *Grundlagen*, required that the existence of the objects of the theory be secured by determining definite truth-values for statements of the theory referring to or quantifying over those objects. It does not appear possible to satisfy this requirement by first specifying what objects are to compose the domain of quantification, and then laying down the interpretations of the primitive predicates and functors with respect to this domain. Such a procedure would presuppose an understanding of what those objects were, and an assurance of their existence; and this would not be a solution of the problem, but a *petitio principii*. Rather, the context principle required that the specification of the domain, in such a way as to guarantee the existence of its elements, should be effected *by* determining truth-values for statements of the theory. It appears, therefore, to follow that Frege's method of meeting the challenge demanded that the *same* stipulations must simultaneously fix the domain and the intended interpretations of the primitives with respect to it. This, at least, is what Frege attempted to do in *Grundgesetze*; given his general principles, it is hard to see how he could have attempted to do any less.

This was the central problem. The status, as logically true, analytic, or otherwise, of the provable statements of a theory for which this problem has been solved is a subordinate matter. Perhaps they would not be analytic according to the definition of 'analytic' given by Frege in *Grundlagen*, though they would certainly not be either a posteriori or synthetic a priori according to his definitions of those notions; but that is not the central issue.

Crispin Wright and I are agreed that Frege's attempt was of the profoundest interest. We are agreed, also, that the problem it was intended to solve is of the utmost importance, and that, if it cannot be circumvented, for instance by holding, with Hartry Field, that the truth of a mathematical theory is of no concern to us, no other solution is in sight. Wright and I differ in that he believes that Frege's attempt essentially succeeded, whereas I think that it failed.

1

Crispin Wright plainly conceives of himself as the champion of Frege's own ideas. He is apt, for example, to refer to a sample proponent of his own interpretation as 'the Fregean'. The Frege of whom he is the champion cannot be identified with the historical Frege, however, neither with the author of *Grundlagen* nor with the author of *Grundgesetze*: he is a Frege reconstructed by Wright.

Having raised the question, in §62 of *Grundlagen*, how numbers are given to us, Frege replied by invoking the context principle, concluding that the question comes down to explaining the sense of a sentence in which a number-word occurs. Numbers are given to us as the meanings of number-words; and we grasp the meanings of number-words by our grasp of the senses of sentences containing them. Since we have agreed to understand number-words as words for self-subsistent objects, we have in particular to fix the senses of statements of identity of the form 'The number of Fs is the same as the number of Gs.' (I here eschew the jargon used by Frege.) In §63 Frege denies that he is defining identity specially for numbers; rather, he is using the concept of identity, taken as already known, as a means of arriving at that which is to be regarded as being identical.

In §64 Frege cites his favourite analogy, that of the direction of a line. Henceforward, the entire discussion, up to the very end of §68, is conducted in terms of this analogy. We need to determine the criterion of identity for directions, that is, for the truth of a statement of identity of the form 'The direction of the line *a* is the same as the direction of the line *b*.' We readily recognize the truth of the biconditional (ED):

The direction of the line *a* is the same as the direction of the line *b* iff *a* is parallel to *b*.

But what is the direction of explanation? Frege argues that it is from right to left, not from left to right: the concept of a direction is to be defined in terms of the relation of being parallel, not the relation of being parallel in terms of the concept of a direction.

In §65 Frege discusses whether the biconditional (ED) can be treated as a definition of phrases of the form 'the direction of *a*', and observes that its legitimacy depends on showing parallelism to be an equivalence relation. In §§66 and 67, however, he raises an objection he considers fatal: the so-called Julius Caesar problem. The purported definition does not handle a statement of identity in which a term of the form 'the direction of *a*' stands on one side, and a term of some quite different form on the other. Although it is on the Julius Caesar problem that Frege concentrates his argument, he leaves it unclear just what is the point at issue. He reasons that it cannot be resolved by supplementing (ED) by some further stipulation: it is useless to define the predicate '*x* is a direction' to mean 'For some line *a*, *x* is the direction of *a*', since, without a solution of the Julius Caesar problem, we cannot say whether or not an arbitrary object *x* is the direction of any line. This makes prominent cases in which the sign of identity is flanked on one side by a term of the form 'the direction of *a*' and on the other by a variable: the purported definition (ED) does not allow us to eliminate the direction-term from such a context. This leaves it ambiguous whether Frege's objection was to the ineliminability or merely to the indeterminacy of reference. The so-called recursive definition of an arithmetical operator such as '+' determines its reference, but does not allow us, in a first-order language, to eliminate it from a context such as '$9 = 3 + x$'. Suppose that Frege were wrong, and some means offered itself of specifying the truth-value of Julius Caesar statements in which the sign of identity connected a direction-term with a specific term of another form. It might then be claimed that this, together with (ED), determined the reference of the direction-operator: but it would probably still not allow the elimination of a direction-term standing on one side of the sign of identity when a variable stood on the other. Such cases are of far greater importance for the definition of the cardinality operator 'the number of Φs' than for that of the direction-operator. On them depends the possibility of applying the cardinality operator to predicates of numbers; and on that in turn depend the definitions of the numbers and the proof that every natural number has a successor.

Having thus rejected (ED) as a *definition* of the direction-operator, Frege (in effect) gives in §68 the explicit definition:

the direction of the line *a* is the class of lines parallel to *a*.

Turning back at last to the cardinality operator with which he is directly concerned, he in effect offers the analogous definition of it:

> the number of Fs is the class of concepts G such that there are just as many Fs as Gs.

From this definition Frege shows in §73 how to derive a biconditional analogous to (ED), namely the biconditional (EN):

> the number of Fs is the same as the number of Gs iff
> there are just as many Fs as Gs.

This is all that Frege uses the definition of the cardinality operator 'the number of Φs' for: all the rest is derived from (EN), together with the definition of 'just as many' in terms of one-one correlation.

The discussion in *Grundlagen* leaves the general problem of vindicating the use of abstract terms and the existence of abstract objects unresolved, because it assumes that we already have at our disposal the means to give an explicit definition of the term-forming operator in question, namely the notion of the extension of a concept. The context principle is invoked only to argue that the definition given *is* in fact correct and adequate: it does not tell us how to proceed before we have such a means at our disposal, or in order to acquire one. This lacuna Frege attempted to fill in *Grundgesetze*.

2

The task Frege undertook in *Grundgesetze* was no longer that of seeking and justifying a definition. Definition in *Grundgesetze* is austerely restricted to the straightforwardly explicit variety, allowing of the elimination from every context of the defined expression, and Frege assumes the justification of the definitions he gives without arguing for it. The task is, instead, to construct a semantics for his formal system; the notion he uses for this purpose is that of *Bedeutung* (as opposed to *Sinn*), so that the objective is an informal characterization, in German as the metalanguage, of the *Bedeutungen* of all well-formed expressions of the formal object-language. This naturally involves the specification of the *Bedeutung* of each of the primitive symbols; such a specification will not amount to, and cannot be transformed into, a definition, and is not intended to provide any means of eliminating the symbol in question. All logical objects, and hence all numbers of whatever kind, are to be represented as value-ranges; unsurprisingly, the task of specifying the *Bedeutungen* of the primitive symbols encounters no difficulties until the abstraction operator, by means of which terms for value-ranges are formed, is reached. The domain of the individual variables is not specified at the outset. The domain is composed of the two truth-values, with which we are assumed to be already acquainted, and otherwise of value-ranges, that is, of the *Bedeutungen* of value-range terms. It is not stated in advance what value-ranges the domain comprises, however; that is tacitly taken to be stipulated simultaneously with the specification of the *Bedeutungen* of well-formed terms for value-ranges.

The task of laying down the *Bedeutungen* of value-range terms is effected by appeal to the context principle. The *Bedeutung* of such a term will be determined by specifying the value for it as argument of every primitive function: in other words, the *Bedeutung* of the term is fixed by fixing the *Bedeutungen* of more complex terms of which it forms part. The task of doing this is reduced to that of determining the truth-value of every statement of identity in which the term figures. Just as with terms for directions or for numbers in *Grundlagen*, there are two cases: that in which the sign of identify connects two value-range terms; and that, of the Julius Caesar type, in which a value-range term stands on one side of the sign of identity and a term of a different form on the other. The first case is handled by means of the criterion of identity for value-ranges formally incorporated into Frege's fatal Axiom V: the value-ranges of functions *f* and *g* coincide just in case *f* and *g* are extensionally equivalent. This time, however, Frege does not confess himself defeated by the Julius Caesar problem, but has a solution to it. The only terms other than value-range terms that can be framed in the formal language are terms for truth-values; the problem is solved by equating each truth-value with its own unit class, thus making every element of the domain a value-range. The uncertainty whether the Julius Caesar problem relates to eliminability, or merely to the determination of reference, is now resolved: eliminability is not in question; we are concerned only to determine the reference or *Bedeutung* of value-range terms, and hence with the assumption of the existence of value-ranges.

Having to his satisfaction determined the *Bedeutungen* of all primitive symbols, Frege sets out to give an inductive proof that every well-formed expression of his formal language has a unique *Bedeutung*; this will in particular involve that every well-formed sentence of the language has a unique truth-value. Such a proof will justify his having directly stipulated the values for value-ranges as arguments only of the *primitive* functions. Even the context principle would not sanction a claim to have fixed the *Bedeutungen* of value-range terms if the values of complex functions for value-ranges as arguments remained undetermined; the natural hope is that this could not happen once the values of the primitive functions had been provided for. Unfortunately, Frege's inductive proof was hopelessly ill-founded. Not only did he fail to prove what he claimed, but, as everyone knows, Axiom V yielded an actual contradiction.

In face of the contradiction, not even the boldest of Frege's defenders could fail to concede, as Frege himself came to acknowledge, that his procedure for introducing value-ranges into the system of *Grundgesetze* was radically faulty.

3

In his *Frege's Conception of Numbers as Objects*, Crispin Wright paid no attention to *Grundgesetze*, but concentrated wholly upon *Grundlagen*. His principal contention was that Frege had gone wrong in allowing the Julius Caesar problem to deflect him from the original plan (or from what Wright took to be his original plan). The original plan, so viewed, was to treat the biconditional (EN), not as a thesis to be derived from the definition eventually given, but as the direct object of a stipulation. In so far as the aim was to find a means of actually eliminating the cardinality operator 'the number of Φs', the Julius Caesar problem admittedly blocked the achievement of *that* aim: but Wright does not consider its elimination essential for the success of the main enterprise. To the extent that Frege thought the Julius Caesar problem an obstacle to determining the reference of terms for cardinal numbers, on the other hand, and thereby to certifying the existence of cardinal numbers, Wright believed him to have been in error: the biconditional (EN), regarded as the object of a stipulation whereby the cardinality operator is to be introduced, *already* solved the Julius Caesar problem, as a problem about reference, by guaranteeing that no cardinal number is to be identified with anything not given as a cardinal number. By following out the original plan, and treating (EN) as the object of direct stipulation, the need for an explicit definition of the cardinality operator, and hence for appeal to the notion of a class (or value-range), would have been obviated and the contradiction avoided. Perhaps the claim to have established the analyticity of number-theoretic propositions would have had to be modified; but the legitimacy of using numerical terms, and the full-fledged referentiality of those terms, would have been conclusively established, and, with it, a respectable version of logicism for number theory.

In his book, Wright boldly maintained that the context principle, as he was interpreting and defending it, would even support our applying a robust notion of reference to terms introduced by an actual contextual definition allowing their elimination without replacing them by other terms with the same denotations. I took this to be a piece of bravado, since it does not bear on the case with which Wright was primarily concerned, that of terms for cardinal numbers. In *Frege: Philosophy of Mathematics*, I devoted much space to arguing with Wright about the notion of reference, and hence the degree of realism, sustainable in the presence of an application of the context principle, with or without a contextual definition. The principal point, however, or at least the prior point, is the legitimacy of the procedure Wright was recommending in place of that actually followed by Frege in *Grundlagen*. About this, my main complaint was Wright's failure to pay any attention to *Grundgesetze*. Since Frege's ambition was to provide a foundation, not only for number theory, but also for analysis, one might complain that Wright had given no indication how, in the absence of classes or value-ranges, he thought Frege's theory of real numbers ought to be reconstructed. Challenged to do so, however, he might have left the work to others by saying, 'Oh, in some similar way'; but that was not my complaint. My complaint was the obvious one. In *Grundgesetze*, value-ranges are introduced in a manner precisely analogous to that in which Wright argued, in his book, that Frege ought to have introduced cardinal numbers (save that it involved a more explicit means of resolving the Julius Caesar problem): and yet it was so far from being justified as to lead to actual contradiction. It therefore *could* not be maintained that this procedure is, in and of itself, legitimate. Possibly some restriction, distinguishing the case of cardinal numbers from that of value-ranges, could be framed. But Wright had put forward no such restriction; and hence his thesis, as it stood, could not be sound.

There is admittedly one large disanalogy between the introduction of cardinal numbers in *Grundlagen* and the introduction of value-ranges in *Grundgesetze*, namely that the settings are quite different. In *Grundlagen*, there is no distinction between object-language and metalanguage: the cardinality operator is being introduced into the only language under consideration, German. There is therefore no question of our having to specify the domain of quantification: unrestricted quantification, within everyday speech, is taken for granted as already understood. In *Grundgesetze*, the matter stands quite otherwise: the abstraction operator is a symbol of a formal language quite distinct from that in which its *Bedeutung* is to be specified, and the domain of the individual variables of that language has therefore to be determined, just as does the interpretation of any of its primitive symbols. This difference is doubtless highly germane to an evaluation of the context principle as applied to the introduction of new terms, either into the language of everyday or into a formalized language governed by an explicit semantics. It is not germane to the present issue, however. (EN) would not give rise to a contradiction when incorporated into a formal system any more than it does informally. If Axiom V were converted into an informal stipulation made in the context of everyday language, it would no longer be possible to deal with the Julius Caesar problem by restricting attention to truth-values as the only objects not given as value-ranges; but the stipulation *would* still generate a contradiction. The question how a procedure endorsed by Wright could prove to yield an inconsistency would remain one he needed to answer.

4

This is a form of objection that Wright has labelled 'Bad Company'. In his 'On the Harmless Impredicativity of $N^=$ ('Hume's Principle')', Wright shows himself unimpressed by it. Now we have seen what claims the historical Frege made in his two books on arithmetic: what claim does Wright make? His thesis relates specifically to a type of biconditional made prominent by Frege in *Grundlagen*, §§62–9, which Wright calls an 'abstraction principle'; the word 'abstraction' is here used to refer to what in *Frege: Philosophy of Mathematics* I called 'logical abstraction', to distinguish it from the psychological abstraction that Frege combated so vigorously. The schema for a biconditional constituting an abstraction principle is as follows. A unary term-forming operator Σ, acting on expressions of some given logical type, is to be introduced: it may act on singular terms, on one-place predicates or on expressions of any one other type. The left-hand side of the biconditional consists of the equation '$\Sigma(\alpha_1) = \Sigma(\alpha_2)$', where α_1 and α_2 are free variables of the type on which Σ acts, together with any bound variables that Σ may be required to govern. The right-hand side of the biconditional is of the form $\mathscr{E}(\alpha_1, \alpha_2)$', where this is any expression, in general complex, which is already understood, in which α_1 and α_2 are the only free variables and which expresses the holding of an equivalence relation between the values of those variables; that, in a given case, it *is* an equivalence relation is a condition that needs to be demonstrated in order to establish that the biconditional is a genuine instance of the abstraction schema. A further condition is probably to be added, namely that there is no antecedent requirement that the referents of Σ-terms be identified with the referents of terms of any other form (nor, perhaps, that they should *not* be so identified). (ED), (EN), and Axiom V are all abstraction principles in this sense.

In *Grundlagen*, Frege indeed advanced a thesis concerning abstraction principles, namely that '$\Sigma(\alpha)$' should be defined as 'the $\mathscr{E}$-equivalence class to which α belongs'; we must except the case in which Σ is itself the abstraction operator. This, naturally, is not at all Wright's doctrine. He holds, rather, that in any such instance we are entitled to *stipulate* outright that the abstraction priniciple shall be taken as holding, and that this stipulation is a legitimate way of introducing the operator Σ as forming genuine singular terms subject to the standard logical laws governing such terms. Let us call this 'Wright's abstraction thesis'.

On the face of it, the claim that we are always free to stipulate the truth of an abstraction principle, and may thereupon treat Σ-terms as genuine terms, introduced in a legitimate manner, must fail in face of the fact that a well-known example of just such a biconditional led straight to contradiction. That would appear to end the matter for the time being: Wright owes us some restrictions with which to hedge his thesis, if it is to be tenable; without them it cannot be sustained.

Wright's attitude to this demand is a mixed one. One the one hand, he says (p. 345) that 'it should not seem at all obvious that I *do* owe any such explanation'. My supposition that he does is due, he thinks, to my incorrect belief that inconsistency is a symptom of some deeper malady. Since, as Wittgenstein taught us, it is not, the only explanation that could be required is one showing that Axiom V did lead to a contradiction, and why (EN) is highly unlikely to do so: and that Wright *had* provided, and proceeds to repeat (pp. 346–7).

Wittgenstein, as I read him, maintained more than that a contradiction does not point to some *underlying* malady: he maintained that it does not signify anything wrong at all. We need cast no doubt upon the assumptions or the forms of reasoning that led to the contradiction; we need merely to avoid deriving the contradiction or inferring anything from it. Wright, by contrast, allows that the contradiction shows Axiom V not to have been a legitimate abstraction principle: a contradiction *does*, in his view, require us to abandon the most plausible culprit among the assumptions that led to it. What the contradiction does *not* do is to call in question the general abstraction thesis as it stands. A contradiction may require us to jettison some underlying assumption, so long as that assumption does not lie *too* deep: it is a matter of getting the depth just right. My mistake was to think that the contradiction impugned the more general assumption upon which the assumption to be rejected had been based.

On the other hand, Wright is disposed to humour those who demand a general restriction on the abstraction thesis by after all providing one (p. 347). The restriction he proposes relates solely to cases in which the operator Σ acts on expressions for concepts. He points out that it will exclude Axiom V, and neatly observes that any abstraction principle that violates it will generate a contradiction. He is, however, very far from showing that any that complies with it will be in order. In fact, it will not even exclude Frege's own modification of Axiom V, which, as we know, was still inconsistent.

Wright fails to perceive what is required of him to justify his abstraction thesis. If he had come up with a restriction that would exclude any abstraction principle leading to inconsistency or any other untoward consequence, he would have a defensible thesis; but he still would not have one for which he had offered any justification. The thesis propounds that an abstraction principle serves as an adequate means of introducing a range of abstract terms into a language. What, at least from Frege's standpoint, is required of a means of introducing a range of singular terms? Plainly, that it suffice to specify the contribution of any of those terms to determining the truth-value of any sentence containing it. According to the usual conception of the semantic role of a singular term, a direct specification of its reference will suffice for this purpose. The context principle says, in effect, that any *other* method of specifying the semantic contribution of a term to any sentence containing it is just as good; indeed, it says, further, that such a method counts as an *indirect* means of laying down its reference.

Frege thought that he had succeeded in determining the truth-values of all sentences of his formal language. (Here 'determining' is to be understood in the semantic or model-theoretic, not the proof-theoretic, sense, as Wright's prime-pair example (p. 363) might suggest.) If he had really succeeded in this, there would have been no question but that he had specified the semantic contribution of every term for a value-range; and so he could legitimately appeal to the context principle to justify his means of introducing such terms. But of course Frege's claim was grossly overblown, and Wright has no thought of claiming so much. An argument is therefore needed to explain *why* it should be accepted that an abstraction principle serves to determine the semantic contribution to all sentences of a term in the range introduced by it. Wright, so far as I can see, has provided no such argument, and seems oblivious of the need for one.

5

Wright repudiates the idea that a single abstraction principle could simultaneously determine the domain of the individual variables and the references, within that domain, of the Σ-terms. In this, he certainly diverges from Frege, who, in *Grundgesetze*, supplied no means independent of his attempted specification of the *Bedeutungen* of value-range terms for forming a conception of a domain over which Axiom V could be satisfied. Rather, Wright appeals to a highly non-Fregean distinction between the meaning of a sentence and the condition for its truth. A sentence containing indexicals has a constant meaning, he says, though its truth-condition varies with occasion of utterance; likewise one involving quantification has a constant meaning, while its truth-condition may vary with the domain assigned to the variables. An abstraction principle determines the meanings of certain terms occurring in a sentence; its truth-condition will change according to the domain, which is to be specified independently.

Now this certainly makes abstraction principles far less interesting than, on Frege's fallacious use of one in *Grundgesetze*, they promised to be. They were meant to solve the problem how we could come by a conception of logical objects, and of a domain comprising sufficiently many of them to accommodate the natural numbers and the real numbers. On Wright's account, however, they will do no more than tell us what numbers of the one (and presumably of the other) kind are, once we have already formed the conception of domains big enough to contain them. And how are we to do this? How, before we have the concept of a number of either kind, can we come by a conception of an at least denumerable domain, or of one at least of the power of the continuum?

Wright's answer is not very clear. He begins (pp. 352–3) by deprecating appeal to the domain of 'all objects' (an indefinitely extensible concept if ever there was one), and expresses sympathy with Russell's (and my) view that very general ranges of objects do not constitute definite totalities capable of being taken as domains of quantification. But he then goes on (p. 353) to recommend the notion of unrestricted generality, arguing that we understand valid formulas of first-order predicate logic as those whose instances hold good in every determinate non-empty domain, without specifying any such domains. There seem to be two confusions here. It might possibly be useful to specify the 'meaning' of a statement involving bound variables independently of the domain over which those variables are to range; that is not at all to say that there is no need to specify that domain. 'What are the means whereby that domain may . . . have been determined?', Wright asks, as if this were a knock-down challenge to his opponent (me). Well, when the domain is that of a fundamental mathematical theory, that is precisely the problem: the problem to which Frege thought that the context principle offered a solution, and to which no one else has offered a satisfactory solution since. It is a sharper problem for Wright than for me. I merely observed that it did not seem possible, as Frege thought, simultaneously to specify the domain and to fix the references within it of a range of terms; it is Wright who, in what is supposed to be a *defence* of Frege, has sundered the task of fixing the meanings of such terms from that of specifying the domain, which, on his account, is a quite distinct task remaining to be accomplished. We are indeed entitled to ask of Wright: what are the means whereby the domain is to be determined? As for first-order logic, there is a difference between recognizing a formula as holding good in any domain, whatever the objects it comprises, and recognizing it as holding good in a domain comprising everything whatever. To confer a determinate truth-value upon a formula, we must impose an interpretation upon it, specifying, among other things, the domain of the variables. This does not cease to be true when it happens to be a valid formula, true under every interpretation; on the contrary, it is part of what is meant by a formula's being true under an interpretation.

Consider someone who (unlike all of us) comes to Frege's *Grundlagen* theory innocent of the concept of natural number. He does not *deny* that there are infinite totalities: rather, he does not have so much as the conception of an infinite totality. What will he make of Frege's theory, and, in particular of his proof that every natural number has a successor? Why, he will probably take it as demonstrating that *natural number* is an indefinitely extensible concept, and *therefore* unsuited to delineate a determinate totality. What will Wright say to him? I cannot guess.

And here we come upon the most astonishing feature of Wright's paper, considered as a reply to me. Twice he mentions the choice between classical and intuitionistic logic as if it were a matter to be decided later, or even one of taste (p. 353: 'we could do with further explanation . . . of whether classical forms of . . . reasoning continue to be appropriate'; p. 362: 'whatever semantics for the quantifiers, classical or constructivist, is found agreeable'). But I had argued that the notion of a *determinate* domain of quantification is a *classical* one: we need it to be determinate just which objects the domain comprises if we are to be able to assume that quantification, applied to a predicate determinately true or false of any given object, will always yield a statement with a determinate truth-value, and will therefore be subject to the laws of classical logic. Generalizing over the (entire) extension of an indefinitely extensible concept is quite meaningful, I argued, and will often yield a statement demonstrably true or demonstrably false: what it will not do is to yield such a statement *in every case*. Both in *Frege: Philosophy of Mathematics* and in 'What is Mathematics About?' (*Mathematics and Mind*, ed. Alexander George, New York, 1994), to which Wright refers, I argued that we can specify the domain of a fundamental mathematical theory by means of an indefinitely extensible concept, but that the consequence of doing so was that we must admit only intuitionistic reasoning; and I cited as a root of Frege's difficulties his unwavering attachment to classical logic. Wright probably disagrees: how he can pass the matter by without a glance perplexes me deeply.

Even if we can find a non-question-begging means of specifying a suitably large domain, it remains unclear how Wright's strategy of separating that task from the abstraction principle is supposed to work. Assume that we have found a means of specifying an at least denumerable domain that is intelligible and acceptable to a subject as yet without the concept of *natural number*. He is now introduced to that concept by means of Frege's theory (considered as founded upon $N^=$, and without the definition in terms of extensions). He has to be shown that $N^=$ is satisfiable in the domain which has been explained to him: how is this to be done? Presumably by associating the N-terms with elements of the domain. These will not have been described as natural numbers, since he did not have that concept when the domain was originally explained. To establish the needed identification of Fregean cardinal numbers (the referents of N-terms) with elements of the domain, the Julius Caesar problem must have been solved, at least to the degree of deciding which such identifications are legitimate. Wright skirts round the Julius Caesar problem, expressing confidence that it can be solved, but not actually offering a solution. Perhaps it is more central to his reconstruction of Frege than he realizes.

(EN) and other abstraction principles are presented as forms of conceptual innovation (n. 39); indeed, I am accused of failing to recognize this aspect of them.[1] In *Frege's Conception of Numbers as Objects*, $N^=$ was more than this: it was a means of justifying the ascription of reference to terms for cardinal numbers, and, thereby, the assumption that such numbers exist; the analogue of the claim Frege made in *Grundgesetze* for his stipulations concerning value-range terms. But in 'Harmless Impredicativity' the role of abstraction principles has been *reduced* to mere means of introducing hitherto unfamiliar concepts. The entire enterprise has become far less ambitious—has been in fact subverted and rendered radically unFregean. $N^=$ now serves merely to fix the *sense* of the cardinality operator, when taken as defined over a sufficiently large domain; it is powerless to show us how to attain the conception of such a domain.

6

The role allotted to Wright's 'Hero' (a male character, despite his feminine name) is less than perspicuous. He begins with no conception of numbers, but with a higher-order language for talking about concrete objects; possibly also about classes (p. 361), but let us ignore this. How he conceives the domain of his individual variables is left unspecified. He is introduced to terms for cardinal numbers by means of $N^=$. He understands that, since N-terms may contain N-terms, this is not a contextual definition enabling him to eliminate them; he is nevertheless determined to appeal to it to help him, step by step, to make sense of some, perhaps of all, sentences containing them.

The first unclarity concerns how he conceives of his efforts. At one point (p. 364) Wright remarks: 'To know the truth-conditions of a numerical identity statement, as determined via $N^=$, need not be . . . to be empowered to conceive it full-bloodedly *as* an identity statement.' This suggests a very interesting interpretation of Hero's labours to attain understanding. According to it, he treats the new N-terms as not intended to have genuine reference, but as virtual terms, to be understood, as far as $N^=$ admits, in the light of that stipulation. This will compel him to interpret quantification in

[1] I should indeed hesitate to say, with Wright, that it is 'of the *essence* of cardinal numbers that facts about identity and distinctness among them are constituted in facts about one–one correspondence among concepts' (p. 361), since I am chary of speaking of the essences of objects; but I should certainly agree that is intrinsic to the *concept* of cardinal numbers, see e.g. *Frege*: *Philosophy of Mathematics*, 130–1, where I explained why Frege was right to make cardinal *equivalence* the fundamental notion.

his now expanded language substitutionally, which is in accordance with Wright's dictum (p. 362) that 'a predicate ought to be regarded as determinate in content whose every completion by a term whose meaning is understood is likewise determinate in content'; this concurs with Frege's condition in *Grundgesetze* for the referentiality of a predicate. (To avoid a mixed interpretation of the quantifiers, let us assume that Hero has in his vocabulary a term for every concrete object.) Likewise, Hero will not treat the sign of identify, when flanked by an N-term, as a genuine sign of identity, but as to be interpreted, whenever it can be, by appeal to $N^=$.

Hero (like Frege in *Grundgesetze*) essays an inductive strategy. The induction is on the *rank* of a term or sentence, meaning essentially the depth of embedding of N-terms in N-terms. It is unclear that rank, as so defined, will serve the purpose. We may reasonably take the term '$Nx: x \neq x$' as of rank 0: but how about '$Nx: x = x$'? This would be intended to represent the number of all objects; not just of all concrete objects, but of cardinal numbers as well. It seems that, to obtain a suitable notion of rank, the individual variables occurring within N-terms ought to be indexed by rank, something not provided for in the expanded language as described by Wright.

Given that Hero understands all sentences and terms of rank 0, how is he to interpret a sentence of the form

$$Nx:A(x) = Ny:(y = Nx:x \neq x),$$

where we may assume the predicate '$A(x)$' to be applicable only to concrete objects and hence of rank 0? By appeal to $N^=$, he knows that this is equivalent to:

> For some R, R maps the x's such that $A(x)$ one-to-one on to the y's such that $y = Nx: x \neq x$.

How is Hero to understand *this*? Since quantification is being interpreted substitutionally, he must allow for the instantiation of '$R(\xi,\zeta)$' by any two-place predicate '$B(\xi,\zeta)$' of the language, of however high a rank; this will of course block the induction. It seems that predicate-variables, too, must be indexed by rank; it is far from clear that the result will tally with the original intention with which $N^=$ was stipulated.

We need hardly pursue Hero's difficulties in his enterprise, so understood, any further; for it is highly unlikely that Wright meant it to be understood in this way. If it could be carried through, Hero would have discovered how to interpret all contexts containing N-terms *without* treating them as denoting genuine objects; and this would give the lie to everything that Wright has argued for. We should therefore take Hero as so understanding $N^=$ as requiring an interpretation of the N-terms as genuine, full-fledged singular terms. He still might do this by thinking of them in either of two ways: as denoting concrete objects of the kind he is accustomed to refer to, or as denoting objects of a sort with which he had previously been unfamiliar. In the first case, the objects are being presented in quite new guises, without his knowing how to identify their old guises ('Even if the number 3 is in fact an apple (the one on the top of the pile)', p. 358). In the second, he understands these new objects as incapable of being identified with any concrete ones ('there is no co-reference between any of his initial stock of terms and any . . . numerical singular terms', p. 361).

There are problems with both these alternatives. If the first be intended, how can Hero be sure that the domain with which he started was large enough for $N^=$ to be satisfiable within it? If the second, on the other hand, Wright would appear to be going back on his declared view that the domain must be fixed *independently* of the abstraction principle: it seems, after all, that the mere stipulation of $N^=$ serves to enlarge the domain with the new objects. Wright offers no solution to either problem: instead, he directs attention to a more restricted problem.

Merely by accepting as legitimate one or other of the two ways of introducing the new terms as having genuine reference, Hero has not yet attained an understanding of all sentences containing them, Wright thinks: the induction on their rank is intended to show how he can do so. There is still a problem about the notion of rank: What rank, for example, should be assigned to the term 'Nx: (for some F, $x = Ny$: Fy)', representing the number of cardinal numbers? The question cannot be brushed aside, since every N-term containing the expression for the successor relation—a fortiori, one containing 'nat(ξ)'—will contain an embedded pseudo-N-term whose N-operator is attached to an expression containing a predicate-variable bound from without.

Wright wishes to prove that:

(1) If Hero understands all numerical identity statements of rank n, then he understands all N-terms of rank n.

From this, he says, it follows that:

(2) If Hero understands all numerical identity statements of rank n, then he understands all sentences of rank n whatever.

From this in turn it follows that:

(3) If Hero understands all numerical identity statements of rank n, then he understands all numerical identity statements of rank $n + 1$.

And from this, by induction, it follows that Hero understands all sentences whatever containing N-terms (at least if we grant, as we may, that he knows that every identity statement connecting an N-term with a term for a concrete object is false). Wright must therefore establish lemma (1) to the induction step (3).

To be said to understand a singular term, Wright tells us (p. 365) one must both know what kind of thing it denotes, and grasp the condition which a particular thing of that kind must meet in order to constitute its referent. This does *not* tally with Frege's condition for the referentiality of a term, which is that the result of inserting it in the argument-place of every referential predicate must have been assigned a reference (truth-value); that this condition is satisfied is, more or less, what Wright is trying to prove. Now, Wright argues, Hero can immediately infer from $N^=$ what cardinal numbers are, and that N-terms stand, in general, for such objects. But, to achieve the second qualification Wright has laid down, Hero must do some work. He must at least derive from $N^=$ what condition a specific finite number **n** must satisfy to be the referent of an N-term 'N*x*: A*x*'—namely that there be exactly **n** objects satisfying 'A*x*'; and this Wright is at pains to show that he can do.

Whatever may be thought of this as a proof of lemma (1), the fact is that Wright's inductive argument is almost as lame as Frege's, and for the same reason, that it is ill-founded. The ill-foundedness also arises for the same reason: a neglect of the range of the quantifiers. This question cannot be evaded: representing the abstraction principle as merely conferring meanings on the new terms, rather than determining the domain of quantification, will not serve Wright's turn any better than it served Frege's to assume that the former effected the latter. Wright has attempted a reformulation of the problem, as follows: *given* that Hero has accepted $N^=$ as a legitimate stipulation, can he be credited with an understanding of N-terms? He thinks that such an understanding can be considered independently of the domain of quantification, which of course it cannot, both because those terms are meant to be construed as denoting elements of that domain, and because they crucially contain both individual variables and predicate-variables defined over it. Frege saw more clearly that the two things went together, but failed to solve the joint problem. Wright has more work yet to do before he can establish his abstraction thesis.

7

Wright expresses 'strenuous disagreement' (p. 343) with my view (cited under (i), pp. 340–1) that the context principle, even when legitimately applied to introduce a new range of terms, can at best sustain a notion of reference too weak to support a realistic interpretation of sentences containing those terms, rather than the robust notion claimed by him to be in place. He does not address this disagreement in his present paper; I wish to plead that, when he comes to do so, he would treat of the criterion I proposed for a notion of reference for singular terms to be robust enough to support a realistic interpretation of them. This was that their reference should be semantically operative. Whether a notion of reference for terms of a given range is semantically operative or semantically idle depends on the sense we attach to a sentence containing a term of that range. In grasping its sense, we have a conception of the way in which its truth-value is determined. If the determination of the truth-value of any such sentence goes through the identification of the referent of the term, the notion of reference, as applied to it, is semantically operative; if it does not, that notion, even though legitimate, is semantically idle. If we follow Frege's stipulations of the *Bedeutungen* of value-range terms, as stated in Part I of *Grundgesetze*, the notion of *Bedeutung*, as applied to them, is semantically idle. In order to determine the truth-value of a sentence containing such a term **t**, we have to invoke Frege's stipulation of the *Bedeutung* of the smallest term (or the truth-value of the smallest sentence) properly containing the term **t**; determining the *Bedeutung* of **t** itself is no part of the process, because no stipulation was made stating what that was to be. The same would hold good of numerical terms if they were introduced by a direct stipulation of $N^{=}$. Even a satisfactory vindication of such a use of an abstraction principle would not justify a realistic interpretation of the terms introduced by means of it.

The point is, however, directly relevant to how the stipulation is to be vindicated, and in particular to Wright's argument in favour of the lemma (1) to his inductive reasoning. Wright wishes to show that if Hero understands all identity-statements involving N-terms of rank *n*, then he must understand all such N-terms, and hence (lemma (2)) must understand all sentences, of whatever kind, in which they occur. He tries to show this by formulating general conditions for being said to understand a singular term, and then arguing that Hero fulfils these in respect of N-terms of rank *n*. But the general conditions seem appropriate only when the semantic role of the singular term is conceived as indicating how its referent is to be identified. In such a case, it is plausible that one who fulfils these conditions will understand any sentence in which the term occurs, provided that he understands the rest of the sentence. For the process of determining the truth or falsity of such a sentence will involve identifying some object as the referent of the term; and, when that has been done, the contribution of the term to the determination of the truth-value of the sentence will have been exhausted. None of this applies when the introduction of the term is justified by appeal to the context principle. The truth-value of a sentence of the form 'Nx: $A(x) = Nx$: $B(x)$' is not determined by the coincidence of the referents of the two N-terms, but by the existence of a one-one map of the As on to the Bs; it is the existence of that map that constitutes the coincidence of their referents, so that the identification of those referents plays no role in the determination of the truth or falsity of the sentence. Wright has given no reason whatever to suppose that it will play a role in the determination of the truth or falsity of *other* sentences, and the presumption is that it will not. Until this very large lacuna is filled, Wright's argument for lemmas (1) and (2) carries no weight.

APPENDIX ON THE TERM 'HUME'S PRINCIPLE'

I wish to add a protest against the use of the name 'Hume's Principle', popularized by George Boolos and adopted from him by Richard Heck, William Demopoulos, Matthias Schirn, and others, though sparingly used by Crispin Wright in the present paper, where he prefers his earlier appellation '$N^=$'. There are two grounds for the protest.

First, Frege's ascription of the principle to Hume in *Grundlagen*, §63, is surely little more than a joke. The sentence he quotes from Hume—'When two numbers are so combined that the one has always an unit answering to every unit of the other, we pronounce them equal'—is very vague. Moreover, it has to do with *units*, not with objects falling under a concept. Hume evidently regarded numbers as collections of units, a conception Frege utterly repudiated.

Secondly, the term 'Hume's Principle' blurs a vital distinction between two quite distinct principles:

(1) the definition of 'just as many . . . as' (of 'equinumerous'):

> There are just as many Fs as Gs iff there is a one-one map of the Fs on to the Gs

and

(2) the equivalence (EN)—the abstraction principle proper:

> The number of Fs = the number of Gs iff there are just as many Fs as Gs.

Admittedly, in §63 of *Grundlagen*, Frege himself blurs the distinction between (1) and (2); but that is no good reason for his commentators to follow him in this, since the two principles are essentially distinct, so that either of them could be accepted while the other was rejected. §63 is a purely preliminary paragraph, and, as soon as Frege treats of the matter in any detail, he keeps (1) and (2) distinct. (1) appears as a definition in §72; (2) as a theorem in §73. (Admittedly, Frege does not keep them so clearly apart in *Grundgesetze*, where he is less concerned with the justification of his definitions.)

When principles (1) and (2) are distinguished, it becomes at once apparent that the sentence Frege quoted from Hume expresses neither of them. It is not, as it stands, about when it is right to say that there are just as many things of one kind as of another, but about equality of *numbers*, implicitly taken to be objects. From it there could no doubt be derived a definition or explanation of 'just as many'. But the most natural way to arrive at it, with Hume's (stated) principle as a base, would be to say that, to discover whether or not there are just as many Fs as Gs, you must first determine the number of Fs and the number of Gs (presumably by counting the Fs and the Gs), and then examine the numbers so arrived at to determine whether or not they are equal, i.e. whether one has always a unit answering to every unit of the other. This is in effect what, in criticizing principle (1), Husserl *does* say: so far as I know, there is nothing whatever in Hume to rule out such an explanation. So, not only does the expression 'Hume's Principle' unwarrantably run together two principles that ought to be clearly distinguished from one another, but it credits Hume with an idea he probably did not have and certainly did not state.

Principles (1) and (2) raise entirely different issues. (1) is a straightforward explicit definition, which Frege adopted but did not invent, which Waismann, following Wittgenstein, and, before him, Husserl both repudiated, and which there is some interest in discussing. (2) is the principle with which, in *Frege's Conception of Numbers as Objects*, Wright was principally concerned. If he had become convinced that (1) is not correct, it would have made very little difference to most of his discussion in that book. For my part, I tried in *Frege: Philosophy of Mathematics* to keep the two principles clearly apart: my expression 'the original equivalence' referred to (2) independently of (1).

(1) and (2) may of course be combined to yield (3), which is $N^{=}$:

> The number of Fs = the number of Gs iff there is a one-one map of the Fs on to the Gs.

(The implication from right to left is theorem 32 of *Grundgesetze*.) We indeed have occasion to refer to (3); but this supplies no ground for suppressing (1) and (2) as distinguishable ingredients of (3). The term 'Hume's Principle' is productive of intellectual as well as historical confusion, and its use should be resolutely avoided.

15

Response to Dummett

CRISPIN WRIGHT

I am grateful to Michael Dummett for his searching comments on my paper. Since he seems to have found so much of it unclear, I shall here try to clarify some of the essentials of the view I was defending, and to respond to his principal criticisms.

I. FOUR NEO-FREGEAN CLAIMS

Loosely expressed, the neo-Fregean thesis about arithmetic is that a knowledge of its fundamental laws (essentially, the Dedekind–Peano axioms)—and hence of the existence of a range of objects which satisfy them—may be based a priori on the explanatory principle, $N^=$. More specifically, the thesis involves four ingredient claims:

(i) that the vocabulary of higher-order logic plus the cardinality operator, 'Nx: . . . x . . .', provides a sufficient definitional basis for a statement of the basic laws of arithmetic;

(ii) that when they are so stated, $N^=$ provides for a derivation of those laws within higher-order logic;

(iii) that someone who understood a higher-order language to which the cardinality operator was to be added would learn, on being told that $N^=$ is analytic of that operator, all that it is necessary to know in order to construe any of the new statements that would then be formulable.[1]

(iv) Finally and crucially, that $N^=$ may be laid down *without significant epistemological obligation*: that it may simply be stipulated as an explanation of the meaning of statements of numerical identity, and that—beyond the issue of the satisfaction of the truth-conditions it thereby lays down for such statements—no competent demand arises for an independent assurance that there *are* objects whose conditions of identity are as it stipulates.

[1] Or anyway, any which corresponded to something finite-arithmetical—there is no need for a neo-Fregean about *arithmetic* to make larger claims about larger cardinals.

The first and third of these claims concern the epistemology of the meaning of arithmetical statements, while the second and fourth concern the recognition of their truth. With which of them does Dummett disagree? Pleasingly, he seems to have no quarrel with either of the first two; pleasingly because it is worth emphasizing that to accept the correctness of the first and second claims is already to acknowledge a substantial Fregean achievement: the analytical reduction of the primitive vocabulary of arithmetic to a base that contains just one non-logical expression, the cardinality operator; and a demonstration—'Frege's Theorem'—that on that basis, the fundamental laws of arithmetic can be reduced to just one: $N^=$ itself.

Dummett's disagreement entirely concerns, rather, the third and fourth claims. Some of his remarks about impredicativity suggest an unqualified rejection of (iii). Elsewhere, when he stresses the putative inability of $N^=$ to solve (the relevant instance of) what he terms the 'central problem'—that of securing the existence of the objects of fundamental mathematical theories—it appears that he regards the neo-Fregean thesis as foundering over (iv). But the line of thought most prominent in his comments in this volume is to the effect that it is not possible to satisfy (iii) and (iv) simultaneously: that to interpret $N^=$ so as to empower it to address, rather than presuppose a solution to the central problem, is to place an unsustainable explanatory burden upon it. He thinks he has demonstrated this dilemma to be inescapable and lethal. But I believe that impression depends upon a failure to confront squarely the version of the neo-Fregean proposal which—although not, it seems, with sufficient clarity—my paper actually advanced.

II. DUMMETT'S DILEMMA

How does the dilemma arise? Assume—what apparently continues to be a fixed point in Dummett's thinking—that, in order for $N^=$ to be usable as an explanation of the meaning of any arithmetical statements, a specification must be provided of a domain over which its objectual variables are to range.[2] Then it may seem to follow that the neo-Fregean has just two options. First, he may give the required specification in advance and independently of the stipulation of $N^=$. But in that case, it may seem that he will have to ensure that it contains sufficient objects to validate $N^=$. And, accordingly, since $N^=$ entails the infinity of the number series, that will demand an assurance that the domain in question contains an at least countable infinity of objects—an assurance we will possess only if we have, in effect, already solved the problem at which Frege's and the neo-Fregean's efforts

[2] This is an idea which was scrutinized in Sect. IV of my paper (henceforward 'Harmless Impredicativity') though Dummett has not, in his comments, directly responded to that discussion.

are directed: the problem of how it may be recognized from scratch that such a countable infinity exists. The alternative is to offer no such independent specification but to attempt to view the very stipulation of $N^=$ as determining *simultaneously* both the meaning of the N-operator—the sole item of non-logical vocabulary which it contains—and the domain over which its objectual variables range. However, Dummett takes it that this latter option is hopeless[3]—that no such stipulation can do both those things—and he takes it that I agree.

It is presumably because he views this dilemma as decisive that Dummett feels impelled to interpret me as making the 'radically unFregean' and etiolated proposal characterized at the end of the fifth section of his remarks. For on neither horn of the dilemma can the stipulation of $N^=$ lead us towards the recognition of a countably infinite domain. On the first horn, that recognition is presupposed. On the second, when no independent characterization of the domain is given, the most that can be claimed is that

> $N^=$ now serves merely to fix the *sense* of the cardinality operator, when taken as defined over a sufficiently large domain; it is powerless to show us how to attain the conception of [that is, I suppose: acquire reason to believe in] such a domain.[4]

Dummett takes me to have accepted this limitation, and thus effectively to have surrendered the Fregean project—of explaining how the infinity of the series of natural numbers can be recognized—altogether. But I do not accept the limitation, and the proposal I actually put forward is of course quite different and designed precisely to address the Fregean project.

To rehearse, briefly. First, there is indeed something which $N^=$ can explain before any consideration is given to the question of the range of its objectual variables. This is what I called the *range unspecific* understanding of the principle: an understanding, for any antecedently understood predicates, F and G, how things have to stand in an arbitrary range of objects if the number of F-objects in that range is to coincide with the number of G-objects in it. Dummett thinks, to repeat, that, constrained by the above dilemma, this point must have been the sum total of my proposal. And of course it's true that, unsupplemented, it would do nothing to explain how we can 'come by a conception' of the infinity of the series of natural numbers. However we *can* come by that conception—in just the way that Frege thought—if we may now legitimately adjoin to the range-unspecific understanding of $N^=$ the interpretation of the range of its first-order quantifiers as *wholly unrestricted*.

The suggestion, in short, is that Dummett's dilemma may be finessed by recourse to unrestricted first-order quantification. We avoid the second

[3] See the passages quoted under n. 25 of 'Harmless Impredicativity'.
[4] This volume p. 381.

horn because we do not require that, just by stipulating that $N^=$ is to hold, we somehow determine both the sense of the cardinality operator and the range of its quantifiers. The range of the quantifiers is indeed determined independently. But this independent determination does not impale us on the first horn because it does not presuppose a *prior* recognition of a countable infinity of objects. Rather, equipped with an antecedent understanding of what it is for a statement to make a claim about everything—about all objects without restriction—we interpret $N^=$ accordingly and are thereby led—via the very proof it provides of the infinity of the number series—to the recognition that there are at least countably infinitely many objects within the range of such a statement.

The proposal is, naturally, hostage to a number of assumptions:

(*a*) that there is indeed an intelligible and, in this context, legitimate use of unrestricted objectual quantification;

(*b*) that this use may be grasped in innocence of many of the kinds exemplified by the objects it encompasses, and

(*c*) that it can be freed from commitment to any notion of an all-encompassing *totality* of the kind jeopardized by the set-theoretic paradoxes—in other words, that the neo-Fregean is free to accept that unrestricted objectual quantification must include indefinitely extensible populations within its range and hence—if Dummett is right—that it can involve no determinate *domain* (if a domain of quantification is essentially something set-like).

Each of these represents a point of possible attack. But Dummett, so far as I can tell, has nowhere mounted such an attack, either in his remarks in the present exchange or elsewhere. True, he has insisted in a number of places that quantification over indefinitely extensible totalities cannot be interpreted classically, and that the logic for statements involving such quantification must be intuitionistic. If he is right about that—I shall not pursue the matter here—we may not assume without further ado that any statement involving unrestricted quantification is determinately either true or false. That might conceivably present a problem for the neo-Fregean, if certain of the proofs involved in the construction of arithmetic on the basis of $N^=$ would not go through except in a classical logic, based on the principle of bivalence. But again, Dummett has not presented an objection on this score.[5]

[5] In correspondence he has written that he regards unrestricted objectual quantification—even without the presupposition of a determinate domain of all objects—as a 'phantasmagoria'. Well, that's the agenda for a further debate. But notice that the minimal requirement for the neo-Fregean's purposes is not that the objectual quantifiers in $N^=$ be unrestricted but only that it be understood that their range embraces, *inter alia*, whatever objects that very principle may lead us to recognize. So even should it transpire that the suggestion of 'phantasmagoria' is right, I see no reason to doubt just on that account that the mode of response to Dummett's dilemma outlined can survive in essentials. But that's a matter for another round.

The question stands, then: How is Dummett's dilemma supposed to engage the neo-Fregean proposal actually canvassed in 'Harmless Impredicativity'? What, really, is his objection?

III. UNDERESTIMATING CAESAR?

A possible clue can be gleaned from Dummett's suggestion[6] that the Caesar problem has an importance for my reconstruction of Frege's project which I have failed to realize. This claim may puzzle some careful readers of 'Harmless Impredicativity' who note the absolutely fundamental role played by the assumption of a solution to the Caesar problem in Hero's transition from arithmetical innocence to an understanding of higher-order arithmetic.[7] Dummett's remark, however, is not made with Hero or his project in mind. He writes

> Even if we can find a non-question-begging means of specifying a suitably large domain, it remains unclear how Wright's strategy of separating that task from the abstraction principle is supposed to work. Assume that we have found a means of specifying an at least denumerable domain that is intelligible and acceptable to a subject as yet without the concept of *natural number*. He is now introduced to that concept by means of Frege's theory (considered as founded upon $N^=$, and without the definition in terms of extensions). He has to be shown that $N^=$ is satisfiable in the domain which has been explained to him: How is this to be done? Presumably by associating the N-terms with elements of the domain. These will not have been described as natural numbers, since he did not have that concept when the domain was originally specified. To establish the needed identification of Fregean cardinal numbers (the referents of N-terms) with elements of the domain, the Julius Caesar problem must have been solved, at least to the degree of deciding which such identifications are legitimate.[8]

Now according to the view I proposed, there *is*, in effect, 'a non-question-begging means of specifying a . . . domain that is intelligible and

[6] This volume p. 381.

[7] Indeed, the relevant point is plausibly independent of the detail of the discussion of Hero. For there can be no way, surely, of representing a general competence in a higher-order language containing the cardinality operator which does not involve ascribing an understanding of open sentences formulable in such a language; and no way of representing understanding of such open sentences save in terms of an understanding of the results of introducing into them any antecedently understood singular term of the language in question. Since these may be terms standing for as diverse a range of entities as you please—arithmetic being applicable to objects of absolutely any kind—a solution to the Caesar problem is not merely an integral part of the particular route that I had Hero follow but an indispensable component in *any* movement to the same result, viz. an understanding of higher-order arithmetic based upon $N^=$ in the context of a higher-order language which may contain terms for objects of absolutely any particular kind. (But for advocacy of the contrary view, that the Fregean may justifiably finesse the Caesar problem, see Richard Heck's 'The Julius Caesar Objection', in his edition, *Language, Thought and Logic*, Oxford, Oxford University Press 1997, pp. 273–308.).

[8] This volume p. 380.

acceptable to a subject as yet without the concept of *natural number*'. That specification is accomplished by stipulating that the range of the objectual variables is to be unrestricted.[9] Of course, it may not be apparent that the 'domain' in question *is* at least denumerable at the point at which the stipulation is made—that fact may emerge only when we recover the proof of the infinity of the number theories from $N^=$. So the whole process may *seem* to be begging a question: the question, roughly, how we are supposed to be sure, just by laying down that the domain of quantification is unrestricted, that we have latched onto a population of objects of which $N^=$ can be true? How can we be sure that all (or enough of) the N-terms which we are empowered to form will actually refer? Dummett apparently assumes that the neo-Fregean's response to this concern will have to be to show how enough items which may be given independently—i.e. not presented as natural numbers—may be found to serve as the referents of the N-terms to ensure that $N^=$ can be satisfied. Hence the need to address the Caesar problem, at least to the extent of showing that the favoured items are eligible so to serve.

Possibly this is the real crux of our disagreement.[10] In any case, I want to say that, so far from posing a good question to which, on my construction of the neo-Fregean project, an answer would have to be given along the lines indicated, the outlined objection merely betrays a flat refusal to entertain the means for addressing the ontological issues that the stipulation of $N^=$, and of abstraction principles generally, is supposed, by the neo-Fregean, to provide. Consider, for instance, the principle Dummett calls ED, stipulating that the directions of a pair of lines are to be the same just in case those lines are parallel. If the stipulation of this principle succeeds, as intended, in fixing the concept of direction, then it does so by determining the truth-conditions of statements about the identity of directions. So there cannot be any space for a doubt that *when those stipulated truth-conditions are met*, the relevant directions are indeed the same—and hence exist. The concept of direction is set up in such a way that no space is left for the possibility that a pair of lines be parallel and yet there be no objects in the

[9] I am now waiving any misgivings one might have about the appropriateness of the notion of a *domain* in the context of unrestricted quantification.

[10] It is, at any rate, a line of objection which has featured in Dummett's thinking for a long time. The following is from his entry on Frege in the *Encyclopaedia of Philosophy* (ed. Paul Edwards, New York, Macmillan, 1967).

> The minimal assumption Frege needed in *Grundlagen* is that there is a mapping by which each concept F is mapped onto a class of concepts containing just those concepts G such that there are just as many F's as G's. When classes are taken, as Frege took them, to be objects—that is, to be in the domain over which the concepts in question are defined—this is tantamount to the assumption that there are at least denumerably many objects. From this point of view it does not much matter whether the numerical operator is taken as primitive or as defined in terms of classes: the existential assumption is the same. (vol. 3, p. 236)

domain in question suitable to serve as their direction. For a pair of lines to be parallel *is* for their directions to be the same. And any language in which this stipulation is made must, accordingly, be interpreted as incorporating directions within the range of its objectual variables.[11]

The point is fundamental. If sortal concepts can indeed legitimately be introduced by Fregean abstraction, then there have to be appropriate objects of the kind for which a particular instance calls on its left-hand side in any case where its right-hand-side condition is met. And if the range of the quantifiers is stipulated as unrestricted, then of course those objects must fall within that range. Accordingly, the objection which Dummett lodges in the passage quoted *just reduces to a bald rejection of Fregean abstraction*. It is tantamount to the insistence that, before any abstraction principle may be accepted, we must independently verify that there are objects which behave as its (non-vacuous) truth would demand. So in no case, it would follow, can such objects be wholly given to us as the referents of a range of terms so introduced; an abstraction principle at best correctly characterizes a range of objects our acceptance of which must be justified by quite other means.

This is nothing but an unargued contradiction of claim (iv) cited at the beginning. And it is worth remembering that it is *not* Dummett's official view. Officially he means, albeit with qualifications about the robustness of the notion of reference that may be applicable to them, to be generally tolerant of abstraction as a method for introducing a range of terms and for justifying ascribing objectual reference to them. The problem that exercises him was supposed to concern specifically the impredicative case. But I believe that, in the objection under consideration, he effectively cuts the ground from under this official tolerance. And the nature of the alleged more specific difficulty remains, after his remarks no less than before, obscure to me.

IV. A PERSISTING MISREPRESENTATION

To other matters. Dummett writes of my proposal about the Caesar problem in *Frege's Conception*:

> To the extent that Frege thought the Julius Caesar problem an obstacle to determining the reference of terms for cardinal numbers, on the other hand, . . . , Wright believed him to be in error: the biconditional [N⁼], regarded as the object of a stipulation whereby the cardinality operator is to be introduced, *already* solved the

[11] For further discussion of the point, see *Frege's Conception*, 146; also my 'Field and Fregean Platonism', in Andrew Irvine (ed.), *Physicalism in Mathematics*, Dordrecht, Kluwer, 1990, 73–93, and sect. I of my 'On the Philosophical Significance of Frege's Theorem', in Richard Heck, ed., *Language, Thought and Logic*, Oxford, Oxford University Press 1997, pp. 201–44.

> Julius Caesar problem, as a problem about reference, by guaranteeing that no cardinal number is to be identified with anything not given as a cardinal number.[12]

This propounds the misinterpretation of the *Frege's Conception* proposal about Julius Caesar which I had hoped to have scotched in 'Harmless Impredicativity', n. 12. So I repeat: it is *not* an implication of the proposal about the Caesar problem in that book that nothing can be a number unless explicitly introduced as such. The *Frege's Conception* proposal, encapsulated in the principle N^d,[13] was that in order for a sortal concept, G, to be apt to include any numbers among its instances, it is necessary that there could be singular terms, 'a' and 'b', overtly purporting to denote instances of G, such that the truth-conditions of the identity-statement, 'a = b', could correctly be explained as the obtaining of a one–one correlation between some suitable pair of concepts.[14] The underlying motive for this proposal was the thought that inclusion among sortal concepts requires overlap in identity criteria. As indicated in 'Harmless Impredicativity', I would no longer wish to stick exactly to the letter of the *Frege's Conception* proposal. But it should be perfectly clear that it does not have the implication which Dummett keeps on suggesting: in particular, it does nothing to prohibit the inclusion of Natural Number within Class.

V. ROTTEN APPLES

Dummett continues to press the Bad Company objection. He sticks to it that the inconsistency of principles, like Basic Law V, which are formally similar to $N^=$, is sufficient just by itself to raise a question about the availability of $N^=$ for the neo-Fregean's purposes. And it is notable that he holds to this line without apparently feeling any need to claim that the consistency of $N^=$, for its part, is open to serious question.

True, he justly observes that the restriction on abstraction principles that I mooted—viz. that the abstractive relation not be, in the sense defined in 'Harmless Impredicativity', *Russellian*—is not by itself sufficient to ensure consistency, and that Frege's ill-judged modification of V, in particular, would at least pass *that* test.[15] But we know anyway that, in a second-order

[12] This volume p. 374.

[13] See *Frege's Conception*, 116.

[14] Note, as stressed in *Frege's Conception* n. 8, p. 178, that the invocation of the idea of *correct explanation* is essential to the purport of the proposal. In a particular context a question of personal identity, for instance—say, that of Jones's assailant with Three-Fingered Jake—might well be settled by a fact of one–one correspondence—say, that of the fingerprints on the cosh with the digits on Jake's left hand. But the truth-conditions of the claim of identity could hardly be adequately thereby *explained*. By contrast, the truth-conditions of claims of identity among a certain kind of equivalence class may—it is plausible—precisely be explained as the obtaining of the relevant equivalence relation.

[15] Justly since it would be pedantic to rejoin that Frege's modification actually involves not an inconsistent—unsatisfiable—principle but one with only singleton models.

context, we are not going to be able to formulate what may be uncontentiously shown to be a general consistency-preserving restriction.[16] And at least *one* of the questions which Dummett wants to press, namely

> the question how a procedure endorsed by Wright could prove to yield an inconsistency[17]

—this question *is* answered, it seems to me, by pointing out that the abstractive relation involved in V is Russellian, and that all principles based on such relations generate contradictions, for a generally statable reason.

Dummett's idea continues to be that the inconsistency of V, and of other abstractions like that for Relation-Numbers, somehow creates a rotten-apple effect, permeating the whole barrel of abstractions. Anyway, he thinks we'd be wise to look askance at what otherwise appear to be perfectly healthy fruit. What is odd about this emerges when, pursuing the image, we find it transpires also to be his view that we should look askance at pieces of fruit which we have cut open and have found—by the best, albeit fallible, assurances that we can gather—to contain no signs of any rot! There really isn't any understanding this view, it seems to me, except as attributable to the idea that inconsistency—the rot—is merely the symptom of an underlying disorder which may infect even rot-free specimens. This is, I take it, exactly Dummett's view: the underlying disorder is impredicativity, and it *is* a disorder, he thinks, even in consistent cases like $N^=$. Well, *that's* the contention which 'Harmless Impredicativity' is concerned to rebut![18,19]

I said above that a proscription on Russellian relations answers one of the questions which Dummett thinks that, in view of the inconsistency of V, the neo-Fregean has to answer. But it is only one of the questions that so arise. Dummett charges that I fail to perceive the extent of my explanatory and justificatory debts and continues:

> Wright fails to perceive what is required of him . . . If he had come up with a restriction that would exclude any abstraction principle leading to inconsistency or any other untoward consequence, he would have a defensible thesis; but he still would not have one for which he had offered any justification. The thesis propounds that

[16] See Richard Heck, 'On the Consistency of Second-Order Contextual Definitions', *Noûs*, 26 (1992), pp. 491–4.

[17] This volume p. 376. Of course, the formulation of the question is tendentious: I have never endorsed a general principle which would license both $N^=$ and V as admissible abstractions.

[18] By the way, I did not, in noting that Dummett's attitude here is of a kind unsympathetically treated by Wittgenstein, intend to suggest—as Dummett supposes—that I regard Wittgenstein as having shown that such an attitude is always incorrect. Rather I intended to signal Dummett's unremarked assumption that Wittgenstein had made no case to answer.

[19] Let me stress again that I do not of course regard the presumable consistency of $N^=$ as settling its suitability for the neo-Fregean's purpose. There remains to be addressed the distinct Bad Company objection, due to Boolos (see 'Harmless Impredicativity', n. 23), which takes centre stage in my 'The Philosophical Significance of Frege's Theorem'. But that is not the objection which Dummett is making.

an abstraction principle serves as an adequate means of introducing a range of abstract terms into a language. What, at least from Frege's standpoint, is required of a means of introducing a range of singular terms? Plainly, that it suffice to specify the contribution of any of those terms to determining the truth-value of any sentence containing it.[20]

Replace, as seems reasonable, 'contribution . . . truth-value' in the last part of that passage by 'contribution . . . truth-*conditions*', and let it be that to *understand* one of such a range of terms is to know what that contribution is. Then it emerges that what Dummett is demanding, in this passage, is that I argue that $N^=$, and legitimate abstractions in general, fix the meanings—furnish an understanding—of the terms for whose introduction they provide. That this is indeed what he intends seems quite clear in the succeeding paragraph:

An argument is . . . needed to explain *why* it should be accepted that an abstraction principle serves to determine the semantic contribution to all sentences of a term in the range introduced by it.[21]

I find it very strange that Dummett should make such a complaint. Of course, I completely agree that such an argument is required. What is at stake is no less than the third of the four neo-Fregean claims cited at the start of this response. Since the requirement is one which is strictly independent of any issues to do with inconsistency or other unfortunate properties that abstraction principles may have, it is no doubt potentially misleading of Dummett to have brought it up in the context of the Bad Company objection. Still, it is fundamental. And it is precisely to satisfy it that the entire last section of 'Harmless Impredicativity' is dedicated! The whole point of that discussion is to show how, equipped only with second-order logic and a range of non-arithmetical concepts, Hero has the means, once he is given $N^=$, to understand any proposition of second-order arithmetic, when formulated on the basis of the vocabulary of second-order logic plus the N-operator in the fashion which Frege invented.

I am accordingly quite at a loss to understand Dummett's immediately succeeding claim:

Wright, so far as I can see, has provided no such argument, and seems oblivious of the need for one.

Presumably he reads the discussion of Hero as directed at some quite different objective. Small wonder, then, that he finds that

The role allotted to Wright's 'Hero' . . . is less than perspicuous.

[20] This volume p. 378. [21] Ibid.

To someone who thought that Hero's role was other than the explanatory one just emphasized, it would, for sure, have to seem obscure indeed what the last section of 'Harmless Impredicativity' is for.

VI. UNRANKED NUMERICAL TERMS

Hero's progress through his recursion makes explicit—if I made no mistake—how he can arrive at an understanding of any higher-order arithmetical statement each of whose ingredient occurrences of the cardinality operator is as the dominant operator in a numerical term assigned a definite rank under the ranking scheme there described. But that, Dummett claims, does not cater for all possible cases:

> What rank, for example [he asks] should be assigned to the term 'Nx:((∃F)x = Ny:Fy)', representing the number of cardinal numbers? The question cannot be brushed aside, since every N-term containing the expression for the successor relation—a fortiori, one containing 'nat(ξ)'—will contain an embedded pseudo-N-term whose N-operator is attached to an expression containing a predicate-variable bound from without.[22]

Let me unpack the objection a little. It is a consequence of the ranking scheme that any statement of determinate rank must somewhere embed, however deeply, an 'unproblematic' predicate. And even in the case when all its primitive predicates are unproblematic, there are, of course, means in any higher-order language for forming predicates which embed no such predicate, to which the cardinality operator can then be applied. The resulting terms will then slip through the ranking scheme. For instance, consider any rank-0 identity:

Nx:Ax = Nx:Bx.

Then we may move to

(∃F)Nx:Ax = Nx:Fx

and thence to the open sentence,

(∃F)y = Nx:Fx, which may in turn be bound by the numerical

operator to generate the term:

Ny:(∃F)y = Nx:Fx,

which Dummett cites. This is a well-formed, prima-facie intelligible term. But the recursion assigns it no rank. So no indication has been provided how Hero can construe (identity-statements containing) it.

[22] This volume p. 383.

True, it is no part of an *arithmetical* competence to be able to construe that particular term. But recall Frege's definition of immediate predecession:

$$Pxy \leftrightarrow (\exists F)(\exists w)\,(y = Nx{:}Fx \;\&\; x = Nz{:}(Fz \;\&\; -(z{=}w))).$$

For particular choices, a, b, and A, for 'x', 'y' and 'F', the sentence

$$(\exists w)(b = Nx{:}Ax \;\&\; a = Nz{:}(Az \;\&\; -(z{=}w)))$$

may well have a determinate rank. But as soon as we generalize on 'A', the definition of rank will have nothing to say about it. And such generalizations have to be intelligible if Hero is to construe the axioms for successor and appreciate the force of the proofs of them which $N^{=}$ facilitates. And indeed, as Dummett says, the predicate, 'natural number' defined as characterizing the items which zero ancestrally precedes, inherits the lacuna.

How can it be filled? Well, perfectly straightforwardly. But I certainly ought explicitly to have indicated how in 'Harmless Impredicativity'. Let me repair the omission now.

There are two kinds of case to deal with—two kinds of way to generate occurrences of the N-operator which may elude the hierarchy of ranks: (i) we can quantify from without on a predicate within the scope of an occurrence of the N-operator—the move from '. . . Nx:(. . . A . . . x . . .)' to '(∃F)(. . . Nx:. . . F . . . x . . .)'. This is what is involved in Frege's definition of immediate predecession, hence in the definition of 'nat'. Note that the result, in this case, is what Dummett dubs a 'pseudo-N-term'. Alternatively (ii) we may generalize within the scope of an occurrence of the N-operator—the move from 'Nx:(. . . A . . . x . . .)' to 'Nx:(∃F)(. . . F . . . x . . .)'. In this case, we form a genuine but unranked numerical term. This is what is involved in forming the term, 'Nx:((∃F) x = Ny : Fy)', for the number of cardinals (if the range of 'F' is unrestricted).

How is Hero equipped to construe these two types of move?

Case (i): Here a very simple reflection suffices. Recall that Hero is hypothesized to be competent in higher-order logic. That requires him to understand higher-order quantifications of any sentences he understands. So the first type of move can never take him from something he understands to something he doesn't.

Case (ii) is only slightly more complex. Let 'Nx : (. . . A . . . x . . .)' be a term which Hero understands. So he understands the open sentence, '. . . A . . . x . . .' and hence—appealing again to his logical competence—its generalization: '(∃F)(. . . F . . . x . . .)'. So he understands the result of replacing the former by the latter in any context which he understands, including contexts of the form:

. . . A . . . x . . . 1–1 Bx,

where Bx is any open sentence which he understands. So—via $N^=$—he understands any identity-statement

$$Nx:(\exists F)(\ldots F\ldots x\ldots) = Nx:Bx$$

for which he understands a corresponding

$$Nx:(\ldots A\ldots x\ldots) = Nx:Bx.$$

From which it follows that he has as effective an understanding of '$Nx:(\exists F)(\ldots F\ldots x\ldots)$' as he has of '$Nx:(\ldots A\ldots x\ldots)$', and hence—now deploying exactly the crucial argument in the recursion[23]—that he understands the result of introducing the former into any open sentence which he understands.

These modes of occurrence of the N-operator thus present no additional difficulty. Note that their treatment relies on no particular construal of the higher-order quantifiers, and no particular view about what it is to understand any of the statements in the range in question. Those are questions that cannot be postponed indefinitely, of course. But they do not need to be taken on before we can be clear how, granted his understanding higher-order logic in some appropriate way—whatever that may prove to be—Hero may use $N^=$ to advance to a construal of any statement in the target range.

Dummett expresses a further, specific misgiving concerning the term, '$Nx:x = x$', standing—if for anything—then for the Universal Number, the number of all objects. I say 'if for anything' because there is, so far as I can see, no pressure intrinsic to the neo-Fregean position to grant a reference to that particular term: one might very well suppose that only stable—non-indefinitely-extensible—totalities can have a cardinal number, and hence that there simply is no such number associated with the condition of self-identity. The neo-Fregean would seem to be free to take that issue on its merits.[24] But suppose that '$Nx:x = x$' *is* granted a reference, and that Hero is consequently obliged to understand it. Dummett's complaint is that it cannot plausibly be assigned to rank 0—since self-identity is not, presumably, an unproblematic predicate—and so, since it embeds no occurrences of the N-operator, it must elude all ranks. So how can Hero construe it?

Well, no new considerations need to be invoked. Recall the two

[23] This volume pp. 364–5.

[24] And the issue doesn't just concern indefinite extensibility and size. The point is familiar that there are many meaningful predicates—'red', 'water', and 'large' respectively illustrate three kinds—which are not suitable arguments for the cardinality operator. However, those which are so suitable—usually termed *sortal*—standardly combine with those which are not to generate compounds—'red letter', 'water bed', 'large woman'—which are potentially suitable (though vagueness may divest the compound of a determinate extension on a particular occasion of use). So it's striking that self-identity never exerts that rehabilitative effect: 'x is a red self-identical' is just as unsuitable as 'x is red'.

platitudes concerning the understanding of descriptive singular terms which were deployed in my argument that an understanding of all numerical identity-statements of rank n provides for an understanding of all numerical terms of rank n. According to those platitudes, in order to understand 'Nx:x = x' it will be sufficient to know that it stands for a cardinal number and to know what would qualify a cardinal number to be its referent. Hero will know the first. As for the second, he will know that the condition for any particular cardinal number to be the referent of 'Nx:x = x' is that the particular numerically definite quantifier associated with that number should generate a truth when applied to 'x = x'. Of course, he may not be able to *tell* of any particular such quantifier that it does indeed generate such a truth (though he will certainly be able to tell of all the finite such quantifiers that they do not). But understanding a statement is one thing and actually being able to determine its truth-value is another. (That is *not* a realist assumption: the mathematical intuitionists do not in general deny meaning to statements which they cannot guarantee to be able to decide.)

VII. UNDERSTANDING NUMERICAL TERMS

It is, however, Dummett's view that the platitudes about the understanding of descriptive singular terms to which I appealed are not appropriately invoked to the neo-Fregean's purpose. This is because, while they—or something of the sort—may justifiably be applied in cases 'when the semantic role of the singular term is conceived as indicating how its referent is to be identified', it remains that

> None of this applies when the introduction of the term is justified by appeal to the context principle. The truth-value of a sentence of the form 'Nx:Ax = Nx:Bx' is not determined by the coincidence of the referents of the two N-terms, but the existence of a one–one map of the As onto the Bs; it is the existence of that map that constitutes the coincidence of their referents, so that the identification of those referents plays no role in the determination of the truth or falsity of the sentence.[25]

Dummett's wording here leaves some play for interpretation. But the main point, I take it, is that the conditions on the understanding of singular terms to which Hero's induction step appeals, are apt only in the case when the notion of reference associated with those terms is, as Dummett has elsewhere termed it, 'realistic', or 'robust'. When the notion of reference is by contrast 'thin'—when the ascription of reference to a range of terms is justified solely by the Context Principle, that is, when reference is imposed on them just by their occurrence in statements which we have to reckon true in the light of some contextual stipulation of their truth-conditions, like $N^=$

[25] This volume pp. 385–6.

or the principle ED—in that case, the platitudes misrepresent the form that the understanding of a singular term assumes.[26]

There are two questions here. The first is whether Dummett has indeed made out a contrast between 'robust' and 'thin' singular reference. The second is whether, if so, it carries the significance that he here ascribes to it—whether it does indeed *not* suffice for the understanding of a merely thinly referential singular term to know what kind of thing it thinly refers to and what specific condition a thing of that kind must meet if it is to qualify as the thin referent of the term. Both questions need a more detailed discussion than I have space to attempt here. But I'll close by saying a little.

Concerning the second, Dummett's position is surely extremely strange: What *more* can the understanding of a descriptive singular term—even a thinly referential descriptive singular term—demand if knowing what kind of thing it stands for and what a thing of that kind has to be like in order to qualify in its referent, is not to be sufficient? Dummett in effect agrees that these conditions do suffice for the understanding of robustly referential terms, but then proceeds to treat that fact as though it were somehow already sufficient to raise a doubt about their sufficiency when reference is—as he conceives it—'thin'. I should have thought, to the contrary, that even granted a distinction between the two kinds of reference, the platitudes go so deep that nothing should count as a notion of reference at all, even a thin one, unless similarly constrained by them.

What Dummett might possibly better have argued, rather, is that what is demanded by the two relevant kinds of knowledge—knowledge of the sort to which the referent of a term would belong, and knowledge of what would qualify a thing of that sort to be its referent—will vary depending on whether the terms in question are robustly or thinly referential, and that my argument misconceives the requirement in the thin case. But he does not essay such an argument; and I shall not now try to anticipate how one might run.

What of the distinction between robust and thin reference itself? The question which notion of reference applies to a given range of terms turns, as Dummett conceives the distinction, on whether it is 'semantically operative' or 'semantically idle'.

> If the determination of the truth-value of [any sentence containing a term of that range] goes through the identification of the referent of the term, the notion of reference, as applied to it, is semantically operative; if it is not, that notion, even though legitimate, is semantically idle.[27]

[26] Note that it doesn't matter for the present purpose whether the principle also provides the resources for a rigorous contextual definition, Dummett here apparently thinks.

[27] This volume p. 385.

Accordingly, he maintains, in any case where the statements containing the terms in question have their truth-conditions stipulated contextually, the notion of reference will be semantically idle because it will be possible to determine the truth-values of such statements by working exclusively on the right-hand sides, so to speak, of the relevant stipulations, without any consideration of the semantic structure of the left-hand sides.

Dummett pleads that I respond to this distinction. So I will. The question is not the validity of the point on which it is based—whether, for instance, it is possible to verify the truth-values of statements about directions without identification of the reference of direction-terms. Obviously it is: someone who verified an appropriate statement about lines and their parallelism, and who conceived of the contextual stipulations merely as introducing notational variants of such statements, would have done just that. But *that* point hardly suffices to establish that the notion of reference associated with direction-terms is properly regarded as semantically idle. To establish that, it ought to be necessary, rather, to show that the assignment of reference to direction-terms plays *no role in determining the content of sentences in which they occur*—it's not enough just to consider sentences in which they *don't* occur and, trading on their stipulated equivalence in truth-conditions to ones in which they do, merely to make the point encapsulated in the datum.

In order to see what force there might be in Dummett's distinction, we ought therefore to ask: Does the *understanding* of statements in which direction-terms occur proceed through a knowledge of the reference of those terms? If we answer affirmatively, there can be no good sense in which their reference could be stigmatized as 'semantically idle'. *And surely we must so answer.* The reason is simply that, while the truth-conditions of such statements may indeed be given, via the contextual stipulations, as those of statements in which no such terms occur, it is necessary, in order to understand statements of the former kind, to know *more* than that their truth-conditions may be so given: you have in addition to follow through the Fregean abstraction—to read the left-hand sides of the appropriate principles not merely as notational variants of the right-hand sides, but in a way which is constrained by their surface syntax and the familiar vocabulary that they contain. In brief, it is of the essence of Fregean abstraction that it demands a notion of statement content *narrower than* a primitive—non-inferential—sameness of truth-conditions. That's what's involved in the notion of 'carving up' of content in a new way.[28] Once this distinction is recognized, the question of whether a certain property of sub-sentential expressions is semantically operative or semantically idle becomes the

[28] *Grundlagen* §64; cf. sect. 4 of Bob Hale's 'Dummett's Critique of Wright's Attempt to Resuscitate Frege', in *Philosophia Mathematica*, (3) vol. 2 (1994), 122–47.

question whether it contributes towards the determination of the identity, in this more exigent sense, of the content of statements containing it.

By this test, the reference of singular terms introduced by Fregean abstraction is fully semantically operative, and no germane disanalogy between them and terms of the kind which Dummett would regard as robustly referential has been established.[29]

[29] Thanks to Michael Dummett, Bob Hale, and Stewart Shapiro.

16

Die Grundlagen der Arithmetik, §§82–3

GEORGE BOOLOS AND RICHARD G. HECK, JNR.

Reductions of arithmetic, whether to set theory or to a theory formulated in a higher-order logic, must prove the infinity of the sequence of natural numbers. In his *Was sind und was sollen die Zahlen?*, Dedekind attempted, in the notorious proof of theorem 66 of that work, to demonstrate the existence of infinite systems by examining the contents of his own mind. The axioms of General Set Theory, a simple set theory to which arithmetic can be reduced, are those of Extensionality, Separation ('Aussonderung'), and Adjunction: $\forall w \forall z \exists y \forall x(x \in y \leftrightarrow x \in z \vee x=w)$. It is Adjunction that guarantees that there are at least two, and indeed infinitely many, natural numbers. The authors of *Principia Mathematica*, after defining zero, the successor function, and the natural numbers in a way that made it easy to show that the successor of any natural number exists and is unique, were obliged to assume an axiom of infinity on those occasions on which they needed the proposition that different natural numbers have different successors.

In §§70–83 of *Die Grundlagen der Arithmetik*, Frege outlines derivations of some familiar laws of the arithmetic of the natural numbers from principles he takes to be 'primitive' truths of a general logical nature. In §§70–81, he explains how to define zero, the natural numbers, and the successor *relation*; in §78 he states that it is to be proved that this relation is one–one and adds that it does not follow that every natural number *has* a successor; thus by the end of §78, the existence, but not the uniqueness, of the successor remains to be shown. Frege sketches, or attempts to sketch, such an existence proof in §§82–3, which would complete his proof that there are infinitely many natural numbers.

§§82–3 offer severe interpretive difficulties. Reluctantly and hesitantly, we have come to the conclusion that Frege was at least somewhat confused in these two sections and that he cannot be said to have outlined, or even to have intended, any correct proof there. We will discuss two (correct) proofs of the statement that every natural number has a successor which might be extracted from §§82–3. The first is quite similar to a proof of this proposition that Frege provides in *Grundgesetze der Arithmetik*, differing

We are grateful to Kathrin Koslicki and Jason Stanley. The first author also wishes to express his thanks to the Alexander von Humboldt Foundation.

from it only in notation and other relatively minor respects. We will argue that fidelity to what Frege wrote in *Die Grundlagen* and in *Grundgesetze* requires us to reject the charitable suggestion that it was this (beautiful) proof that he had in mind when he wrote *Die Grundlagen*. The second proof we discuss conforms to the outline Frege gives in §§82–3 more closely than does the first. But if it had been the one he had in mind, the proof-sketch in these two sections would have contained a remarkably large gap that was never filled by any argument found in *Grundgesetze*. In any case, it is certain that Frege did not know of this proof.

We begin by discussing §§70–81.

In §70, Frege begins the definition of equinumerosity by explaining the notion of a relation, arguing that like (simple) concepts, relational concepts belong to the province of pure logic. In §71, he defines 'the objects falling under F and G are correlated with each other by the relation ϕ'. Using modern notation, but strictly following Frege's wording, we would write:

$$\forall a \neg(Fa \wedge \neg\exists b(a\phi b \wedge Gb)) \wedge \forall a \neg(Ga \wedge \neg\exists b(Fb \wedge b\phi a)).$$

To put the definition slightly more transparently, the objects falling under F and G are correlated by ϕ iff

$$\forall x(Fx \rightarrow \exists y(Gy \wedge x\phi y)) \wedge \forall y(Gy \rightarrow \exists x(Fx \wedge x\phi y)).$$

In §72 Frege defines what it is for the relation ϕ to be one–one ('beiderseits eindeutig', 'single-valued in both directions'): it is for it, as we should say, to be a function, i.e. $\forall d\forall a\forall e(d\phi a \wedge d\phi e \rightarrow a=e)$, that is one–one, i.e. $\forall d\forall b\forall a(d\phi a \wedge b\phi a \rightarrow d=b)$. Frege then defines 'equinumerous' ('gleichzahlig'): F is equinumerous with G iff there is a relation that correlates the objects falling under F one–one with those falling under G:

$$\exists\phi[\forall x(Fx \rightarrow \exists y(Gy \wedge x\phi y)) \wedge \forall y(Gy \rightarrow \exists x(Fx \wedge x\phi y)) \wedge$$
$$\forall d\forall a\forall e(d\phi a \wedge d\phi e \rightarrow a=e) \wedge \forall d\forall b\forall a(d\phi a \wedge b\phi a \rightarrow d=b)].$$

We abbreviate this formula: $F\approx G$.

At the end of §72, Frege defines the number that belongs to F as the extension of the concept 'equinumerous with the concept F'. He also defines 'n is a (cardinal) number': there is a concept such that n is the number that belongs to F. His next task, attempted in §73, is to prove a principle that Crispin Wright once called $N^{=}$ (for numerical equality),[1] Michael Dummett calls 'the original equivalence',[2] and we call 'Hume's Principle': the number belonging to F is identical with that belonging to G iff F is equinumerous with G.

[1] Crispin Wright, *Frege's Conception of Numbers as Objects*, Scots Philosophical Monographs Number Two, Aberdeen University Press, 1983.

[2] Michael Dummett, *Frege: Philosophy of Mathematics*, Harvard University Press, 1991.

The trouble with the definition of number given in §72 and the proof of Hume's Principle given in §73 is that they implicitly appeal[3] to an inconsistent theory of extensions of second-level concepts. Russell of course demonstrated the inconsistency of Frege's theory, presented in *Grundgesetze der Arithmetik*, of extensions of first-level concepts; a routine jacking-up of Russell's argument shows that of the theory Frege tacitly appeals to in *Die Grundlagen.*[4] It is by now well known, however, that Frege Arithmetic, i.e. the result of adjoining a suitable formalization of Hume's Principle to axiomatic second-order logic, is consistent if second-order arithmetic is, and is strong enough to interpret second-order arithmetic (as of course Frege can be seen as attempting to prove in *Grundlagen*). Indeed, Frege Arithmetic and second-order arithmetic are equi-interpretable: in Appendix 2 we show how to interpret Frege Arithmetic in second-order arithmetic.

Writing: $^{\#}F$ to mean: the number belonging to the concept F, we may symbolize Hume's Principle: $^{\#}F = {}^{\#}G \leftrightarrow F{\approx}G$.

The development of arithmetic sketched in §§74–81 makes use only of Frege Arithmetic and can thus be reconstructed in a consistent theory (or one we believe to be so!). Nothing will be lost and much gained if we henceforth suppose that Frege's background theory is Frege Arithmetic.

In §74, Frege defines 0 as the number belonging to 'not identical with itself': $0 = {}^{\#}[x{:}\ x{\neq}x]$. ([x: . . . x . . .] is the concept *being an object* x *such that* . . . x . . . .) Frege notes that it can be shown on logical grounds alone that nothing falls under [x: x ≠ x]. In §75, he states that $\forall x\neg Fx \rightarrow (\forall x\neg Gx \leftrightarrow F{\approx}G)$ has to be proved, from which $\forall x\neg Fx \leftrightarrow 0 = {}^{\#}[x{:}\ Fx]$ follows. These have easy proofs; Frege outlines that of the former in detail.

§76 contains the definition of 'n follows immediately after m in the "natürliche Zahlenreihe" ':

$$\exists F \exists x(Fx \wedge {}^{\#}F{=}n \wedge {}^{\#}[y{:}Fy \wedge y{\neq}x] = m).$$

It is advisable, we think, to regard the relation defined in this section as going from m to n, despite the order of 'n' and 'm' in both the definiens and definiendum of 'n immediately follows m in the natural series of numbers'. Thus we shall symbolize this relation: mPn ('P' for '(immediately) precedes').

[3] The appeal is made when Frege writes 'In other words:' at the end of the second paragraph of §73.

[4] Let ($\mathbb{V}$) be $\forall\mathcal{F}\forall\mathcal{G}({'\mathcal{F}} = {'\mathcal{G}} \leftrightarrow \forall X(\mathcal{F}X \leftrightarrow \mathcal{G}X))$. Then ($\mathbb{V}$) is inconsistent (in third-order logic). For let $\mathcal{F}$ be $[X{:}\ \forall\mathcal{H}(\forall x(Xx \leftrightarrow x = {'\mathcal{H}}) \rightarrow \neg\mathcal{H}X)]$ and let X be $[x{:}\ x = {'\mathcal{F}}]$. Suppose $\mathcal{F}X$. Then $\forall\mathcal{H}(\forall x(Xx \leftrightarrow x = {'\mathcal{H}}) \rightarrow \neg\mathcal{H}X)]$. So $\forall x(Xx \leftrightarrow x = {'\mathcal{F}}) \rightarrow \neg\mathcal{F}X$, whence $\neg\mathcal{F}X$ by the definition of X. Thus $\neg\mathcal{F}X$. So for some $\mathcal{H}$, $\forall x(Xx \leftrightarrow x = {'\mathcal{H}})$ and $\mathcal{H}X$, and then $X{'\mathcal{F}} \leftrightarrow {'\mathcal{F}} = {'\mathcal{H}}$. By the definition of X again, ${'\mathcal{F}} = {'\mathcal{F}} \leftrightarrow {'\mathcal{F}} = {'\mathcal{H}}$, ${'\mathcal{F}} = {'\mathcal{H}}$, and by ($\mathbb{V}$), $\forall X(\mathcal{F}X \leftrightarrow \mathcal{H}X)$, contra $\neg\mathcal{F}X$ and $\mathcal{H}X$. (We use ' ' ' to mean 'the extension of' and '[. : . . .]' to denote concepts (of whatever level).)

Call a concept *Dedekind infinite* if it is equinumerous with a proper subconcept of itself; equivalently, if it has a subconcept equinumerous with the concept *being a natural number*. With the aid of the equivalence of these definitions of Dedekind infinity, it is not difficult to see that nPn if and only if n is the number belonging to a Dedekind infinite concept. Thus the number of finite numbers, which Frege designates: ∞_1, but which we shall as usual denote: $\aleph_0$, follows itself in the 'natürliche Zahlenreihe', in symbols, $\aleph_0 P \aleph_0$. Since $\aleph_0$ is not a finite, i.e. natural, number, we shall translate 'in der natürlichen Zahlenreihe' as 'in the natural sequence of numbers'.[5]

§77 contains the definition of 1, as $^{\#}$[x: x=0], and a proof that 0P1. In §78, Frege lists a number of propositions to be proved:

(1) 0Pa → a=1;
(2) 1 = $^{\#}$F → ∃xFx;
(3) 1 = $^{\#}$F → (Fx ∧ Fy → x=y);
(4) ∃xFx ∧ ∀x∀y(Fx ∧ Fy → x=y) → 1 = $^{\#}$F;
(5) P is one–one ('beiderseits eindeutig'), i.e. mPn ∧ m′Pn′ → (m=m′ ↔ n=n′).[6]

Frege observes that so far it has not yet been stated that every number immediately follows or is followed by another. He then states

(6) Every number except 0 immediately follows a number in the natural sequence of numbers.

It is clear from §44 of *Grundgesetze*[7] that Frege did not take (6) to imply that 0 does not immediately follow a number, that is, that ¬zP0. This proposition is proved separately in *Grundgesetze*, as proposition 108, and will be used later on here, at a key point in the argument.

[5] Timothy Smiley observed that 'in the natural series of numbers' is to be preferred as a translation of 'in der natürlichen Zahlenreihe' to Austin's 'in the series of natural numbers'. T. J. Smiley, 'Frege's "series of natural numbers"', *Mind*, 97 (1988), 388–9. We have substituted 'sequence' for 'series' throughout.

[6] Frege does not indicate what proof of 78.5 he might have intended. Here is an obvious one that he might have had in mind.

Suppose mPn and m′Pn′. Then for some F, F′, x, x′, Fx, F′x′, $^{\#}$F = n, $^{\#}$F′ = n′, $^{\#}$[y: Fy ∧ y≠x] = m, and $^{\#}$[y′: F′y′ ∧ y′≠x′] = m′.

Assume m = m′. Then by Hume's Principle, there is a one-one correspondence φ between the objects y such that Fy and y≠x and the objects y′ such that F′y′ and y′≠x′. We may assume that if yφy′, then Fy, y≠x, F′y′, and y′≠x′. Let yψy′ iff (yφy′ ∨ [y=x ∧ y′=x′]). Then ψ is a one-one correspondence between the objects falling under F and those falling under F′, and so by Hume's Principle, n = n′.

Assume n = n′. By Hume's Principle, let ψ be a one-one correspondence between the objects falling under F and those falling under F′. We may assume that if yψy′, then Fy and F′y′. Let yφy′ iff (Fy ∧ y≠x ∧ F′y′ ∧ y′≠x′ ∧ [yψy′ ∨ (yψx′ ∧ xψy′)]). Then φ is a one-one correspondence between the objects y such that Fy and y≠x and the objects y′ such that F′y′ and y′≠x′, and so by Hume's Principle, m = m′.

[7] All references to sections of *Grundgesetze der Arithmetik* are to sections of vol. I of that work. The numbering of sections starts over again in vol. II.

§79 contains the definition of the strong ancestral of ϕ, 'x precedes y in the ϕ-sequence' or 'y follows x in the ϕ-sequence':

$$\forall F(\forall a(x\phi a \to Fa) \to \forall d\forall a(Fd \to d\phi a \to Fa) \to Fy),$$

which was definition (76) of *Begriffsschrift*. Frege will use this definition in §81 to define 'member of the natural sequence of numbers ending with n'. We shall use the standard abbreviation: $x\phi^*y$ for the strong ancestral. To prove that if $x\phi^*y$, then . . . y . . . , it suffices, by the comprehension schema $\exists F\forall a$ (Fa $\leftrightarrow$ . . . a . . .) of second-order logic, to show that $\forall a(x\phi a$ $\to$ . . . a . . .) and $\forall d\forall a$ (. . . d . . . $\to d\phi a \to$ . . . a . . .). We call this method of proof *induction 1*. (Induction 2 and induction 3 will be defined below.)

Here and below, we associate iterated conditionals to the right. Thus, e.g. '$A \to B \to C$' abbreviates '$(A \to (B \to C))$'. This convention provides an easy way to reproduce in a linear symbolism one major notational device of both *Begriffsschrift* and *Grundgesetze*.

Frege mentions in §80 that it can be deduced from the definition of 'follows' that if b follows a in the ϕ-sequence and c follows b, then c follows a; the transitivity of the strong ancestral is proposition (98) of *Begriffsschrift*. The proof Frege gives there can be formalized in second-order logic only with the aid of the comprehension schema (or something to the same effect); however, there is an easier proof that makes use only of the ordinary quantifier rules, applied to the universal second-order quantifier in the definition of ϕ^*. For the proof in §82–3, Frege will also need proposition (95) of *Begriffsschrift*: if $x\phi y$, then $x\phi^*y$, which easily follows from the definition of ϕ^*.

At the very end of §80 Frege states that only by means of the definition of following in a sequence is it possible to reduce the method of inference ('Schlussweise', which Austin mistranslates as 'argument') from n to n + 1, to the general laws of logic. Of course, the method of inference from n to n + 1 is what we call mathematical induction; Frege's remark may be taken to be a claim that mathematical induction can be proved with the aid of the definition of the ancestral of P.

In §81 Frege defines the weak ancestral: 'y is a member of the ϕ-sequence beginning with x' and 'x is a member of the ϕ-sequence ending with y' are to mean: $x\phi^*y \vee y{=}x$. We shall use the abbreviation: $x\phi^{*=}y$. He states at the beginning of the section that if ϕ is P, then he will use the term 'natural sequence of numbers' instead of 'ϕ-sequence'. We thus have five terms: 'y follows x in the natural sequence of numbers', 'x precedes y . . .', 'y immediately follows x . . .', 'x is a member of the natural sequence of numbers ending with y', and 'y is a member . . . beginning with x'. We shall abbreviate these as: xP^*y, xP^*y, xPy, $xP^{*=}y$, and $xP^{*=}y$, respectively.

Induction 2 is the following method of proof, in which weak ancestrals occur as hypotheses: To prove that if $x\phi^{*=}y$, then . . . y . . . , it suffices to prove

(i) . . . x . . .

and

(ii) $\forall d \forall a(\ldots d \ldots \rightarrow d\phi a \rightarrow \ldots a \ldots)$.

Induction 2 quickly follows from induction 1: if (i) and (ii) hold, then so does $\forall a(x\phi a \rightarrow \ldots a \ldots)$; thus if $x\phi^*y$, then . . . y . . . , by induction 1. But if $x = y$, then by (i), . . . y . . . again. Frege proves induction 2 as proposition 144 of *Grundgesetze*.

A basic fact about the weak ancestral, to which we shall repeatedly appeal, is that $x\phi^*a$ and thus $x\phi^{*=}a$, provided that $x\phi^{*=}d$ and $d\phi a$, for then either $x\phi^*d\phi a$, $x\phi^*d\phi^*a$, and $x\phi^*a$, or $x = d\phi a$, $x\phi a$, and $x\phi^*a$, by (95) and (98) of *Begriffsschrift*. That $x\phi^*a$ if $x\phi^{*=}d$ and $d\phi a$ is proposition 134 of *Grundgesetze der Arithmetik*; that $x\phi^{*=}a$ if $x\phi^*a$ is proposition 136.

Frege has not yet defined finite, or natural, number. He will do so only at the end of §83, where 'n is a finite number' is defined as 'n is a member of the natural sequence of numbers beginning with 0', i.e. as: $0P^{*=}n$. By induction 2, to prove that . . . n . . . if n is finite, it suffices to prove . . . 0 . . . and $\forall d \forall a(\ldots d \ldots \rightarrow dPa \rightarrow \ldots a \ldots)$.

In the formalism in which we are supposing Frege to be working the existence and uniqueness of 0, defined in §74 as $^{\#}[x: x \neq x]$, are given by the comprehension scheme for second-order logic and the standard convention of logic that function signs denote *total* functions. Thus $^{\#}$ denotes a total function from second- to first-order entities and the existence of $^{\#}[x: x=x]$, that of $^{\#}[x: x \neq x]$, and that of $^{\#}[x:x = {}^{\#}[x: x \neq x]\ ]$ will count as truths of logic. The propositions that 0 is a natural number and that any successor of a natural number is a natural number follow immediately from the definition of 'natural number'; 78.5 says that P is functional and one–one. So apart from the easily demonstrated statement that nothing precedes zero, by the end of §81 Frege can be taken to have established the Dedekind–Peano axioms for the natural numbers, *except for the statement that every natural number has a successor*.

Using the notation we have introduced, we may condense §§82–3 as follows:

§82. It is now to be shown that—subject to a condition still to be specified—(0) $nP\ ^{\#}[x: xP^{*=}n]$. And in thus proving that there exists a number k such that nPk, we shall have proved at the same time that there is no last member of the natural sequence. . . .

. . . It is to be proved that (1) $dPa \rightarrow dP\ ^{\#}[x: xP^{*=}d] \rightarrow aP\ ^{\#}[x: xP^{*=}a]$.

It is then to be proved, secondly, that (2) $0P\ ^{\#}[x: xP^{*=}0]$. And finally it is to be deduced that (0′) $0P^{*=}n \rightarrow nP\ ^{\#}[x: xP^{*=}n]$. The method of inference ('Schlussweise') here is an application of the definition of the expression 'y follows x in the natural sequence of numbers', taking $[y: yP\ ^{\#}[x: xP^{*=}y]\ ]$ for our concept F.[8]

[8] This sentence seems to throw Austin. But we take its last half to mean: when one takes for the concept F what is common to the statements about d and about a, about 0 and about

§83. In order to prove (1), we must show that $a = {}^{\#}[x: xP^{*=}a \wedge x \neq a]$ (3). And for this again it is necessary to prove that $[x: xP^{*=}a \wedge x \neq a]$ has the same extension as[9] $[x: xP^{*=}d]$ (4*). For this we need the proposition that $\forall a(0P^{*=}a \rightarrow \neg aP^{*}a)$ (5′). And this must once again be proved by means of our definition of following in a sequence, on the lines indicated above.

We are obliged hereby to attach to the proposition that $nP\ {}^{\#}[x: xP^{*=}n]$, the condition that $0P^{*=}n$. For this there is a convenient abbreviation . . . : n is a finite number. We can thus formulate (5′) as follows: no finite number follows itself in the natural sequence of numbers.

(We have added some reference numbers; (1) is Frege's own; primes indicate the presence of a finiteness condition in the antecedent; the asterisk * in (4*) indicates (what at least appears to be) a reference to extensions.)

It might appear that Frege proposes in these two sections to prove, not (0), but (0′), as follows: First prove (5′) by an appeal to the definition of P*. Then derive (4*) from (5′) and (3) from (4*). From (3) derive (1). Prove (2). Then, finally, infer (0′) from (2) and (1), by a similar appeal to the definition of P*.

However, it will turn out that this precise strategy cannot succeed. It cannot be (4*) and (3) that Frege wishes to derive—(3) e.g. is false if $a = \aleph_0$, as we shall see—but rather certain conditionals (4′) and (3′), whose consequents are (4*) (or rather an equivalent of it) and (3).

We do not, of course, know how Frege might have tried to fill in the details of this proof-sketch at the time of composition of *Die Grundlagen*. In particular, we do not exactly know how he would have proved (5′). (We can be reasonably certain that his proof of (2), however, would have been at least roughly like the proof we shall give below.) But, since he later proved a version of the following lemma as proposition 141 of *Grundgesetze*, it seems plausible to us to speculate that he might have intended to appeal to it or something rather like it in his proof of (5′). The lemma is a logicized version of the arithmetical truth: if $i < k$, then for some j, $j+1 = k$ and $i \leq j$.

Lemma. $xP^{*}y \rightarrow \exists z(zPy \wedge xP^{*=}z)$.

Proof. Let $Fa \equiv \exists z(zPa \wedge xP^{*=}z)$. Then $xPa \rightarrow Fa$, for if xPa, then certainly Fa: take z=x. And $Fd \wedge dPa \rightarrow Fa$: Suppose Fd and dPa. Then for some z, zPd and $xP^{*=}z$. By the basic fact about the weak ancestral, $xP^{*=}d$. But since dPa, Fa. The lemma follows by induction 1.

With the aid of the lemma, we can now use induction 2 to prove (5′).

n, and thus that the concept in question is $[x: xP\ {}^{\#}[y: yP^{*=}x]\ ]$. (Austin's translation makes it sound as if some binary relation holding between d and a and also between 0 and n were meant. However good his German and English may have been, Austin was no logician. It is time for a reliable English translation of *Grundlagen*.)

[9] Austin's translation here reads 'has an extension identical with that of'.

Proof. 0 = $^{\#}$[x: x≠x]. By Hume's Principle and the definition of P, ∀z¬zP0, and therefore by the lemma, ¬0P*0.

Now suppose dPa and aP*a. Then by the lemma, for some z, zPa and aP*⁼z, i.e. either aP*z or a=z, and therefore either zPaP*z or zPa=z. In either case zP*z. Since dPa and zPa, z=d by 78.5 and so dP*d. Thus ¬dP*d ∧ dPa → ¬aP*a.

(5′) now follows by induction 2.

(5′) merits a digression. The part of *Die Grundlagen der Arithmetik* entitled 'Our definition completed and its worth proved' begins with §70 and ends with §83; the concluding sentence of §83 reads, 'We can thus formulate the last proposition above as follows: No finite number follows itself in the natural sequence of numbers.' Apart from its position in the book and the fact that Frege mentions it in both the table of contents and the recapitulation of the book's argument at the end of *Die Grundlagen*, there are a number of reasons for thinking that Frege regarded this proposition as especially significant.

First, there is, according to Frege, an interesting connection with *counting*. When we count, he points out in §108 of *Grundgesetze*, we correlate the objects falling under a concept Φ(ξ) with the number words in their normal order from 'one' up to a certain one, 'N'; N is then the number of objects falling under Φ(ξ). Since correlating relations between concepts are not in general unique,

> the question arises whether one might arrive at a different number word 'M' with a different choice of this relation. By our stipulations, M would then be the same number as N, but at the same time one of the two number words would follow after the other, e.g. 'N' after 'M'. Then N would follow in the series of numbers after M, i.e. after itself. Our proposition [(5′)] excludes this for finite numbers.

We find this argument of considerable interest, but will not enter into a discussion of its correctness here.

Secondly, one of Frege's major philosophical aims, as is well known, was to show that reason, under the aspect of logic, could yield conclusions for which many philosophers of his day might have supposed some sort of Kantian intuition to be necessary. The proof of (5′) is a paradigm illustration of how the role of intuition in delivering knowledge can be played by logic instead.

One might think that the truth of (5′) could be seen by the following mixture of reason and intuition: (5′) says that there is no (non-null) loop of P-steps leading from a back to a whenever a is a finite number. So if a is finite but not zero and there is a loop from a to a, then within the loop, there is some number x that (immediately) precedes a, and therefore there is a loop from x (through a, back) to x. But since a is finite, there is a finite sequence of P-steps from zero to some number d preceding a; since *pre-*

cedes is one-one, d = x, and therefore there is a loop from d to d. Thus a loop 'rolls back' from a to d, and then all the way back to zero. But there is no loop from zero to zero; otherwise some number would precede zero, and that is impossible.

Of course, Frege's proof of proposition 145 avoids any appeals to intuition like those found in the foregoing argument.

Finally, in proposition 263 of *Grundgesetze*, Frege shows that any structure satisfying a certain set of four conditions is isomorphic to that of the natural numbers. We find it quite plausible to think that Frege realized that the statements that the natural numbers satisfy these conditions constitute an axiomatization of arithmetic and regarded them as *the* basic laws of arithmetic, the *Grundgesetze* of the title of his book.[10] Since one of these conditions is the one that (5′) shows to be satisfied, there is considerable reason to think that Frege regarded (5′) as one of the basic laws of arithmetic. End of digression.

(4*) at least appears to mention extensions of (first-level) concepts and may well do so. But (4*) is unlike the definition of cardinal number and proof of Hume's Principle in that any mention of extensions it contains is readily eliminable without loss: Frege could have written to exactly the same point, 'a member of the natural sequence of numbers ending with a, but not identical with a, is a member of the natural sequence of numbers ending with d, and vice versa.'

It is evident that Frege cannot in fact be proposing to derive (4*) or the equivalent

$$(4)\ \forall x([xP^{*=}a \wedge x \neq a] \leftrightarrow xP^{*=}d)$$

from (5′) since both of these contain free occurrences of 'd'. Since the supposition of §82 that dPa is clearly still in force, it might be thought that Frege wishes to derive

$$(4^{\dagger})\ dPa \rightarrow \forall x([xP^{*=}a \wedge x \neq a] \leftrightarrow xP^{*=}d)$$

from (5′).

However, if d = a = $\aleph_0$, then, as we have observed, dPa; and then, since $aP^{*=}a$, (4†) has a true antecedent and false consequent. Thus it cannot be (4†) that Frege is proposing to derive from (5′).

We may note, though, that $\forall x([xP^{*=}a \wedge x \neq a] \leftrightarrow xP^{*=}d)$ can be derived from dPa and $\neg aP^{*}a$. So we may take it that Frege is proposing to derive

$$(4')\ 0P^{*=}a \rightarrow dPa \rightarrow \forall x([xP^{*=}a \wedge x \neq a] \leftrightarrow xP^{*=}d)$$

from (5′).

[10] For elaboration of this suggestion, see the second author's 'The Development of Arithmetic in Frege's *Grundgesetze der Arithmetik*', *Journal of Symbolic Logic*, 58/2 (1993), 579–601.

Proof. Suppose $0P^{*=}a$ and dPa. Assume $xP^{*=}a \wedge x \neq a$. Then xP^*a. By the lemma, for some c, cPa and $xP^{*=}c$. By 78.5, c = d. Thus $xP^{*=}d$. Conversely, assume $xP^{*=}d$. Since dPa, xP^*a by the basic fact about the weak ancestral, and so $xP^{*=}a$. If $\neg aP^*a$, then also $x \neq a$. But since $0P^{*=}a$, it follows from (5′) that indeed $\neg aP^*a$. Thus (4′) is proved.

Nor could Frege be proposing to derive (3) $a = {}^{\#}[x: xP^{*=}a \wedge x \neq a]$ from any proposition he takes himself to have demonstrated. For (3) is false if 'a' has $\aleph_0$ as value. In fact, ${}^{\#}[x: xP^{*=}\aleph_0 \wedge x \neq \aleph_0] = 0$. For if $xP\aleph_0$, then since $\aleph_0 P \aleph_0$, $x = \aleph_0$, by 78.5. Let S be the converse of P. Then if $\aleph_0$ Sx, $x = \aleph_0$. Thus if $\aleph_0 S^*x$, $x = \aleph_0$. (Let $Fa \equiv a = \aleph_0$ in the definition of S^*.) But the ancestral is the converse of the ancestral of the converse. So if $xP^*\aleph_0$, then $x = \aleph_0$. Thus $xP^{*=}\aleph_0$ iff $x = \aleph_0$, and therefore for *no* x, $xP^{*=}\aleph_0 \wedge x \neq \aleph_0$. By a proposition given in §75, ${}^{\#}[x: xP^{*=}\aleph_0 \wedge x \neq \aleph_0] = 0$.

However, it is important to observe that at this point it is not only the conjunct dPa of the antecedent of (1) that is assumed to be in force; the *other* conjunct $dP\,{}^{\#}[x: xP^{*=}d]$ is also assumed to hold. (It is easy to be oblivious to this further assumption since '$a = {}^{\#}[x: xP^{*=}a \wedge x \neq a]$' does not mention d. But it is supposed at this point that a is such that dPa, and it is likewise *also* supposed that d is such that $dP\,{}^{\#}[x: xP^{*=}d]$.) Since $a = {}^{\#}[x: xP^{*=}a \wedge x \neq a]$ follows from these two conjuncts and the consequent of (4′), we may take it that Frege wishes to prove (3′):

$$(3')\ \ 0P^{*=}a \rightarrow dPa \rightarrow dP\ {}^{\#}[x: xP^{*=}d] \rightarrow a = {}^{\#}[x: xP^{*=}a \wedge x \neq a].$$

Proof. Suppose dPa and $dP\,{}^{\#}[x: xP^{*=}d]$. Then by 78.5 (the other way), $a = {}^{\#}[x: xP^{*=}d]$. Suppose further that $0P^{*=}a$. Then by (4′), $\forall x([xP^{*=}a \wedge x \neq a] \leftrightarrow xP^{*=}d)$. By Hume's Principle, ${}^{\#}[x: xP^{*=}a \wedge x \neq a] = {}^{\#}[x: xP^{*=}d]$. Thus $a = {}^{\#}[x: xP^{*=}a \wedge x \neq a]$.

We come now to the difficult question how Frege proposes to derive (1) from (3′). Frege tells us that to prove (1), we must show (3). But (3) is not unconditionally true. However, (3′), whose consequent is (3) and whose antecedent contains a conjunct stating that the value of 'a' satisfies the condition of finiteness, can be proved. Thus it might seem reasonable to think that Frege may be proposing, as in the case of (4) and (3), not to derive (1) from (3), but some conditional instead whose antecedent expresses a finiteness condition and whose consequent is (1). Moreover, since dPa is one of the clauses of the antecedent, if we take $0P^{*=}d$ as another conjunct of the antecedent, we need not also take $0P^{*=}a$. So we let (1′) be

$$0P^{*=}d \rightarrow dPa \rightarrow dP\ {}^{\#}[x: xP^{*=}d] \rightarrow aP\ {}^{\#}[x: xP^{*=}a]).$$

(1′) follows readily from (3′):

Proof. Suppose that $0P^{*=}d$, dPa and $dP\,{}^{\#}[x: xP^{*=}d]$. By the basic fact about the weak ancestral, $0P^{*=}a$. By (3′), $a = {}^{\#}[x: xP^{*=}a \wedge x \neq a]$. Since

$a{=}a$, $aP^{*=}a$. By the definition of P, $^{\#}[x{:}\ xP^{*=}a \wedge x{\neq}a]\ P\ ^{\#}[x{:}\ xP^{*=}a]$. Thus $aP\ ^{\#}[x{:}\ xP^{*=}a]$.

It may be useful to recapitulate here our (somewhat intricate) derivation of (1′) from (5′) and the other propositions to which Frege appeals:

Suppose $0P^{*=}d$, dPa and $dP\ ^{\#}[x{:}\ xP^{*=}d]$. By the basic fact about the weak ancestral, $0P^{*=}a$, and thus by (5′), $\neg aP^{*}a$. If $xP^{*=}a$ and $x \neq a$, then $xP^{*}a$, and so by the lemma, for some z, $xP^{*=}z$ and zPa. By one half of 78.5, $z{=}d$, and so $xP^{*=}d$; conversely if $xP^{*=}d$, then by the basic fact, $xP^{*}a$, whence $x{\neq}a$ (since $\neg aP^{*}a$) and $xP^{*=}a$. Thus $\forall x(xP^{*=}a \wedge x{\neq}a \leftrightarrow xP^{*=}d)$, which is (4), and so by Hume's Principle, $^{\#}[x{:}\ xP^{*=}a \wedge x{\neq}a] = {}^{\#}[x{:}\ xP^{*=}d]$, and therefore $dP\ ^{\#}[x{:}\ xP^{*=}a \wedge x{\neq}a]$. By the other half of 78.5, $a = {}^{\#}[x{:}\ xP^{*=}a \wedge x{\neq}a]$, which is (3). Since $aP^{*=}a$ (trivially), by the definition of P, $^{\#}[x{:}\ xP^{*=}a \wedge x{\neq}a]\ P\ ^{\#}\ [x{:}\ xP^{*=}a]$, and therefore $aP\ ^{\#}[x{:}\ xP^{*=}a]$.

(2) is proved much more easily.

$$\text{(2)}\quad 0P\ ^{\#}[x{:}\ xP^{*=}0].$$

Proof. $0 = {}^{\#}[x{:}\ x{\neq}x]$. By Hume's Principle and the definition of P, $\forall z\neg zP0$. By the lemma, $\forall x\neg xP^{*}0$, and so $\forall x\neg(xP^{*=}0 \wedge x{\neq}0)$. By a result of §75 mentioned above, $^{\#}[x{:}\ xP^{*=}0 \wedge x{\neq}0] = 0$. But $0P^{*=}0$, whence $^{\#}[x{:}\ xP^{*=}0 \wedge x{\neq}0]\ P\ ^{\#}[x{:}\ xP^{*=}0]$, and therefore $0P\ ^{\#}[x{:}\ xP^{*=}0]$.

(0′) must now be derived from (1′) and (2). It is not possible to appeal to induction 2 because of the presence of '$0P^{*=}d$' in the antecedent of (1′). But, it might be supposed, Frege can appeal here to induction 3, which he explicitly demonstrated in *Grundgesetze* as proposition 152: To prove that if $x\phi^{*=}\ y$, then . . . y . . . , it suffices to prove

(i) . . . x . . .

and

(ii) $\forall d\forall a(x\phi^{*=}d \rightarrow \ldots d \ldots \rightarrow d\phi a \rightarrow \ldots a \ldots)$.

Note the formula '$x\phi^{*=}d$', whose presence weakens (ii) and thereby strengthens the method. The derivation of induction 3 from induction 2 is significantly more interesting than that of induction 2 from induction 1. It appeals to the comprehension scheme of second-order logic and uses a technique sometimes called 'loading the inductive hypothesis'. (At the beginning of §116 of *Grundgesetze*, Frege writes, 'To prove proposition (γ) of §114, we replace the function mark "F(ξ)" with "$\neg(aP^{*=}\xi \rightarrow \neg F(\xi)$)".')

Proof of induction 3: Suppose (i) and (ii). Let $Ga \equiv \ldots a \ldots \wedge x\phi^{*=}a$ (second-order comprehension). Now, $x\phi^{*=}x$ trivially; thus by (i), Gx. We now show $\forall d\forall a(Gd \rightarrow d\phi a \rightarrow Ga)$: Suppose $d\phi a$ and Gd, i.e. . . . d . . . and $x\phi^{*=}d$. By (ii), . . . a. . . . By the basic fact about the weak ancestral, $x\phi^{*=}a$. Thus $\forall d\forall a(Gd \rightarrow d\phi a \rightarrow Ga)$. By induction 2, Gy, whence . . . y. . . .

(0′) is now immediate from (2) and (1′) by induction 3: let x = 0, ϕ = P, y = n, and . . . y . . . iff $yP^{\#}[x: xP^{*=}y]$.

We believe that no one will seriously dispute that this proof of (0′), which features a derivation of (1′) from (5′) and an appeal to induction 3, is Fregean in spirit, ingenious, and of a structure that fits the proof-sketch found in §§82–3 rather well. But there are a number of strong reasons for doubting that Frege had *it* in mind while writing these two sections. Accordingly, we shall refer to it as the *conjectural* proof.[11]

First of all, Frege twice *says* that (1) is to be proved, once in §82 and again in §83. He says, moreover, 'The method of inference here is an application of the definition of the expression "y follows x in the natural sequence of numbers", taking $[y: yP^{\#}[x: xP^{*=}y]\,]$ for our concept F.' It would thus seem natural to take Frege as arguing by appeal to induction 1 or induction 2 (with P as ϕ). Frege mentions the condition that n be finite, but does not also mention, as he might easily have done, the need to assume that d (or a) is finite as well. Thus it would seem overly charitable to assume that the argument he really intended proceeds via induction 3.

Secondly, notice that Frege says in §83 that (5′), which Frege proved in *Grundgesetze* by appeal to induction 2, 'must likewise ("ebenfalls") be proved by means of our definition of following in a series, as indicated above.' It seems plain that Frege does not intend to use induction 3 to prove (5′); 'ebenfalls' suggests that the induction used to prove (0′) would be like the one used for (5′).

The most telling objections to the suggestion that Frege was intending to sketch the conjectural proof in *Die Grundlagen*, however, arise from a close reading of Section H (eta) of Part II of *Grundgesetze*. We quote and comment upon part of Section H.[12]

H. Proof of the proposition $(0P^{*=}b \rightarrow bP^{\#}[x: xP^{*=}b])$

§114. Analysis

We wish to prove the proposition that the Number that belongs to the concept

member of the number-series ending with b

follows after b in the number series if b is a finite number. Herewith, the conclusion that the number-series is infinite follows at once; i.e. it follows at once that there is, for each finite number, one immediately following after it.

We first attempt to carry out the proof with the aid of proposition (144) {viz. $aq^{*=}b \rightarrow \forall d(Fd \rightarrow \forall a(dqa \rightarrow Fa) \rightarrow (Fa \rightarrow Fb)$}, replacing the function-mark '$F(\xi)$' with '$\xi P^{\#}[x: xP^{*=}\xi]$'. For this we need the proposition '$dP^{\#}[x: xP^{*=}d] \rightarrow dPa \rightarrow aP^{\#}[x: xP^{*=}a]$'.

[11] Of course, what is conjectural is whether the proof is Frege's, not whether it is a (correct) proof.

[12] The present translation is based upon one due to the second author and Jason Stanley. We have changed Frege's notation to our own and added some material in braces.

That is to say, one's 'first' idea might be to prove (0′) by applying induction 2 to the concept [y: yP #[x: xP*⁼y]], which would, among other things, require a proof of (1). (A footnote, to which we shall return, is attached to this last sentence.)

> Substituting . . . in (102) {viz. #[x: Fx ∧ x≠b] = c → Fb → cP #F}, . . . we thus obtain '#[x: xP*⁼a ∧ x≠a] = a → aP*⁼a → aP #[x: xP*⁼a]', from which we can remove the subcomponent 'aP*⁼a' by means of (140) {viz. aq*⁼a}. The question arises whether the subcomponent '#[x: xP*⁼a ∧ x≠a] = a' can be established as a consequence of 'dPa' and 'dP #[x: xP*⁼d]'.

Put differently, the problem reduces to that of proving

$$(3^{\dagger})\ \text{dPa} \rightarrow \text{dP}\ {}^{\#}[\text{x: xP}^{*=}\text{d}] \rightarrow \text{a} = {}^{\#}[\text{x: xP}^{*=}\text{a} \wedge \text{x} \neq \text{a}],$$

which is (3′) without the finiteness condition 0P*⁼a, and which, together with the relevant instance of Frege's proposition 102, implies (1).

> By the functionality of progression in the number series . . . , we have 'dP #[x: xP*⁼d] → dPa → #[x: xP*⁼d] = a'. . . . We thus attempt to determine whether '#[x: xP*⁼a ∧ x≠a] = #[x: xP*⁼d]' can be shown to be a consequence of 'dPa'. . . . For this it is necessary to establish '([bP*⁼a ∧ b≠a] ↔ bP*⁼d)' as a consequence of 'dPa'. . . .

Or, (3†) will follow from

$$(4^{\dagger})\ \text{dPa} \rightarrow \forall \text{x}([\text{xP}^{*=}\text{a} \wedge \text{x} \neq \text{a}] \leftrightarrow \text{xP}^{*=}\text{d}),$$

an easy consequence of Hume's Principle and the one–oneness of P.

> For this it is necessary to establish '(bP*⁼d → [bP*⁼a ∧ b≠a])' and '([bP*⁼a ∧ b≠a] → bP*⁼d)' as consequences of 'dPa'. But it turns out that another condition must be added if 'b≠a' is to be shown to be a consequence of 'bP*⁼d' and 'dPa'. By (134) we have 'bP*⁼d → dPa → bP*a'. If b coincided with a, then the main component would transform into 'aP*a'. By (145) {our (5′)}, this is impossible if a is a finite number. Thus the subcomponent '0P*⁼a' is also added.
>
> Admittedly, the desired application of (144) thereby becomes impossible; but with (137) {viz. aq*⁼e → eqm → aq*⁼m}, we can replace this subcomponent with '0P*⁼d' and derive from (144) the proposition [152], '(aq*⁼b → ∀d(Fd → aq*⁼d → ∀a(dqa → Fa)) → (Fa → Fb))' . . . , which takes us to our goal.

That is, to establish the first half of (4†), we need to know that ¬aP*a; this will follow from (5′) and the additional assumption that a is finite. However, this new assumption must then be carried along throughout the proof, transforming (4†) into (4′), (3†) into (3′), and (1) into '0P*⁼a → dP #[x: xP*⁼d] → dPa → aP #[x: xP*⁼a]', from which (1′) easily follows. The attempt to prove (0′) via induction 2 then fails, since we simply have not proved (1), though we can still complete the proof by making use of induction 3 instead.

It is, we think, difficult to read these paragraphs without supposing that they reveal Frege's *second thoughts* about his idea in *Die Grundlagen* of applying induction 2 to prove $(0P^{*=}n \rightarrow nP\ {}^{\#}[x\colon xP^{*=}n])$, by substituting $[y\colon yP\ {}^{\#}[x\colon xP^{*=}y]\ ]$ for F. The attempt won't work, he says, because we need the hypothesis that a is finite in order to derive $\neg aP^{*}a$, which is needed for $(bP^{*=}d \rightarrow dPa \rightarrow b{\neq}a)$, which is in turn necessary for the rest of the proof. Read side by side with §§82–3 of *Die Grundlagen*, Frege's discussion in these paragraphs strikes one as penetrating and direct criticism of his earlier work. Moreover, the criticism suggests a way in which the conjectural proof can be regarded as Frege's after all: it is the proof obtained on amending the proof-sketch of §§82–3 in the way suggested in this section of *Grundgesetze*.

It is striking that the formal proof Frege actually gives in *Grundgesetze*, though closely related to the conjectural proof, is not quite the same proof. The formal proof,[13] given in §§115, 117, and 119, does proceed by deriving (0′) by means of induction 3 (Frege's proposition 152), from (1′) {= ε150}[14] and (2) {= 154}. And the proof of (1′) does begin with a derivation of (4′) {= α149}, from (5′) {= 145}. But (1′) is not derived from (4′) *via* (3′); the argument is slightly different.

This part of the *Grundgesetze* proof, translated into English plus our notation, runs as follows: By the basic fact about the weak ancestral it suffices to show that if $0P^{*=}a$, dPa, and $dP\ {}^{\#}[x\colon xP^{*=}d]$, then $aP\ {}^{\#}[x\colon xP^{*=}a]$. By (4′) and (an easy consequence of) Hume's Principle, we have that ${}^{\#}[x\colon xP^{*=}a \wedge x{\neq}a] = {}^{\#}[x\colon xP^{*=}d]$ (cf. 149). But substituting into proposition (102) quoted above, we have ${}^{\#}[x\colon xP^{*=}a \wedge x{\neq}a] = {}^{\#}[x\colon xP^{*=}d] \rightarrow aP^{*=}a \rightarrow {}^{\#}[x\colon xP^{*=}d]P\ {}^{\#}[x\colon xP^{*=}a]$. Hence, by (140), ${}^{\#}[x\colon xP^{*=}d]P\ {}^{\#}[x\colon xP^{*=}a]$ (cf. β150). Since dPa and $dP\ {}^{\#}[x\colon xP^{*=}d]$, $a = {}^{\#}[x\colon xP^{*=}d]$ (cf. γ150), whence $aP\ {}^{\#}[x\colon xP^{*=}a]$ (cf. δ150) and we are done.

Comparing this argument with the relevant portion of the conjectural proof, one sees immediately how little they differ from each other; one might therefore overlook (or ignore) the fact that (3′) does not actually appear in the proof given in *Grundgesetze*. But the omission of (3′) is significant, since the 'proof' discussed in §114 explicitly highlights ($3^{\dagger}$) as what must be proved if (1) is to be derived from ($4^{\dagger}$). The typical point of a section of *Grundgesetze* headed 'analysis' is to describe a formal proof found in 'construction' sections that follow it. Thus on reading §114, one would naturally expect the following proof to include, not just proofs of the results of adding a finiteness condition to ($4^{\dagger}$) and to (1), but also, as part of the derivation of the latter from the former, a proof of a proposition similarly related to ($3^{\dagger}$). As we said, however, the derivation of (1′) from (4′) in §115

[13] For a fuller account, see the second author's 'The Development of Arithmetic in Frege's *Grundgesetze der Arithmetik*', loc. cit. For the benefit of interested readers, Appendix 3 of the present paper contains translations into our notation of relevant parts of Frege's proofs.

[14] By proposition xn we mean the proposition labelled with Greek letter x which occurs *during*, as opposed to *after*, the proof of proposition number n.

does not go via (3′). That ($3^\dagger$) is so much as mentioned in §114 is therefore bound to seem mysterious unless one reads it as we have suggested, i.e. as criticism of Frege's *own* 'first attempt' to prove (0′) in §§82–3 of *Die Grundlagen*, for ($3^\dagger$) or (3′) is indeed an intermediate step in *that* proof.

This observation concerning how the *Grundgesetze* proof differs from the conjectural proof also suggests a plausible explanation of the origin of the mistake of which we have accused Frege. Consider the two lists of propositions:

(1′) $0P^{*=}d \rightarrow dPa \rightarrow dP\ {}^{\#}[x: xP^{*=}d] \rightarrow aP\ {}^{\#}[x: xP^{*=}a]$
(3′) $0P^{*=}a \rightarrow dPa \rightarrow dP\ {}^{\#}[x: xP^{*=}d] \rightarrow a = {}^{\#}[x: xP^{*=}a \wedge x{\neq}a]$
(4′) $0P^{*=}a \rightarrow dPa \rightarrow \forall x([xP^{*=}a \wedge x{\neq}a] \leftrightarrow xP^{*=}d)$
(5′) $0P^{*=}a \rightarrow \neg aP^{*}a$

(1) $dPa \rightarrow dP\ {}^{\#}[x: xP^{*=}d] \rightarrow aP\ {}^{\#}[x: xP^{*=}a]$
($3^\dagger$) $dPa \rightarrow dP\ {}^{\#}[x: xP^{*=}d] \rightarrow a = {}^{\#}[x: xP^{*=}a \wedge x{\neq}a]$
($4^\dagger$) $dPa \rightarrow \forall x([xP^{*=}a \wedge x{\neq}a] \leftrightarrow xP^{*=}d)$
($5^\dagger$) $\neg aP^{*}a$

As we have seen, (4′) follows from (5′), (3′) from (4′), and (1′) from (3′). But notice also that ($4^\dagger$) follows from ($5^\dagger$), ($3^\dagger$) from ($4^\dagger$), and (1) from ($3^\dagger$), as obvious modifications of our proofs show. Frege, able to prove (5′) and desirous of proving (1), may well have lost sight of the need for a finiteness condition somewhere in the middle of his argument—perhaps he had not yet fully written out the argument in *Begriffsschrift*—and mistakenly concluded that he could deduce (1) from (5′). If forced to guess, we would suppose that it was between (4′) and (1), i.e. at ($3^\dagger$) or (3′), that the finiteness condition vanished, for it is there that the *Grundgesetze* proof differs from the conjectural proof.

The first sentence of the second paragraph of §83 calls for some discussion. Frege writes there that we are obliged 'hereby' ('hierdurch') to attach to the proposition that $nP\ {}^{\#}[x: xP^{*=}n]$ the condition that $0P^{*=}n$. One might be forgiven for thinking that, in so stating, Frege is indicating that this condition is required by the presence of the finiteness condition in (5′), since it is with an indication of how (5′) is to be proven that the previous paragraph ends. But this thought cannot be right. Frege says in §82 that, once (1) and (2) are proved, 'it is to be deduced that $0P^{*=}n \rightarrow nP^{\#}[x: xP^{*=}n]$' by means of induction 2. Thus what subjects n in (0) to a finiteness condition is not the presence of such a condition in (5′), but the kind of proof of (0′) being given in the first place. 'Hierdurch' refers to the use in the proof of (0′) of the 'definition of following in a series, on the lines indicated above', that is, as was discussed in §82.

There is one final piece of textual evidence to which we should like to draw attention. As we said earlier, a footnote is attached to proposition (1) when it is first mentioned in §114: 'This proposition is, as it seems, unprovable, but it is not here being asserted as true, since it stands in quotation marks.' The natural explanation for this remark of Frege's is that he once *did* believe (1) to be provable, namely when he wrote *Die Grundlagen*, and any defender of the view that Frege was outlining the conjectural proof in §§82–3 will have the occurrence of this remark to explain away.

Apart from the light it may throw on the question whether Frege made a reparable error, the footnote is astonishing. Note that Frege says, not that (1) seems to be false, but that it 'seems to be *unprovable*' (itals. ours). There is, moreover, reason to suppose Frege believed (1) to be not false, but *true*. For one thing, had Frege believed it to be false, he presumably would have said so. Furthermore, Frege's difficulty was probably not that he did not know how to prove (1), but rather that he did not know how to prove it *in his formal system*. There is a very simple proof of (1) which depends only upon (1′), Dedekind's claim that every infinite number is (the number of a concept that is) Dedekind infinite,[15] and the observation, made earlier, that 'd is Dedekind infinite' is equivalent to 'dPd'. We may take (1′) to be one half of a dilemma, the other half of which is:

$$\neg 0P^{*=}d \rightarrow dPa \rightarrow dP\ {}^{\#}[x: xP^{*=}d] \rightarrow aP\ {}^{\#}[x: xP^{*=}a].$$

This proposition may be proved as follows: suppose the subcomponents. Since d is not finite, it is Dedekind infinite. So dPd, and since dPa, d=a and (1) follows immediately.

This proof is one Frege might well have known. It is not at all difficult and once (1′) has been proved, a proof of (1) by dilemma immediately suggests itself. Moreover, Frege was familiar with Dedekind's claim and, at least while he was working on Part II of *Grundgesetze*, believed it to be true.[16] As for the observation, not only is it easily proved, it is natural, in Frege's system, just to use 'dPd' as a *definition* of 'd is Dedekind infinite' (cf. *Grundgesetze*, proposition 426). We conclude that Frege believed (1) to be a true but unprovable formula of Frege Arithmetic.

Frege's belief that (1) is unprovable in Frege Arithmetic was mistaken, however. A proof of (1) can be given that makes use of techniques that are different from any found in §§82–3 of *Die Grundlagen* or in relevant sections of *Grundgesetze*, but with which Frege was familiar. What we shall prove is that the hypothesis $0P^{*=}d$ of (1′), that d is finite, is dispensable. More precisely, we shall prove that if $dP\ {}^{\#}[x: xP^{*=}d]$, then ${}^{\#}[x: xP^{*=}d]$ is finite,

[15] Of course in set theory without the axiom of choice, i.e. ZF as opposed to ZFC, this claim cannot be proved.

[16] See Frege's review, published in 1892, of Cantor's 'Zur Lehre vom Transfiniten'. Of course, if Frege did know of this proof and believed (1) to be unprovable, then he must have believed Dedekind's result too to be unprovable, which he (rightly) did. For further discussion, see the second author's 'The Finite and the Infinite in Frege's *Grundgesetze der Arithmetik*', elsewhere in this volume.

from which it follows that d is finite, since by proposition (143) of *Grundgesetze* {viz. $dPb \rightarrow aP^{*}b \rightarrow aP^{*=}d$}, any predecessor of a finite number is finite.

Theorem (FA).[17] Suppose $dP\,{}^{\#}[x: xP^{*=}d]$. Then ${}^{\#}[x: xP^{*=}d]$ is finite.

Proof. In FA, define $h: [x: 0P^{*=}x] \rightarrow [x: xP^{*=}d]$ by:

$h(0) = d;\ h(n{+}1) = y$ if $yPh(n)$ and

$= h(n)$ if $\neg\exists y\, yPh(n)$.

The definition is OK since P is one-one.

Since in general $yR^{*}z \leftrightarrow z(R^{\cup})^{*}y$,[18] $\forall x(xP^{*=}d \leftrightarrow d(P^{\cup})^{*=}x)$, and so h is onto. Therefore $[x: xP^{*=}d]$ is countable, i.e. either finite or countably infinite. If the latter, then ${}^{\#}[x: xP^{*=}d] = \aleph_0$ and by the supposition of the theorem, $dP\aleph_0$. But as we saw just after the proof of (4′), $xP^{*=}\aleph_0 \leftrightarrow x = \aleph_0$. Since $dP\aleph_0$, $dP^{*=}\aleph_0$, $d = \aleph_0$, and ${}^{\#}[x: xP^{*=}d] = 1$, contra ${}^{\#}[x: xP^{*=}d] = \aleph_0$. Therefore ${}^{\#}[x: xP^{*=}d]$ is finite.

Thus Frege could have proved (1) after all and thus appealed to induction 2 to prove (0′). Of course the technology borrowed from second-order arithmetic used in the proof just given, particularly the inductive definition of h, is considerably more elaborate than that needed to derive induction 3 from induction 2. The conjectural proof is unquestionably to be preferred to this new one on almost any conceivable grounds.

So. Frege erred in §§82–3 of *Die Grundlagen*, where an oversight marred the proof he outlined of the existence of the successor. Mistakes of that sort are hardly unusual, though, there are four or five ways the proof can be patched up, and Frege's way of repairing it cannot be improved on. But even if one ought not to make too much of Frege's mistake, there is *lots* to be made of his belief that (1) was true but unprovable in his system. One question that must have struck Frege is: If there are truths about numbers unprovable in the system, what becomes of the claim that the truths of arithmetic rest solely upon definitions and general logical laws? Another that may have occurred to him is: Can the notion of a truth of logic be explained otherwise than via the notion of provability?

APPENDIX 1. COUNTERPARTS IN *GRUNDGESETZE* OF SOME PROPOSITIONS OF *DIE GRUNDLAGEN*

Proposition of this paper	Proposition of *Grundgesetze*
Hume's Principle	32, 49
$\forall x\neg Fx \leftrightarrow 0 = {}^{\#}[x: Fx]$	94, 97
78.1	114
78.2	113

[17] This result is due to the second author.

[18] $R^{\cup}$ is the converse of R.

Proposition of this paper	Proposition of *Grundgesetze*
78.3	117
78.4	122
78.5	90
78.6	107
¬zP0	108
Induction 1	123
The basic fact about the weak ancestral	134, 136
The lemma	141
Induction 2	144
(5′)	145
(4′)	α149
(1′)	150
Induction 3	152
(2)	154
(0′)	155

APPENDIX 2. INTERPRETING FREGE ARITHMETIC IN SECOND-ORDER ARITHMETIC

The language (of second-order arithmetic) contains variables x, y, z, . . . over natural numbers; variables α, β, γ, . . . over sets of numbers; and variables ρ, σ, . . . over binary relations of numbers. (We do not need variables over n-place relations for n>2.) Its non-logical symbols are 0, s, +, ×, <. Terms t are built up out of 0, s, +, × as usual; the atomic formulas are t=t′, t<t′, αt, ρtt′; formulas are then built up as usual.

The axioms of second-order arithmetic are induction: $(\alpha 0 \wedge \forall x(\alpha x \rightarrow \alpha sx) \rightarrow \alpha x)$ [a single formula]; the recursion axioms for successor, plus and times and the definition of less-than:

$$0 \neq sx,\ sx = sy \rightarrow x = y,\ x+0 = x,\ x+sy = s(x+y),\ x \times 0 = 0,$$
$$x \times sy = (x \times y)+x,\ x<y \leftrightarrow \exists z\ sz+x = y;$$

and the comprehension axioms (which are axioms of standard second-order logic):

$\exists\alpha\forall x(\alpha x \leftrightarrow A)$, $\exists\rho\forall x\forall y(\rho xy \leftrightarrow B)$, A a formula in which α is not free and B a formula in which ρ is not free.

Since we have + and ×, we could have dispensed with binary relation variables; and since we have binary relation variables, we could have dispensed with + and × and set variables:

$J, = \lambda xy\iota z(x^2+2xy+y^2+3x+y = 2z)$, is an onto pairing function.

Thus if we have + and ×, we have J, and so we can replace ρtt′ by αJ (t,t′). And we can define x+y = z and x×y = z from 0 and s using binary relation variables: x+y = z iff $\forall\rho(\rho 0x \wedge \forall u\forall v(\rho uv \rightarrow \rho susv) \rightarrow \rho yz)$; $x{\times}y = z$ iff $\forall\rho(\rho 00 \wedge \forall u\forall v\ (\rho uv \rightarrow \rho su\,(v{+}x)\,) \rightarrow \rho yz)$. And, uninterestingly, we can replace αx by ρxx.

The least number principle can be proved from the induction axioms as usual.

Introduce the notation: $\{x{:}\ A\}$ as usual: $\{x{:}\ A\}t$ abbreviates
$\exists\alpha\ (\forall y[\alpha y \leftrightarrow \exists x(x{=}y \wedge A)] \wedge \alpha t)$, or, equivalently,
$\forall\alpha\ (\forall y[\alpha y \leftrightarrow \exists x\ (x{=}y \wedge A)] \rightarrow \alpha t)$. $[\alpha, y$ new$]$.[19]

Def. $\alpha{\approx}\beta$ via $\rho \equiv \forall x(\alpha x \leftrightarrow \exists y(\beta y \wedge \rho xy)\,) \wedge \forall y(\beta y \leftrightarrow \exists x(\alpha x \wedge \rho xy)\,) \wedge \forall x\forall y\forall a\forall b(\rho xy \wedge \rho ab \rightarrow (x{=}a \leftrightarrow y{=}b)\,)$
Def. $\alpha{\approx}\beta \equiv \exists\rho(\alpha{\approx}\beta$ via $\rho)$.
Def. $N = \{i{:}\ i{=}i\}$.

1. $\vdash \approx$ is an equivalence relation.
2. $\vdash \{i{:}\ i{<}m\} \approx \{i{:}\ i{<}n\} \rightarrow m{=}n$.
3. $\vdash \neg(N \approx \{i{:}\ i{<}n\})$.
4. $\vdash N{\approx}\alpha \vee \exists n\ (\{i{:}\ i{<}n\}{\approx}\alpha)$.

1. will be used without citation below; it is evident how to prove it. Sketches of proofs of 2, 3, 4 in analysis are given below.

5. $\vdash \exists!x[\exists n(\{i{:}\ i{<}n\}{\approx}\alpha \wedge x{=}n{+}1) \vee (N{\approx}\alpha \wedge x{=}0)]$.

Pf. By 2, 3, 4.
Def. ${}^{\#}\alpha = \iota x[\exists n(\{i{:}\ i{<}n\}{\approx}\alpha \wedge x{=}n{+}1) \vee (N{\approx}\alpha \wedge x{=}0)]$.

Theorem. $\vdash {}^{\#}\alpha{=}{}^{\#}\beta \leftrightarrow \alpha{\approx}\beta$.
Pf. Case 1. ${}^{\#}\alpha{=}0{=}{}^{\#}\beta$. Then $\alpha{\approx}N{\approx}\beta$.
Case 2. ${}^{\#}\alpha{=}0$, ${}^{\#}\beta{=}n{+}1$. Then $\alpha{\approx}N$, $\beta{\approx}\{i{:}\ i{<}n\}$, and $\neg(\alpha{\approx}\beta)$, by 3.
Case 3. ${}^{\#}\alpha{=}m{+}1$, ${}^{\#}\beta{=}0$. Similar to case 2.
Case 4. ${}^{\#}\alpha{=}n{+}1{=}{}^{\#}\beta$. Then $\alpha \approx \{i{:}\ i{<}n\}{\approx}\beta$.
Case 5. ${}^{\#}\alpha{=}m{+}1$, ${}^{\#}\beta{=}n{+}1$, $m{\neq}n$. Then $\alpha{\approx}\{i{:}\ i{<}m\}$, $\beta{\approx}\{i{:}\ i{<}n\}$, and by 2, $\neg(\alpha{\approx}\beta)$.

Proof of 2. Suppose not. By the least number principle, let n be least such that for some least $m{<}n$, $\{i{:}\ i{<}m\} \approx \{i{:}\ i{<}n\}$, via some ρ. $m, n{\neq}0$, hence for some j, k, $m = sj$, $n = sk$. So $j{<}k{<}n$. Let $b = \rho(j)$; $a = \rho^{-1}(k)$; let $\sigma = (\,(\rho - \{\langle j, b\rangle, \langle a, k\rangle\}) \cup \{\langle a, b\rangle\}) - \{\langle j, k\rangle\}$. Then $\{i{:}\ i{<}j\} \approx \{i{:}\ i{<}k\}$ via σ, contra leastness of n.

Proof of 3. Suppose $N \approx \{i{:}\ i{<}n\}$, n least. Let $N \approx \{i{:}\ i{<}n\}$ via ρ. Clearly $n{=}m{+}1$, some m. $m{<}n$. Let i be such that $\rho(i) = m$. Let $\sigma(j) = \rho(j)$ if $j{<}i$ and let $\sigma(j) = \rho(j{+}1)$ if $j{\geq}i$. Then $N \approx \{i{:}\ i{<}m\}$ via σ, contra leastness of n.

Proof of 4. Suppose that for no n, $\{i{:}\ i{<}n\}{\approx}\alpha$. We prove the existence of a one–one function ρ mapping N onto α.

Say that σ is good to n if σ is a function; domain $(\sigma) = \{i{:}\ i{<}n\}$; for all $i{<}n$, $\sigma(i) \in \alpha$; for all $j{<}i{<}n$, $\sigma(j){<}\sigma(i)$; and for all $i{<}n$, if $k \in \alpha$ and $k{<}\sigma(i)$, then for some $j{<}i$, $k{=}\sigma(j)$. [Officially: σ is good to n iff
$\forall i\forall p\forall q(\sigma ip \wedge \sigma iq \rightarrow p = q) \wedge \forall i(\exists p\sigma ip \leftrightarrow i{<}n) \wedge$
$\forall i\forall p(\sigma ip \wedge \rightarrow \alpha p) \wedge \forall j\forall i\forall p\forall q(j{<}i \wedge \sigma jq \wedge \sigma ip \rightarrow q{<}p) \wedge$
$\forall i\forall k\forall p(\alpha k \wedge \sigma ip \wedge k{<}p \rightarrow \exists j(j{<}i \wedge \sigma jk)\,)$.]

[19] Thanks here to Roy Dyckhoff.

$\emptyset$ is good to 0. If σ is good to n, then $\sigma \cup \{<$ n, least member of α not in the range of $\sigma>\}$ is good to sn (the existence of the least member of α not in the range of σ follows from our supposition). By induction, for every n, some σ is good to n. If σ is good to n, σ' is good to n′, and $n \leq n'$, then by induction for every $i<n$, $\sigma(i) = \sigma'(i)$. By comprehension, let $\rho = \{\langle n, k\rangle: \exists\sigma\ (\sigma$ is good to sn and $\sigma(n) = k)\}$. It is sufficiently clear that ρ is a function; domain $(\rho) = N$; for all i, $\rho(i) \in \alpha$; if $j<i$, $\rho(j)<\rho(i)$, and if $k \in \alpha$, $k < \rho(i)$, then for some $j<i$, $k = \rho(j)$. We must show that ρ is onto α. Since $\rho(j)<\rho(i)$ whenever $j<i$, for every k, $k \leq \rho(k)$; otherwise for some least k, $\rho(k)<k$, and then $\rho(\rho(k))< \rho(k)$, contra leastness of k. Thus if $k \in \alpha$, then $k\leq\rho(k)$, and so for some $j\leq k$, $k = \rho(j)$. Thus ρ is onto α.

APPENDIX 3. TRANSLATIONS INTO PRESENT-DAY NOTATION OF §§115, 117, 119 AND PART OF §§113 OF *GRUNDGESETZE*

Below we use boldface instead of Fraktur and notation introduced above in place of Frege's own, we omit the signs indicating which rule of inference is applied and reference numbers to axioms of Frege's system, and we utilize certain easy equivalences (e.g. $p \wedge q$ for $\neg(p \rightarrow \neg q)$).

§113 (part)

$aq^*b \rightarrow \exists\mathbf{e}(\mathbf{e}qb \wedge aq^{*=}\mathbf{e})$ (141)

$\forall\mathbf{e}(\mathbf{e}qb \rightarrow \neg aq^{*=}\mathbf{e}) \rightarrow \neg aq^*b$ (142)

$\neg aP^{*=}d \rightarrow d{=}c \rightarrow \neg aP^{*=}c$

(88): $\neg aP^{*=}d \rightarrow dPb \rightarrow cPb \rightarrow \neg aP^{*=}c$

$\neg aP^{*=}d \rightarrow dPb \rightarrow \forall\mathbf{e}(\mathbf{e}Pb \rightarrow \neg aP^{*=}\mathbf{e})$

(142): $\neg aP^{*=}d \rightarrow dPb \rightarrow \neg aP^*b$

$aP^*b \rightarrow dPb \rightarrow aP^{*=}d$ (143)

§115

130 $bq^{*=}a \rightarrow \neg bq^*a \rightarrow a{=}b$

$\neg bq^*a \rightarrow bq^{*=}a \rightarrow b{=}a$ (146)

$bq^{*=}a \wedge b{\neq}a \rightarrow bq^*a$ (147)

147 $bP^{*=}a \wedge b{\neq}a \rightarrow bP^*a$

(143): $dPa \rightarrow (bP^{*=}a \wedge b{\neq}a) \rightarrow bP^{*=}d$

$(bP^{*=}d \rightarrow (bP^{*=}a \wedge b{\neq}a)\ ) \rightarrow dPa \rightarrow (\ (bP^{*=}a \wedge b{\neq}a) \leftrightarrow bP^{*=}d)$ (β)

134 $bP^{*=}d \rightarrow dPa \rightarrow bP^*a$

$\neg aP^*a \rightarrow bP^{*=}d \rightarrow dPa \rightarrow b{\neq}a$

(145): $0P^{*=}a \rightarrow bP^{*=}d \rightarrow dPa \rightarrow b{\neq}a$

$bP^{*=}a \rightarrow 0P^{*=}a \rightarrow bP^{*=}d \rightarrow dPa \rightarrow (bP^{*=}a \wedge b{\neq}a)$

(137): $dPa \rightarrow 0P^{*=}a \rightarrow bP^*{=}d \rightarrow (bP^{*=}a \wedge b{\neq}a)$

(β): $dPa \rightarrow 0P^{*=}a \rightarrow (\ (bP^{*=}a \wedge b{\neq}a) \leftrightarrow bP^{*=}d)$

(77): $dPa \rightarrow 0P^{*=}a \rightarrow ([x{:}\ xP^{*=}a \wedge x{\neq}a]b \leftrightarrow bP^{*=}d)$ (148)

$dPa \rightarrow 0P^{*=}a \rightarrow \forall\mathbf{a}([x{:}\ xP^{*=}a \wedge x{\neq}a]\mathbf{a} \leftrightarrow \mathbf{a}P^{*=}d)$

(96): $dPa \rightarrow 0P^{*=}a \rightarrow {}^{\#}[x{:}\ xP^{*=}a \wedge x{\neq}a] = {}^{\#}[x{:}\ xP^{*=}d]$ (149)

(102): dPa → 0P*=a → aP*=a → #[x: xP*=d]P #[x: xP*=a]
(140): dPa → 0P*=a → #[x: xP*=d]P #[x: xP*=a]
#[x: xP*=d] = a → dPa → 0P*=a → aP #[x: xP*=a]
(70): dP #[x: xP*=d] → dPa → 0P*=a → aP #[x: xP*=a]
(137) dP #[x: xP*=d] → 0P*=d → dPa → aP #[x: xP*=a]
∀**d**(**d**P #[x: xP*=**d**] → 0P*=**d** → ∀**a**(**dPa** → **aP** #[x: xP*=**a**])) (150)

§117
∀**d**(F**d** → aq*=**d** → ∀**a**(**dqa** → F**a**)) → (Fd → aq*=d → ∀**a**(dq**a** → F**a**))
∀**d**(F**d** → aq*=**d** → ∀**a**(**dqa** → F**a**)) → (Fd → aq*=d → dqb → Fb)
∀**d**(F**d** → aq*=**d** → ∀**a**(**dqa** → F**a**)) → (¬Fb → aq*=d → dqb → ¬Fd)
∀**d**(F**d** → aq*=**d** → ∀**a**(**dqa** → F**a**)) →
((aq*=b → ¬Fb) → aq*=b → aq*=d → dqb → ¬Fd)
(137) ∀**d**(F**d** → aq*=**d** → ∀**a**(**dqa** → F**a**)) →
((aq*=b → ¬Fb) → dqb → (aq*=d → ¬Fd))
∀**d**(F**d** → aq*=**d** → ∀**a**(**dqa** → F**a**)) →
((aq* = d ∧ Fd) → dqb → (aq*=b ∧ Fb))
∀**d**(F**d** → aq*=**d** → ∀**a**(**dqa** → F**a**)) →
∀**d**((aq*=**d** ∧ F**d**) → ∀**a**(**dqa** → (aq*=**a** ∧ F**a**)) (151)
140 aq*=a
Fa → (aq*=a ∧ Fa)
(144): aq*=b → ∀**d**((aq*=**d** ∧ F**d**) → ∀**a**(**dqa** → (aq*=**a** ∧ F**a**))) →
(Fa → (aq*=b∧ Fb))
(151): aq*=b → ∀**d**(F**d** → aq*=**d** → ∀**a**(**dqa** → F**a**)) →
(Fa → (aq*=b ∧ Fb))
aq*=b → ∀**d**(F**d** → aq*=**d** → ∀**a**(**dqa** → F**a**)) → (Fa → Fb) (152)
150 ∀**d**(**d**P #[x: xP*=**d**] → 0P*=**d** → ∀**a**(**dPa** → **aP** #[x: xP*=**a**]))
(152) 0P*=b → 0P #[x: xP*=0] → bP #[x: xP*=b] (153)

§119
126. ¬aP*0
(130) aP*=0 → 0=a
aP*=0 → a=0
(58): ¬[x: xP*=0 ∧ x≠0]a
∀**a**¬[x: xP*=0 ∧ x≠0]**a**
(97): #[x: xP*=0 ∧ x≠0] = 0
(102): 0P*=0 → 0P #[x: xP*=0]
(140): 0P #[x: xP*=0] (154)
(153): 0P*=b → bP #[x: xP*=b] (155)

Propositions cited but not proved above:
(58) F(fa) → F(¬[x: ¬fx]a)
(70) ePd → ePa → d=a
(77) F(fa) → F([x: fx]a)
(88) dPe → aPe → d=a
(96) ∀**a**(u**a** ↔ v**a**) → #u = #v
(97) ∀**a**¬u**a** → #u = 0
(102) #[x: ux ∧ x≠c] = m → uc → mP #u

(130) $F(aq^{*=}c) \rightarrow F(aq^{*}c \vee c^{=}a)$
(137) $aq^{*=}e \rightarrow eqm \rightarrow aq^{*=}e$
(140) $aq^{*=}q$
(144) $aq^{*=}b \rightarrow \forall \mathbf{d}(F\mathbf{d} \rightarrow \forall \mathbf{a}(\mathbf{d}q\mathbf{a} \rightarrow F\mathbf{a})) \rightarrow (Fa \rightarrow Fb)$
(145) $0P^{*=}b \rightarrow \neg bP^{*}b$

17

The Finite and the Infinite in Frege's *Grundgesetze der Arithmetik*

RICHARD G. HECK, JNR.

1. OPENING

Some of the best-known work on the foundations of mathematics done in the late nineteenth century was concerned with the concepts of the finite and the infinite, rigorous characterizations of which were famously given by both Cantor and Dedekind. It is little known, however, that, in his *Grundgesetze der Arithmetik*,[1] Frege also studied the notions of finitude and infinity. His work in *Grundgesetze* had little influence upon that of his contemporaries because, quite simply, almost no one appears to have read it. One obstacle was Frege's notation, which is notoriously difficult to learn to read (though easy enough to read, actually); following Russell's discovery that the formal system of *Grundgesetze* is inconsistent, mathematicians might justifiably have thought themselves to have had little reason to bother. Indeed, one might wonder why, given his system's inconsistency, the proofs Frege gives in it merit any serious study at all.

The answer, as I have argued elsewhere,[2] is that Frege's proofs depend far less upon the axiom responsible for the inconsistency than one might have supposed. For present purposes, the relevant axiom, Axiom V, may be taken to be:

$$\grave{\varepsilon}.F\varepsilon = \grave{\varepsilon}.G\varepsilon \text{ iff } \forall x(Fx \equiv Gx).$$

That is: the value-range (or extension) of the concept $F\xi$ is the same as that of the concept $G\xi$ just in case the Fs are exactly the Gs. Now, Frege uses value-ranges, and so Axiom V, throughout *Grundgesetze*, but he makes *essential* appeal to Axiom V only in the course of his proof of what may be

[1] Gottlob Frege, *Grundgesetze der Arithmetik*, Hildesheim, Georg Olms Verlagsbuchhandlung, 1966. Further references are in the text, marked '*Gg*' with a roman numeral, for the volume number, and a section number. The first volume was published in 1893; the second, in 1903.

[2] See my 'The Development of Arithmetic in Frege's *Grundgesetze der Arithmetik*', *Journal of Symbolic Logic*, 58 (1993), 579–601, esp. §1, repr. with a postscript in W. Demopoulos (ed.), *Frege's Philosophy of Mathematics*, Cambridge, Mass., Harvard University Press, 1995, 257–94.

called 'Hume's Principle', which states that the number of Fs is the same as the number of Gs just in case the Fs can be correlated one-to-one with the Gs. Hume's Principle may be formulated, in second-order logic, thus:

$$\begin{aligned} Nx{:}Fx = Nx{:}Gx \text{ iff } (\exists R)[&(\forall x)(\forall y)(\forall z)(Rxy \ \& \ Rxz \rightarrow y{=}z) \ \& \\ &(\forall x)(\forall y)(\forall z)(Rxz \ \& \ Ryz \rightarrow x{=}y) \ \& \\ &(\forall x)(Fx \rightarrow (\exists y)(Gy \ \& \ Rxy)) \ \& \\ &(\forall y)(Gy \rightarrow (\exists x)(Fx \ \& \ Rxy))]. \end{aligned}$$

With the exception of Hume's Principle itself, the arithmetical theorems Frege proves in *Grundgesetze* can be, *and effectively are*, derived from Hume's Principle within second-order logic: that is to say, they are, in effect, proven in *Frege Arithmetic*, that (standard, axiomatic) second-order theory whose sole non-logical axiom is Hume's Principle. Since Frege Arithmetic is equi-consistent with second-order arithmetic,[3] all the important theorems concerning arithmetic which Frege proves in *Grundgesetze* are proven in a consistent sub-theory of his formal theory.

A large part of *Grundgesetze* is devoted to proofs of axioms for arithmetic, not the Dedekind–Peano Axioms—though each of them is indeed proven—but the axioms of Frege's own, somewhat different, axiomatization.[4] Frege's investigation of arithmetic does not, however, end with his proofs of these axioms. He goes on to prove a number of theorems concerning, for example, the infinite cardinal aleph-null, which he calls by the name 'Endlos'. Perhaps the most powerful of these results is that all structures satisfying Frege's axioms for arithmetic are isomorphic, a result analogous to one for which Dedekind's *Was sind und was sollen die Zahlen?* is celebrated.[5] Frege's efforts are also directed, however, towards the pro-

[3] See George Boolos, 'The Consistency of Frege's *Foundations of Arithmetic*', in J. J. Thomson (ed.), *On Being and Saying: Essays for Richard Cartwright*, Cambridge, Mass., MIT Press, 1987, 3–20, repr. in W. Demopoulos (ed.), op. cit., 211–33.

[4] Frege's axioms are:

1. $\forall x \forall y \forall z(Sxy \ \& \ Sxz \rightarrow y{=}z)$.
2. $\neg\exists x[Nx \ \& \ \mathscr{F}(S)(x,x)]$.
3. $\forall x[Nx \rightarrow \exists y.Sxy]$.
4. $Nx \equiv \mathscr{F}^{=}(S)(0,x)$.

See below for the definitions of '$\mathscr{F}$' and '$\mathscr{F}^{=}$', which are the strong and weak ancestral, respectively. For an argument that these are Frege's axioms for arithmetic, see my 'Development of Arithmetic', 598–9, and 'Definition by Induction in Frege's *Grundgesetze der Arithmetik*', in M. Schirn (ed.), *Frege: Importance and Legacy*, New York, De Gruyter, 1996, 200–33, repr. in W. Demopoulos (ed.), op. cit, 295–333.

[5] See Richard Dedekind, 'The Nature and Meaning of Numbers', in his *Essays on the Theory of Numbers*, tr. by W. W. Beman, New York, Dover Publications, 1961, 29–115, theorem 132. For discussion of Frege's proofs, see my 'Definition by Induction'.

vision of characterizations of the finite and the infinite, characterizations which, as we shall see, are given in pure second-order logic, which do not rely even upon Hume's Principle.

The main goal of this paper is to present Frege's proofs in a form in which they will be more accessible to potential readers. We shall also be trying to understand *why* he proves what he proves. The theorems proven in Part II are intended as proofs of what Frege calls, in its title, 'The Basic Laws of Cardinal Number'. The theorems we shall discuss do not, at first sight, look much like basic laws of cardinal number, and Frege does not spend much time, in *Grundgesetze*, explaining the importance of the results he proves. That does not, however, mean that his theorems do not each have some purpose, some significance, philosophical or mathematical, and it is implausible that the theorems included in what was intended to be the definitive statement and proof of Frege's logicism should be without such significance. Thus, though much of the present paper amounts to speculation about Frege's ulterior motives in *Grundgesetze*, it is well-motivated speculation.

We shall see that paying attention to what Frege proves, to how he proves it, and even to what he finds himself unable to prove, can be fruitful. We shall see, for example, that Frege probably discovered, around 1892, that Dedekind's proof that every infinite set is Dedekind-infinite depends upon an axiom of countable choice. This fact would, in itself, be of but passing interest were it not that Frege's interest in the axiom of choice vividly illuminates certain controversial aspects of his views, namely his conception of logic and his views about the significance of formalization and axiomatization. For, *c*. 1897, Frege appears to have been thinking not only about the truth-value of the axiom he had discovered but also about its epistemological status. That Frege regarded it as an intelligible question whether the axiom of countable choice was a law of logic is inconsistent with the claim, advanced by some commentators, that he regarded all 'meta-logical' discourse as nonsensical.

2. DEFINITIONS OF CENTRAL NOTIONS

In his proofs, Frege makes use of a number of defined notions. As most of the definitions are straightforward, and I have discussed them elsewhere, I shall present them with minimal commentary. Frege's definitions are not given in pure second-order logic, but instead make use of value-ranges. Here, however, they will silently be transformed into definitions in second-order logic (or in Frege Arithmetic, in such cases as the operator '$Nx{:}\Phi x$' is needed). Later, when we discuss Frege's proofs, they too will silently be transformed into proofs in Frege Arithmetic.

The first of Frege's definitions is that of the *converse of a relation* (*Gg* I §39):[6]

$$\text{Conv}_{\alpha\varepsilon}(R\alpha\varepsilon)(x,y) \equiv\text{df } Ryx.$$

Thus, x stands in the converse of the relation Rξη to y if y stands in the relation Rξη to x. The second definition is that of a relation's being *functional*, i.e. function-like (*Gg* I §37):

$$\text{Func}_{\alpha\varepsilon}(R\alpha\varepsilon) \equiv\text{df } \forall x\forall y\forall z(Rxy \;\&\; Rxz \rightarrow y{=}z).$$

The third is that of a relation's *mapping* one concept into another (*Gg* I §38):

$$\text{Map}_{\alpha\varepsilon xy}(R\alpha\varepsilon)(Fx,Gy) \equiv\text{df } \text{Func}_{\alpha\varepsilon}(R\alpha\varepsilon) \;\& \\ \forall x[Fx \rightarrow \exists y(Rxy \;\&\; Gy)].$$

Thus, the relation Rξη maps the concept Fξ into the concept Gξ just in case Rξη is functional and every F stands in it to some (hence exactly one) G.[7] Additionally, we need Frege's famous definition of the *strong ancestral* of a relation (*Gg* I §45):

$$\mathscr{F}_{\alpha\varepsilon}(Q\alpha\varepsilon)(a,b) \equiv\text{df } \forall F[\forall x(Qax \rightarrow Fx) \;\&\; \forall x\forall y(Fx \;\&\; Qxy \rightarrow Fy) \\ \rightarrow Fb].$$

That is, *b follows after a in the Q-series* if, and only if, b falls under every concept (i) under which all objects to which a stands in the Q-relation fall and (ii) which is *hereditary in the Q-series*, i.e. under which every object to which an F stands in the Q-relation falls. Frege defines the *weak ancestral* as follows (*Gg* I §46):

$$\mathscr{F}^{=}{}_{\alpha\varepsilon}(Q\alpha\varepsilon)(a,b) \equiv\text{df } \mathscr{F}_{\alpha\varepsilon}(Q\alpha\varepsilon)(a,b) \vee a{=}b.$$

Thus, *b belongs to* (or *is a member of*) *the Q-series beginning with a* if, and only if, either b follows after a in the Q-series or b is identical with a.

Finally, we need Frege's definitions of specifically arithmetical notions. The number zero is defined as the number of objects which are not self-identical (*Gg* I §41; see *Gl* §74):[8]

$$0 = Nx{:}x{\neq}x.$$

[6] I insert the bound variables in the definitions but will drop them when it causes no confusion to do so, for readability.

[7] Note that the definition does *not* say that Rξη is functional and is *onto* the Gs, but that Rξη is functional and that every F is mapped, by Rξη, to some G. Indeed, one might well write 'Map(R)(F,G)' as 'R: F → G'.

[8] References to Gottlob Frege, *The Foundations of Arithmetic*, tr. by J. L. Austin (Evanston, Ill., Northwestern University Press, 1980), are in the text, marked by '*Gl*' and a section number.

We say that *m (immediately) precedes n in the number-series* if, and only if, there is a concept $F\xi$, whose number is n, and an object y falling under $F\xi$, such that m is the number of Fs other than y (*Gg* I §43; see *Gl* §74):

$$\text{Pred}(m,n) \equiv\text{df } \exists F\exists y[n = Nx{:}Fx \ \& \ Fy \ \& \ m = Nx{:}(Fx \ \& \ x\neq y)].$$

The concept of a *finite* or *natural number* may then be defined as follows:

$$\mathbb{N}x \equiv\text{df } \mathscr{F}^{=}(\text{Pred})(0,x).$$

A number is finite, is a natural number, if and only if it belongs to the Pred-series (the number-series) beginning with zero. And the first transfinite cardinal may be defined thus (*Gg* I §122):

$$\infty = \text{df } Nx{:}\mathscr{F}^{=}(\text{Pred})(0,x).$$

'Endlos' is the number of natural numbers.

As was mentioned above, after completing the proofs of axioms for arithmetic, Frege turns to the proofs of theorems concerning the number Endlos. The most interesting of these constitute, together, a characterization of concepts whose number is Endlos. Frege's Theorem 207 is:

$$\exists Q[\text{Func}(Q) \ \& \ \neg\exists x.\mathscr{F}(Q)(x,x) \ \& \ \forall x(Gx \rightarrow \exists y.Qxy) \ \& \\ \exists x\forall y(Gy \equiv \mathscr{F}^{=}(Q)(x,y))] \rightarrow Nx{:}Gx = \infty$$

Following Frege, we say that the Q-series is *simple* if (1) $Q\xi\eta$ is functional and (2) no object follows after itself in the Q-series. We say that the Q-series beginning with a is *endless* if, whenever x belongs to the Q-series beginning with a, there is an object onto which $Q\xi\eta$ maps x. Then Theorem 207 amounts to the claim that the number of Gs is Endlos if the Gs can be ordered as a series which is both simple and endless (or *simply endless*). Frege's Theorem 263 is the converse of Theorem 207. Putting these together, we thus have:

$$Nx{:}Gx = \infty \equiv \exists Q[\text{Func}(Q) \ \& \ \neg\exists x.\mathscr{F}(Q)(x,x) \ \& \ \forall x(Gx \rightarrow \exists y.Qxy) \\ \& \ \exists x\forall y(Gy \equiv \mathscr{F}^{=}(Q)(x,y))].$$

That is: the number of Gs is Endlos *if and only if* the Gs can be ordered as a simply endless series. Note that the concept of a simply endless series, and so that of a concept the objects falling under which may be ordered as a simply endless series, is definable (because here defined) in second-order logic.

As we shall see, Frege's two main theorems concerning concepts whose number is finite are analogues of Theorems 207 and 263.

3. FREGE'S CHARACTERIZATION OF FINITUDE

We turn now to Frege's characterization of finitude. To state it, we need an additional definition:

$$\text{Btw}_{\alpha\varepsilon}(Q\alpha\varepsilon; a, b)(x) \equiv\text{df Func}(Q) \ \& \ \neg\mathscr{F}(Q)(b,b) \ \& \ \mathscr{F}^{=}(Q)(a,x) \ \& \ \mathscr{F}^{=}(Q)(x,b).$$

Frege reads 'Btw(Q;a,b)(x)' as 'x is a member of the Q-series running from a to b' (see *Gg* I §158). However, I prefer to read it as 'x is between a and b in the Q-series' or, better, 'x is Q-between a and b'. So we say that x is Q-between a and b if, and only if: Qξη is functional; b does not follow itself in the Q-series; and x belongs both to the Q-series beginning with a and to that ending with b.

The theorems of interest at present are Theorems 327 and 348. Theorem 327 is:

$$\exists Q\exists x\exists y\forall z[Fz \equiv \text{Btw}(Q; x, y)(z)] \rightarrow \mathscr{F}^{=}(\text{Pred})(0, \text{Nx:Fx}).$$

Theorem 348 is the converse of Theorem 327, so we have:

$$\exists Q\exists x\exists y\forall z[Fz \equiv \text{Btw}(Q; x, y)(z)] \equiv \mathscr{F}^{=}(\text{Pred})(0, \text{Nx:Fx}).$$

As we shall see below, the content of this theorem is, essentially, that the number of Fs is finite just in case the Fs can be ordered as a simple series which *ends*. Let us say that a concept is *Frege finite* in these circumstances. Note again that Frege finitude is definable (because so defined) in pure second-order logic.

To understand the point of Frege's so characterizing finitude, it is essential to look at his proofs of these two theorems. We begin our discussion, then, with his proof of Theorem 327, by far the more difficult of the two. Frege derives it from his Theorem 325:

$$\mathscr{F}^{=}(\text{Pred})[0, \text{Nz:Btw}(Q; x, y)(z)].$$

That is: the number of objects Q-between x and y is a natural number. Theorem 327 follows easily: If, for some Qξη, x, and y, the Fs just *are* the objects Q-between x and y, then certainly the number of Fs is a natural number.

Theorem 325 is a simple consequence of two lemmas, the first of which is Theorem 321:

$$\mathscr{F}^{=}(Q)(x,y) \ \& \ \text{Func}(Q) \ \& \ \neg\mathscr{F}(Q)(y,y) \rightarrow \mathscr{F}^{=}(\text{Pred})[0, \text{Nz:Btw}(Q; x, y)(z)].$$

The other lemma is Theorem 325, γ:

$$\neg\{\mathscr{F}^{=}(Q)(x,y)\ \&\ \mathrm{Func}(Q)\ \&\ \neg\mathscr{F}(Q)(y,y)\} \rightarrow$$
$$\mathscr{F}^{=}(\mathrm{Pred})[0,\mathrm{N}z{:}\mathrm{Btw}(Q;x,y)(z)].$$

Theorem 325 follows by propositional logic. The proof of Theorem 325, γ is easy: If the antecedent holds, then *nothing* is Q-between x and y (by definition); so, by Frege's Theorem 97, the number of objects Q-between x and y is zero, which is a natural number.

The main work in the proof of Theorem 327 thus lies in the proof of Theorem 321. Frege's proof of this result is extremely interesting. One can perhaps best understand its significance by comparing it to a different proof, which is less complex, and rather shorter, than Frege's. This proof of Theorem 321 is, one should note, simple, utterly straightforward, and entirely obvious; indeed, since the antecedent of Theorem 321 includes a conjunct of the form '$\mathscr{F}^{=}(Q)(x,y)$', a proof by (logical) induction is what immediately suggests itself.[9] The induction is justified by Frege's Theorem 152 (of which mathematical induction is an instance):

$$\mathscr{F}^{=}(Q)(x,y)\ \&\ Fx\ \&$$
$$\forall z\forall w[\mathscr{F}^{=}(Q)(x,z)\ \&\ Fz\ \&\ Qzw \rightarrow Fw] \rightarrow Fy.$$

That is: if y belongs to the Q-series beginning with x, if x is itself F, and if Fξ is hereditary, as one might say, in the Q-series *beginning with x*, then y

[9] To simplify the exposition of Frege's proofs, I shall make free use of what might be called *the basic facts* about the weak and strong ancestrals; these basic facts may be thought of as manifestations of the ancestrals' transitivity. (Special thanks here to George Boolos.) First, we have the following, which are easily derived from the definitions of the strong and weak ancestral:

$$\mathscr{F}^{=}(Q)(a,b)\ \&\ Qbc \rightarrow \mathscr{F}(Q)(a,c)\ \&\ \mathscr{F}^{=}(Q)(a,c).$$
$$Qab\ \&\ \mathscr{F}^{=}(Q)(b,c) \rightarrow \mathscr{F}(Q)(a,c)\ \&\ \mathscr{F}^{=}(Q)(a,c).$$

Note that the second conjunct of the consequent simply follows from the first. Moreover, for this same reason, we also have:

$$\mathscr{F}(Q)(a,b)\ \&\ Qbc \rightarrow \mathscr{F}(Q)(a,c)\ \&\ \mathscr{F}^{=}(Q)(a,c).$$
$$Qab\ \&\ \mathscr{F}(Q)(b,c) \rightarrow \mathscr{F}(Q)(a,c)\ \&\ \mathscr{F}^{=}(Q)(a,c).$$

All four of these also hold if 'Q$\xi\eta$' is weakened to '$\mathscr{F}(Q)(\xi,\eta)$', so we have:

$$[\mathscr{F}(Q)(a,b) \vee \mathscr{F}^{=}(Q)(a,b)]\ \&\ \mathscr{F}(Q)(b,c) \rightarrow \mathscr{F}(Q)(a,c)\ \&\ \mathscr{F}^{=}(Q)(a,c).$$
$$\mathscr{F}(Q)(a,b)\ \&\ [\mathscr{F}(Q)(b,c) \vee \mathscr{F}^{=}(Q)(b,c)] \rightarrow \mathscr{F}(Q)(a,c)\ \&\ \mathscr{F}^{=}(Q)(a,c).$$

Note that the transitivity of the strong ancestral is a consequence of either of these. Finally, we have the transitivity of the weak ancestral:

$$\mathscr{F}^{=}(Q)(a,b)\ \&\ \mathscr{F}^{=}(Q)(b,c) \rightarrow \mathscr{F}^{=}(Q)(a,c).$$

For discussions of these results, and other results about the ancestral, see my 'Development of Arithmetic' and 'Definition by Induction'.

too is F. To prove Theorem 321, then, suppose that $Q\xi\eta$ is functional. For $F\xi$ in Theorem 152, we take: $\neg\mathcal{F}(Q)(\xi,\xi) \rightarrow \mathcal{F}^{=}(Pred)[0,Nz:Btw(Q;x,\xi)(z)]$. We must show that x falls under this concept, i.e. that

$$\neg\mathcal{F}(Q)(x,x) \rightarrow \mathcal{F}^{=}(Pred)[0,Nz:Btw(Q;x,x)(z)].$$

and that it is hereditary in the Q-series beginning with x, i.e. that

$$\mathcal{F}^{=}(Q)(x,a)\ \&\ [\neg\mathcal{F}(Q)(a,a) \rightarrow \mathcal{F}^{=}(Pred)[0,Nz:Btw(Q;x,a)(z)]]$$
$$\&\ Qab \rightarrow [\neg\mathcal{F}(Q)(b,b) \rightarrow \mathcal{F}^{=}(Pred)[0,Nz:Btw(Q;x,b)(z)]].$$

Theorem 321 will then follow from Theorem 152.

For the former, we must show that the number of objects Q-between x and itself is a natural number. By Frege's Theorem 282, no object other than x is Q-between x and itself:[10]

$$Btw(Q;x,x)(z) \rightarrow z=x.$$

Hence, the number of objects Q-between x and itself is going to be a natural number: namely either zero or one, depending upon whether x is Q-between x and itself (which, actually, it is, given that $Q\xi\eta$ is functional and x does not follow itself in the Q-series).

We now show that the concept $\neg\mathcal{F}(Q)(\xi,\xi) \rightarrow \mathcal{F}^{=}(Pred)[0,Nz:Btw(Q;x,\xi)(z)]$ is hereditary in the Q-series beginning with x. That is, we prove:

$$\mathcal{F}^{=}(Q)(x,a)\ \&\ [\neg\mathcal{F}(Q)(a,a) \rightarrow \mathcal{F}^{=}(Pred)(0,Nz:Btw(Q;x,a)(z))]$$
$$\&\ Qab \rightarrow [\neg\mathcal{F}(Q)(b,b) \rightarrow \mathcal{F}^{=}(Pred)(0,Nz:Btw(Q;x,b)(z))].$$

Suppose the antecedent. Suppose further that $\neg\mathcal{F}(Q)(b,b)$. Suppose, for reductio, that $\mathcal{F}(Q)(a,a)$. Then, we have Frege's Theorem 242, which is closely related to Theorem 124 of *Begriffsschrift*:[11]

$$Func(Q)\ \&\ \mathcal{F}(Q)(a,x)\ \&\ Qab \rightarrow \mathcal{F}^{=}(Q)(b,x).$$

[10] For the proof of Theorem 282, suppose that z is between x and itself in the Q-series and that $z \neq x$. By the definition of betweenness, $\mathcal{F}^{=}(Q)(x,z)$ and $\mathcal{F}^{=}(Q)(z,x)$; but $z \neq x$, so $\mathcal{F}(Q)(x,z)$, by the definition of the weak ancestral. But then $\mathcal{F}^{=}(Q)(x,z)$ and $\mathcal{F}(Q)(z,x)$, whence $\mathcal{F}(Q)(x,x)$. But that contradicts the supposition that z is Q-between x and itself, since, if Btw(Q;x,x)(z), then $\neg\mathcal{F}(Q)(x,x)$.

I shall confine the proofs of similarly unspectacular results to the footnotes.

[11] Gottlob Frege, *Begriffsschrift, a Formula Language, Modeled upon that of Arithmetic, for Pure Thought*, tr. by J. van Heijenoort, in J. van Heijenoort (ed.), *Frege and Gödel: Two Fundamental Texts in Mathematical Logic*, Cambridge, Mass., Harvard University Press, 1970.

Since $\mathscr{F}(Q)(a,a)$ and Qab, Theorem 242 implies that $\mathscr{F}^{=}(Q)(b,a)$. But then $\mathscr{F}^{=}(Q)(b,a)$ and Qab, so $\mathscr{F}(Q)(b,b)$. Contradiction. Hence, $\neg\mathscr{F}(Q)(a,a)$.

By *modus ponens*, then, the number of objects Q-between x and a is a natural number. To show that the number of objects Q-between x and b is a natural number, we show that the number of objects between x and b is the immediate successor of the number of objects between x and a, whence the former is a natural number, since the latter is. That is, we show that, on the assumptions made so far:

$$\text{Pred}[\text{Nz:Btw}(Q;x,a)(z),\ \text{Nz:Btw}(Q;x,b)(z)].$$

To prove this, we employ Frege's Theorem 102

$$Fb\ \&\ m = \text{Nx:}(Fx\ \&\ x \neq b) \rightarrow \text{Pred}(m, \text{Nx:}Fx),$$

which follows from the definition of 'Pred(ξ,η)'. Instantiating the variable 'Fξ' with 'Btw(Q;x,b)(ξ)', and 'm' with 'Nz:Btw(Q;x,a)(z)', we have:

$$\text{Btw}(Q;x,b)(b)\ \&\ \text{Nz:Btw}(Q;x,a)(z) = \text{Nz:}[\text{Btw}(Q;x,b)(z)\ \&\ z \neq b] \rightarrow \text{Pred}[\text{Nz:Btw}(Q;x,a)(z),\ \text{Nz:Btw}(Q;x,b)(z)].$$

We thus need only to establish the antecedent. The first conjunct is fairly obvious.[12] To prove the second conjunct, we show not only that the number of objects falling under Btw(Q;x,b)(ξ) & ξ≠b is the same as the number of objects falling under Btw(Q;x,a)(ξ), but that the objects falling under the former concept *just are* the objects falling under the latter, whence, by Frege's Theorem 96, the concepts have the same number.

For one direction, suppose that Btw(Q;x,a)(z); we must show that Btw(Q;x,b)(z) & z≠b. Since Btw(Q;x,a)(z), we have that $\mathscr{F}^{=}(Q)(x,z)$ and $\mathscr{F}^{=}(Q)(z,a)$, by definition. But Qab, so $\mathscr{F}^{=}(Q)(z,b)$. Hence, z is between x and b in the Q-series (since $\neg\mathscr{F}(Q)(b,b)$ and Func(Q), by hypothesis). So, suppose z=b. Then $\mathscr{F}^{=}(Q)(b,a)$ and Qab; hence, $\mathscr{F}(Q)(b,b)$, contradicting the assumption that $\neg\mathscr{F}(Q)(b,b)$.

For the other direction, we show that, if Btw(Q;x,b)(z) and z≠b, then Btw(Q;x,a)(z). If Btw(Q;x,b)(z), then $\mathscr{F}^{=}(Q)(z,b)$ and $\mathscr{F}^{=}(Q)(x,z)$. Since, as argued above, $\neg\mathscr{F}(Q)(a,a)$, we need only show that $\mathscr{F}^{=}(Q)(z,a)$. We then have Frege's Theorem 243, which is closely related to Theorem 133 of *Begriffsschrift*:

$$\text{Func}(Q)\ \&\ \mathscr{F}^{=}(Q)(x,a)\ \&\ \mathscr{F}^{=}(Q)(x,z) \rightarrow [\mathscr{F}(Q)(a,z) \vee \mathscr{F}^{=}(Q)(z,a)].$$

[12] By definition: $\mathscr{F}^{=}(Q)(x,b)\ \&\ \mathscr{F}^{=}(Q)(b,b)\ \&\ \text{Func}(Q)\ \&\ \neg\mathscr{F}(Q)(b,b) \rightarrow \text{Btw}(Q;x,b)(b)$.

Since $\mathscr{F}^{=}(Q)(x,a)$, by hypothesis, and $\mathscr{F}^{=}(Q)(x,z)$, either $\mathscr{F}(Q)(a,z)$ or $\mathscr{F}^{=}(Q)(z,a)$. Suppose that $\mathscr{F}(Q)(a,z)$. Since Qab and Qξη is functional, $\mathscr{F}^{=}(Q)(b,z)$, by Theorem 242. By hypothesis, b≠z, so $\mathscr{F}(Q)(b,z)$. But then $\mathscr{F}(Q)(b,z)$ and $\mathscr{F}^{=}(Q)(z,b)$, so $\mathscr{F}(Q)(b,b)$, contradicting the assumption that $\neg\mathscr{F}(Q)(b,b)$. Hence, $\mathscr{F}^{=}(Q)(z,a)$, and we are done.

That, then, completes the alternative proof of Theorem 321. We have made use of none of the theorems which Frege proves in the course of his proof of Theorem 321, other than certain very general facts about the ancestral and the notion of betweenness. Most importantly, however, we have not had to involve ourselves in the definition of any relations by induction, and we have not had to define any one–one correlations between objects falling under various concepts. As will shortly become clear, however, this proof, being a proof by brute force, has none of the interest of Frege's proof.

4. FREGE'S METHOD OF DEFINING FUNCTIONS BY INDUCTION

In order to discuss Frege's proof of Theorem 321, we need a few additional definitions, those required for the recursive definition of functions, *à la* Frege. Consider the definition of a function, by induction, on the natural numbers. Suppose given an object *a* and a function g(ξ); we wish to define a function φ(ξ), on the natural numbers, satisfying the recursion equations:

$$\varphi(0) = a.$$
$$\varphi(Sn) = g(\varphi(n)).$$

Consider, then, the following relation, which we may call Φξη:

$$\exists x \exists y \exists z \exists w[\xi = \langle x,z\rangle \ \& \ \eta=\langle y,w\rangle \ \& \ y=Sx \ \& \ w=g(y)].$$

An ordered pair stands in this relation to another just in case the first member of the latter is the successor of the first member of the former, and the second member of the latter is the value of g(ξ) for argument the second member of the former. Consider, then, the members of the Φ-series beginning with ⟨0,*a*⟩: These are the ordered pairs ⟨0,*a*⟩, ⟨*S*0,g(*a*)⟩, ⟨*SS*0,g(g(*a*))⟩, and so on. Prima facie, it would appear that the members of this series are exactly the ordered pairs we need to have in the extension of φ(ξ). And this is how Frege defines the (functional) relation η=φ(ξ), by means of the ancestral of the relation Φξη, viz. $\mathscr{F}^{=}(\Phi)(\langle 0,a\rangle,\langle\xi,\eta\rangle)$.

More generally, Frege defines the *coupling* of relations Rξη and Qξη as follows:

$$(R\pi Q)(a,b) \equiv\text{df}\ \exists x\exists y\exists z\exists w[a=\langle x,y\rangle \ \& \ b=\langle z,w\rangle \ \& \ Rxz \ \& \ Qyw].$$

Thus, the coupling of Rξη and Qξη relates one ordered pair to another just in case the first member of the former bears Rξη to the first member of the latter, and the second member of the former bears Qξη to the second member of the latter. Obviously, in the definition of coupling, Frege makes use of ordered pairs, so these must be introduced into his system. For present purposes, we may take them as primitive, terms standing for ordered pairs being governed by the ordered pair axiom:[13]

$$\langle x,y\rangle = \langle z,w\rangle \text{ iff } x=z \ \& \ y=w.$$

The central theorem on the recursive definition of functions on the natural numbers is Frege's Theorem 256, which is essentially:[14]

$$\begin{aligned}&\text{Func}(Q)\ \&\ \forall x[\mathscr{F}^{=}(Q)(a,x) \rightarrow \exists y.Qxy] \rightarrow\\ &\quad \text{Func}[\mathscr{F}^{=}(\text{Pred}\ \pi\ Q)(\langle 0,a\rangle,\langle \xi,\eta\rangle)]\ \&\\ &\quad \forall x\{\mathscr{F}^{=}(\text{Pred})(0,x) \rightarrow \exists y[\mathscr{F}^{=}(Q)(a,y)\ \&\\ &\quad \mathscr{F}^{=}(\text{Pred}\ \pi\ Q)(\langle m,a\rangle,\langle x,y\rangle)]\}.\end{aligned}$$

That is: if Qξη is functional, and if the Q-series beginning with a is endless, then the relation $\mathscr{F}^{=}$(Pred π Q)(⟨0,a⟩,⟨ξ,η⟩) is functional and, strengthening the result slightly,[15] its *domain* is exactly the natural numbers and its *range* is wholly contained in the Q-series beginning with a. It is not difficult to see that, if Qξη is taken to be η=g(ξ), the relation so defined satisfies the recursion equations mentioned above.

5. FREGE'S PROOF OF THEOREM 321

Because Frege's proof of Theorem 321 is both long and complex, I shall merely outline it here, noting those aspects of the proof which are of particular interest. A large portion of the proof will be confined to the appendix.

[13] Frege actually defines ordered pairs by means of value-ranges. It does not appear to be possible to define a *total* pairing function in Frege Arithmetic: It is, of course, possible to define such a function on the natural numbers, but we have no assurance that the domain contains only natural numbers.

In any event, the uses Frege actually *makes* of ordered pairs may be eliminated, and he knew as much. For details, see 'Definition by Induction'. We need not concern ourselves with such complexities here, however.

[14] This theorem follows from a more general result, Frege's Theorem 254, on the recursive definition of functions on simple series in general. For discussion, see 'Definition by Induction'.

[15] The strengthening is justified by Frege's Theorem 232:

$$\mathscr{F}^{=}(R\pi Q)(\langle m,a\rangle,\langle x,y\rangle) \rightarrow \mathscr{F}^{=}(R)(m,x)\ \&\ \mathscr{F}^{=}(Q)(a,y)$$

Recall that Theorem 321 is:

$$\text{Func}(Q)\ \&\ \neg\mathcal{F}(Q)(y,y)\ \&\ \mathcal{F}^{=}(Q)(x,y)\rightarrow$$
$$\mathcal{F}^{=}(\text{Pred})(0,\text{Nz:Btw}(Q;x,y)(z)).$$

That is: if $Q\xi\eta$ is functional, and if y is a member of the Q-series beginning with x which does not follow after itself, then the number of objects Q-between x and y is finite. The proof requires three lemmas, the first of which is Theorem 315:

$$\mathcal{F}^{=}(\text{Pred})(0,n)\ \&\ \mathcal{F}^{=}(Q\ \pi\ \text{Pred})(\langle x,1\rangle,\langle y,n\rangle)\ \&\ \text{Func}(Q)\ \&$$
$$\neg\mathcal{F}(Q)(y,y)\rightarrow\mathcal{F}^{=}(\text{Pred})[0,\text{Nz:Btw}(Q;x,y)(z)].$$

That is: if n is finite, if ⟨y,n⟩ is a member of the (Q π Pred)-series beginning with ⟨x,1⟩, if $Q\xi\eta$ is functional, and if y does not follow itself in the Q-series, then the number of objects Q-between x and y is finite. The second lemma is Theorem 317:[16]

$$\mathcal{F}^{=}(Q\ \pi\ \text{Pred})(\langle x,1\rangle,\langle y,n\rangle)\ \rightarrow\mathcal{F}^{=}(\text{Pred})(0,n).$$

The third is Theorem 319:[17]

$$\mathcal{F}^{=}(Q)(x,y)\rightarrow\exists n.\mathcal{F}^{=}(Q\ \pi\ \text{Pred})(\langle x,1\rangle,\langle y,n\rangle).$$

Theorem 321 is then an immediate consequence of these lemmas. Theorem 317 allows us to drop the assumption that n is finite from the antecedent of Theorem 315; Theorem 319 guarantees that an appropriate n exists, if y belongs to the Q-series beginning with x.

It is worth considering Theorem 319, whose point is not obvious, a little more carefully. Theorem 319 states that, if y belongs to the Q-series beginning with x, then, for some n, ⟨y,n⟩ belongs to the (Q π Pred)-series beginning with ⟨x,1⟩. But what relation, exactly, is this relation $\mathcal{F}^{=}$(Q π Pred)(⟨x,1⟩,⟨ξ,η⟩)? Let us abbreviate it: P(ξ,η). Consider, first, the case in which the Q-series beginning with x is simple. In this case, P(y,n) just in case y is the nth member of the Q-series beginning with x. For let x_1=x. Then certainly, $\mathcal{F}^{=}$(Q π Pred)(⟨x_1,1⟩,⟨x_1,1⟩), so P(x_1,1), and x_1 is indeed the first member of the Q-series beginning with x_1. Suppose then that Qx_1x_2. Since Qx_1x_2 and Pred(1,2), we have, by definition: (Q π Pred)(⟨x_1,1⟩,⟨x_2,2⟩). So $\mathcal{F}^{=}$(Q π Pred)(⟨x_1,1⟩,⟨x_2,2⟩); hence, P(x_2,2). And so on. Now, if the Q-series is not simple, then, firstly, we can not say that P(y,n) just in case y is *the* n^{th}

[16] For the proof of this theorem, we need only Frege's Theorem 232, mentioned in the preceding note.

[17] The proof of Theorem 319 is by induction, the relevant concept being:

$$\exists n.\mathcal{F}^{=}(Q\ \pi\ \text{Pred})(\langle x,1\rangle,\langle \xi,n\rangle)$$

member of the Q-series beginning with x; at most, we can say that P(y,n) if y is *an* n^{th} member.[18] Moreover, in this case, members of the Q-series beginning with x need not be uniquely an nth member, rather than an mth member. Rather, we will have that P(y,n) just in case it is possible to get from x to y in n-1 'steps', that is, if, and only if, there is some sequence $x=x_1, x_2, \ldots, x_n=y$, where Qx_ix_{i+1}. This justifies our reading '$\mathscr{F}^{=}(Q\ \pi\ \text{Pred})(\langle x,1\rangle,\langle y,n\rangle)$' as 'y is an nth member of the Q-series beginning with x' or, when justifiable, as 'y is the nth member of the Q-series beginning with x'.[19]

Thus, what Theorems 319 and 317 together say is just this: if y belongs to the Q-series beginning with x, then, for some natural number n, y is an nth member of the Q-series beginning with x. This theorem should reinforce one's confidence in Frege's definition of the ancestral, for y is, intuitively, an 'ancestor' of x just in case it is some finite number of 'steps' from x.[20]

Let us now return to the main line of argument. The main lemma in the proof of Theorem 321 is Theorem 315, which is, recall:

$$\mathscr{F}^{=}(\text{Pred})(0,n)\ \&\ \mathscr{F}^{=}(Q\ \pi\ \text{Pred})(\langle x,1\rangle,\langle y,n\rangle)\ \&\ \text{Func}(Q)\ \&$$
$$\neg\mathscr{F}(Q)(y,y) \rightarrow \mathscr{F}^{=}(\text{Pred})[0, \text{Nz:Btw}(Q;x,y)(z)].$$

That is: if n is a natural number, if y is the nth member of the Q-series beginning with x, if $Q\xi\eta$ is functional, and if y does not follow after itself in the Q-series, then the number of objects Q-between x and y is finite. Theorem 315 itself follows from two lemmas, the first of which is Frege's Theorem 314:

$$\mathscr{F}^{=}(\text{Pred})(0,n) \rightarrow n = \text{Nz:Btw}(\text{Pred};1,n)(z).$$

That is: every natural number is the number of numbers Pred-between 1 and itself. The proof of Theorem 314 poses little difficulty.[21] The second lemma is Theorem 298:

[18] That $P(\xi,\eta)$ is functional if the Q-series beginning with x is simple follows from Frege's Theorem 253: $\text{Func}(Q)\ \&\ \neg\exists y[\mathscr{F}^{=}(Q)(x,y)\ \&\ \mathscr{F}(Q)(y,y)]\ \&\ \text{Func}(R) \rightarrow \text{Func}[\mathscr{F}^{=}(Q\ \pi\ R)(\langle x,z\rangle,\langle\xi,\eta\rangle)]$.

[19] The foregoing can be formalized and proven, but these intuitive remarks should suffice for present purposes. One of the differences between Frege's proof of the validity of definition by induction and Dedekind's is that Frege's proves the validity of recursive definitions of *relations*. The relation $\mathscr{F}^{=}(Q\ \pi\ \text{Pred})(\langle x,1\rangle,\langle\xi,\eta\rangle)$ is a wonderful example of a useful and important *non-functional* relation naturally definable by Frege's methods. See 'Definition by Induction', §4.

[20] Cf. Crispin Wright, *Frege's Conception of Numbers as Objects*, Aberdeen, Aberdeen University Press, 1983, 159–60: 'Does [Frege's definition] capture the intuitive meaning? . . . [D]oes satisfaction of Frege's condition guarantee the accessibility of y from x by a series of [Q]steps?' Yes, and provably so, though the reasoning required to show that it does is admittedly impredicative.

[21] The proof appeals to Frege's Theorem 160, which is:

$$\mathscr{F}^{=}(\text{Pred})(0,n) \rightarrow n = \text{Nx:}[\mathscr{F}^{=}(\text{Pred})(x,n)\ \&\ x{\neq}0]$$

$$\mathscr{F}^{=}(Q\,\pi\,\mathrm{Pred})(\langle x,1\rangle,\langle y,n\rangle)\ \&\ \mathrm{Func}(Q)\ \&\ \neg\mathscr{F}(Q)(y,y)\rightarrow$$
$$\mathrm{Nz{:}Btw}(Q;x,y)(z)=\mathrm{Nz{:}Btw}(\mathrm{Pred})(1,n)(z).$$

That is: if y is the nth member of the Q-series beginning with x, if Qξη is functional, and if y does not follow itself in the Q-series, then the number of objects Q-between x and y is the same as the number of numbers Pred-between 1 and n.

The derivation of Theorem 315 is then straightforward. For suppose that n is finite, that y is the nth member of the Q-series beginning with x, that Qξη is functional, and that y does not follow itself in the Q-series. Then, by Theorem 298, the number of objects Q-between x and y is the same as the number of numbers Pred-between 1 and n. But, by Theorem 314, this number is n, which is, by hypothesis, finite.

The proof of Theorem 298, in turn, requires a number of lemmas, but the idea behind the proof is relatively straightforward. The main lemma is Theorem 288:

$$\mathscr{F}^{=}(R\,\pi\,\mathrm{Pred})(\langle x,1\rangle,\langle y,n\rangle)\ \&\ \mathrm{Func}(R)\ \&\ \neg\exists z.\mathscr{F}(R)(z,z)\rightarrow$$
$$\mathrm{Nz{:}Btw}(R;x,y)(z)=\mathrm{Nz{:}Btw}(\mathrm{Pred})(1,n)(z).$$

Except for the change of variable, this result differs from Theorem 298 in only one respect: the condition, in Theorem 288, that no object follows itself is replaced, in Theorem 298, by the weaker condition that y not follow itself.

Theorem 298 is derived from 288 as follows. Suppose, as in the antecedent of Theorem 298, that y is an nth member of the Q-series beginning with x, that Qξη is functional, and that y does not follow itself in the Q-series. We now define, in terms of Qξη, a new relation, Rξη, which will satisfy the antecedent of Theorem 288. The relation Rξη is to be Qξη & $\mathscr{F}^{=}(Q)(\eta,y)$, the *restriction* of Qξη to the Q-series ending with y. To show that this relation satisfies the antecedent of Theorem 288, we thus have to prove:[22]

This in turn is a relatively easy consequence of the fact that n precedes $Nx{:}\mathscr{F}^{=}(\mathrm{Pred})(x,n)$, if n is finite, which is Theorem 155. For $Nx{:}[\mathscr{F}^{=}(\mathrm{Pred}(x,n)\ \&\ x\neq 0]$ precedes $Nx{:}\mathscr{F}^{=}(\mathrm{Pred})(x,n)$, by Theorem 102, and $Nx{:}\mathscr{F}^{=}(\mathrm{Pred})(x,n)$, like every number, has at most one predecessor.

Theorem 314 then follows from Theorem 313,ζ:

$$\mathscr{F}^{=}(\mathrm{Pred})(0,n)\rightarrow\forall x\{\mathrm{Btw}(\mathrm{Pred})(1,n)(x)\equiv[\mathscr{F}^{=}(\mathrm{Pred})(x,n)\ \&\ x\neq 0]\}$$

This, however, is an easy consequence of the definition of betweenness, the axioms of arithmetic, and Frege's Theorem 242, mentioned earlier.

[22] To simplify the exposition, I have here taken certain liberties. Theorem 189* actually follows from the more general result that, if Qξη is functional, then, for any F, Qξη & Fη is functional. Also, Frege's Theorem 290 is more general. '1' being replaced by 'm' and 'Pred' by 'R'.

$$189^*: \text{Func}(Q) \rightarrow \text{Func}[Q\xi\eta \;\&\; \mathcal{F}^{=}(Q)(\eta,y)].$$

$$290^*: \mathcal{F}^{=}(Q\ \pi\ \text{Pred})(\langle x,1\rangle,\langle y,n\rangle) \rightarrow \mathcal{F}^{=}[(Q\xi\eta \;\&\; \mathcal{F}^{=}(Q)(\eta,y))\ \pi\ \text{Pred}](\langle x,1\rangle,\langle y,n\rangle).$$

$$297: \text{Func}(Q) \;\&\; \neg\mathcal{F}(Q)(y,y) \rightarrow \neg\exists z.\mathcal{F}[Q\xi\eta \;\&\; \mathcal{F}^{=}(Q)(\eta,y)](z,z).$$

From these results and Theorem 288, it then follows that the number of objects $[Q\xi\eta \;\&\; \mathcal{F}^{=}(Q)(\eta,y)]$-between x and y is the same as the number of numbers Pred-between 1 and n. To complete the proof of Theorem 298, we thus need only show that the number of objects $[Q\xi\eta \;\&\; \mathcal{F}^{=}(Q)(\eta,y)]$-between x and y is the same as the number of Q-between x and y:

$$295: \text{Func}(Q) \;\&\; \neg\mathcal{F}(Q)(y,y) \rightarrow \text{Nz:Btw}(Q;x,y)(z) = \text{Nz:Btw}[Q\xi\eta \;\&\; \mathcal{F}^{=}(Q)(\eta,y);x,y](z).$$

As the proofs of these four lemmas are neither difficult nor terribly interesting, I leave them to the notes.[23]

[23] Theorem 189* is completely obvious. Any restriction of a functional relation is functional. Theorem 290* is proven by induction, using Frege's Theorem 257, which is mentioned in the appendix, and the definition of coupling.

Theorem 297 follows from these two:

$$193: \mathcal{F}[Q\xi\eta \;\&\; F\eta](x,y) \rightarrow Fy$$

$$194: \mathcal{F}[Q\xi\eta \;\&\; F\eta](x,y) \rightarrow \mathcal{F}(Q)(x,y)$$

For suppose that z follows itself in the $[Q\xi\eta \;\&\; \mathcal{F}^{=}(Q)(\eta,y)]$-series. Then, by (193), $\mathcal{F}^{=}(Q)(z,y)$; and, by (194), $\mathcal{F}(Q)(z,z)$. But following oneself is hereditary in any functional series (see Theorem 278), so y must also follow itself. Contradiction.

Finally, Theorem 295 follows from Theorem 295, ε (which is the theorem, marked 'ε', proved *during* the proof of Theorem 295):

$$295,\varepsilon: \text{Func}(Q) \;\&\; \neg\mathcal{F}(Q)(y,y) \rightarrow \forall z\{\text{Btw}(Q;x,y)(z) \equiv \text{Btw}[Q\xi\eta \;\&\; \mathcal{F}^{=}(Q)(\eta,y);x,y](z)\}$$

The proof, from right-to-left, depends, for the most part, upon the following analogue of (194):

$$201: \mathcal{F}^{=}[Q\xi\eta \;\&\; F\eta](x,y) \rightarrow \mathcal{F}^{=}(Q)(x,y)$$

From left to right, it is a little more difficult. The crucial result is

$$292: \mathcal{F}^{=}(Q)(x,d) \;\&\; \mathcal{F}^{=}(Q)(d,y) \rightarrow \mathcal{F}^{=}[Q\xi\eta \;\&\; \mathcal{F}^{=}(Q)(\eta,y)](x,d)$$

Suppose that d is Q-between x and y. Then $Q\xi\eta$ is functional; hence, $Q\xi\eta \;\&\; \mathcal{F}^{=}(Q)(\eta,y)$ is. Moreover, $\neg\mathcal{F}(Q)(y,y)$, so $\neg\mathcal{F}[Q\xi\eta \;\&\; \mathcal{F}^{=}(Q)(\eta,y)](y,y)$, by (194). The antecedent of (292) holds, so we have that $\mathcal{F}^{=}[Q\xi\eta \;\&\; \mathcal{F}^{=}(Q)(\eta,y)](x,d)$; and, substituting into (292), we have

$$\mathcal{F}^{=}(Q)(d,y) \;\&\; \mathcal{F}^{=}(Q)(y,y) \rightarrow \mathcal{F}^{=}[Q\xi\eta \;\&\; \mathcal{F}^{=}(Q)(\eta,y)](d,y)$$

whose antecedent again holds. Done.

It remains to prove Theorem 288. Frege must show that, if the R-series is simple and if y is the nth member of the R-series beginning with x, then some relation correlates the objects R-between x and y one–one with the numbers between 1 and n; one such relation is $\mathcal{F}^{=}(R\ \pi\ \text{Pred})(\langle x,1\rangle,\langle \xi,\eta\rangle)$. What Frege proves is thus that, if the R-series is simple, and if y is the nth member of the R-series beginning with x, then the objects R-between x and y are correlated one–one with the numbers from 1 to n by the relation: ξ is the ηth member of the R-series beginning with x. That this is so should seem overwhelmingly plausible. The proof is somewhat tortuous, however, and the details, while interesting, are not relevant to the central concerns of the present paper. I therefore leave Frege's proof of Theorem 288 to the Appendix.

6. THE SIGNIFICANCE OF THEOREMS 327 AND 348

The differences between Frege's proof of Theorem 321 and that considered earlier are, in my judgement, not due to ignorance on Frege's part that there is a proof of Theorem 321 which is easier than his. Rather, Frege proves Theorem 321 as he does because he wants to establish a result somewhat stronger than Theorem 321 itself, a result which reveals the *intuitive* basis of his characterization of finitude.

Recall that the Q-series is *simple* if $Q\xi\eta$ is functional and if no object follows after itself in the Q-series. Now, the content of Theorems 207 and 263, as mentioned earlier, is that the number of a concept is Endlos just in case the objects falling under that concept can be ordered as a simple series which does not end. The content of Theorems 327 and 348, which Frege introduces as analogues, for the concept of finitude, is that the number of a concept is a natural number just in case 'the objects falling under it can be ordered in a simple series which begins with a certain object and ends with a certain object', i.e. which contains a (unique) object after which no member of the series follows (*Gg* I §158).[24]

[24] This statement of the result is slightly stronger than what Theorems 327 and 348 themselves say, but the stronger claim follows easily from the weaker. To see this, recall that Theorems 327 and 348 together imply:

$$\exists Q\exists x\exists y\forall z[Fz \equiv \text{Btw}(Q;x,y)(z)] \equiv \mathcal{F}^{=}(\text{Pred})(0,\text{N}z{:}Fz)$$

Plainly, if there is a concept $Q\xi\eta$ of the appropriate sort, there is such a concept whose domain and range consist only of Fs. (Consider $R\xi\eta \equiv_{df} Q\xi\eta\ \&\ F\xi\ \&\ F\eta$.) To simplify the exposition, we restrict attention to such concepts. If there is an F, then Func(Q) and $\neg\mathcal{F}(Q)(y,y)$, by the definition of 'Btw'. Moreover, that no object which precedes y in Q-series (so no F, so no object) can follow itself in the Q-series, since following oneself in the Q-series is hereditary if $Q\xi\eta$ is functional: hence, the Q-series is simple.

Finally, we show that no object follows after y in the Q-series. For *reductio*, suppose that $\mathcal{F}(Q)(y,z)$. By Frege's Theorem 242, mentioned earlier, z is in the range of $Q\xi\eta$ and so, by the

Given such an ordering of the Fs, we can show, essentially by brute force, that the number of Fs (the number of members of such an ordering) is a natural number, as we did in §2. That proof, however, does nothing to show, so to speak, *which* natural number is the number of objects so ordered. Frege's proof, on the other hand, rests upon the insight that, given such an ordering of the Fs, the Fs can then be mapped one–one onto an initial segment of the natural numbers: that is to say, it can be shown that the number of Fs is the same as the number of numbers between 0 and *n*, for some natural number *n*. It will then follow, from Frege's theorem on the infinity of the number-series (Theorem 155), that the number of Fs is the successor of *n*.

As we have seen, however, Frege does not proceed in *quite* this way. His not so proceeding is our best indication of the role Theorems 327 and 348 play in *Grundgesetze* and were intended to play in his philosophy of mathematics. What Frege does, instead, is to produce a one–one mapping between the Fs and the numbers between *1* and a natural number *n*. On reflection, it is clear why Frege proceeds in this way. *One* way of producing a simple ordering of the Fs is to *enumerate* or *count* them, that is, to associate each of them, in turn, with a number, beginning with one and ending with some number *n*, which is then the number of Fs. Indeed, as was mentioned, the relation which Frege shows correlates the members of the relevant series with the numbers between 1 and n is the relation ξ is the ηth member of that series, so this correlation itself amounts to a 'counting' of the Fs. The intuitive content of Theorems 327 and 348 is thus that the number of Fs is a natural number just in case the Fs can be counted and that, in that case, the number of Fs is the natural number one reaches *by* counting.

This is especially clear from the proof of Theorem 348. Recall that Theorem 348 is:

$$\mathscr{F}^{=}(\text{Pred})(0,\text{Nx:Gx}) \rightarrow \exists Q\exists x\exists y\forall z[Gx = \text{Btw}(Q;x,y)(z)].$$

That is: if the number of Gs is a natural number, then the Gs can be ordered as a simple series which ends. This Theorem, like Theorem 321, admits of a direct and utterly uninteresting proof by induction.[25] Again, however, Frege's proof is more complicated than it needs to be and, therefore, more illuminating than it might have been. He derives Theorem 348 from two lemmas, the first of which is Theorem 347:

simplifying assumption, is Q-between x and y; hence, $\mathscr{F}^{=}(Q)(z,y)$. But then $\mathscr{F}(Q)(y,z)$ and $\mathscr{F}^{=}(Q)(z,y)$, so $\mathscr{F}(Q)(y,y)$, contrary to hypothesis.

[25] For the base case, in which Nx:Gx = 0, take $Q\xi\eta$ to be any non-functional relation. For the induction step, suppose that, whenever the number of Fs is n, the Fs can be ordered as a simple series which ends; suppose, further, that the number of Gs is n+1. Let a be a G. The number of Gs other than a is n; so the Gs other than a can be ordered as a simple series which ends. Tack a onto the end. Done.

$$Nz{:}Gz = Nz{:}Btw(R;a,b)(z) \rightarrow \exists Q \exists x \exists y \forall z[Gx \equiv Btw(Q;x,y)(z)].$$

That is: if the number of Gs is the same as the number of objects R-between x and y, for some R$\xi\eta$, x, and y, then the Gs can be ordered as a simple series which ends. The other lemma is the previously mentioned Theorem 314:

$$\mathscr{F}^{=}(Pred)(0,n) \rightarrow n = Nz{:}Btw(Pred;1,n)(z).$$

Theorem 347 follows immediately. By Theorem 314, if the number of Gs is a natural number, then the number of Gs is the same as the number of numbers between 1 and the number of Gs; substitute into Theorem 347.

Now, Theorem 347 is a generalization of the following:

$$Nz{:}Gz = Nz{:}Btw(Pred;1,n)(z) \rightarrow \exists Q \exists x \exists y \forall z[Gx \equiv Btw(Q;x,y)(z)].$$

It is this which is really Frege's main goal in his proof of Theorem 348, not Theorem 347 (but Frege rarely passes up an opportunity to prove the most general result possible). The proof proceeds by showing that, if the numbers between 1 and n can be mapped one–one onto the Gs, then the Gs can be ordered as a simple series which ends; the relation which so orders the Gs will be the image of the predecessor relation under the relevant mapping. To correlate the Gs one–one with the numbers between 1 and n is, for all intents and purposes, to count them. So Theorem 348 amounts to this: if the number of Gs is finite, then they can be counted, that is, ordered as a simple series which ends.

Frege's proofs of Theorems 327 and 348 may thus be understood as falling into two parts. The hard work in the two proofs goes into establishing that the objects falling under a concept can be put in one-to-one correspondence with an initial segment of the counting numbers (the non-zero natural numbers) if, and only if, they can be ordered as a simple series which ends. (From left to right, this is Theorem 347; from right to left, it is—roughly—Theorem 288.) From this intermediate result, Theorems 327 and 348 follow via Theorem 314, an immediate consequence of which is that the number of a given concept is finite if, and only if, the objects falling under it can be put in one-to-one correspondence with an initial segment of the counting numbers. Now, I do not want to deny that Theorems 327 and 348 have a purely *technical* point: they yield a nice characterization of finite sets. But the Theorems also have an epistemological point: they show that the concept of (Frege) finitude is a concept of *logic*; Frege would have regarded the characterization as *uncontroversially* logical, because purely second-order. And they have yet another point, for they reveal that the notion of finitude thus shown to be logical is compatible with, or is a rigorous version of, our *intuitive* conception of finitude: a finite set is one which can be counted. That is why Frege proves these theorems as he does, why both proofs go through Theorem 314.

Every philosophy of arithmetic must deal with the natural numbers' two aspects, their use as finite cardinals, and their use as finite ordinals. What is distinctive about Frege's development of arithmetic in *Grundgesetze* is that he takes the natural numbers fundamentally to be finite *cardinals*. The developments of arithmetic with which we are most familiar, on the other hand, take the natural numbers fundamentally to be finite ordinals. (It is for this reason that Dedekind's numbers begin, not with zero, but with one: there is no 'zeroth' position in an ordering and no ordinal 'zeroth', though of course there is a null order-type.) We are familiar with how the use of natural numbers as finite cardinals can be reconstructed from their use as finite ordinals. Indeed, Cantor, whose development of the theory of transfinite numbers was made possible by his taking the notion of an ordinal to be fundamental, showed how (at least part of) the theory of cardinal numbers could be reduced to the theory of ordinals: in the finite case, we say that the number of Fs is that finite ordinal n the ordinals less than which can be put into one–one correspondence with the Fs (here we are assuming that there is an ordinal number zero, but this is inessential). Theorems 327 and 348 are a kind of converse of this definition for finite cardinals: in these two theorems, Frege shows how the use of natural numbers as finite *ordinals* can be reconstructed from their use as finite *cardinals*.

The question of the relationship between finite cardinals and finite ordinals is, to my mind, of great importance, but it is not very often discussed. In part, this may be because it is thought, first, that Cantor showed that the cardinals could be reduced to the ordinals and, secondly, that the converse reduction can not be carried out. The reduction of finite cardinals to finite ordinals mentioned above is easily generalized: in set-theory, the cardinal number of a set x is typically taken to be the smallest ordinal number the ordinals less than which can be correlated one–one with the members of x. And there is no doubt that this provides for a reduction of (at least some) cardinals to ordinals. Frege's reduction of the finite ordinals to the finite cardinals, on the other hand, can not be so generalized. It is this which leads Dummett to remark that 'if Frege had paid more attention to Cantor's work, he would have understood what it revealed, that the notion of an ordinal number is more fundamental than that of a cardinal number'.[26]

Dummett's view is thus that the ordinals are *more* fundamental than the cardinals. And if the foregoing discussion of the significance of Theorems 327 and 348 is correct, Frege must have held the contrary view. But there appears to be no obvious reason to believe either the ordinals or the

[26] Michael Dummett, *Frege: Philosophy of Mathematics*, Cambridge, Mass., Harvard University Press, 1992, 293. What Frege failed to understand may have been, more generally, the importance of the theory of ordinals.

cardinals to be more fundamental. Indeed, Frege ought not to have held either of these views. According to him, a philosophical account of mathematical objects of a certain sort must not only be compatible with, but must from the outset *embody*, an account of the application of our theory of those objects. As Dummett puts the point, the foundations of the theory 'must be so constructed as to display the most general form of [its] applications.'[27] It is on this sort of ground that Frege objects to Cantor's theory of the real numbers; presumably, Frege would have objected to the reduction of cardinals to ordinals on similar grounds, that it treats the use of natural numbers as cardinals not as 'one of their distinguishing characteristics, which ought therefore to figure in their definition', but rather explains this use as one which could be made of *any* simply infinite system.[28] But the use of ordinals to answer questions of the form, 'The whichth member of the ordering?' is as much one of *their* distinguishing characteristics as the use of cardinals to answer questions of the form 'How many?' is one of theirs. One might therefore suppose that Frege was committed to developing the theory of ordinals on the basis of a principle, like Hume's Principle, which embodied this aspect of the ordinal numbers (see *Gl* §86). Unfortunately, the most obvious way of doing so leads to the Burali–Forti Paradox, and it is a nice question whether a decent neo-Fregean theory of the ordinals can be constructed at all.

One should not, however, conclude from this that our only alternative is to reduce the cardinals to the ordinals (set-theoretically understood). For one might wonder whether the familiar reduction of cardinals to ordinals succeeds even on its own terms. I said earlier that one can reduce *at least some* cardinals to ordinals in the familiar way. What I had in mind was not that, without choice, not all sets can be shown to be equinumerous with the ordinals less than some ordinal (though one might, indeed, pursue that point): rather, it seems to me that there is a cardinal number which is provably *not* the number of ordinals less than any ordinal, namely the number of all the ordinals there are. Admittedly, the claim that there is such a number is widely rejected; indeed, Hume's Principle has been questioned on the ground that it entails that there is such a number.[29] But I know of no argument for the claim that there is no such number which does not beg the question—by implicitly appealing to the very reduction of ordinals to cardinals here under discussion (or to some similar claim).[30] There is nothing 'self-contradictory' about the claim that there is a number of all

[27] Michael Dummett, *Frege: Philosophy of Mathematics*, 293.

[28] See here ibid. 47–51. The quotation is from 51.

[29] George Boolos has frequently raised this worry in conversation; it is mentioned, though not endorsed, on p. 4 of his 'Consistency'.

[30] I should emphasize that the status of Zermelo-Fraenkel set-theory, or any other set-theory, as the proper response to the set-theoretic paradoxes is not presently in question;

ordinals, or all sets, or all objects: that Frege Arithmetic, and higher-order ZF+Hume's Principle, are consistent *implies* that there is nothing contradictory about it.

I conclude that, though Frege's reduction of ordinals to cardinals cannot be generalized, his development of a theory of cardinals *independent* of the theory of ordinals is nevertheless worth our continued attention.

7. FINITUDE, THE LEAST NUMBER PRINCIPLE, AND WELL-ORDERING

We have seen that Frege presents a purely second-order characterization of finitude in *Grundgesetze*. According to this characterization, a concept is Frege finite if, and only if, the objects falling under it can be ordered as a simple series which ends. Frege proves, in Frege Arithmetic (*modulo* certain eliminable uses of ordered pairs) that a concept is Frege finite just in case the objects falling under it can be put in one-to-one correspondence with some initial segment of the natural numbers. His characterization is thus equivalent to Cantor's, not to Dedekind's. What I wish to discuss at this point is the importance, both to this characterization and to Frege's characterization of concepts whose number is Endlos, of the condition that the objects falling under the concept be ordered as a *simple* series of a certain form. This should help further to illuminate the relationship between Frege's characterization of finitude and more familiar ones.

Frege's most sustained discussion of the importance of the concept of simplicity occurs in a discussion of his Theorem 145, which states that no natural number follows after itself in the number-series. It is worth quoting this little-known discussion at length:[31]

> The importance of [Theorem 145] will be made more evident by the following considerations. If we determine the number belonging to a concept $\Phi(\xi)$, or, as one normally says, if we count the objects falling under the concept $\Phi(\xi)$, then we successively associate these objects with the number-words from 'one' up to a number-word '*N*', which will be determined by the associating relation's mapping the concept $\Phi(\xi)$ into the concept 'member of the series of number-words from "one" to "*N*"' and the converse relation's mapping the latter concept into the former. '*N*' then denotes the sought number; i.e. *N* is this number. This process of counting may be carried out in various ways, since the associating relation is not completely determined.

rather, the point is that *there are no set-theoretic paradoxes specifically concerning cardinal numbers.*

[31] The translation of this passage is due to myself and Jason Stanley. I discuss this fascinating passage further in 'Definition by Induction'. It is, by the way, well worth reading this passage in conjunction with §56 of *Grundlagen*, on which it throws a great deal of light.

The question therefore arises whether, by another choice of this relation, one could reach another number-word '*M*'. Then, by our assumptions, *M* would be the same number as *N*, but, at the same time, one of these two number-words would follow after the other, e.g. '*N*' would follow '*M*'. Then *N* would follow in the number-series after *M*, which means that it would follow after itself. That is excluded by our proposition concerning finite numbers. (*Gg* I §108)

Thus, it is, according to Frege, because the series of natural numbers is simple that the result of the process of counting is determinate.

Compare with this, then, Theorem 288:

$$\begin{aligned}&\text{Func}(Q)\ \&\ \neg\exists z.\mathscr{F}(Q)(z,z)\ \&\ \mathscr{F}^{=}(Q\ \pi\ \text{Pred})(\langle x,1\rangle,\langle y,n\rangle)\rightarrow\\&\quad \text{Nz:Btw}(Q;x,y)(z)=\text{Nz:Btw}(\text{Pred})(1,n)(z).\end{aligned}$$

As said earlier, the first three conjuncts together imply that y is the nth member of the Q-series beginning with x; it is essential to this that the Q-series be simple. Were the Q-series not simple, y might be not only an nth but also an mth member of the Q-series beginning with x: that is to say, there might be more than one way of correlating objects Q-'between' (though not in Frege's sense, of course) x and y with an initial segment of the counting numbers. Frege, in his discussion of Theorem 145, is concerned with the simplicity of the number-series *itself*. But the similarity between the importance he attaches to that fact and the importance of the simplicity of the Q-series in Theorem 288, is striking: it is because the Q-series is simple that its members have a well-defined ordinal position in it.

These reflections suggest a connection between the concept of a *simple* series and that of a *well-ordered* series, and such a connection there is. Eliminating a few definitions and contraposing, Frege's Theorem 359 is:

$$\begin{aligned}&\text{Func}(Q)\ \&\ \neg\exists x.\mathscr{F}(Q)(x,x)\ \&\ \exists x[\mathscr{F}^{=}(Q)(a,x)\ \&\ Fx]\rightarrow\\&\quad \exists x\{\mathscr{F}^{=}(Q)(a,x)\ \&\ Fx\ \&\ \neg\exists y[\mathscr{F}^{=}(Q)(a,y)\ \&\ Fy\ \&\ \mathscr{F}(Q)(y,x)]\}.\end{aligned}$$

That is: if the Q-series is simple and if some F belongs to the Q-series beginning with a, then there is an F which belongs to the Q-series beginning with a which is ancestrally preceded in that series by no F; that is, there is some F in the Q-series beginning with a which is $\mathscr{F}$(Q)-*minimal*. Theorem 359 is thus a generalization, for all simple series, of the *least number principle*. Substituting 'Pred(ξ,η)' for 'Qξη' and '0' for 'a', we have:

$$\begin{aligned}&\text{Func}(\text{Pred})\ \&\ \neg\exists x.\mathscr{F}(\text{Pred})(x,x)\ \&\ \exists x[\mathscr{F}^{=}(\text{Pred})(0,x)\ \&\ Fx]\rightarrow\\&\quad \exists x\{\mathscr{F}^{=}(\text{Pred})(0,x)\ \&\ Fx\ \&\ \neg\exists y[\mathscr{F}^{=}(\text{Pred})(0,y)\ \&\ Fy\ \&\\&\quad \mathscr{F}(\text{Pred})(y,x)]\}.\end{aligned}$$

But Pred(ξ,η) is functional (Theorem 71), and no natural number follows after itself in the Pred-series (Theorem 145). Thus, writing '$\mathbb{N}\xi$' for '$\mathcal{F}^{=}$(Pred)(0,ξ)' and 'ξ<η' for '$\mathbb{N}\eta$ & $\mathcal{F}$(Pred)(ξ,η)',[32] we have:

$$\exists x(\mathbb{N}x \ \& \ Fx) \rightarrow \exists x[\mathbb{N}x \ \& \ Fx \ \& \ \forall y(y<x \rightarrow \neg Fy)].$$

And that is the least number principle.

Strengthening it slightly, Theorem 359, together with two other theorems, implies that, whatever Qξη and a may be, if the Q-series beginning with a is simple, it is well ordered by the ancestral of Qξη. The first of these other theorems is the previously mentioned Theorem 243, which states that the strong ancestral is connected on the Q-series beginning with a:

$$\text{Func}(Q) \ \& \ \mathcal{F}^{=}(Q)(a,b) \ \& \ \mathcal{F}^{=}(Q)(a,c) \rightarrow$$
$$[\mathcal{F}(Q)(b,c) \vee b=c \vee \mathcal{F}(Q)(c,b)].$$

The second auxiliary result is Theorem 275, which states the strong ancestral is transitive:[33]

$$\mathcal{F}(Q)(x,y) \ \& \ \mathcal{F}(Q)(y,z) \rightarrow \mathcal{F}(Q)(x,z).$$

Now, the Fs are well ordered by $\mathcal{F}$(Q)(ξ,η) just in case (i) $\mathcal{F}$(Q)(ξ,η) linearly orders the Fs[34] and, (ii) whenever some of the Fs are G, there is an $\mathcal{F}$(Q)(ξ,η)-minimal G which is F. But Theorems 243 and 275 together imply that, if the Q-series is simple, the members of the Q-series beginning with a are linearly ordered by $\mathcal{F}$(Q)(ξ,η): for $\mathcal{F}$(Q)(ξ,η) is transitive, by Theorem 275; irreflexive, since the Q-series is simple; and connected on the Q-series beginning with a, by Theorem 243. Moreover, by Theorem 359, every simple series satisfies the second condition. Thus, to repeat: Theorem 359, together with Theorems 243 and 275, implies that, if the Q-series beginning with a is simple, it is well ordered by $\mathcal{F}$(Q)(ξ,η).

There is thus, as was said, a close relationship between the concept of a simple series and that of a well-ordered series. For this reason, of the various characterizations of finitude given in the last hundred years, that most reminiscent of Frege's is one due to Zermelo.[35] Say that the Fs are Zermelo

[32] In general, '$\mathcal{F}$(Pred)(ξ,η)' does *not* mean that ξ is less than η, and it is highly confusing to read it that way, since $\mathcal{F}$(Pred)(∞,∞) (and similarly for any Dedekind infinite cardinal—in fact, '$\mathcal{F}$(Pred)(ξ,ξ)' *is equivalent to* 'ξ is Dedekind infinite'). However, '$\mathcal{F}$(Pred)(ξ,η)' is co-extensional with 'ξ < η', if these relations are restricted to natural numbers.

[33] It is perhaps worth mentioning that these two theorems are the two main theorems of Part III of *Begriffsschrift*, Theorems 133 and 98, respectively. That these two results should become so important at this point is very suggestive.

[34] Qξη linearly orders the Fs if, and only if, Qξη is transitive and irreflexive and connected on the Fs (whenever Fx and Fy, either Qxy or Qyx or x=y).

[35] For a useful discussion of the history of definitions of finitude and infinity, see Charles Parsons, 'Developing Arithmetic in Set Theory without Infinity: Some Historical Remarks',

finite if, and only if, the Fs can be well ordered by a relation whose converse also well-orders them (if the Fs can, as is said, be *doubly well ordered*). Frege's characterization is fairly easily seen to be equivalent to Zermelo's. The proof that, if the Fs are Zermelo finite, they are Frege finite is of no special interest at present.[36] For the other direction, it suffices to show that, if the Q-series beginning with a is simple and ends with b, then it is doubly well ordered by $\mathscr{F}(Q)(\xi,\eta)$.

To simplify the exposition, let us assume the domain and range of $Q\xi\eta$ to contain only members of the Q-series beginning with a. By the foregoing, the Q-series beginning with a is well ordered by $\mathscr{F}(Q)(\xi,\eta)$. Hence, we need only show that the members of the Q-series beginning with a are well ordered by the converse of the ancestral of $Q\xi\eta$, i.e. by $\text{Conv}[\mathscr{F}(Q)](\xi,\eta)$. Now, the converse of $\mathscr{F}(Q)(\xi,\eta)$ is just $\mathscr{F}(\text{Conv}\ Q)(\xi,\eta)$, i.e. the ancestral of the converse of $Q\xi\eta$.[37] It will therefore suffice to show, first, that $\mathscr{F}(\text{Conv}\ Q)(\xi,\eta)$ well-orders the Conv(Q)-series beginning with b and, secondly, that the members of the Conv(Q)-series beginning with b are the members of the Q-series beginning with a. To prove the latter claim, note that $\mathscr{F}^{=}(\text{Conv}\ Q)(b,x)$ iff $\text{Conv}[\mathscr{F}^{=}(Q)](b,x)$ iff $\mathscr{F}^{=}(Q)(x,b)$ iff $\mathscr{F}^{=}(Q)(a,x)$. To prove this last equivalence, note that, if $\mathscr{F}^{=}(Q)(x,b)$, then x is in the domain of $Q\xi\eta$, so $\mathscr{F}^{=}(Q)(a,x)$, by our simplifying assumption; and if $\mathscr{F}^{=}(Q)(a,x)$, then, since $\mathscr{F}^{=}(Q)(a,b)$, either $\mathscr{F}^{=}(Q)(x,b)$ or $\mathscr{F}(Q)(b,x)$, by Theorem 243; but, if $\mathscr{F}(Q)(b,x)$, then the Q-series beginning with a cannot end with b.

To prove the former claim, note first that $\mathscr{F}(\text{Conv}\ Q)(\xi,\eta)$ is certainly a linear order. To show that it satisfies the well-ordering condition, we use Theorem 359 and need only show that the Conv(Q)-series is simple. Plainly, no object follows after itself in the Conv(Q)-series, since $\mathscr{F}(\text{Conv}\ Q)(x,x)$ iff $\text{Conv}[\mathscr{F}(Q)](x,x)$ iff $\mathscr{F}(Q)(x,x)$. That $\text{Conv}(Q)(\xi,\eta)$ is functional follows from the following very general result about the ancestral, whose proof is confined to the notes:[38]

History and Philosophy of Logic, 8 (1987), 201–13. My understanding of Zermelo's work is largely due to this paper.

[36] Suppose the Fs are doubly well ordered by $\xi<\eta$. Consider the relation $Q\xi\eta$ which holds just in case η is a <-minimal object greater than ξ. $Q\xi\eta$ is functional, by linearity. No object follows itself in the Q-series, since, as can be established by induction, if $\mathscr{F}(Q)(x,y)$, then $x<y$, and $\xi<\eta$ is irreflexive. To complete the proof, one must show that all the Fs are Q-between the minimal element, say, a, of the <-ordering and the minimal element, say, b, of the >-ordering (the converse). Consider now the concept: $\forall x[x\leq\xi \rightarrow \text{Btw}(Q;a,\xi)(x)]$; i.e. everything $\leq \xi$ is Q-between a and ξ. Call it $G\xi$. Plainly, a falls under $G\xi$. Since $\xi<\eta$ is a *double* well-ordering, there is a <-maximal object which is G, say, x. If $x\neq b$, then there is some minimal object >x, say, y. Hence, Qxy. But then y falls under $G\xi$, since everything <y is $\leq$x and so Q-between a and x, and so Q-between a and y; and, of course, y itself is near-trivially Q-between a and y. So x = b. So b falls under $G\xi$. So everything $\leq$b is Q-between a and b. So every F is Q-between a and b. Done.

[37] The two directions of this non-trivial result are Frege's Theorems 299 and 302, each of which is proven by (logical) induction.

[38] Suppose that the Q-series is simple, that x and y belong to the Q-series beginning with a,

$$\text{Func}(Q)\ \&\ \neg\exists x.\mathcal{F}(Q)(x,x) \rightarrow \forall x\forall y\forall z[\mathcal{F}^{=}(Q)(a,x)\ \&\ \mathcal{F}^{=}(Q)(a,y)\ \&\ Qxz\ \&\ Qyz \rightarrow x{=}y].$$

That is: if the Q-series is simple, then the converse of $Q\xi\eta$, restricted to members of the Q-series beginning with any given object a, is functional. Since, by our simplifying assumption, the domain and range of $Q\xi\eta$ contain only members of the Q-series beginning with a, the converse of $Q\xi\eta$ is functional.

Frege's characterization of finitude is thus reasonably close to Zermelo's.

8. FREGE AND THE AXIOM OF CHOICE

Frege offered no very complete account of the notions of finitude and infinity, though he did prove a number of important, preliminary results about them. Had Russell's Paradox not disrupted his work, Frege might well have continued his discussion of such matters later. I should like to speculate a bit about what, precisely, he might have discussed and argue that, probably no later than 1892, Frege had formulated at least the axiom of *countable* choice, if not the full axiom of choice, and that he was thinking both about its truth-value and about its epistemological status around 1897.[39]

About 1892, Frege became aware of Dedekind's proof that every concept which is not finite, in Frege's sense, is Dedekind-infinite. He writes, in his review of Cantor's *Zur Lehre vom Transfiniten*:[40]

> Mr. Dedekind gives as the characteristic mark of the infinite that it is similar to a proper part of itself . . . , after which the finite is defined as the non-infinite, whereas Mr. Cantor tries to do what I have done: first to define the finite, after which the infinite appears as the non-finite. Either plan can be carried through correctly, and it can be proved that the infinite systems of Mr. Dedekind are not finite in my sense.[41] This proposition is convertible; but the proof of it is rather difficult. . . .

and that both Qxz and Qyz. By Theorem 243, either $\mathcal{F}(Q)(y,x)$, or $x = y$, or $\mathcal{F}(Q)(x,y)$. Suppose that $\mathcal{F}(Q)(y,x)$. Since Qyz, we have, by the previously mentioned Theorem 242, that $\mathcal{F}^{=}(Q)(z,x)$. But then $\mathcal{F}^{=}(Q)(z,x)$ and Qxz, so $\mathcal{F}(Q)(z,z)$, contrary to the simplicity of the Q-series. Similarly, if $\mathcal{F}(Q)(x,y)$, then $\mathcal{F}(Q)(z,z)$. So $x = y$, and we are done.

[39] It is in the course of such speculations that one most laments the loss of Frege's *Nachlass* and, more specifically, Scholz's decision not to transcribe any of the formal material. This decision was, at the time, understandable, but still regrettable.

[40] Gottlob Frege, 'Review of Georg Cantor, *Zur Lehre vom Transfiniten*', in his *Collected Papers* (ed.) by Brian McGuinness, Oxford, Blackwell, 1984, 178–81, at 180, originally 271. The review was published in 1892, so that sets an upper bound on when Frege encountered *Was sind?* And Frege writes, in the Introduction to *Grundgesetze*, composed in 1893, that the book had 'lately come to [his] notice' (*Gg* I p. vii).

[41] This can be proven in Frege's system in much the same way as in *Was sind?* In fact, it is the final theorem of Part II of *Grundgesetze*, Theorem 484. Indeed, since '$\mathcal{F}(\text{Pred})(\xi,\xi)$' is

This result would, indeed, have been of great interest to Frege. On the one hand, he had shown that a concept is of finite number if, and only if, the objects falling under it can be ordered as a simple series which *ends*; on the other, that a concept is of countably infinite number just in case the objects falling under it can be ordered as a simple series which does *not* end. It would be natural to wonder whether, and perhaps even more natural to *conjecture that*, every concept is either of finite number or has a countably infinite *subconcept*. (This is equivalent to the claim that every infinite concept is Dedekind-infinite.) One might reason, informally, thus. Suppose we set out to build a simple series from objects falling under Fξ, successively choosing distinct Fs and adding them to the series so far constructed. Either we should at some point exhaust the Fs or we should not. If the former, the Fs would have been ordered as a simple series which ends. If the latter, would not *some* of the Fs *eventually* be ordered as a simple series which does not end? This proof would have suggested itself all the more readily to Frege, since it is by this *sort* of method that he establishes his Theorem 428, which states that, if every F is G, and if the Gs are countably infinite, then the Fs are either finite or countably infinite.[42]

The informal proof just given is, of course, hardly one with which Frege would have been prepared to settle. (Here we have exactly the sort of mix of reason and intuition which it was a large part of Frege's purpose to disentangle.) We know that he was not terribly happy with Dedekind's proof either, for what completes the above quotation from the review of Cantor are the words 'and [the proof] is hardly executed with sufficient rigour in Mr. Dedekind's paper'. Frege was right about this, for the theorem cannot be proven without the axiom of (countable) choice, to which Dedekind tacitly appeals.

Dedekind's proof is as follows.[43] Since the number of Fs is not a natural number, there is, for each natural number n, a one–one relation, call it $R_n(\xi,\eta)$, which correlates the numbers between 1 and n with some of the Fs. Now, say that φ(x) just in case, for some n and y, $R_n(y,x)$. Every φ is obviously F, and the number of φs must surely be Endlos.

The problem with this proof certainly does not lie at the first step. It is not difficult to show, in Frege Arithmetic, that[44]

equivalent to 'ξ is Dedekind infinite', Frege's Theorem 145 itself implies that no finite number is Dedekind infinite.

[42] Of course, the proof depends essentially upon the fact that countably infinite sets can be well ordered. Theorem 359 is proved as a central lemma in the proof of Theorem 428.

[43] See 'The Nature and Meaning of Numbers', Theorem 159. Dedekind himself makes reference to a one–one *function*.

[44] The proof is, of course, by induction. For the case n=0, we may take Gξ just to be ξ≠ξ. Suppose then that every G is F, that the number of Gs is n, and that n is a natural number. Then some F is not G, since otherwise the number of Fs would be n, and so finite, contrary to hypothesis. Let a be such an F which is not G. Let Hξ ≡df Gξ ∨ ξ=a. Every H is F, and the number of Hs is n+l. Done.

$$\neg\mathcal{F}^{=}(\text{Pred})(0, \text{Nx:Fx}) \rightarrow \forall n\{\mathcal{F}^{=}(\text{Pred})(0,n) \rightarrow \exists G[\forall x(Gx \rightarrow Fx) \ \& \ n = \text{Nx:Gx}]\}.$$

Given Frege's Theorem 314, mentioned earlier, it will then be easy to complete the first step by proving that

$$\neg\mathcal{F}^{=}(\text{Pred})(0, \text{Nx : Fx}) \rightarrow \forall n\{\mathcal{F}^{=}(\text{Pred})(0,n) \rightarrow \exists R[\text{Func}(R) \ \& \ \text{Func}(\text{Conv } R) \ \& \ \forall x(\text{Btw}(\text{Pred})(1,n)(x) \rightarrow \exists y(Rxy \ \& \ Fy))]\}.$$

The difficulty lies in the next step, and it is Dedekind's terminology which misleads. We can, indeed, prove that, for each n, there is a relation which correlates the numbers between 1 and n with some of the Fs, and we may call this relation '$R\xi\eta$', '$R_n(\xi,\eta)$', 'ξ is greater than η', or whatever we wish. What we may not do is treat the subscript 'n' in '$R_n(\xi,\eta)$'—any more than the 'n' in 'ξ is greater than η'—as if it were a *variable*. More precisely, we may not suppose that 'n' here occupies an *argument-place*, and that is just what Dedekind tacitly supposes: in defining '$\varphi(x)$' to hold just in case, *for some n* and y, $R_n(y,x)$, he treats '$R_n(\xi,\eta)$' as if it were, not a two-place, but a three-place, predicate, 'n' filling the third argument-place; to put it a different way, he supposes that '$R_\zeta(\xi,\eta)$' is a functional expression, denoting a function from natural numbers to binary relations. A tacit, and crucial, inference is thus being made from

$$\forall n\{\mathcal{F}^{=}(\text{Pred})(0,n) \rightarrow \exists R[\text{Func}(R\xi\eta) \ \& \ \text{Func}(\text{Conv } R\xi\eta) \ \& \ \forall x(\text{Btw}(\text{Pred})(1,n)(x) \rightarrow \exists y(Rxy \ \& \ Fy))]\}.$$

to

$$\forall n\{\mathcal{F}^{=}(\text{Pred})(0,n) \rightarrow \exists \Phi[\text{Func}(n\Phi\xi\eta) \ \& \ \text{Func}(\text{Conv } n\Phi\xi\eta) \ \& \ \forall x(\text{Btw}(\text{Pred})(1,n)(x) \rightarrow \exists y(n\Phi xy \ \& \ Fy))]\}.$$

(Here, the variable 'Φ' ranges over ternary relations and is written as it is simply to emphasize that one should think of the ternary relation as if it were a function from objects to binary relations.) Given the validity of this inference, the remainder of Dedekind's proof can be carried out in Frege Arithmetic.[45]

[45] Showing that the number of φs is Endlos is not, as far as I can tell, as easy as one might have thought. The problem is that we need to define a one–one correlation between the φs and the natural numbers. One might just try mapping $\iota x.R_0(1,x)$ to 0, $\iota x.R_1(1,x)$ to 1, $\iota x.R_2(2,x)$ to 2, and so forth. But this may fail to be one–one. Still, the techniques for producing such a

Frege's remark that Dedekind's proof 'is hardly executed with sufficient rigour' strongly suggests that he had attempted to reproduce Dedekind's proof in his formal system and had failed. In formalizing the proof, Frege would quickly have discovered the gap, the inference upon whose validity Dedekind was tacitly relying: its premiss is what one can actually, and easily, prove in second-order arithmetic by following Dedekind's argument; its conclusion, what Dedekind, because of the way he speaks, takes himself to have proven. Now, Frege would probably have thought the inference plausible enough, but he would hardly have been content to let matters rest there. One would have expected him to investigate further, for example, to have attempted to establish the inference as a derived rule within his formal system. And so he did, for what appear to be fragments of this attempt survive in section O(micron) of Part II of *Grundgesetze*. This is a hodgepodge of results of the sort one might prove in attempting to make some progress towards a proof that every infinite set is Dedekind-infinite. The very last theorem of Part II, Theorem 484, is the converse; Theorem 476 states that, if the number of Gs is Dedekind-infinite, and every G is F, then the number of Fs is Dedekind-infinite; and the remainder appear to form part of an attempt to prove Dedekind's theorem in something like the way sketched a few paragraphs back.

In the course of this investigation, Frege would naturally have considered the problematic inference in its general form. This form is obtained by generalizing the specific inference with respect to what is demanded of the relation whose existence is asserted, thus: if, for each natural number n, there is a relation Rξη such that . . . Rξη . . . , then there is some ternary relation (so to speak, some function, from natural numbers to binary relations) which maps each number n to a relation nΦξη such that . . . nΦξη. . . . Formally:

$$\forall n[\mathscr{F}^{=}(\text{Pred})(0,n) \rightarrow \exists R.M_{xy}(Rxy,n)] \rightarrow$$
$$\exists \Phi \forall n[\mathscr{F}^{=}(\text{Pred})(0,n) \rightarrow M_{xy}(n\Phi xy,n)].$$

('M' is a third-order variable indicating relations between binary relations and objects; the subscripts indicate what (first-order) variables it binds.)[46] This is an axiom of countable choice, for binary relations, in Frege Arithmetic.[47] The main reason to suppose Frege formulated (something like) it

mapping were well known to Frege and are employed in his proof of Theorem 428, mentioned previously, which states that if the number of Fs is Endlos and every G is F, then the number of Gs is either finite or Endlos. Indeed, that the number of φs is Endlos can be derived from Theorem 428.

[46] Frege was perfectly familiar with such uses of higher-order variables. See *Gg* I §25.

[47] One can simplify the exposition here a little by considering an attempted proof simpler than Dedekind's. I have not done so in the text, because I have been attempting to reproduce a train of thought which would have been available to anyone reading Dedekind who had

is, to put the point as directly as possible, that one cannot but do so if one investigates Dedekind's proof with the sorts of resources Frege had at his disposal. *That*, one might say, is the sort of thing formalization did for mathematics.

9. CLOSING: THE VALUE OF FORMALIZATION AND FREGE'S CONCEPTION OF LOGICAL TRUTH

The supposition that Frege discovered the axiom of countable choice raises, in a very powerful way, a difficulty for those who would deny him the resources to discuss its truth or its epistemological status, that is, to discuss the question whether it is a law of logic.[48] To understand the sort of problem it raises, it is worth considering a more familiar argument for the claim that Frege regarded the question whether a given thought is a law of logic as intelligible.

Frege set out to derive the basic laws of number within a certain formal system, second-order logic plus Axiom V (it would not have mattered, for present purposes, had he set out to derive them within Frege Arithmetic or within some other formal system). But he was not interested in formalization only for its mathematical benefits (which he rightly thought would be substantial). He, like his philosophical descendants, hoped that formalization would shed light upon epistemological questions. Indeed, Frege

access to Frege's formal theory and understood its workings. Still, it is worth sketching this alternative approach.

One argues that, for each natural number n, there is a *subconcept* $\varphi_n\xi$ of $F\xi$, whose number is n, defines $\varphi\xi$ as $\exists n.\varphi_n\xi$, and then argues that the number of φs is Endlos. But, once again, while we can easily prove

$$\forall n\{\mathscr{F}^{=}(\mathrm{Pred})(0,n) \rightarrow \exists G[\forall x(Gx \rightarrow Fx)\ \&\ n = Nx{:}Gx]\}$$

we have no way to infer

$$\exists R\forall n\{\mathscr{F}^{=}(\mathrm{Pred})(0,n) \rightarrow [\forall x(nRx \rightarrow Fx)\ \&\ n = Nx{:}nRx]\}$$

without an axiom of countable choice. What we need, this time, is to be able, quite generally, to infer

$$\forall n[\mathscr{F}^{=}(\mathrm{Pred})(0,n) \rightarrow \exists G.M_x(Gx,n)] \rightarrow \exists R\forall n[\mathscr{F}^{=}(\mathrm{Pred})(0,n) \rightarrow M_x(nRx,n)]$$

And this is an axiom of countable choice, for concepts.

Interestingly enough, it would appear that showing that the number of φs is Endlos is *still* going to require an appeal to countable choice for binary relations. It may be for this reason that Dedekind proves the theorem as he does, given that the argument just mentioned seems in many ways more intuitive than Dedekind's own.

[48] Perhaps the best-known exponent of this view is Thomas Ricketts. See e.g. his 'Frege, the *Tractatus*, and the Logocentric Predicament', *Noûs*, 19 (1985), 3–16; others are Burton Dreben, Warren Goldfarb, and Joan Weiner. Its first published exponents were probably Jan van Heijenoort and Jaakko Hintikka. For an important critical discussion of Ricketts's views in particular, see Jason Stanley, 'Truth and Meta-theory in Frege', *Pacific Philosophical Quarterly* (1996), 172–95.

is admirably clear about the relation between formalization and epistemology:[49]

> I became aware of the need for a conceptual notation when I was looking for the fundamental principles or axioms upon which the whole of mathematics rests. Only *after* this question is answered can it be hoped to trace successfully the springs of knowledge on which this science thrives. (my emphasis)

The epistemological status of arithmetic is not decided by formalization alone; it is only *after* its axioms have been isolated that the question of arithmetic's epistemological status becomes tractable. But if logicism is not established by a 'reduction' of arithmetic to the formal system of *Grundgesetze*, or Frege Arithmetic, or any other formal system, it simply must be an intelligible question whether the axioms of that system are logical laws. What other question could remain at that point?[50]

This sort of argument concentrates upon Frege's attitude towards those thoughts he accepts as axioms of logic, those rules of inference he accepts as logically valid. I myself regard it as conclusive, but it nevertheless seems not to have carried conviction. A more powerful argument emerges, though, if we concentrate upon Frege's attitude toward a different sort of thought, a good example of which is the axiom of countable choice. In so far as one sets out to derive all the truths of some 'branch of learning' from a fixed set of axioms, one is immediately confronted with the possibility that one's axioms might not be 'complete',[51] might not suffice: there may be apparent truths, propositions of whose truth we had previously been convinced by informal argument, which cannot be proven within the system as it stands. More interestingly, there may be a certain *kind* of inference which, though commonly made, cannot be replicated within the system. Frege writes:[52]

> By [resolving inferences into their simple components] we shall arrive at just a few modes of inference, with which we must then attempt to make do at all times. And if at some point this attempt fails, then we shall have to ask whether we have hit upon a truth issuing from a non-logical source of cognition, whether a new mode of inference has to be acknowledged, or whether perhaps the intended step ought not to be taken at all.

[49] Gottlob Frege, 'On Mr. Peano's Conceptual Notation and My Own', in *Collected Papers*, 234–48, at 235, op. 362.

[50] Famously, the only principle to which Frege appealed about whose status as a law of logic Frege had any doubt was Axiom V. See *Gg* I p. vii. In my opinion, he probably had *no* real doubt that it was *true*. On this, see my 'Frege and Semantics', forthcoming in T. Ricketts (ed.), *The Cambridge Companion to Frege*.

[51] This is a term which Frege uses in this context (see 'Peano', 235, op. 362), though not quite in its modern sense.

I am not, it is worth emphasizing, suggesting that Frege anticipated Gödel. There is no indication that Frege ever considered the possibility that there might be *no* complete axiomatization of arithmetic; it is this, of course, which makes Gödel's result so important. If the incompleteness of *Principia Mathematica* could have been remedied, its incompleteness would have been uninteresting.

[52] 'Peano', 235, op. 363.

Now, it is of course possible that Frege is here speaking completely hypothetically. But, I suggest, he may well be speaking from experience; he may have in mind something like the version of countable choice mentioned above. In any event, the interest of Frege's discussion does not depend upon the correctness of this speculation,[53] which merely serves to make vivid the problem that concerns him, which is one about the epistemological status of such principles. Frege is saying, quite reasonably, that the question whether e.g. the axiom of countable choice is true at all, and if so, whether it is a logical law, is not only intelligible but *important*.

Unable to formalize Dedekind's proof in the formal theory of *Grundgesetze*, or in Frege Arithmetic, and having identified the proposition whose truth it assumes, we have, according to Frege, essentially three options: first, we may reject the proposition in question, as not being true, and reject the proof. If, instead, we accept the proposition's truth, we have two further options: to accept it as a truth peculiar to some special science, say, a part of mathematics not reducible to logic (a part similar, in that respect, to geometry); or to accept it as a truth of logic and to acknowledge some new logical axiom from which it follows or some rule of inference by means of which it can be proven. The presence of the first option shows that Frege did not think of himself merely as formalizing *accepted* mathematical practice: his project had, in his view, potentially revisionary consequences. The presence of the other two options shows that Frege regarded it as an intelligible question whether e.g. the axiom of countable choice was a law of logic. I thus regard it as demonstrable that Frege believed such questions to be intelligible.

Whatever the larger import of this point, however, I make it in the present context for purposes of illustration. It is sad enough that a work as great as *Grundgesetze der Arithmetik*, a work fully worthy of comparison with Dedekind's classic *Was sind und was sollen die Zahlen*?, should have been ignored for one hundred years. Our failure to study it is all the more regrettable, however, if an understanding of *Grundgesetze* promises a non-negligible improvement of our understanding of Frege's philosophy more generally. I admit, indeed, I insist that the offered interpretation of Frege's attitude towards proposed new axioms does not depend upon the speculation inspired by our discussion of *Grundgesetze*. But that does not imply that our understanding of Frege is not improved by an understanding of the kinds of mathematical problems with which he grappled, the sorts of solutions upon which he settled, and the philosophical significance he took them to have.[54]

[53] It is worth noting, though, that it is demonstrable that Frege believed there were arithmetical truths which were unprovable within his system as it stood: see George Boolos and Richard G. Heck, Jnr., '*Die Grundlagen der Arithmetick* §§82–3', elsewhere in this volume.

[54] I should like to thank George Boolos, Burton Dreben, Warren Goldfarb, Mathieu Marion,

APPENDIX 1. FREGE'S PROOF OF THEOREM 288, AND REFLECTIONS ON THE LEAST NUMBER PRINCIPLE

As was said above, the details of Frege's proof of Theorem 288 are not relevant to the central concerns of the present paper. But, as the proof is somewhat difficult to follow, and as certain features of it are interesting for other reasons, it is worth explaining here. I shall discuss only those theorems which Frege proves in the relevant sections of *Grundgesetze*, omitting the proofs of any and all theorems proven earlier in the book, i.e. all theorems earlier than Theorem 264.[55]

Before we begin, we need one additional definition:

$$\mathscr{F}^{\wedge}(T;\langle a,x\rangle)(b,y) \equiv \mathrm{df}\, \mathscr{F}^{=}(T)(\langle a,x\rangle,\langle b,y\rangle).$$

Thus, $\mathscr{F}^{\wedge}(Q\pi R)(\langle a,x\rangle)(\xi,\eta)$ is a relation, defined by induction, according to Frege's method: it is the relation which holds between ξ and η just in case $\langle\xi,\eta\rangle$ belongs to the $(Q\pi R)$-series beginning with $\langle a,x\rangle$.

Recall that Theorem 288 is:

$$\begin{aligned}&\mathrm{Func}(Q)\ \&\ \neg\exists z.\mathscr{F}(Q)(z,z)\ \&\ \mathscr{F}^{=}(Q\ \pi\ \mathrm{Pred})(\langle x, 1\rangle,\langle y,n\rangle)\rightarrow\\ &\quad \mathrm{Nz{:}Btw}(Q;x,y)(z)=\mathrm{Nz{:}Btw}(\mathrm{Pred})(1,n)(z).\end{aligned}$$

That is: If the Q-series is simple and if $\langle y,n\rangle$ is a member of the $(Q\ \pi\ \mathrm{Pred})$-series beginning with $\langle x,1\rangle$, then the number of objects Q-between x and y is the same as the number of objects Pred-between 1 and n.

The proof of Theorem 288 requires two lemmas. The first of these is Theorem 287, whose proof is given in the notes:[56]

$$\neg\exists z[\mathscr{F}^{=}(\mathrm{Pred})(1,z)\ \&\ \mathscr{F}(\mathrm{Pred})(z,z)].$$

The second lemma is Theorem 284:

$$\begin{aligned}&\mathrm{Func}(R)\ \&\ \neg\exists z[\mathscr{F}^{=}(R)(m,z)\ \&\ \mathscr{F}(R)(z,z)]\ \&\ \mathrm{Func}(Q)\ \&\\ &\quad\neg\exists z[\mathscr{F}^{=}(Q)(x,z)\ \&\ \mathscr{F}(Q)(z,z)]\ \&\ \mathscr{F}^{=}(Q\pi R)(\langle x,m\rangle,\langle y,n\rangle)\rightarrow\\ &\quad\mathrm{Nz{:}Btw}(Q;x,y)(z)=\mathrm{Nz{:}Btw}(R;m,n)(z).\end{aligned}$$

Charles Parsons, Bill Tait, and Jason Stanley for discussions which have contributed greatly to this paper. I should also like to thank the members of the conference 'Philosophy of Mathematics Today' whose proceedings this volume is. The paper I read there was not this one but a hybrid of 'The Development of Arithmetic' and 'Definition by Induction'; nevertheless, the discussion which followed—indeed, all our discussions at the conference—had great effect on my thought about these and other issues. Many thanks are due to Matthias for organizing the conference.

This paper is dedicated to the participants in a seminar, informally known as 'Frege's Three Books', which George Boolos and I taught together in the Spring of 1993. Regular members, other than myself and George, were Emily Carson, Janet Folina, Michael Glanzberg, Delia Graff, David Hunter, Darryl Jung, Josep Macia-Fabrega, Ofra Rechter, Lisa Sereno, and Jason Stanley. I shall be a happy man if, one more time in my life, I have the opportunity to teach a seminar which is half as much fun as that one was.

[55] Many of these are discussed in 'Definition by Induction'.

[56] If $\mathscr{F}^{=}(\mathrm{Pred})(1,n)$, then, since $\mathrm{Pred}(0,1)$, we have $\mathscr{F}^{=}(\mathrm{Pred})(0,n)$, by Theorem 285. But then $\neg\mathscr{F}(\mathrm{Pred})(n,n)$, by Theorem 145.

That is: if both the R-series beginning with m and the Q-series beginning with x are simple, and if $\langle y,n\rangle$ is a member of the $(Q\pi R)$-series beginning with $\langle x,m\rangle$, then the number of objects Q-between x and y is the same as the number of objects R-between m and n.

Theorem 288 follows quickly from these two lemmas. For, taking $R\xi\eta$ as $\mathrm{Pred}(\xi,\eta)$ and m as l, the following is a substitution instance of Theorem 284:

$$\begin{aligned}&\mathrm{Func}(\mathrm{Pred})\ \&\ \neg\exists z\big[\mathcal{F}^{=}(\mathrm{Pred})(1,z)\ \&\ \mathcal{F}(\mathrm{Pred})(z,z)\big]\ \&\ \mathrm{Func}(Q)\ \&\\ &\neg\exists z[\mathcal{F}^{=}(Q)(x,z)\ \&\ \mathcal{F}(Q)(z,z)]\ \&\ \mathcal{F}^{=}(Q\,\pi\,\mathrm{Pred})(\langle x,1\rangle,\langle y,n\rangle)\rightarrow\\ &Nz{:}\mathrm{Btw}(Q;x,y)(z)=Nz{:}\mathrm{Btw}(\mathrm{Pred};1,n)(z).\end{aligned}$$

$\mathrm{Pred}(\xi,\eta)$ is functional, by Theorem 71, and the second conjunct is Theorem 287. Moreover, if $\neg\exists z.\mathcal{F}(Q)(z,z)$, then certainly $\neg\exists z[\mathcal{F}^{=}(Q)(x,z)\ \&\ \mathcal{F}(Q)(z,z)]$. So we are done.

We turn then to the proof of Theorem 284, which is derived from Theorem 283:

$$\begin{aligned}&\mathrm{Func}(R)\ \&\ \neg\exists z\big[\mathcal{F}^{=}(R)(m,z)\ \&\ \mathcal{F}(R)(z,z)\big]\ \&\ \mathrm{Func}(Q)\ \&\\ &\neg\exists z[\mathcal{F}^{=}(Q)(x,z)\ \&\ \mathcal{F}(Q)(z,z)]\ \&\ \mathcal{F}^{=}(Q\pi R)(\langle x,m\rangle,\langle y,n\rangle)\rightarrow\\ &\mathrm{Map}\big[\mathcal{F}^{\wedge}(Q\pi R,\langle x,m\rangle)\big][\mathrm{Btw}(Q;x,y)(\xi),\ \mathrm{Btw}(R;m,n)(\xi)].\end{aligned}$$

That is: if the R-series beginning with m and the Q-series beginning with x are simple, and if $\langle y,n\rangle$ is a member of the $(Q\pi R)$-series beginning with $\langle x,m\rangle$, then $\mathcal{F}^{\wedge}(Q\pi R,\langle x,m\rangle)(\xi,\eta)$ maps the objects Q-between x and y into the objects R-between m and n.

The method by which Theorem 284 is derived from Theorem 283 is very common in *Grundgesetze*. Theorem 283 asserts that, under certain conditions, the relation $\mathcal{F}^{\wedge}(Q\pi R,\langle x,m\rangle)(\xi,\eta)$ will map the objects Q-between x any y into the objects R-between m and n. By exchanging 'R' and 'Q', 'x', and 'm', and 'y' and 'n' in Theorem 283, one can show that, under the same conditions,[57] $\mathcal{F}^{\wedge}(R\pi Q,\langle m,x\rangle)(\xi,\eta)$ will map the objects R between m and n into those Q-between x and y. But by Frege's Theorem 259, $\mathcal{F}^{\wedge}(Q\pi R,\langle x,m\rangle)(\xi,\eta)$ is the converse of $\mathcal{F}^{\wedge}(R\pi Q,\langle m,x\rangle)](\xi,\eta)$, so it follows that $\mathcal{F}^{\wedge}(R\pi Q,\langle m,x\rangle)](\xi,\eta)$ correlates the objects Q-between x and y one-to-one with those R-between m and n, whence the numbers of these concepts are the same, by Hume's Principle.

Most of the work in the proof of Theorem 288 thus lies in the proof of Theorem 283. To prove Theorem 283, suppose its antecedent. To show that $\mathcal{F}^{\wedge}(Q\pi R,\langle x,m\rangle)(\xi,\eta)$ maps the objects Q-between x and y into the objects R-between m and n, we must show (i) that $\mathcal{F}^{\wedge}(Q\pi R,\langle x,m\rangle)(\xi,\eta)$ is functional and (ii) that, if z is Q-between x and y, there is an object R-between m and n onto which $\mathcal{F}^{\wedge}(Q\pi R,\langle x,m\rangle)(\xi,\eta)$ maps it. Now, (i) follows immediately from Frege's Theorem 253:

$$\begin{aligned}&\mathrm{Func}(R)\ \&\ \mathrm{Func}(Q)\ \&\ \neg\exists w\big[\mathcal{F}^{=}(Q)(m,w)\ \&\ \mathcal{F}(Q)(w,w)\big]\rightarrow\\ &\mathrm{Func}\big[\mathcal{F}^{\wedge}(Q\pi R,\langle x,m\rangle)(\xi,\eta)\big]\end{aligned}$$

[57] This follows from Frege's Theorem 258: $\mathcal{F}^{=}(Q\pi R)(\langle x,m\rangle,\langle y,n\rangle)\rightarrow\mathcal{F}^{=}(R\pi Q)(\langle m,x\rangle,\langle n,y\rangle)$. In most such proofs, the conditions are not the same but merely corresponding.

And (ii) is the consequent of Frege's Theorem 277, the first of two important lemmas:

$$\begin{aligned}&\mathrm{Func}(R)\ \&\ \mathscr{F}^{=}(Q\pi R,\langle x,m\rangle,\langle y,n\rangle)\ \&\ \neg\mathscr{F}(R)(n,n)\ \&\\ &\quad \exists w.\mathscr{F}^{\wedge}(Q\pi R,\langle x,m\rangle)(z,w) \to \{\mathrm{Btw}(Q;x,y)(z) \to\\ &\quad \exists w[\mathscr{F}^{\wedge}(Q\pi R,\langle x,m\rangle)(z,w)\ \&\ \mathrm{Btw}(R;m,n)(w)]\}\end{aligned}$$

That is: if $R\xi\eta$ is functional, if $\langle y,n\rangle$ belongs to the $(Q\pi R)$-series beginning with $\langle x,m\rangle$, if n does not follow itself in the R-series, and if z is in the domain of $\mathscr{F}^{\wedge}(Q\pi R,\langle x,m\rangle)(\xi,\eta)$, then, if z is Q-between x and y, there is an object R-between m and n onto which $\mathscr{F}^{\wedge}(Q\pi R,\langle x,m\rangle)(\xi,\eta)$ maps it. It will suffice to establish, on the assumption that the antecedent of Theorem 283 holds, the antecedent of

$$\begin{aligned}&\mathrm{Func}(R)\ \&\ \mathscr{F}^{=}(Q\pi R,\langle x,m\rangle,\langle y,n\rangle)\ \&\ \neg\mathscr{F}(R)(n,n)\ \&\\ &\quad \{\mathrm{Btw}(Q;x,y)(z) \to \exists w.\mathscr{F}^{\wedge}(Q\pi R,\langle x,m\rangle)(z,w)\} \to\\ &\quad \{\mathrm{Btw}(Q;x,y)(z) \to \exists w[\mathscr{F}^{\wedge}(Q\pi R,\langle x,m\rangle)(z,w)\ \&\ \mathrm{Btw}(R;m,n)(w)]\}\end{aligned}$$

which is equivalent to Theorem 277. We have assumed the first two conjuncts. Moreover, since $\mathscr{F}^{=}(Q\pi R)(\langle x,m\rangle,\langle y,n\rangle)$, $\mathscr{F}^{=}(R)(m,n)$, by Frege's Theorem 232; so, since no member of the R-series beginning with m follows after itself, n does not. The fourth conjunct is, in turn, the consequent of the second important lemma, Frege's Theorem 283, ζ:

$$\mathscr{F}^{=}(Q\pi R)(\langle x,m\rangle,\langle y,n\rangle) \to \{\mathrm{Btw}(Q;x,y)(z) \to \exists w.\mathscr{F}^{\wedge}(Q\pi R,\langle x,m\rangle)(z,w)\}$$

That is: if $\langle y,n\rangle$ belongs to the $(Q\pi R)$-series beginning with $\langle x,m\rangle$ and if z is Q-between x and y, then z is in the domain of $\mathscr{F}^{\wedge}(Q\pi R,\langle x,m\rangle)(\xi,\eta)$. (Since 'n' appears only in the antecedent, which is equivalent to '$\mathscr{F}^{\wedge}(Q\pi R,\langle x,m\rangle)(y,n)$', we may also say: if y is in the domain of $\mathscr{F}^{\wedge}(Q\pi R,\langle x,m\rangle)(\xi,\eta)$, then everything Q-between x and y is also in its domain.) But we have also supposed its antecedent, so Theorem 283 is proven.

We have now to prove Theorems 277 and 283, ζ. We begin with Theorem 277. Suppose that $R\xi\eta$ is functional, that $\langle y,n\rangle$ belongs to the $(Q\pi R)$-series beginning with $\langle x,m\rangle$, that z is in the domain of $\mathscr{F}^{\wedge}(Q\pi R,\langle x,m\rangle)(\xi,\eta)$—say, $\mathscr{F}^{\wedge}(Q\pi R,\langle x,m\rangle)(z,w)$—that n does not follow itself in the R-series, and that z is Q-between x and y. We must show that $\mathscr{F}^{\wedge}(Q\pi R,\langle x,m\rangle)(\xi,\eta)$ maps z to an object R-between m and n. Indeed, w itself is R-between m and n. For $\mathscr{F}^{=}(Q\pi R)(\langle x,m\rangle,\langle z,w\rangle)$, whence, by Theorem 232, $\mathscr{F}^{=}(R)(m,w)$. Thus, Func(R), $\neg\mathscr{F}(R)(n,n)$, and $\mathscr{F}^{=}(R)(m,w)$. So, if $\mathscr{F}^{=}(R)(w,n)$, then w is between m and n in the R-series and we are done. Therefore, by Theorem 232, it will suffice to show that $\mathscr{F}^{=}(Q\pi R)(\langle z,w\rangle,\langle y,n\rangle)$.

To show this, we employ Theorem 243, an instance of which is:

$$\begin{aligned}&\mathrm{Func}(Q\pi R)\ \&\ \mathscr{F}^{=}(Q\pi R)(\langle x,m\rangle,\langle y,n\rangle)\ \&\ \mathscr{F}^{=}(Q\pi R)(\langle x,m\rangle,\langle z,w\rangle) \to\\ &\quad \mathscr{F}^{=}(Q\pi R)(\langle z,w\rangle,\langle y,n\rangle) \vee \mathscr{F}(Q\pi R)(\langle y,n\rangle,\langle z,w\rangle)\end{aligned}$$

Now, we know that $\mathscr{F}^{=}(Q\pi R)(\langle x,m\rangle,\langle y,n\rangle)$ and $\mathscr{F}^{=}(Q\pi R)(\langle x,m\rangle,\langle z,w\rangle)$. But we also know that Func(R); and, since Btw(Q;x,y)(d), Func(Q), by definition. But, then, Frege's Theorem 252 is:

$$\text{Func}(Q)\ \&\ \text{Func}(R) \rightarrow \text{Func}(Q\pi R)$$

So, either $\mathscr{F}^{=}(Q\pi R)(\langle z,w\rangle,\langle y,n\rangle)$ or $\mathscr{F}(Q\pi R)(\langle y,n\rangle,\langle z,w\rangle)$. But if the latter, $\mathscr{F}(Q)(y,z)$, contradicting the assumption that z is Q-between x and y—since then $\mathscr{F}(Q)(y,z)$ and $\mathscr{F}^{=}(Q)(z,y)$, so $\mathscr{F}(Q)(y,y)$—and we are done.

To complete the proof of Theorem 283, and so of Theorem 288, we need now only prove Theorem 283, ζ, which is:

$$\mathscr{F}^{=}(Q\pi R)(\langle x,m\rangle,\langle y,n\rangle) \rightarrow \\ \{\text{Btw}(Q;x,y)(z) \rightarrow \exists w.\mathscr{F}^{\wedge}(Q\pi R,\langle x,m\rangle)(z,w)\}$$

The proof is by induction, the induction justified by Frege's Theorem 257, which is:[58]

$$\mathscr{F}^{=}(Q\pi R)(\langle x,m\rangle,\langle y,n\rangle)\ \&\ Fxm\ \& \\ \forall t\forall s\{Fts \rightarrow \forall n\forall v[Qtu\ \&\ Rsv \rightarrow Fuv]\} \rightarrow Fyn$$

For the proof, we take Fξη to be:

$$\mathscr{F}^{=}(Q\pi R)(\langle x,m\rangle,\langle \xi,\eta\rangle)\ \&\ \forall z[\text{Btw}(Q;x,\xi)(z) \rightarrow \exists w.\mathscr{F}^{\wedge}(Q\pi R,\langle x,m\rangle)(z,w)]$$

Let us call this 'the relevant relation'. To prove Theorem 283, ξ, we must therefore prove the base case, Theorem 283, ε:

$$\mathscr{F}^{=}(Q\pi R)(\langle x,m\rangle,\langle x,m\rangle)\ \&\ \forall z[\text{Btw}(Q;x,x)(z) \rightarrow \exists w.\mathscr{F}^{\wedge}(Q\pi R,\langle x,m\rangle)(z,w)]$$

and the induction step, Theorem 281, ρ:

$$\forall t\forall s\{\mathscr{F}^{=}(Q\pi R)(\langle x,m\rangle,\langle t,s\rangle)\ \& \\ \forall z[\text{Btw}(Q;x,t)(z) \rightarrow \exists w.\mathscr{F}^{\wedge}(Q\pi R,\langle x,m\rangle)(z,w)] \rightarrow \\ \forall u\forall v[Qtu\ \&\ Rsv \rightarrow \mathscr{F}^{=}(Q\pi R)(\langle x,m\rangle,\langle u,v\rangle)\ \& \\ \forall z[\text{Btw}(Q;x,u)(z) \rightarrow \exists w.\mathscr{F}^{\wedge}(Q\pi R,\langle x,m\rangle)(z,w)]]\}$$

Since only x is Q-between x and x, Theorem 283, ε, follows from the trivial fact that $\mathscr{F}^{=}(Q\pi R)(\langle x,m\rangle,\langle x,m\rangle)$ and so $\mathscr{F}^{\wedge}(Q\pi R, \langle x,m\rangle)$ (x,m).

For the proof of Theorem 281, ρ, suppose that ⟨t,s⟩ belongs to the (QπR)-series beginning with ⟨x,m⟩ and that everything Q-between x and t is in the domain of $\mathscr{F}^{\wedge}(Q\pi R,\langle x,m\rangle)(\xi,\eta)$; suppose further than Qtu and Rsv. We then have Theorem 281, η:

$$\mathscr{F}^{=}(Q)(x,t)\ \&\ Qtu \rightarrow [\text{Btw}(Q;x,u)(z) \rightarrow \text{Btw}(Q;x,t)(z) \vee z=u]$$

So, everything Q-between x and u is either Q-between x and t or is equal to u. But, by hypothesis, everything Q-between x and t is in the domain of

[58] For a discussion of the significance of Theorem 257, see 'Definition by Induction'.

$\mathcal{F}^\wedge(Q\pi R,\langle x,m\rangle)(\xi,\eta)$. Now, $\mathcal{F}^=(Q\pi R)(\langle x,m\rangle,\langle t,s\rangle)$; but since Qtu and Rsv, we have $(Q\pi R)(\langle t,s\rangle,\langle u,v\rangle)$, by definition. So $\mathcal{F}^=(Q\pi R)(\langle x,m\rangle,\langle u,v\rangle)$, whence u is in the domain of $\mathcal{F}^\wedge(Q\pi R,\langle x,m\rangle)(\xi,\eta)$, and we are done.

The proof of Theorem 281, η, is not particularly difficult.[59] So that completes our discussion of Frege's proof of Theorem 288. I hope it is now clear that, when I said that the proof of Theorem 321 given in §2 is simpler, and shorter, than Frege's, I spoke truly.

It is worth discussing Theorem 281, η, a bit further, though, because it is rather more important than it might at first appear. It is one of two results which Frege mentions, in the Foreword to Volume II of *Grundgesetze*, as results he intends to use but which had not previously been marked for later use. In fact, he employs Theorem 281, η, frequently and at important points in the proofs in Volume II. The reason for this is that Theorem 281, η, has the force of an *induction* principle, one very closely related to (Frege's logicized version of) the least number principle. More precisely, suppose that the Q-series beginning with a is simple, and that, whenever everything Q-between a and some member, b, of the Q-series beginning with a, is F, b is itself F. Then Theorem 281, η, very easily implies that every member of the Q-series beginning with a is F. That is, Theorem 281, η, implies what one might call *induction for betweenness*:

$$\text{(IB)}\ \mathcal{F}^=(Q)(a,b)\ \&\ \text{Func}(Q)\ \&\ \neg\exists z[\mathcal{F}^=(Q)(a,z)\ \&\ \mathcal{F}(Q)(z,z)]\ \&\\ \forall x\{\mathcal{F}^=(Q)(a,x)\ \&\ \forall z[\text{Btw}(Q;a,x)(z)\ \&\ x\neq z\rightarrow Fz]\rightarrow Fx\}\rightarrow Fb$$

Frege does not prove this theorem. His logicized version of the least number principle is his Theorem 359, which (eliminating a couple definitions, again, and strengthening slightly) is:[60]

$$\text{Func}(Q)\ \&\neg\exists z[\mathcal{F}^=(Q)(a,z)\ \&\ \mathcal{F}(Q)(z,z)]\ \&\ \forall x\{\mathcal{F}^=(Q)(a,x)\ \&\\ \forall z[\mathcal{F}^=(Q)(a,z)\ \&\ \mathcal{F}(Q)(z,x)\rightarrow\neg Fz]\rightarrow\neg Fx\}\rightarrow\forall x[\mathcal{F}^=(Q)(a,x)\rightarrow\neg Fx]$$

Now, if we substitute '¬Fξ' for 'Fξ', we have:

$$\text{Func}(Q)\ \&\neg\exists z[\mathcal{F}^=(Q)(a,z)\ \&\ \mathcal{F}(Q)(z,z)]\ \&\ \forall x\{\mathcal{F}^=(Q)(a,x)\ \&\\ \forall z[\mathcal{F}^=(Q)(a,z)\ \&\ \mathcal{F}(Q)(z,x)\rightarrow Fz]\rightarrow Fx\}\rightarrow\forall x[\mathcal{F}^=(Q)(a,x)\rightarrow Fx]$$

This is easily seen to be equivalent to induction for betweenness. It differs from induction for betweenness only in that it has, as its third conjunct:

$$\forall x\{\mathcal{F}^=(Q)(a,x)\ \&\ \forall z[\mathcal{F}^=(Q)(a,x)\ \&\mathcal{F}(Q)(z,x)\rightarrow Fz]\rightarrow Fx\}$$

[59] Suppose that $\mathcal{F}^=(Q)(x,t)$, that Qtu, and that Btw(Q;x,u)(z). Then Func(Q), $\neg\mathcal{F}(Q)(u,u)$, $\mathcal{F}^=(Q)(x,z)$, and $\mathcal{F}^=(Q)(z,u)$. Suppose then that z ≠ u. To show that Btw(Q;x,t)(z), we need only show that $\mathcal{F}^=(Q)(z,t)$. Since $\mathcal{F}^=(Q)(x,z)$ and $\mathcal{F}^=(Q)(x,t)$, either $\mathcal{F}^=(Q)(z,t)$ or $\mathcal{F}(Q)(t,z)$, by Theorem 243. And since Func(Q), $\mathcal{F}(Q)(t,z)$, and Qtu, $\mathcal{F}^=(Q)(u,z)$, by Theorem 242. Now, u ≠ z, so $\mathcal{F}(Q)(u,z)$. But then $\mathcal{F}(Q)(u,u)$, since $\mathcal{F}^=(Q)(z,u)$. Contradiction.

[60] Frege's proof of Theorem 359 is, of course, different from that given here of induction for betweenness, but it is strikingly similar.

whereas, in induction for betweenness, we have:

$$\forall x\{\mathcal{F}^{=}(Q)(a,x)\ \&\ \forall z[Btw(Q;a,x)(z)\ \&\ x\neq z \rightarrow Fz] \rightarrow Fx\}$$

But, in the presence of the other conjuncts of the antecedent, these are equivalent. Indeed, the proof of Theorem 281, η, just is a demonstration that they are.

APPENDIX 2. TREE OF IMPORTANT THEOREMS IN FREGE'S PROOF OF THEOREM 327

The following is a graphical representation of the dependencies among various results used by Frege in his proof of Theorem 327. I have omitted references to

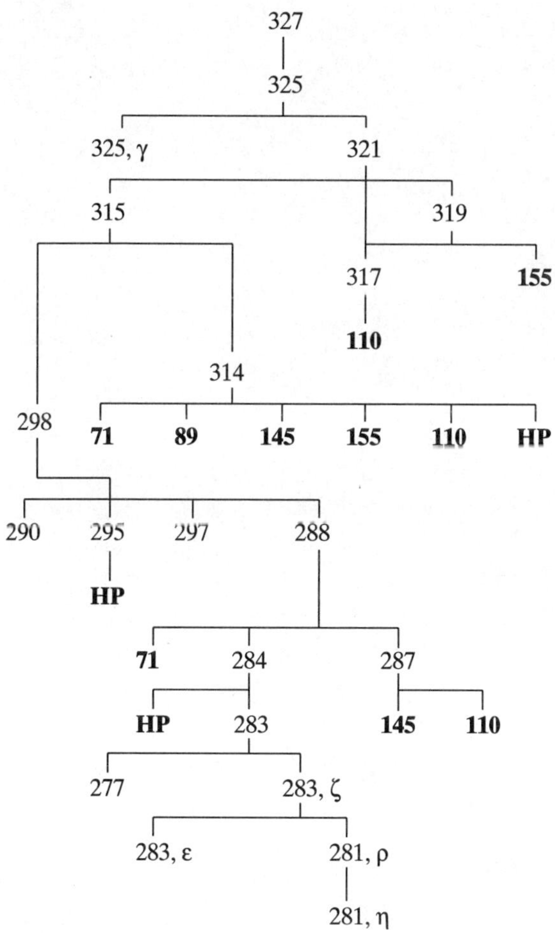

general results about the ancestral, uses of definitions, and theorems which are immediate consequences thereof. I have also omitted reference to theorems proven in sections A-I of *Grundgesetze*, except for those to specifically arithmetical truths and to Hume's Principle, all of these being marked in boldface.[61]

[61] The arithmetical theorems in question are: (71) Func(Pred); (89) Func(Conv Pred); (110) Pred(0,l); (145) $\mathscr{F}^{=}$(Pred)(0,x) → ¬$\mathscr{F}$(Pred)(x,x); (155) $\mathscr{F}^{=}$(Pred)(0,n) → Pred[n,Nx:$\mathscr{F}^{=}$(Pred)(x,n)].

V. Sets, Structure, and Abstraction

18

Zermelo's Conception of Set Theory and Reflection Principles

W. W. TAIT

1. ZERMELO 1930

Zermelo (1930) is concerned with impredicative second-order set theory. He treats the general case of set theory with urelements, but it will be enough to consider only the case of pure set theory, i.e. without urelements. In this context, Zermelo's theory is the axiomatic second-order theory T_2 in the language of pure set theory whose axioms are *Extensionality*, *Regularity*, *Null Set*, *Unordered Pairs*, *Union*, *Power Set*, *Second-order Separation*, *Second-order Replacement*, and *Choice*.[1] Thus T_2 with the Axiom of Infinity added is the full system T_3 of Morse–Kelley (also sometimes called Bernays–Morse) set theory with the Axiom of Choice, which Zermelo regards as a principle of logic. Since the theory is second-order and we are concerned with genuine models of the second-order theory, which Zermelo calls *normal domains*, and not with arbitrary Henkin models, the power-set operation is absolute and the von Neumann ordinals in a model are well ordered by $\in$. So with each model M is associated an ordinal $o(M)$, called the *ordinal of* M, which is the order type of its set of von Neuman ordinals; and a model M is characterized to within isomorphism by $o(M)$. More generally, the substructures of any two models M and N consisting of the sets

This is a modified version of the talk that I gave at the conference 'Philosophy of Mathematics Today'.

[1] Second-order Separation is a consequence of the other axioms, but it will be useful to retain it as an axiom. There is some question about what Zermelo intended. In fn. 1 of (1930), Zermelo asserts that the Axioms of Separation, yielding $\{x \in t \mid F(x)\}$, and Replacement, yielding $\{G(x) \mid x \in t\}$, for any set t, are to hold for arbitrary propositional function F and function G. But then he goes on to refer to the debate over the notion of 'definiteness' and to Zermelo (1929) in which he characterizes the definite functions in effect as those definable by second-order formulas of set theory. If he intended his arbitrary F and G to be restricted in this way to those which are definite, then there are Henkin models of T_2 (as John Burgess pointed out to me), but not genuine models. In this case, his arguments for the results summarized below are invalid. So I will assume that he was referring to *arbitrary* F and G in genuine models and that the reference to Zermelo (1929) was simply a mistake. For example, how could an isomorphism from a model M to a model N with $o(M) = o(N)$ be defined when Replacement in M is restricted to functions that are second-order definable on M?

of rank $< \alpha$ are isomorphic provided α is no greater than the ordinals of either M or N. Again, since we are concerned with genuine models M of the second-order theory, $o(M)$ is a regular strong-limit ordinal—a *Grenzzahl*, as Zermelo calls it. In other words, it is either ω or a (strongly) inaccessible cardinal.

Concerning the question of the existence of models of T_2, Zermelo appeals to criteria set by Hilbert: if an extension T of T_2 is consistent and categorical (i.e. all of its models are isomorphic), then we are entitled to introduce a model of T. He speaks of this as 'mathematical' or '*ideelle*' existence. His concept of consistency seems to be that of the syntactical consistency of T in second-order logic. But this is of course inadequate for the results summarized in the last paragraph, since by Gödel's first incompleteness theorem there are $T \supset T_2$ which are syntactically consistent in second-order logic (assuming that T_2 is) but have no models.[2] Zermelo himself simply left aside the question of proving consistency; but that was in 1930, and we now know that not only would a syntactical consistency proof for any particular formal system be inadequate for the purpose of guaranteeing the existence of models, but (by Gödel's second incompleteness theorem) that in no reasonable sense are such proofs forthcoming. So, although one would certainly insist that when T is inconsistent it has no model, the establishment of consistency cannot profitably be required in developing set theory. We must seek to carry out this development on other grounds, but agnostically so to speak, accepting the possibility of Descartes' demon, that the development of set theory or even of first-order number theory leads to contradiction. On the other hand, the condition for the existence of a model that it be the model of a categorical extension T makes very good sense, especially in the light of the incompleteness of higher-order logic. To introduce a structure simply by the condition that it be a model of T should certainly require that the set of sentences true in the structure be determined from T. But for sentences of more than first-order, how else is this completeness property of T to be guaranteed otherwise than by categoricity? Actually, Zermelo appeals to a weaker condition when he speaks of categoricity. In his admittedly tentative sketch of how his view should be developed, he argues for the adoption of an axiom asserting that, for any given ordinal α, there is a Grenzzahl $> \alpha$. The argument is that the conjunction $\phi(\alpha)$ of the axioms of T_2 together with the assertion that there is no Grenzahl $> \alpha$ is 'categorical'. If M is a model of it, then $o(M)$ is the required Grenzzahl. But $\phi(\alpha)$ is not simply a sentence of second-order set theory when it contains the parameter α, and in Zermelo's argument α must

[2] Once Henkin models are admitted, there are no categorical theories aside from the trivial case of theories with only finite models. In the case of consistent extensions of T_2 in second-order logic, there are, for each infinite cardinal number κ, non-isomorphic Henkin models of cardinality κ.

be understood as denoting itself or, more precisely, the corresponding von Neumann ordinal, which consequently must be in the model. Accordingly, when Zermelo speaks of this sentence as being categorical, he is extending that notion from its usual application. $\phi(\alpha)$ is categorical if and only if any two models of it in which the set of predecessors of the von Neumann ordinal denoted by α have the order type α are isomorphic. So the structure M is given not just by the sentence $\phi(\alpha)$, but by this together with the order type α. This extended notion of categoricity can also be applied to sentences $\phi(t)$, where t is an arbitrary set in some model. In this case t determines a well-founded tree, called its *$\in$-tree*, rather than simply a well-ordered set and it is the isomorphism type of this tree together with the sentence $\phi(t)$ that determines the structure.[3] It is sometimes charged that categoricity of second-order theories is a sham. Thus, all models of T_3 together with the axiom that there are no inaccessible cardinals are isomorphic. But that does not tell us whether or not the continuum hypothesis is true—only that if it holds in one of these models, then it holds in all of them. However, in view of the second-order Axiom of Separation, the question of whether or not it holds is a question of what second-order classes, in particular subclasses of $\mathcal{P}(\omega)$, there are. Thus, the incompleteness involved is that of second-order logic. If one accepts the view, stated by Gödel and advanced in Section 3 below, that the notion of set involved in second-order logic, called the *logical* notion of set in Shapiro (1991), is distinct from and prior to the notion of set as it occurs in set theory, then categoricity, in the strict sense at least, reduces the question of truth in a model of set theory to that of logical truth in second-order logic.[4] On the other hand, in Zermelo's extended sense of the term 'categoricity', this is no longer so.

As noted, we must drop as an active principle the condition of consistency in developing Zermelo's idea. But obviously something more is required than the extended notion of categoricity to motivate extensions of T_2. Some idea must underlie the choice of these extensions. Zermelo's argument, cited above, for the existence of an unbounded sequence of inaccessible cardinals, is the germ of such an idea.

2. FIRST- AND SECOND-ORDER REFLECTION

Zermelo's argument leads to the following more general principle

(1) $\forall x[\phi(x) \rightarrow \exists\beta\phi^{\beta}(x)]$

[3] The nodes of the tree are the elements of the transitive closure of t and a node r is immediately above a node s just in case r is an element of s.

[4] There is an interesting discussion of this issue in Shapiro (1991).

for any formula $\phi(x)$ of set theory with just x free. We shall want to consider formulas not only of second-order set theory, but of set theory of finite type. The *finite types* are inductively defined by the condition that, if $n \geq 0$ and $\tau_1 \ldots \tau_n$ are finite types, then $\tau = (\tau_1, \ldots, \tau_n)$ is a finite type. The *order* of τ is defined to be 1 greater than the maximum order of the τ_i. When $n = 0$, then variables of type $\tau = ()$ range over sets. When $n > 0$, then variables of type τ range over relations whose elements have the form $(t_1, \ldots, t_n)$, where t_i is of type τ_i for $i = 1, \ldots, n$. Objects of type $((\ldots(\,)\ldots))$ of order n (i.e. with n pairs of parentheses) will be called *classes of order n*. The *formulas* that we shall consider are built up by means of the propositional connectives and quantifiers of arbitrary finite type from atoms of the form $x \in y$, where x and y are of type (), or $(X_1, \ldots, X_n) \in Y$, where $n > 0$, Y is of type $(\tau_1, \ldots, \tau_n)$ and X_i is of type τ_i for $i = 1, \ldots, n$. In the case of $n = 1$, we write $X \in Y$ for $(X) \in Y$. When ϕ is a formula, then ϕ^β is the result of restricting the bound variables of type () in ϕ to the set $R(\beta, ()) = R(\beta)$ of sets of rank $<\beta$ and, for $n > 0$, the bound variables of type $\tau = (\tau_1, \ldots, \tau_n)$ to the set $R(\beta, \tau) = \mathcal{P}(R(\beta, \tau_1) \times \ldots \times R(\beta, \tau_n))$. So we are taking the logical framework of set theory to be, not just second-order predicate logic, but predicate logic of finite type.[5] The *order* of a formula is the maximum order of its bound variables.

Notice that the Axiom of Infinity and so the system T_3 follows from T_2 together with (1) when $\phi(x)$ is the first-order sentence 'There is a null set and, for every set y, there is a set whose elements are those of y together with y itself.' Conversely, every instance of (1) for a first-order formula ϕ is derivable in T_3—indeed, in ZF. On the other hand, when second-order formulas $\phi(x)$ are admitted, then Zermelo's proposed axiom that, for each ordinal α, there is a Grenzzahl $> \alpha$ follows from (1) by taking $\phi(\alpha)$ to be the conjunction of $\alpha \in V$, which is an abbreviation for $\exists x\, [x = \alpha]$, and the axioms of T_2.

(1) has been called a *reflection principle* because it expresses that truth in the universe of all sets reflects down to truth in some initial substructure $R(\beta)$ of it. In terms of Zermelo's idea that the models are given by categorical extensions of T_2, the argument might be that if $\phi(t)$ is true in the universe for some set t and (1) is false, then $\phi(t) \wedge \forall\beta[t \in R(\beta \rightarrow \neg\, \phi^\beta(t)]$ would be true. But this sentence together with T_2 is categorical and so the universe of sets would itself constitute a model M, so that $o(M)$ would be an ordinal β for which $t \in R(\beta) \wedge \phi^\beta(t)$ holds after all. The requirement of consistency could be thought to be replaced in this argument by truth in the universe of sets. The same line of argument would seem to support the stronger reflection principle

[5] It is usual in set theory to restrict the finite types $\tau = (\tau_1, \ldots, \tau_n)$ to the case $n = 0$ or $n = 1$. The reason is that relations can always be coded as classes. But in Sect. 5 below, we shall need to consider a class of formulas, the *positive* formulas, which is not closed under this coding.

(2) $\forall X[\phi(X) \rightarrow \exists\beta\phi^{\beta}(X^{\beta})]$

where X is a variable of type (()), ranging over second-order classes, and X^{β} is $X \cap R(\beta)$. In this case the categorical determination of such a β requires not only the formula $\phi(A) \wedge \forall\beta\, \neg\phi^{\beta}(A^{\beta})$, but also the specific class A. It is easy to see that the second-order Axiom of Replacement is an instance of (2) with $\phi(X)$ a first-order formula.

But there is a difficulty with regarding either (1) or (2) as a reflection principle in this way. The argument is that if A is a second-order class and $\phi(A)$ is true in the universe of all sets, then it should be true in some $R(\beta)$. But how are we to understand truth in the universe of all sets? Cantor (1883), in initiating the mathematical theory of infinite totalities, distinguished between those determinate totalities which can be mathematically treated, namely those which have a cardinal number, and the 'absolute infinite', which includes the totalities of all ordinals or of all cardinals and, of course, the totality of all sets. Zermelo is very explicit about this: set theory has many models, determined by certain categorical extensions of T_2. Among these models is *not* the universe of all sets, since every model has proper extensions. He sees Cantor's distinction between the absolute infinite and the determinate infinite totalities in an interesting way, as the distinction between the theory T_2 and its models (given by certain categorical extensions of T_2). For him, the paradoxes of set theory rest on the neglect of this distinction. We can speak of a universe of sets as a model of set theory only if we have a categorical determination of it; but then it would not be the universe of all sets. We may 'define' the totality V of all sets or the totality $\Omega = \{x \mid x \text{ is a von Neumann ordinal}\}$ of all ordinals, etc.; but these are just formal terms which have no meaning in themselves: what is 'x' in the expression $\{x \mid \ldots\}$ supposed to range over? It is only with respect to a given structure $\langle M, \in_M \rangle$ that they have meaning. In that case, 'x' ranges over M, V denotes M and Ω denotes the totality of von Neumann ordinals in M; and these are legitimate sets in models of higher ordinal than M. From this viewpoint the 'paradoxes of set theory' vanish. But also from this viewpoint the picture of (1) and (2) reflection principles, reflecting *down* from the universe of all sets, becomes problematic. For it seems to require that we know what it means to say that a sentence $\phi(t)$ is true in the universe of all sets. For example, even to apply (1) on this interpretation to obtain Zermelo's principle that there are Grenzzahlen greater than any ordinal, we must assume that the axioms of T_2 are true in the universe of all sets. But whatever is the relevant notion of truth, it is not our usual notion of truth in a model.[6]

[6] Hellman (1989, 89–91) notes the difficulty from the Zermelian point of view with the top-down picture of (1) and (2) as reflection principles. Hellman works out a foundation for certain instances of (2) based on modal ideas.

3. THE ITERATIVE CONCEPTION OF SET THEORY

In preparation for discussing this problem of truth, we turn to an idea which is first explicit in (Gödel 1947), but which derives from Zermelo's paper. First, the type-free notion of set which is formalized in ZF or in T_2 should be distinguished from the notion of set as it occurs commonly in many parts of mathematics and is formalized in second-order logic. In the latter sense, we have some universe U of objects—the integers or the real numbers or real-valued functions of real numbers, etc.—and we then pass to the universe $\mathscr{P}(U)$ of arbitrary sets of objects from U. This notion, Shapiro's *logical* notion of set, is implied by the very framework of second-order logic.[7] By a *domain* we shall mean the result of iterating the operation $\mathscr{P}$, starting with the null domain of objects. Secondly, the former, type-free, notion of set may be founded on the latter, type-theoretic, notion in that we may understand by a set in the former sense an object occurring in some domain. For sets s and t, $s \in t$ (in the type-free sense) means that for some domain D, s is in D, t is in $\mathscr{P}(D)$, and s is an element of t in the type-theoretic sense. In particular, this yields the standard type-free meaning for the symbol $\in$ in any domain. So we may speak about the truth of a sentence of set theory in a domain D, meaning its truth in $\langle D, \in_D\rangle$, where $\in_D$ is the restriction of the type-free elementhood relation to elements of D.[8] Models, i.e. normal domains, are a special case of domains and Zermelo's argument mentioned in Section 1 shows that if D and D' are domains with order types $o(D) \leq o(D')$,then D is to within isomorphism the subdomain of D' consisting of the elements of rank $< o(D)$. Let T_1 denote the system in the language of set theory with the unary function constant *ran* (for the rank function) added and with the following axioms: *Extensionality*, *Regularity* (in the second-order form that there are no classes which form infinite descending $\in$-chains), *Union*, *Second-order Separation*, *Choice*,

$$ran(x) = \bigcup\{ran(y) \cup \{ran(y)\} \mid y \in x\}$$

(formulated so as to avoid the general assumption of the Axiom of Unordered Pairs) and, with α ranging over the von Neumann ordinals,

$$\forall\alpha[\{x \mid ran(x) \in \alpha\} \text{ is a set}].$$

It is easy to see that the models of T_1 are, to within isomorphism, precisely the non-empty domains (with *ran* denoting the rank function). For α a von

[7] Gödel (1947, in fn. 13 on p. 519) writes of the operation $\mathscr{P}$ that 'as opposed to the concept of set in general (if considered as primitive) we have a clear notion of this operation.'

[8] Zermelo is not explicit about what he means by a domain; but when he assumes that 'every categorically determined domain whatsoever can also be regarded as a "set", i.e. can appear as an element of a (suitably chosen) normal domain', it is clear from the consequences that he draws that the domain in question is not simply the abstract set M, but the structure $\langle M, \in\rangle$ and its occurrence in the normal domain, the model, is as an initial substructure.

Neumann ordinal in such a domain D, $\{x \mid \text{ran}(x) \in \alpha\}$ defines a domain included in D, namely the domain $R(\alpha)$ of all sets of rank $<\alpha$. We shall denote that domain by M_α. On the iterative conception, sets are the objects obtained in domains that arise from iterating the operation $\mathcal{P}$ 'as far as possible', obtaining bigger and bigger domains. So T_1 is the natural system with which to begin, rather than T_2.

There are two issues in connection with Zermelo (1930) and especially in connection with T_1 that should be mentioned. The first concerns the logical basis of T_1 itself, in particular of the Axiom of Choice and the Comprehension Axiom as principles of second-order logic. This is discussed in Tait (1994), where it is shown that from what I shall call the *type-theoretic* point of view, these *are* principles of logic. The second issue, raised already in Skolem (1923) and immediately apparent to the reader of Zermelo (1930), concerns the notion of a domain. What do we mean by a 'domain', if not a set? Again, I want to answer this from the type-theoretic point of view. On this view, mathematical objects are always objects of some *type*. A type is given by effective introduction and elimination rules and in this way is *proposition-like*. The question of whether or not a given object is of a given type is always a trivial question: an object is given *as* an object of a certain unique type. In contrast to the question of whether or not an object is of a given type, the question of whether or not an object of type A is in a given set of type $\mathcal{P}(A)$ is in general a non-trivial question. Given a type A, there is always its *power type* $\mathcal{P}(A)$, which we may take to be the 'proposition' $A \rightarrow \mathbf{2}$, $\mathbf{2}$ being the two-element type. Strictly speaking, we should distinguish the operation of passing from a set to its power-set from the operation of passing from a type to its power-type; although if a is the set of all objects of type A, then its power-set is the set of all objects of the power-type $\mathcal{P}(A)$. Now, the view that we are adopting here is that *domains are types*: we introduce a domain by introducing a new way of constructing objects and reasoning about them. The null domain is of course just the absurd proposition, and we have already noted the construction of the power domain $\mathcal{P}(D)$ from the domain D. The representation of domains as types does raise a small point. We generally regard M_α as a subdomain of M_β when $\alpha < \beta$. But types are disjoint, and so instead of the inclusion map of M_α into M_β, we have to speak in terms of the natural embedding of M_α into M_β. In particular then, to construct *limit* domains, i.e. domains M_γ with γ a limit ordinal, we must take, not the union of the M_α with $\alpha < \gamma$, but their *direct limit*. This construction is discussed in more detail in (Tait, n. d.).

There still remains the question of what limit ordinals are to be admitted under the iterative conception. The extensions of T_1 that we envision as a development of the iterative conception will be obtained by introducing further operations for constructing ordinals, operations that arise from the analysis of the idea of iteration 'as far as possible'. In particular, we may introduce new operations which are the closures of previously introduced operations. In this way we shall obtain ever stronger extensions T of T_1. As Gödel wrote concerning the axioms of ZF

> the axioms of set theory by no means form a system closed in itself, but, quite on the contrary, the very concept of set on which they are based suggests their extension by new axioms which assert the existence of still further iterations of the operation 'set of'. The simplest of these strong axioms of infinity assert the existence of inaccessible numbers (and of numbers inaccessible in the stronger sense) > $\aleph_0$. The latter axiom, roughly speaking, means nothing else but that the totality of sets obtainable by exclusive use of the processes of formation of sets expressed in the other axioms forms again a set (and, therefore, a new basis for a further application of these processes). (Gödel 1947, 520)

In a revised note, footnote 20, of the 1964 version of his paper, Gödel expresses the view that (1) and (2) follow from the iterative conception of set, at least when ϕ is a second-order formula, and on p. 265 speaks, by implication at least, of their *intrinsic necessity*. In this respect he contrasts these principles with axioms expressing, for example, the existence of measurable cardinals, which do not follow—or at least not obviously—from the iterative conception. Notice that, in the above quote, he already takes the axioms of set theory, and, in particular, the Axioms of Infinity and of Replacement, to be based on the iterative conception. We have noted that Infinity follows from (1). Clearly so does the Axiom of Unordered Pairs; and Replacement and so all of T_2 follows from (2).[9]

So far in this section we have been discussing Gödel's view rather than Zermelo's. To what extent can the iterative conception be attributed to Zermelo? Gödel himself refers to (Zermelo 1930) at the end of the passage quoted above. It has been claimed that the iterative conception arose with Gödel and was no part of Zermelo's conception. The argument is that Zermelo did not regard the Axiom of Foundation as following from the

[9] It has been written that the iterative conception does not imply the Axiom of Replacement. The argument for this is that the iterative conception is based upon having a prior system of stages, essentially ordinals, along which the iterations are to be carried out; and there is no reason to suppose that the order type of the system of stages is regular, as would be implied by the Axiom of Replacement. Of course on these grounds we should also reject the Axiom of Infinity, since there is no reason to suppose that the given system of stages contains an infinite stage. But in any case, one can see that this is neither Zermelo's nor Gödel's conception of set theory. The iterations are not given by a prior system of ordinals, but rather the ordinals and sets are generated autonomously by means of operations which follow from the idea of as far as possible. In particular, this leads to the principles (1) and (2) and so to the Axioms of Infinity and Replacement.

It is also worth noting that the view that the Axiom of Replacement is based on the principle of limitation of size is at least misleading from the point of view of the iterative conception. Replacement is a special case of Reflection, and surely not all cases of the latter can be accounted for by the principle of limitation of size.

notion of set. But this is true of Gödel as well; neither of them makes the claim that the iterative conception exhausts our concept of a totality: for both the introduction of the Axiom of Foundation marks out a particular conception of set, and Gödel's reference to Zermelo's paper is a recognition of the fact that Zermelo saw that this conception leads to further axioms and, in particular, to the existence of an unbounded sequence of inaccessible cardinals.

4. THE UNIVERSE OF SETS

There is one point on which Zermelo and Gödel may be in conflict: as we have noted, for Zermelo there is no one universe of all sets which constitutes a model for the axioms of set theory. On this is based his dissolution of the 'paradoxes of set theory'. In a footnote to a draft of the 1964 version of his paper, Gödel refers to (Zermelo 1930) as containing 'substantially the same solution of the paradoxes' as his; but the reference is dropped in the final version. One explanation for his dropping it (although others are also possible) would be that he did not accept Zermelo's view that there is no universe of all sets forming a model of the axioms of set theory. There is independent evidence that Gödel did indeed differ from Zermelo on this point. For example, in footnote 17 of the 1947 paper, he mentions the possibility of extending first-order set theory to higher-order set theory, but says that its 'consequences for propositions referring to limited domains of sets (such as the continuum hypothesis) are contained already in the axioms depending on the concept of set', i.e. first-order set theory. Presumably he has in mind that if ϕ expresses the proposition referring to limited domains and ψ is the conjunction of axioms in higher-order set theory which implies ϕ, then for a sufficiently large cardinal κ, ψ^κ implies ϕ^κ, which is just ϕ, and, by the principle (1), there is such a κ for which ψ^κ is true. But in ψ^κ the *n*th-order quantifiers range over $R(\kappa + n - 1)$ and so the reference to higher-order objects is eliminated. But what is significant in the present context about Gödel's remark is the implication, on its most natural reading, that there are also propositions which do *not* refer to limited domains, i.e. which are expressed by formulas ϕ with unbounded quantifiers, and that, if ϕ is a consequence of the axioms of higher-order set theory, then it is true of the universe of all sets. This implies that the axioms of set theory are themselves true of the universe of all sets. For Zermelo, on the other hand, ϕ does not express a single proposition at all; rather, in each normal domain M_κ it expresses the proposition ϕ^κ and, if ϕ is a consequence of the axioms, then ϕ^κ is true.

Whether or not Gödel really held the view that set theory is the theory of the universe of all (pure) sets in the way that Euclidean geometry is the theory of Euclidean space, it is certainly a commonly expressed view. But in as much as it implies that the universe of sets is a well-defined totality, it seems to be a difficult view to defend. On the conception we are discussing, all sets are obtained by iterating the power-set operation as far as possible. But this is not a mathematical characterization; rather, it is an informal idea which leads us to successively stronger precise principles of iteration, such as the reflection principles above. Indeed, a mathematical characterization is precluded on the iterative conception by the fact that any mathematically definable totality P of such principles would lead to a new one, namely closure under all the principles in P. Moreover, it is not at all clear that, as we try to develop the idea of iteration as far as possible, there is only one direction that this would take us. It is conceivable that there is some development of this idea that leads to the construction, for example, of measurable cardinals; but that there would also be grounds for rejecting the construction. So it does not seem possible to speak of the universe of sets as a well-defined extension, even in the sense of speaking of those sets which we construct or might have constructed on the basis of the iterative conception. Even if one believed that, by the nature of the human psyche, there is only one way in which the idea could be developed, this single path itself would still be incapable of a mathematical characterization: we could then only say of it that it is the path we are destined to take—an anthropological characterization, not a mathematical one.[10]

So when we affirm the truth of a proposition about the universe of all sets, we must speak of it on the basis of those operations for constructing sets that have already been accepted, while admitting that further principles might also be admitted. For this reason, the logic that applies to arbitrary formulas of set theory, when these are interpreted in the universe of all sets, should be constructive, not classical logic. For there is no such well-defined universe; there is just the possibility of stronger and stronger operations which we might introduce for constructing sets. So, when x ranges over all sets, there are no grounds for assuming $\forall x\phi(x) \vee \exists x\neg\phi(x)$ unless it can be derived on the basis of operations that we have already introduced. Indeed, it is precisely in connection with the absolute infinite, the universe of all sets, that the ideas of constructive mathematics become compelling, even for those who reject the restrictions to constructive methods in analysis. In the case of analysis we are dealing with sets of . . . of sets of finite ordinals and the means for constructing finite ordinals are fixed once and

[10] When one thinks of the development of Euclidean geometry from earliest civilizations through the classical Greeks and finally to Hilbert's *Grundlagen der Geometrie*, one is tempted perhaps to believe that there is indeed some such thing in the nature of the human psyche. But set theory is in one way different from geometry. The quest for foundations of Euclidean geometry came to an end with a categorical axiomatization. In the nature of the case, the quest for foundations of set theory can never come to an end in this way.

for all: 0 is a finite ordinal and, whenever n is a finite ordinal, so is its successor $n + 1$. The situation is essentially different for the totality of all ordinals: there is the further operation of taking the least upper bound of totalities X of ordinals. But there can be no complete specification of those totalities X for which the least upper bound can be taken. The idea of *potential infinity* makes very good sense in connection with the totality of all ordinals or sets, even for those who fail to understand it in the case of the finite ordinals.

Of course, when ϕ is interpreted in a particular domain, then $\phi \vee \neg\phi$ is valid. In this respect, the role of a finite domain in constructive mathematics is played by arbitrary domain in the present setting. More generally, if ϕ is a bounded formula, i.e. a first-order formula in which all the quantifiers occur in contexts $\forall x \in t$ or $\exists x \in t$, then the meaning of ϕ does not depend on the particular domain in which it is interpreted and so the Law of Excluded Middle will be admitted. So to say that $\phi \vee \neg\phi$ is valid when referred to any particular domain is just to say that the bounded formula $\phi^\beta \vee \neg\phi^\beta$ is valid for all β.

Adopting this conception of the logic of propositions referring to the universe of all sets, when $\phi(x)$ is a first-order formula, then Gödel's grounds for accepting (1) on the basis of the iterative conception remain valid. But in the case of (2) and even in the case of (1) with $\phi(x)$ of order > 1, there is an immediate difficulty; namely what are we to mean by a higher-order object? If we cannot regard the 'universe of sets' as a well-defined totality, then neither can we regard, for example, second-order classes in general as well-defined extensions. It seems clear that we must regard such classes and, in general, objects of order > 1 *intensionally*, as given by defining properties (but not, of course, with the definition restricted to any particular extension of T_1). But notice that when we relativize a second-order class A, for example, to $R(\beta)$ for some β, we do not relativize the *definition* of A, only its extension: for elements t of $R(\beta)$, $t \in A^\beta$ just in case $t \in A$. The definition acts only as an oracle, giving a sense to the assertion $t \in A$ for any set t in any domain that we might construct.

There remains a complication. Suppose that $\phi(A)$ and that A is some second-order class $\{y \mid \psi(y)\}$ for which $\psi(0)$ is undecidable, in the sense that no decision about its truth is possible on the iterative conception of set. For example, as we mentioned above, it could happen that there is no clear sense in which question of the existence of a measurable cardinal is decided on the iterative conception. So if $\psi(0)$ asserts the existence of a measurable cardinal, then $0 \in A$ is undecidable. Now suppose that $\phi(A)$ holds and implies the existence of 0. Then by (2) there should be a β such that $\phi^\beta(A^\beta)$ and $0 \in M_\beta$. Therefore $(0 \in A)^\beta$, i.e. $0 \in A \cap M_\beta$, is equivalent to $0 \in A$. But the former is a bounded formula and so should, according to the ideology we have been developing, be decidable, whereas the latter is undecidable.

For this reason, as a valid assertion about the totality of all sets, (2) must be weakened. A second-order class A will be called *decidable* iff $\forall x(x \in A \vee x \notin A)$; and similarly for other second-order objects. In place of (2) we should have

(3) $\forall X[\phi(X) \wedge X$ is decidable $\rightarrow \exists\beta\phi^{\beta}(X^{\beta})]$.

Moreover, in the case of (1), when the true sentence $\phi(t)$ is second-order, its truth may involve the truth of existential sentences of the form $\exists X\psi(X,t)$ and so of instances $\psi(A,t)$ of them. Hence the truth of $\phi^{\beta}(t)$ will involve that of $\psi^{\beta}(A^{\beta},t)$. So in both (1) and (2) when ϕ is second-order, all of the second-order quantifiers in ϕ must be restricted to decidable classes. But with this restriction, (1) and (3) are justified for second-order ϕ on the iterative conception in the same way that (1) is for first-order ϕ.

However, having accepted the restricted instances of (1) or (3), we may apply (1) to those instances themselves, ensuring that they have models. So, in keeping with Zermeo's and Gödel's conception, we may add the restricted (1) and (3) as axiom schemas to T_1. But in this theory, which is in the framework of classical logic (and whose models are domains, not the universe of all sets), the clause 'X is decidable' is provable and so drops out, and we obtain (1) and (2) without the restriction to decidable objects. For this reason, we shall ignore the question of decidability from now on. Also, since (1) is reducible to (2), from now on we shall just speak of (2).

5. HIGHER-ORDER REFLECTION

Notice that we have not derived (2) from the iterative conception when ϕ is of order > 2. But even adopting it for formulas ϕ of arbitrary finite order does not yield very large cardinals. When (2) holds in M_{κ} for a given ϕ, we say that the cardinal κ is *ϕ-indescribable.* κ is *totally indescribable* if it is ϕ-indescribable for all formulas ϕ of finite order. But it turns out that, in the universe of large cardinals, total indescribability is a not very strong property: let κ be a cardinal. $C \subseteq \kappa$ is called *unbounded* if for every $\alpha < \kappa$, there is a $\beta \in C$ with $\alpha < \beta$. C is called *closed* if for every $\beta < \kappa$, if $C \cap \beta$ is unbounded in β, then $\beta \in C$. $S \subseteq \kappa$ is called *stationary* in κ if $S \cap C$ is non-empty for every closed unbounded class C in κ. Let $[C]^n$ denote the set of all n-element subsets of C. We denote by $P(n)$ the following partition property: for any function $f: [\kappa]^n \rightarrow 2$ there is a stationary subset S of κ which is *f-homogeneous*, i.e. is such that f is a constant on $[S]^n$. Now, when κ satisfies even the relatively weak principle $P(2)$[11] then the set of totally indescribable cardinals < κ is stationary in κ. So if the iterative conception

[11] A weaker condition, that κ be *subtle* suffices.

of set theory is to develop significantly, it must entail stronger principles for constructing ordinals.

The extension of (2) to higher-order classes X is a natural candidate;[12] but it seems to pose a problem. For example, if we take the analogue of $X^\beta = X \cap R(\beta)$ for a third-order class X to be $X \cap R(\beta+1)$, then let U be the totality of unbounded second-order classes (i.e. not included in $R(\beta)$ for any β) and let $\phi(U)$ be the true proposition that U is not empty. For all β, $\phi^\beta(U \cap R(\beta+1))$ is false. On the other hand, $X \cap R(\beta+1)$ is not the right relativization of the third-order class X anyway. Relative to $R(\beta)$, the second-order class A is $A^\beta = A \cap R(\beta)$, since two such classes, A and B, are equal from the point of view of $R(\beta)$ just in case $\forall x \in R(\beta)[x \in A \leftrightarrow x \in Y]$. So the relativization X^β of X should be the the class of these. In general, when X is of type $(\tau_1, \ldots, \tau_n)$, where $n > 0$, then we should set

$$X^\beta = \{(Y_1^\beta, \ldots, Y_n^\beta) \mid Y_1^\beta, \ldots, Y_n^\beta \text{ are defined and } (Y_1, \ldots, Y_n) \in X$$

When $\tau_i = ()$, then Y_i^β is just Y_i when Y_i is in $R(\beta)$ and otherwise is undefined. Notice that (2) now has meaning for X of arbitrary finite type. On the same grounds that we gave for replacing (2) by (3) and restricting the second-order quantifiers in ϕ to decidable objects, a similar restriction to 'decidable' objects is required when we allow X or $\phi(X)$ to be of order >2. I will omit the required definition of decidability in this case, since, as in the second-order case, once an instance of (3) is itself reflected down to some $R(\beta)$, the decidability restriction is trivialized and we are left simply with an instance of (2).

But there still remains a problem with the generalized (2), even for X a third-order class: when U is the class of bounded or of unbounded second-order classes, we have the true sentence $\phi(U)$ that every class in U is bounded or unbounded, respectively; whereas for every β, $\phi^\beta(U^\beta)$ is false since U^β is just $R(\beta+1)$ and, in particular, contains both $R(\beta)$) and the null set. So the generalization of (2) to third order is still problematic. Presumably it was in view of such considerations that William Reinhardt (1974) suggested that there is no genuine extension of (1) and (2) to higher orders; at least he challenges the reader to find one.

The problem is not merely with generalizing (2) to cases in which X is of order greater than 2, but with (2) itself or even (1) when $\phi(X)$ contains quantifiers of order greater than 2. For example, $\phi(A) = \exists Y \psi(A,Y)$ may be true, where A is a second-order class or a set and Y is of order greater than

[12] One proposal for developing the iterative conception beyond (2) is given by the notion of β-*reflection* in (Tait 1990). The formulation of the principle in that paper is not quite what was intended; but in the case in which it is applied, namely to derive the existence of a cardinal satisfying $P(2)$, the formulation is correct. The real problem with it (when properly formulated) is that it fails to follow from the iterative conception of set: it requires a 'top-down' conception of reflection.

two, because $\psi(A,B)$ is true for some B. If $\psi^\beta(A^\beta,B^\beta)$ is false for all β, then on what grounds do we infer the existence of a β such that $\phi^\beta(A^\beta)$? In order to understand the difference between reflecting formulas containing parameters or bound variables of order at most two and the reflection of formulas containing parameters or bound variables of higher order, notice this difference: in the former case, if ϕ is a basic (i.e. atomic or negated atomic) sentence containing parameters and it is true, then its relativization to $R(\beta)$ is true for any β such that the set parameters in ϕ are of rank less than β. This is clearly so of $s = t, s \neq t, s \in t, s \notin t, (s, \ldots, t) \in S$, and $(s, \ldots, t) \notin S$, when S is second-order. Moreover, it remains true for the higher-order basic sentences $(S, \ldots, T) \in R$; but it is not in general true for their negations. Thus, the restriction of (2) to sentences containing no quantifiers of order greater than 2 hides the need for a restriction that is required in the more general case: call a formula is *positive* iff it is built up by means of the operations $\wedge, \vee, \forall$ and $\exists$ from atoms of the form $x = y, x \neq y, x \in y, x \notin y, x \in Y, x \notin Y, X = Y$ and $(X, \ldots, Y) \in Z$. The required restriction on (2) in the general case is that $\phi(X)$ must be a positive formula. With this restriction, $\phi(X)$ may be any formula of set theory of finite type and X may be of any finite type. We shall refer to the general case of (2) with this restriction simply as the *Reflection Principle*.

My aim in this paper has been to show that the iterative conception of set is a foundation, not only for the principle (2) in second-order set theory, i.e. when X and $\phi(X)$ are at most second order, but also for the general Reflection Principle. In conclusion, I want to indicate something of the strength of this principle. Let $A, \ldots, B$ be of any order, but assume that $\phi(X, \ldots, Y)$ is first order. If $\phi(A, \ldots, B)$ is true in M_κ, where κ is an uncountable regular cardinal, then the set of $\beta < \kappa$ such that $\phi^\beta(A^\beta, \ldots, B^\beta)$ is true contains a closed unbounded subset of κ. So in this case, the Reflection Principle is weak, no matter what the orders of $A, \ldots, B$. However, let Γ_n denote the set of all positive formulas of the form

$$\forall X_1 \exists Y_1 \ldots \forall X_n \exists Y_n \psi.$$

where ψ is first order, the X_i are all second order and the $\exists Y_j$ are sequences $\exists Z_{j,1} \ldots \exists Z_{j,m}$ for some m, and the $Z_{j,i}$ are variables of arbitrary types. A cardinal κ is called *n-reflective* iff M_κ satisfies (2) for all $\phi(X)$ in Γ_n with X of any order > 1. Although, as we have just noted, the property of being 0-reflective is not very strong, the property of being 1-reflective already goes beyond total indescribability: κ is 1-reflective precisely if it has the property $P(2)$. More generally, we have

Theorem 1. For $n > 0$, if κ is n-reflective, then it has the property $P(n + 1)$.

This result holds even when Γ_n is further restricted to formulas $\phi(X)$ in which X is of order at most $n + 2$. When $n > 1$, it is unknown whether or not the converse of Theorem 1 also holds. For an indication of the proofs of the above theorem and of the following theorem, see (Tait, n.d.). In fact, the best bound on the strength of the Reflection Principle restricted to formulas in Γ_n that is known at the moment may not be too reassuring to everyone:

Theorem 2. Every measurable cardinal is n-reflective for all n.

Of course, this is not an argument that the Reflection Principle is valid on the iterative conception. We have already argued that this principle is a particular explication of that conception and, so far, *no* explication of that conception has led to the existence of a measurable cardinal. Rather we should say that the result in question gives some evidence for the consistency of the iterative concept as we have explicated it. Concerning the strength of (2) when $\phi(X)$ is a positive formula containing universal quantifiers of order greater than 2, nothing is known.

REFERENCES

Cantor, G. (1883), *Grundlagen einer allgemeinen Mannigfaltigkeitslehre. Ein mathematisch-philosophischer Versuch in der Lehre des Unendlichen*, Leipzig, Teubner.

Gödel, K. (1947), 'What is Cantor's continuum problem?', *American Mathematical Monthly*, 54: 515–25.

Hellman, G. (1989), *Mathematics without Numbers: Towards a Modal-Structural Interpretation*, Oxford, Clarendon Press.

Reinhardt, W. (1974), 'Remarks on reflection principles, large cardinals, and elementary embeddings', *Proceedings of Symposia in Pure Mathematics 10. Part II*, 189–205.

Shapiro, S. (1991), *Foundations without Foundationalism: A Case for Second-Order Logic*, Oxford, Clarendon Press.

Skolem, T. (1923), 'Einige Bemerkungen zur axiomatischen Begründung der Mengenlehre', *Matematikerkongressen i Helsingfors den 4–7 Juli 1922, Den femte skandinaviska matematikerkongressen, Redogörelse*, 217–232.

Tait, W. (1990), 'The iterative hierarchy of sets', *Iyyun: A Jerusalem Philosophical Quarterly*, 39: 65–79.

—— (1994), 'The law of excluded middle and the axiom of choice', *in* A. George (ed.), *Mathematics and Mind*, Oxford, Oxford University Press, 45–70.

—— (n.d.), 'Foundations of set theory', to appear in the proceedings of the conference 'Truth in Mathematics', held in Mussomeli in Sept. 1995.

Zermelo, E. (1929), 'Über den Begriff der Definitheit in der Axiomatik', *Fundamenta Mathematicae*, 14: 339–344.

—— (1930), 'Über Grenzzahlen und Mengenbereiche. Neue Untersuchungen über die Grundlagen der Mengenlehre', *Fundamenta Mathematicae*, 16: 29–47.

19

Structure and Abstraction

PETER SIMONS

Ave Caesar! Morituri te salutant.

1. INTRODUCTION

My ontological inclinations are nominalistic, and nominalism traditionally faces a number of stern challenges. One is from the theory of predication and universals. I think this challenge is, for a philosophical one, fairly easy for a nominalist to face. A far more daunting prospect is to reconcile the nominalist's denial of the existence of abstract objects with the acceptance of a vast body of mathematical propositions by a large number of highly intelligent people. Taken at face value, the mathematical propositions these people hold true entail the existence of untold infinities of abstract mathematical objects.

However, rather than presuppose or work within a nominalist framework I shall be exploring aspects of the kind of Platonistic realism of mathematical objects which I consider the most promising and plausible, namely structuralism. I shall not attempt to argue for structuralism against other kinds of Platonistic mathematical ontology. Rather I wish to canvass the possibility of a union—some might consider it a shotgun wedding—between the philosophies of mathematics of Dedekind and Frege. This is not in order to rescue the logicism they held in common, but because each emphasizes in his philosophy of mathematics something the other can profitably use, namely structure (Dedekind) and abstraction (Frege).

2. ABSTRACTION

It may seem anachronous to take abstraction as the Fregean component of the proposed synthesis because after all Frege vociferously rejected and indeed poured scorn on what other mathematicians such as Dedekind and Cantor and various empiricists called 'abstraction'. By 'abstraction' I shall intend the procedure of contextual introduction of singular terms and their purported denotata via an equivalence relation, a procedure that Frege discussed but eventually rejected in *Grundlagen* §§63–7. The singular terms in question, purporting to denote various kinds of abstract object, I shall call simply *abstract singular terms* and their denotata *abstracta*. The reason why Frege rejected such contextual introductions has come to be known as the 'Caesar Problem'. Suppose one introduces abstract singular terms via identities containing them, as in the example

[H] the number of *F*s = the number of *G*s ⇔ there is a bijection between the *F*s and the *G*s

Then for all one knows about the identity of the object which is the number of *F*s, from this introduction, 'we can never . . . decide by means of our definitions whether any concept has the number JULIUS CAESAR belonging to it, or whether that same familiar conqueror of Gaul is a number or is not.'[1] In order to have a convenient label, I shall call the procedure *Frege abstraction*. If the general Caesar Problem were solved, then I think the ontological credentials of abstract objects introduced via Frege abstraction would be considerably enhanced, but I shall not comment here on the solubility or otherwise of the problem. In a certain sense the Caesar Problem must be welcome to a structuralist account of mathematical objects because it provides at least a partial account of the source, nature, and limits of that indeterminacy of mathematical singular terms that structuralism alleges there to be.

Let us try to get a little clearer about what Frege abstraction is and what it is not. What empiricists called abstraction was a psychological process of focusing one's mental attention on some characteristics of objects and ignoring or overlooking others. There are in fact interesting structural parallels between this process (which, properly understood, is nothing disreputable, but a psychological necessity) and Frege abstraction, and also parallels to the Husserlian method of eidetic or essential abstraction. But Frege abstraction is neither empiricist nor phenomenological abstraction. I shall also not be using the term 'abstraction' to stand for variable-binding logical operators like set-abstraction, lambda-abstraction, or Frege's own course-of-values-abstraction. Again, such operators are not irrelevant but they open up considerations better avoided here. There are of course many other mathematical examples of such abstraction operators, such as the definite and indefinite integrals of analysis or the ∇ operator in the theory of partial differential equations. The effect of any such variable-binding operator is, as perhaps Ajdukiewicz was the first to recognize, the product of

[1] Frege (*Grundlagen* §56). This might also be called the England Problem, since Frege's example against contextual introduction is whether England = the direction of the Earth's axis, but Caesar is now safely dead and England has other problems so I shall retain Michael Dummett's appellation.

two distinct operations: a pure binding operation using a lambda or (as in Russell and Ajdukiewicz) a circumflex sign to stand for it, followed by the application of a functor. The former, variable-binding operation is a purely syntactic gathering mechanism, which in itself makes no difference to the meaning, whereas it is in the functor that the semantic 'meat' is always to be found: it carries us from one meaning to another, whereas the operator is a systematic mechanism for obtaining a single connected functorial expression out of any complex expression of the same effective grammatical category. In all the examples below I shall for simplicity accordingly use functors, not variable-binding operators.

The reason for naming the abstraction procedure I shall consider after Frege, despite his rejection of it, is that he was the first person to give anything like an adequate account of it, and furthermore he only rejected it with some reluctance after deciding it would not solve his difficulties. Even after rejecting Frege abstraction, Frege practised it. The courses of values of *Grundgesetze* are contextually introduced and face the Caesar Problem. In §10 of that work Frege thought he had managed to get round that problem by stipulations, but Russell's Paradox proved he had not.

Granted that a general comprehension principle, as expressed in Theorem I of *Grundgesetze*, is inadmissible; granted also, as George Boolos has shown, that adding Hume's Principle (the universal closure of [H] above) to second-order logic is consistent relative to second-order arithmetic, so that some contextual definitions admit the serpent into Eden, as Dummett says, while others do not, could we not then, as Crispin Wright has put it, opportunistically open the hunting season for other acceptable introduction principles? And indeed the hunting prospects are not bad. However, what I prefer to do is go back to the initial rejected *Grundlagen* theory and try to salvage enough of it to work with, both because Frege felt its attraction and because it seems to me still to have certain attractions over restricted *Grundgesetze*-style explicit introduction principles.

Let us recall first of all the form of Frege abstraction in *Grundlagen*. We have in fact two examples:

[D] $\mathrm{Dir}(a) = \mathrm{Dir}(b) \Leftrightarrow a \parallel b$.
[H] $\#(F) = \#(G) \Leftrightarrow F$ equinum G.

The *Grundgesetze* add a third member of the species

[V] $\mathrm{CV}(f) = \mathrm{CV}(g) \Leftrightarrow f$ eq g, where f eq $g =$ Df. $\forall x(Fx = Gx)$,

which has as a special or restricted version for concepts

[C] $\mathrm{Ext}(F) = \mathrm{Ext}(G) \Leftrightarrow F$ coext G.

It is important to understand the strength of the double arrow here. It signifies neither material nor strict equivalence but a tighter equivalence relation, sameness of content, which is the object-language equivalent of the metalinguistic relation of synonymy. We may call it *S-equivalence*, by analogy with Carnap's L-equivalence. In 1884 Frege describes the introduced identities which contain two occurrences of a new singular-term-building functor (the direction of, the number of, the course-of-values of) as involving a *recarving of the content* of a sentence. Although it was still six or seven years before Frege introduced and systematized the distinction between sense and reference, it is fairly clear that 'content' here is much closer to what later became sense than what became reference. That this is so is reinforced by Frege's avowal in the 1891 essay *Funktion und Begriff* that both sides of Basic Law V (which is [V] above) express the same sense. Whether this is warrantable Frege exegesis or not, I shall simply assume from now on that the two sides of any such *Recarving Scheme*, whose general form is

[R] $\wedge(R)(a) = \wedge(R)(b) \Leftrightarrow a\ R\ b$,

are synonymous.

Leaving the Caesar Problem aside for a moment, and leaving aside also the fact that the scheme as here presented is typically ambiguous (the arguments can be functions, concepts, relations, as well as individuals), let us consider where to go from here. Instances of the recarving scheme only introduce new terms, but nothing else. They do not introduce any new predicates to go along with these, or tell us anything about which predicates (old or new) truly apply to the new terms. Here one may rely on the tradition of abstraction theory stemming from Peano, Weyl, and Lorenzen, and consider predicates *invariant* under the equivalence. For 1- and *n*-place predicates invariance with respect to *R* are defined respectively as

[I1] $\mathrm{inv}(P,R) = \mathrm{Df.}\ \forall xy(x\ R\ y \supset (Px \equiv Py))$.
[I*n*] $\mathrm{inv}(P^n,R) = \mathrm{Df.}\ \forall x_1x_2 \ldots x_ny_1y_2 \ldots y_n((x_1\ R\ y_1 \wedge x_2\ R\ y_2 \wedge \ldots \wedge x_n\ R\ y_n) \supset (P^nx_1x_2 \ldots x_n \equiv P^ny_1y_2 \ldots y_n))$.

For example, on the plane the line $y = 1$ intersects the parabola $y = x^2$ but the parallel line $y = -1$ does not. Hence the property of intersecting the parabola is not invariant under the equivalence of being parallel, whereas the relational property of intersecting a given straight line *L* orthogonally is so invariant. An example of a two-place invariant predicate (from the same domain) is that of a pair of lines intersecting at a given angle, say 45°. As a last example for consideration later, consider the curve C of the cube function, satisfying $y = x^3$: this intersects every line $y = ax + b$ for any *a* and *b*, so the property of intersecting C is (trivially) invariant under the equivalence of being parallel.

Ordinary language and to some extent scientific language support the convenient but inexact practice of simply carrying the same words (predicates) over from talking about the concreta to talking about their abstracta (I call this *transcriptive abstraction*), but this practice is to be heartily deplored and condemned, despite its occasional advocacy by Lorenzen's Erlangen school. If a is a concretum and $°(a)$ is its abstractum under an equivalence R, and if P is a predicate invariant under R, then if $P(a)$, the predicate truly predicable of $°(a)$ is not P but a closely related predicate $*P$, whose sense is obtained from that of P by compensatorially *adjusting for* the change in sense between a and $°(a)$ brought about by the recarving of $a\ R\ a$ into $°(a) = °(a)$: this same sense-component which $°(a)$ *gains* over a is to be *deducted* from the sense of P to obtain that of $*P$. I call this, following Geo Siegwart, *adjustive abstraction*.[2]

Several simple examples show up the inadequacy of transcriptive abstraction, and at the same time show how uncertain ordinary language is here as an analytic tool. If the equivalence relation is *is as heavy as* and if John is 10kg heavier than Mary, then while we might on occasion feel tempted to say John's weight is 10kg heavier than Mary's, we should strictly say it is 10kg *greater* than Mary's weight. Weighty objects, not weights, are heavier and lighter. But if this strikes one as unnecessarily fussy, consider the cases in Table 19.1.

Adjustive abstraction accounts for certain felt meaning-shifts, e.g. the move in generics from *Shamu is a mammal* said of an individual killer whale, to *Orca orca is a mammal* said of the species. As John Burgess has pointed out,[3] since individual spatio-temporal location and causal influence tend to vary over concreta, abstracta derived from them cannot truly be said to have a spatio-temporal location or causal powers. On the other hand *is located somewhere* and *has some causal powers* are true of all cubic objects, but the Cube is not located somewhere, nor does it have causal powers. Only adjustive, not transumptive abstraction can retain and account for this intuition.[4]

It also becomes clear on this account how new concepts and predicates can be introduced on the basis of existing ones, on the coat-tails of abstractive introduction. We also seem to have a rough measure of *how* abstract certain abstract objects are. If we use the abstracta obtained in one round as concreta for a second round, we can say the second abstracta are in this limited sense 'more abstract' than the first. Charles Parsons has spoken in this context of quasi-concrete objects, and of objects which are more or less abstract.[5] But more abstract abstracta can typically be obtained from original concreta in a single move, so the rank of abstractness is dependent on

[2] My original expression was *transumptive* but it was reasonably objected that this is too like *transcriptive* in both sound and meaning for easy discrimination.

[3] In discussion.

[4] I do not however consider biological species to be abstracta.

[5] Parsons (1990).

TABLE 19.1

Equivalence	*Invariant predicate*	*Bad (transcriptive)*	*Good (adjustive)*
equiform (tokens)	X is a token	the type of X is a token	the type of X is a type
parallel (lines)	L intersects C	the direction of L intersects C	the direction of L is such that it is of lines which all intersect C
parallel (lines)	L is a line, not a direction	the direction of L is a line, not a direction	the direction of L is a direction, not a line
equimassive (bodies)	M has non-zero mass	the mass of M has non-zero mass	the mass of M is non-zero
equimassive (bodies)	M is spatially extended	the mass of M is spatially extended	the mass of M is such that it is of things which are all spatially extended
co-specific (organisms)	died before 1800	the Dodo died before 1800	the Dodo was extinct before 1800

the abstractions actually made. Iterated abstraction is not well understood, though it is clearly an important phenomenon. Since abstraction takes place within mathematics itself, e.g. in defining quotient structures or defining groups on existing mathematical structures, there is good evidence from mathematics that such iterations are not confined to the lowest levels of mathematics but are endemic within it. That one may start out from concreta and obtain abstracta at more than one level of proximity goes a long way towards explaining the applicability of mathematics. Michael Resnik has correctly noted this as an advantage for the structuralist view, though his account works from the top down rather than the bottom up: abstract structures are realized in concrete configurations. He also notes that if we can recognize abstract structures in concrete configurations, then the basis is laid for a non-occult epistemology of abstract objects.[6]

There is one thing over which adjustive abstraction leaves us in the dark, namely the sense in which abstracta may be said to exist. This was already

[6] Resnik (1981).

a problem in Lorenzen's early paper on abstraction.[7] By 'exist' I mean the first-order predicate of singular existence, usually captured in free logic by the definition

$$\mathrm{E!}a \equiv_{\mathrm{Df.}} \exists x(x = a).$$

It would seem as though we can easily assert that abstract objects exist. Consider a particular line l. Then l is parallel to l so $\mathrm{Dir}(l) = \mathrm{Dir}(l)$ so by existential instantiation $\exists x(x = \mathrm{Dir}(l))$. This is too quick, however. That '$\exists x(x = a)$' may be used to mean 'a exists' in free logic depends on the *existential* reading of the particular quantifier, and the inference in free logic from $F(a)$ to $\exists xF(x)$ is only licensed if a exists, but that is precisely what is at stake here. In so-called positive free logics '$a = a$' is true even when '$\exists x(x = a)$' is false. On the other hand if we allow an unrestricted particular quantifier Σ so that the inference from $F(a)$ to $\Sigma xF(x)$ is valid, then the inference from '$a = a$' to '$\Sigma x(x = a)$' is valid but we are no longer entitled to understand the latter as meaning 'a exists' because the identity predicate is existentially neutral. One may indeed introduce a non-reflexive identity predicate '$\approx$' with existential import, so that existence may then be defined as

$$\mathrm{E!}a \equiv_{\mathrm{Df.}} \sum x(x \approx a),$$

but then there is no non-question-begging assurance that we are entitled to use the predicate '$\approx$' in adjustive abstraction.

It is unclear whether this impasse is a disadvantage for adjustive abstraction. Here are three somewhat inchoate advantages. The first is that in view of the general doubts as to what existence is for abstract objects, whether it is the same concept as existence for concrete objects or a different concept, it is fitting that there is the same uncertainty in the adjustive theory, and further that it is connected with the interpretation of the particular quantifier, since one of the time-honoured ways to try and gain the ideological advantages of Platonism without paying the ontological price has been to look for an ontologically non-committal way to interpret the quantifiers.

The second point is that whether or not we grant the possibility of a shift in meaning of the term 'exists', there will be no *additional* shift under any subsequent adjustive abstraction, so all abstracta will have the same mode of being (or non-being): there will be no grades of ontological tenuousness among abstracta. What differences there are (and what may motivate Parsons' terminology of 'quasi-concrete' objects) will be differences of materiality. The 1926 Model T Ford, Mozart's *Cosi fan tutte*, and the lower-case 12 point Linotype Times Bembo Italic '*a*' are not purely formal objects,

[7] Lorenzen (1962).

whereas the number 2 is, but all are equally abstract. The more 'earthy' abstracta are not 'pure': they still have (adjusted) predicates pertaining to them which have to do with particular domains, such as music, typesetting, and automotive engineering, whereas cardinal numbers are universal and topic-neutral. If mathematical objects are those abstracta which are pure structures, what then are pure structures? I shall return to that question below.

Another attractive feature of bottom-up adjustive abstraction is that it gives full rein to what Wittgenstein called the 'motley' of mathematics. Mathematical structures crop up in disorderly variety in manifold domains in response to a variety of contexts of human activity. Counting, measurement, land surveying, weighing, water waves, plucked strings, the construction of buildings, telling the time, predicting eclipses, the reflection of light in mirrors, ballistic warfare, dice games, and much more have all given rise to their own areas of mathematics. This is not only historically a more enlightening view in terms of the genesis of such topics and their subsequent and increasing unification, simplification, and reduction in mathematics, it also emphasizes how applications frequently drive mathematics. Pure structuralism as a philosophy of mathematics is a view of (mature) hindsight, and fits the attitudes of mathematicians of this century much better than those of most of their predecessors.

3. STRUCTURALISM

Structuralism as a philosophy of mathematics comes in at least two versions: *quantificational* structuralism, associated with Benacerraf, Shapiro, and in a modalized form with Hellman;[8] and *referential* structuralism, associated with Resnik.[9] In referential structuralism the apparent constant singular terms of mathematical theories are genuinely referential, but refer to objects which are schematic or indeterminate in various ways. In quantificational structuralism the apparent constant singular terms are to be construed as really free variables; restoring their missing quantifiers makes it clear that they do not refer to anything but range over all structures (or positions in structures) of a certain kind, whether these structures are abstract or concrete.[10]

Hellman and Resnik have enumerated advantages of their respective versions of structuralism. Hellman cites the adherence as far as possible to

[8] Benacerraf (1965), Shapiro (1991), Hellman (1989).

[9] Resnik (1981).

[10] Charles Parsons' distinction between eliminative and non-eliminative structuralism (Parsons 1990) may be coextensional with the distinction between quantificational and referential structuralism.

classical logic, a non-mystical epistemology, recognition of the primacy of proof in mathematics, and satisfying accounts of applicability and analyticity. Resnik effectively agrees with the first four of these, adds that we also have a satisfying account of the abstract–concrete distinction, and an explanation for mathematicians' interest in mathematical objects only 'up to isomorphism'. Each kind of structuralism has its own peculiarities and it is difficult to balance their pros and cons: what may be an advantage of one is a disadvantage of another. The issues are too complex to enter into here; I shall simply state without further justification that referential structuralism seems to cohere better with the role of adjustive abstraction than quantificational structuralism, especially in the modalized form offered by Hellman. Being blatantly Platonist, referential structuralism quantifies from the start over abstract structures and so does not run the risk of 'running out of objects', there being no limit to the size of structures it canvasses. Concretistic tendencies on the other hand tend to dampen the enthusiasm one feels for structures of large size. If the cosmos turns out to be finite in extent, divisibility, and cardinality, only modal structuralism will grant the truth of theories requiring infinite domains, but it may be doubted whether even an infinite cosmos will grant enough concreta to satisfy the more elevated regions of large cardinal theory.

It may be that the two versions of structuralism could be reconciled using Kit Fine's notion of *arbitrary objects*.[11] Arbitrary objects are like Resnik's structures in being in various respects indeterminate, but they are genuinely referred to, while on the other hand they are used to explicate the way in which we understand quantification, and indeed the way we naturally reason within mathematics. Such a suggestion for reconciliation however ignores the modal dimension of Hellman's version of structuralism.

In *Frege: Philosophy of Mathematics* Michael Dummett gives a more elaborate classification of kinds of structuralism.[12] Structuralism as a philosophical theory about the subject-matter of mathematics divides into weak and strong. In weak structuralism, which we can ascribe to Frege, applications enter into the definition of mathematical objects, but only in that *anything* of suitable structure is catered for. For example real numbers apply for Frege to any set of relations forming what he calls a *magnitude domain*: the conditions such a domain needs to satisfy are simply laid down, but the objects defined (real numbers) are defined as extensions covering all such domains. Strong structuralism divides for Dummett into hard-headed and mystical. Hard-headed structuralism allows that structures do have non-structural properties, whereas mystical structuralism claims they have no non-structural properties.

[11] Fine (1985). [12] Dummett (1991, 295–7).

A bottom-up or abstractionist approach to structures takes some of the venom out of the oppositions between the various approaches. One may grant that there are structures with (adjusted) non-structural properties, while denying that they are mathematical objects: the IBM extended computer keyboard has non-structural properties but is not a mathematical object, yet it has many purely mathematical structures associated with it and applicable to it by being instantiated in it.

4. STRUCTURES OF STRUCTURES

Resnik does not allow structures of structures, since he takes all structures to consist of structureless points. But I believe we can make sense of this notion. (Once we allow it once, we may iterate the move.) The following figures will illustrate. A structure is marked by a closed curve, the positions within it as points (dots). The relations among the points may sometimes be partially represented graphically, for instance left-to-right positioning may represent an ordering. Here to begin with is a simple example, the structure ω of the Dedekind–Peano numbers (Fig. 19.1). The leftmost point is the first number, the point to the right of a point is its successor, the three dots are a fudge to show we have not depicted all the numbers.

Unlike most set theorists I take ordinals to be distinct from cardinals, and both of these to be distinct from the Dedekind–Peano numbers. Ordinals I understand just as Cantor did, as the *order-types* of well-ordered sets. Cardinals on the other hand are the *powers* of sets, they measure how many elements a set has. One *can* get from ordinals to cardinals by literally abstracting from order, as Cantor (wrongly) thought one *must*. Fig. 19.2

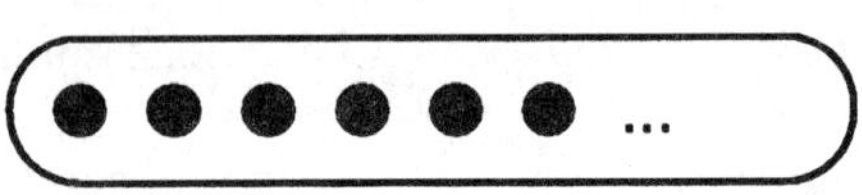

FIG. 19.1

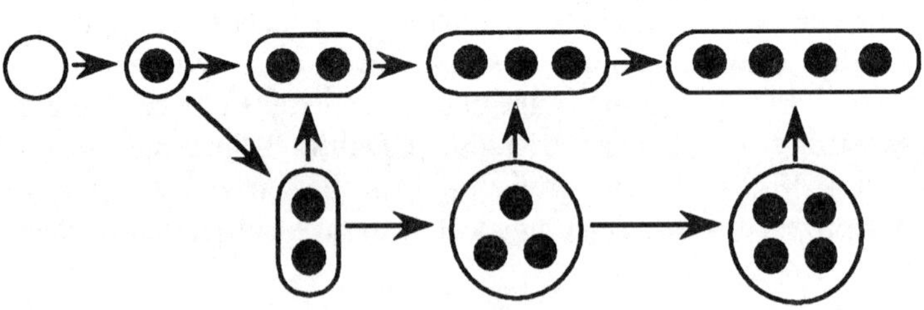

FIG. 19.2

shows the first few ordinals and cardinals: the ordinals are above and the cardinals below. Consider first the ordinals: each ordinal is itself a structure, the structure of the ordered sets whose order-type it represents.

The leftmost structure is here the (unique) empty structure, the one with no points. Because each ordinal is a structure, they cannot be the same as the structureless points of the Dedekind–Peano structure ω. But there is a intimate relation between the finite ordinals and the Dedekind–Peano numbers. In the diagram, an arrow between one structure and another indicates that the first is an immediate substructure of the second, or occurs immediately in it. The relation *occurs immediately in* is an internal relation among the structures making up the class **Ord** of ordinal numbers, and if we *abstract from* the *internal* structure of each ordinal and consider only its relation to the ordinal it occurs immediately in, we obtain an ordered structure of points: *occurs immediately in* becomes the successor relation. The poignancy of the example consists among other things in the fact that ω itself occurs in the class **Ord** as the first transfinite ordinal, but it cannot be reached from the finite ones by finite iteration of the *occurs immediately in* relation, only by a 'leap', a new procedure, and it is precisely the ordinals themselves that give expression to this structural relation, the limit ordinals representing the 'leaps'. Burali–Forti's Paradox may now be understood as proving that there is no member of **Ord** which is obtained from **Ord** by ignoring internal structure. **Ord** is too big to have a structure of its own which can be one structure among the many making up **Ord** itself. Clearly though there are *parts* of **Ord** which *do* have their own structures, the sequences of finite ordinals for instance, whose structures are precisely the finite ordinals and ω. The finite ordinals, ordered by *occurs immediately in*, form a structure of structures. We might call all the ordinals a *hyperstructure* of structures.

The lower part of the example shows the first few cardinals. The first two cardinals (zero and one) are actually identical with the first two ordinals, thereafter the cardinals are different structures because they lack internal ordering. From this point of view, with the exception of zero and one, the ordinal so-called numbers are strictly no more numbers than any other structures whose identity depends on more than just how many points they contain (how many positions they have). But the ordinals and the cardinals are again intimately related. Take any ordinal and abstract from the ordering relation among its points, and one obtains its corresponding cardinal: the cardinal occurs as a structure within the ordinal, but its inclusion is not like that of the ordinals in other ordinals, rather it is inclusion by loss of internal relational structure. Cardinal 7 is included in ordinal 7, it is also included in cardinal 8, and thus in ordinal 8, etc., but ordinal 7 is not included in any cardinal, no matter how big, since cardinals lack internal order: their points are 'merely distinct'. The inclusion relation among

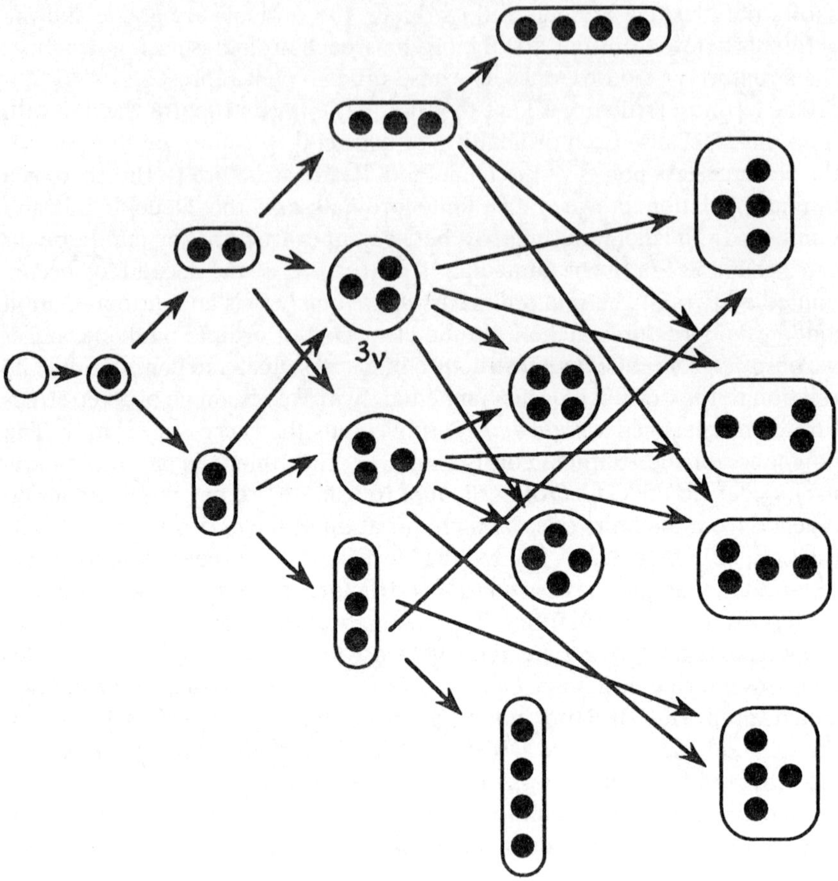

FIG. 19.3

cardinals looks initially just like that for the ordinals, and the structure of the finite cardinals (ignoring their internal structure, i.e. the fact that they have different numbers of points) is precisely ω again. A question mark—the continuum problem—hangs over the issue of how the transfinite cardinals are related by cardinality, and Cantor's Paradox shows again that the class **Card** does not have a structure which is a member of **Card**.

Fig. 19.3 shows the beginning of the structures making up the class **POrd** of order-types of partially ordered sets (pordinals): these include **Ord** as a subclass. Again the internal order of the structures in **POrd** is represented by the left–right ordering: incomparable points are in a vertical line, but up–down has no significance: incomparable points are again 'merely dis-

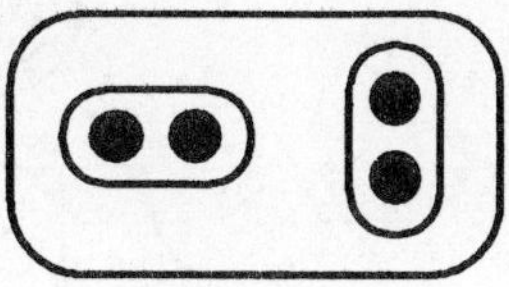

FIG. 19.4

tinct'. But completely 'flat' orderings, where all points are incomparable, are not cardinals, since they are still ordered structures and order types. If the 'flat' 4 were cardinal 4, then it would be included in ordinal 4 and other partial-order types of four points, but clearly it is quite distinct from these. No cardinal (except zero and one again) is a pordinal, but each pordinal has its own cardinal, the one giving its cardinality.

While we have seen structures related by inclusion, we have not yet seen structures literally inside other structures. Fig. 19.4 is a very simple example: two two-point structures making up a complex structure, in fact the structure consisting of simply the two pordinals of cardinality two. Here the ordinal 2 and the flat 2 are contained in the whole structure in a different way from that in which ordinal 2 is in ordinal 3 or cardinal 2 is in ordinal 2: they are *elements* of the whole structure, not substructures of another. The term 'element' is in my view just right here: when one thing is immediately in another in this way we are looking at the set-membership relation.

Fig. 19.5 then is a more complex example of structures within structures, where here the arrows represent membership. This is a depiction of the very beginning of a very interesting structure of structures, a pure set-theoretic hierarchy. It is pure because there are no urelements, no black dots: it starts with the empty set (empty structure, zero in every sense). The labelled structures are further objects which have been or might be called 'numbers': 3_Z is Zermelo's 3, while 3_N is von Neumann's 3. Zermelo's 3 does not have three elements, von Neumann's does. Each sequence, the Zermelo 'numbers' and the von Neumann 'numbers', forms a subcollection of the whole hierarchy whose ordering is that of the ordinals. We could have another kind of 'number': rather than take the smallest set at each iteration as Zermelo does, we could take the largest, the power-set. So 3_P is the power 3, and it has four elements, being the power-set of all its predecessors in the cumulative hierarchy. Von Neumann's 3 is elegant in containing not all predecessors in the set-theoretic hierarchy but only those in the 'number' hierarchy, but as tokens for counting the Zermelo and power 'numbers' are no less effective than von Neumann's.

This picture offers the glimmering of a structuralist understanding of set theory, which has tended to fit uneasily hitherto into the structuralist

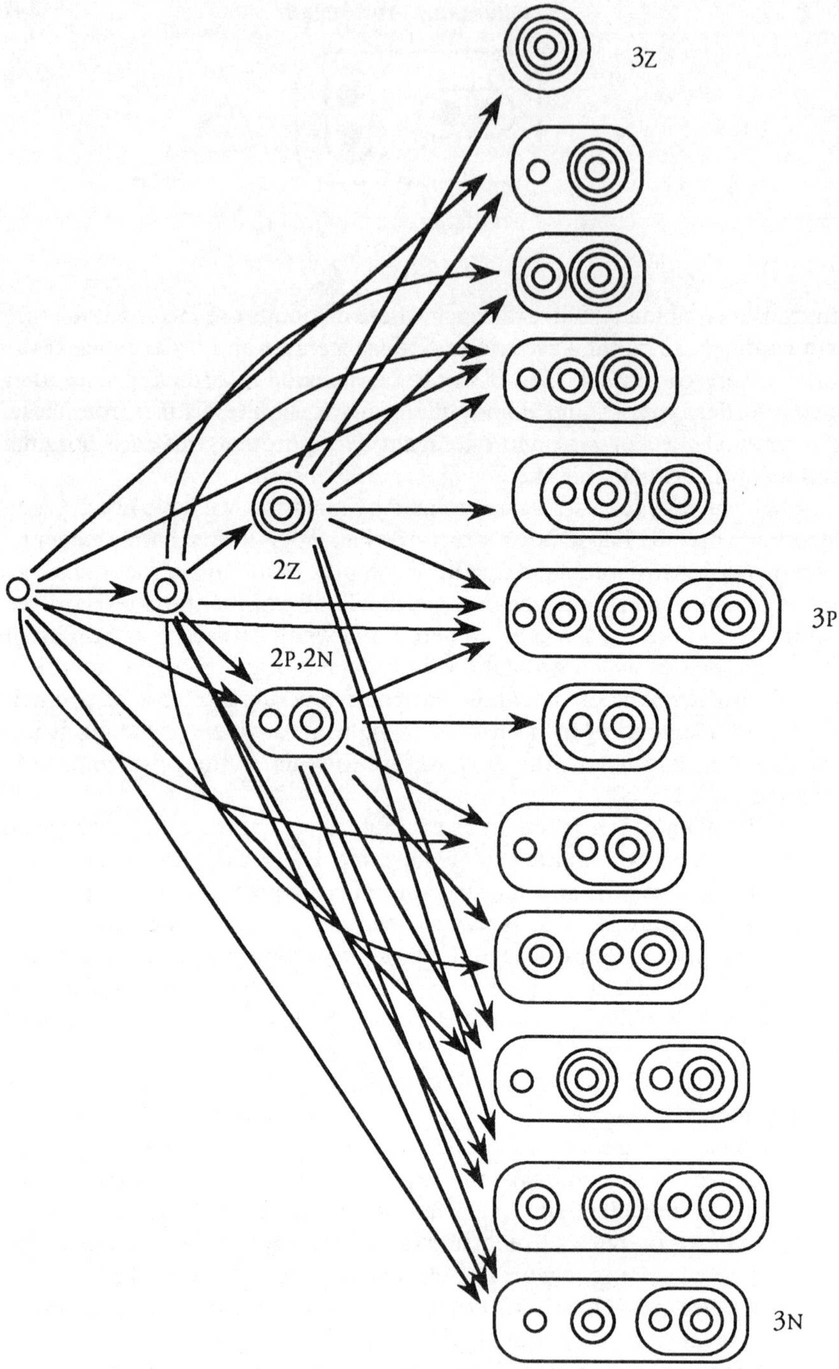

FIG. 19.5

picture. Resnik for instance sees set theory as portraying just one structure among others, but this does not do justice to the special status of set theory within mathematics, whereas to see set theory as the structural theory of structural containment does give it a very special role. The imponderables about how to manage the iterative continuation into the transfinite of set-theoretic hierarchies whose very beginnings are depicted here are substantive issues within set theory and not material for philosophical comment: it is up to mathematicians to discover rather than philosophers to legislate how the continuations can or must go.

The prospects for generating new structures out of old through inclusion and containment *lighten* a problem faced by would-be abstractionist nominalists, the shortage of objects problem. No basis of relative concreta for an abstractive step can be of lower cardinality than the equivalence types defined over it, so it would seem there can be no more abstracta defined for a given collection of concreta than there are partitions of these. This is the same problem which led Hellman to introduce the modal component into his quantificational structuralism, just in case the actual world is too sparing with concreta to generate sufficient mathematical objects (in the worst case, only finitely many). But if we can rope in the abstracta themselves into providing new bases for abstraction the prospects are brighter. Indeed provided the equivalence type of certain objects is always distinct from each of these objects (what the Erlanger followers of Paul Lorenzen call the 'Difference Postulate') we can start from one abstractum. Frege explicitly designed his cardinal numbers so that no empirical facts about how many concreta there are have any bearing on what numbers exist.

Concatenating structures in the way I have sketched illustrates another way to do this, and the cumulative hierarchy shows exactly where abstractive 'innocence' is lost, namely in Zermelo's Axiom of Infinity. This suggests however that there appears to be no way from a purely abstractionist point of view, according to which only abstraction and not the combination of structures is permissible, in which one can generate enough structures for mathematics.

5. PURITY

A piece of unfinished business is the question what makes a structure pure, since it was hinted earlier that mathematics is the theory of *pure* structures. Consider first a simple structure, that is, one having only points as elements. Its identity is determined only by the cardinality of its points and by the *logical* properties of the relations which hold among these in the structure. It is important to be able to state what a logical property is without invoking the notion of a pure structure, otherwise circularity will result.

Fortunately Tarski has shown when a relation or property is logical, namely when it is invariant under all permutations of any domain of individuals over which it is defined.[13] Transitivity is a logical property of relations for example, and it applies to many relations that are not themselves logical, such as the relation *acquired later than* defined over the books in my library. That an ordering relation on a linearly ordered set—whatever the relation is in concreto—is an ordering may be an empirical fact, but to be a linear ordering is to have that logical property, so the order type of a linearly ordered set is a structure, where only the cardinality and the logical properties of being a linear ordering are invariant. Consider the pordinal 3_V. What makes it a pure structure is that it is a partial ordering on three elements such that there is one element preceding the other two, neither of which precedes the other, and these two obviously have no successors. These are all logical properties.

So we have an idea what it means to be a pure *simple* structure. Now consider a complex structure $, whose immediate elements are either points or themselves pure structures. Then if the structure obtained from $ by replacing each element by a point is itself a pure simple structure then $ is itself pure. The finite structure of structures (suppressing the dots) in Fig. 19.3, showing the inclusion of pordinals in others, is a pure structure because each pordinal is a pure simple structure and their inclusion structure is itself a pordinal somewhat higher up the pordinal hierarchy, and likewise pure. There is a question mark over how this iteration, determining what purity is, may be continued into the transfinite.

6. SUMMARY

I indicated how Frege-abstraction delivers lowish-level abstracta, not just mathematical ones, and that mathematical structures may be obtained by judicious choice of equivalence relation. The idea that mathematics is the theory of pure structures possibly only works for pure mathematics, but an attempt was made to elucidate purity of structure. I suggested how the adjustive conception of abstraction, and being given leeway to concatenate and combine structures, suggest a discriminating and powerful ontology of structures, and noted several imponderables as to how far structures may be generated by iterative combination: well-known paradoxes indicate there are limits to the procedure. Pure abstraction on its own appears too cardinally impoverished to deliver the structures mathematicians find interesting. One of the important issues in structuralism which must be faced concerns the identity conditions for structures and points, of which Resnik

[13] Tarski (1986).

has made much. This I deliberately skirted. Likewise I left the question whether structures really exist or not where I seemed to find it, namely up in the air.

REFERENCES

Benacerraf, P. (1965), 'What Numbers Could Not Be', *Philosophical Review*, 74: 47–73.

Dummett, M. (1991), *Frege: Philosophy of Mathematics*, London, Duckworth.

Fine, K. (1985), *Reasoning with Arbitrary Objects*, Oxford, Blackwell.

Hellman, G. (1989), *Mathematics without Numbers*, Oxford, Clarendon.

Parsons, C. (1990), 'The Structuralist View of Mathematical Objects', *Synthese*, 84: 303–46.

Resnik, M. (1981), 'Mathematics as a Science of Patterns: Ontology and Reference', *Noûs*, 15: 529–50.

Shapiro, S. (1991), *Foundations without Foundationalism: A Case for Second-Order Logic*, Oxford, Clarendon Press.

Tarski, A. (1986), 'What are Logical Notions?', *History and Philosophy of Logic*, 7: 143–54.

20

The Limits of Abstraction

KIT FINE

This paper has been written more from a sense of curiosity than commitment. I was fortunate enough to attend the Munich Conference on the Philosophy of Mathematics in the Summer of 1993 and to overhear a discussion of recent work on Frege's approach to the foundations of mathematics. This led me to investigate certain technical problems connected with the approach; and these led me, in their turn, to reflect on certain philosophical aspects of the subject. I was concerned to see to what extent a Fregean theory of abstraction could be developed and used as a foundation for mathematics and to place the development of such a theory within a general framework for dealing with questions of abstraction. My conclusions were somewhat mixed: a theory of abstraction could be developed somewhat along the lines that Frege has envisaged; and it could indeed be used as a basis for both arithmetic and analysis. When wedded to a suitable version of the context principle, the theory was capable of accounting for our reference to numbers and other abstract objects. But without the support of the principle, it was not clear that the theory had any great advantage over its rivals. Thus my results would be congenial to someone already committed to the Fregean approach though unconvincing to someone who was not. I therefore present them in somewhat the same spirit as someone who sends off a message in a bottle. I have no desire to announce my communication to the world; but if someone stumbles across it and finds it to be of interest, I shall be pleased.

The paper is in three parts. The first is devoted to philosophical matters, which help explain the motivation for the subsequent technical work and also its significance. It is centred on three main questions: What are the correct principles of abstraction? In what sense do they serve to define the

I am much indebted to the participants in that discussion—principally, Boolos, Clark, Hale, Heck, and Wright. Preliminary versions of the paper were given at the third Austrian philosophy conference in Salzburg, at a talk at the City University of New York, at a philosophy of mathematics workshop at UCLA, and at a workshop on abstraction in St Andrews; and I am grateful for the comments that I received on those different occasions. I have been greatly influenced by the writings of Michael Dummett and Crispin Wright and have greatly benefited from the comments of Tony Martin.

abstracts with which they deal? And to what extent can they provide a foundation for mathematics? The second part proposes and investigates a set of necessary and sufficient conditions for an abstraction principle to be acceptable. The acceptable principles, according to this criterion, are precisely determined and it is shown, in particular, that there is a strongest such principle. The third part attempts to develop a general theory of abstraction within the technical limitations set out by the second part; the theory is equipped with a natural class of models; and it is shown to provide a foundation for both arithmetic and analysis.

The original version of the paper contained a lengthy section on the context principle. But this acquired a life of its own (just as reference does under the principle); and it has therefore been omitted. I hope to be able to present the material elsewhere.

PART I: PHILOSOPHICAL INTRODUCTION

An abstraction principle associates objects with the items from a given domain, the objects associated with two items being the same when the items are suitably related and the objects being distinct when the items are not so related. For example, we may abstract on concepts in accordance with the principle that concepts between which there exists a one–one correspondence are to be associated with the same number; or we may abstract on lines in accordance with the principle that parallel lines are to be associated with the same direction.

A principle of abstraction is said to be *conceptual* when the items upon which it abstracts are concepts and it is said to be *objectual* when the items in question are objects. Thus of the two examples above, the first is a principle of conceptual abstraction and the second a principle of objectual abstraction. The two kinds of principle are fundamentally different; for the conceptual principles involve a 'projection' of the larger domain of concepts into the smaller domain of objects, where these objects may themselves fall under the concepts upon which an abstraction is made. It is this special reflexive feature of the conceptual principles that makes them so powerful—and also so dangerous.

The serious study of conceptual abstraction, as it is understood here, began with Frege. He was the first to provide a clear statement of its principles and the first to provide a cogent account of how such principles might be of use in determining an ontology of abstract objects and a foundation for mathematics. Two principles of conceptual abstraction are prominent in his work. The first is the principle for abstracting to numbers, which was mentioned above and will be called 'Hume's Law'. The second is the principle of extensional abstraction, Law V of the *Grundgesetze*; it associates extensions with concepts, the extensions being the same when the objects falling under the concepts are the same.

Frege attempted to derive the whole of arithmetic and analysis from Law V with the help of other, less problematic, logical principles. But the attempt failed if for no other reason than that Law V leads, by the argument of Russell's Paradox, to a contradiction. One might have thought, prior to the discovery of the paradoxes, that extensional abstraction was innocuous. For what is to prevent one from picking out the extension of each concept? Indeed, the appeal of extensional abstraction in this regard would seem to extend to all forms of conceptual abstraction. For just as there would appear to be nothing to prevent one from picking out the extension of each concept, so there would appear to be nothing to prevent one from picking out its counter-extension or its number or any other aspect of it.

But the appeal of conceptual abstraction in this regard depends upon overlooking the critical respect in which it differs from objectual abstraction. For what makes it appear so innocuous is the view of concepts as being given independently of the objects which are abstracted from them. Once it is recognized that the abstracts themselves may play a role in the determination of concepts, it is no longer so clear that one can simultaneously abstract on the concepts and 'conceptualize' over the resulting objects.

Russell's Paradox has, until recently, tended to deflect attention away from the topic of conceptual abstraction. Philosophers and logicians alike have followed Frege's own lead in considering other approaches to the foundation of mathematics and have not been so interested in the topic of conceptual abstraction as such. The prevalent view is that abstracts should just be treated as equivalence classes (or perhaps we should say equivalence *sets*). The theory of abstraction thereby becomes a part of the much more comprehensive theory of sets or classes.

But conceptual abstraction, as so conceived, represents a double departure from the Fregean conception. In the first place, it is not conceptual, i.e. abstraction on concepts, since the members of the classes are objects, not concepts; in the second place, it is not even Fregean abstraction, since the classes in question will not arise from a principle of abstraction of the sort envisaged by Frege.

Maybe the Fregean idea of conceptual abstraction should be given up; and maybe the current conception of abstraction should take its place. But it is surely premature to dismiss the Fregean conception on the grounds of a single failure.

In developing the Fregean approach, there are three main questions that need to be considered. The first is which, if any, of the principles of conceptual abstraction are true (or otherwise acceptable); the second is, given that an abstraction principle is true, what kind of truth is it and how, in particular, does it relate to the objects with which it deals; and the third is whether abstraction principles can serve as a foundation for mathematics or, at least, a significant part thereof. Frege, in his state of pre-Russellian innocence, could provide oversimplified answers to all three questions. He could embrace all abstraction principles indiscriminately by way of their reduction to extensional abstraction; he could treat Law V as determinative of its subject-matter; and, with the help of the Law, he could provide a foundation for arithmetic and analysis. On all these matters, he was mistaken. But we, in our state of post-Russellian sophistication, can still profitably consider the questions that his work raises even if we cannot accept his answers.

1. TRUTH

Of our three questions the most important is certainly the first; for it is only when one has settled on the truth of a principle of abstraction that one can sensibly raise the question of its status as a truth or its role in providing a foundation for mathematics.

Any abstraction principle can be stated in the form:

the abstract of α = the abstract of β iff . . . α . . . β . . . ,

where the variables α and β range over the items on which the abstraction is to be performed, the phrase 'the abstract of' stands in for an abstraction operator, such as 'the number of' or 'the direction of', and the clause '. . . α . . . β . . .' represents a criterion of identity, such as equinumerosity or coextensiveness. Two necessary conditions for the truth of an abstraction principle hold as a matter of logic (see Section II.4). In the first place, it follows from the truth of an abstraction principle that its underlying criterion of identity on concepts should be an equivalence relation (reflexive, symmetric, and transitive). Thus each abstraction principle will induce a partition of the domain of items into mutually exclusive 'equivalence classes'.

Secondly, it follows from the truth of an abstraction principle that the identity criterion should not be inflationary, the number of equivalence classes must not outstrip the number of objects. There must, that is to say, be a one–one correspondence between all of the equivalence classes, or their representatives, on the one hand, and some or all of the objects, on the other. It is, of course, exactly on this score that Law V proves unacceptable; for where there are n objects, it demands that there be 2^n abstracts.

The question naturally arises as to whether these two necessary conditions are jointly sufficient for the truth of a principle of conceptual abstraction. In considering this question, it is important to distinguish between circular and non-circular criteria of identity. A criterion is circular or not according as to whether it involves the very notion of abstraction that is in question.[1] The standard criteria of identity, such as equinumerosity or coextensionality, are non-circular; while the identity criterion behind Frege's proposed amendation to Law V is not, since coextensionality is restricted to the objects that are distinct from the given abstracts.

It is clear from the consideration of examples that our proposed conditions are not sufficient for the truth of abstraction principles whose identity criterion is circular. The general difficulty is that how one associates objects with equivalence classes of concepts will help determine what those equivalence classes are. Thus there may be no consistent way of simultaneously fixing the equivalence classes and associating them with objects (see the comments to lemma II.4.1).

In the non-circular case, however, there is no such impediment to the truth of an abstraction principle; the concepts can be associated with objects in such a way as to render the principle true (lemma II.4.1). It is therefore tempting to suppose, on the grounds of a general principle of plenitude for abstract objects, that the abstraction principle, in such a case, is actually true. Since there *can* be abstract objects of the sort in question, it is supposed that there *are* such objects.

It is important in considering this proposal to distinguish the question of whether a given principle is correct from the question of whether it is genuinely a principle of abstraction. Whether something is a principle of abstraction has been for us a matter of form; the principle must state an equivalence between an appropriate form of identity statement, on the one side, and an appropriate form of identity criterion, on the other. But from among the abstraction principles in this formal sense, only some are abstraction principles in a genuine or 'material' sense. This point can be made especially vivid if we bear in mind that to any statement whatever there will correspond an abstraction principle; for we may take its criterion of identity to be one which identifies two concepts just in case the statement holds (c.f. Heck 1992). The statement and the principle will then be equivalent (at least granted the existence of a universal method of abstraction, one which identifies all concepts). But a principle obtained in this way will not in general be a genuine principle of abstraction; for its truth will not rest entirely, or even principally, on the existence of abstracts.

I assume it would be very difficult to say what makes a true principle a genuine principle of abstraction. However, our present concern is not with picking out the genuine principles from among those that are true but in

[1] A non-circular criterion can involve the notion of abstraction associated with some other criterion of identity as long as there is no circularity in the system of principles as a whole. The possibility of such non-circular systems of principles is countenanced in the technical discussion of Parts II and III but, for purposes of simplicity, I have here assumed that each non-circular criterion will not involve any other criteria.

picking out the true principles from among those that are of the required form; and, regarded in this way, the proposed condition of sufficiency is far more plausible.

Unfortunately, it is still subject to difficulties. One, which we dub *additivity*, is familiar from the literature (see e.g. Boolos (1990), 273–4). Two principles may make incompatible demands on the size of the universe, even when each alone makes a consistent demand on its size. We may imagine a principle, for example, which inflates on finite universes but not on infinite ones and another principle which inflates on infinite universes but not on finite ones. Thus the two principles, when taken together, will inflate on any universe; and so our proposed conditions cannot be taken to guarantee the truth of the principles.

Such a difficulty could be overcome if the domain of objects were taken to be given in advance of the adoption of any principle of abstraction. The non-inflationary principles could then be taken to be those that were non-inflationary on this particular domain rather than on some domain or other; and the clash between the different non-inflationary principles would then be avoided. The natural way to fix the domain in advance is to let it be the universal domain, i.e. inclusive of all objects whatever. The idea is that somehow, prior to the adoption of any principle of abstraction, we can determine the size of the universe and that, once this is done, we can determine which of the principles are true.

But even if we overlook the difficulty of fixing the size of the universe prior to the adoption of any principles of abstraction, there is another, less familiar, difficulty which can arise. Let us suppose that there are more principles than objects or, to be more exact, that they give rise to more equivalence classes of concepts than there are objects. Then, on the assumption that the abstractions associated with distinct equivalence classes of concepts are distinct, there will again be more abstracts than objects. Thus although no principle alone inflates on the given universe, taken together they do inflate. We have what might be called the problem of *hyper*inflation.

This problem can arise even with abstraction principles that merely serve to divide the universe of concepts into two. For given any concept, we may divide the universe of concepts into those that are coextensive with the given concept and those that are not. Each such division then yields 2 abstracts; but, taken together, they yield 2^n abstracts, where n is the size of the universe.[2]

[2] There is a corresponding problem of hyperinflation for objectual abstraction even though there is no corresponding problem of inflation. For given any concept F, we may divide the universe of objects into those that F and those that do not. Thus once we allow there to be multiple principles of abstraction, the difference between the objectual and conceptual cases is not so great.

Although the problem has been stated in the metalanguage using standard set-theoretic machinery, it is readily reproduced in the object-language. Thus, in the case of objectual

I propose to solve this problem by imposing a further condition on the non-circular criteria of identity. For some are purely logical in the sense that they can be formulated without the aid of non-logical concepts. Thus the familiar criteria of coextensiveness and of equinumerosity are logical; but the divisor criteria above (with the sole exceptions of division by the universal and empty concepts) are not. Each logical criterion corresponds to an invariant equivalence relation, one which can take no account of the specific identity of the different objects (or, more technically, to one which is invariant under any permutation).

Under a certain assumption, it can be shown that the principles, as so restricted, will not hyperinflate. Let us say that a transfinite cardinal **c** is *unsurpassable* if $2^{\mathbf{d}} \leq \mathbf{c}$, where **d** is the number of cardinals less than **c**. Then, by means of the analysis of invariance in II.7, it may be demonstrated that the non-inflationary and logical abstraction principles will fail to hyperinflate just in case the cardinality of the domain is unsurpassable. Thus the unsurpassable cardinals are a kind of analogue, in our theory, to the inaccessible cardinals of ZF; and, just as with the inaccessibles, the question of their existence cannot be settled within ZF (Section II.8).

The proposal can be taken further. For let us say that an abstraction principle is *predominantly logical* if its identity criterion involves only a 'small' number of objects in relation to the number of objects in the universe as a whole.[3] For example: given a small class of objects, the agreement of two concepts with respect to the objects in the class is predominantly logical; and any criterion which uses the concept of being an individual will be predominantly logical as long as there is a small number of individuals in relation to the number of objects as a whole. The result can then be extended to the principles that are predominantly logical; they too will not hyperinflate as long as the domain is of unsurpassable cardinality.

abstraction, let [F] be the divisor method of abstraction associated with the concept F. Subject [F] to the principle:

$$\forall F \forall x, y([F]x = [F]y \leftrightarrow (Fx \leftrightarrow Fy));$$

lay down the identity postulate:

$$\forall F, G \forall x, y([F]x = [G]y \rightarrow \forall z((Fz \leftrightarrow Fx) \leftrightarrow (Gz \leftrightarrow Gy)));$$

and define $\in$ by:

$$x \in y =_{df} \exists F(Y = [F]x \,\&\, Fx).$$

Then we may show:

$$\forall F(\exists x Fx \rightarrow \exists y \forall x(x \in y \leftrightarrow Fx);$$

and from this $\forall x,y(x = y)$ follows by the reasoning of Russell's Paradox.

[3] The notion of being small, as used here, is not the usual one. A subset *C* of cardinality **c** is said to be small relative to a domain *D* of cardinality **d** if $\mathbf{d}^{\mathbf{c}} \leq \mathbf{d}$, i.e. if the number of subsets of the same cardinality as the given subset does not exceed the cardinality of the domain itself. Thus a predominantly logical criterion is one whose extension is invariant under any permutation that is an identity on some small predetermined set of objects.

A kind of converse result also holds. For if we allow abstractions to be defined on a large number n of objects, then hyperinflation will result. For corresponding to any set of n objects, there will then be a unique abstract; and there are more such sets, and hence more such abstracts, than there are objects in the universe. Thus the restriction to those identity criteria that are predominantly invariant is, in a sense, the best possible.

It follows from the preceding considerations that the two necessary conditions, being an equivalence on concepts and being non-inflationary, can be taken to be jointly sufficient for the truth of any principle of abstraction that is non-circular and predominantly logical. Furthermore, it can with some plausibility be supposed that any principle of abstraction should be predominantly logical; for if one non-inflationary principle involving a large number of objects is permitted it is hard to see why all such principles should not also be permitted. It is also plausible to require that abstraction principles should be like definitions in general in being non-circular. Under these two suppositions, we then obtain a necessary and sufficient condition for the truth of an abstraction principle; for an abstraction principle will be true just in case its identity criterion and yields a non-inflationary and predominantly logical equivalence on concepts.

The two requirements provide a condition of truth. But how useful are they in providing a *test* for truth? To what extent can they be used in *establishing* that a given abstraction principle is true? In the special case in which an abstraction principle is non-circular, i.e. innocent of reference to the notion of abstraction, the requirements for its truth will likewise be innocent of such reference. Indeed, it will be possible to formulate the requirements within the language of higher-order logic—and, under certain simplifying assumptions, within second-order logic itself (Section II.4). It might therefore appear that we could settle which abstraction principles are to hold on the basis of the corresponding higher-order or second-order conditions and thereby avoid any appeal to intuitions regarding the abstracts themselves.

However, the epistemological advantages that such a reduction appears to provide are largely illusory. For the satisfaction of the corresponding higher-order conditions is itself to be settled on the basis of substantive non-logical considerations. Consider Hume's Law, for example. The satisfaction of the corresponding non-inflation condition will require the existence of infinitely many objects. And how is this to be ascertained without presupposing that there are infinitely many numbers, or sets, or abstract objects of some other sort?

Still, the proposed conditions are not altogether useless as a test. Let us distinguish between objects of abstraction, which are given by a principle of abstraction, and abstract objects in a broader sense of the term. Thus points in abstract Euclidean space are abstract objects and yet are not objects of abstraction, since they are not introduced, and cannot sensibly be taken to be introduced, through a principle of abstraction of the sort envisaged by Frege. The conditions then enable us to settle the question of the existence of objects of abstraction of a given kind either on the basis of the existence of objects of abstraction of some other kind or kinds or on the basis of the existence of abstract objects that are not given through abstraction. An example of the first kind of justification is illustrated by our own axiomatic theory of abstraction; for within that theory, an infinity of divisor abstractions (ones that divide the universe of concepts into two) will enable us to justify number abstraction. A powerful example of the second kind is provided by standard set theory; for its huge ontology will bring a host of non-inflationary conditions in its wake.

Our conditions have so far been stated by appeal to an informal concept of truth. I want now to consider two model-theoretic criteria of acceptability and see how they compare with the informal criterion.

We can think of each model-theoretic criterion as being obtained from the informal criterion by adopting an appropriate model-theoretic criterion of truth. The first, which we call *tenability*, results from taking the universe of objects to be given as a set. This set will determine a standard model, one containing a concept for each subset of objects; and truth can then be taken to be truth relative to the model.

Under the simplifying assumption that the abstraction principle is logical or 'invariant', the truth of the equivalence and non-inflation conditions in a given model will depend only upon its cardinality. We may therefore say that an abstraction principle is *tenable on* a cardinal if its identity criterion determines a non-inflationary equivalence on the concepts in a standard model of that cardinality (Section II.4).

The second criterion, which we call stability, is obtained by treating truth as a limit concept; a statement is taken to be true when it is true in all models of sufficiently large cardinality. Adopting this criterion of truth, we may then say that an abstraction principle is *stable* if, for some cardinal, the principle is tenable on all greater cardinals (cf. Heck 1992, n. 4).

How faithful are the model-theoretic criteria to our informal conception of truth? In considering this question, we must take account of the attitude of the abstractionist to standard set theory (as embodied in ZF or ZFC). He can either be compromising or uncompromising. The uncompromising abstractionist rejects set theory. He therefore sees the theory of abstractions as an alternative, rather than as a supplement, to the standard theory of sets.

There are, of course, various grounds upon which ZF might be rejected. But there is one directly related to the abstractionist's position. For he may adopt an imperialistic stand and see all abstract objects as arising from abstraction. For him, to be an abstract object is simply to be an object of abstraction, one that is introduced by means of an appropriate principle.

The attractions of the thoroughgoing position are manifest; for it provides us with a single and simple unified point of view from which the various philosophical problems concerning abstract objects may be considered. But the difficulties in the position are enormous. For many kinds of abstract object, from both within and without mathematics, do not appear to fit within the abstractionist mould. They must therefore be rejected or, at best, accepted in a distorted form.

What is the uncompromising abstractionist to make of the tenability criterion? It appears as if it must be rejected—presumably as unintelligible, though possibly as false. For the criterion—with its reference to models, cardinals, satisfaction, and the like—is stated within the language of ZF, and so must be rejected along with ZF. Perhaps there is some way of stating the criterion, or something like it, in terms acceptable to the abstractionist. But if there is a such way, it is not at all clear what it is.

The advocate of ZF, on the other hand, is in a position to recognize the criterion as a correct account of truth for the uncompromising abstractionist. For he can acknowledge that, for certain cardinals, the principles tenable on those cardinals might be exactly those that would be true were his opponent correct. We have therefore what might be called an *externalist* characterization of a given position, one which can only be regarded as correct, or even as intelligible, by someone who does not hold the position. (A similar, and more familiar, case is provided by the classical characterizations of the constructivist conception of logical validity.)

Given that we adopt the externalist perspective, it is natural to ask: What is the size of our opponent's universe? In response to this question, it seems that the best one can do is to see which cardinals will render true what our opponent takes to be true. Thus if he adopts the principle of plenitude, as stated above, then we will know that the cardinal will be unsurpassable and hence uncountable. But this will not enable us to distinguish between his acceptance of one unsurpassable as opposed to another.

In the face of this difficulty, one might be tempted to suppose that our abstractionist should endorse any abstraction principle that stabilizes on some cardinal. For if his thinking is governed by a general principle of plenitude, then should he not recognize as true anything that would be true in a sufficiently large universe? But this is to import too much of the externalist standpoint into his own way of thinking. For how large can a universe be? For us, the possible size of the universe can be determined on the basis of set theory. But, for him, it must be determined in some other way; and if he is thoroughgoing, it must be determined on the basis of the methods of abstraction themselves. There is no way of seeing 'from above' which methods of abstraction can be accepted and which not.

Consider, for example, the restricted forms of the principle of extensional abstraction. They allow one to form the extension of a concept that holds of a sufficiently small number of objects—finitely many, or countably many, or what have you. We would know that such a principle was non-inflationary if the universe were sufficiently large. But from the standpoint of a thoroughgoing abstractionist, this is not something that can simply be assumed. It must itself be established on the basis of principles that are independently plausible. In fact, I shall later suggest that there are such principles and that they enable us to determine the cardinality of the universe of the uncompromising abstractionist as being that of the first unsurpassable.

We now turn to the case of the abstractionist who is prepared to endorse set theory. In considering this case, it is important to bear in mind that the proper setting for the theory of conceptual abstraction is second-order logic; for it is this which provides us with the logic of the concepts upon which the abstraction is made. We shall suppose that the second-order logic contains a general comprehension axiom, one to the effect that each condition determines a concept. It then follows, in particular, that there is a universal concept, one holding of every object in the domain. We must now ask: What is the attitude of our advocate of set theory to second-order logic? How does he conceive of the domain of objects from which the concepts are to be drawn? Two basically different attitudes may be distinguished. He might, on the one hand, be prepared to accept second-order logic as applying to the whole universe of sets and urelements. In this case, he will be willing to endorse the second-order version of ZF, with concepts having a far wider range of application than sets.

On the other hand, he might think of the domain of objects from which the concepts are drawn as itself a set. Indeed, this attitude is forced upon him if he thinks of sets as the most general method of collection; for given that there were a universal concept, one holding of every object in the underlying domain, there would have to be a universal set, one to which every object in that domain belongs. In this case, our theorist will merely endorse first-order ZF and will most naturally think of second-order logic as being interpreted within it. Thus, in the one case, our theorist will think of second-order logic as lying *alongside* ZF; while, in the other, he will think of it as lying *within* ZF.

Consider the second case first. Since there is no fixed domain for second-order logic, there is no fixed domain for the theory of abstraction and hence no fixed conception of truth. The proper setting for a theory of abstraction on such a view is not really second-order logic but the first-order theory of sets. Thus the resulting theory will be of a quite different sort from the one envisaged by Frege.

All the same, we may ask: Which second-order principles of abstraction can be made true on a domain of given cardinality? And presumably, the answer is: those principles that are tenable on the cardinal. For it seems reasonable to suppose that the abstracts exist; and given that they exist, they can be included within the domain without danger of inflation.

But it would also be desirable to have an absolute criterion of acceptability, one that is not relative to a given domain or cardinal. We would like the acceptable principles to be true in domains of arbitrarily large size, so that they can be applied regardless of the size of the underlying domain; and we would like singly acceptable principles to be jointly acceptable, so that the application of one such principle does not exclude the application of another. We are therefore led to propose the following requirement: all acceptable principles should be jointly tenable on arbitrarily large domains.

In attempting to meet this requirement, it will not do to define acceptability as tenability on arbitrarily large cardinals. For one principle may be tenable on exactly the successor cardinals while the other may be tenable on exactly the limit cardinals. Both will therefore be acceptable according to the criterion; and yet they will not be jointly tenable.

A natural solution to this problem is to use stability as the criterion of acceptability. For any set of stable principles can be made jointly true in a domain of sufficiently large cardinality (even allowing that different principles invariably yield different abstracts). But the solution is not altogether satisfactory. For a proper class of stable principles cannot, in general, be rendered jointly true. Given any cardinal, for example, consider an abstraction principle that is untenable up to the cardinal and tenable thereafter. Then no set-domain can render all such principles true.

Nor is the proposal optimal (under the more limited requirement that any *set* of acceptable principles can be made jointly true). For the requirement may be satisfied by taking a principle to be acceptable when the class of cardinals upon which the principle is tenable contains an unbounded and closed subclass of cardinals.[4] And even if there were an optimal solution, it could not be unique. For instead of appealing to the given acceptability-inducing classes of cardinals, we could appeal instead to the classes of their successors.

We turn, finally, to the case of a compromising abstractionist, one who is prepared to endorse second-order set theory. It might appear from our previous discussion that such a position is impossible. For set theory requires a domain (if we think of it as a set) whose cardinality is inaccessible,

[4] I owe this suggestion to Tony Martin. But note that even this proposal may not be optimal; for in a model of set theory whose cardinality is weakly Mahlo, the class of regular cardinals will be stationary and hence can be added to the given acceptability-inducing classes without violation of the requirement.

whereas abstraction theory, under a principle of plenitude, requires a domain whose cardinality is unsurpassable. But no inaccessible cardinal can be unsurpassable; and it therefore appears as if no set-domain can be taken to provide a model both for set theory (i.e. ZFC) and for our theory of abstraction.

But this line of thought is in error, even granted that it is appropriate to conceive of the domains as sets. For the proper framework for a joint theory is ZFI, i.e. ZF (or ZFC) with urelements. The abstracts can then figure among the urelements; and there is nothing to prevent the domain of urelements being of greater cardinality than the domain of sets.

What can be shown to be impossible, though, is the non-circular representation of abstracts by means of sets. For within second-order ZFI with abstracts, there will be a principle of abstraction corresponding to any condition on cardinals. Each such condition will yield a 'generalized cardinal' that is the abstract of the concepts whose cardinal conforms to the condition. It may then be proved, within the theory itself, that there is no one–one correspondence between these abstracts and the pure sets (theorem 1.2 of Part III). But any non-circular method of representing abstracts by means of sets should yield such a one–one correspondence; and so no such method is possible. The means of abstraction are therefore seen to extend beyond the reach of set theory.

We have here the ultimate *coup de grâce* for the identificatory standpoint. In has been customary to identify abstracts with equivalence classes. But sometimes, as with the cardinal numbers, this has not been possible; and the abstracts have then been identified with representatives from the equivalence-classes. But the present result shows that, even allowing for arbitrary methods of identification, it will not be possible to find a place for every abstract within the set-theoretic universe. Abstracts, on grounds of cardinality alone, must be treated as objects in their own right.

What attitude should the compromising abstractionist adopt towards the two formal criteria? The situation here is not so very different from that which prevails in set theory. We can hope that some cardinal will reflect the behaviour of the whole universe, with the principles tenable on that cardinal exactly coinciding with those that are true. But if there are such cardinals, we cannot, in the light of the semantic paradoxes, expect to be able to identify them independently of the concept of truth.

We might therefore attempt to *approximate* to the truth. One rather crude way is to appeal to the notion of stability. Thus granted that any true abstraction principle is true in domains of arbitrarily large size, stability will provide a sufficient condition for truth. But a more sophisticated way is to identify cardinals that reflect more and more of the properties that we perceive the universe to have. It will be natural on the generative conception of abstractions developed below to suppose that when the universe of pure sets has cardinality **c** the universe of abstracts will have as its cardinality the smallest unsurpassable cardinal greater than **c**. In this case, the present problem will reduce to the analogous problem in set theory of identifying the cardinals that reflect more and more of the properties of the universe of pure sets.

2. DEFINITION

It is natural to suppose that we can gain an understanding of number through Hume's Law, that it can somehow serve to say what number is. In the same way, it might be thought that the abstraction principle for directions can serve in defining direction and that, in general, any acceptable abstraction principle can serve in defining the sort of abstract with which it deals.

My aim in this and the next section is to examine some of the different ways in which Hume's Law might help in defining number or in which an abstraction principle might help in defining some particular kind of abstract. The focus in the present section is on definitions of an orthodox kind, while the focus in the succeeding section is on definitions that are in some way unorthodox.

It will be helpful, before considering Hume's Law itself, to outline what I take the orthodox conception of definition to be. A definition, we shall suppose, is of something linguistic by means of something linguistic. That which is defined is called the *definiendum* and that by which it is defined the *definiens*.

We may also talk of defining a non-linguistic item by means of something linguistic. We may say, for example, that 'the successor of 1' is a definition of the *number* 2 rather than of the *numeral* '2'. But in such cases we define the object by defining, or by providing the means for defining, an expression for the object. We shall later have occasion to consider cases of non-linguistic definition that are not straightforwardly reducible, in such a way, to the linguistic case.

We may distinguish between explicit and implicit definition. In an explicit definition, the definiendum will be a term (of a given grammatical type) and the definiens will be a term of the same grammatical type. The aim of such a definition is then to set up an appropriate equivalence between the two terms. To take two familiar examples of this sort: 'bachelor' may be defined as 'unmarried man'; or 'God' as 'the being no greater than which can be conceived'. In an implicit definition, on the other hand, the definiens will be a statement or condition that involves the term to be defined. Thus '+' may be defined by the condition '$n + 0 = 0$ & $n + m' = (n + m)'$'. The purpose of the definition is then not, of course, to set up an equivalence between the term and the defining condition but somehow to constrain the interpretation of the term by means of the condition.

We shall find it convenient to treat the explicit form of definition as a special case of the implicit form. Thus the explicit definition of 'bachelor' above will be treated as equivalent to the implicit definition of 'bachelor' by means of the condition, 'All and only bachelors are unmarried men'; and, similarly, the explicit definition of 'God' will be treated as equivalent to its implicit definition by means of the condition 'God = the being no greater than which can be conceived'.

An implicit definition may be simultaneously of several terms; and the terms, whether there be one or many, may be defined by several conditions. These conditions can perhaps be regarded as one by forming their (possibly infinite) conjunction; and the terms can perhaps be regarded as one by refashioning the definition to be of a term for a sequence and then defining the original terms as the members of the sequence. Be that as it may, our focus will be on the cases in which there is but a single term and a single condition; and we shall only consider the complications of there being several terms or conditions as they arise.

The purpose of a definition is to assign an interpretation to an hitherto uninterpreted expression. We may take the interpretation to consist either in the assignment of a referent or of a sense or both. Thus prior to the definition, the definiendum may be taken to be without a referent or sense. After the definition, if it is successful, the expression will have been assigned a referent or sense. One may, of course, talk about defining locutions that are already understood; and it is presumably in this way that Hume's Law serves to define number. But what we are saying, in such cases, is that if the expression were devoid of meaning, then the interpretation assigned by the definition would match, or correspond to, the interpretation that it actually has.[5]

Let us say that a definition has *referential import* if it is intended to result in the assignment of a referent to the term t that is to be defined. The assignment of the referent should, of course, be in conformity with the defining condition D(t). It is characteristic of what I have called the orthodox conception of definition that conformity amount to no more than the requirement that D(t) should be true; the reference of t is simply to be determined by the truth of D(t). When a definition with referential import is successful, i.e. when it succeeds in appropriately assigning a referent to its term in conformity with its defining condition, we shall say that it is *referentially effective*.

[5] We may also take a definition to *redefine* a previously interpreted expression. In some cases (an example is the assignment of n + 1 to 'n' in certain programming languages), a previously interpreted expression may be used in setting up a new interpretation for that very expression.

A definition may result in the assignment of more than one referent to its term—not in the sense that there is the simultaneous assignment of several referents to the term, but in the sense that there are several assignments, each of a single referent, to the term. A definition with referential import is said to be *deterministic* if it is meant to result in the single assignment of a referent; and it is said to be *indeterministic* if it is meant to result in an assignment of a referent but not necessarily a single assignment. For a deterministic definition to be referentially effective it must succeed in assigning a single referent to its term and for an indeterministic definition to be referentially effective it must succeed in assigning at least one referent to its term.[6]

There are semantical analogues to these referential notions of definition. We may say that a definition has *semantical import* if it is intended to result in the assignment of a sense to the term to be defined, where the sense is to be in conformity with the defining condition. Likewise, we may draw a distinction between the determinate and non-determinate forms of definition. In the semantical case, however, we do not want to take seriously the idea that the definition can result in the assignment of different senses to the term to be defined. The distinction must therefore be drawn along somewhat different lines.

Accordingly, we may say that a definition with semantical import is *deterministic* if it is meant to assign a sense that purports to be of a single object and that it is *indeterministic* if it is meant to assign a sense that purports to be of a range of objects. Thus the definition of 'God' by the condition 'God = the being no greater than which can be conceived' will be deterministic, while the definition of 'n' by the condition 'n is a number' will be indeterministic.

But what should the sense be taken to be in such cases? What is it for the sense of the term t, whether it be of a deterministic or indeterministic sort, to be 'in conformity' with the defining condition D(t)? Our answer to the corresponding question concerning reference makes natural an answer to the question concerning sense. When the definition is deterministic, the sense assigned to the term t should be the sense of the description 'the x such that D(x)'; and when the definition is indeterministic, the sense assigned to the term t should be the sense of a variable whose range is determined by the predicate D(x).[7] (I leave open the question of how the idea

[6] The definition of a constant, such as 'Let 2 be the successor of 1', is naturally taken to be deterministic, while the definition of a variable, such as 'Let n be a number', is naturally taken to be indeterministic. I have made it a requirement on the successful definition of a variable that it should be assigned at least one value. But this requirement might be dropped.

[7] There are certain cases in which we might want to give a somewhat different account of the sense. Consider the definition of 'God' by the condition 'God = the being no greater than which can be conceived'. Then rather than take the sense of 'God' to be that of 'the x such that x = the being no greater than which can be conceived', we might take it to be the sense

of the sense of a variable should be made to fit in with the rest of Frege's views on sense.) Given that the sense is assigned in this manner, the referents assigned by the definition will just be those determined by the sense.

We see that there are four different aims a definition might have: it might attempt to assign either a referent or a sense to the term to be defined; and in each case, it might attempt to do so either deterministicly or indeterministicly. On a Fregean view, at least, it is impossible for a definition to assign a referent without also assigning a sense. Any definition with referential import should therefore be taken to have semantical import. However, not all definitions need assign a referent or even be taken to have referential import. The definition of 'God' as 'the being no greater than which can be conceived', for example, might merely be intended to fix the sense of 'God' ; and it will not have a referent unless, of course, there is such a being.

On the other hand, even if a definition fails to assign a referent, it will always yield a sense; for whatever the defining condition D(t), there will always be a sense attaching to the description 'the x such the D(x)' or the sense attaching to the variable whose range is determined by D(x). Indeed, on a Fregean conception of language, it is hard to see how a definition could serve its purpose without assigning a sense to the term or terms that are to be defined.

In the light of these general remarks, let us now consider the possible definitional role of Hume's Law and other such principles. Given that Hume's Law is to be construed as a definition, then what should it be taken to define?

The most natural answer is that it defines the operator 'the number of'. Indeed, if one takes the syntax of the Law at its face value, then this would appear to be the only possible answer. For the other terms occurring in the Law are all of a logical character; and it would clearly be absurd to suppose that the Law somehow served to define them by means of the other terms.

There is a way, however, in which the Law (or whatever principle is in question) might be taken to do more. Any definition of a term, such as the number operator, presupposes a domain of discourse over which the term is to have application. Now under the standard forms of definition, this domain is taken to be given along with the meaning or interpretation of the undefined terms. The purpose of the definition is then to assign an interpretation to the undefined term from within the given domain. Consider, for example, the recursive definition of addition as given by the equations:

of the embedded description 'the being no greater than which can be conceived'. In this way, the sense determined by the implicit definition could be made to be the same as the sense determined by the corresponding explicit definition.

$m + 0 = 0$; and $m + n' = (m + n)'$. It is here understood that the domain of discourse is to be the set of natural numbers; and it is not required that the definition supply an interpretation of the symbol '+' over the rationals or the reals or some other kind of number or object.

Now one might adopt a similar approach in the case of abstraction. One might take the intended domain to be the universe of all objects and let the number operator have application to the concepts over that domain and the objects within it. However, even from this standpoint, it would be natural to suppose that the principle is capable of securing the interpretation of the operator for various other domains. For the success of the principle in determining an interpretation of the operator should not depend upon the individuals being what they are; and so the principle should also have application to those universes which would arise under a different disposition of what individuals there are.

But one might be more ambitious still and take the principle itself to be of help in determining the different domains of discourse. We no longer regard the domains of objects from which the values for the abstraction operation are to be drawn as simply given. Rather, we think of the principle itself as helping to determine what those domains should be. It is somehow meant to succeed both in telling us what objects there are and in assigning them as abstracts to the concepts.

We may be more precise. Before we only required that the principle determine an operation F for each appropriate choice of the domain M. Now we require the principle should determine M on the basis of the underlying subdomain I of individuals (or non-abstracts). Thus the principle must not only determine the operator F as a function of M, it must also determine M as a function of I. It must, that is to say, also determine a method of 'generation', a function g which will take us from a given domain I of individuals into an appropriate domain $M = g(I)$ of individuals and abstracts.[8]

The generation of the domain M from I may be compared to the generation of sets under the cumulative hierarchy from a given domain of urelements. Just as there are different universes of sets for different domains of urelements, so there may be different universes of numbers for different domains of individuals. For different domains of individuals of finite size, the corresponding universes of numbers will be the same (in contrast to the case for sets). But it is plausible to suppose that for larger and larger domains of individuals (which may include abstracts of other kinds), new numbers of ever greater size will be added to the universe.

[8] In the more general case there will be several different abstraction principles. Each principle will then be associated with a function g. The total domain (if it exists) will be obtained by successively applying the generation functions to the initial subdomain of non-abstracts until no further applications are possible.

Now there are two rather different ways in which the Law may be taken to perform this second task. One rests upon treating the definition as having a creative capacity, as having the power to introduce new objects into the domain of discourse. This is an approach that we shall consider in the next section. But here we are interested in definitions of an orthodox sort; and when treated in this way, the definition is incapable, as it stands, of determining the domain M on the basis of the subdomain I of individuals. For it contains no predicate either for M or for I and hence nothing that could be taken to determine M or to be determined by I.

But this defect is readily remedied. For we may suppose that the variables for concepts and objects that appear in the statement of the Law are relativized to a predicate D for the domain. Thus instead of saying 'for all concepts', we say 'for all concepts defined over D'; and similarly for the other locutions. We may also introduce a predicate I for being an individual and add to Hume's Law the assertion that every individual is in the domain (i.e. every I is a D). With these changes, Hume's Law may not only serve to determine a number operation F_0 on the given domain M_0 of all objects; it may also serve to determine a domain $M = g(I)$ on the basis of the subdomain I and a number operation F_M of the basis of M.

We therefore see that Hume's Law, and other such abstraction principles, may be taken to perform two related definitional tasks. But how well do they perform them? To what extent do they help in determining the domain of objects or the operation of abstraction?

The answer is 'very badly'. The abstraction principle, even when modified, is unable to determine a unique expansion M of any given subdomain of individuals I. For given that it has determined one expansion, it will be incapable of excluding any other expansion with the same number of abstracts (or abstract-surrogates).

The principle is also generally incapable of determining a unique abstraction operation on a given domain. For suppose that Φ is an abstraction principle concerning a certain abstraction operator §; and let us replace the operator § in Φ by an appropriate variable F to obtain the general condition Φ(F). Then Φ(F) is never uniquely satisfied over a given domain (except in the case in which the domain contains only one object). For let F be one solution to Φ(F). Let $\vee$ be the universal abstract, i.e. the abstract $F(\bigvee)$ where $\bigvee$ is the universal concept. Let x be any other object in the domain (possibly even an abstract). Permute $\vee$ and x. The resulting operation F' will then be distinct from F and yet still be a solution to Φ(F).[9]

In light of the general importance of this phenomenon, it will be helpful to pinpoint more exactly how and where the indeterminacy arises. One issue

[9] The more usual method of proving non-uniqueness uses Frege's switching argument (cf. lemma 3.7, 8 of Part II). But the proof breaks down when the identity criterion is circular.

concerns existence. Is there, for any subdomain of individuals, an expansion of the subdomain with abstracts and an interpretation of the abstraction operator for which the abstraction principle is true? The answer, at least for the non-circular principles, is clear; there will exist such an expansion and interpretation as long as the principle is non-inflationary over arbitrarily large domains.

Second, there is the question of uniqueness (aside from existence). Are there different ways of expanding a subdomain of individuals? And are there different interpretations of the abstraction operator that can be imposed over a given domain?

Uniqueness, in either case, can fail for two rather different reasons, one structural and the other 'material'. There can be structurally different, i.e. non-isomorphic, solutions. But even if all of the solutions are structurally the same, there may still be solutions which differ in how the structure is realized. The one is a failure of categoricity, and the other a failure of determinacy within a given isomorphism-type.

Let us see how each source of failure might manifest itself. Consider the structural question first. In regard to the problem of domain expansion, the question is whether the number of abstracts is determined by the number of individuals. To this the answer is no. Hume's Principle, for example, is compatible, in the absence of any individuals, with there being a countable infinity of numbers or with there being an inaccessible number of numbers.

Let us suppose that the structural problem of domain expansion has been solved; it is known how many abstracts go with each predetermined number of individuals. Then in regard to the problem of interpreting the operator, the structural question that remains is whether, given the number of individuals and the number of abstracts, the interpretation is determined up to isomorphism.

The answer is again no. Suppose, for example, that Φ is a principle which takes each finite concept into its extension and every other concept into the same abstract ('Infinity'). Then such a principle is compatible with the finite extensions being well founded over a given domain and also compatible with them not being well founded (see the comment to theorem II.6.5). It turns out that the answer is yes for number abstraction and for the various weaker forms of abstraction (theorem II.6.6); and this is perhaps the reason why this particular source of indeterminacy has been overlooked.

Consider now the question of identity. In regard to the problem of domain expansion, this is partly a question of ascertaining which objects are abstracts of the kind in question and which are individuals. We can then be sure that the expansion is always from individuals to abstracts of the right sort. But the question is also partly a matter of getting the abstracts right in any given case. For example, in the case of Hume's Principle, we only want to include a number in the expanded domain as long as all smaller numbers are included.

A principle of abstraction is powerless to settle the question. For take any solution *g* to the problem of domain expansion. Then we may replace abstracts and individuals at will, always taking care to replace distinct objects with distinct objects, and we will still have a solution.

Let us now suppose that the identity problem for domain expansion has been solved; we know which abstracts go with which individuals. Then in regard to the problem of interpreting the operator, the identity question that remains is the question of determining which abstracts go with which concepts. Thus whereas the first identity question is a matter of getting the range of the abstraction operation right, the second is a matter of getting the values within the range right, of correctly assigning them to the concepts.

Again, a principle of abstraction is powerless to settle the question. For, as we have seen, we may permute the universal abstract with any other and still retain a satisfactory interpretation of the abstraction operator.

The general question of identity corresponds to what has been called the 'Caesar problem'. This is the problem of determining the truth-values of sentences of the form 'the number of Fs = t', where t itself is not of the form 'the number of—' (and similarly for other forms of abstraction). The exact relationship between the two questions depends upon what terms are admitted as substituends for t. Suppose that each object in the domain is taken to have a name whose designation remains the same regardless of the interpretation of the abstraction operator. Then the two issues are essentially the same. For example, if '0' is a 'rigid' designator of the number 0, then the truth of 'the number of non-self-identicals = 0' will guarantee that the correct object is assigned to the empty concept.

However, in discussions of the Caesar problem it is usually assumed, if only implicitly, that any designator for a number will be reducible to one of the form 'the number of—'. This assumption is, in fact, quite controversial. It requires a logicist account of arithmetical operations, for an arithmetical expression like '3 + 5' must be understood in some such way as ' the number of the disjunction of two exclusive concepts of which 3 and 5 are the respective numbers'; and in the case of an expression like 'Caesar's favourite number', it becomes even more problematic whether a reduction to the required form can be given.

In any case, as long as this assumption is made, the two problems will be inequivalent. For the 'Caesar' term t must then always designate an individual; and so a solution to the Caesar problem will merely guarantee that the abstraction terms refer to abstracts, not that they refer to the right abstracts. We only have a solution to what might be called the *external*, as opposed to the *internal*, Caesar problem.

In the face of these various forms of indeterminacy, what kind of definitional status should we accord to Hume's Law and the various other abstraction principles? One possible response to the indeterminacy is to say that it is no worse than the indeterminacy that besets our actual conception of number or of some other kind of abstract. The definitions provided by the various abstraction principles may therefore reasonably be taken to be indeterministic in character. Thus Hume's Law may be taken to determine a variable domain of objects on the basis of the individuals, one which will allow different choices of objects for the numbers; and it may be taken to determine a variable number operation for a given domain, one which will allow different assignments of the objects from the domain to the concepts. It may then be maintained that our understanding of the numerical locutions, when so defined, exactly matches our ordinary understanding of those locutions.

This is not the view I am myself inclined to take. For there seems to be a striking difference between our conception of numbers, on the one hand, and our conception of an axiomatically defined class of elements, such as a group, on the other. The indeterminacy that infects our conception of the unit element *e*, for example, seems to be of a quite different order from that, if any, which infects our conception of the number 1. Thus whereas it is acceptable to let *e* 'be' 1 under a suitable interpretation of the group operator, it makes no sense to suppose that 1 might be the undetermined group element e under a suitable interpretation of the numbering operator. Any view must therefore be judged unsatisfactory if it does not explain the asymmetry between the two cases.

However, it is not my aim to debate the merits of the indeterministic conception here.[10] I shall merely take for granted that some form of deterministic conception is correct. In this case, it must be conceded that the mere requirement of conformity to an abstraction principle is incapable of yielding our ordinary conception of the kind of abstract in question; and it is therefore incumbent upon the advocate of the abstractionist approach to provide an account of what else might be relevant in determining our understanding of that kind.

There are two main forms of response that he might pursue. One is to supplement the principle with further conditions. The principle would then be only part of the full definition; it would be the definitional analogue of

[10] However, let me note that the indeterministic conception of number may be used to evade the dilemma that Crispin Wright poses for Field on pp. 450–1 of his (1988) (and that is taken from his (1983), n. 20, 148–52). He asks: How is it possible to explain number using Hume's Law and yet deny the existence of numbers? To which the answer is: by treating Hume's Law as an explanation of a *variable* number operator. The existence of numbers may then intelligibly be denied since the denial simply amounts to the claim that there is no operator that conforms to the Law. If we regard Hume's Law as part of a 'scientific' theory, then this response is equivalent to a Ramsey-style treatment of the theoretical terms.

an enthythematic argument. Under such an approach, it may be asked why the principle, rather than the supplementary material or the full condition, should be taken to define the operator in question. But, as we shall see, the principle may provide what is distinctive about the full defining condition and there may be a uniform method for going from the principle to what else is in the condition.

The other line of response is to strengthen the requirement of conformity. Instead of requiring mere conformity to the principle, we require conformity of a special sort. The principle alone then serves to define, but in a way that goes beyond its mere satisfaction.

The distinction between the two forms of strengthening can be illustrated with the case of inductive definition. Consider the inductive definition of *even* number: 0 is even; and for each natural number n, n + 2 is even if n is. Mere conformity to these rules is not sufficient to determine the required interpretation of 'even'. All the same, the rules do provide a basis for a satisfactory definition of the term in either of the indicated ways. For we can either stick to conformity and supplement the given rules with the rule that any even number belongs to all classes which are closed under the given rules; or we can stick to the given rules and constrain conformity by the requirement that the extension for 'even' should be minimal.

As this example shows, it may be possible to strengthen a given definition in either of the two ways. And, in general, either kind of strengthening can be traded in for the other: for conformity to additional conditions can be regarded as a special kind of conformity; and conformity to the given conditions in a special kind of way can be regarded as an additional condition. Usually, however, it will be more natural to regard a definition in one way rather than another.

How then might an abstraction principle be supplemented or the manner of conformity strengthened so as to yield a determinate conception of the domain of abstracts and of the abstraction operation?

A solution of sorts to the structural problem may be given in terms of the generative model of abstraction that is developed in Section II.5. Suppose that we are given a domain of individuals of fixed cardinality. Then we may successively generate the abstracts over that domain in a natural manner. In the case of number abstraction, for example, we may first generate the numbers of the various concepts of individuals. We may then add these numbers to the given domain and generate the numbers of the various concepts over the augmented domain. Continuing in this way, we obtain a domain and a partial interpretation of number abstraction at each finite stage. By taking the limit of these partially interpreted domains, we may take the construction into the transfinite, stopping at the first stage at which no new numbers are introduced.

What such a construction rules out is the possibility of an 'ungrounded' abstract. Any number, for example, must be the number of some concept. Moreover, it must be possible so to choose the concept that it can be specified without reference, either directly or indirectly, to the number in question. Continuing in this way, the identity of every number will ultimately be grounded in the identity of the individuals.

The construction will give us a categorical interpretation of the number operator over the given domain; any two interpretations which are generated in this way will be isomorphic as long as the cardinalities of their respective domains of individuals are the same. Moreover, the result will generalize. The condition of being in one–one correspondence which figures as the identity criterion in Hume's Principle has the property of being *absolute*: whether two concepts are related by the condition depends only upon the objects which fall under the concepts in question, not on those which fall outside of the concepts. And it is merely the absoluteness of the identity criterion that is required for the resulting interpretation of the abstraction principle to be categorical.[11]

This solution may be characterized under each of the two heads for completing a definition. On the one hand, we may suppose that Hume's Principle, to continue with the example, is supplemented by the principle that there is no proper subdomain that contains all individuals and contains the number of any concept of objects in the subdomain. Or we may require that the solution be minimal in the sense of not containing a smaller solution. Thus the way the completion goes is not, in this case, essentially different from the case of inductive definition.

The above generative model can be extended to apply simultaneously to several abstraction operators. We might imagine, for example, that numbers and finite extensions are generated in tandem. But, more significantly, the means of abstraction can themselves be taken to be subject to generation. Let us suppose that we start off with a given domain. We may then generate the acceptable means of abstraction over that domain, i.e. the ones that are non-inflationary and predominantly logical. We may then apply these means of abstraction to the concepts over the domain to obtain new abstracts. These abstracts are then added to the domain and the whole process repeated *ad infinitum*.

Such a construction yields not only a determinate conception of what abstracts there are, given the means of abstraction, it also yields a determinate conception of what means of abstraction there are. It therefore solves

[11] Our account here is reminiscent of the generative model of understanding that is given in Wright (1983, 142–5) and has been elaborated by Wright (1983) and Hale (1994) in response to criticisms of Dummett (1991*a*, 236). However, as far as I can see, Wright's account is not intended as a way of generating the domain of all numbers but as a way of showing how we can understand terms for all of the natural numbers. Wright gestures towards the notion of absoluteness at the close of Sect. II of his (1997), but in connection with the question of acceptability, not of impredicativity or categoricity.

the problem of determining what the cardinality of the universe of abstracts should be: for the cardinality is determined by the point at which the process yields nothing new (Section III.2).

The above solution is very natural and, indeed, it is hard to think of reasonable alternatives. The problem of identity, though, is not so tractable. Even if it is agreed that the abstracts are generated from the individuals in the manner proposed, we still need to say what the abstracts are. One solution to this problem is implicit in the general theory of abstraction that is developed in Part III. The basic notion of this theory is that of an object being the abstract *of* a concept *with respect to* a relation on concepts. Thus the notion relates three items: an object, which is the result of the abstraction; a first-order concept, which is what is abstracted; and a second-order relation on concepts, which is the means of abstraction. In the case of numbers, for example, we may say that 0 is the abstract of the empty concept with respect to the relation of one–one correspondence.

In order to solve the identity problem, we can simply declare that the number of Fs is the abstract of F with respect to the relation of equinumerosity, that the extension of F is the abstract of F with respect to coextensiveness, and similarly for the other cases of abstraction. Such an account then provides us with an explanation of the particular kinds of abstract in terms of the general notion of an abstract.

This solution can be regarded as a generalization of Frege's account of the number of Cs as the extension of the concept of being equinumerous with C. For upon treating the abstract of a concept with respect to a given equivalence relation, in our account, as the class of all concepts equivalent to the given concept, we obtain the Fregean abstract. However, our viewpoint is quite different from Frege's. For he took there to be a privileged form of abstract, one to which all others were reducible, whereas we take the different forms of abstract to be *sui generis*.

Indeed, from the standpoint of a general theory of abstracts, the Fregean view is incoherent. For he identifies each conceptual abstract, with respect to a given means of abstraction, with the corresponding extension or class of concepts. But the extensions are themselves among the abstracts. So, by the same general rule, each extension should be identified with the class of all the concepts which have that extension, which is absurd.

We have obtained solutions to the problems of identity and of structure. However, the two solutions do not fit well together. For given our solution to the problem of identity, we already have a complete explicit definition of number as a certain sort of abstract; and in the presence of an explicit definition of an operator, it is hard to see what further role might be served by an implicit definition. Our definition of number in terms of abstracts, like Frege's in terms of extensions, should be regarded as an alternative to the definition in terms of Hume's Law, not as an addition to it.

This is not to deny the analyticity of Hume's Law, or of whatever other kind of abstraction principle might be in question. For even when the principle does not follow directly from the explicit definition of the abstraction operator, it may still follow from the definition and the analytic principles governing the notions in terms of which it has been defined. (Indeed, this is just what Frege hoped for in the case of Hume's Law.) Thus an abstraction principle may still be indirectly definitive of its operator even though it forms no part of its explicit definition.

But is there any reasonable way in which Hume's Law could be taken to contribute directly, as part of an explicit definition, to our understanding of the number operator? One possibility is to appeal to the essential properties of numbering and of the numbers. Call an operation *quasi-numerical* if it conforms to Hume's Law. As we have observed, many different operations are quasi-numerical. By adding further conditions to Hume's Law, we want to distinguish the (genuinely) numerical operation, the one that takes each concept into its number, from the others. Once we have successfully defined what it is to be a numerical operation, we can take the number of Cs to be the result of applying this operation to C.

Now one way the genuinely numerical operation is distinguished from some of the others is that it is *essentially* quasi-numerical; it is, by its nature, quasi-numerical. Consider, for example, an operation of being a 'qumber', defined by the condition:

> the qumber of Fs is the number of Fs if that number is not 9 and otherwise is the number of the planets.

Then although the qumber operation conforms to Hume's Law it only does so accidentally.

This condition is not enough, on its own, to distinguish the genuinely numerical operator from all of the others. Let us define the operation of being a schumber by the condition:

> the schumber of Fs is 0 if F is singleton, it is 1 if F is empty, and it is the number of Fs otherwise.

Thus schumber is like number, but with the roles of 0 and 1 reversed. It may then plausibly be maintained that the operation of schumbering is also essentially quasi-numerical.

However, the genuinely numerical operation has a further distinguishing feature. For each value of the operator, i.e. each number, is essentially the value of the operation for suitable choices of the concept. Thus the number 0 is essentially the number of the concept of being non-self-identical, the number 1 is essentially the number of the concept of being identical to 0, and similarly for the other cases. This further feature distinguishes numbering from schumbering. For 0 is not essentially the schumber of any singleton concept.

Moreover, this feature is itself essential to numbering. I therefore propose to define a genuine numerical operation as one that essentially conforms to Hume's Law and is essentially definitive of its values in the way described.

It is of the utmost importance, in evaluating this proposal, to distinguish between the definition of a linguistic expression and the definition of the non-linguistic item for which it stands. Thus the expression 'the number of objects that are not self-identical' may be taken either as a definition of the numeral '0' or as a definition of the number 0. In the first case, the identity '0 = the number of objects that are not self-identical' is taken to be true because of the meaning of the numeral '0'; and in the second case it is taken to be true because of the nature or essence of the number 0.

What we are proposing is that each number (not numeral) has a definition of the form 'the number of Cs' (it is essentially of that form for suitable C). We are also proposing that the number operation (not operator) has a definition part of whose content includes the definability of the numbers. The predicate (not property) 'is a numerical operator' is then defined by reference to these non-linguistic forms of definition.[12]

I am not sure that the present definition of number in terms of its essential properties should be regarded as being at odds with the previous definition of number as a certain kind of abstract. It might even be regarded as providing a deeper account of the notion of abstract. For we may ask: What is it for x to be an abstract of a concept C with respect to a means of abstraction $\boldsymbol{R}$? Our essentialist form of definition then provides an answer: for an abstraction operation can be singled out in terms of its essentialist relationship to $\boldsymbol{R}$ and the abstract x can then be taken to be the value of this operation for the concept C as value.

I turn finally to the question of how a definition of number in terms of Hume's Law might be of significance for the philosophy of mathematics. There are two major problems in the area that such a definition might be thought to be of help in solving. The first is semantical: it is to show how we can make determinate reference to numbers. The second is

[12] Let me make a few further comments on this proposal (a full discussion is out of place). (1) I do not presuppose an account of essence in modal terms and, indeed, am inclined to reject such an account for the reasons stated in Fine (1994). (2) For the purposes of the account, operations are best taken in an intensional sense though concepts may be taken in either an extensional or intensional sense. (3) There are other views as to what the definability of the numbers might consist in. One might maintain, for example, that the number 0 is essentially the number of any empty concept, that the number 1 is essentially the number of any singleton concept, and similarly for the other cases. Another possibility is that for *any* singleton concept, except one whose object presupposes the number 1 itself, the number 1 may be defined as the number of that concept. Of the three possibilities it is the third that is the most plausible. However, it involves a distinction between definitive and essential properties which is explained in Fine (1995) but which I have here preferred to ignore.

epistemological: it is to show how we can have knowledge of number-theoretic truths.

Consider the semantical problem first. In order for a definition of number to resolve philosophical doubts over determinacy, it must satisfy three desiderata. First, it must satisfy the formal requirement of having a unique solution. Second, the terms by which the number locutions are defined must not themselves suffer from a similar indeterminacy. Third, the definition must be correct and, in particular, must not rest upon an arbitrary identification of numbers with one kind of object as opposed to another.

Frege's definition of number as an extension of concepts is based upon an incoherent conception of extension and fails, in any case, to satisfy the third desideratum. Our alternative definition of number in terms of abstracts satisfies the first desideratum and arguably satisfies the third. For there is nothing arbitrary in the identification of numbers with a certain sort of abstract. However, the definition fails to satisfy the second desideratum. For the notion of an abstract is subject to the same sort of indeterminacy as the notion of a number. The problem of determinacy is merely pushed back.

Hume's Law on its own fails to satisfy the first desideratum; for, as we have seen, many different operators will conform to the Law. It must therefore be supplemented in some way if it is to be of any use. We have suggested a supplementation in terms of the essentialist role of the Law; and it is arguable that the resulting definition satisfies all three desiderata. It satisfies the second; for even though the notion of essence may be subject to indeterminacy, it does not seem to be of a kind that is relevant to the indeterminacy of number. It satisfies the critical part of the third; for the definition does not even attempt to identify number with some broader category of object. Therefore the only room for doubt lies in the essentialist intuitions upon which the correctness and uniqueness of the definition rest. One might think, for example, that 0 can with equal legitimacy be defined as the number of an empty concept or as the schumber of a singleton concept. But in the absence of any such doubt, it is hard to see how determinacy could reasonably be denied.

Consider now the epistemological problem: How can arithmetical truths be known? One natural approach to the problem is to provide an explicit definition of the number operator in terms of other notions which, in the special case of logicism, will be logical notions. Arithmetic can then be derived from Hume's Law; and Hume's Law can be derived from principles governing the notions in terms of which number is defined. The problem of our knowledge of arithmetic therefore reduces to the problem of our knowledge of logic and of those other principles. When the principles are themselves logical, we have the familiar form of logicism and arithmetic is reduced to logic.

The possibility of implicitly defining number by means of Hume's Law might appear to give us an epistemological edge over the more usual forms of logicism. For Hume's Law is then secured by definition; and so we only need the logic to take us from the Law to arithmetic. But this is too simple. For whereas an explicit definition stands in no need of justification, an implicit definition does. An epistemological cost must therefore still be borne; but it lies now in the justification of the definition rather than in the justification of the Law on the basis of the definition.

Since the point is of some importance, it will be worth a more thorough discussion. To this end, let us ignore the question of whether the proposed definition is faithful to our ordinary conception of number and whether, in particular, it is referentially determinate. Our aim may be only to provide what might be called *reinterpretative* foundations; we aim to provide an interpretation of number that possibly deviates from the usual interpretation but will serve the same purposes and, in particular, will provide a basis for our knowledge of what goes for the usual arithmetical principles. Thus for these purposes, Hume's Law itself can be adopted, without supplementation, as an indeterministic definition of the number operator.

Now in order to derive arithmetic from such a definition, we must not only be able to endorse Hume's Law as a definition of number, we must also be able to make the transition from the endorsement of the Law as a definition to its endorsement as a truth. The interpretation assigned to the number operator by means of the definition must be such as to render the Law true.

Call a definition *materially effective* when this transition is warranted, i.e. when the making of the definition will result in its defining condition being true. Not all definitions are materially effective even when they are successful. For in case the definition has only semantical import, an interpretation will be assigned regardless of whether or not the defining condition is thereby rendered true. Should the definition be deterministic, the sense assigned may be that of an empty description; and should the definition be indeterministic, the sense assigned may be that of an empty variable. In neither case is it plausible to suppose that the resulting interpreted condition will be true.

There are, of course, various methods for evaluating a statement that contains an empty description or an empty variable; and some may result in such statements being true. For example, under the generality interpretation of the variable (according to which a statement containing a variable is true if it is true for all of its values), all such statements will be vacuously true. But such a possibility is of no help in the present case. For we require that Hume's Law should be true under a conception of truth that more or less agrees with our ordinary conception of what is true; it should not render all arithmetical statements true or, in some other radical manner, cut across the usual distinction between what is true and what is false.

Thus in order to be justified in using the definition of Hume's Law as a premiss in a demonstration of arithmetic we must know that it is materially effective, i.e. that it results in a truth (according to our ordinary conception of what is true). But how can we know this? There appears to be only one answer. To know that the definition is materially effective we must know that it is referentially effective, i.e. that it succeeds in assigning a referent to the number operator in conformity with the Law (or its supplementation). But how can we know this further fact? Again, there appears to be only one answer. To know that the definition is referentially successful we must know that there exist objects that play the role required of the referents. What this means, in case we take the definition to be deterministic, is that we must know that there exists a unique solution to Hume's Law (or its supplementation); and what it means, in case we take the definition to be indeterministic, is that we must know that there exists at least one solution to Hume's Law.

The use of Hume's Law as a definition is therefore not without epistemological cost; and whether it can secure an epistemological advantage and, in particular, whether it can be used to vindicate a form of logicism, is far from clear. For the only obvious way of justifying the existential generalization of Hume's Law (with respect to the number operation) is by means of an instance. We see that the generalization is true by taking the operation in question to be the number operation or some other, specific, mathematically defined operation. Thus even though the existential generalization is stated in purely logical terms, it is not clear that it has a purely logical justification or one that is in some other way less problematic than that of the Law itself.

3. RECONCEPTUALIZATION

I want to consider a non-classical approach to the question of defining number by means of Hume's Law. It might be called *definition by reconceptualization* and rests on the idea that new senses may emerge from the reanalysis of a given sense. The idea derives from Sections 63–4 of the *Grundlagen*. However, it has not been my aim to be faithful to Frege's thought; I have been content, in this matter, to follow the exposition of others, principally Dummett (1991) and Wright (1983). My interest, rather, has been to see whether the ideas themselves can be sustained.

There is another non-classical approach to using Hume's Law as a definition of number. This is the context principle; and it rests on the idea that certain truths may be used to fix the reference of the terms that they contain. Thus the first approach rests on the adoption of an unorthodox mechanism for the determination of sense, and the second on an unorthodox mechanism for the determination of reference. I hope to discuss the context principle and its relevance to the definition of number elsewhere.[13]

It is worth noting, however, that it is highly implausible that we might altogether avoid the epistemic cost involved in adopting Hume's Law as a classical implicit definition of number by adopting a less orthodox form of definition in its place. For whatever the mechanism by which the sense or reference of the defined terms is secured, we surely require assurance that the adoption of the definition is consistent. This problem becomes especially acute in view of the fact that the adoption of a formally similar principle, Law V, is inconsistent. Surely we require assurance that the two laws are not alike in this respect if we are to have confidence in the conclusions obtained from the one as opposed to the other.

But how is such assurance to be obtained? A *post facto* justification, one that already rests upon the acceptance of the definition in question, is useless; for an inconsistent definition is as capable of justifying its own consistency as a consistent definition. We could, of course, appeal to the corresponding existential claim (that there exists an operation in conformity with the Law). But, as we have seen, it is quite unclear whether this provides us with any epistemic advantage; and it would not, in any case, provide us with any advantage over the adoption of a straight implicit definition.

If we really are to provide a distinctive foundation for arithmetic on the basis of one of the novel forms of definition, then some mark of consistency must be stated and justified without implicit appeal to the abstract objects in question. We are therefore in much the same position as a nominalist who is required to defend the consistency of a given branch of mathematics. Perhaps this can be done. After all, one might follow the lead of Field (1989) by taking consistency to be a modal primitive and establishing the claim of consistency by some form of non-mathematical induction. But we should not then pretend that the definitionalist has any epistemological advantage over his nominalistic rival. Whatever advantage he gains will derive, not from his having any special access to the putative truths of arithmetic, but from his being able to endow their terms with reference.

We should note that this problem may be peculiar to the mathematical applications of abstraction. For granted the existence of the appropriate mathematical objects, the non-mathematical applications will give rise to no special difficulty; for consistency, or whatever is required to justify the form of definition in question, can be established in the usual way. Thus it is the non-mathematical, rather than the mathematical, applications of abstraction that are most plausibly taken to provide us with a special epistemic access to their objects.

[13] The reader might like to consult C. Parsons (1965), Wright (1983), Hodes (1984), and Dummett (1981*a*, 1981*b*, 1991*a*, 1991*b*) for discussion on the topic.

Let us now turn to the question of whether number can be defined through reconceptualization. The basic idea behind this approach is that we may understand the left-hand side of Hume's Law as another way of saying what is said on the right and in such a way that the number operator is thereby endowed with sense.

In discussing this approach, it will be important to distinguish between a universal and a schematic formulation of Hume's Law. A universal formulation is given by a closed sentence:

> For all concepts F and G, the number of Fs = the number of Gs iff the Fs and the Gs are equinumerous.

The schematic formulation is given by a scheme:

> the number of Fs = the number of Gs iff the Fs and the Gs are equinumerous,

where 'F' and 'G' are now schematic letters for predicates, not variables for concepts.

What is meant by talk of the 'left' and 'right' sides of Hume's Law? Under the schematic formulation, the left side is presumably a (closed) identity statement of the form 'the number of Fs = the number of Gs'; and similarly for the right. But under the universal formulation, there is no *sentence* on the right or left. Presumably, then, the left-hand side is the *open* sentence 'the number of Fs = the number of Gs'; and similarly for the right.

How should reconceptualization be understood, as relating to the schematic or to the universal formulation? It might be thought not to make much difference, since the opportunities for reconceptualizing what is expressed on the right are much the same whether we have a closed or an open sentence. But this is not in general correct. Consider Frege's paradigm of objectual abstraction:

> the direction of L = the direction of M iff L and M are parallel.

If the lines here are the lines of abstract Euclidean space, it is not clear that any reference to them is possible. (It is idle to fix an origin and axes, since that already requires that we be able to single out a point and some lines.)[14] But if that is so, then there are no relevant closed sentences whose senses can be reconceptualized. It is therefore the open sentences that should be taken to be the target of reconceptualization. Usually, when philosophers have talked of reconceptualization they have supposed that the sense of the right- and left-hand sides is a thought, i.e. the sense of a sentence. But if I

[14] I doubt that there is even *in principle* a way of referring to one of the lines, or one of the points, as opposed to the others. I note, by the way, that this raises a serious problem for the usual development of Fregean intensional logic. For it means that we cannot assume that for each object there is a sense that picks it out.

am right, it would be better to take the sense to be a schematic thought, i.e. the sense of an open sentence or formula.

We may divide the process of conferring a sense on the number operator into two stages: first, the formula on the left is taken to have the same sense as the formula on the right; second, this sense is subject to reanalysis in such a way as to provide a sense for the number operator. In a way, the method is a combination of the methods of explicit and implicit definition. For if the structure of the left-hand side is ignored (apart from the presence of the variables 'F' and 'G'), we can use the method of explicit definition to endow it with the same sense as the formula on the right. On the other hand, if we do not require the synonymy of the left and right sides, we can use the method of implicit definition to confer a sense on the number operator. The idea behind reconceptualization is to get the implicitly defined sense, not directly from the principle, but indirectly from the explicitly defined sense.

Does the method work? Various questions should be distinguished here. Can we obtain a unique sense for the number operator in this way? Can we obtain at least one sense, even if not a unique sense? And, in each case, what is required by way of justifying the definition?

It should be noted that if the definition succeeds in conferring a unique sense on the number operator, then the definition is referentially effective and, indeed, referentially determinate. For if the sense did not determine a referent or if the referent were an operation on concepts that were not always defined, the formula 'the number of Fs = the number of Fs' would lack a truth-value for certain concepts as values for 'F' and so its sense could not be the same as that of 'the Fs are equinumerous with the Fs'. Thus the sense determines a referent; and there seems to be nothing in the sense which would allow for any indeterminacy in the referent that was so determined.

This point is of some importance, because it means that the method is more powerful than it is commonly taken to be. Several authors, I suspect, have thought of definition by reconceptualization and the context principle as linked in the defence of logicism. Perhaps the thought is that the one provides a route to sense and the other a route to reference. But since reconceptualization already yields reference, it can be allowed to stand on its own.

A related point holds with regard to Sections 64–6 of the *Grundlagen*. Frege there considers three objections to the proposed method of defining the number operator. The first two he dismisses, but the third he upholds. It is that the definition is unable to settle the question of whether Caesar is a number. But his objection appears to be misguided. For granted that the sense of the number operator has been fixed, that the recarving of content has been successful, the operation that is the referent of the operator will have been fixed; and so it is determined whether or not Caesar is a value of this operation. Of course, at this point Frege did not have the distinction between sense and reference; but the point remains that the 'concept' of the number operation can hardly have been fixed without its values also having been fixed. It is possible that in posing this objection Frege is here reverting to the model of implicit definition; for an implicit definition of number by Hume's Law will, of course, fail to settle the question of whether Caesar is a number. Alternatively, it is possible that the objection is meant to show that the proposed recarving will not work. Since it is evident that the Caesar question is not resolved, there must be something wrong with the proposed attempt to recarve, even if it is not clear what it is.[15]

Does definition by reconceptualization, then, deliver a unique sense? I am not sure that there is anything in principle to be said against the possibility of such a definition fixing sense. To take an example of Dummett's (1991*a*, 173), suppose we attempt to define the predicate 'commits suicide' by means of the condition:

$$\forall x(x \text{ commits suicide iff } x \text{ kills } x).$$

Then it might reasonably be thought that the definition works by reconceptualization: there is a sense expressed on the right, which is the result of linking the argument-places in the sense of the predicate 'kills'; and this sense is then reconceptualized as that of a one-place predicate, with a single argument-place rather than two linked argument-places.[16]

There are, however, reasons of detail against supposing that anything analogous could work in the case of Hume's Law. Consider Law V, its inconsistency being irrelevant for present purposes. Suppose the Law is given in the formulation $\S xCx = \S xDx \leftrightarrow \forall x(Cx \leftrightarrow Dx)$; and let us adopt the following abbreviations:

(A) for $\forall x(Fx \leftrightarrow Gx)$;

(B) for $\S xFx = \S xGx$;

(C) for $\forall x(x = \S xFx \leftrightarrow x = \S xGx)$; and

(D) for $\S x(x = \S xFx) = \S x(x = \S xGx)$.

[15] This is how Dummett appears to construe him on p. 126 of (1991*a*), although on pp. 168–9 he also takes him to be making a claim of synonymy. I might note that if Frege had an implicit definition in mind, he might also have objected that it fails to serve his epistemological purposes, for its existential presupposition would stand in need of justification.

[16] Dummett's account (1991*a*, 174–5) of the way the definition works is somewhat different from mine. He supposes that there is a 'linguistic device' which converts the LHS into the RHS and two corresponding notions of sense, one coarse and the other more refined. The coarse sense is only assigned to an expression after the linguistic device has been applied; and so the coarse sense of the two sides is the same. The refined sense of the LHS also includes 'the functioning of the device'; and so the refined senses of the two sides will be different. Thus he does not posit a single sense that is subject to two different analyses.

Then on the view under consideration, (B) should have the same sense as (A) and (D) the same sense as (C). Now presumably (C) should also have the same sense as (B). Indeed, it is hard to see how a criterion of synonymy could be so tolerant as to allow the synonymy of (B) to (A) (or of (D) to (C)) and yet not tolerant enough to allow the synonymy of (C) to (B). If this is right, it follows that (A) has the same sense as (D). But then if the sense of an operator that is given through reconcept ualization is unique, the sense of §xFx will be the same as the sense of §x(x = §xFx), which presumably is not so.

A similar, though more complicated, argument can be constructed for the case of number. Let F^+ be the concept of falling under F or of being the least natural number not to fall under F; and similarly for G^+. Let (A), (B), (C), and (D) now be as follows:

(A) for 'the Fs are equinumerous with the Gs';
(B) for 'the number of Fs = the number of Gs';
(C) for 'the F^+s are equinumerous with the G^+s';
(D) for 'the number of F^+s = the number of G^+s'.

It is then plausible to suppose, especially given the synonymy of (A) to (B), that (A) is synonymous with (C). But from this it follows that 'the number of Fs' and 'the number of F^+s' will be synonymous, which is impossible since, for any finite concept F, the number of F^+s will be the successor of the number of Fs.

In the light of such examples, it is difficult to see what general principles governing the identity of sense might permit its reconceptualization and yet prevent its proliferation. Nor does it help to weaken the relationship under which reconceptualization takes place to something like sameness of truth-conditions, as is proposed by Wright (1988, 459) and Hale (1994). For if we cannot recover the sense under one relationship, nor can we recover it under a weaker relationship.

But let us ask whether there exists at least one sense of the number operator for which the two sides of the Law will be synonymous or otherwise equivalent. In this case, it will certainly help to weaken the relationship. For suppose we adopt an extensional conception of sense, under which sense and reference are to be identified. Then any of the many operations for which Hume's Law holds will provide a 'sense' for which the two sides are 'synonyms'. It might even be possible to maintain the existence of a sense under the more reasonable construal of synonymy as analytic equivalence. For if logicism is true, it will provide a sense of the number operator for which Hume's Law is analytic; and, for this sense, the two sides of the Law will, in the required way, be synonyms.

But the difficulty now concerns not the actual existence of the sense, but the grounds that we have for supposing it to exist. It is worth reminding ourselves that Law V is inconsistent. Thus it cannot be on the basis of any simple formal considerations that the existence of a sense is to be ascertained. But then what grounds can there be for supposing the required sense to exist? One should not be dogmatic on such a matter, but it is hard to believe that the existence of a sense for which the two sides are, in some strict sense, equivalent could be any less problematic than the existence of an operation for which the two sides are materially equivalent. Thus it seems that the possibility of definition by reconceptualization can take us no further than an implicit definition of a standard sort.

4. FOUNDATIONS

Frege attempted to provide a logical basis for arithmetic and analysis. All of the notions of arithmetic and analysis were to be defined in logical terms; and all of the theorems of arithmetic and analysis were to be derived, with the help of the definitions, from logical axioms. The attempt failed if for no other reason than the inconsistency of the critical axiom, Law V.

Recently, Crispin Wright (1983) has attempted to revive the Fregean logicist programme. He observes that the Fregean derivation of arithmetic may be broken down into two steps: first, Hume's Law may be derived from Law V; and then, arithmetic may be derived from Hume's Law without any help from Law V.[17] He therefore suggests that Hume's Law be used instead of Law V as a logical basis for arithmetic.

One difficulty with the approach is that it is unclear, in the absence of Law V, how one is to obtain a logical foundation for (higher-order) analysis. We can no longer treat reals as extensions of number concepts or the like; and nor is it clear what other form of abstract should be used in place of extensions. One might try experimenting with restricted forms of extensionality; but they appear either to be inconsistent or too weak or *ad hoc*.

Perhaps the closest we come in the existing literature to a satisfactory formulation is with Boolos (1986–7, 1993). In (1986–7), he proposes that extensional abstraction be permitted on *small* concepts, i.e. on those that cannot be put into one to one correspondence with the universal concept; and in (1993), he extends the proposal to those concepts whose complements are small even when they themselves are not. He shows how the natural numbers can be identified with representative finite extensions, such as those considered by Zermelo or von Neumann. However, he cannot

[17] Heck (1993) suggests that Frege himself was probably aware of the possibility of separating the two steps.

adopt Frege's definition of number as the extension of a second level concept; and nor is it clear that he can provide any other definition of number that is capable of yielding Hume's Law in its full generality. The system is in a certain respect too weak; since the reals can only be obtained by adding some *ad hoc* axiom, such as that the concept of being a natural number is small. And the system is in another respect too strong; since it only allows a standard model of cardinality **c** when **c** is a regular strong limit cardinal or a successor cardinal $\mathbf{d}^+$ for which $2^{\mathbf{d}} = \mathbf{d}^+$.

Given the general acceptance of the 'context principle', the reals might simply be 'posited' as the limits of rationals. Thus we might add to the language a limit-operator on concepts and a restricted criterion of identity to the effect that when the concepts C and D of rationals have rational upper bounds the limit of the Cs is to be identical to the limit of the Ds just in case the rational upper bounds of the one are the same as the rational upper bounds of the other. However, such a principle seems entirely *ad hoc*; and it would be desirable to have a more general basis upon which such a principle could be established.

In Part III, I propose what I hope is a natural solution to these problems. I first present a general theory of abstraction, one which is independently plausible and which, in fact, was developed without regard to foundational considerations. I then show that higher-order analysis can be developed within the theory under a suitable identification of reals with abstracts.

The reals are not identified with extensions but with what one might call *generalized numbers*. With each concept of numbers, one may associate a criterion of identity, one by which two concepts are related when their numbers either both fall under the concept or both fall outside it. The abstracts corresponding to these criteria of identity can be shown to exist within the theory; and it is these abstracts, rather than the corresponding sets of numbers, that do duty for the reals.

A distinctive feature of the theory is its generality. Instead of attempting to account for this or that form of abstraction, we attempt to account for them all. Thus the basic notion of the theory is that of an object being an abstract of a concept with respect to a given possible means of abstraction; and the aim of the theory is to provide an axiomatic characterization of all those possible means of abstraction which will actually give rise to abstracts.

There are two basic postulates of the theory: one concerning existence, and the other concerning identity. The existence postulate says, roughly, that as long as the means of abstraction is non-inflationary and predominantly logical it will give rise to abstracts; and the identity postulate, in its strong form, says that two abstracts are the same if and only if they are associated, through their respective means of abstraction, with the same 'equivalence classes' of concepts. For the purposes of foundations, however, the two postulates can be weakened: the existence axiom can be confined to the logical criteria; and the identity can be restricted to its left-right direction.

The question of the identity of abstracts will be discussed in the next section; and the question of their existence has already been taken up. But we should note, in connection with our foundational aims, that the instances of the existence axiom are conditional in form. An instance of an abstraction principle, for a given method of abstraction, is only asserted under the condition that the method is not inflationary, i.e. does not result in more abstracts than objects. Thus Hume's Law cannot be asserted in its categorical form but only under the condition that the universe is already infinite.

This means that we must reverse the procedure adopted by Frege. Instead of using Hume's Law to show that the universe is infinite, we must use the fact that the universe is infinite to establish Hume's Law. But how can this be done, given the absence of any strong categorical forms of abstraction?

The answer lies in the generality of the theory and in the additional assumption, which we make, of the existence of at least two objects. For we can then prove the legitimacy of an infinite number of means of abstraction, each of which produces at least two abstracts. By the identity postulate, it can then be shown that all of these different abstracts are distinct. Thus it is from the diversity in the allowable means of abstraction, rather than from the power of any one of them, that the theory derives its strength.

If the diversity of means is genuinely to yield a diversity in abstracts, however, it is essential that abstracts which derive from different means of abstraction not be indiscriminately identified. One and the same abstract is not to wear different 'faces' in connection with different means of abstraction. Thus a satisfactory solution to the Caesar problem is essential to our foundational programme and is not, as it is for others, something merely required for philosophical purposes.

There is another approach to the question of foundations that is suggested by our work and is more in line with modern thinking on the subject. For we can treat the objects of our theory as classes rather than as abstracts and dispense with any reference to an underlying criterion of identity. Thus we no longer regard an 'abstract' as being associated to various items by means of a method of abstraction but as simply being a collection of those items. On this view, although every abstract is taken to be a class, not every class is taken to be an abstract with respect to the coextensionality of concepts. Thus our previous difficulty over identifying abstracts with classes does not arise.

There are somewhat different ways in which this alternative approach might be presented. Perhaps the most orthodox, by current standards, is to take the notion of membership as primitive. The setting for the theory is then second-order logic and comprehension for classes of concepts takes the form:

$$\exists y \forall F(F \in y \leftrightarrow \phi(F)), \text{ where } \phi(F) \text{ is predominantly invariant.}$$

(The condition that ϕ (F) be predominantly invariant is something that can be expressed within the theory itself.) One can formulate a similar comprehension principle for classes of objects:

$$\exists y \forall x(x \in y \leftrightarrow \phi(x)), \text{ where } \phi(x) \text{ is predominantly invariant.}$$

However, in order to obtain a theory of reasonable strength, it is essential to countenance the application of the principle to classes of concepts and not just to classes of objects.

One might see these principles as arising from a certain natural approach to the problem of impredicativity. Consider the universal class V. One may define it as the class of all objects x for which x = x. Now there is a sense in which this definition is impredicative, for the universal class V is already included within the range of the variable x. But there is also a sense in which it is not. For one can understand what the condition is without presupposing any knowledge of the class V. Indeed, our understanding of the condition x = x is not something that requires the understanding of any particular objects at all. Once we have defined the class V, we may then define other classes with its help, such as the unit class of V, or the class of all objects not identical to V; and so on. In general, it seems reasonable to suppose that we can define any class by means of a condition as long as our understanding of it does not involve too many objects. We can think of our comprehension principle as a way of making this view precise: the idea of a condition involving an object is explained in terms of invariance; and the idea of 'too many' objects is explained in terms of (exponential) smallness.

An alternative method of presentation, more in keeping with Frege's own, is to take an extension-forming operator § as primitive and then to adopt the following restricted form of abstraction for extensions:

$$\S F\phi(F) = \S F\psi(F) \leftrightarrow \forall F(\phi(F) \leftrightarrow \psi(F)), \text{ where } \phi(F) \text{ and } \psi(F) \text{ are both predominantly invariant.}$$

Similarly, in the case of objectual extensions, the principle would take the form:

$$\S x\phi(x) = \S F\psi(x) \leftrightarrow \forall x(\phi(x) \leftrightarrow \psi(x)), \text{ where } \phi(x) \text{ and } \psi(x) \text{ are both predominantly invariant.}$$

Within such a theory, one can then provide a foundation for both arithmetic and analysis. Hume's Law can be derived in much the same way Frege derived it from Law V, using the very same definition of the number operator; and analysis can then be derived along the same lines as for the general theory of abstracts.

I do not regard these theories as providing support for Frege's overall logicist aims. If one adheres to the context principle, then one can conceive of the various abstraction principles as providing a definition of the notion of natural number and perhaps also of the notion of real number. Moreover, if one takes the notions of second-order logic to be logical, then these definitions will be stated entirely in logical terms; and this makes it very reasonable to suppose that the defined notions will themselves be logical. But the logical status of the notions will not provide the defining principles with any special epistemological status; for, even though they are logical definitions, they will still stand in need of an extra-logical justification.

Perhaps the best that can be said on their behalf is that they provide us with some sort of theoretical insight into the nature of the various kinds of mathematical object. Mathematics is like the natural world in displaying a wide variety of different kinds of object—natural numbers, the reals, points, lines, figures, determinants, groups and so on. It is therefore natural to seek some kind of uniformity within this diversity. What our theories show is how many of the different kinds of mathematical object can be regarded as forms of abstraction and how their various special theories can be seen to have their basis in the general theory of abstraction. What we achieve is an ontologically significant reduction of certain branches of mathematics to the theory of abstraction, but not an epistemologically significant understanding of the theory of abstraction as a branch of logic.

5. THE IDENTITY OF ABSTRACTS

An abstraction principle yields a necessary and sufficient condition for abstracts of the same sort to be identical. But what of abstracts of different sorts? Given an abstract obtained from an item by one means of abstraction and an abstract obtained from an item by another means of abstraction, under what conditions are the two abstracts the same?

The left-to-right direction of the identity axiom provides a very plausible necessary condition: for the two abstracts to be the same they must be associated, through their respective means of abstraction, with the same equivalence class of items. It seems absurd to suppose that the extension of the concept of being an odd number, for example, could be the number 0; and, in general, it is hard to see what reasonably systematic view could associate the same abstract with two different equivalence classes of items. Perhaps the only view of this sort with any degree of plausibility is the one that identifies higher-order abstracts, i.e. abstracts of abstracts, with lower-order abstracts. It might be supposed, for example, that the non-negative integers mod 2 (obtained from the non-negative integers by means of the equivalence 'leaving the same remainder on division by 2') are the same as the generalized numbers *Odd* and *Even* (which are obtained directly from the finite concepts by means of a suitable generalization of equinumerosity). But even this view will not yield the cross-sortal identity of *conceptual* abstracts.

Many philosophers have been attracted by the idea that numbers are classes—that a number might be the class of its predecessors, for example, or the class of all concepts that are of that number. If this idea is combined with the view that classes are conceptual abstracts, we would obtain another counter-example to the necessity of our condition. For the very same abstract would, *qua* number, be associated with a range of concepts equinumerous with one another and would, *qua* class, be associated with a range of concepts coextensive with one another; and with possibly the single exception of the number 0, the two ranges of concepts would not be the same. On what basis, then, could an abstract *qua* number and *qua* class be judged the same? The only reasonable view to suggest itself is that *any* abstract, associated through a means of abstraction with certain items, is to be identified with the class of those items. But, as we have seen, such a view leads to the absurd conclusion that any class C of items is identical to the class of the concept(s) whose extension is C.[18]

The other direction of the identity axiom, though, is much more problematic. Wright (1983, 116–17, 122) and Hale (1987, 206) have discussed the question of sufficiency for cross-sortal identity, but their focus is different from ours. For they want to leave room for the view that numbers are classes; they do not want to reject it out of hand simply on the grounds that numbers and classes are associated with different ranges of concepts. The focus of their discussion is, accordingly, on the question of how two abstracts might be the same even though they are associated, through their respective means of abstraction, with different concepts.

As I have indicated, this focus appears to be misplaced; for there seems to be no coherent view that will let both numbers be classes and classes be

[18] A related objection in Hale (1987, 186) is directed against the general thesis that abstracts are to be identified with equivalence classes. He later (p. 187) seems to endorse the view, proposed here, that 'abstract objects are differentiated by their association with distinctive [*sic*] equivalence classes'. I therefore find it puzzling that later still, in Sect. 8.2, he should envisage the possibility of numbers being classes even though the equivalence classes associated with a number and with the corresponding class are not the same. It may be that he wants there to allow for the possibility that classes are not a form of conceptual abstract.

abstracts.[19] Our focus, on the other hand, is on the question of how two abstracts might be distinct even though they are associated, through their respective means of abstraction, with the same concepts. What might contribute to their being different?

The obvious answer is: the identity criteria themselves. On this view, then, two abstracts will be distinct if they are associated with different means of abstraction. But even if one does not believe in the possibility of numbers being classes, this might still appear to be too strong. Suppose one defines *natural number* using equinumerosity as the criterion of identity, but only in application to finite concepts, and that one defines *cardinal number* again using equinumerosity as the criterion of identity, but now in application to all concepts. Does one want to say that the natural number 0 and the cardinal number 0 are not the same on account of their criteria of identity not being the same?

Of course, one might insist in the face of such an example that a conceptual criterion of identity should have application to all concepts. But suppose that the proposed criterion of identity for natural number is extended to all concepts by treating it as a universal relation on any two infinite concepts. We then obtain a single infinite number ∞, just as in pre-Cantorian mathematics. Do we still want to say that the respective 0s are not the same?

These examples force us to face the possibility that the criteria of identity might be different in a way that is not relevant to the identities of the abstracts in question. And this might lead one to shift to the opposite extreme and take two abstracts to be the same when their associated equivalence classes are the same, regardless of the means of abstraction by which they were obtained. But this view is also subject to difficulties.

[19] I might add that, in my view, neither Wright nor Hale succeed in formulating a *selective* criterion, one that will let in certain identifications and not others. Thus Zermelo numbers are not excluded on Wright's criterion because one can always specify a Zermelo number as the Zermelo number corresponding to a given NBG number (cf. Wright (1983), 123); and Frege numbers are not included on Hale's criterion since one can always specify the Frege number of Cs as the class of Ds for which there exists a many–one correspondence from C onto D and a many–one correspondence from D on to C (cf. Hale (1987), 208). It is important, in this connection, not to confuse the standard terms for certain classes with the classes themselves. Thus Wright (1983, 122) advises us 'to pick on classes among which all questions of identity and distinctness . . . are logically equivalent to questions concerning 1–1 correlation among concepts.' But it is only terms for the classes that can yield the required logical equivalents. Or again, Hale (1987, 208) talks of 'membership of a class' being 'specifiable in terms of 1–1 correlation'. But the membership of any class whatever might be specified by a formula that uses an expression for 1–1 correspondence and also by a formula that uses no such expression. There is nothing intrinsic to a class, as opposed to a term, that involves 1–1 correspondence.

There are two other aspects of their discussion that I find unfortunate. One is that they propose a criterion for when all Fs are Gs (e.g. numbers are classes), but one would prefer a criterion for when a particular F is a G. It is perfectly possible for the Fs and Gs to overlap. The other is that they state the criterion in cognitive terms such as 'explanation' or 'understanding'. But one would prefer a purely objective formulation.

For could not two abstracts just *happen* to be associated with the same equivalence class? Perhaps there is a certain weight and a certain density, both unique to a certain thing. Under the proposed criterion, the weight and the density are the same. But they do not appear to be. Indeed, we seem to have a proof that they are not. For surely it is possible for that very weight and density to belong to different things; and if they do, then they will not be the same (cf. Hale 1987, 185–6).

It may be that there does not exist a similar possibility in the case of logical abstracts such as numbers; for all of the possibilities for two such abstracts to be associated with different ranges may, in all essential respects, already be realized. But, be that as it may, one would still like to have a general account of identity, one which has application to all forms of abstraction and from which the sufficiency of our original condition, in the case of logical abstracts, might then be proved rather than simply assumed.

In the light of these difficulties, it is natural to adopt a modal form of the criterion: two abstracts are to be the same iff they are *necessarily* associated with the same items under their respective means of abstraction. The criterion in effect reduces the intra-world question of cross-sortal identity to the cross-world question of intra-sortal identity. For we can determine whether abstracts of different sorts are the same by 'tracking' each from one world to another and 'seeing' whether the items with which they are associated are the same.

This question of cross-world identity is clearly important to the development of a modal theory of abstracts. But it is also important to the ontology of abstracts. For we might ask: What lengths are there? If one takes lengths to be the lengths of actual things, then there is the danger of missing lengths—a length between two others, perhaps, which we think should exist but which is not the length of anything. On the other hand, if we take there to be lengths that are not the length of anything, we appear to give up on the idea of lengths as a form of abstract.

One natural way out of this dilemma is to take the lengths there are to be the lengths of possible as well as actual objects: given the possibility of a length we suppose there already is that length. But this solution raises the problem (among others) of how the unrealized lengths are related to the realized lengths. On what basis do we say that they are the same or distinct, shorter or longer? Clearly any answer must again rest upon an account of cross-world identity.

So when do we say that an abstract of an item *a* with respect to a given means of abstraction in one world is the same as an abstract of an item *b* with respect to the same means of abstraction in another world? One natural line of response is to make a cross-world application of the criterion of identity: the two abstracts are the same if and only if item *a* in the one world is related by the criterion of identity to item *b* in the other world.

Unfortunately, a criterion of identity has, in itself, no meaningful application from one world to another; it cannot, Colossus-like, bestride the two worlds. In some cases, though, it is completely clear how the application is to be made. We are in no doubt, for example, about the cross-world identity of numbers. The number of *C*s in one world will be the same as the number of *D*s in another world just in case the *C*s and *D*s are equinumerous in a world (or domain) in which the respective extensions of *C* and *D* are both preserved; and similarly for any other criterion of identity that is logical and absolute in the sense defined above.

Other cases, though, are more problematic. When do we say that something in one world is of the same length as something in another world? Or of the same colour? It seems that any general answer to such a question should be given in terms of the possession of a common property: an item in one world is to be taken to be related by the criterion to an item in another world just in case the one has, in its world, a property of the required sort that is also had by the other in the other world.

Even the idea that we have an alternative explanation of cross-world equinumerosity may be something of an illusion. For we need to make sense of the idea that there is a one–one correspondence whose domain is the extension of the concept *C* in the one world and whose range is the extension of the concept *D* in the other world. But if there is no world that contains the two extensions, then why should there be a world which contains the required 1–1 correspondence? Indeed, on any plausible actualist view of the matter, no such world would exist.

Given the need to ground the cross-world identities in properties, what might those properties plausibly be taken to be? A natural suggestion is that they are the relational properties that correspond to the criterion of identity—the properties of being related by the criterion to *a*, for any item *a*. Now it will be true that two items will be related by the criterion just in case they share such a property. But these properties are not, in general, the ones we require. Common possession of such a property is not sufficient, in the cross-world case, for the abstracts to be the same. Two sticks may, in their respective worlds, be of the same length as a given stick or even of the very same things; but that does not mean that they themselves have the same length. Nor is common possession necessary. For one stick in one world may have the same length as another stick in another world and yet the associated equivalence classes in the respective worlds be disjoint.

The relational properties will not even work too well in the case of numbers. If concepts are allowed to have a variable extension, then two distinct numbers may be of the same concept in their respective worlds—they might both, for example, be the number of planets; while if concepts are required to have a rigid extension, there is the problem that they may not exist in every world and so there is the possibility that a large number, such as $\aleph_1$, may not share a concept (even a concept of numbers) in a particular pair of worlds.[20]

Another property that might be associated with an abstract is the property of yielding that abstract. With a given length, for example, we might associate the property of being of that length. Common possession of such a property would, of course, be necessary and sufficient for the cross-world identity of the abstracts; but it would provide no explanation of that identity. Granted that such an explanation is possible, we must therefore have an understanding of the common properties that is not directly explicable either in terms of the criterion of identity or in terms of the abstracts.

In particular cases, we seem to be able to say what these properties are. In the case of numbers, they will be what one might call the *numerical* properties—the property, for example, of there being an object that falls under the given concept and to which every object that falls under the concept is identical; and in the case of extensions, they will be what one might call the *enumerative* properties—the property, for example, of an object falling under the given concept just in case it is either Caesar or Brutus. The case of length is more difficult. But one might suppose that significant overlap between paradigm equivalence classes in the two worlds could be used to determine that their items were of the same length. Once a cross-world comparison between these items had been established, they could be used as a 'yardstick' to make the other cross-world comparisons.

As these, and other, examples suggest, there appears to be no uniform method by which the relevant properties can be derived from the criterion of identity. Indeed, it might even be argued that the same intra-world criterion of identity might be associated with distinct cross-world criteria and hence with different kinds of abstract. Consider, for example, the contrast between the conceptions of temperature as given by the Celsius scale and by the thermodynamical account. The concept of same temperature is the same under the two conceptions. But it might be argued that the temperature at which water freezes (under normal atmospheric conditions) must remain the same under the Celsius scale but, because of possible changes in the behaviour of water, could be different under the thermodynamical account.

Even when there appears to be only one property type associated with a given criterion of identity, its adoption is by no means forced upon us. In the case of numbers, for example, we could take the property associated with a number to be the property possessed by a concept when it is empty,

[20] I might add that the possibility of the concepts of a given number being different is yet further evidence against the view that a number is the class of concepts which are of that number, for the membership of the class cannot change.

or singleton, or what have you, *and* when the universe of all objects is of the particular size that it is. Equinumerosity would then still serve as the criterion of identity within a world; but in any possible world of a different size from our own, the numbers would all be different.[21]

If this is right, there must be more to our understanding of a form of abstraction than is contained in the adoption of a specific criterion of identity; we must have an understanding of the common properties in virtue of which the identity criterion holds that goes beyond our understanding of the criterion itself. Although it has become standard to identify a form of abstraction with its criterion of identity, it should in general be identified with a type of property, common possession of which provides the basis for the criterion to hold.

In the simplest case, and perhaps in general, the properties will be mutually exclusive and exhaustive over the domain of items on which they are to be defined. Thus with each item x, we may associate a unique property P_x of the type in question. If $\boldsymbol{R}$ is the corresponding criterion of identity, then the two should conform to the condition:

(*) $x\boldsymbol{R}y$ iff $P_x = P_y$

for any items x and y. If we think of the relation $\boldsymbol{R}$ and the properties P_x in extensional terms, then the choice of P_x, for any given x, is uniquely determined by $\boldsymbol{R}$; and it is perhaps this fact that has made it so easy to overlook the independent role of the properties P_x. However, if we think of the relation and the properties in intensional terms and require that (*) hold of necessity for any items x and y, then the choice of P_x is by no means uniquely determined by $\boldsymbol{R}$. For each way of drawing the transworld-lines between the equivalence classes in different worlds will correspond to different choices of P_x.

In developing the theory below, I have only been interested in the extensional properties of abstracts and therefore have had no reason to depart from the conventional approach. However, it seems to me that, within a fuller account, the means of abstraction should be specified in terms of property types rather than equivalence relations. Perhaps the theory of classes, as developed above on the basis of the theory of abstracts, can be seen as a step in this direction—though, of course, the Extensionality axiom must be given up.[22]

[21] Tennant (1997), in a different way, has stressed the independent importance of numerical properties in understanding what numbers are; and Peacocke (1991) has stressed the general importance of properties in providing a modal account of our understanding of abstracts.

[22] Under an intensional conception of abstracts as envisaged here, there is a new danger of inflation; for the possibilities of distinguishing abstracts associated with the same actual range of items might exceed the number of items in the universe. This difficulty must therefore also be resolved under a satisfactory intensional account.

Let me mention, in conclusion, that one might envisage a criterion of identity that was intermediate in strength between the present modal criterion and the earlier criterion in terms of the means of abstraction. Under the modal criterion, the number 0 and the empty class will be the same; for in each possible world they will be associated with the same concept(s), viz. the empty one(s). However, one might wish to maintain that the two are distinct and that therefore something more is required for the identity of two abstracts than their necessarily being associated with the same range of items.

I myself would be inclined to question the intuition that lies behind this view. For to the extent that we wish to conceive of a class as an abstract on concepts, it is not clear that the null class is to be distinguished from the number 0. But if the intuition is upheld, it is hard to see how it might coherently be developed. For the relevant difference between the number and the class seems to be tied to the type of abstract in question. What is common to the empty concepts can either be regarded as a form of number or as a form of extension. But if the two can be distinguished in this way, then what is to stop us distinguishing between the natural numbers that are tied to a single pre-Cantorian infinity and the ones that are tied to the usual range of infinite numbers? To draw the line, we seem to require an, as yet, undeveloped theory of basic classificatory types or forms.

PART II: THE ANALYSIS OF ACCEPTABILITY

The present part centres on several questions concerning the acceptability of a theory of abstraction. Sections 1–3 are devoted to preliminaries: Section 1 describes the basic languages and systems of interest to us; Section 2 specifies the semantics for the languages; and Section 3 establishes some basic proof- and model-theoretic results. Sections 4–6 deal with various requirements for acceptability: Section 4 contains some preliminary results on tenability, which will later be refined in the section on invariance; Section 5 introduces the notion of a generated model; and Section 6 shows how an abstraction theory may be categorical with respect to its generated models. Sections 7–9 are devoted to the topic of invariance: Section 7 provides an intrinsic analysis of those criteria of identity that are invariant and noninflationary; Section 8 builds on these results to determine when the class of all invariant methods of abstraction will not 'hyperflate' over a given domain; and Section 9 shows how the results on invariance may be internalized to a theory of abstraction and limits thereby set on which abstraction principles can consistently be assumed.

1. LANGUAGE AND LOGIC

The underlying language of our investigations is the language L^2 of second-order logic. This is obtained from the usual language of first-order logic by adding variables for relations of arbitrary finite non-zero degree. The relational variables apply to the first-order objectual terms in the same way as the relational constants; and the quantifiers apply to both the first- and the second-order variables. We suppose that the language contains a symbol for identity but only allow it to flank terms of first order.

We shall occasionally appeal to the language L^3 of third-order logic. This is obtained from L^2 by adding relational variables, and possibly also some relational constants, of arbitrary third-order degree. In particular, it will contain a binary relational variable of third order whose first term is of second order and whose second term is of first order.

Both of the languages L^2 and L^3 may have non-logical constants where they have variables; and so, strictly speaking, there are different languages for different choices of constants. But our main interest is in languages without non-logical constants. These languages and their associated formulas, when they need to be distinguished, will be called *(purely) logical*.

For the most part, our notation will either be standard or intelligible from the context. We reserve: C, D, E, etc. for unary relational symbols; R, S, T, etc. for the binary ones; and P, Q, etc. for those of arbitrary degree. Relational constants and variables of third order will be distinguished by being displayed in bold face. We follow Frege in thinking of the first-order terms as standing for *objects* and the unary second-order terms as standing for *concepts*.

The language $L^§$ of *abstraction* is obtained from L^2 by adding an operator § of second-order abstraction. This applies to a concept-term C and results in an object-term §C. Where **§** is a set of abstraction operators, we may also form the result $L^{\mathbf{§}}$ of adding all of the members § of **§** to L^2. We should perhaps have treated § as a variable-binding device that applies to a variable x and a formula ϕ to form an object-term §xϕ; what we now express as §C would then be expressed as §xCx. Whatever the philosophical reasons for preferring the one notation to the other, nothing of great technical significance hangs on the choice; and it will turn out that the variable-free notation is more convenient for our purposes.

The following abbreviations within the language L^2 will be useful (the type of the various terms occurring in the definienda being evident from the context):

Domain: $x \in \text{Dm}(R)$ for $\exists y Rxy$;
Range: $x \in \text{Rg}(R)$ for $\exists y Ryx$;

Field: $x \in \mathrm{Fld}(P)$ for $\exists x_2 \ldots x_n(Pxx_2x_3 \ldots x_n \vee Px_2xx_3 \ldots x_n \vee Px_2x_3 \ldots x_nx)$;

Reflexivity: Refl(R) for $\forall xRxx$;

Symmetry: Sym(R) for $\forall xy\,(Rxy \rightarrow Ryx)$;

Transitivity: Trans(R) for $\forall xyz\,(Rxy \,\&\, Ryz \rightarrow Rxz)$;

Equivalence: Eq(R) for Refl(R) & Sym(R) & Trans(R);

Inclusion: $P \subseteq Q$ for $\forall x_1 \ldots x_n(Px_1 \ldots x_n \rightarrow Qx_1 \ldots x_n)$;

Coextensiveness: $C \equiv D$ for $\forall x(Cx \leftrightarrow Dx)$;

Complementation: D compl C for $\forall x(Dx \leftrightarrow \neg Cx)$;

Mapping: $P \rightarrow_R Q$ for $\forall x_1 \ldots x_n \forall y_1 \ldots y_n\ (\bigwedge Rx_iy_i \rightarrow (Px_1 \ldots x_n \leftrightarrow Qy_1 \ldots y_n)$;

One–one: 1–1(R) for $\forall xyz(((Rxy \,\&\, Rxz) \vee (Ryx \,\&\, Rzx)) \rightarrow y = z)$;

Equinumerosity via R: C eq_R D for $C \rightarrow_R D$ & 1–1(R) & $\forall x((x \in \mathrm{Dm}(R) \leftrightarrow Cx) \,\&\, (x \in \mathrm{Rg}(R) \leftrightarrow Dx))$;

Equinumerosity: C eq D for $\exists R$ (C eq_R D);

Inequality: $C \leq D$ for $\exists D'(D' \subseteq D$ & C eq D$')$;

Bi-equinumerosity: C beq D for $C \approx D \,\&\, \forall C', D' \forall x([(C'$ compl C) & (D$'$ compl D)] $\rightarrow C' \approx D')$;

Permutation: Perm(R) for 1–1(R) & $\forall x(x \in \mathrm{Dm}(R) \,\&\, x \in \mathrm{Rg}(R))$.

The logic L^2 of second-order is based on the second-order language L^2. (So, strictly speaking, we obtain different logics for different choices of the non-logical constants.) The logic contains the usual truth functional and quantificational principles for first- and second-order terms. However, I assume that these principles are formulated so as to be tolerant of an empty domain. The logic L^2 contains, in addition, the *Comprehension Scheme* for relations of arbitrary degree: where P is a relational variable of degree n which does not occur free in ϕ, $\exists P \forall x_1 \ldots x_n(Px_1 \ldots x_n \leftrightarrow \phi)$ is to be an axiom.

L^3 is the language of the third-order logic L^3, which is obtained from L^2 by extending the truth-functional and quantificational principles in the obvious manner and by adding a comprehension scheme for third-order relations. For example, in the case of a relation between a concept and an object, it would contain axioms of the form $\exists \mathbf{F} \forall C \forall x(\mathbf{F}(C,x) \leftrightarrow \phi)$.

$L^{\S}$ is the language for the various theories of abstraction. Let ϕ be a formula whose only free variables are the concept variables C and D. Φ is then taken to be the formula $\S C = \S D \leftrightarrow \phi(C,D)$; and the *abstraction theory* T^{ϕ} is taken to be the result of adding Φ as an axiom to L^2 (as defined over the language which results from adding § to the language of ϕ). Similarly, when $L^{\S}$ is a language containing several abstraction operators and ϕ associates a formula $\phi_{\S}$, meeting the specifications above, with each § in **§**, the *generalized* abstraction theory T^{ϕ} is the result of adding all of the formulas $\Phi_{\S}$ as axioms to the logic L^2 (based on $L^{\S}$).

The formula ϕ is called an *identity criterion* since it provides a criterion for two abstracts to be the same, and the function ϕ is called a *system* of identity criteria. The associated formula Φ is called *Abstraction* or the *Abstraction Principle* (*for* ϕ).

The identity criterion ϕ (the associated principle Φ, the theory T^ϕ, and the operator § in T^ϕ) are said to be *logical* if ϕ contains no occurrence of non-logical constants or of §; and they are said to be *grounded* if ϕ contains no occurrence of §. Logical criteria are also called *L-criteria.* Within a logical language, logical and grounded criteria will, of course, coincide.

These definitions can be extended to generalized theories T^ϕ and even to theories in which the identity conditions are permitted to be infinitary. Let us say that § *precedes* §′ in the generalizd theory T^ϕ—in symbols, § < §′—if § occurs in $\phi_§$. Then T^ϕ is said to be *logical* if all of its identity conditions $\phi_§$ are logical; and it is said to be *grounded* if the associated relation < of precedence is well founded, i.e. if there is no infinite sequence $§_1, §_2, \ldots$ of abstraction operators and of corresponding identity conditions $\phi_1, \phi_2, \ldots$ with the property that $§_{i+1}$ occurs in ϕ_i for $i = 1, 2, \ldots$. Regarding abstraction principles as forms of definition, the requirement that a system of such principles be grounded can be seen as part of the general requirement that definitions be grounded.

For any well-founded relation < on the operators in **§**, there exists a compatible well-ordered sequence $<\xi_\varphi: \xi < \alpha>$, i.e. one with the property that $§_\nu < §_\xi$ only if $\nu < \xi$. Accordingly, let us say that a generalized theory T^ϕ is *definable over* a well-ordered sequence $<§_\xi: \xi < \alpha>$ of its operators **§**, or is a $<§_\xi: \xi < \alpha>$-theory, if the well ordering of the operators **§** is compatible with the relation of precedence for T^ϕ. Thus in any $<§_\xi: \xi <\alpha>$-theory T^ϕ, $\phi_§$, for $§ = §_\xi$, will only contain operators $§_\nu$ for $\nu < \xi$.

The following abbreviations within $L^§$ will be useful:

Abstract: Ab(x) for $\exists C(x = §C)$;
Individual: I(x) for $\neg Ab(x)$;
Null abstract: NAb(x) for $\exists C(\forall x \neg Cx \ \& \ x = §C)$;
Universal abstract: UAb(x) for $\exists C(\forall x Cx \ \& \ x = §C)$.

We shall also be interested in *restricted* abstraction theories. Let ϕ be an identity criterion, as before, and let $\psi = \psi(C)$ be a formula whose sole free variable is C. Then the *restricted abstraction principle* Φ_ψ is the formula $\psi(C) \ \& \ \psi(D) \rightarrow (§C = §D \leftrightarrow (C,D))$ and the *restricted abstraction theory* $T^{\phi,\psi}$ is the result of adding Φ_ψ to the logic L^2 based on the language of Φ_ψ. Thus in a restricted principle, the application of the identity criterion is limited to those concepts which conform to a certain constraint.

We use $\vdash^2$, $\vdash^3$, $\vdash^\phi$, and $\vdash^\phi$ to indicate provablity in L^2, L^3, T^ϕ, and T^ϕ, respectively.

2. MODELS

We outline the model theory for the various languages and define some basic model-theoretic notions.

A *mode* $\boldsymbol{M}$ *for* the language L^2 is a quadruple $(M, \boldsymbol{R}, e. v)$, where M (the domain of *objects*) is a (possibly empty) set, $\boldsymbol{R}$ is a function which takes each n, for $n > 0$, into a set R_n (the domain of n-ary *relations*), e (*extension*) is a function which takes each member of $\boldsymbol{R}_n$ into a subset of M^n of n-tuples from M, and v (*valuation*) is a function which takes each object constant into a member of M and each n-ary relation constant into a member of $\boldsymbol{R}_n$.

A *model* $\boldsymbol{M}^§$ *for* $L^§$ (also called a §-*model*) is a quintuple $(M, \boldsymbol{R}, e, \mathcal{S}, v)$, with $(M, \boldsymbol{R}, e, v)$ as before and with § (the *abstractor*) a function from the domain $\boldsymbol{R}_1$ of concepts into the domain M of objects. The model $\boldsymbol{M}^§ = (M, \boldsymbol{R}, e, \mathcal{S}, v)$ is said, in such a case, to be a §-*expansion of* the model $\boldsymbol{M} = (M, \boldsymbol{R}, e, v)$. Similar definitions can be given for the other languages. But, of course, when the language contains several abstraction operators §, the model will contain an abstractor $\mathcal{S}$ for each of the operators. If $\boldsymbol{M}$ is a §-model then the *reduction* $\boldsymbol{M}_{\mathcal{S}}$ *of* $\boldsymbol{M}$, for § ∈ **§**, is the §-model with the one abstractor § in place of the many abstractors in $\boldsymbol{M}$.

Given a model $\boldsymbol{M}^§ = (M, \boldsymbol{R}, e, \mathcal{S}, v)$, the objects within the range of § are called *abstracta* and the remaining objects of M are called *concreta* or *individuals*. Similarly, in a model with several abstractors, the abstracta are the objects lying in their range and the individuals are the remaining objects. Note that individuals should only be taken to be individuals in a relative sense; even if they cannot be obtained as a result of the methods of abstraction represented within the model, they may still result, intuitively speaking, from some other kind of abstraction.

We use I for the set of individuals and A for the set of abstracta; and we may employ an index $\boldsymbol{M}$, both here and elsewhere, to indicate dependence upon the underlying model. The model $\boldsymbol{M}^§$ is said to be *pure* when its domain I of individuals is empty.

The model $\boldsymbol{M} = (M, \boldsymbol{R}, e, v)$ (or its expansion $\mathbf{M}^§$) is said to be *extensional* if extensionally equivalent relations are the same, i.e. if $e(R) = e(S)$ implies $R = S$ for any R and S in $\boldsymbol{R}_n$, $n > 0$. A subset of M^n, $n > 0$, is said to be *represented in* $\boldsymbol{M}$ if it is the extension of some relation in $\boldsymbol{R}_n$; and the model $\boldsymbol{M}$ (or $\boldsymbol{M}^§$) is said to be *full* if every subset of M^n, for $n = 1, 2, \ldots,$ is represented.

The model $\boldsymbol{M}$ (or $\boldsymbol{M}^§$) is said to be *set-theoretic* if M consists entirely of urelements (i.e. non-sets) and e is an identity function on $\cup \boldsymbol{R}_n$; and $\boldsymbol{M}$ (or $\boldsymbol{M}^§$) is said to be *standard* if it is both set-theoretic and full. Clearly, any set-theoretic model is extensional. We shall assume, for convenience, that there is a proper class U of urelements and that all of the urelements that we need are drawn from U.

Any standard model $\boldsymbol{M}$ of a language L^2 without non-logical constants is uniquely determined by its domain M and hence is uniquely determined up to isomorphism by the cardinality $\mathbf{c}$ of its domain. We denote this model (as given up to isomorphism) by $\boldsymbol{M}_c$.

Three kinds of abstractor will be of special importance in what follows. These are the *cardinal* abstractors §, which give the value $§(C) = \text{card}(C)$ for each subset C of the domain M, the *bicardinal* abstractors, which give the value $§(C) = \langle \text{card}(C), \text{card}(M - C)\rangle$ for each subset C of M, and the *divisor* abstractors, those with a range of cardinality ≤ 2.

Let $\boldsymbol{M}$ be a model for $L^§$ (or for $L^§$). Then with each abstractor § in $\boldsymbol{M}$ we may associate an equivalence relation $\equiv_§$ on concepts defined by: $C \equiv_§ D$ iff $§(C) = §(D)$. The associated equivalence classes $|C|_§$ and partition $\boldsymbol{P}_§$ may then be defined in the usual way.

We may distinguish three ways in which a model can be separated. The model $\boldsymbol{M}$ for $L^§$ is said to be *strictly separated* if the ranges of the different $§_i$'s are disjoint. No two methods of abstraction yield the same abstract. The model $\boldsymbol{M}$ for $L^§$ is said to be *separated* (simpliciter) if, for any abstractors § and §′ in $\boldsymbol{M}$, $§(C) = §'(D)$ holds iff $|C|_§ = |D|_{§'}$. Two abstracts, of different types, are identified when and only when their associated equivalence classes are the same. The model $\boldsymbol{M}$ for $L^§$ is said to be *weakly separated* if, for any abstractors § and §′ in $\boldsymbol{M}$, $§(C) = §'(D)$ holds only if $|C|_§ = |D|_{§'}$. Two abstracts, of different types, are the same only when their associated equivalence classes are the same.

Satisfaction and truth for the models of our various languages are defined in the usual way. In particular, §C will denote $§(C)$, where C is the concept from $\boldsymbol{R}_1$ assigned to the variable C. With truth as so defined, any model will be a model of the theorems of second-order logic without Comprehension. From henceforth, we shall take our models also to be models for Comprehension unless there is an explicit rider to the contrary.

When $\boldsymbol{M}$ is a model for one of our languages L and ϕ is a formula of L with free variables $x_1, \ldots x_m, P_1, \ldots, P_m$ (in that order of appearance), we take the *extension* $\boldsymbol{E}_{\phi,\mathbf{M}}$ *of* ϕ *in* $\boldsymbol{M}$ to be the set $\{\langle x_1, \ldots, x_m, P_1, \ldots, P_n\rangle : \boldsymbol{M} \models \phi[x_1, \ldots, x_m, P_1, \ldots, P_n]\}$. (The subscript $\boldsymbol{M}$ may be omitted from $\boldsymbol{E}_{\phi,M}$, and elsewhere, when it is obvious from the context.)

We are especially interested in the case in which $\phi = \phi(C,D)$ is an identity criterion (with free variables C and D). In this case, E_ϕ is a relation on concepts, i.e. a subset of $\boldsymbol{R}_1 \times \boldsymbol{R}_1$, and is called the *criterial relation in* $\boldsymbol{M}$. Should E_ϕ be an equivalence relation, it will induce a partition P_ϕ over $\boldsymbol{R}_1$ in the obvious way; and when $\boldsymbol{M}$ is a standard model, R_ϕ and P_ϕ will be an equivalence and a partition over the subsets $\wp(M)$ of M.

Given a set $\boldsymbol{M}$ of urelements, we take a *(local) set relation over* M to be a subset of $\wp(M) \times \wp(M)$. A *global set relation* is then taken to be a (proper class) function which takes each set M of urelements into a local set relation over M. Each logical identity criterion ϕ will determine a global set relation, also denoted by $\boldsymbol{E}_\phi$, which takes each set M of urelements into $\boldsymbol{E}_{\phi,\boldsymbol{M}}$, where $\boldsymbol{M}$ is the standard model with domain M.

3. PRELIMINARY RESULTS

We establish various elementary model- and proof-theoretic results for our systems. We begin with the proof theory.

Where $\phi(C)$ is a formula containing free occurrences of the concept variable C, let $\phi(D)$ be the result of replacing any free occurrence of C in $\phi(C)$ with a free occurrence of D. Then within second-order logic, we may show:

Lemma 1. (Extensionality). $\vdash^2 C \equiv D \rightarrow (\phi(C) \leftrightarrow \phi(D))$.

Proof. By a straightforward induction on the complexity of the formula $\phi(C)$.

This result fails to hold in third-order logic or if we allow identity on concepts. For in the first case, $C \equiv D \rightarrow (\mathbf{F}(C) \leftrightarrow \mathbf{F}(D))$ is unprovable for $\mathbf{F}$ a third-order concept term; and in the second case, $C \equiv D \rightarrow (C = C) \leftrightarrow (C = D)$ is unprovable. The result also fails to hold in an arbitrary abstraction theory. For let the theory be the degenerate one T_ϕ in which ϕ is the formula $\S C = \S D$. Then $C \equiv D \rightarrow \S C = \S D$ is unprovable.

We may establish the fixed point theorem within the confines of second-order logic. For $\phi(C,D)$ a formula (which we treat as defining a function with argument C and value D), we adopt the following abbreviations:

Functionality: Func (ϕ) for $\forall C \exists D(\phi(C,D) \ \& \ \forall E(\phi(C,E) \rightarrow E \equiv D))$;

Monotonicity: Mon(ϕ) for $\forall C,D,C',D'(\phi(C,D) \ \& \ \phi(C',D') \ \& \ C \subseteq C' \rightarrow D \subseteq D')$;

Fixed Point over B: $FP_{\phi,B}(F)$ for $B \subseteq F \ \& \ \phi(F,F)$;

Least Fixed Point over B: $LFP_{\phi,B}(F)$ for $FP_{\phi,B}(F) \ \& \ \forall E(FP_{\phi,B}(E) \rightarrow F \subseteq E)$.

The fixed point theorem now takes the following form:

Theorem 2. (Least Fixed Point) $\vdash^2$ Func(ϕ) & Mon$(\phi) \rightarrow \exists F(LFP_{\phi,B}(F))$.

Proof. We present an informal proof, but in a way that makes clear how the proof is to be formalized. We first consider the case in which B is empty. Define a concept C to be *determining* if $\forall D(\phi(C,D) \rightarrow D \subseteq C)$. Now say that an object is *determined* if it falls under every determining concept; and let F be the concept of being a determined object. We may then show in the usual way that F is a 'least fixed point'.

In the general case, let ϕ^+ be the formula: $\exists C^+[\forall x(C^+x \leftrightarrow (Bx \vee Cx))$ & $\phi(C^+,D)]$. Applying the first case to this formula then yields the desired result.

We turn now to abstraction theories and show first that the relation on concepts determined by the identity criterion $\phi(C,D)$ is an equivalence:

Lemma 3 (Equivalence). In the abstraction theory T^ϕ, the formulas $\phi(C,C)$, $\phi(C,D) \rightarrow \phi(D,C)$ and $[\phi(C,D)$ & $\phi(D,E)] \rightarrow \phi(C,E)]$ are all theorems.

Proof. Using the abstraction principle, the above formulas 'translate' into theorems of the first-order logic of identity. For example, $\phi(C,C)$ translates into §C = §C.

Note that the above proof makes no appeal to the Comprehension Scheme in T_ϕ.

Using the above lemma, we can show that coextensive concepts must be identified in any abstraction theory T_ϕ in which ϕ is grounded (i.e. §-free):

Lemma 4 (Inclusion). $\vdash^\phi C \equiv D \rightarrow \phi(C,D)$ for grounded ϕ.

Proof. Given that ϕ is §-free, it follows by the extensionality lemma that $\vdash C \equiv D \rightarrow (\phi(C,C) \rightarrow \phi(C,D))$. By the equivalence lemma, $\vdash \phi(C,C)$. Hence $\vdash C \equiv D \rightarrow \phi(C,D)$.

The extensionality lemma can now be extended to abstraction theories:

Lemma 5 (Extended Extensionality). $\vdash^\phi C \equiv D \rightarrow (\psi(C) \leftrightarrow \psi(D))$ for grounded ϕ and any ψ.

Proof. As before by induction. But to take care of the presence of § in the language, we need to show that C ≡ D → §C = §D is a theorem. But C ≡ D → $\phi(C,D)$ is a theorem by the inclusion lemma; and $\phi(C,D) \rightarrow$ §C = §D follows from the abstraction principle Φ.

The inclusion and extensionality lemmas can be extended by means of the obvious transfinite induction to generalized abstraction theories which are grounded.

Invariance under permutations will play a major role in our attempt to understand what can be conveyed by a logical criterion of identity. It is the formal counterpart to Frege's idea of the generality of logic (see e.g. §3 of the *Grundlagen*). We have, in the first place, the following syntactic formulation:

Lemma 6 (Permutation). Suppose that $\psi = \psi(x_1, \ldots, x_m, P_1, \ldots, P_n)$ is a logical formula whose only free variables are the distinct variables $x_1, \ldots, x_m, P_1, \ldots, P_n$. Let $y_1, \ldots, y_m$ and $Q_1, \ldots, Q_n$ be distinct variables which do not occur in ψ, with each Q_j of the same arity as P_j. Then $\vdash^2$ (Perm(R) & $\wedge_i Rx_iy_i$ & $\bigwedge_j P_j \rightarrow_R Q_j) \rightarrow [\psi(x_1, \ldots, x_m, P_1, \ldots, P_n) \rightarrow \psi(y_1, \ldots, y_m, Q_1, \ldots, Q_n)]$.

Proof. By induction on the complexity of the formula. For the case that ψ is a second-order quantification we need to appeal to comprehension.

As a special case of the above result, we have:

Corollary 7 (Invariance). Suppose that $\phi = \phi(C,D)$ is a logical identity criterion. Then $\vdash^2$ Perm(R) & $C \rightarrow_R C'$ & $D \rightarrow_R D' \rightarrow (\phi(C,D) \rightarrow \phi(C',D'))$.

Thus we may prove within second-order logic that logical identity criteria are invariant under permutations of the universe of objects: the images of concepts related by a criterion are also related by the criterion. Note that the lemma may fail when the formula $\psi(x_1, \ldots, x_m, P_1, \ldots, P_n)$ contains non-logical constants or the operator §. For from the fact that Perm(R) & $C \rightarrow_R D$ we cannot infer $Ca \rightarrow Da$ for a an object constant (and similarly for relation constants); and from the fact that Perm(R) & $C \rightarrow_R D$ we cannot infer §C = §D. On the other hand, we should note that a version of this result will continue to hold when the language is enriched with further logical constants, be they finitary or infinitary in character.

We now consider the model-theoretic results. The first tells us that Abstraction does nothing to determine the identity of abstracts (beyond their being objects). Let us note, for the purposes of the proof, that an §-model $\boldsymbol{M}$ is a model for the abstraction principle Φ just in case for every C, $D \in \boldsymbol{R}_1$, $\S(C) = \S(D)$ iff $\langle C,D\rangle \in \boldsymbol{E}_{\phi,\boldsymbol{M}}$.

Lemma 8 (Switching). Suppose that $\boldsymbol{M}$ is a model for grounded Φ and that f is a one–one function from M into M. Let $\boldsymbol{M}'$ be the result of replacing the abstractor § in $\boldsymbol{M}$ with the compositional map $\S' = f \circ \S$. Then $\boldsymbol{M}'$ is also a model for Φ.

Proof. $\boldsymbol{M}$ is a model for Φ iff for every $C, D \in \boldsymbol{R}^1$, $\S(C) = \S(D)$ iff $\langle C,D\rangle \in \boldsymbol{E}_{\phi,\boldsymbol{M}}$; while $\boldsymbol{M}'$ is a model for Φ iff for every $C, D \in \boldsymbol{R}^1$, $\S'(C) = \S'(D)$ iff $\langle C,D\rangle \in \boldsymbol{E}_{\phi,\boldsymbol{M}'}$. But given that f is one–one, $\S(C) = \S(D)$ iff $\S'(C) = \S'(D)$; and given that ϕ is §-free, $\boldsymbol{E}_{\phi,\boldsymbol{M}} = \boldsymbol{E}_{\phi,\boldsymbol{M}'}$.

This proof is related to Frege's argument in section 10 of the *Grundgesetze.* (See Schroeder-Heister (1987) and T. Parsons (1987) for a careful discussion of his argument.) Our proof fails in case Φ is ungrounded. For example, let Φ be the principle §C = §D $\leftrightarrow \exists x$(UAb(x) & $Cx \leftrightarrow Dx$), in which concepts are identified when they agree on the universal abstract; and let $\boldsymbol{M}$ be a model for Φ with two elements. Switching the universal abstract with any other object then gives a stucture $\boldsymbol{M}'$ which is not a model for Φ.

Call a model $\boldsymbol{M}$ for $L^{\S}$ *§-extensional* if $\S(C) = \S(D)$ whenever $C, D \in \boldsymbol{R}_1$, and $e(C) = e(D)$. Thus a §-extensional model is one which conforms to the *extensionality principle*: $C \equiv D \rightarrow \S C = \S D$. With any §-extensional model $\boldsymbol{M} = (M, \boldsymbol{R}, e, \S, v)$ whose domain M consists of urelements, we may associate a set-theoretic model $\boldsymbol{M}^s = (M, \boldsymbol{R}', e', \S', v')$ by setting $\boldsymbol{R}'_n = \{e(R): R \in \boldsymbol{R}_n\}$, $e'(R) = R$ for $R \in \boldsymbol{R}'_n$, $\S'(e(C)) = \S(C)$ for $e(C) \in \boldsymbol{R}'_1$, $v'(\mathrm{R}) = e(v(\mathrm{R}))$ for R a relational constant, and $v'(\mathrm{a}) = v(\mathrm{a})$ for a an object constant. In effect, extensionally equivalent relations are identified with their extensions. It is then readily shown that:

Lemma 9. Suppose that $\boldsymbol{M}$ is an §-extensional model $(M, \boldsymbol{R}, e, \S, v)$ with associated set-theoretic model $\boldsymbol{M}^s$. Then $\boldsymbol{M} \models \phi[P_1, \ldots, P_m, x_1, \ldots, x_n]$ iff $\boldsymbol{M}^s \models \phi[e(P_1), \ldots, e(P_m), x_1, \ldots, x_n]$.

This lemma will justify us in confining our attention to the set-theoretic models of an abstraction theory T^{ϕ} in case ϕ is §-free.

We now provide a model-theoretic account of invariance under permutation. Say that a global set relation $\boldsymbol{R}$ is *invariant* if, for each non-empty subset M of urelements, one–one map f from M onto N, and subsets C and D of M, $\boldsymbol{R}_M(C,D)$ implies $\boldsymbol{R}_N(f(C), f(D))$; and say that a local set relation R over M is *invariant* if, for any subsets C and D of M and permutation f on M, $R(C, D)$ implies $R(f(C), f(D))$. Clearly, if the global set relation $\boldsymbol{R}$ is invariant then so are each of the local set relations $\boldsymbol{R}_M$.

Lemma 10 (Outer Invariance). For ϕ a logical criterion, the corresponding global set relation $\boldsymbol{E}_{\phi}$ is invariant.

Proof. Suppose that M and N are non-empty sets of urelements, C and D are subsets of M, and f is a one–one map from M onto N. Let $\boldsymbol{M}$ and $\boldsymbol{N}$ be standard models (for a purely logical language) with respective domains M and N. Suppose that $\boldsymbol{R}_M(C,D)$ for $\boldsymbol{R}$ the global relation $\boldsymbol{E}_{\phi}$ associated with ϕ. Then $\boldsymbol{M} \models \phi[C,D]$. Now f can be extended to an isomorphism from $\boldsymbol{M}$ onto $\boldsymbol{N}$; and so, $\boldsymbol{N} \models \phi[f(C), f(D)]$. But then $\boldsymbol{R}_N(f(C), f(D))$.

We finally consider those identity criteria whose behaviour is independent of the underlying domain. Such criteria play an imnportant role in the study of generated abstraction models. We provide a syntactic and semantic criterion for such domain independence and connect the two. Given the sequence S of relational and objectual terms $P_1, \ldots, P_m, t_1, \ldots, t_n$, say that the formula ϕ is *relativized to* S if each (universally) bound objectual variable x is relativized to the formula $x \in \mathrm{Fld}(P_1) \vee \ldots \vee x \in \mathrm{Fld}(P_m) \vee x = t_1 \vee \ldots \vee x = t_n$ and each (universally) bound relational variable P is relativized to the formula $\forall x(x \in \mathrm{Fld}(P) \rightarrow x \in \mathrm{Fld}(P_1) \vee \ldots \vee x \in \mathrm{Fld}(P_m) \vee x = t_1 \vee \ldots \vee x = t_n)$. When S is the sequence of all of the objectual and relational terms which occur—either as free variables or as non-logical constants—in the very formula ϕ, we say that ϕ is *restricted*.

On the semantic side, let $\boldsymbol{M} = (M, \boldsymbol{R}, e, v)$ be a second-order model and N a subset represented in M. We let the *restriction* or *submodel* $\boldsymbol{N} = \boldsymbol{M}/N$ *of* $\boldsymbol{M}$ *induced by* N be the model $(N, \boldsymbol{R}', e', v)$, where $\boldsymbol{R}'_n = \{P \in \boldsymbol{R}_n\colon e(P) \subseteq N^n\}$, and e' is the restriction of e to $\cup_n \boldsymbol{R}'_n$. For a §-model $\boldsymbol{M}^§ = (M, \boldsymbol{R}, e, §, v)$, we also require that §′ in the restriction $\boldsymbol{M}^§/N$ should be the restriction of § in $\boldsymbol{M}$ to $\boldsymbol{R}'_1$. The values of the non-logical constants must be defined in N for the restriction to be well defined. However, we shall allow §′(C) to be undefined when §(C) does not belong to N but C belongs to $\boldsymbol{R}'_1$. In this case the restriction $\boldsymbol{M}^§/N$ will be said to be a *partial* model, i.e. one whose abstraction operation is not always defined; and otherwise it will be said to be *total*. For the most part, our interest is in standard models of L^2; and, in this case, every subset N of the domain M of the model $\boldsymbol{M}$ will determine a restriction of $\boldsymbol{M}$.

It is readily shown that the restriction $\boldsymbol{N}$ verifies the comprehension principle $\exists P \forall x_1 \ldots x_n(Px_1 \ldots x_n \leftrightarrow \phi)$ given that $\boldsymbol{M}$ does. For $\boldsymbol{M} \models \exists P \forall x_1 \ldots x_n(Px_1 \ldots x_n \leftrightarrow \phi^C)$, where ϕ^C, for C a new concept variable, is in the obvious sense the relativization of ϕ to C. Letting C 'be' a concept whose extension is N then yields in $\boldsymbol{M}$ a concept 'P' which, in $\boldsymbol{N}$, will correspond to ϕ.

Let $\phi = \phi(P_1, \ldots, P_m, x_1, \ldots, x_n)$ be a formula of L^2 whose free variables are as displayed. Then ϕ is said to be *absolute* if for any model $\boldsymbol{M}$ for the language of ϕ and restriction $\boldsymbol{N}$ of $\boldsymbol{M}$, $\boldsymbol{M} \models \phi[P_1, \ldots, P_m, x_1, \ldots, x_n]$ iff $\boldsymbol{N} \models \phi[P_1, \ldots, P_m, x_1, \ldots, x_n]$ for any relations $P_1, \ldots, P_m$ (of appropriate type) and objects $x_1, \ldots, x_n$ from N. Thus whether the formula holds of certain objects and relations is unaffected by what in the model lies beyond those objects and the extensions of the relations.

Suppose that $\boldsymbol{M}$ is a model for a language L^2 with finitely many non-logical constants; and suppose that it verifies Comprehension. Let $P_1, \ldots, P_m, x_1, \ldots, x_n$ be relations and objects from $\boldsymbol{M}$. There will then be a smallest restriction $\boldsymbol{N}$ of $\boldsymbol{M}$ to contain $P_1, \ldots, P_m, x_1, \ldots, x_n$. If there are no non-logical constants in the language, the domain of $\boldsymbol{N}$ will be the union of the fields of $e(P_1), \ldots, e(P_m)$ with $\{x_1, \ldots, x_n\}$; while if there are nonlogical constants in the language, the objects from their values must also be included. In each case, the domain N will be definable in $\boldsymbol{M}$ by an appropriate disjunction; and it should be clear that a formula $\phi = \phi(P_1, \ldots, P_m, x_1, \ldots, x_n)$ will be absolute just in case satisfaction in $\boldsymbol{M}$ is equivalent to satisfaction in the smallest restriction of $\boldsymbol{M}$ for each choice of relations and objects $P_1, \ldots, P_m, x_1, \ldots, x_n$.

It may be shown by a straightforward induction:

Lemma 11 (Absoluteness). Any restricted second-order formula is absolute.

Clearly, the result also holds for any formula which is provably equivalent to a restricted formula. For example: the formula $C \equiv D$ is equivalent to the restricted formula ∀x(Cx ∨ Dx → Cx ↔ Dx) and is therefore absolute. Similarly, for the formula C eq D. On the other hand, the formula C beq D is not absolute and therefore is not provably equivalent to any restricted formula.

Absolute identity criteria define an especially simple form of global relation. Say that a global relation $\boldsymbol{R}$ is *internal* if, for any subsets X and Y of urelements and subsets C and D of $X \cap Y$, $\boldsymbol{R}_x(C,D)$ implies $\boldsymbol{R}_Y(C,D)$. Equivalently, $\boldsymbol{R}$ will be internal if $\boldsymbol{R}_{C \cup D}(C,D)$ holds iff $\boldsymbol{R}_Y(C,D)$ holds for any $Y \supseteq C \cup D$.

Corresponding to any global set relation $\boldsymbol{R}$ is an ordinary (unindexed) relation R which holds between any two sets C and D just in case $\boldsymbol{R}_x(C,D)$ holds for some set X of urelements. In general, a global relation cannot be recovered from the corresponding unindexed relation. But in case the global relation is internal, it can be: for $\boldsymbol{R}_x(C,D)$ holds, for $C, D \subseteq X$, just in case $R(C,D)$ holds.

We say that the global relation $\boldsymbol{R}$ is *internally invariant* (or *I-invariant*, for short) if it is both internal and invariant; and we say that a local relation R over M is *internally invariant* (*I-invariant*) if $R(C, D)$ implies $R(f(C), f(D))$ for any subsets C and D of M and any one–one function f from $C \cup D$ into M. Clearly, when $\boldsymbol{R}$ is internally invariant, then so are each of the corresponding local relations $\boldsymbol{R}_M$. It is readily shown from lemmas 10 and 11 that:

Lemma 12. For ϕ a restricted L-criterion, $\boldsymbol{E}_\phi$ is internally invariant.

4. TENABILITY

We consider various formal conditions that might be proposed for the truth or correctness of a principle of abstraction. The philosophical significance of the conditions has been discussed towards the end of section I.1; and further conditions are considered in some of the later sections.

Given a §-free model $\boldsymbol{M}$, we say that the identity criterion ϕ (or the associated abstraction principle Φ) is *tenable on* $\boldsymbol{M}$ if some expansion of $\boldsymbol{M}$ is a model of Φ and, for $\mathbf{d}$ a cardinal, we say that ϕ is $\mathbf{d}$-*tenable on* $\boldsymbol{M}$ if some expansion of $\boldsymbol{M}$ is a model of Φ and contains exactly $\mathbf{d}$ individuals. When the model $\boldsymbol{M}$ is standard and the criterion ϕ is logical, all that matters about $\boldsymbol{M}$ is its cardinality. We may therefore say that the L-criterion ϕ is *tenable on the cardinal* $\mathbf{c}$ if it is tenable on a standard model $\boldsymbol{M}$ with domain of cardinality $\mathbf{c}$ and that it is $\boldsymbol{d}$-*tenable on* $\mathbf{c}$ if it is $\mathbf{d}$-tenable on a standard model $\boldsymbol{M}$ whose domain is of cardinality $\mathbf{c}$.

The above condition is relative to a model or a cardinal. We may obtain an absolute condition by declaring an L-criterion ϕ to be *stable* if, for some cardinal $\mathbf{c}$, ϕ is tenable on $\mathbf{d}$ for any cardinal $\mathbf{d} \geq \mathbf{c}$. I have here presupposed that the underlying model is full; but the account could readily be accommodated to other, more restrictive, conceptions of what concepts and relations are admitted into a model. If ϕ is stable, it will be said to *stabilize* on the smallest cardinal **c** for which ϕ is tenable on every $\mathbf{d} \geq \mathbf{c}$.

Other absolute notions might also be distinguished. For example, we might say that an L-criterion ϕ is *generally* tenable if it is tenable on each infinite cardinal, or that ϕ is *indefinitely* tenable if there is no greatest cardinal upon which it is tenable.

It might be thought that the requirement on cardinality in the definition of stability is too weak. For given a model whose domain of objects is of transfinite cardinality **c**, one might want to insist that the domain of individuals (relative to the given abstractor) should be capable of being of any reasonable size $\mathbf{d} \leq \mathbf{c}$. However, it may be argued that this further requirement is automatically met. For let us suppose that the cardinality of the abstracts is $\mathbf{c}' \leq \mathbf{c}$. If $\mathbf{c}' = \mathbf{c}$, we may—by means of the switching lemma 3.8 above—allow **d** to be of any cardinality $\leq \mathbf{c}$. On the other hand, if $\mathbf{c}' < \mathbf{c}$, it may plausibly be maintained that **d** should be equal to **c**. For there will be at least as many abstracts of some sort or another as individuals (in the absolute sense) and so if there are only $\mathbf{c}' < \mathbf{c}$ abstracts of the given sort there must be at least **c** abstracts of another sort, which will, in the required relative sense, be individuals. But given $\mathbf{d} = \mathbf{c}$, we may again by means of the switching lemma allow **d** to be the cardinality of the domain of individuals (in the relative sense).

We now provide a simple characterization of tenability. A deeper account will later be provided in §7.

Lemma 1 (Anti-inflation). Suppose that $\boldsymbol{M}$ is a standard §-free model and that ϕ is a §-free identity criterion. There is then an §-expansion $\boldsymbol{M}^+$ of $\boldsymbol{M}$ which is a model of Φ iff:

(i) $E_{\phi,\boldsymbol{M}}$ is an equivalence relation; and
(ii) $\mathrm{card}(\boldsymbol{P}_{\phi,\boldsymbol{M}}) \leq \mathrm{card}(M)$.

Proof. We first prove the direction from left to right. Suppose that the §-expansion $\boldsymbol{M}^+$ of $\boldsymbol{M}$ is a model of Φ. By the Equivalence Lemma 3.3, E_ϕ is an equivalence relation. For each C in $\boldsymbol{R}^1$, let $|C| = \{D: \langle C,D\rangle \in E_\phi\}$ (so $P_{\phi,\boldsymbol{M}} = \{|C|: C \in \boldsymbol{R}^1\}$). Since $\boldsymbol{M}$ is a model for Φ, $\S(C) = \S(D)$ iff $\langle C,D\rangle \in E_\phi$; and so § induces a one–one map f from P_ϕ into M (with $f(|C|) = \S(C)$). But then $\mathrm{card}(P_{\phi,\boldsymbol{M}}) \leq \mathrm{card}(M)$.

For the other direction, suppose that (i) and (ii) hold. Let $\mathbf{c} = \mathrm{card}(P_\phi)$; and pick a subset M' of M of cardinality **c**. There is then a one–one map f from P_ϕ onto M'. Define § by $\S(C) = f(|C|)$ for each $C \in \boldsymbol{R}^1$. Then adding § as the abstractor to $\boldsymbol{M}$ yields an §-expansion $\boldsymbol{M}^+$ which is a model of Φ.

The above proof does not rest on any special assumptions concerning the model $\boldsymbol{M}$, although its formulation does require that ϕ be §-free. If we were to drop the requirement that $\boldsymbol{M}$ and ϕ be §-free, conditions (i) and (ii) would provide neither a necessary nor a sufficient condition for a variant of $\boldsymbol{M}$ (with possibly different §) to be a model of ϕ. They would not be jointly necessary since ϕ might be the trivial condition §C = §D. Nor would they be sufficient. For let ϕ be the formula $(\forall C,D(\S C = \S D)\ \&\ (\forall xCx \leftrightarrow \forall xDx))\vee(\neg\forall C,D(\S C = \S D)\ \&\ C \equiv D)$. If we let card($M$) > 1 and let *§* in $\boldsymbol{M}$ have a singleton range, conditions (i) and (ii) will be satisfied; and yet there is no model for Φ.

An analogous result can be proved for the existence of a model for Φ with exactly **d** individuals. In this case, condition (ii) should be replaced with:

(ii)′ $\mathbf{d} + \text{card}(\boldsymbol{P}_{\phi,M}) = \text{card}(M)$.

If $\mathbf{m} = \text{card}(M)$ is infinite, then this condition is satisfied iff $\max\{\mathbf{d}, \text{card}(\boldsymbol{P}_{\phi,\mathbf{M}})\} = \mathbf{m}$.

Suppose that ϕ is an L-criterion. Then we say that the cardinal **c** is *non-inflationary* (*wrt* ϕ) if the conditions (i) and (ii) are satisfied when $\boldsymbol{M}$ is the standard model $\boldsymbol{M}_\mathbf{c}$. We now have the following immediate consequences of the lemma:

Corollary 2. For ϕ an L-criterion:
(i) ϕ is tenable on **c** iff **c** is non-inflationary;
(ii) ϕ is stable iff for some cardinal **c**, every $\mathbf{d} \geq \mathbf{c}$ is non-inflationary.

An analogue to (i) holds for **d**-tenability. Say that the cardinal **c** is ***d**-static* if condition (i) of lemma 1 and the condition (ii)′ above are satisfied for the case in which $\boldsymbol{M}$ is the standard model $\boldsymbol{M}_\mathbf{c}$. Then ϕ will be **d**-tenable on **c** iff **c** is **d**-static. The matter might be put in terms of fixed points. For each cardinal **c**, let the *partition cardinal* $p(\mathbf{c}) = \text{card}(\boldsymbol{P}_{\phi,M})$ (for $\boldsymbol{M} = \boldsymbol{M}_\text{c}$ and $\boldsymbol{P}_{\phi,M}$ defined) and let $p_\mathbf{d}(\mathbf{c}) = \mathbf{d} + p(\mathbf{c})$. Then ϕ will be **d**-tenable on **c** just in case **c** is a fixed point of the function $p_\mathbf{d}$.

Given that ϕ is absolute, $p_\mathbf{d}$ will be a monotonic function—or, more exactly, $\mathbf{e} \leq \mathbf{f}$ and $p_\mathbf{d}(\mathbf{f})$ defined will imply that $p_\mathbf{d}(\mathbf{e})$ is defined and $p_\mathbf{d}(\mathbf{e}) \leq p_\mathbf{d}(\mathbf{f})$. From this it follows that, if $p_\mathbf{d}$ has a fixed point, then its least fixed point can be reached by iteration from below. Thus if we set $\mathbf{c}^0 = p_\mathbf{d}(0)$, $\mathbf{c}^{\alpha+1} = p_\mathbf{d}(\mathbf{c}^\alpha)$ and $\mathbf{c}^\lambda = \sup\{\mathbf{c}^\xi{:}\xi < \lambda\}$, then the least fixed point of $p_\mathbf{d}$, if it exists, will be of the form $\mathbf{c}^\alpha$, for α the least ordinal for which $\mathbf{c}^\alpha = \mathbf{c}^{\alpha+1}$.

The definitions and results can be carried over to systems containing several abstraction operators. For example, when it is required that the model be separated, lemma 1 takes the form:

Lemma 3. (Anti-inflation). Suppose that $\phi = \langle\phi_§ : § \in §\rangle$ is a system of §-free identity criteria and that $\boldsymbol{M}$ is a standard §-free model. There is then a separated §-expansion $\boldsymbol{M}^+$ of $\boldsymbol{M}$ which is a model of the $\Phi_§$ iff:

(i) each $\boldsymbol{E}_{\phi,M}$, for $\phi = \phi_§$, is an equivalence relation;
(ii) $\text{card}(\cup\{\boldsymbol{P}_{\phi,M}: \phi = \phi_§\}) \leq \text{card}(M)$.

Let us say that a system $\phi = \langle\phi_§ : § \in \mathbf{§}\rangle$ of identity criteria is tenable on **c** if some separated expansion of the standard model $\boldsymbol{M}_{\mathbf{c}}$ is a model of all the $\phi_§$ and that the system is *stable* if, for some cardinal **c**, it is tenable on every $\mathbf{d} \geq \mathbf{c}$. We then have the corollary in the same form as before, but with non-inflation defined by means of the new conditions (i) and (ii) and with the system of criteria ϕ in place of the single criterion ϕ.

The tenability conditions can be expressed within second-order logic itself. This point will be important both for proving certain technical results and for setting up the general theory of abstraction in Part III. In the first place, we may note that it follows from lemma 3.3 that the abstraction principle Φ, corresponding to an identity criterion ϕ, will deductively imply the formula:

$$\text{Eq}_\phi\text{: } \forall C,D,E[\phi(C,C) \;\&\; (\phi(C,D) \rightarrow \phi(D,C)) \;\&\; (\phi(C,D) \;\&\; \phi(D,E) \rightarrow \phi(C,E)].$$

Thus we may show within second-order logic that it is necessary for an abstraction principle to hold that its identity criterion determine an equivalence relation on concepts.

We also wish to prove that it is necessary for an abstraction principle to hold that it not be 'inflationary'. Suppose again that the criterion of identity is ϕ. Then the condition that the principle not be inflationary can be expressed by means of the following third-order formula (where **R** is a variable ranging over second-order relations between concepts and objects):

$$(*)\ \exists\mathbf{R}[\forall C\exists!x(\mathbf{R}(C,x) \;\&\; \forall C,D\forall x,y(\mathbf{R}(C,x) \;\&\; \mathbf{R}(D,y) \rightarrow (x = y \leftrightarrow \phi(C,D)))].$$

It is clear that (*) follows within third-order logic from the abstraction principle Φ (i.e. $\forall C,D(§C = §D \leftrightarrow \phi(C,D))$); for we may take $\mathbf{R}(C,x)$ to be given by the condition $§C = x$. Moreover, if ϕ is itself §-free, then the formula (*) will also be §-free.

Given an appropriate version of the axiom of choice for objects, (*) will be equivalent to:

$$(**)\ \exists\mathbf{R}[\forall C\exists x(\mathbf{R}(C,x) \;\&\; \forall C,D\forall x(\mathbf{R}(C,x) \;\&\; \mathbf{R}(D,x) \rightarrow \phi(C,D))].$$

Moreover, given an appropriate version of the axiom of choice of concepts, (*) will, in its turn, be equivalent to the following second-order formula (where R is now a binary relation on objects):

$$\text{Noninfl}_\phi\text{: } \exists R\forall C\exists D\exists y[\phi(C,D) \ \&\ \forall x(R(x,y) \leftrightarrow Dx)].$$

R in effect picks out a representative D from each equivalence class |C|.

Given the 'meaning' of Eq_ϕ and Noninfl_ϕ in a standard model $\boldsymbol{M}_\mathbf{c}$, it follows that:

Lemma 4. The L-criterion ϕ is tenable on **c** iff (Eq_ϕ & Noninfl_ϕ) is true in the model $\boldsymbol{M}_\mathbf{c}$.

This result can be used to determine the least cardinal at which all stable criteria stabilize. Given an L-criterion ϕ that stabilizes, let $\mathbf{c}_\phi$ be the cardinal at which it stabilizes; and let the *stability number* (of pure second-order logic) be the least upper bound of the $\mathbf{c}_\phi$. Recall that the *Hanf number* (of pure second-order logic) is the least cardinal such that any sentence with a standard model of that cardinality or greater has standard models of arbitrarily large cardinality. Let us extend the notion of stabilizing and say that an arbitrary *sentence* ψ of L^2 *stabilizes on* a cardinal **c** if **c** is the smallest cardinal with the property that ψ is true on all models of equal or greater cardinality. Then the Hanf number will be the least upper bound of the cardinals upon which a sentence of L^2 stabilizes. We now have:

Theorem 5. The Hanf number of second-order logic is identical to the stability number.

Proof. Let us use **c** for the Hanf number and **c*** for the stability number. First suppose that **c** < **c***. Since **c** is less than the stability number, there is an L-criterion ϕ that stabilizes on some cardinal **d** > **c**. Let ψ be the sentence (Eq_ϕ & Noninfl_ϕ). By lemma 4, ψ stabilizes on **d**, contrary to assumption that **c** is the Hanf number.

Now suppose **c*** < **c**. Since **c*** is less than the Hanf number, it follows that there is a sentence ψ of L^2 that stabilizes on some cardinal **d** > **c***. Define the identity criterion ϕ as: $(\psi \vee (C \equiv D))$. Then it is readily shown that ϕ stabilizes on **d** > **c***, contrary to assumption that **c*** is the stability number.

5. GENERATION

We deal here with the notion of a generative model, one in which the abstracts are generated in stepwise fashion from the individuals. The philosophical significance of the notion is discussed in Sections I.2 and I.3.

Let $\boldsymbol{M}$ be a standard §-model (with I, recall, as its subdomain of individuals). Define g on $\wp(M)$ by:

$$g(K) = \{\S(C): C \subseteq K\}$$

Thus $g(K)$ consists of the abstracts which can be directly generated from the concepts that are defined over the objects of K. For each ordinal ξ, we define the *generation function* G_ξ by:

$$G_0 = I;$$
$$G_{\alpha+1} = I \cup g(G_\alpha);$$
$$G_\lambda = \cup\{G_\xi: \xi < \lambda\}.$$

We also set $g^+(X) = g(X) \cup X$. $G_{\alpha+1}$ can then be alternatively defined as $g^+(G_\alpha)$.

We prove in the usual way that there is a least ordinal α such that $G_{\alpha+1} = G_\alpha$ and call α the *critical* ordinal *for* $\boldsymbol{M}$. We use G—without danger of confusion, I hope—for this G_α. Thus G is the set of objects which can be generated, either directly or indirectly, from the individuals of $\boldsymbol{M}$. The requirement that the objects be generable then amounts to the condition that, in the model $\boldsymbol{M}$, $M = G$. We call a model conforming to this condition *minimal*.

The condition that $M = G$ might be compared to the axiom of constructibility, $V = L$, or to the principle $V = \cup V_\alpha$, where the V_α's are the sets in the cumulative hierarchy. But the construction is in a way broader. For it relates to any kind of abstract, not just to sets; and the abstracts are generated by means of concepts, which may or may not be definable. One could, of course, carry out an analogous construction in a non-standard model $\boldsymbol{M}$.

As an example of a minimal model, consider the standard model $\boldsymbol{M}$ in which the domain M is taken to consist of the natural numbers with $\aleph_0$ and § is the cardinal abstractor defined by: $\S(C) = \text{card}(C)$. Then $\boldsymbol{M}$ is generative. For: $G_0 = I = \varnothing$; $G_{n+1} = \{0, 1, \ldots, n\}$; $G_\omega = \{0, 1, \ldots\}$; and $G_{\omega+1} = \{0, 1, \ldots, \aleph_0\} = M$ and is therefore a fixed point. On the other hand, the standard model $\boldsymbol{M}$, in which M consists of all cardinals $\leq$ some inaccessible cardinal and § is the cardinal abstractor, is not generative. For, again, I will be $\varnothing$ and the least fixed point G will be $\{0, 1, \ldots, \aleph_0\}$, which is a proper subset of M. In such a model, there will be no non-circular account of any of the cardinals $> \aleph_0$, for each of them must be explained by reference to a concept which holds of some of those very cardinals.

The definitions can be generalized. Say that a subset K of M is *downward-closed* if each element c of K is either an individual or is of the form $\S(C)$ for some subset C of K; say that K is *upward-closed* if, for any

$C \subseteq K$, $\S(C) \in K$; and say that K is *closed* simpliciter if it is both downward- and upward-closed. The construction can then be relativized with an arbitrary downward-closed subset K of M in place of I. Each of the sets $G_{K,\alpha}$ obtained by means of the construction will then be downward-closed. Let G_K be the resulting least fixed point. Then G_K will also be upward-closed and hence closed. We say that the model $\boldsymbol{M}$ is *K-minimal* if $M = G_K$. Note that if a model is to be K-minimal then K must contain I.

Similar definitions can also be given with generalized §-models in place of the singular §-models. The set g(K) must then contain $\S(C)$ for each subset C of K and abstractor $\S$ for which $\S \in \S$.

We have the following alternative characterization of the sets G_K (proved in the usual way):

Lemma 1. Let $\boldsymbol{M}$ be a standard §-model and K a downward-closed subset of M. Then G_k is the smallest upward-closed subset of M to contain K.

It follows that a model $\boldsymbol{M}$ is minimal if M is the smallest upward-closed subset of M to contain I. This alternative formulation of minimality can itself be expressed within the language of the abstraction theory. For let us use:

Closed(C) for $\forall$x(Ix $\rightarrow$ Cx) & $\forall$D(D $\subseteq$ C $\rightarrow$ C§(D)); and
Min for $\neg\exists$C($\exists$x$\neg$Cx & Closed(C)).

We then see that:

Lemma 2. A standard model $\boldsymbol{M}$ is minimal iff $\boldsymbol{M} \models$ Min.

Even in the absence of Min, we may define the smallest closed concept and show that it exists. For let $\psi = \psi$(C,D) be the formula $\forall$x(Dx $\leftrightarrow$ $\exists$C′(C′ $\subseteq$ C & x = §C′). Then, in any abstraction theory T^{ϕ}, we can prove the functionality and monotonicity conditions Func (ψ) and Mon (ψ) required for the application of the Fixed Point Theorem (theorem 3.2); and so, by the Fixed Point Theorem itself, we can prove $\exists$Gχ(G), where χ(G) is the formula $\exists$C($\forall$x(Cx $\leftrightarrow$ Ix) & $LFP_{c,\psi}$(G)).

We may also express the generative account of closure within the system by using the following sequence of definitions:

Well-foundedness: WF(R) for $\forall$C($\exists$yCy $\rightarrow$ $\exists$y(Cy & $-\exists$x(Cx & xRy));
Generating relation: Gen(R) for [WF(R) & $\forall$y(y $\in$ Fld(R) & $\exists$C(y = §C) $\rightarrow$ $\exists$C(y = §C & $\forall$x(Cx $\leftrightarrow$ xRy)))];
Generable objects: Gen(x) for $\exists$R(Gen(R) & x $\in$ Fld(R)).

It may then be shown that the generable objects constitute the smallest closed domain (containing all individuals). But to this end, we need a form of the axiom of choice.

Another requirement that is natural from the generative standpoint is that each identity criterion ϕ be absolute in the sense of Section 3. If it were not, then we would not properly know which abstract was being introduced by a definition of the form §C; for which abstract was denoted would depend not only upon the extension of the concept *C* but also upon what the underlying domain would ultimately turn out to be. Suppose, for example, that § were interpreted as the bicardinality operator, i.e. as an operator which took each concept into the pair consisting of the cardinality of its extension and the cardinality of its counter-extension. Then it would not be known which 'bicardinal' resulted from the application of § to an empty concept until it had been determined how many objects there would be in the domain.

The requirement of absoluteness is not as restrictive as it would appear. For suppose we are dealing with an abstract, like bicardinality, whose identity depends upon the underlying domain. We can then trade it in for an abstract on two concepts, where one concept corresponds to the original concept and the other to the domain. The new abstract of C and D will then correspond to the old abstract of C in the domain D. But of course this approach, once it is generalized, requires that we consider abstraction on any finite number of concepts.

In the case of absolute identity criteria, the existence of a minimal submodel is always guaranteed:

Theorem 3. For ϕ an absolute L-criterion, the minimal submodel $\boldsymbol{N}$ of the standard model $\boldsymbol{M}$ will be a model of the theory T^{ϕ} as long as $\boldsymbol{M}$ itself is.

Proof. Let *C* and *D* be any two subsets of *N*. Then:

$$\S_N(C) = \S_N(D) \text{ iff } \S_M(C) = \S_M(D) \text{ (since } \boldsymbol{N} \text{ is a submodel of } \boldsymbol{M})$$

$$\text{iff } \boldsymbol{M} \models \phi[C, D] \text{ (given that } \boldsymbol{M} \text{ is a model for } T^{\phi})$$

$$\text{iff } \boldsymbol{N} \models \phi[C, D] \text{ (given that } \phi \text{ is absolute).}$$

A syntactic form of this result may also be established. For where $\phi = \phi(C,D)$ is an identity criterion, let Absolute (ϕ) be the formula $\forall C\{\forall D,E(D,E \subseteq C \rightarrow (\phi^C(D,E) \leftrightarrow \phi(D,E)))\}$. Thus Absolute ($\phi$) states that ϕ is absolute (though not absolutely absolute, only absolute relative to the underlying domain!). We may now prove within T^{ϕ} the formula: Absolute(ϕ) $\rightarrow \forall C(\text{Closed}(C) \rightarrow \Phi^C)$. Let Min(C) be the formula (Closed(C) & $\forall D(\text{Closed}(D) \rightarrow C \subseteq D)$. Then we may also prove $\exists C\text{Min}(C)$. Hence given a proof of Absolute(ϕ), we may show $\exists C(\text{Min}(C) \;\&\; \Phi^C)$. The formula that defines C may be regarded as an 'inner model' for the theory T^{ϕ}.

The theorem can be extended to generalized theories T^{ϕ} in which each $\phi_{\S}$ is an absolute L-criterion. A version of the result can also be proved for generalized theories T^{ϕ} defined on a well-ordered sequence $\langle \S_\xi : \xi < \alpha \rangle$ of abstraction operators (i.e. theories in which $\phi_{\S}$, for $\S = \S_{\xi}$, only contains the operators $\S_{\nu}$ for $\nu < \xi$). In this case, we require that $\boldsymbol{M}$ and ϕ conform to the condition that, for any $K \subseteq M$ and any $C,D \subseteq K$, $\boldsymbol{M} \models \phi_{\xi}[C,D]$ iff $\boldsymbol{M}/K \models \phi_{\xi}[C,D]$ as long as $\boldsymbol{M}/K$ is a model of each Φ_{ν} for $\nu < \xi$.

The condition that ϕ be absolute is required for the theorem to hold. Suppose, for example, that ϕ is the criterion of bi-equinumerosity and that $\boldsymbol{M}$ is the model for T^{ϕ} in which $M = \{\langle \mathbf{c},\mathbf{d}\rangle : \mathbf{c} + \mathbf{d} = \aleph_0\}$ and $\S(C) = \langle \text{card}(C), \text{card}(M - C)\rangle$ for each $C \subseteq M$. Then the minimal submodel $\boldsymbol{N}$ will be the restriction of $\boldsymbol{M}$ to $\{\langle \mathbf{c},\aleph_0\rangle : \mathbf{c} \leq \aleph_0\}$, which is not a model for T^{ϕ}. Indeed, in this case it can be shown that the theory T^{ϕ} has no standard minimal model.

6. CATEGORICITY

We prove a basic categoricity result and then outline various ways in which it might be extended. The relevance of categoricity to the question of definition has been discussed in Sections I.2 and I.3. We assume throughout this section that the underlying languages contain no non-logical symbols apart from the abstraction operators.

Our basic categoricity result will depend upon two lemmas concerning extensions and chains. We say that two standard §-models $\boldsymbol{M}$ and $\boldsymbol{N}$ are *internally similar* if, for any subsets C and D of M, subsets C' and D' of N, and one–one map f for which $f[C] = C'$ and $f[D] = D'$, $\S_M(C) = \S_M(D)$ iff $\S_N(C') = \S_N(D')$. Thus whether two subsets are identified in either model depends upon their possessing the same internal structural relationships.

Lemma 1 (Extension). Let $\boldsymbol{M}$ and $\boldsymbol{N}$ be two internally similar standard models; and let K and L be two downward-closed subsets of the respective domains M and N. Suppose that f is an isomorphism from $\boldsymbol{M}/K$ onto $\boldsymbol{N}/L$. Then f can then be extended to an isomorphism f^+ from $\boldsymbol{M}/g^+(K)$ onto $\boldsymbol{N}/g^+(L)$.

Proof. It should be noted that the various restrictions $\boldsymbol{M}/K$, $\boldsymbol{N}/g^+(L)$, etc., will in general be partial models. We extend f to a function f^+ on $g^+(K)$ by letting $f^+(\S_M(C))$ be $\S_N(f[C])$ for each subset C of K for which $\S_M(C)$ is not in K. We first need to show that f^+ is well defined, which is to say that $\S_N(f[C])$ and $\S_N(f[D])$ are the same whenever $\S_M(C)$ and $\S_M(D)$ are the same (and not in K). But this follows from the fact that f is one–one and the models are internally similar.

We now show that f^+ is one–one. To this end, we show that $f^+(c)$ and $f^+(d)$ are distinct for distinct $c, d \in g^+(K)$. We distinguish three cases:

(a) $c, d \in K$. Then $f^+(c) = f(c)$ and $f^+(d) = f(d)$, and $f(c)$ and $f(d)$ are distinct by f an isomorphism.

(b) $c, d \in g(K)$. Then c is of the form $\S_M(C)$ and d is of the form $\S_M(D)$ for some subsets C and D of K. By the definition of f^+ and the fact that f is an isomorphism, $f^+(\S_M(C)) = \S_N(f[C])$ and $f^+(\S_M(D)) = \S_N(f[D])$; and given that $c = \S_M(C)$ and $d = \S_M(D)$ are distinct, it follows from internal similarity that $\S_N(f[C])$ and $\S_N(f[D])$ are distinct.

(c) $c \in K - g(K)$ and $d \in g(K)—K$ (and similarly when $d \in K—g(K)$ and $c \in g(K) - K$). Then $f^+(d)$ is of the form $\S_N(D')$ for some subset D' of L, where D' is itself of the form $f(D)$ for some subset D of K. Suppose, for reductio, that $f^+(c) = \S_N(D')$. Then since $f^+(c) = f(c)$, it follows from f an isomorphism that $c = \S_M(D)$; and hence $c \in g(K)$ after all.

We next show that f^+ is onto. Suppose $c' \in g^+(L)$. If $c' \in L$, then c' is in the range of f and hence of f^+. So suppose $c' \in g(L) - L$. Then c' is of the form $\S_N(C')$ for some subset C' of L. But then $f^+(\S_M(f^{-1}[C'])) = c'$.

Finally we show that, for any subset C of $g^+(K)$, $\S_M(C)$ is defined iff $\S_N(f^+(C))$ is defined and that, in case they are both defined, $f^+(\S_M(C)) = \S_N(f^+(C))$. Let us suppose that $\S_M(C)$ is defined and identical to c (the other direction is similar). Then:

(*) $\S_M(C) = \S_M(C')$ for some subset C' of K.

For either $c \in g^+(K) - K$, in which case (*) holds by the definition of g^+, or $c \in K$, in which case (*) holds by the fact that K is downward-closed. From (*), $f^+(\S_M(C)) = f^+(\S_M(C')) = \S_N(f[C']) = \S_N(f^+[C'])$; and, by internal similarity, $\S_N(f^+[C]) = \S_N(f^+[C'])$.

Note, for the subsequent application of this lemma, that $g^+(K)$ will be downward-closed whenever K is.

Lemma 2 (Chains). Let $\langle \boldsymbol{M}/K_\xi : \xi < \lambda\rangle$ and $\langle \boldsymbol{N}/L_\xi : \xi < \lambda\rangle$ be two increasing chains of partial §-models and $\langle f_\xi : \xi < \lambda\rangle$ an increasing chain of functions (i.e. $K_\xi \subseteq K_{\xi'}$, $L_\xi \subseteq L_{\xi'}$, and $f_\xi \subseteq f_{\xi'}$ for $\xi < \xi' < \lambda$). Suppose that f_ξ is an isomorphism from $\boldsymbol{M}/K_\xi$ onto $\boldsymbol{N}/L_\xi$ for each $\xi < \lambda$. Then $f = \cup f_\xi$ is an isomorphism from $\boldsymbol{M}/\cup K_\xi$ onto $\boldsymbol{N}/\cup L_\xi$.

Proof. A straightforward verification.

Also note, for purposes of subsequent application, that the limit $\langle \boldsymbol{M}/K_\xi : \xi < \lambda\rangle$ will be downward-closed whenever each of the models $\boldsymbol{M}/K_\xi$ is downward-closed.

Theorem 3 (Categorical Extension). Suppose that $\boldsymbol{M}$ is a standard K-minimal §-model for K a downward-closed subset of M and that $\boldsymbol{N}$ is an internally similar standard L-mimimal §-model for L a downward-closed subset of N. Suppose that f is an isomorphism from $\boldsymbol{M}/K$ onto $\boldsymbol{N}/L$. Then there exists a unique extension f^+ of f to an isomorphism from $\boldsymbol{M}$ onto $\boldsymbol{N}$.

Proof. Let G and H be the respective generation functions for $\boldsymbol{M}$ and $\boldsymbol{N}$ (relativized to K and L respectively). For each ordinal ξ, we define an isomorphism f_ξ from $\boldsymbol{M}/G_\xi$ onto $\boldsymbol{N}/H_\xi$ as follows:

(i) For $\xi = 0$, we let f_ξ be f. f_ξ is then by supposition an isomorphism from $\boldsymbol{M}/G_0$ onto $\boldsymbol{N}/H_0$.

(ii) For $\xi = \alpha + 1$, we let f_ξ be an extension of f_α which is an isomorphism from $\boldsymbol{M}/G_\xi$ onto $\boldsymbol{N}/H_\xi$. Since (by inductive hypothesis) we may assume that f_α is an isomorphism from $\boldsymbol{M}/G_\alpha$ onto $\boldsymbol{N}/H_\alpha$ and, given that G_α is downward-closed, such an isomorphism will exist by the extension lemma.

(iii) For $\xi = \lambda$, we let f_λ be the union of the f_ξ for $\xi < \lambda$. By the chain lemma, f_λ will again be an isomorphism from $\boldsymbol{M}/G_\lambda$ onto $\boldsymbol{N}/H_\lambda$.

Since both $\boldsymbol{M}$ and $\boldsymbol{N}$ are minimal, we may choose an ordinal α for which $G_\alpha = M$ and $H_\alpha = N$. But then $f^+ = f_\alpha$ is an isomorphism from $\boldsymbol{M}$ onto $\boldsymbol{N}$.

To show that the extension f^+ is unique, suppose that there is another extension f^* of f which is an isomorphism from $\boldsymbol{M}$ onto $\boldsymbol{N}$. Choose the least α for which there exists an element c in G_α upon which f^+ and f^* disagree. Clearly $\alpha > 0$. But then c is of the form $\S_M(C)$ for C a subset of G_β for some $\beta < \alpha$. Now $f^+(\S_M(C)) = \S_N(f^+[C])$ and $f^*(\S_M(C)) = \S_N(f^*[C])$. Since $\beta < \alpha$, $f^+[C] = f^*[D]$; and so $f^+(c) = f^*(c)$ after all.

As a special case of the theorem, we obtain:

Corollary 4. Suppose that $\boldsymbol{M}$ is a standard minimal §-model with I-invariant $\equiv_\S$. There are then no proper automorphisms on $\boldsymbol{M}$ that are fixed on I_M.

Proof. Since $\equiv_\S$ is I-invariant, $\boldsymbol{M}$ is internally similar with itself. Suppose that f is the identity on I_M. Then, by the theorem, there exists a unique extension f^+ of f that is an automorphism on $\boldsymbol{M}$.

Thus, once the identity of the individuals is taken to be given, each abstract will have its own structural 'position'.

As a further special case of the theorem, we have:

Corollary 5. Suppose that $\boldsymbol{M}$ and $\boldsymbol{N}$ are internally similar standard minimal §-models. Then they are isomorphic as long as $\text{card}(I_M) = \text{card}(I_N)$.

Proof. Since $\text{card}(I_M) = \text{card}(I_N)$, there is a one–one map f from I_M onto I_N. Clearly, I_M and I_N are both downward-closed and f is an isomorphism from $\boldsymbol{M}/I_M$ onto $\boldsymbol{N}/I_N$. But then $\boldsymbol{M}$ and $\boldsymbol{N}$ are isomorphic by the theorem.

We do not have this result without the requirement that both $\boldsymbol{M}$ and $\boldsymbol{N}$ be minimal, even when we fix the cardinality of the subdomains of individuals and of abstracts. For let $\boldsymbol{M}$ be a standard model with domain $\{a^0,$

$a^1, \ldots\}$, $\S(\{a^n\}) = a^{n+1}$ for $n \geq 0$, and $\S(C) = a^0$ for C non-singleton; and let $\boldsymbol{N}$ be a standard model with domain $\{a^0, a^1, \ldots\} \cup \{\ldots, b^{-1}, b^0, b^1, \ldots\}$, $\S(\{a^n\}) = a^{n+1}$ for $n \geq 0$, $\S(C) = a^0$ for C non-singleton, and $\S(b^n) = b^{n+1}$ for any integer n. Then $\boldsymbol{M}$ and $\boldsymbol{N}$ are internally similar; and both have zero individuals and denumerably many abstracts. But they are not isomorphic. Indeed, $\boldsymbol{M}$ is minimal; whereas $\boldsymbol{N}$ is not, since it contains a 'backwards' regress with respect to §.

There is, however, a circumstance in which cardinality alone is sufficient to secure isomorphism. Let us say that a §-model $\boldsymbol{M}$ is *numerical* if $\S(C) = \S(D)$ whenever C and D are two subsets of M of the same cardinality.

Theorem 6. Suppose that $\boldsymbol{M}$ and $\boldsymbol{N}$ are internally similar standard numerical §-models of the same cardinality. Then they are isomorphic as long as $\text{card}(I_M) = \text{card}(I_N)$.

Proof. In this case we may define the isomorphism f directly by letting f/I_M be a one–one map onto I_N and by letting $f(\S_M(C)) = \S_N(C')$ for C' a subset of N of the same cardinality as C.

The proof of theorem 6 actually only requires that the models be binumerical in the sense of identifying concepts with the same bicardinality and that they be exactly similar in the sense that, for any one–one map f from M onto N, $\S_M(C) = \S_M(D)$ iff $\S_N(f[C]) = \S_M(f[D])$.

Corollary 5 has an immediate consequence for theories:

Corollary 7 (Categoricity). Suppose that ϕ is an absolute L-criterion. Then any two standard models for T^ϕ + Min with the same number of individuals are isomorphic.

Proof. Given that ϕ is absolute, it follows that any two standard models $\boldsymbol{M}$ and $\boldsymbol{N}$ of T^ϕ will be internally similar. It therefore follows from the corollary that any two such models will be isomorphic as long as the cardinalities of I_M and I_N are the same.

Thus the combination of absoluteness and minimality is able to secure categoricity relative to the cardinality of the individuals. The cardinality of the minimal model with **d** individuals will be the least fixed point **c** of the function $p_{\mathbf{d}}$ defined in Section 4. For Φ will have a model with exactly **c** objects and **d** individuals, and so will have a generated submodel, which has at most **c** objects and exactly **d** individuals, and hence which has exactly **c** objects given that **c** is the least fixed point.

In the case of numerical identity criteria (those which determine numerical §-models), any formula that fixes the cardinality of the abstracts will do in place of Min. Suppose, for example, that we define *Natural Number* in the abstraction theory with Hume's Law. Then any axiom to the effect that there is a one–one correspondence between the universe and the natural numbers will be sufficient to secure (relative) categoricity, as will any axiom to the effect that there is a one–one correspondence between the universe and all sets of natural numbers. Boolos (1987, 16) and Dummett (1991*a*, 227) have both commented on the failure of Hume's Law to settle questions of cardinality. But this, we see, is the only respect in which it fails to achieve completeness or even, categoricity.

The generation of the minimal model, in these cases proceeds in an especially simple way. Given $\nu < \xi < \alpha$, where α is the critical ordinal, $\text{card}(G_\nu) < \text{card}(G_\xi)$. For clearly the result will hold if it holds for $\xi = \nu + 1$. So suppose, for reductio, that $\text{card}(G_\nu) = \text{card}(G_{\nu+1})$. Then $\nu + 1$ must be the critical ordinal. For take an abstract $\S(C)$ in $G_{\nu+2}$, with $C \subseteq G_{\nu+1}$. Choose a subset C' of G_ν of the same bicardinality as C. Then $\S(C') = \S(C)$; and hence no new abstracts are introduced at stage $\nu + 1$.

Let us now consider some extensions of the basic result to generalized theories of abstraction. Different results are obtained according as to whether or not we insist on the identity criteria being logical or on the models being strictly separated. In each case, the abstractors are taken to be denoted by an abstraction operator; and it is only in Section III.2 that we consider the problem of categoricity for a theory in which we are allowed to quantify over the methods of abstraction themselves.

The first extension is to generalized models in which the abstractors are taken to be logical but separated. Let $\boldsymbol{M}$ and $\boldsymbol{N}$ be two generalized §-models. We say that they are *internally similar* if the §-models $\boldsymbol{M}_\S$ and $\boldsymbol{N}_\S$ are internally similar for each $\S \in \boldsymbol{\S}$.

Theorem 8. Suppose that $\boldsymbol{M}$ is a strictly separated and standard K-minimal $\boldsymbol{\S}$-model, with K a downward-closed subset of M, and that $\boldsymbol{N}$ is a strictly separated and standard L-minimal $\boldsymbol{\S}$-model, with L a downward-closed subset of N. Suppose that $\boldsymbol{M}$ and $\boldsymbol{N}$ are internally similar and that f is an isomorphism from $\boldsymbol{M}/K$ onto $\boldsymbol{N}/L$. Then there exists a unique extension f^+ of f to an isomorphism from $\boldsymbol{M}$ onto $\boldsymbol{N}$. In particular, two internally similar models $\boldsymbol{M}$ and $\boldsymbol{N}$ will be isomorphic as long as $\text{card}(I_{\boldsymbol{M}}) = \text{card}(I_{\boldsymbol{N}})$.

Proof. Form an ordered set $\langle \S_\zeta : \zeta < \alpha_0 \rangle$ of the operators $\S \in \boldsymbol{\S}$. For each ordinal ξ, define an isomorphism f_ξ from $\boldsymbol{M}_\xi = \boldsymbol{M}/K_\xi$ onto $\boldsymbol{N}_\xi = \boldsymbol{N}/L_\xi$ as follows:

(i) $\xi = 0$. Let $K_0 = K$, $L_0 = L$, and $f_\xi = f$. Then by supposition, f is an isomorphism from $\boldsymbol{M}_\xi$ onto $\boldsymbol{N}_\xi$.

(ii) ξ of the form $\alpha_0\beta + \zeta'$ for $\zeta < \alpha_0$ (where ζ' is the successor of ζ). Let ν be the immediate predecessor $\alpha_0\beta + \zeta$ of ξ; and temporarily set $\S = \S_\zeta$. Recall that $\boldsymbol{M}_\S$ and $\boldsymbol{N}_\S$ are the respective reducts of $\boldsymbol{M}$ and $\boldsymbol{N}$ wrt §. By theorem 3 and IH, there is an extension $(f_\nu)^+$ of f_ν which is an isomorphism from a closed submodel $M_§/K^*$ of $M_§$ onto a closed submodel $N_§/L^*$ of $N_§$. We then set $K_\xi = K^*$, $L_\xi = L^*$, and $f_\xi = (f_\nu)^+$.

(iii) ξ of the form $\alpha_0\beta + \lambda$ for λ a limit ordinal $<\alpha_0$. We let K_ξ, L_ξ and f_λ be the respective unions of the K_ν, L_ν and f_ν for all $\nu < \xi$.

Using this construction, the results may then be proved in much the same way as before. Note that the models M and N are required to be strictly separated in order to guarantee that no new §′-abstracts, for §′ ≠ §, will be introduced at step (ii).

We turn next to the case in which the operators are allowed to be defined in terms of one another. Let $\langle §_\zeta: \zeta < \alpha\rangle$ be a well-ordered sequence of the operators § ∈ **§** (which we regard as respecting their order of definition). Given such a sequence, an ordinal $\zeta < \alpha$, and a **§**-model M, we let $M_{<\zeta}$ be the reduct of M to the operators $§_\zeta$, with $\zeta' < \zeta$, and we say that the subset K of M is *closed*$_{<\zeta}$ if it is closed wrt each operator $§_{\zeta'}$ for $\zeta' < \zeta$. We say that the **§**-models M and N are *internally similar* (*relative to* $\langle §_\zeta: \zeta < \alpha\rangle$) if, for each $\zeta < \alpha$, for any isomorphism f from $M_{<\zeta}/K$ onto $N_{<\zeta}/L$, with K and L both closed$_{<\zeta}$, and for any subsets $C, D \subseteq K$ and $C', D' \subseteq L$ for which $f[C] = C'$ and $f[D] = D'$, $§_{\zeta,M}(C) = §_{\zeta,M}(D)$ iff $§_{\zeta,N}(C') = §_{\zeta,N}(D')$. Thus the map f is now required to be a full isomorphism on the preceding abstractors.

Theorem 9. Suppose that M is a strictly separated and standard K-minimal §-model, with K a downward-closed subset of M, and that N is a strictly separated and standard L-mimimal §-model, with L a downward-closed subset of N. Suppose that M and N are internally similar relative to the well-ordered sequence $\langle §_\zeta: \zeta < \alpha\rangle$ of the operators § ∈ **§** and that f is an isomorphism from M/K onto N/L. There then exists a unique extension f^+ of f to an isomorphism from M onto N. In particular, the two models M and N will be isomorphic as long as $\text{card}(I_M) = \text{card}(I_N)$.

Proof. By transfinite induction on α. For $\alpha = 0$, the result is trivial. So suppose α is a successor ordinal $\beta + 1$. We then prove the extension lemma in the form:

(*) Let M and N be two strictly separated standard models that are internally similar relative to the sequence $\langle §_\zeta: \zeta < \alpha\rangle$; and let K and L be subsets of the respective domains M and N, closed wrt each $§_\xi$, for $\xi < \beta$, and downward-closed wrt $§_\beta$. Suppose that f is an isomorphism from M/K onto N/L. Then f can be extended to an isomorphism f^+ from $M/g^+(K)$ onto $N/g^+(L)$ (with g^+ defined in terms of $§_\beta$).

The proof follows that of lemma 1. Note that in the former proof an appeal to internal similarity is made at two places: the first is at (b) under the verification of one-oneness; and the second is at the end under the verification of the condition of isomorphism. The first gives rise to no difficulty; but the second does, since the application of the internal similarity condition in its relativized form now requires that we have a suitable underlying isomorphism. To overcome this difficulty, we appeal to the inductive hypothesis. For the f^+, as defined under (*), can be extended to an isomorphism from $\boldsymbol{M}_{<\beta}/K^+$ onto $\boldsymbol{N}_{<\beta}/L^+$, where K^+ and L^+ respectively contain $g^+(K)$ and $g^+(L)$ and are closed$_{<\beta}$.

Having established the extension lemma, we show, along the lines of the proof of theorem 3, that there is an extension f^+ of f (in the statement of the present theorem) which is an isomorphism from $\boldsymbol{M}_\beta/K^+$ onto $\boldsymbol{N}_\beta/L^+$, where K^+ and L^+ respectively contain K and L and, by strict separation, contain no new $\S_\zeta$-abstracts, for $\zeta < \beta$. The inductive hypothesis may now be applied, with K^+ and L^+ in place of K and L. This may, in its turn, be followed by an extension to take care of the operator $\S_\beta$—and so on, until a fixed point is obtained.

Finally, suppose that α is a limit ordinal λ. For each $\xi < \lambda$, let f_ξ be the unique extension of f which is an isomorphism from $\boldsymbol{M}_{<\xi}/K_\xi$ onto $\boldsymbol{N}_{<\xi}/L_\xi$ with K_ξ and L_ξ closed$_{<\xi}$ supersets of K and L. Then it is clear that the f_ξ's, for $\xi < \lambda$, form an increasing chain; and so $f^+ = \cup f_\xi$ will be an isomorphism from $\boldsymbol{M} = \boldsymbol{M}/\cup K_\xi$ onto $\boldsymbol{N} = \boldsymbol{N}/\cup L_\xi$.

We can apply this result to particular generalized theories by extending the notion of a restricted axiom. Instead of requiring that the 'meaning' of the identity criterion ϕ(C,D) remain the same under the relativization to any concept that subsumes C and D, we now require that the meaning remain the same under the relativization to any concept that subsumes C and D and is closed under the preceding abstraction operations.

We turn finally to the question of categoricity for non-strictly separated §-models. In this case, the analogue of theorem 8 does not hold. For consider a § with two members $\S_1$ and $\S_2$. Let $\boldsymbol{M}$ be a standard §-model with $M = \{c\}$ and $\S_{1,\boldsymbol{M}}(C) = \S_{2,\boldsymbol{M}}(C) = c$ for all subsets C of M; and let $\boldsymbol{N}$ be a standard model with $N = \{c, d, e\}$, $\S_{1,\boldsymbol{N}}(C) = c$ for all subsets C of N, $\S_{2,\boldsymbol{N}}(C) = d$ for empty and singleton subsets C of N, and $\S_{2,\boldsymbol{N}}(C) = e$ for doubleton subsets C of N. Then $\boldsymbol{M}$ and $\boldsymbol{N}$ are separated minimal models without individuals, satisfy the internal similarity condition, and yet are not isomorphic.

The discrepancy in the generation of $\boldsymbol{M}$ and $\boldsymbol{N}$ emerges at the first stage. For whereas $\S_{1,\boldsymbol{M}}$ and $\S_{2,\boldsymbol{M}}$ yield the same abstract c in application to the empty concept, $\S_{1,\boldsymbol{N}}$ and $\S_{2,\boldsymbol{N}}$ yield the different abstracts c and d. There is no violation of separation in the case of $\boldsymbol{N}$ since the abstracts c and d can be distinguished in terms of concepts defined with the help of those very abstracts.

It is reasonable to think that the model $\boldsymbol{N}$ should not be allowed under a proper conception of the generative method; for at the first stage, we have no basis in terms of the objects already generated, for distinguishing between the abstracts *c* and *d*. Accordingly, let us redefine the notions of generation and minimality.

Let *K* be a subset of the domain *M* of a standard **§**-model ***M***. For §, §′ ∈ **§**, we say that the abstract §′(*C*) is *indistinguishable from* the abstract §(*C*) *in K* if $C \subseteq K$ and $\{D \subseteq K$: §′(D) = §′$(C)\}$ = $\{D \subseteq K$: §(D) = §$(C)\}$. (Note that the indistinguishability of the two abstracts depends upon their respective representation as §-and §′-abstracts). We now say that the subset *K* of *M* is *quasi-upward-closed* if it contains *I* and if for any subset *C* of *K* and abstractor §, with § ∈ **§**, there is an abstractor §′, with §′ ∈ **§**, such that §(*C*) is indistinguishable from §′(*C*) in *K* and §′(C) ∈ *K*. Thus we no longer require that § (*C*) itself be in *K* but only that some indistinguishable abstract be in *K*. We do not insist on the stronger condition of upward-closure since there is no basis from within *K* for distinguishing between indistinguishable abstracts.

The generation of a minimal upward-closed subset of a domain *M* is no longer straightforward since, at a given stage in the generation of the subset, we may have a choice as to which of two currently indistinguishable abstracts should be introduced. Accordingly, the previous generation function *g* should be replaced by a system of alternative generation functions. Accordingly, let us say that *g* is *a generation function for* the standard **§**-model ***M*** if, for every subset *K* of *M*, $K^+ = K \cup g(K)$ is a set with the properties:

(i) for each $C \subseteq K$ and § ∈ **§**, there is a §′ ∈ **§** such that §(C) is indistinguishable from §′(*C*) in *K* and §′(C) ∈ $g(K)$;

(ii) any object $c \in g(K)$ is of the form §′(*C*) for $C \subseteq K$ and is indistinguishable in *K* from any other abstract §(*C*) in $K \cup g(K)$.

Thus the $g(K)$ may be obtained by first forming the indistinguishability classes of the §(*C*), for § ∈ **§** and $C \subseteq K$, and then picking a representative from those classes which do not already have a representative in *K*.

The generation of the upward-closed subset may now be relativized to a generation function *g* in the usual manner. Thus for any ordinal ξ:

$$G_0 = I;$$
$$G_{\alpha+1} = I \cup g(G_\alpha);$$
$$G_\lambda = \bigcup\{G_\xi : \xi < \lambda\}.$$

By considerations of cardinality, there will be a least ordinal α for which $G_{\alpha+1} = G_\alpha$. We again call α the *critical* ordinal *for **M*** and again use *G* for G_α. It is evident that *G* is quasi-upward-closed and that it is separated in the sense that any two abstracts §(*C*) and §′(*C*) in *G* are distinguishable in *G*.

Let us say that the **§**-model ***M*** is *strictly minimal* if it is minimal and if

M properly contains no quasi-upward-closed subset. As is evident from the example ***N*** above, there is no guarantee that even an internally invariant §-model will contain a strictly minimal submodel. However, should such a submodel exist, it may be generated by means of any one of the generation functions *g*. In particular, a strictly minimal model may itself be generated by means of such a function and hence the objects of such a model may be generated in such a manner that the abstracts introduced at any given stage are always distinguishable.

Using the condition of strict minimality in place of minimality, we may extend our previous categoricity results to separated §-models. The proofs are much the same as before, but instead of considering an arbitrary generation of the objects of the model we consider a generation that is in accord with one of the generation functions *g*.

It is natural to think of each generated abstract as being represented by a term. Given such a representation, one may then provide a truth-definition for the sentences of an abstraction theory in purely nominalistic terms, without reference to an underlying domain of objects. Such a truth-definition is very much in accord with the kind of truth-conditions that the proponent of the context principle might be thought to have in mind. However, he will be unable to offer the truth-definition that is stated here since it requires mathematical resources that go beyond those at his disposal.

We suppose given a set of *individual constants* $\{a_0, a_1, \ldots\}$ of arbitrary cardinality, which we regard as denoting the individuals—one constant for each individual. We define the *object* and the *relation* terms by the following rules:

(i) each individual constant is an object-term;

(ii) each set of n-tuples of object-terms is a relation term (in particular, a set of object-terms is said to be a *concept* term);

(iii) If C is a concept-term, then §C is an object-term. Note that the object- and the concept-terms will each form a proper class.

We think of the concept-term $\{t_1, t_2, \ldots\}$ as denoting the concept whose extension consists of the denotations of $t_1, t_2, \ldots$ (and similarly for the relation terms). We have represented a concept-term by a set, but we could equally well have represented it by something more linguistic in character, such as the 'expression' $\lambda x\ (x = t_1 \vee x = t_2 \vee \ldots)$. The infinitary character of the concept-terms reflects the fact that we think of the concepts as being given platonistically.

We think of the object-term §C as denoting the abstract on the concept denoted by C. There will, of course, be no assurance that different object-terms §C denote different abstracts and hence no assurance that the different concept-terms denote different concepts. To each object term t, we associate a *rank* rk(t) according to the rules:

(i) rk(a) = 0 for a an individual constant;
(ii) rk(§C) = sup{rk(s) + 1: s ∈ C}.

Similarly, the rank of a set T of object-terms is taken to be sup{rk(t) + 1: t ∈ T}.

The new object- and relation-terms may be used to expand the language $L^{\S}$ of abstraction theory. To this end, we define the notions of *objectual term*, *relational term*, and *formula*:

(i) any objectual variable or object-term is an objectual term;
(ii) if C is a conceptual term, then §C is an objectual term;
(iii) any relational variable or relation-term is a relational term;
(iv) s = t is a formula when s and t are objectual terms;
(v) $Pt_1 \ldots t_n$ is a formula if P is an n-ary relational term and $t_1, \ldots, t_n$ are objectual terms;
(vi) $\neg\phi$ and $(\phi \vee \psi)$ are formulas if ϕ and ψ are formulas;
(vii) $\forall x\phi$ is a formula when x is an objectual variable and ϕ is a formula; and
(viii) $\forall P\phi$ is a formula when P is a relational variable and ϕ is a formula.

A set of object-terms T is said to be a (*term*) *domain* if s ∈ T whenever §C ∈ T and s ∈ C. Thus a domain must contain the object-terms which figure in the concept-terms from which the object-terms in the domain are formed. Given any set C of object-terms, there will be a smallest domain T_C to contain C. With each closed sentence ϕ of the expanded language, we may associate the set $L(\phi)$ of object-terms that occur in ϕ and the corresponding domain $T_\phi = T_{L(\phi)}$.

We define truth for the (closed) sentences of the expanded language. In fact, for the purposes of the induction, we need to define truth relative to a term domain T. It is assumed that this notion of relative truth only has application to a sentence ϕ and a domain T when $T_\phi \subseteq T$.

We suppose that we have been supplied with a specific identity criterion $\phi(C,D)$ for the operator §. The various definitions and results are then relative to that specific choice of $\phi(C,D)$. The relative truth-definition is as follows:

(i) a = b is true in T iff a = b, where a and b are individual constants;
(ii) a = §C for a an individual constant is never true in any domain T;
(iii) §C = §D is true in T iff $\phi(C,D)$ is true in $T_{C \cup D}$;
(iv) $Pt_1 \ldots t_n$ is true in T iff for some $s_1, \ldots, s_n$, $t_1 = s_1, \ldots, t_n = s_n$ are true in T and $\langle s_1, \ldots, s_n \rangle \in P$, for P an n-ary relation term;
(v) $\neg\phi$ is true in T iff ϕ is not true in T;
(vi) $(\phi \vee \psi)$ is true in T iff ϕ is true in T or ψ is true in T;
(vii) $\forall x\phi(x)$ is true in T iff $\phi(t)$ is true for every object-term t in T;
(viii) $\forall P^n\phi(P^n)$ is true in T iff $\phi(Q)$ is true for every relation-term $Q \subseteq T^n$.

The clauses are analogous to the clauses of a standard substitutional semantics, with the sole exception of clause (iii). Here, instead of adopting:

(iii)′ §C = §D is true in T iff $\phi(C,D)$ is true in T,

we restrict the domain on the right-hand side. In effect, we are assuming that §C = §D is true in the domain T iff it is true in the subdomain $D_{C\cup D}$. We shall say that an identity s = t is *true* (simpliciter) if it is true in the domain $T_{\{s,t\}}$.

Without this modification to (iii)′, the truth-definition would not be well founded. But with the modification, it is. For let the *complexity of* a pair $\langle T, \phi\rangle$, for ϕ a formula and T a domain containing T_ϕ, be the pair $\langle \alpha, n\rangle$, where α is the rank of T and n is the number of occurrences of logical symbols in ϕ. Let us adopt a lexicographic ordering on measures of complexity; and let us suppose that clause (iv) is rewritten so that applications of clauses (i)–(iii) are already built into the evaluation of the right-hand side. Then it may be shown that the application of each clause to a formula results in a drop in complexity or a direct assignment of truth-value.

Let $Eq_=$ be the formula $\forall x(x = x)$ & $\forall xy(x = y \rightarrow y = x)$ & $\forall xyz(x = y$ & $y = z \rightarrow z = z)$ (contrast with the formula Eq_ϕ, which states that the criterion $\phi = \phi(C,D)$ is an equivalence). Given the clauses for = in the truth-definition, there is no guarantee that $Eq_=$ will be true in any given term domain T. But let us assume that it is, and define $\approx_T$ by: $s \approx_T t$ iff s and t are terms in T and s = t is true. Then $\approx_T$ will be an equivalence relation. So with each term $t \in T$, we may associate an equivalence class $|t| = \{u: u \approx_T t\}$. Let $M_T = \{|t|: t \in T\}$; for each $Q \subseteq T^n$, let $|Q| = \{\langle |t_1|, \ldots, |t_n|\rangle: \langle t_1, \ldots, t_n\rangle \in Q\}$; and, for $T = T_C$, let $\boldsymbol{M}_C$ designate the standard second-order model with domain M_T. We use the subscripted notation $|t|_C$ and $|Q|_C$, where necessary, to indicate that the equivalence classes are with respect to the relation $\approx_T$.

Lemma 10. Suppose that $Eq_=$ is true in a term domain T. Let $\psi = \psi(x_1, \ldots, x_m, P_1, \ldots, P_n)$ be a formula of L^2 with free variables as displayed; and let $t_1, \ldots, t_m$ be object-terms of T and $Q_1, \ldots, Q_n$ be relation-terms of the same arity, respectively, as $P_1, \ldots, P_n$. Then $\psi\{t_1, \ldots, t_m, Q_1, \ldots, Q_n)$ is true in T iff $\psi[|t_1|, \ldots, |t_m|, |Q_1|, \ldots, |Q_n|]$ is true in $\boldsymbol{M}_T$.

Proof. By a straightforward induction on the truth-definition treating the truth-values of the identity sentences as given.

It should be noted, from the proof of this result, that nothing turns upon the rules of evaluation for the identity sentences.

Lemma 11. Suppose that $Eq_=$ is true in the domain T. Let S be a subdomain of T; let $\boldsymbol{N}$ be the restriction of $\boldsymbol{M}$ to $\{|s|_T: s \in S\}$; and, for each $s \in S$, let $f(|s|_T) = |s|_T \cap S$. Then f is an isomorphism from $\boldsymbol{N}$ onto $\boldsymbol{M}_S$.

Proof. A straightforward verification.

When can we expect Eq_T to be true in a term domain? Call an L-criterion $\phi = \phi(C,D)$ *regular* if it is restricted and if Eq_ϕ is valid (i.e. true in any second-order model).

Lemma 12. Suppose that the L-criterion ϕ is regular. Then $Eq_=$ is true in any term domain T.

Proof. We consider the T-truth of $\forall xyz\ (x = y \ \&\ y = z \rightarrow x = z)$ (the other cases being similar). The proof is by induction on the rank α of T. If $\alpha = 0$, then T is empty and the result is immediate. So suppose $\alpha > 0$. Let s, t, and u be terms from T. Then we need to show that $s = t \ \&\ t = u \rightarrow s = u$ is true in T. So suppose that $s = t$ and $t = u$ are true (in T). If one of s or t or u is an individual constant, then all of them are by clause (ii) in the truth-definition. But then $s = t$ and $t = u$ by clause (i); so $s = u$; and so $s = u$ is true in T by clause (ii) again.

We may therefore suppose that all of s, t, and u are abstract terms. So put $s = §C$, $t = §D$, and $u = §E$. Since $s = t$ is true (in T), $\phi(C,D)$ is true in $T_{C \cup D}$. But $rk(T_{C \cup D \cup E}) < rk(T)$; and so by IH, $Eq_=$ is true in $T_{C \cup D}$. By lemma 10, $\phi[|C|, |D|]$ is true in $\boldsymbol{M}_{C \cup D}$; so, by lemma 11, $\phi[|C|_U, |D|_U]$, with $U = C \cup D \cup E$, is true in the restriction of $\boldsymbol{M}_U$ to $\{|t|_U: t \in C \cup D\}$; and so, by lemma 3.11 and the fact that ϕ is restricted, $\phi[|C|_U, |D|_U]$ is true in $\boldsymbol{M}_U$. Similarly, $\phi[|D|_U, |E|_U]$ is true in $\boldsymbol{M}_U$. But then from the validity of Eq_ϕ, $\phi[|C|_U, |E|_U]$ is true in $\boldsymbol{M}_U$. By the use of lemma 11 and lemma 3.11 again, $\phi[|C|_{C \cup E}, |E|_{C \cup E}]$ is true in $\boldsymbol{M}_{C \cup E}$; by lemma 10, $\phi(C,E)$ is true in the term domain $T_{C \cup E}$; and hence, by the truth-definition $§C = §E$ is true in T.

How might a §-model be formed from a term domain? We cannot in general expect that *§* will be defined in $\boldsymbol{M}_T$ for every concept |C| with $C \subseteq T$. The object-term §T itself, for example, will not belong to T. Accordingly, let us say that the domain T is *representative* if for each object-term s there is an object-term t in T such that $s = t$ is true. Thus a representative domain represents the totality of objects that can be denoted by terms.

Given a representative domain T and a regular L-criterion ϕ, Let $\boldsymbol{M}_T$ be a standard §-model, with domain M_T as before and with *§* defined, for any $C \subseteq T$, by:

$$§(|C|) = |t|, \text{ for } t \in T \text{ and } §C = t \text{ true.}$$

Each subset $C \subseteq M_T$ will be of the form |C| for $C \subseteq T$; and hence *§* will, putatively, be defined on every concept of $\boldsymbol{M}_T$. If the domain is representative, there will be a t conforming to the definiens for each $C \subseteq T$; and hence *§* will have a value for each argument |C|. Given that ϕ is regular, $Eq_=$ will be true in T by lemma 12, and so the value |t| will not depend upon the choice of t. Nor will the value depend upon the choice of C. For suppose $|C| = |D|$.

Then $\phi[|C|, |D|]$ is true in the second-order model $\boldsymbol{M}_T$ and hence, by ϕ restricted, is also true in the restriction $\boldsymbol{M}_{CUD}$. Given that $Eq_=$ is true in T, it follows from lemmas 10 and 11 that $\phi(C,D)$ is true in T_{CUD} and hence that §C = §D is true.

We can now extend lemma 10 to the language $L^§$:

Lemma 13. Suppose that ϕ is a regular L-criterion and that T is representative term domain. Let $\psi = \psi(x_1, \ldots, x_m, P_1, \ldots, P_n)$ be a formula of $L^§$ with free variables as displayed; and let $t_1, \ldots, t_m$ be object-terms of T and $Q_1, \ldots, Q_n$ be relation-terms of the same arity, respectively, as $P_1, \ldots, P_n$. Then $\psi(t_1, \ldots, t_m, Q_1, \ldots, Q_n)$ is true in T iff $\psi[|t_1|, \ldots, |t_m|, |Q_1|, \ldots, |Q_n|]$ is true in $\boldsymbol{M}_T$.

Proof. Given the proof of lemma 10, it suffices to show that the formula $\psi(C,D) = (§C = §D)$ is true (in T) iff $\psi\,[§|C|, §|D|]$ is true in $\boldsymbol{M}_T$, i.e. iff $\mathcal{S}(|C|) = \mathcal{S}(|D|)$. But the latter holds iff for some t and u in T, §C = t, §D = u are true and |t| = |u|, i.e. iff for some t in T, §C = t = §D. But given that T is representative, this holds, iff §C = §D is true.

We finally obtain our central result, that the truth-definition, relativized to a representative domain of terms, will yield the same truth-values as the corresponding minimal model.

Theorem 14. Suppose that ϕ is a regular L-criterion and that T is representative term domain. Then:

(i) the corresponding §-model $\boldsymbol{M}_T$ is a minimal model of Φ; and
(ii) a sentence ϕ of $L^§$ is true in T iff it is true in $\boldsymbol{M}_T$.

Proof. (i) may be straightforwardly established by using an induction on the rank of $t \in T$ in order to show that |t| must be a member of the generated subdomain *G* of $\boldsymbol{M}_T$. (ii) follows from lemma 13.

The existence of a representative domain with **d** individual constants will imply the existence of a standard model for Φ (indeed, of a minimal model) with **d** individuals. The converse also holds. For if there exists a standard model with **d** individuals, it will have a minimal submodel with **d** individuals. Suppose now that α is the critical ordinal for the generating function G_ξ. Then it is readily shown that the set of terms of rank $\leq \alpha$ (or of rank $< \alpha$ for α a limit ordinal) will be representative domain.

We might compare the above proof of soundness to the attempted demonstration of the soundness of Law V in section 31 of Frege's *Grundgesetze*. The proof there is infected by two sources of impredicativity: one to the predicates and the other to the abstraction operator (see Dummett (1991*a*), ch. 17). The first is removed here by adopting a Platonic conception of relations and concepts; the second is removed by assuming the existence of a representative domain of object-terms. Whether such a domain exists depends, of course, on the identity criterion in question. In the case of coextensionality, there will be no such domain; while in the case of equinumerosity, there will.

Although my interest has principally been in a Platonic conception of concepts, it is worth pointing out it is also possible to obtain models of the abstraction theories in which all objects and all concepts are denoted by closed terms from the language itself (assuming that we allow the use of λ to form complex predicates). Thus from the results of Boolos and Putnam (1968), we can obtain a term model for Hume's Law in which the concepts are taken, in effect, to be the predicatively defined sets of numbers. Also, as Tony Martin has pointed out, by assuming the axiom of projective determinacy, we can obtain term models which are elementary submodels of the standard model (conceived now in a first-order way): and it seems likely that these methods could be extended to various other forms of abstraction. We would thereby obtain something closer to the kind of model that Frege seemed to have in mind, though without the straightforward inductive determination of the truth-conditions.

7. INVARIANCE

We shall attempt to characterize the non-inflationary invariant equivalence relations on a given power-set. The analysis will then be extended to other forms of invariance.

Given an equivalence relation $\approx$ on the power set $\wp(M)$ of a set M, let $P_{\approx}$ be the partition of $\wp$ (M) induced by $\approx$ and call the equivalence relation $\approx$ *non-inflationary* if card($P_{\approx}$) $\leq$ card(M). Each L-criterion ϕ induces an invariant relation $\approx$ on the power-set $\wp(M)$ of a set M of urelements; and ϕ will be tenable on the corresponding standard model $\boldsymbol{M}$ just in case $\approx$ is an equivalence relation and the induced partition $P_{\approx}$ is non-inflationary. Thus determining the non-inflationary equivalences on $\wp(M)$ will help us determine which L-criteria are tenable.

In the ensuing discussion, we take M to be a set of urelements; and, unless otherwise indicated, we shall assume that M is infinite. Recall that a *bicardinal* is an ordered pair $\langle \mathbf{c}, \mathbf{d} \rangle$ of cardinals; and given a subset C of M, we set bicard(C) = $\langle$card(C), card($M - C$)$\rangle$. Suppose that $\mathbf{m}$ is the cardinality of M. If card(C) $< \mathbf{m}$, then bicard(C) = $\langle$card(C), $\mathbf{m}\rangle$. Thus bicardinality, in this case, is a function of cardinality: card(C) = card(D) implies bicard(C) = bicard(C). On the other hand, if card(C) = $\mathbf{m}$, then bicard(C) can take any of the values $\langle \mathbf{m},\mathbf{d} \rangle$ for $\mathbf{d} \leq \mathbf{m}$.

Let C and D be any two subsets of M. Then the *cardinality distribution* cdstr (C,D) *of* C *and* D is the ordered quadruple

⟨card($C-D$), card($C\cap D$), card($D-C$), card($M-(C\cup D)$)⟩. Thus the cardinality distribution of two sets identifies the cardinalities of the smallest demarcated areas in their Venn diagram.

We first consider the question: When does identification of one pair of subsets of M under an invariant equivalence relation imply identification of another pair? Say that the pair $\langle C,D\rangle$ of subsets of M *yields* the pair $\langle E,F\rangle$ if, for any invariant equivalence $\approx$ over $\wp(M)$, $C\approx D$ implies $E\approx F$.

Cardinality distribution provides a sufficient condition for one pair of sets to yield another:

Lemma 1. $\langle C,D\rangle$ yields $\langle E,F\rangle$ if cdstr (C,D) = cdstr(E,F).

Proof. Suppose that cdstr(C,D) = cdstr(E,F), i.e. card$(C\cap D)$ = card$(E\cap F)$, card$(C-D)$ = card$(E-F)$, card$(D-C)$ = card$(F-E)$, and card$(M-(C\cup D))$ = card$(M-(E\cup F))$. Let f_1 be a one–one map from $C\cap D$ onto $E\cap F$, f_2 a one–one map from $C-D$ onto $E-F$, f_3 a one–one map from $D-C$ onto $F-E$, and f_4 a one–one map from $M-(C\cup D)$ onto $M-(E\cup F)$. Then $f=f_1\cup f_2\cup f_3\cup f_4$ is a permutation on M with $f[C]=E$ and $f[D]=F$. Suppose now that $C\approx D$. Given that $\approx$ is invariant, $f[C]\approx f[D]$, i.e. $E\approx F$.

Use of this lemma will often be implicit. Once it has been shown that $C\approx D$ for sets C and D of given cardinality distribution, it will be taken for granted that $E\approx F$ holds for any sets E and F of the same cardinality distribution.

By a *combination* is meant a pair $\langle C,D\rangle$ of subsets of M of the same bicardinality. A combination $\langle C,D\rangle$ of subsets of M (of bicardinality $\langle \mathbf{c},\mathbf{d}\rangle$) is said to be *representative* if it yields all combinations of subsets of M of the same bicardinality, i.e. if, for any invariant equivalence relation $\approx$ on $\wp(M)$, $C\approx D$ implies $E\approx F$ whenever bicard(E) = bicard(F) = $\langle \mathbf{c},\mathbf{d}\rangle$.

We wish to determine which combinations of sets are representative. The relation of *being almost the same* will be especially significant in this regard. Given two sets C and D, define their *symmetric difference* $C\sim D$ to be $(C-D)\cup(D-C)$. We say that two subsets C and D of M are *almost the same*—in symbols, $C\equiv' D$—if either (i) they are finite and $C=D$ or (ii) they are infinite, card(C) = card(D) and card$(C\sim D)$ < card(C); and we say that the subsets C and D are *very different* if they are not almost the same.

Lemma 2. $\equiv'$ is an equivalence relation on the subsets of M.

Proof. The relation $\equiv'$ is clearly reflexive and symmetric. To establish transitivity, suppose that $C\equiv' D$ and $D\equiv' E$. We wish to show $C\equiv' E$. If one of C, D, or E is finite, then they are all finite and the result is obvious. So suppose that they are all infinite. Then card(C) = card(D) = card(E). So it remains to show that card$(C-E)<\mathbf{c}$ and that card$(E-C)<\mathbf{c}$, where

$\mathbf{c} = \text{card}(C)$. By symmetry, it suffices to establish the first of the two inequalities. Now $C - E \subseteq ((C - D) \cup (D - E))$. So $\text{card}(C - E) \leq \text{card}((C - D) \cup (D - E)) < \mathbf{c} + \mathbf{c} = \mathbf{c}$ (given that $\mathbf{c}$ is infinite).

A subset C of M is said to be *small* (*relative to* M) if $\text{card}(C) < \text{card}(M)$ and otherwise is said to be *large* (*relative to* M). (Later, another notion of smallness and largeness will be introduced.) When we are dealing with small subsets C and D of a set M of infinite cardinality $\mathbf{c}$, the cardinality of $M - (C \cup D)$ will always be $\mathbf{c}$ and hence may be ignored in assessing cardinality distribution.

We divide the results on repesentative combination into three parts:

Lemma 3 (part 1). A combination $\langle C,D \rangle$ of small infinite sets is representative when they are very different.

Proof. Suppose $\text{card}(C) = \text{card}(D) = \mathbf{c} < \mathbf{m}$. We distinguish two subcases:

(1) C and D are disjoint. Take any two sets E and F of cardinality $\mathbf{c}$ (so that $\text{card}(M - E) = \text{card}(M - F) = \mathbf{m}$). Given that $C \approx D$ for invariant $\approx$, we wish to show that $E \approx F$. Clearly, we can pick a subset B of M of cardinality $\mathbf{c}$ disjoint from both E and F. (The reader might like at this and similar points in the proofs to draw a Venn diagram.) Since $\text{cdstr}(C,D) = \text{cdstr}(E,B)$, it follows from lemma 1 that $E \approx B$. Similarly, it follows from the fact that $\text{cdstr}(C,D) = \text{cdstr}(F,B)$ that $F \approx B$. But then $E \approx F$.

(2) C and D overlap. Since C and D are very different, either $C - D$ or $D - C$ is of cardinality $\mathbf{c}$. Without loss of generality, suppose that $C - D$ is. Let us also suppose that $\text{card}(C \cap D) = \mathbf{d}$ and that $\text{card}(D - C) = \mathbf{e}$. Pick a subset A of $C - D$ for which $\text{card}(A) = \mathbf{d}$ and $\text{card}(C - (D \cup \text{A}) = \mathbf{c}$; and pick a subset B of $M - (C \cup D)$ of cardinality $\mathbf{e}$ (note that at least one of $\mathbf{d}$ or $\mathbf{e}$ is $\mathbf{c}$). Let $D^+ = A \cup B$. Then $\text{card}(D^+) = \mathbf{d} + \mathbf{e} = \mathbf{c}$, $\text{card}(C - D^+) = \mathbf{c}$, and $\text{card}(D^+ - C) = \mathbf{e}$. Thus $\text{cdstr}(C,D^+) = \text{cdstr}(C,D)$. So $C \approx D^+$ and, given $C \approx D$, $D^+ \approx D$. Now D^+ and D are disjoint. By (1) above, the combination $\langle D^+,D \rangle$ is representative; and therefore so is the combination $\langle C,D \rangle$.

Say that a subset C of the infinite set M is *almost universal* (*relative to* M) if $\text{card}(M - C) < \text{card}(C) = \text{card}(M)$. Thus an almost universal set is one which is almost the same as the universal set M.

Lemma 3 (part 2). A combination $\langle C,D \rangle$ of almost universal subsets of M is representative when their complements $M - C$ and $M - D$ are very different.

Proof. We reduce this case to the other. Given any equivalence relation $\approx$ on $\wp(M)$, we define the dual relation $\approx'$ by:

$$C \approx' D \text{ iff } M - C \approx M - D.$$

It is readily established that $\approx'$ is an equivalence relation and that it is invariant if $\approx$ is.

Now suppose that $C \approx D$ for $\langle C,D\rangle$ a combination of almost universal sets. Then $M - C \approx' M - D$. Given that C and D are almost universal, $M - C$ and $M - D$ are small; and given that C and D are of the same bicardinality, so are $M - C$ and $M - D$. Since $M - C$ and $M - D$ are very different, it follows from the first part of the lemma that $C' \approx' D'$ for any sets C' and D' of the same bicardinality as $M - C$. But then $M - C' \approx M - D'$ for any such sets C' and D'; and so $E \approx F$ for any sets E and F of the same bicardinality as C.

Say that the set C *bifurcates* D if C is a subset of D and card(C) = card $(D - C)$ and that the subset C of M is a *bifurcator* if it bifurcates M. A bifurcator divides the universe M into parts of equal size; and it will, for an infinite universe, be of the same size as the universe. From among the large subsets of M, the bifurcators are those which are not almost universal.

Lemma 3 (part 3). A combination $\langle C,D\rangle$ of bifurcators C and D is representative when C is very different both from D and from $M - D$.

Proof. We distinguish three subcases:

(1) C and D are disjoint.

Proof. Clearly, $M - (C \cup D)$ must be large, for otherwise C would be almost the same as $M - D$. We first show that $E \approx F$ must hold for any bifurcators E and F whenever $M - (E \cup F)$ is large. For choose a bifurcator B of $M - (E \cup F)$. Then cdstr(E,B) = cdstr(C,D) = cdstr(F,B). So $E \approx B$ and $F \approx B$; and consequently, $E \approx F$. Now consider the case in which $M - (E \cup F)$ is small. Then both $E - F$ and $F - E$ must be large. Choose a bifurcator B of $E - F$. Then all of $B \approx E$, $F \approx F - E$ and $B \approx F - E$ hold by what was shown earlier; and so $E \approx F$.

(2) $C \cap D$ is small.

Proof. In this case, $C - D$ and $D - C$ must both be large. But also $M - (C \cup D)$ must be large, for otherwise C would be almost the same as $M - D$. Suppose card$(C \cap D) = \mathbf{c}$. Choose a subset A of $C - D$ of cardinality $\mathbf{c}$ and a bifurcator B of $M - (C \cup D)$. Let $D^+ = A \cup B$. Then it is readily verified that cdstr(C,D) = cdstr(C,D^+). So $C \approx D^+$; and hence $D^+ \approx D$. But D and D^+ are disjoint. By case 1, the combination $\langle D,D^+\rangle$ is representative; and therefore so is the combination $\langle C, D\rangle$.

(3) $C \cap D$ is large.

Proof. Split $C \cap D$ into two disjoint parts B_1 and B_2, each of cardinality $\mathbf{m}$. Let $C' = C - B_2$ and $D' = D - B_1$. Then cdstr(C',D) = cdstr(C,D) and cdstr(C,D') = cdstr(C,D). Given $C \approx D$, $C' \approx D$ and $C \approx D'$; and hence $C' \approx D'$. So we again have a reduction to the case of disjoint sets.

It should be noted that, for bifurcators C and D, $C \equiv' D$ iff $M - C \equiv' M - D$; for the symmetric differences of the sets C and D and of their complements are the same and the cardinalities of the sets and their complements are also the same. Summing up:

Theorem 4 (Representative Combinations). A combination $\langle C,D \rangle$ of infinite subsets of M is representative if (i) C and D are small but very different or (ii) C and D are almost universal but with very different complements or (iii) C and D are bifurcatory with one very different from the other and from its complement.

Another way to present our results (and others like them) is in terms of what one might call a calculus of cardinality distributions. Suppose that Γ is a set of possible cardinality distributions and that τ is a particular such distribution (for a given set M). We might say that Γ *entails* τ if any invariant equivalence on $\wp(M)$ that identifies sets C and D for which cdstr(C, D) $\in \Gamma$ will identify sets C and D for which cdstr(C,D) = τ; and we might then attempt to determine a sound and complete 'calculus' for the relation of entailment as so defined. However, this is not the approach we shall pursue.

We now attempt to take account of the non-inflationary character of an equivalence $\approx$. To this end, we need further information on the relation $\equiv'$ of approximate identity. For each subset C of M, let $|C|_{\equiv'} = \{D \subseteq M: C \equiv' D\}$, let $Q = \{|C|_{\equiv'}: C \subseteq M\}$, and let $Q_{\mathbf{c}} = \{|C|_{\equiv'}: C \subseteq M$ and C is of cardinality $\mathbf{c}\}$. Recall that $[M]^{\mathbf{c}}$ is the set of all subsets of M of cardinality $\mathbf{c}$. We may then show that the identification of approximately identical sets does not reduce their cardinality.

Lemma 5. card($Q_{\mathbf{c}}$) = $\mathbf{m}^{\mathbf{c}}$ for $\mathbf{m}$ transfinite and $\mathbf{c} \leq \mathbf{m}$.

Proof. For $\mathbf{m}$ transfinite and $\mathbf{c} \leq \mathbf{m}$, $\mathbf{m}^{\mathbf{c}}$ is the cardinality of $[M]^{\mathbf{c}}$; and since card($Q_{\mathbf{c}}$) $\leq$ card($[M]^{\mathbf{c}}$), card($Q_{\mathbf{c}}$) $\leq \mathbf{m}^{\mathbf{c}}$.

The other direction is proved by 'magnifying' the differences between distinct sets. To this end, take $\mathbf{c}$ disjoint copies M_ζ of M. Clearly, $M^+ = \cup \{M_\zeta: \zeta < \mathbf{c}\}$ is of cardinality $\mathbf{m}$. For each $\zeta < \mathbf{c}$, let f_ζ be a one–one map from M onto M_ζ. For each subset C of M of cardinality $\mathbf{c}$, let $f(C) = \cup \{f_\zeta[C]: \zeta < \mathbf{c}\}$. Thus f is a map from $[M]^{\mathbf{c}}$ into $[M^+]^{\mathbf{c}}$. Moreover, if $C \neq D$, then not $f(C) \equiv' f(D)$. For if $C - D$, let us say, contains an element a, then $f(C) - f(D)$ contains $\mathbf{c}$ copies of a and hence is of cardinality $\mathbf{c}$. So if $C \neq D$, then $|f(C)|_{\equiv'} \neq |f(D)|_{\equiv'}$ (wrt M^+); and hence card($Q_{\mathbf{c}}$) $\geq$ card($[M]^{\mathbf{c}}$) $= \mathbf{m}^{\mathbf{c}}$.

Let us call an equivalence $\approx$ ℘(M) *strictly acceptable* if it is invariant and non-inflationary. Our characterization of the strictly acceptable equivalences will be in terms of a minimal such relation. A cardinal $\mathbf{c} \leq \mathbf{m}$ is said to be *exponentially small* (*relative to* $\mathbf{m}$) if $\mathbf{m}^{\mathbf{c}} \leq \mathbf{m}$ and *exponentially large* (*relative to* $\mathbf{m}$) if $\mathbf{m}^{\mathbf{c}} > \mathbf{m}$. The subset C of M is said to be *exp. small* if card$\{D \subseteq M$: card$(D) \leq$ card$(C)\} \leq$ card(M); and otherwise it is said to be *exp. large*. If M is finite, then C is an exp. small subset of M iff C is empty and M is non-empty. If M is infinite, then C is an exp. small subset of M iff card(C) is exp. small relative to card(M). We note that, by Cantor's Theorem, $\mathbf{m} > 1$ is itself exponentially large (relative to $\mathbf{m}$) and that if the generalized continuum hypothesis holds then every transfinite cardinal less than $\mathbf{m}$ is exponentially small (relative to $\mathbf{m}$). It is important to observe that a cardinal $\mathbf{c}$ may be exp. small relative to a cardinal $\mathbf{m}$ and yet exp. large relative to a larger cardinal n. For example, given GCH, $\aleph_1^{\aleph_0} = \aleph_1$, whereas $\aleph_\omega^{\aleph_0} = \aleph_{\omega+1}$.

We define the *basal* relation $\approx_0$ on ℘(M) by:

$C \approx_0 D$ iff either (i) card(C) and card(D) are exp. small and $C = D$ or (ii) card$(M - C)$ and card$(M - D)$ are exp. small and $C = D$ or (iii) card(C), card(D), card$(M - C)$ and card$(M - D)$ are exp. large and bicard(C) = bicard(D).

Thus $\approx_0$ is a kind of hybrid relation—behaving like identity on sets that are sufficiently small or that have sufficiently small complements and behaving like bi-equinumerosity on the remaining sets. By the basal relation *for* the cardinal $\mathbf{m}$, we mean the basal relation on ℘(M) for M a set of urelements of cardinality $\mathbf{m}$. The basal relation for a cardinal is clearly unique up to isomorphism.

Say that the relation R is *as refined as* the relation S, or that S *subsumes* T, if S holds between any two objects between which R holds. We then have:

Theorem 6 (Characterization). Suppose that M is infinite. Then the basal relation $\approx_0$ is the most refined strictly acceptable equivalence relation on ℘(M).

Proof. Let us first show that $\approx_0$ is a strictly acceptable equivalence. It is readily verified to be an equivalence and, from its characterization, is clearly invariant.

To show that $\approx_0$ is non-inflationary, let us count the equivalence classes $|C|$ in the partition P_0 induced by $\approx_0$. Consider first the C for which card(C) $= \mathbf{c}$ is exp. small. There are $\mathbf{m}^{\mathbf{c}}$ subsets C of M of cardinality $\mathbf{c}$ and hence $\mathbf{m}^{\mathbf{c}}$ corresponding equivalence classes $|C|$ in P_0. But given that $\mathbf{c}$ is exp. small, $\mathbf{m}^{\mathbf{c}} \leq \mathbf{m}$. Consider now the C for which card$(M - C)$ is exp. small. Then by similar reasoning, there are again $\mathbf{m}^{\mathbf{c}}$ corresponding equivalence classes $|C|$ in P_0. Finally, consider the C for which both C and $M - C$ are exp. large. Let

n be the cardinality of the set of cardinals ≤ **m**. Then **n** ≤ **m** and there are at most 2.**n** = **n** equivalence classes $|C|$.

We see that in each case there are at most **m** equivalences classes. But there are at most **n** cases; and hence card(P_0) ≤ **m**.

Let us now show that $\approx_0$ is the most refined of the strictly acceptable equivalences. Consider an arbitrary non-inflationary and invariant equivalence ≈. Suppose that $C \approx_0 D$. We then wish to show $C \approx D$.

Let **c** = card(C) = card(D), and **d** = card($M - C$) = card($M - D$). If either **c** or **d** is exp. small, it follows from the definition of $\approx_0$ that $C = D$; and so it is evident that $C \approx D$. So we may suppose that **c** and **d** are both exp. large, i.e. $\mathbf{m}^{\mathbf{c}} > \mathbf{m}$ and $\mathbf{m}^{\mathbf{d}} > \mathbf{m}$. It then follows from the definition of $\approx_0$ that bicard(C) = bicard(D).

Suppose, for reductio, that not $C \approx D$. By theorem 4 on representative combinations (TRC for short), ≈ cannot hold between any representative pair of subsets of M of the same bicardinality as C and D. There are three cases, each of which will require ≈ to be inflationary.

Case 1 **c** < **m**. Then card($M - C$) = card($M - D$) = **m**; and so, by TRC(part 1), $E \approx F$ implies $E \equiv' F$ for any subsets of E and F of M of cardinality **c**. But then:

$$\begin{aligned} \text{card}(P_{\approx}) &\geq \text{card}(Q_{\mathbf{c}}) \\ &= \mathbf{m}^{\mathbf{c}} \text{ (by lemma 5)} \\ &> \mathbf{m} \text{ given that } \mathbf{c} \text{ is exp. large.} \end{aligned}$$

Case 2 **c** = **m** and **d** < **m**. In a similar way to case 1, it follows by TRC (part 2) that $E \approx F$ implies $(M - E) \equiv' (M - F)$ for any subsets $M - E$ and $M - F$ of cardinality **d**. But then:

$$\begin{aligned} \text{card}(P_{\approx}) &\geq \text{card}(\{|(M-E)|_{\equiv'} : \text{card}(M-E) = \mathbf{d}\} \\ &= \text{card}(Q_{\mathbf{d}}) \\ &= \mathbf{m}^{\mathbf{d}} \text{ (by lemma 5)} \\ &> \mathbf{m}, \text{ given that } \mathbf{d} \text{ is exp. large.} \end{aligned}$$

Case 3 **c** = **d** = **m**. By TRC, $E \approx F$ implies that either $E \equiv' F$ or $E \equiv' (M - F)$. Define $\equiv''$ on bifurcators by: $E \equiv'' F$ iff $E \equiv F$ or $E \equiv (M - F)$. It is readily shown that $\equiv''$ is an equivalence relation. Define a map f from $P_{\equiv'}$ (as restricted to bifurcators) into $P_{\equiv''}$ by: f $(|C|_{\equiv'}) = |C|_{\equiv''}$. Then it should be clear that f maps exactly two equivalence classes from $P_{\equiv'}$ into each of the equivalence classes from $P_{\equiv''}$; and so card($P_{\equiv'}$) = card(Q_{m}) = $\mathbf{m}^{\mathbf{m}}$ > **m**.

We now consider the case in which the equivalence relation is required to be internally invariant. We say that an equivalence on ℘(M) is *strictly I-*

acceptable if it is I-invariant and non-inflationary. Define the *super-basal* relation $\approx_1$ on $\wp(M)$ by:

$C \approx_1 D$ iff either (i) card(C) and card(D) are exp. small and $C = D$ or (ii) card(C) and card(D) are exp. large and card(C) = card(D).

We then have:

Corollary 7. Suppose that M is infinite. Then the relation $\approx_1$ is the most refined strictly I-acceptable equivalence on $\wp(M)$.

Proof. Given the theorem, it suffices to show that $\approx_1$ is the smallest I-invariant equivalence to contain $\approx_0$. It is readily verified that $\approx_1$ is an I-invariant equivalence. Now consider any I-invariant equivalence $\approx$ containing $\approx_0$; and suppose that $C \approx_1 D$. There are two cases: (i) card(C) and card(D) are exp. small and $C = D$. But then it is evident that $C \approx D$. (ii) card(C) and card(D) are exp. large and card(C) = card(D). Suppose that card($C \cap D$) = **e**. Clearly, we can find subsets C' and D' of M such that card(C') = card(D') = card(C), card($C' \cap D'$) = **e**, and card($M - C'$) = card($M - D'$) = **m**. There is therefore a one–one map f from $C \cup D$ onto $C' \cup D'$ with $f[C] = C'$ and $f[D] = D'$. So by the internal invariance of $\approx_1$, $C' \approx_1 D'$. But card(C') (= card(D')) and card($M - C'$) (= card($M - D'$)) are both exp. large; and so $C' \approx_0 D'$. Given that $\approx$ contains $\approx_0$, $C' \approx D'$; and so by the internal invariance of $\approx$, $C \approx D$.

According to the theorem, the most refined form of invariant abstraction, compatible with full second-order logic, is one which yields the extensions of concepts whose extensions or counter-extensions are exp. small and yields the bicardinalities of the concepts otherwise. According to the corollary, if it is further required that the identity criterion be absolute, then the most refined form of abstraction is one that yields extensions of concepts whose extensions are exp. small and yields the cardinalities of the concepts otherwise. Thus bi-equinumerosity and equinumerosity represent the 'outer limits' of invariant abstraction; any acceptable method of invariant abstraction must eventually degenerate into one of the equinumerosity relations or something coarser.

There are various strengthenings of our basic results, some of which will later prove useful. We concentrate on the case in which invariance is assumed to be internal, although related results can be established for the cases in which it is not. Suppose that M is a set of urelements of (finite or infinite) cardinality **m**. For $\mathbf{c} \leq \mathbf{m}$ and $\mathbf{d} \leq \mathbf{c}^+$ (the successor of **c**), define the relation $\equiv_\mathbf{d}$ on the members of $[M]^\mathbf{c}$ by: $C \equiv_\mathbf{d} D$ if card($C \sim D$) < **d**. For **d** = **c**, $\equiv_\mathbf{d}$ is coincident with the earlier relation $\equiv'$ of being almost the same on infinite sets of cardinality **c**; and for $\mathbf{d} = (2.\mathbf{c})^+$, $\equiv_\mathbf{d}$ is coincident with the relation of equinumerosity on sets of cardinality **c**. In the same way as for lemma 2, it may be shown that the relation $\equiv_\mathbf{d}$ is an equivalence for transfinite **d**.

It will be helpful to classify the consequences of $C \approx D$ holding in relationship to the cardinality of $C \sim D$. When the cardinality is infinite, we have:

Lemma 8. Let $\approx$ be an I-invariant equivalence over the set M; and suppose that $C \approx D$ holds with card($C \sim D$) = **d** infinite and card(C) = card(D) = **c**. Then $E \equiv_{\mathbf{d}+} F$ implies $E \approx F$ for any sets E and F of cardinality *c*.

Proof. We may assume that card($M - (C \cup D)$) = card(M). For C and D may be mapped one–one onto subsets C' and D' for which card($M - (C' \cup D')$) = card(M); since Icdstr(C,D) = Icdstr(C',D'), $C' \approx D'$; and so it suffices to establish the result for C' and D'.

Let card($C - D$) = $\mathbf{d}_1$ and card($D - C$) = $\mathbf{d}_2$. The result will then follow from the following two facts:

(1) If $\mathbf{d} = \mathbf{d}_1 > \mathbf{d}_2$, then $E \approx F$ whenever Icdstr(E,F) = $\langle \mathbf{d}_1, \mathbf{c}, \mathbf{d}_1 \rangle$ or Icdstr (E, F) = $\langle \mathbf{d}_2, \mathbf{c}, \mathbf{d}_2 \rangle$.

Pf. Pick a subset B of $M - (C \cup D)$ of cardinality **d**. Let $C' = C \cup B$. Then by considerations of cardinality distribution (CCD, for short), $C' \approx D$. Hence $C \approx C'$. But Icdstr(C,C') = $\langle \mathbf{d}, \mathbf{c}, \mathbf{d} \rangle$. The proof for the other case is similar.

(2) If $\mathbf{d}_1 = \mathbf{d}_2 = \mathbf{d}$, then $E \approx F$ whenever Icdstr(E,F) = $\langle \mathbf{d}, \mathbf{c}, \mathbf{e} \rangle$ for any $\mathbf{e} < \mathbf{d}$.

Pf. Pick a subset B of $M - (C \cup D)$ of cardinality **e**. Let $C' = C \cup C$. By CCD, $C' \approx D$. Hence $C \approx C'$. But Icdstr(C, C') = $\langle \mathbf{d}, \mathbf{c}, \mathbf{e} \rangle$.

When the cardinality of $C \sim D$ is finite, we have:

Lemma 9. Let $\approx$ be an I-invariant equivalence on $\wp(M)$ (for M either finite or infinite). Suppose that $C \approx D$ holds for card($C \sim D$) finite but non-zero and card(C) = **c**. Then $E \equiv_{\omega} F$ implies $E \approx F$ for any subsets E and F of M of cardinality **c**.

Proof. Suppose that $\approx$ is an invariant equivalence on $\wp(M)$; and, for the purposes of the proof, let us define Ey/x by $(E - \{x\}) \cup \{y\}$. If C is empty, the result is trivial; and so we may assume that C is non-empty. In case M is finite, we may assume that $M - C$ is non-empty, since otherwise the result is again trivial; and in case M is infinite, we may assume that card($M - (C \cup D)$) = card(M) for the same reasons as before. In each of these cases, it then suffices to show that $C' \approx C$ for any subset C' of the same cardinality as C.

We distinguish three cases:

(i) D is a proper subset of C. Suppose that $x \in C - D$. Let y be a member of $M - C$; and set $D' = Cy/x$. By considerations of cardinality distribution (CCD, for short), $D' \approx D$. But then $C \approx D'$; and so we have a reduction to case (iii) below.

(ii) C is a proper subset of D. Suppose that $x \in C$ and that $y \in D - C$. Let $D' = Cy/x$. By CCD, $D' \approx D$. But then $C \approx D'$; and we have a reduction to case (iii) below.

(iii) C and D properly overlap. Suppose $z \in D - C$. Let n = card ($C \sim C'$), where C' is a subset of M of the same cardinality as C. We prove $C \approx C'$ by induction on n (since M may be finite, we need to take some special care in the reasoning). For n = 0, the result is trivial. So suppose the result holds for n ≤ k; and let us establish it for n = k + 1. Pick an element x in $C - C'$ and an element y in $C' - C$. Let $C^* = Cy/x$. It then suffices to show $C^* \approx C$: for card($C^* \sim C'$) = (k – 1) and C^* and C properly overlap; so given $C^* \approx C$, it follows by IH that $C^* \approx C'$; and hence $C \approx C'$.

To show $C \approx C^*$, we distinguish three subcases:

(a) $y \in M - D$. Let $D' = Dy/z$. Then by CCD, $C \approx D'$ and we have a reduction to case (b) or (c) below.

(b) $y \in D - C$ and $x \in M - D$. Suppose $u \in C \cap D$. Let $D' = Dx/u$. By CCD, $C \approx D'$ and we have a reduction to case (c).

(c) $y \in D - C$ and $x \in C \cap D$. But then $C^* = Cy/x \approx D$ by CCD; and so $C \approx C^*$.

As a special case of the lemma, note the following:

Corollary 10. Let $\approx$ be an I-invariant equivalence on $\wp(M)$. Suppose that $C \approx D$ holds for distinct finite sets C and D. Then $E \approx F$ holds for any subsets E and F of the same cardinality as C.

We consider finally the identification of sets of different cardinality. Say that an equivalence $\approx$ over M is *numeric on* the cardinal **c** if $C \approx D$ for any subsets C and D of cardinality **c**; and say that it is *numeric* simpliciter if it is numeric on every cardinal ≤ card(M).

Lemma 11. Suppose that $\approx$ is an I-invariant equivalence over an infinite set M. If $C \approx D$ for sets C and D of distinct cardinality, then $\approx$ is numeric on the cardinality of C.

Proof. For C and D finite, the result follows from corollary 10. So suppose that card(C) < card(D) with D infinite. We may then find C' and D' such that Icdstr(C',D') = Icdstr(C,D) and yet $M - (C' \cup D')$ is of the same cardinality as M. We may therefore find a C^* such that Icdstr(C^*,D') = Icdstr(C',D') and C' is disjoint from C^*. So by theorem 4, $\langle C,C^* \rangle$ is a representative combination, and the required result follows.

Now suppose that card(D) < card(C) with C infinite. By the same reasoning as before, we may suppose that $M - (C \cup D)$ is of the same cardinality as M. Choose a subset C' of $M - (C \cup D)$ which is of the same cardinality as C and for which $M - (C \cup D \cup C')$ is of the same cardinality as M. Then we may find a D' such that $\mathrm{Icdstr}(C,D) = \mathrm{Icdstr}(C',D')$. By the previous case, $D \approx D'$; by the invariance of $\approx$, $C' \approx D'$; and so $C \approx C'$. So again by theorem 4, $\langle C,C' \rangle$ is a representative combination; and the required result follows.

Summing up lemmas 9 and 11, we obtain:

Theorem 12. Let $\approx$ be an I-invariant equivalence over the set M; and suppose that $C \approx D$ holds with $\mathrm{card}(C \sim D)$ non-zero and $\mathrm{card}(C) = \mathbf{c}$. Then $E \equiv_{\mathrm{d}} F$ implies $E \approx F$ for any sets E and F of cardinality $\mathbf{c}$, where $\mathbf{d} = \max\{\aleph_0, \mathrm{card}(C \sim D)^+\}$.

The theorem provides a good picture of the structure of I-invariant equivalences. For subsets of a given finite cardinality, either no two sets of that cardinality are identified or all are; for sets of a given infinite cardinality $\mathbf{c}$, the sets are identified according to the relationship $\equiv_{\mathrm{d}}$ for some cardinal $\mathbf{d}$ ($\mathbf{d}$ will be the successor of $\sup\{\mathbf{e}: C \approx D, \mathrm{card}(C) = \mathrm{card}(D) = \mathbf{c}$, and $\mathrm{card}(C \sim D) = \mathbf{e}\}$); and if two sets of distinct cardinality are identified, then all sets of those cardinalities are identified.

The lemmas have some further consequences of interest. From theorem 4 and lemmas 10 and 11, we readily obtain:

Corollary 13. Suppose that $\approx$ is an I-invariant equivalence over infinite M. If $C \approx D$ for very different sets C and D, then $\approx$ is numeric on $\mathrm{card}(C)$.

If we insist that the equivalence should be non-inflationary, then we also obtain:

Theorem 14. Suppose that $\approx$ is a non-inflationary I-invariant equivalence on $\wp(M)$, with M finite. Then $\approx$ is numeric.

Proof. Suppose that $C \approx D$ fails to hold for C and D of the same cardinality k. Then C and D must be distinct and non-empty, and so $0 < \mathrm{k} < \mathrm{m} = \mathrm{card}(M)$. But then by corollary 10, no two subsets of M of cardinality k can be related by $\approx$. Now the number of such subsets is at least m; and so, given that $\approx$ is non-inflationary, $M \approx C$ for some such set C. But then, by CCD, $M \approx C$ for every such set C; and the different Cs are related by $\approx$ after all.

Thus we see that the possibilities for non-inflationary identification are much more limited in the finite than in the infinite case. Any strictly I-acceptable equivalence must contain the relation of equinumerosity. But this relation is itself inflationary, of course, on finite domains, since for a domain M of n elements the relation yields n + 1 equivalence classes. Indeed, it is easy to see that there can be no smallest non-inflationary equivalence on $\wp(M)$ should $\mathrm{m} = \mathrm{card}(M) > 1$, since we are free to distinguish between the k-membered and the (k + 1)-membered subsets for *any* k = 0, 1, . . . , (m – 1), but we are not free to distinguish between them for *all* k.

With each numeric equivalence ≈ on ℘(*M*), we may associate an equivalence $\approx_{num}$ on all cardinals according to the rule:

c $\approx_{num}$ **d** iff either (i) for some subsets *C* and *D* of *M*, *C* ≈ *D*, card(*C*) = **c** and card(*D*) = **d** or (ii) for no *C* does **c** = card(*C*) and for no *D* does **d** = card(*D*).

Thus *C* ≈ *D* iff *C* and *D* are subsets of *M* and card(*C*) $\approx_{num}$ card(*D*). The equivalence classes induced by $\approx_{num}$ may be regarded as a form of generalized number.

There is a rather different way in which we may extend our basic results; for we need not require that the underlying relation be an equivalence. Call the relation ↔ on an arbitrary set *X* a *bi-equivalence* if *x* ↔ *y* & *x'* ↔ *y* & *x'* ↔ *y'* implies *x* ↔ *y'*. Given a bi-equivalence ↔ on a set *X*, we say *x*, *y* ↔ (*x* and *y* are *connected on the left*) if *x* ↔ *z* and *y* ↔ *z* for some *z*; and, similarly, we say ↔ *x*, *y* (*x* and *y* are *connected on the right*) if *z* ↔ *x* and *z* ↔ *y* for some *z*. We also denote these two relations by $\leftrightarrow_L$ and $\leftrightarrow_R$. It is readily shown that the relation $\leftrightarrow_L$ is an equivalence relation on the domain of ↔ and that $\leftrightarrow_R$ is an equivalence relation on the range of ↔. Let us use $|x|_L$ and $|y|_R$ for the corresponding equivalence classes, and P_L and P_R for the corresponding partitions. For $X \in P_L$ and $Y \in P_R$, say that *X* ↔ *Y* if *x* ↔ *y* for some *x* ∈ *X* and some *y* ∈ *Y*. Then it is readily shown that ↔ is a one–one relation between P_L and P_R and that *x* ↔ *y* iff $|x|_L \leftrightarrow |y|_R$ for any *x* in the domain D_L of ↔ and any *y* in the range D_R of ↔. Conversely, given a one–one relation ↔ between partitions P_L and P_R of $\wp(D_L)$ and $\wp(D_R)$ respectively, we may determine the relation ↔ between D_L and D_R to which it corresponds by means of the definition: *x* ↔ *y* if $|x|_L \leftrightarrow |y|_R$.

Our interest is in the case in which ↔ is a relation on the power set ℘(*M*). For this case, we have the following analogue of theorem 12:

Theorem 15. Let ↔ be an I-invariant bi-equivalence on ℘(*M*); and suppose that *C* ↔ *D* holds with card(*C* ~ *D*) = non-zero and card(*C*) = **c**. Then $E \equiv_{\mathbf{d}+} F$ implies *E* ≈ *F* for any sets *E* and *F* of cardinality **c**, where $\mathbf{d} = \max\{\aleph_0, \text{card}(C \sim D)^+\}$.

Proof. By a straightforward modification of the proof of theorem 12. Details are omitted.

It is also possible to extend the above results to relations that are *relatively* invariant, i.e. to relations that ignore the identity of all but a fixed set of individuals. Say that a function *f* from *X* into *Y* is *fixed on* a subset *X'* of *X* if for all *x* ∈ *X'*, *f*(*x*) = *x*. For *R* a relation on ℘(*M*) and *K* a subset of *M*, we then say that *R* is *K-invariant* if *R*(*C*,*D*) implies *R*(*f*(*C*), *f*(*D*)) for any permutation *f* of *M* that is fixed on *K* and we say that *R* is *internally K-*

invariant if $R(C, D)$ implies $R(f(C), f(D))$ for any one–one map from f from $C \cup D$ into M that is fixed on $K \cap (C \cup D)$.

For M an infinite set and K a subset of M, define the equivalence $\approx_{0,K}$ on $\wp(M)$ by:

$$C \approx_{0,K} D \text{ iff } C \cap K = D \cap K \text{ and } C - K \approx_0 D - K.$$

And similarly for $\approx_{1,K}$. Let us say that an equivalence $\approx$ on $\wp(M)$ is (*internally*) *K-acceptable* if it (internally) K-invariant and non-inflationary. We may then show:

Corollary 16. For M an infinite set and K an exp. small subset of M, $\approx_{0,K}$ (resp. $\approx_{1,K}$) is the most refined (internally) K-acceptable equivalence on $\wp(M)$.

Proof. We prove the result for the relation $\approx_{0,K}$, the proof for the relation $\approx_{1,K}$ being exactly similar.

Given that $\approx_0$ is an equivalence relation, it is readily shown that $\approx_{0,K}$ is an equivalence relation; and it is evident that $\approx_{0,K}$ is internally K-invariant. Let us use the unardorned $|C|$ for the equivalence classes wrt $\approx_{0,K}$. For L a subset of M, let $M_L = \{C \subseteq M\colon C \cap K = L\}$. Then, for each $C, D \in M_L$, $C \approx_{0,K} D$ iff $C - K \approx_0 D - K$. So, by theorem 6, the number of equivalence classes $|C|$ for $C \in M_L$ is at most $\mathbf{m} = \text{card}(M)$. But there are at most $2^{\mathbf{k}}$ subsets of K, where $\mathbf{k} = \text{card}(K)$. So the total number of equivalence classes $|C|$ is at most $\mathbf{m}.2^{\mathbf{k}} \leq \mathbf{m}.\mathbf{m}^{\mathbf{k}} \leq \mathbf{m}$, given that $\mathbf{k}$ is exp. small; and hence $\approx_{0,k}$ is non-inflationary.

To show that $\approx_{0,K}$ is the most refined internally K-acceptable equivalence, let $\approx$ be an arbitrary internally K-acceptable equivalence on M. For each subset L of K, define $\approx_L$ on $\wp(M - K)$ by:

$$C \approx_L D \text{ iff } C \cup L \approx D \cap L.$$

Then $\approx_L$ is a strictly acceptable equivalence on $\wp(M - K)$. So by theorem 6, it contains $\approx_0$ as defined on $\wp(M - K)$ and hence is readily shown to contain the restriction of $\approx_0$, as defined on $\wp(M)$, to $\wp(M - K)$. Now suppose $C \approx_{0,K} D$ for $C, D \subseteq M$. Then $C - K \approx_0 D - K$ and $C \cap K = D \cap K$. Hence $C - K \approx_{C \cap K} D - K$; and so $C = (C - K) \cup (C \cap D) \approx (D - K) \cup (D \cap K) = D$.

It should be noted that the various minimal relations that we have appealed to can all be expressed within second-order logic. In order to express the absolute relations $\approx_0$ and $\approx_1$, we use the definitions:

Exponential Smallness: Expsmall(C) for $\exists R \forall D(D \leq C \rightarrow \exists y \forall x(xRy \leftrightarrow Dx))$;

Exponential Largeness: Explarge(C) for $\neg$Expsmall(C);

Suitability: Stb(C) for (Expsmall(C) $\vee$ $\exists$D(D compl C & Expsmall(D));

Unsuitability: Unstb(C) for $\neg$Stb(C).

The basal relation $\approx_0$ is then expressed by the formula ϕ_0(C,D):

(Stb(C) & Stb(D) & C ≡ D) ∨ (Unstb(C) & Unstb(D) & C beqD).

Thus if M is an infinite set and $\approx_0$ is the basal relation on $\wp(M)$ then, for any subsets C and D of M, C $\approx_0$ D iff $\phi_0[C, D]$ is true in the standard model $\boldsymbol{M}$ based on M. Similarly, the super-basal relation $\approx_1$ is expressed by the formula $\phi_1(C,D)$:

(Expsmall(C) & Expsmall(D) & C ≡ D) ∨ (Explarge (C) & Explarge(D) & C eq D).

The relativized relations $\approx_{0,K}$ and $\approx_{1,K}$ are then expressed by relativizing the formulas ϕ_0 and ϕ_1 to a predicate K in the obvious way.

Using the above results, we can provide a deeper analysis of the various forms of acceptability. We limit our attention to the case of invariant (as opposed to K-invariant) criteria.

Theorem 17. Suppose that $\boldsymbol{M}$ is an infinite standard model. Then:

(i) An (absolute) L-criterion ϕ is tenable on $\boldsymbol{M}$ iff $E_{\phi,\boldsymbol{M}}$ is an equivalence which subsumes the basal (resp. the super-basal) equivalence on $\wp(M)$;

(ii) An L-criterion ϕ is stable if, for each sufficiently large M, $E_{\phi,\boldsymbol{M}}$ is an equivalence that subsumes the basal equivalence on $\wp(M)$;

(ii)′ An absolute L-criterion ϕ is stable iff the corresponding global relation E_ϕ is an equivalence that subsumes equinumerosity for all sets of sufficiently large cardinality;

(iii) An L-criterion ϕ is generally tenable iff $E_{\phi,\boldsymbol{M}}$ is an equivalence that subsumes the relation $\approx_{\text{be}}$ of bi-equinumerosity on $\wp(M)$ for each infinite model $\boldsymbol{M}$;

(iii)′ An absolute L-criterion ϕ is generally tenable iff the corresponding global relation $\boldsymbol{E}_{\phi,M}$ is an equivalence that subsumes equinumerosity on $\wp(M)$ for each infinite set M of urelements.

Proof. (i) Since ϕ is an L-criterion, $\boldsymbol{E}_{\phi,\boldsymbol{M}}$ is invariant by lemma 3.10. Now ϕ is tenable over $\boldsymbol{M}$ iff $\boldsymbol{E}_{\phi,\boldsymbol{M}}$ is an equivalence and non-inflationary, by corollary 4.2. But given that $\boldsymbol{E}_{\phi,\boldsymbol{M}}$ is invariant, it follows from theorem 6 above that it is a non-inflationary equivalence iff it contains the basal relation on $\wp(M)$.

The other results are proved similarly.

From (i) it follows that if the criterion $\phi = \phi$ (C,D) is tenable on M then the formula $\phi_0(C,D) \rightarrow \phi(C,D)$ will be true in the standard model $\boldsymbol{M}$ and, should ϕ be absolute, the formula $\phi_1(C,D) \rightarrow \phi(C,D)$ will also be true in $\boldsymbol{M}$. But note that the criterion $\phi_0(C,D)$ is not itself absolute.

There is a most refined (global) relation that is generally tenable. It is defined, for sets C and D of urelements, by:

$C \approx_2 D$ iff (i) C and D are both finite and $C = D$ or (ii) C and D are both infinite and $\text{card}(C) = \text{card}(D)$.

This relation can be expressed by an absolute (indeed, by a restricted) formula. For let Fin(C) be a restricted expression of finitude. Then $\approx_2$ can be expressed by the formula $\phi_2 = \phi_2(\text{C}, \text{D})$:

$$(\text{Fin(C)} \,\&\, \text{Fin(D)} \,\&\, \text{C} \equiv \text{D}) \vee (\neg\text{Fin(C)} \,\&\, \neg\text{Fin(D)} \,\&\, \text{C eq D}).$$

Thus if we are to remain neutral about the size of the universe (apart from its being infinite), then ϕ_2 represents the best we can do using absolute means of expression. If we also insist on a 'uniform meaning' from one domain to another, then it would appear to be equinumerosity, simpliciter, that represents the best we can do; and this fact may go some way towards explaining the privileged role of Hume's Law in discussions of abstraction.

Our results can be extended to restricted theories of extensional abstraction, i.e. theories of the form $\text{T}^{\phi,\psi}$ where ϕ is the criterion $\text{C} \equiv \text{D}$.

Corollary 18. Suppose that $\text{T} = \text{T}^{\phi,\psi}$ is a restricted theory of extensional abstraction, with ψ logical, and that $\boldsymbol{M}$ is an infinite standard §-free model. Then some expansion of $\boldsymbol{M}$ is a model of T iff $\boldsymbol{E}_{\phi,\boldsymbol{M}}$ is an equivalence on $\boldsymbol{E}_{\psi,\boldsymbol{M}}$ and $\boldsymbol{E}_{\psi,\boldsymbol{M}}$ contains only exp. small subsets of M.

Proof. Associated with the restricted principle $\Phi_\psi = [\psi(\text{C}) \,\&\, \psi(\text{D}) \rightarrow (\S\text{C} \equiv \S\text{D} \leftrightarrow \text{C} \equiv \text{D})]$ is the unrestricted principle Φ': $[\S\text{C} = \S\text{D} \leftrightarrow ((\psi(\text{C}) \,\&\, \psi(\text{D}) \,\&\, \text{C} \equiv \text{D}) \vee (\neg\psi(\text{C}) \,\&\, \neg\psi(\text{D})))$. Thus in the associated unrestricted principle, all concepts which fail to conform to the given restriction are identified. It is then easy to show that some expansion of $\boldsymbol{M}$ is a model for Φ_ψ iff ϕ' is tenable on $\boldsymbol{M}$. By theorem 17(i) above, ϕ' is tenable on $\boldsymbol{M}$ iff $\boldsymbol{E}_{\phi',\boldsymbol{M}}$ is an equivalence that subsumes the basal equivalence $\approx_0$ on $\wp(M)$. But $E_{\phi',M}$ is an equivalence iff $E_{\phi,\text{M}}$ is an equivalence on $E_{\psi,M}$; and $E_{\phi',M}$ subsumes $\approx_0$ iff $E_{\psi,M}$ contains only exp. small subsets of M.

Thus the limits of extensional abstraction are set by concepts with exp. small extensions. When the domain M is countably infinite, the exp. small sets are, of course, the same as those that are small in the usual sense of having a cardinality less than that of the universe. But for larger domains M the two notions may diverge—some small sets in the usual sense may not be exp. small.

Similar results can be established for abstraction theories T^Φ with several abstraction operators. For example, in analogy to theorem 17(i), we have:

Corollary 19. If $\boldsymbol{M}$ is an infinite standard model and T^{Φ} is a generalized abstraction theory, containing only logical abstraction principles, then some §-expansion of $\boldsymbol{M}$ is a model for T^{Φ} iff $\boldsymbol{E}_{\phi,M}$ is an equivalence containing the basal equivalence on $\wp(M)$ for each of the identity criteria $\phi = \phi_{\S}$.

Proof. We may show that some §-expansion of $\boldsymbol{M}$ is a model of the generalized theory T^{Φ} iff for each § ∈ §, some §-expansion of $\boldsymbol{M}$ is a model of the individual theory T^{Φ} with $\phi = \phi_{\S}$. The left-to-right direction is trivial; and for the right-to-left direction, we may simply 'piece together' all of the component §-expansions to obtain the §-expansion. The result then follows from theorem 17(i).

It should be noted that there is no guarantee that the §-expansion of $\boldsymbol{M}$ obtained under this corollary will be separated; for, in piecing together the component §-expansions, different abstracts may be identified. In Section 8, we consider the possibilities for obtaining separated models.

8. HYPERINFLATION

We build on the previous analysis of invariance to determine the conditions under which there will be no hyperinflation among the non-inflationary invariant principles. The analysis is then extended to other forms of invariance.

By a *cell* is meant a non-empty collection of sets. A cell X is said to be *over* a set M if each member of X is a subset of M. A cell X is said to be *strictly admissible in* M if it is over M and if it is a member of $P_{\approx}$ for some strictly acceptable equivalence $\approx$ on $\wp(M)$. Let $\boldsymbol{C}_M$ (or simply $\boldsymbol{C}$ if M is understood) be the set of strictly admissible cells in M; and let $\boldsymbol{E}_M = \{\approx : \approx$ is a strictly acceptable equivalence on $\wp(M)\}$. Our aim is to determine when card($\boldsymbol{C}_M$) and card($\boldsymbol{E}_M$) are ≤ card(M).

For $\approx$ an equivalence on $\wp(M)$, let cdstr($\approx$) = {cdstr(C,D): $C \approx D$}.

Lemma 1. $\approx$ and $\approx'$ are the same invariant equivalences on $\wp(M)$ if cdstr($\approx$) = cdstr($\approx'$).

Proof. Assume that the condition holds; and suppose that $C \approx D$. Then cdstr(C,D) ∈ cdstr($\approx$) and hence cdstr(C,D) ∈ cdstr($\approx'$). So for some C' and D', $C' \approx' D'$ and cdstr(C',D') = cdstr(C,D). But then for some permutation p on M, $p[C'] = C$ and $p[D'] = D$; and so, by $\approx'$ invariant, $C \approx' D$.

For each cardinal $\mathbf{c}$, let cp($\mathbf{c}$) = card({$\mathbf{d}$: $\mathbf{d}$ a cardinal ≤ $\mathbf{c}$}). ('cp' here stands for cardinality of predecessors and is meant to be reminiscent of 'cf' for cofinality.) The notion will play an important role in our consideration of hyperinflation; and it is therefore of interest that the related finitary notion plays an important role in Frege's proof of the infinity of the natural numbers.

Lemma 2. For $\mathbf{m} = \text{card}(M)$ transfinite:

(i) $\text{card}(\boldsymbol{E}_M) = 2^{\text{cp}(\mathbf{m})}$;
(ii) $\text{card}(\boldsymbol{C}_M) = \max(\mathbf{m}, 2^{\text{cp}(\mathbf{m})})$.

Proof. We first show that $\text{card}(\boldsymbol{E}_M) \leq 2^{\text{cp}(\mathbf{m})}$. By lemma 1, cdstr is a one–one map from the invariant equivalences on $\wp(M)$ into sets of quadruples of cardinals $\leq\mathbf{m}$. But the number of such quadruples is at most $2^{\text{cp}(\mathbf{m})}$; and hence $\text{card}(\boldsymbol{E}_M) \leq 2^{\text{cp}(\mathbf{m})}$.

We next show that $\text{card}(\boldsymbol{E}_M), \text{card}(\boldsymbol{C}_M) \geq 2^{\text{cp}(\mathbf{m})}$. Given a set Γ of cardinals $\leq \mathbf{m}$, define $\equiv_\Gamma$ on $\wp(M)$ by: $C \equiv_\Gamma D$ iff card(C) and card(D) both belong to Γ or both fail to belong to Γ. Clearly, for each Γ, $\equiv_\Gamma$ is an invariant equivalence; and, clearly, it is non-inflationary. The relation $\equiv_\Gamma$ divides $\wp(M)$ into two cells, one containing all subsets of M of cardinality in Γ and the other containing all other subsets of M. Thus for each Γ, $X_\Gamma = \{C \subseteq M: \text{card}(C) \in \Gamma)$ is a strictly admissible cell. But $\text{card}(\{\Gamma: \Gamma \text{ a set of cardinals} \leq \mathbf{m}\} = 2^{\text{cp}(\mathbf{m})}$; and for distinct Γ, the X_Γ are distinct. So $\text{card}(\boldsymbol{E}_\text{M}), \text{card}(\boldsymbol{C}_M) \geq 2^{\text{cp}(\mathbf{m})}$. This establishes (i) (and part of (ii)).

We now show that $\text{card}(\boldsymbol{C}_M) \geq \mathbf{m}$. Define $\approx_s$ on $\wp(M)$ by: $C \approx_s D$ iff either C and D are singleton and the same or C and D are not singleton. Clearly, $\approx_s$ is invariant and non-inflationary; and so each $\{\{x\}\}$ for $x \in M$ is a strictly admissible cell. But there are $\mathbf{m}$ such cells.

We finally show that $\text{card}(\boldsymbol{C}_M) \leq \max(\mathbf{m}, 2^{\text{cp}(\mathbf{m})})$. Since each equivalence in $\boldsymbol{E}_M$ induces a partition with at most $\mathbf{m}$ cells, $\text{card}(\boldsymbol{C}_M) \leq \text{card}(\boldsymbol{E}_M)$. $\mathbf{m} = 2^{\text{cp}(\mathbf{m})}$. $\mathbf{m} = \max(2^{\text{cp}(\mathbf{m})}, \mathbf{m})$. This establishes (ii).

We say that a cardinal $\mathbf{c}$ is *unsurpassable* if $2^{\text{cp}(\mathbf{c})} \leq \mathbf{c}$. The unsurpassable cardinals are the counterpart in our general theory of abstraction to the inaccessible cardinals of ZF. $\aleph_0$ is not unsurpassable since $\text{cp}(\aleph_0) = \aleph_0$. On the other hand, given CH, $2^{\text{cp}(\aleph_1)} = 2^{\aleph_0} = \aleph_1$ and hence $\aleph_1$ will be unsurpassable.

From lemma 2, we immediately obtain:

Theorem 3. For $\mathbf{m} = \text{card}(M)$ transfinite, the following three conditions are equivalent:

(i) $\mathbf{m}$ is unsurpassable;
(ii) $\text{card}(\boldsymbol{E}_M) \leq \mathbf{m}$;
(iii) $\text{card}(\boldsymbol{C}_M) \leq \mathbf{m}$.

Similar results can be proved using the notion of internal invariance instead of invariance simpliciter. We define the *internal cardinality distribution* Icdstr(C,D) *of* two subsets C and D of M by $\langle\text{card}(D - C), \text{card}(C \cap D), \text{card}(C - D)\rangle$; and similarly for the related notions. We use $\boldsymbol{IE}_M$ and $\boldsymbol{IC}_M$ for the corresponding sets of the strictly I-acceptable equivalences and of the strictly I-admissible cells that belong to their partitions. Then clearly the previous upper bounds on card($\boldsymbol{E}_M$) and card($\boldsymbol{C}_M$) still apply; and the lower bounds also apply given that the equivalences that were used in the constructions were all I-invariant.

Call a §-model $\boldsymbol{M}$ *replete* (*I-replete*) if, for each invariant (resp. I-invariant) non-inflationary equivalence $\approx$ on $\wp(M)$ there is a § $\in$ § such that $\equiv_§$ is $\approx$. From the theorem we immediately obtain:

Corollary 4. Let M be an infinite domain of urelements and let $\mathbf{m}$ = card(M). Then the following conditions are equivalent:

(i) $\mathbf{m}$ is unsurpassable;
(ii) there is an weakly replete and separated model $\boldsymbol{M}$ with domain M;
(iii) there is a replete and strictly separated model $\boldsymbol{M}$ with domain M.

In the light of this result it is of interest to ascertain which cardinals are unsurpassable. Note that cp($\mathbf{c}$) = $\mathbf{c}$ iff $\mathbf{c}$ is a regular limit cardinal. From this it follows that no regular limit cardinal is unsurpassable. But what of the singular or the successor cardinals? There are two extreme possibilities. One is that they are all unsurpassable; the other is that none is. The first possibility is consistent with ZFC since it follows from GCH; and the second is consistent with ZFC by a result of Woodin and Foreman (1991) according to which it is consistent with ZFC + the existence of certain strong cardinals to assume that $2^{\mathbf{c}}$ is (weakly) inaccessible for each transfinite cardinal $\mathbf{c}$.

Relativized versions of these results, with elements from a given subset K of M kept fixed, can also be proved. We say that a subset of $\wp(M)$ is a K-*cell* if it is a member of $P_{\approx}$ for some K-invariant equivalence $\approx$ on M and that a K-cell is *K-admissible* if it is a member of $P_{\approx}$ for some K-acceptable equivalence $\approx$ over M. The *cardinality distribution* cdstr$_K$(C,D) *of C and D relative to* K is taken to be the sextuple (K_1, K_2, $\mathbf{c}$, $\mathbf{d}$, $\mathbf{e}$, $\mathbf{f}$), where $K_1 = C \cap K$, $K_2 = D \cap K$, and ($\mathbf{c}$, $\mathbf{d}$, $\mathbf{e}$, $\mathbf{f}$) = cdstr($C \cap (M - K)$, $D \cap (M - K)$) (relative to $(M - K)$ as domain). The cdstr$_K$(X) for X a K-cell is defined as before. We then have, in analogy to lemma 1, that K-invariant equivalences with the same K-cardinality distributions will be the same.

Let $\boldsymbol{C}_{M;K}$ be the set of K-admissible cells over M and let $\boldsymbol{E}_{M;K}$ be the set of non-inflationary K-invariant equivalences on $\wp(M)$. Then in analogy to lemma 2 and theorem 3, we have:

Theorem 5. For $\mathbf{m}$ = card(M) transfinite and K a subset of M of cardinality $\mathbf{k} < \mathbf{m}$:

(i) card($\boldsymbol{E}_{M;K}$) = max($2^{\mathrm{cp}(\mathbf{m})}$, $2^{\mathbf{k}}$);
(ii) card($\boldsymbol{C}_{M;K}$) = max($\mathbf{m}$, $2^{\mathrm{cp}(\mathbf{m})}$, $2^{\mathbf{k}}$).

Proof. Much as for lemma 2. To show that card($\boldsymbol{E}_{M;K}$), card($\boldsymbol{C}_{M;K}$) $\geq 2^{\mathbf{k}}$, we let $\equiv_L$, for each subset L of K, be the relation defined by: $C \equiv_L D$ iff $C = D = L$.

Corollary 6. Suppose that $\mathbf{m} = \text{card}(M)$ is transfinite and K is a subset of M of cardinality $\mathbf{k} < \mathbf{m}$. Then the following three conditions are equivalent:

(i) $\max(2^{\text{cp}(\mathbf{m})}, 2^{\mathbf{k}}) \leq \mathbf{m}$;
(ii) $\text{card}(\boldsymbol{E}_{M;K}) \leq \mathbf{m}$;
(iii) $\text{card}(\boldsymbol{C}_{M;K}) \leq \mathbf{m}$.

Proof. Straightforwardly from the theorem.

Recall that, for any two cardinals $\mathbf{c}$ and $\mathbf{d}$, $\mathbf{c}^{<\mathbf{d}} = \sup\{\mathbf{c}^{\mathbf{e}}: \mathbf{e} < \mathbf{d}\}$. Where $\mathbf{k}$ is a cardinal $\leq\mathbf{c}$, call a §-model $\boldsymbol{M}$ ***k**-replete* (*internally* ***k**-replete*) if, for each K-invariant (resp. internally K-invariant) non-inflationary equivalence $\approx$ on $\wp(M)$, where K is a subset of M of cardinality $<\mathbf{k}$, there is a $\S \in \boldsymbol{\S}$ for which $\equiv_{\S}$ is $\approx$. In analogy to corollary 4, we have:

Corollary 7. For $\mathbf{k}$ a cardinal $\leq\mathbf{m} = \text{card}(M)$, there is an (internally) $\mathbf{k}$-replete and (weakly, strictly) separated model $\boldsymbol{M}$ with domain M iff $\max(\mathbf{m}^{<\mathbf{k}}, 2^{\text{cp}(\mathbf{m})}) \leq \mathbf{m}$.

A particular case of this result is of special interest. Let $\mathbf{k}$ (which we call the *exponential bound* exbd($\mathbf{m}$) *for m*) be $\sup\{\mathbf{c}: \mathbf{m}^{\mathbf{c}} \leq \mathbf{m}\}$. Then the condition $\mathbf{m}^{<\mathbf{k}} \leq \mathbf{m}$ is automatically satisfied; and so there will be a $\mathbf{k}$-replete and separated model $\boldsymbol{M}$ just in case $\mathbf{m}$ is unsurpassable, as before. We say that an equivalence $\approx$ on $\wp(M)$ is *predominantly (I-) invariant* if it is (*internally*) K-invariant for some exp. small subset K of M; and we say that it is *broadly (I-) acceptable* (or *(I-) acceptable*, simpliciter) if it is non-inflationary and predominantly (I-) invariant. There is, of course, no most refined broadly acceptable equivalence $\approx$; since, for whichever exp. small K $\approx$ was K-acceptable, there would be a more refined K^{+}-acceptable equivalence for any exp. small superset K^{+} of K. A cell is said to be *broadly (I-) admissible* (or *(I-) admissible*, simpliciter) if it is induced by an (I-) acceptable equivalence. Thus when $\mathbf{k} = \text{exbd}(\mathbf{m})$, each admissible cell will be represented by an abstract in the $\mathbf{k}$-replete model $\boldsymbol{M}$.

The I-admissible cells have the following interesting property:

Theorem 8. A cell on the infinite domain M is I-admissible iff it is predominantly I-invariant, i.e. stays the same under all internally K-invariant permutations on M for some exp. small subset K of M.

Proof. Suppose first that the cell X is internally K-admissible for K an exp. small subset of M. Thus for some internally K-acceptable equivalence $\approx$ over M, X is a member of the partition induced by $\approx$. For any subset P of K, let $X_P = \{C \subseteq (M - K): C \cup P \in X\}$ and let $\approx_P$ be the relation on $M - K$ defined by: $C \approx_P D$ if $C \cup P \approx D \cup P$. For any subsets P and Q of K, define $\leftrightarrow_{P,Q}$ on subsets C and D of $M - K$ by: $C \leftrightarrow_{P,Q} D$ if $C \cup P \approx D \cup Q$. It is readily verified that $\leftrightarrow_{P,Q}$ is a bi-equivalence and that the relation $\leftrightarrow_{\text{L}}$ of connection on the left is coincident with $\approx_P$ and that the relation $\leftrightarrow_{\text{R}}$ of connection on the right is coincident with $\approx_Q$.

We establish the following facts (for P and Q subsets of K):

(1) If $C \approx_P D$, for C very different from D, then $\approx_P$ is numeric on card(C).

Proof. It is clear that the equivalence $\approx_P$ is an I-invariant equivalence over $M - K$. The result therefore follows from corollary 7.13.

(2) If $C \leftrightarrow_{P,Q} D$, for C very different from D, then $\approx_P$ is numeric on card(C).

Proof. From theorem 7.15.

(3) If X_P contains an exp. large set C then it contains a set D very different from C.

Proof. $\approx_P$ is a strictly I-acceptable equivalence over $M - K$. So by corollary 7.7, it is numeric on card(C). But then we may readily find a D that is equinumerous with C and yet very different from C.

(4) If $X^- = \cup\{X_P: P \subseteq K\}$ contains an exp. large set or two sets that are very different, then each X_P is numeric, i.e. is closed under equinumerosity.

Proof. If X^- contains an exp. large set C then it follows from (3) that it contains two sets that are very different; and so we need only consider the one case in which X^- contains two sets C and D that are very different. Take now any set E in X_P for $P \subseteq K$. Then E is either very different from C or from D. Without loss of generality, assume the former. If $D \in X_P$, then the desired result follows from (1). If $D \in X_Q$, for Q distinct from P, then the desired result follows from (2).

We are now in a position to establish the direction from left to right. By (4), we need only consider the following two cases:

(i) Each X_P is numeric. Let $X^P = \{C \in X: X \cap K = P\}$. Then each X^P is internally K-invariant; and so $X = \cup\{X^P: P \subseteq K\}$ is K-invariant.

(ii) X^- contains only exp. small sets and any two are almost the same. We may show that the members of any two non-empty sets X_P are the same. For suppose $C \in X_P$ and $D \in X_Q$, with C distinct from D. Then $C \leftrightarrow_{P,Q} D$. Let $\mathbf{d}$ = card($C \sim D$). Then by theorem 15, $C \approx_P E$ for any E for which $E \equiv_{\mathbf{d}+} C$; and so, since $D \equiv_{\mathbf{d}+} C$, $D \in X_P$.

We now distinguish two subcases. (a) X^- is the singleton $\{C\}$. But then X will be $K \cup C$ invariant, with $K \cup C$ exp. small. (b) X^- is not singleton. Given that X^- is a cell induced by any of the $\approx_P$ for which X_P is non-empty, it follows from theorem 12 that X^- will be an equivalence class $|C|$ wrt to one of the relations $\equiv_{\mathbf{d}}$. But then, again, X will be $K \cup C$ invariant, with $K \cup C$ exp. small.

For the other direction, which is much more straightforward, suppose that X is an internally K-invariant cell for some exp. small subset K of M. Define $\approx$ on $\wp(M)$ by: $C \approx D$ iff C and D both belong or both fail to belong to X. Then it is clear that $\approx$ is non-inflationary and internally K-invariant.

An analogous result does not hold for strict admissibility. For when C is exp. small, $\{C\}$ will in general be strictly I-admissible and yet not I-invariant.

9. INTERNALIZED PROOFS

For the purposes of the present section, we take ourselves to be working within a particular abstraction theory T^{ϕ}, with ϕ §-free. We will suppose that the theory contains a well-ordering principle, WO, which states that there is a well ordering of the whole universe, and an axiom of infinity, Inf, which states that there exists an ordering without any last element.

We show how the main negative results of Section 7 can be formalized within the resulting theory. This means that the results will have a claim on any abstraction theorist who accepts the underlying second-order logic and that they do not rest upon his adopting a 'full' Platonic stand on concepts.

Theorem 1. In any logical abstraction theory T^{ϕ} + WO + Inf, Unstb(C) & Unstb(D) & (C beq D) → §C = §D is a theorem.

Proof. Let me briefly sketch how the proof of theorem 7.6 may be formalized within T^{ϕ} + WO + Inf. We need to establish appropriate internal analogues of lemmas 7.3 and 7.5. For simplicity, let us concentrate on the case of bifurcators. We then need to show:

(1) If two bifurcatory concepts C and D are identified by ϕ, with C very different from both D and its complement, then all bifurcators are identified; and

(2) If no two bifurcatory concepts C and D, with C very different from both D and its complement, are ever identified, then there is definable one–one correspondence taking the bifurcatory concepts into objects.

The condition that the bifurcatory concepts C and D are very different can be expressed as C-D or D-C being equinumerous with C. We cannot say in the second-order language L^{ϕ} that there exists a one–one correspondence between concepts and objects; but we will be able to produce, on the basis of a proof, a formula $\chi(C,x)$ (containing parameters) with the properties of a one–one correspondence.

To formalize the proofs of (1) and (2), we may use lemma 3.6 to obtain the 'internalized' counterpart to the claim that the relation defined by ϕ is invariant. We then need to imitate the reasoning about cardinals and the various constructions on sets, some of which takes place outside of the underlying domain *M*. This will require the Schroeder-Bernstein Theorem in the form: if there is a one–one correspondence from C into D and a one–one correspondence of D into C, then there is a one–one correspondence of C onto D. The theorem in this form can be proved along the lines of the proof on p. 49 of Drake (1974). For as we have seen, the Fixed Point Theorem can be proved within L^2 (theorem 3.2); and it is then an easy matter to define the relevant 'functions' on concepts and to establish the relevant properties (cf. Shapiro 1991, 102–3).

For some of the proofs, we will need Well Ordering. Given this assumption, the ordinals can be identified with the objects under the posited well-ordering and the cardinals can be identified with the surrogates for the initial orderings. It can be then be shown that the objects falling under any concept can be put into one–one correspondence with an initial segment of the 'ordinals' and hence assigned a 'cardinal'.

To divide the objects falling under an infinite concept into two equal-sized parts, we take advantage of the one–one correspondence onto an initial segment of the 'ordinals'. A definition by transfinite induction can then be used to divide the ordinals into two equal-sized parts; and this then induces a corresponding division of the objects.

The original proof of (2) requires that we take **m** copies of the universe *M*. Clearly, this cannot be done in L^2. But what we can do instead is to set up a one–one correspondence from 'pairs' of object from the domain into the whole domain. Given Well Ordering, there is no difficulty in defining such a correspondence. A 'copy' of *M* is then obtained by taking the objects which correspond to all the pairs (x_0,y) for a fixed x_0.

Recall that $\phi_0(C,D)$ is the object-language version of the basal equivalence $\approx_0$. An equivalent form of the result is:

Corollary 2. In any theory $T^\phi + WO + Inf$, with logical ϕ, the formula $\phi_0(C, D) \rightarrow \phi(C,D)$ is a theorem.

Let Bf(C) be the formula in L^2 which says that there is a one–one correspondence between the objects which fall under C and those which fail to fall under C. Since Bf(C) → Unstb(C) and Bf(C) & Bf(D) → C beq D are theorems of L^2, as a special case of the theorem we obtain:

Corollary 3. In any logical abstraction theory $T^\phi + WO + Inf$, the formula Bf(C) & Bf(D) → §B = §D is a theorem.

These results have consequences for restricted theories of extensional abstraction. A natural response to the paradoxes is to restrict the concepts that are capable of having an extension. Recall that, when $\psi = \psi(C)$ is a formula containing C as its sole free variable, $T^{\psi,=}$ is the result of adding to L^2 the axiom:

$$\text{Ext}_{\psi}.\ \psi(C) \,\&\, \psi(D) \rightarrow (\S C = \S D \leftrightarrow C \equiv D).$$

We may now prove within the theory of restricted abstraction that no unsuitable concepts can be subject to extensional abstraction:

Corollary 4. In any theory $T^{\psi,=}$ + WO + Inf, with logical ψ, $\psi(C) \rightarrow \text{St}(C)$ (and hence $\psi(C) \rightarrow \neg\text{Bf}(C)$) is a theorem.

Proof. Let $\phi = \phi(C,D)$ be the formula: $(\psi(C) \,\&\, \psi(D) \,\&\, C \equiv D) \vee (\neg\psi(C) \,\&\, \neg\psi(D))$. By corollary 2, we can prove the formula Unstbt(C) & Unstbt(D) & C beq D $\rightarrow \phi(C,D)$ within T^{ϕ} + WO + Inf. But the § of L^{ϕ} can be 'defined' within $L^{\psi,=}$ by 'identifying' §C, for the C which do not conform to ψ, with an arbitrary object from the domain (which can always be 'detached' from the rest of the universe by means of WO and Inf). Theorems of T^{ϕ} then translate into theorems of $T^{\psi,=}$; and hence Unstb(C) & Unstb(D) & C beq D $\rightarrow \phi(C,D)$ can be proved within $T^{\psi,=}$ + WO + Inf. We can then prove $\psi(C) \rightarrow \text{Stb}(C)$ within $T^{\psi,=}$ by transcribing the following informal proof:

Suppose there were an unsuitable C which conformed to ψ. Choose a D that is bi-equinumerous with but not coextensive with C. Now D is unsuitable; and by the invariance of ψ, it also conforms to ψ. But then from Unstb(C) & Unstb(D) & C beq D $\rightarrow \phi(C,D)$, it follows that C and D are coextensive. A contradiction.

The only property of ψ required for the above proof is that it be invariant. Hence if we add a third-order monadic predicate constant $\boldsymbol{S}$ to the language $L^{\S}$, $\boldsymbol{S}(C) \rightarrow \text{Stb}(C)$ can be proved in the theory with the axioms: WO, Inf, $\boldsymbol{S}(C) \,\&\, \boldsymbol{S}(D) \rightarrow (\S C = \S D \leftrightarrow C \equiv D)$, and $(\boldsymbol{S}(C) \,\&\, C \text{ beq } D) \rightarrow \boldsymbol{S}(D)$ as axioms. Thus whatever logical restriction be imposed on the concepts subject to extensional abstraction, it can be *proved* within the resulting theory that they must all be suitable; either their extensions or their counter-extensions must be exp. small.

The above results can be used to establish the inconsistency of a large number of logical abstraction principles; for all that need be shown is that the principles in question fail to identify unsuitable concepts which are bi-equinumerous. They can therefore be regarded as providing a highly general form of the paradoxical reasoning, though subject to the limitation that the criterion of identity be logical and that Well Ordering and Infinity be assumed. It would be desirable to produce versions of these results that did not depend upon these assumptions.

PART III. THE GENERAL THEORY OF ABSTRACTION

We develop a general theory of abstraction, one which is intended to account for the existence and behaviour of abstracts in general, and not of any kind of abstract in particular. The material is in three sections: first we outline the systems; then we look at their models; and finally we show how they serve to provide a foundation for both arithmetic and analysis.

1. THE SYSTEMS

Since we want to talk about all equivalence relations on concepts, the most natural framework for our theory is third-order rather than second-order logic. Most of our previous definitions for second-order logic will extend in the obvious way to third-order logic; but to avoid confusion, we will use bold face to distinguish the third-order terms from the others. Later it will be shown how a version of the theory can be given within the framework of second-order logic.

The sole extra-logical primitive of our theory is the three-place predicate Abstr, which applies to terms for an object, a concept, and a relation, in that order. The predications formed from Abstr may be written in the form t $\text{Abstr}_{\mathbf{R}}$ C; and the intended reading is that t is an abstract of C with respect to **R**. Note that in contrast to a functional notation, such as $\S_{\mathbf{R}}(C)$, the predications formed from Abstr can be used without any commitment to the existence of abstracts.

The following definitions will be useful (some have already been given but are here restated in an appropriately modified form):

Unrelativized Abstraction: x Abstr C for $\exists\mathbf{R}(x\ \text{Abstr}_{\mathbf{R}}\ C)$;
R-Abstractable concepts: $\text{Abstr}_{\mathbf{R}}(C)$ for $\exists x(x\ \text{Abstr}_{\mathbf{R}}\ C)$;
R-Abstracts: $\text{Abstr}_{\mathbf{R}}(x)$ for $\exists C(x\ \text{Abstr}_{\mathbf{R}}\ C)$;
Abstract: Ab(x) for $\exists\mathbf{R}\exists C(x\ \text{Abstr}_{\mathbf{R}}\ C)$;
Applicability: App(**R**) for $\forall C(\text{Abstr}_{\mathbf{R}}(C))$;
Invariance: Invar(**R**) for $\forall P,C,D,C',D'\ [\text{Perm}(P)\ \&\ \mathbf{R}(C,D)\ \&\ \forall x,x'\ (Pxx' \rightarrow (Cx \leftrightarrow C'x')\ \&\ (Dx \leftrightarrow D'x'))\rightarrow \mathbf{R}(C',D')]$;
Internal Invariance: I-invar(**R**) for $\forall P,C,D,C',D'\ [\text{1-1}(P)\ \&\ \forall x((x \in \text{Dm}(P) \leftrightarrow Cx \vee Dx)\ \&\ \mathbf{R}(C,D)\ \&\ \forall x,x'(Pxx' \rightarrow (Cx \leftrightarrow C'x')\ \&\ (Dx \leftrightarrow D'x'))\rightarrow \mathbf{R}(C',D')]$;
K-invariance: $\text{Inv}_K(\mathbf{R})$ for $\forall P,C,D,C',D'\ [\text{Perm}(P)\ \&\ \forall x(Kx \rightarrow Pxx)\ \&\ \mathbf{R}(C,D)\ \&\ \forall x,x'\ (Pxx' \rightarrow (Cx \leftrightarrow C'x')\ \&\ (Dx \leftrightarrow D'x'))\rightarrow \mathbf{R}(C',D')]$.
Internal K-invariance: $\text{I-invar}_K(\mathbf{R})$ for $\forall P,C,D,C',D'[\text{1-1}(P)\ \&\ \forall x((x \in \text{Dm}(P) \leftrightarrow Cx \vee Dx \vee Kx)\ \&\ \forall x(Kx \rightarrow Pxx)\ \&\ \mathbf{R}(C,D)\ \&\ \forall x,x'(Pxx' \rightarrow (Cx \leftrightarrow C'x')\ \&\ (Dx \leftrightarrow D'x'))\rightarrow \mathbf{R}(C',D')]$;

Predominant invariance: Prdinv(**R**) for $\exists$K(Expsmall(K) & Inv_K(**R**));

Predominant internal invariance: I-Prdinv(**R**) for $\exists$K(Expsmall(K) & I-inv_K(**R**));

Non-inflationary: Noninfl(**R**) for $\exists$**P**[$\forall$C$\exists$x**P**(C,x) & $\forall$C,D,x(**P**(C,x) & **P**(D,x) $\rightarrow$ **R**(C,D)].

Related notions (e.g. the invariance of a second-level concept) may be defined in an analogous way; and their formal definitions will not always be given.

A concept is abstractable when it yields an abstract (wrt a given method of abstraction); a third-order relation is invariant when its extension remains the same under any permutation; and a third-order relation (which we conceive as a method for identifying concepts) is non-inflationary when the non-identified concepts can be mapped one–one into objects.

Let me first give a formulation of the basic theory. I shall then consider various ways in which it may be modified. There are three axioms in all. The first gives necessary and sufficient conditions for two abstracts to be the same:

Identity: (x $\text{Abstr}_{\mathbf{R}}$ C & y $\text{Abstr}_{\mathbf{S}}$ D) $\rightarrow$ [x = y $\leftrightarrow$ (**R**(C,D) & $\forall$E(**R**(C,E) $\leftrightarrow$ **S**(C,E)))].

Thus this axiom says that two abstracts, possibly associated with different methods of abstraction **R** and **S** and different concepts C and D, will be the same just in case each method identifies the two concepts and each method identifies the same concepts with one of the given concepts (given the second conjunct there is no need to state ***S***(C,D) since it follows from the first conjunct).

The second axiom gives a sufficient condition for abstractions to exist:

Existence: Eq(**R**) & Noninf(**R**) & Inv(**R**) $\rightarrow$ App(**R**).

Thus according to this axiom, any invariant and non-inflationary equivalence is an applicable form of abstraction; it will yield abstracts for each first-order concept.

None of the conditions in the antecedent can sensibly be dropped. This is obvious for the first. If we drop the second, the requirement that **R** be non-inflationary, then a contradiction can be proved in the manner of Russell's Paradox by letting **R** be the relation of coextensiveness. If we drop the third conjunct, the requirement that **R** be invariant, then a contradiction can be proved by a hyper-version of the Russellian reasoning. For with each concept *C* we may associate a non-inflationary equivalence $\equiv_C$ according to the condition that $D \equiv_C E$ iff $D \equiv C \leftrightarrow E \equiv C$. (This equivalence divides the universe of concepts into those that are coextensive with *C* and those that are not.) With respect to each such equivalence $\equiv_C$, there will exist an abstract ex(C) of C. We may define a concept C_r by the condition that C_rx iff x is an object of the form ex(C) which does not fall under C. Considering the question of whether or not ex(C_r) falls under C_r then allows us to derive a contradiction.

The third axiom states that either no concepts are abstractable (with respect to a given method of abstraction) or all are:

Application: $\text{Abstrs}_{\mathbf{R}}(\text{C}) \rightarrow \text{Abstrs}_{\mathbf{R}}(\text{D})$.

The various axioms can be weakened or strengthened in various respects. The identity axiom implies a necessary and sufficient condition for an abstract x of one kind **R** to be the same as an abstract y of another kind ***S***; and this is that each method of abstract associate the same concepts with the two objects. It is important for the deductive development of the theory that the condition be accepted as necessary; if the associated concepts are distinct then the abstracts themselves must be distinct. The requirement of sufficiency, however, can be dropped; and the question of what might plausibly take its place has been considered in Section I.4.

The existence axiom only provides for the existence of invariant methods of abstraction. It would be desirable to extend it to various non-invariant methods even though, on pain of contradiction, we cannot admit them all. Our earlier results on K-invariance suggest a natural way in which this might be done:

Broad Existence: $\text{Eq}(\mathbf{R})$ & $\text{Noninf}(\mathbf{R})$ & $\text{Prdinv}(\mathbf{R}) \rightarrow \text{App}(\mathbf{R})$.

Under a generative conception of abstraction, we would want to replace the existence axiom by the weaker principle:

Internal Existence: $\text{Eq}(\mathbf{R})$ & $\text{Noninf}(\mathbf{R})$ & $\text{I-inv}(\mathbf{R}) \rightarrow \text{App}(\mathbf{R})$.

Similarly, the axiom of broad existence might be replaced with:

Broad Internal Existence: $\text{Eq}(\mathbf{R})$ & $\text{Noninf}(\mathbf{R})$ & $\text{I-Prdinv}(\mathbf{R}) \rightarrow \text{App}(\mathbf{R})$.

We might also insist that all methods of abstraction are to be predominantly invariant. Assuming internality, the required axiom would then take the form:

Exclusivity: $\text{App}(\mathbf{R}) \rightarrow \text{I-Prdinv}(\mathbf{R})$.

Using this axiom, the arrow in the preceding existence axiom can then be reversed.

A principle of minimality might also be adopted. In the light of the discussion at the end of Section II.6, this could be stated in the form:

Minimality: $\neg\exists\text{D}[\exists\text{x}\neg\text{Dx}$ & $\forall\text{x}(\text{I}(\text{x}) \rightarrow \text{Dx})$ & $\forall\mathbf{R},\text{C},\text{x}(\text{C} \subseteq \text{D}$ & $\text{x Abstr}_{\mathbf{R}}\text{ C} \rightarrow \exists\text{y}[\text{Dy}$ & $\forall\text{C}\subseteq\text{D}(\text{y Abstr C} \leftrightarrow \text{x Abstr}_{\mathbf{R}}\text{ C}]$.

However, we may, in the present context, state the axiom in a stronger form. Let us note, in this regard, that the definitions of being exponentially small and being non-inflationary can be relativized to a subdomain, as given by a concept D. We use $\text{Expsmall}^D(C)$ and $\text{Non-inflationary}^D(\mathbf{R})$ for these relativized notions. We define the relativized notions of K-acceptability, acceptability, and strict acceptability by:

$\text{K-Acc}^D(\mathbf{R})$ for $[\text{Eq}(\mathbf{R})$ & $\text{Non-inflationary}^D(\mathbf{R})$ & $(K \subseteq D)$ & $\forall B_1,B_2,C_1,C_2,P(1\text{-}1(P)$ & $\forall x(Kx \rightarrow Pxx)$ & $\mathbf{R}(B_1,B_2)$ & $B_1 \rightarrow_p C_1$ & $B_2 \rightarrow_P C_2 \rightarrow \mathbf{R}(C_1,C_2)]$;

$$\text{Acc}^D(\mathbf{R}) \text{ for } \exists K(\text{Expsmall}^D(K) \,\&\, K\text{–}\text{Acc}^D(\mathbf{R}));$$

$$S - \text{Acc}^D(\mathbf{R}) \text{ for } \exists K(\forall x \neg Kx \,\&\, K\text{–}\text{Acc}^D(\mathbf{R})).$$

Then the alternative formulation is:

Strong Minimality. $\neg\exists D[\exists x \neg Dx$ & $\forall x(I(x) \rightarrow Dx)$ & $\exists x \exists y(x \neq y)$ & $\forall \mathbf{R},C,x(\text{Acc}^D(\mathbf{R})$ & $C \subseteq D$ & $x \text{ Abstr}_{\mathbf{R}} C \rightarrow \exists y[Dy$ & $\forall C \subseteq D(y \text{ Abstr } C \leftrightarrow x \text{ Abstr}_{\mathbf{R}} C]$.

Thus closure is here required only with respect to the means of abstraction that are acceptable within the given subdomain. In case only logical definitions of the abstracts are allowed, the axiom should take the same form but with $\text{Acc}^D(\mathbf{R})$ replaced by $\text{S-Acc}^D(\mathbf{R})$.

The Application axiom is rather strong and might be weakened by allowing restricted methods of abstraction, i.e. ones that are only defined on the concepts in the field of their relation. The principle should then take the form:

Weak Application: $(\text{Abtrs}_{\mathbf{R}}(C)$ & $C \in \text{Fld}(\mathbf{R})) \rightarrow \forall D(D \in \text{Fld}(\mathbf{R}) \rightarrow \text{Abstr}_{\mathbf{R}}(D)$.

The existence axiom must also then be modified so as to allow for the existence of restricted methods of abstraction. However, there is no great disadvantage in working with the unrestricted principles. For any restricted or partially defined method of abstraction can be identified with an unrestricted method that identifies all of the unabstractable concepts with one of the abstractable concepts.

The original application axiom or its variants are not really essential for the intended deductive development of the system, although they do make it simpler.

We turn to the second-order formulation of the theory. This uses schemes in place of universal principles. The primitive non-logical predications of the theory now take the form t $\text{Abstr}_{[C,D]\phi}$ E, where t is an object term, C and D are distinct concept variables, ϕ is a formula, and E is a concept term. In the resulting formula, the variables C and D are taken to be bound.

The intuitive meaning of the formula t $\text{Abstr}_{[C,D]\phi}$ E is given by t $\text{Abstr}_{\mathbf{R}}$ E, where **R** is the relation $(\lambda C,D)\phi$. However, in the formula t $\text{Abstr}_{[C,D]\phi}$ E, we treat the complex $[C,D]\phi$ as a part of the whole notation and not as a meaningful constituent in its own right. Thus the only second-order relations we can talk about are those that can be specified by a formula (with or without parameters).

The axioms are the same as before but with **R** now taken to be a complex of the form $[C,D]\phi$. Thus whereas the axioms, under the previous formulation, could be taken to be single sentences universally quantified by a third-order variable **R**, they must now be taken to be schemes.

The retrenchment from the third- to the second-order formulation involves some complications. First, the application **R**[E,F] of **R** to the concept terms E and F must now be understood as the simultaneous substitution of E and F for C and D in ϕ. Second, the original definition of Noninfl(**R**) cannot be used since it is third order. Instead of existentially quantifying over the third-order correlation **P** between concepts and objects, we may suppose that **P** is specified by a formula (possibly with parameters). Third, the Minimality axiom, which involves an ineliminable reference to third-order relations, must be dropped. It is not clear which second-order axioms, if any, should be put in its place.

Let us denote the basic third-order theory, consisting of the axioms Identity, Broad Internal Existence, and Application, by GA; and let us denote the result of adding Exclusivity and Strong Minimality to GA by GA^+. We use GA^2 for the second-order counterpart of GA; and we use SGA and SGA^2 for the systems that result from replacing the broad existence axiom in GA and GA^2, respectively, by its normal invariant form. For many purposes, the differences between the systems will not matter; and, in particular, the derivation of arithmetic and analysis given in Section 3 can be carried out equally well in GA^+ and in SGA^2.

It follows from theorem 8.7 of Part II that GA will have a standard model of cardinality **c** just in case **c** is an unsurpassable cardinal. Since $\aleph_0$ is surpassable, GA will not have a standard model of cardinality $\aleph_0$. But given CH, it will have a standard model of cardinality $\aleph_1$; and so from the consistency of CH will follow the consistency of GA. In the next section, we shall construct some natural models for GA on any domain of an unsurpassable cardinality and thereby show that the various additional axioms can also be satisfied.

We state the following elementary results:

Lemma 1. The following are theorems of GA (and GA^2):

(i) x $\text{Abstr}_{\mathbf{R}}$ C & y $\text{Abstr}_{\mathbf{R}}$ D $\rightarrow$ (x = y $\leftrightarrow$ $\mathbf{R}$(C,D));

(ii) $\exists C(\text{Abstrs}_{\mathbf{R}}(C)) \to \text{Eq}(\mathbf{R})$;
(iii) $x\ \text{Abstr}_{\mathbf{R}}\ C \to (\mathbf{R}(C,D) \leftrightarrow x\ \text{Abstr}_{\mathbf{R}}\ D)$;
(iv) $\text{Abstr}_{\mathbf{R}}(x)\ \&\ \text{Abstr}_{\mathbf{S}}(y) \to [x = y \leftrightarrow \forall E(x\ \text{Abstr}_{\mathbf{R}}\ E \leftrightarrow y\ \text{Abstr}_{\mathbf{S}}\ E)]$;

Proof. The proofs are stated informally but are readily formalized.

(i) Let **R** and **S** in Identity be the same. The second conjunct on the right is then trivially satisfied.

(ii) Given $\exists C(\text{Abstr}_{\mathbf{R}}(C))$, it follows by Application that App(**R**). Now to prove Sym(**R**), suppose **R**(C,D). By App(**R**), x $\text{Abstr}_{\mathbf{R}}$ C and y $\text{Abstr}_{\mathbf{R}}$ D for some x and y. By (i) and the fact that **R**(C,D), x = y. But then by another application of (i), **R**(D,C). Trans(**R**) and Refl(**R**) are proved similarly. (Note that the assumption of App(**R**) is essential to the proof; for there is no need for inapplicable relations to be equivalence relations.)

(iii) For the left-to-right direction, assume x $\text{Abstr}_{\mathbf{R}}$ C and **R**(C,D). It then follows by Application that y $\text{Abstr}_{\mathbf{R}}$D for some y. But x = y by (i); and hence x $\text{Abstr}_{\mathbf{R}}$ D.

For the right-to-left direction, use (i).

(iv) We may write the antecedent in the form x $\text{Abstr}_{\mathbf{R}}$ C & y $\text{Abstr}_{\boldsymbol{S}}$ D. Assume its truth in this form. Then by (iii), the final consequent of (iv) is equivalent to $\forall E(\mathbf{R}(C,E) \leftrightarrow \boldsymbol{S}(D,E))$. Hence given Identity, it suffices to show the equivalence of $\forall E(\mathbf{R}(C,E) \leftrightarrow \boldsymbol{S}(D,E))$ to $\mathbf{R}(C,D)\ \&\ \forall E(\mathbf{R}(C,E) \leftrightarrow \boldsymbol{S}(C,E)))$. Assume the first formula. Now $\boldsymbol{S}$(D,D) by (ii); and so **R**(C,D). Also **R**(C,C), by (ii) again; and so $\boldsymbol{S}$(D,C). But then by transitivity, $\boldsymbol{S}(C,E) \leftrightarrow \boldsymbol{S}(D,E)$; and so $\forall E(\mathbf{R}(C,E) \leftrightarrow \boldsymbol{S}(C,E))$. Now assume the second formula. We obtain $\boldsymbol{S}$(C,D) by instantiation and $\boldsymbol{S}$(D,C), from this, by Symmetry. But then $\boldsymbol{S}(D,E) \leftrightarrow \boldsymbol{S}(C,E)$ and hence $\forall E(\mathbf{R}(C,E) \leftrightarrow \boldsymbol{S}(D,E))$.

The principle under (i) provides a criterion for intra-sortal identity while the principle under (iv) provides a criterion for cross-sortal identity, stated directly in terms of abstraction and not the underlying relation. These two principles are combined together in our original Identity axiom.

Suppose that we were to use $\S_{\mathbf{R}}C$ to designate the abstract of C with respect to the equivalence relation **R**. Then the abstraction principle for **R**-abstracts could be stated in the categorical form:

$$\S_{\mathbf{R}}C = \S_{\mathbf{R}}D \leftrightarrow \mathbf{R}(C,D).$$

From this principle, within a standard logical system, the existence of the **R**-abstract $\S_{\mathbf{R}}C$ of each concept C would then follow. The principle under (i) is a kind of conditional version of the principle in its categorical form. It states that *if* there are **R**-abstracts of C and D, then they will behave in accord with that principle. The existential implications of the categorical principle has led many authors to doubt its analyticity. But the principle in its conditional form is free of any such implications and hence has a much better claim to be considered analytic.

It is of interest to consider the theory that results from combining the general theory of abstraction with ZF. This is the theory that might be adopted by someone who believed both in sets (as given by the cumulative hierarchy) and in abstracts. Let us assume a standard formulation of third-order ZFI (with Choice). There will be a special predicate I for the individuals, but it will not be assumed that the individuals constitute a set. On this is then imposed the third-order theory GA (or a variant thereof). Let us call the result ZFA.

Given the existence of an inaccessible cardinal $\mathbf{c}_0$ and of a greater unsurpassable cardinal $\mathbf{c}_1$, it may be shown that the theory ZFA has a 'standard' model. For we may let the domain I of individuals be of cardinality $\mathbf{c}_1$; and over this domain we may construct a cumulative hierarchy U_α of sets, for $\alpha < \mathbf{c}_0$, by requiring the sets generated at each stage to be of cardinality less than $\mathbf{c}_0$. Over the resulting domain U of individuals and sets, we may then construct a standard model of GA.

In this model, the size of I is greater than that of U. It can actually be proved within ZFA that this must hold, i.e. that the abstracts must outnumber the pure sets. This result is of some philosophical significance since it shows that it is impossible for our proponent of sets and abstracts to identify the abstracts, in any reasonable manner, with the sets.

Let us use Pset(x) to mean that x is a pure set (one whose transitive closure contains no urelements). We then have:

Theorem 2. Within the theory ZFA, we may prove

$$\neg(\exists R)(1-1(R) \,\&\, \forall x(Ab(x) \rightarrow x \in Dm(R)) \,\&\, \forall y(y \in Rg(R) \rightarrow PSet(y))).$$

Proof. We sketch an informal version of the proof. Given a non-empty concept C of (ZF) cardinals we may form the sequence s_C of its members in increasing order of magnitude. (s_C in general will, of course, not be a set but a concept or class. It could, for example, be represented by a one–one relation from ordinals to cardinals.) We may then order such concepts by the rule:

$C \leq D$ iff either (i) C and D are coextensive or (ii) s_C is of smaller length than s_D or (iii) they are of the same length and, at the first point at which they differ, the value of s_C is smaller than the value of s_D.

It is readily shown that $\leq$ is a transitive and connected relation and that the corresponding strict relation $<$ is well founded. (Note that to express that $<$ is well founded we need to go to the third order.)

With each non-empty concept of cardinals C we may associate the equivalence relation $\equiv_C$ on concepts, where:

D $\equiv_C$ E iff D and E both constitute sets of a cardinal that falls under C or both fail to constitute sets of a cardinal that falls under C.

With respect to each such equivalence relation $\equiv_C$, we may associate the abstract x_C of a concept D whose extension is of a cardinal that falls under C. (The abstract is the same whichever concept D we choose.) When C < D, the concepts C and D will be distinct and hence so will the associated abstracts x_C and x_D.

Suppose, for reductio, that there is a one–one map from abstracts to pure sets. Then, in particular, there is a one–one map from the abstracts x_C to the pure sets (call these sets the *surrogates*). The well-founded ordering < on the concepts C therefore induces a well-founded ordering $<_s$ on the surrogates.

Now this ordering $<_s$ can be used to construct a one–one map from surrogates to ordinals. For we may reorder the surrogates according to the rule:

$$x < y \text{ iff either } \text{rank}(x) < \text{rank}(y) \text{ or } \text{rank}(x) = \text{rank}(y) \text{ and } x <_s y.$$

Each surrogate y will then be preceded by a *set* of surrogates according to this ordering; and so the resulting ordering will induce a one–one map from surrogates to ordinals.

We may therefore obtain a corresponding one–one map from surrogates to cardinals. But since surrogates are associated with concepts of cardinals, we have a one–one map from concepts of cardinals to cardinals; and the reasoning of Russell's Paradox may be applied to obtain a contradiction.

There is another, rather different, way in which the theory GA might be related to set theory; for, as mentioned in Section I.4, the abstracts themselves might be regarded as classes (or extensions) of concepts. Thus given an abstract x, we may take $C \in x$ to hold just in case $\exists \mathbf{R}(x \text{ Abstr}_{\mathbf{R}} C)$. Given the Identity axiom, we may then establish:

Extensionality: $\forall x,y(\forall C(C \in x \leftrightarrow C \in y) \rightarrow x = y)$;

and by formalizing the proof of the easy direction of theorem 8.8 of Part II, we may also establish:

Comprehension: $\text{I-Prdinv}(\mathbf{E}) \rightarrow \exists x \forall C(C \in x \leftrightarrow \mathbf{E}(C))$.

Similarly, within the strict theory SGA, we may show:

Strict Comprehension: $\text{I-inv}(\mathbf{E}) \rightarrow \exists x \forall C(C \in x \leftrightarrow \mathbf{E}(C))$;

and within the corresponding schematic formulations of GA and SGA, we may show:

Schematic Comprehension: $\text{I-Prdinv}(\psi(C)) \rightarrow \exists x \forall C(C \in x \leftrightarrow \psi(C))$;
Strict Schematic Comprehension: $\text{I-inv}(\psi(C)) \rightarrow \exists x \forall C(C \in x \leftrightarrow \psi(C))$.

On the other hand, the predicate $\in$ could itself be taken as a primitive and Extensionality, along with one of the forms of Comprehension, be adopted as axioms. Let us use STE^2 (theory of extensions) for the weak second-order theory consisting of Extensionality and Strict Schematic Comprehension; and let us use TE for the strong third-order theory consisting of Extensionality and Comprehension in its broad third-order formulation.

It is natural to extend the resulting theories to allow for classes of objects as well of concepts. Comprehension would then take the form:

$$\text{I-Prdinv}(\psi(z)) \rightarrow \exists x \forall z(z \in x \leftrightarrow \psi(z)).$$

However, such an extension would not significantly increase the power of the theory, since objects could always be identified with their singleton concepts and the corresponding form of comprehension on sets of concepts be used in place of a given instance of comprehension on sets of objects.

Within the strong theory TE, one might define what it is for a relation **R** on concepts to be broadly I-acceptable. By formalizing the proof of the hard direction of theorem 8.8 (using the appropriate form of the axiom of choice), one could then prove within the theory itself:

$$(\forall \mathbf{R})\,(\mathbf{R} \text{ is broadly I-acceptable } \rightarrow (\forall C)\,(\exists y)\,\forall D(D \in y \leftrightarrow \mathbf{R}(C,D));$$

and similarly for the schematic versions of the theory.

By adding an appropriate form of the axiom of choice, it should also be possible to establish the 'equivalence' (or, more exactly, the mutual interpretability) between different versions of the theory of classes and corresponding versions of the theory of abstraction. But this is not a matter that we shall investigate.

2. SEMANTICS

We give what might plausibly be regarded as the standard models for our general theory of abstraction. We concentrate on the case in which individuals are allowed in the definition of the methods of abstraction, although we shall also consider the case in which individuals are not allowed. In either case, the models are constructed according to the generative standpoint described in Section I.2. Thus starting off with a given class *I* of individuals, each construction yields a cumulative sequence of domains (denoted respectively by $U_{I,\alpha}$ and $V_{I,\alpha}$); and the intended model (either U_I or V_I) is given by a fixed point of the construction.

We collect together some results on cardinals, which will later be of use. On unsurpassables, we have:

Lemma 1. (i) If $\mathbf{u}$ is unsurpassable, then so is $2^{\mathrm{cp}(\mathbf{u})}$;
(ii) If $\mathbf{u}$ is the smallest unsurpassable cardinal greater than a surpassable cardinal $\mathbf{c}$, then $2^{\mathrm{cp}(\mathbf{u})} = \mathbf{u}$.

Proof. For the purposes of the proof, let us use $\mathbf{u}^*$ for $2^{\mathrm{cp}(\mathbf{u})}$.

(i) Given that $\mathbf{u}$ is unsurpassable, $\mathbf{u}^* \leq \mathbf{u}$; so $\mathrm{cp}(\mathbf{u}^*) \leq \mathrm{cp}(\mathbf{u})$; and so $2^{\mathrm{cp}(\mathbf{u}^*)} \leq 2^{\mathrm{cp}(\mathbf{u})} = \mathbf{u}^*$.

(ii) If $\mathbf{u}^* < \mathbf{c}$, then $2^{\mathrm{cp}(\mathbf{c})} \leq 2^{\mathrm{cp}(\mathbf{u})} = \mathbf{u}^* < \mathbf{c}$; and so $\mathbf{c}$ is unsurpassable, contrary to supposition. So $\mathbf{u}^* \geq \mathbf{c}$. But then $\mathbf{u}^* = \mathbf{u}$, since otherwise $\mathbf{u}^*$, by (i), will be a smaller unsurpassable than $\mathbf{u}$ that is greater than $\mathbf{c}$.

We call an unsurpassable $\mathbf{u}$ *steady* if $\mathbf{u}^* = \mathbf{u}$. Let us note that, under the assumption of GCH, the steady unsurpassables are exactly the successors of the regular limit cardinals. Thus the first is $\aleph_1$ and the next (if it exists) is the successor of the least inaccessible.

On steady unsurpassables, we have:

Lemma 2. (i) No successor $\mathbf{u}^+$ of a steady cardinal $\mathbf{u}$ is steady;
(ii) no limit cardinal $\mathbf{u}$ is steady;
(iii) for any steady unsurpassable $\mathbf{u}$, $\mathrm{cp}(\mathbf{u}) \leq \mathrm{exbd}(\mathbf{u}) < \mathbf{u}$.

Proof. (i) $\mathrm{cp}(\mathbf{u}) = \mathrm{cp}(\mathbf{u}^+)$; and hence $2^{\mathrm{cp}(\mathbf{u}+)} = 2^{\mathrm{cp}(\mathbf{u})} = \mathbf{u} < \mathbf{u}^+$.

(ii) Suppose $\mathbf{u}$ is steady limit cardinal. Then:

$$\begin{aligned}\mathbf{u}^{\mathrm{cf}(\mathbf{u})} &\leq \mathbf{u}^{\mathrm{cp}(\mathbf{u})}, \text{ since for limit cardinals } \mathrm{cf}(\mathbf{u}) \leq \mathrm{cp}(\mathbf{u}),\\ &= (2^{\mathrm{cp}(\mathbf{u})})^{\mathrm{cp}(\mathbf{u})}, \text{ since } \mathbf{u} \text{ is steady,}\\ &= 2^{\mathrm{cp}(\mathbf{u}).\mathrm{cp}(\mathbf{u})}, \text{ by cardinal arithmetic,}\\ &= 2^{\mathrm{cp}(\mathbf{u})}\\ &= \mathbf{u}.\end{aligned}$$

But $\mathbf{u}^{\mathrm{cf}(\mathbf{u})} > \mathbf{u}$ by cardinal arithmetic. A contradiction.

(iii) $\mathrm{cp}(\mathbf{u})$ is exp. small (relative to $\mathbf{u}$) – for $\mathbf{u}^{\mathrm{cp}(\mathbf{u})} = (2^{\mathrm{cp}(\mathbf{u})})^{\mathrm{cp}(\mathbf{u})} = 2^{\mathrm{cp}(\mathbf{u})} = \mathbf{u}$; and so $\mathrm{cp}(\mathbf{u}) \leq \mathrm{exbd}(\mathbf{u})$. By (ii), $\mathbf{u}$ is a successor cardinal; and so if $\mathrm{exbd}(\mathbf{u}) = \mathbf{u}$, $\mathbf{u}$ itself must be exp. small. But this contradicts Cantor's theorem.

Given a cardinal $\mathbf{c}$, we let $\mathbf{u}_c$ (or $\mathbf{u}/\mathbf{c}$) be the smallest unsurpassable greater than or equal to $\mathbf{c}$ (assuming that such a cardinal exists). Thus $\mathbf{u}_0$ is the smallest unsurpassable cardinal and hence, by lemma (ii), is steady.

If $\mathbf{u}_c$ exists, it may be constructed 'from below' in a natural manner. For each ordinal ξ, define the cardinals $\mathbf{u}_{c,\xi}$ by:

(i) $\mathbf{u}_{c,0} = \mathbf{c}$;
(ii) $\mathbf{u}_{c,\alpha+1} = \max(\mathbf{u}_{c,\alpha}\ 2^{cp(\mathbf{u}/c,\alpha)})$; and
(iii) $\mathbf{u}_{c,\lambda} = \lim_{\zeta<\lambda}\mathbf{u}_{\zeta}$.

It is clear that $\mathbf{u}_{c,\xi} \leq \mathbf{u}_c$ for each ξ. Moreover, $\mathbf{u}_{c,\alpha} \leq \mathbf{u}_{c,\beta}$ for any $\alpha \leq \beta$. On the assumption that $\mathbf{u}_c$ exists, there will therefore be an $\alpha \leq \mathbf{u}_c$ for which $\mathbf{u}_{c,\alpha} = \mathbf{u}_{c,\alpha+1} = \mathbf{u}_c$. As long as $\mathbf{c}$ is not an unsteady unsurpassable cardinal, we will always have $2^{cp(\mathbf{u}/c,\alpha)} \geq \mathbf{u}_{c,\alpha}$; and hence clause (ii) can be written in the simpler form:

(ii)′ $\mathbf{u}_{c,\alpha+1} = 2^{cp(\mathbf{u}/c,\alpha)}$.

We conclude with a result on exponential smallness. Say that a cardinal $\mathbf{p}$ is *exponential* if it is of the form $2^{\mathbf{q}}$ for some $\mathbf{q}$. If $\mathbf{k}$ is exp. small relative to $\mathbf{m}$, i.e. if $\mathbf{m}^{\mathbf{k}} \leq \mathbf{m}$, then $\mathbf{k}$ might not be exp. small relative to a greater cardinal $\mathbf{p}$. However, this result will hold in case $\mathbf{p}$ is exponential:

Lemma 3. If $\mathbf{m}^{\mathbf{k}} \leq \mathbf{m}$, then $\mathbf{p}^{\mathbf{k}} \leq \mathbf{p}$ for $\mathbf{p}$ an exponential cardinal $\geq \mathbf{m}$.

Proof. The result is evident in case $\mathbf{m}$ is finite or $\mathbf{m} = \mathbf{p}$. So let us assume that $\mathbf{m}$ is infinite and that $\mathbf{p} = 2^{\mathbf{q}} > \mathbf{m}$. Now $\mathbf{q} \geq \mathbf{k}$. For assume otherwise, i.e. $\mathbf{k} > \mathbf{q}$. Then $\mathbf{m} = \mathbf{m}^{\mathbf{k}} \geq 2^{\mathbf{k}} \geq 2^{\mathbf{q}} = \mathbf{p}$, contrary to assumption. But then $\mathbf{p}^{\mathbf{k}} = (2^{\mathbf{q}})^{\mathbf{k}} = 2^{\mathbf{q}.\mathbf{k}} = 2^{\mathbf{q}.\mathbf{k}} = 2^{\mathbf{q}} = \mathbf{p}$.

We now embark on the construction of the domain of the intended model. This proceeds by adding an abstract to a given domain if it can be produced by applying an acceptable means of abstraction to a concept that is defined over the given domain. In our account of acceptability, we need to modify the condition of being non-inflationary slightly in order to allow for the possibility of there being two objects even when they are not already in the domain. Accordingly, let us say that an equivalence $\approx$ on $\wp(M)$ is *non-inflating* if $\mathrm{card}(P_{\approx}) \leq \max\{2, \mathrm{card}(M)\}$. Thus it is only in case $\mathrm{card}(M) < 2$ that there can be non-inflating equivalences that are not also non-inflationary. We now say that the equivalence relation $\approx$ on $\wp(M)$ is *K-acceptable* if $\approx$ is non-inflating and internally K-invariant for K an exp. small subset of M, that it is (*broadly*) *acceptable* if it is K-acceptable for some K, and that it is *strictly acceptable* if it is $\varnothing$-acceptable. The strictly acceptable equivalences correspond to those acceptable means of abstraction that can be given a purely logical definition. (For $\mathrm{card}(M) \leq 1$, the definitions differ slightly from those previously given; and since we are only interested in internal invariance, we have dropped the I-suffix.)

We shall identify abstracts with cells. However, it will be important to indicate the domain from which the cell originated. Accordingly, we take an *indexed cell* to be an ordered pair $\langle X,M\rangle$ for X a non-empty subset of $\wp(M)$. An indexed cell $\langle X,M\rangle$ can be legitimately introduced on the basis of the domain M in the following ways: it is *K-admissible* if X is induced by some K-acceptable equivalence $\approx$ on $\wp(M)$; it is *broadly admissible* (or *admissible*, simpliciter) if it is K-admissible for some subset K of M; and it is (*strictly*) *admissible* if it is $\varnothing$-admissible. It should be noted that, by theorem 8.8 of Part II, the admissible cells $\langle X,M\rangle$ may be characterized, without reference to an underlying equivalence, as those for which X is predominantly I-invariant.

There is, however, a difficulty in representing an abstract x by means of an indexed cell $\langle X,M\rangle$. For under an expansion M^+ of the domain, the abstract x might be associated with further concepts from $\wp(M^+)$. We therefore need a method for determining which concepts will be associated with the cell on the basis of the concepts which already are associated with it.

One obvious solution to this problem is to represent an abstract, not by a single cell, but by a sequence of cells, one for each of the expansions of the original domain. However, the details of this approach are extremely messy; and we prefer instead to fix upon a canonical way in which a cell can grow from one domain to another. The cell is taken to contain the 'seed' from which its subsequent growth is determined. We describe one natural way in which this can be done—though there are other possibilities, all giving rise to the same end-result, that might also be considered.

Suppose that M^+ is a superset of M. Then given an equivalence $\approx$ on $\wp(M)$, we call $\approx'$ an *expansion of* $\approx$ *from M to* M^+ if it is an equivalence on $\wp(M^+)$ that coincides with $\approx$ on $\wp(M)$; and given an indexed cell $\langle X, M\rangle$, we say that the indexed cell $\langle X^+,M^+\rangle$ is an *expansion of* $\langle X, M\rangle$ if $X^+ \supseteq X$ and both X^+ and X agree on subsets of M. The cell $\langle X^+,M^+\rangle$ is said to be a *K-legitimate* expansion of $\langle X,M\rangle$ if X is induced by some K-acceptable equivalence $\approx$ on $\wp(M)$ and X^+ is induced by some K-acceptable expansion $\approx'$ of $\approx$ on $\wp(M^+)$. The expansion $\langle X,M\rangle$ is *broadly legitimate* (or *legitimate*, simpliciter) if it is K-legitimate for some K; and it is *strictly legitimate* if it is $\varnothing$-legitimate.

The legitimate expansions of a cell correspond to the different possible ways the abstract represented by the cell could behave in an extended domain. Given a set C (corresponding to a given concept), we must now somehow determine whether C should be associated with the cell $\langle X,M\rangle$—or, as we shall put it, whether the cell *admits* the set—on the basis of the legitimate expansions $\langle X^+,M^+\rangle$ of the cell. We make two decisions in this regard: we look only at the expansions of the form $\langle X^+, M \cup C\rangle$; and we require that the set C should belong to every such expansion. Thus we say that $\langle X,M\rangle$ *K-admits* the set C if (a) there is a K-legitimate expansion of $\langle X,M\rangle$ of the form $\langle Y, M \cup C\rangle$ and (b) $C \in Y$ for any K-legitimate expansion of the form $\langle Y, M \cup C\rangle$; and similarly for the definitions of *broad* and *strict admittance*.

We are able to give an intrinsic characterization of admittance. But first we need some results on the expansions of equivalences.

Lemma 4. Suppose that $\approx$ is a K-acceptable equivalence on $\wp(M)$ for K an exp. small subset of M; and let N be a superset of M. There is then a K-legitimate expansion $\approx'$ of $\approx$ to N as long as M is finite or N is of the same cardinality as M or N is of exponential cardinality.

Proof. We distinguish between the finite and infinite case:

(i) M finite. Then K is empty and $\approx$ is strictly invariant. So it follows from theorem 7.14 of Part II that $\approx$ is a numeric equivalence (with associated equivalence $\approx_{\text{num}}$ on the cardinals). Let us define $\approx'$ on subsets of N by:

$$C \approx' D \text{ iff } \text{card}(C) \approx_{\text{num}} \text{card}(D).$$

Then it is readily verified that $\approx'$ is non-inflationary, internally invariant, and an expansion of $\approx$.

(ii) M infinite. Set $\mathbf{m} = \text{card}(M)$, and define $\approx'$ on subsets C and D of N by:

$$C \approx' D \text{ iff either } \text{Icdstr}_K(C, D) \,\varepsilon\, \text{Icdstr}_K(\approx) \text{ or}$$
$$\text{Icdstr}_K(C,C),\ \text{Icdstr}_K(D,D) \notin \text{Icdstr}_K(\approx).$$

Note that the condition $\text{Icdstr}_K(C,C) \in \text{Icdstr}_K(\approx)$ is satisfied just in case $\text{card}(C) \leq \mathbf{m}$.

It is readily verified that $\approx'$ is internally K-invariant. We show that $\approx'$ is an equivalence. It is clearly symmetric. As for reflexivity, first consider the case in which $\text{card}(C) > \mathbf{m}$, for C a subset of N. It is then evident that $C \approx' C$. Now consider the case in which $\text{card}(C) \leq \mathbf{m}$. We may then find a one–one function f, fixed over K, from C into M. Since $f[C] \approx f[C]$, it follows that $C \approx' C$. For transitivity, suppose that $C \approx' D$ and $D \approx' E$. If any of C, D, or E have cardinality $> \mathbf{m}$, then they all do; and so it is clear that $C \approx' E$. So suppose that they all have cardinality $\leq \mathbf{m}$. By thinning out members of M from C, D, and E, we may find sets C', D', and E' for which $M - (C \cup D \cup E)$ is of cardinality $\mathbf{m}$ and yet $\text{Icdstr}_K(C',D') = \text{Icdstr}_K(C,D)$, $\text{Icdstr}_K(D',E') = \text{Icdstr}_K(D,E)$, and $\text{Icdstr}_K(C',E') = \text{Icdstr}_K(C,E)$. By the K-invariance of $\approx'$, $C' \approx' D'$ and $D' \approx' E'$. Let $B = (C' \cup D' \cup E') - M$. Choose a subset B^* of $M - (C' \cup D' \cup E' \cup K)$ of the same cardinality as B; and let $C^* = (C' - B) \cup B^*$, $D^* = (D' - B) \cup B^*$, and $E^* = (E' - B) \cup B^*$. It is readily verified that $\text{Icdstr}_K(C',D') = \text{Icdstr}_K(C^*,D^*)$, $\text{Icdstr}_K(D',E') = \text{Icdstr}_K(D^*,E^*)$, and $\text{Icdstr}_K(C',E') = \text{Icdstr}_K(C^*,E^*)$. But then $C^* \approx D^*$, $D^* \approx E^*$. Hence $C^* \approx E^*$ by the transitivity of $\approx$; so $C' \approx' E'$ by the definition of $\approx'$; and so $C \approx' E$ by the K-invariance of $\approx'$.

To show that $\approx'$ is non-inflationary it suffices, by corollary 7.16 of Part II, to show that it contains the equivalence $\approx_{1,K}$ as defined on N. So suppose $C \approx_{1,K} D$, where C and D are exp. large in N and of the same cardinality

p. There are two cases: (i) **p** > **m**. Then $C \approx' D$ by the definition of $\approx'$; (ii) **p** ≤ **m**. By lemma 3 above, **p** is exp. large relative to card(M); and so $\approx$ will relate any two subsets of M of cardinality **p**. But then we may readily find subsets C' and D' of M of cardinality **p** for which $\text{Icdstr}_K(C',D') = \text{Icdstr}_K(C,D)$; and so $C' \approx' D'$.

It should be noted that the expansion $\approx'$ is defined differently for the finite and for the infinite case. Neither definition will work for the other case. Suppose, for example, that we apply the definition for the infinite case to the finite case. Take a domain M with three elements x, y, z and an acceptable equivalence $\approx$ that relates any two membered subsets of M. Then $\approx'$ will not even be an equivalence on any proper superset N of M. For take an element w in N but not in M. We will then have $\{x,y\} \approx' \{yz\}$ and $\{y,z\} \approx' \{z,w\}$, but not $\{x,y\} \approx' \{z,w\}$.

Let us use $\approx_+$ and $\approx_{K'}$ respectively, for the expansion $\approx'$ as defined in the finite and in the infinite case, but not restricted to any particular domain. Thus for *any* sets C and D:

$$C \approx_+ D \text{ iff } \text{card}(C) \approx_{\text{num}} \text{card}(D); \text{ and}$$

$$C \approx_K D \text{ iff either } \text{Icdstr}_K(C,D) \in \text{Icdstr}_K(\approx) \text{ or}$$
$$\text{Icdstr}_K(C,C),\ \text{Icdstr}_K(D,D) \notin \text{Icdstr}_K(\approx).$$

We now characterize admittance, dealing first with the finite and then with the infinite case:

Lemma 5. For M finite, $\approx$ an acceptable equivalence on $\wp(M)$, X a cell induced by $\approx$, and $B \in X$, the following conditions are equivalent:

(i) $\langle X,M\rangle$ admits C:
(ii) $C \approx_+ B$;
(iii) card(C) = card(D) for some $D \in X$.

Proof. (i) implies (ii). Suppose $\langle X,M\rangle$ admits C. Then some acceptable equivalence on $\wp(M)$ induces the cell X. By the proof of lemma 4, the restriction of $\approx_+$ to $N = M \cup C$ is an acceptable equivalence on $\wp(N)$. But then, by the definition of admittance, $C \approx_+ B$.

(ii) implies (iii). Suppose $C \approx_+ B$. Then for some subsets D and E of M, $D \approx E$ with card(C) = card(D) and card(B) = card(E). By theorem 7.14 of Part II, the equivalence $\approx$ is numeric; and so $E \approx B$. Given $D \approx E$ and $D \approx B$, $D \approx B$; and so $D \in X$.

(iii) implies (i). Suppose card(C) = card(D) for some D in X. Then card(C) is finite and hence so is $N = M \cup C$. Since $\langle X,M\rangle$ is acceptable, there is an acceptable equivalence $\approx$ on $\wp(M)$ that induces X. By lemma 4, there is an acceptable expansion of $\approx$ to $\wp(N)$. Take now any acceptable expansion $\approx'$ of $\approx$ to $\wp(N)$. By theorem 7.14 of Part II, $\approx'$ is numeric; and so $D \approx' C$.

Lemma 6. For M infinite, $\approx$ a K-acceptable equivalence on $\wp(M)$, X a cell induced by $\approx$, and B **ε** X, the following conditions are equivalent:

(i) $\langle X,M \rangle$ K-admits C;
(ii) $C \approx_K B$ (i.e. $\text{Icdstr}_K\,(C, B)$ **ε** $\text{Icdstr}_K(\approx)$);
(iii) $\text{Icdstr}_K\,(C, B)$ **ε** $\text{Icdstr}_K\,(X)$.

Proof. (i) → (ii). Suppose $\langle X,M \rangle$ K-admits C. Then some K-acceptable equivalence $\approx'$ on $\wp(N)$, for $N = M \cup C$, induces an expansion $\langle Y,N \rangle$ of $\langle X,M \rangle$ that contains C. Now $\text{card}(C) \leq \text{card}(M)$. For suppose $\text{card}(C) = \mathbf{n} > \text{card}(C)$, and define a new equivalence on $\wp(N)$ by:

$$D \approx'' E \text{ iff either } \text{card}(D) = \text{card}(E) = \mathbf{n} \text{ or } \text{card}(D), \text{card}(E) \neq \mathbf{n}$$

and $D \approx' E$.

Then it is readily verified that $\approx'$ is K-acceptable and that it induces an expansion $\langle Y,N \rangle$ of $\langle X,M \rangle$ that does not contain C.

Now since $\text{card}(N) = \text{card}(M)$, it follows by the proof of lemma 4 and the definition of K-admittance that the restriction $\approx^*$ of $\approx_K$ to $\wp(N)$ is K-acceptable. But then $C \approx^* B$; and hence $C \approx_K B$.

(ii) → (iii). Suppose $C \approx_K B$, i.e. $\text{Icdstr}_K\,(C, B) = \tau = \text{Icdstr}_K(E,F)$ for some E and F for which $E \approx F$. If B is exp. large (in M) then, by corollary 7.16 of Part II, we may 'thin' B and thereby obtain a B' for which $B \approx B'$, $\text{Icdstr}_K(C,B') = \tau$ and $M - (B' \cup K)$ is large. Clearly it then suffices to prove the result by setting $B = B'$. If B is small, then $M - (B \cup K)$ is already large. So in either case, we may assume that $M - (B \cup K)$ is large.

Given that $\text{Icdstr}_K(C,B) = \tau$, we can now find a subset C' of M for which $\text{Icdstr}_K(C',B) = \text{Icdstr}_K(C,B) = \tau$. So $C' \approx B$; and hence $\text{Icdstr}_K(C,B)$ **ε** $\text{Icdstr}_K(X)$.

(iii) → (ii). Trivial.

(ii) → (i). Suppose that $\text{Icdstr}_K(C,B) \in \text{Icdstr}_K(\approx)$, i.e. $\text{Icdstr}_K(C,B) = \text{Icdstr}_K(E, F)$ for $E \approx F$. Given that $\langle X,M \rangle$ is K-admissible, there is a K-acceptable equivalence $\approx$ on $\wp(M)$ that induces X. Given that $\text{Icdstr}_K(C,B) \in \text{Icdstr}_K(\approx)$, $\text{card}(C) \leq \text{card}(M)$. So $\text{card}(N) = \text{card}(M)$ (for $N = M \cup C$). But then, by lemma 4, the restriction of $\approx_K$ to N is a K-legitimate expansion of $\approx$. Take now any K-legitimate expansion $\approx'$ of $\approx$ on $\wp(N)$. Since $E \approx F$, $E \approx' F$; and since $\text{Icdstr}_K(C,B) = \text{Icdstr}_K(E,F)$, $C \approx' B$. But then C **ε** $|B|_{\approx'} \supseteq X$.

What $\langle X,M \rangle$ broadly or strictly admits can depend upon M. For suppose that $X = \{\{x\}\}$, M_1 is any finite set containing the doubleton $\{x,y\}$, and M_2 is any infinite set containing $\{x,y\}$. Then $\langle X, M_1 \rangle$ admits any singleton subset of M_1 while $\langle X,M_2 \rangle$ admits only the set $\{x\}$ itself. However, note that condition (ii) (or condition (iii)) in both lemmas 5 and lemma 6 makes no reference to M. This means that it is only the status of M as finite or infinite that can make a difference to what $\langle X,M \rangle$ admits. We call $\langle X,M \rangle$ *finitary* if M if finite and *infinitary* otherwise.

Given the two lemmas, we may prove:

Lemma 7. Suppose that the cell $\langle X,M \rangle$ is admissible. Then:

(i) If it admits C, it L-admits C for any L for which $\langle X,M \rangle$ is L-admissible;

(ii) It admits a subset C of M iff $C \in X$.

Proof. (i) There are two cases:

(a) M finite. The result is then trivial since only $\varnothing$ is exp. small.

(b) M infinite. Suppose that $\langle X,M \rangle$ admits C, i.e. K-admits C for K exp. small. So $\langle X,M \rangle$ is K-admissible and, by lemma 6 (iii), $\tau = \text{Icdstr}_K(C,B) \,\varepsilon\, \text{Icdstr}_K(X)$ for some B in X. We may suppose that $M - B$ is large. For assume otherwise. Then B is large; and so, by corollary 7.16 of Part II, X must contain any set B' that is equinumerous with B and agrees with B on K. But then, given that K is small, we may, by appropriate 'thinning' of B, find a B' in X for which $M - B'$ is large and $\text{Icdstr}_K(C,B) = \text{Icdstr}_K(C, B')$.

Let $D = C - M$. Note that in order for $\tau \in \text{Icdstr}_K(X)$ we must have $\text{card}(D) \leq \text{card}(M)$. We distinguish two subcases. (1) $\text{card}(C - (B \cup D)) \geq \text{card}(D)$. We then let $C' = C - D$. (2) $\text{card}(C - (B \cup D)) < \text{card}(D)$. We then pick a subset of D' of $M - (B \cup K \cup L)$ of the same size as D and let $C' = (C - D) \cup D'$. It is then readily ascertained, in either case, that $\text{Icdstr}_{K \cup L}(C',B) = \text{Icdstr}_{K \cup L}(C,B)$. Since $\text{Icdstr}_K(C',B) - \tau \,\varepsilon\, X$, $C' \in X$; and since $\text{Icdstr}_L(C',B) = \text{Icdstr}_L(C',B)$, X L-admits C by lemma 6.

(ii) Again there are two cases. (a) M finite. Given that $\langle X,M \rangle$ admits a subset C of M, $\text{card}(C) = \text{card}(B)$ for some $B \in X$ by lemma 5. So since any acceptable equivalence on $\wp(M)$ must be numeric, $C \in X$. Now suppose that $C \,\varepsilon\, X$. Then trivially C is of the same cardinality as some set in X; and so again by lemma 5, $\langle X,M \rangle$ admits C:

(b) M infinite. Assume that $\langle X,M \rangle$ is admissible. Then, for some K exp. small in M and some K-acceptable equivalence $\approx$ on $\wp(M)$, $X \in P_{\approx}$. Suppose now that $\langle X,M \rangle$ admits the subset C of M. Then, for some cells D, E, F in X, $\text{Icdstr}_K(C,D) = \text{Icdstr}_K(E,F)$. So $D \approx E \approx F$; and, by $\approx$ internally K-invariant, $C \approx D$. But then $C \in X$. Next suppose that $C \in X$. Then $\text{Icdstr}_K(C,C) \in \text{Icdstr}_K(X)$; and hence X admits C in M.

Two indexed cells $\langle X,M \rangle$ and $\langle Y,N \rangle$ are said to be *indistinguishable* if they admit the same subsets of $M \cup N$ and to be *absolutely indistinguishable* if they admit all the same sets. Similar notions can be defined with strict in place of broad admittance. We establish two sufficient conditions for absolute indistinguishability.

Lemma 8. Suppose that the cells $\langle X,M\rangle$ and $\langle Y,N\rangle$ are admissible, $M \subseteq N$, and either M is finite or N is of the same cardinality as M or N is of exponential cardinality. Then the two cells are indistinguishable iff they are absolutely indistinguishable.

Proof. The direction from right to left is trivial. Suppose now that $\langle X,M\rangle$ and $\langle Y,N\rangle$ are indistinguishable. The case in which M is finite is straightforward. So let us suppose that M is infinite. Given that $\langle X,M\rangle$ is admissible, X is induced by some K-acceptable equivalence $\approx$ on $\wp(M)$, for K an exp. small subset of M. Take any subset C of N. Then:

$C \in Y$ iff Y admits C, by lemma 7 (ii)
iff X admits C, by supposition
iff $C \approx_K B$ for some $B \in X$.

By the proof of lemma 4 and the conditions on N, it follows that the restriction $\approx'$ of $\approx_K$ to $\wp(N)$ is an acceptable equivalence that induces Y. Take now any C whatever. Then, by lemma 6, $\langle Y,N\rangle$ admits C iff $C \approx_K B$; and, by another application of lemma 6, $C \approx_K B$ iff $\langle X,M\rangle$ admits C.

Lemma 9. Suppose that $\langle X,M\rangle$ and $\langle Y,N\rangle$ are infinitary K-legitimate cells. Then they are absolutely indistinguishable if X and Y have a member in common and $\text{Icdstr}_K(X) = \text{Icdstr}_K(Y)$.

Proof. Let B be the common member of X and Y. Suppose that $\langle X,M\rangle$ admits C. Then by lemma 7 (i), it K-admits C. So by lemma 6, $\text{Icdstr}_K(C,B) \in \text{Icdstr}_K(X)$. So $\text{Icdstr}_K(C,B) \in \text{Icdstr}_K(Y)$; and hence, by lemma 6, again, $\langle Y,L\rangle$ K-admits C.

The construction of the domain U can now be described. Let I be a set of urelements (which we regard as the individuals, i.e. the non-abstracts). We make no assumption regarding I beyond the fact that card(I) is not an unsteady unsurpassable. The construction then proceeds relative to a fixed choice of I. For each ordinal ξ, we define $U_{I,\xi}$ by:

(i) $U_{I,0} = I$;
(ii) $U_{I,\alpha+1} = U_{I,\alpha} \cup \{\langle X,U_{I,\alpha}\rangle : \langle X,U_{I,\alpha}\rangle$ is an admissible cell distinguishable from every cell in $U_{I,\alpha}\}$;
(iii) $U_{I,\lambda} = \bigcup_{\zeta<\lambda} U_{I,\zeta}$.

For the purposes of the strict construction, the U's should be changed to V's and clause (ii) should be replaced with:

(ii) $V_{I,\alpha+1} = V_{I,\alpha} \cup \{\langle X,V_{I,\alpha}\rangle : \langle X,V_{I,\alpha}\rangle$ is a strictly admissible cell and is strictly distinguishable from every cell in $V_{I,\alpha}\}$.

We say that the cell $\langle X,M\rangle$ is *introduced at* the ordinal α if $\langle X,M\rangle$ belongs to $U_{I,\alpha}$ but to no $U_{I,\beta}$ for $\beta < \alpha$ (and similarly for the strict construction).

We need to know that the construction will stabilize, i.e. that $U_{I,\alpha} = U_{I,\alpha+1}$ for some ordinal α. To this end, we compute the cardinalities of the domains U_ξ.

Lemma 10. (i) $\text{card}(U_{I,\alpha}) \leq \text{card}(U_{I,\beta})$ for $\alpha \leq \beta$;
(ii) $\text{card}(U_{I,\alpha+1}) \geq 2^{\text{cp}(\text{card}(U/I,\alpha))}$ for $U_{I,\alpha}$ infinite;
(iii) if $U_{I,\alpha}$ is finite, then $\text{card}(U_{I,\alpha+1})$ is finite and $> \text{card}(U_{I,\alpha})$;
(iv) if $U_{I,\alpha}$ is infinite, then $\text{card}(U_{I,\alpha+1}) \leq \max(\text{card}(U_{I,\alpha}), 2^{\text{cp}(\text{card}(U/\alpha))})$;
(v) $\text{card}(U_{I,\lambda}) = \lim_{\zeta<\lambda}\text{card}(U_{I,\zeta})$.

Proof. (i) Evident from the definition.

(ii) Set $\mathbf{u} = \text{card}(U_{I,\alpha})$; and let Γ be a non-empty set of cardinals $\leq \text{cp}(\mathbf{u})$. Define $\equiv_\Gamma$ on $U_{I,\alpha}$ by: $C \equiv_\Gamma D$ iff card(C) and card(D) both belong to Γ or neither do. Then the equivalence $\equiv_\Gamma$ is clearly internally invariant; and since the associated partition contains at most two equivalence classes, it is non-inflating on $U_{I,\alpha}$. So if X_Γ is the equivalence class of the form $|C|$ for card(C) $\varepsilon\ \Gamma$, $\langle X_\Gamma, U_{I,\alpha}\rangle$ is admissible. For distinct Γ, the X_Γ's are distinct; and so, given that $\mathbf{u}$ is transfinite, there are $2^{\text{cp}(\mathbf{u})}$ such cells.

(iii) Let $n = \text{card}(U_{I,\alpha+1})$ and $m = \text{card}(U_{I,\alpha})$. Then $n \leq 2^p$, for $p = 2^m$, and hence is finite given that m is finite. We see that $n > m$ from the construction of $\equiv_\Gamma$ above. For this shows that $n \geq (2^{m+1} - 1) \geq m$.

(iv) Fix on a particular exp. small subset K of $U_{I,\alpha}$ of cardinality $\mathbf{k} < \mathbf{u} = \text{card}(U_{I,\alpha})$. Then, by theorem 7.5 of Part II, the number of internally K-invariant cells over $U_{I,\alpha+1}$ is at most $\max(\mathbf{u}, 2^{\text{cp}(\mathbf{u})}, 2^{\mathbf{k}}) = \max\{\mathbf{u}, 2^{\text{cp}(\mathbf{u})})$, given that $\mathbf{k}$ is exp. small. But there are at most $\sup\{\mathbf{u}^{\mathbf{k}}: \mathbf{k}$ exp. small$\} \leq \mathbf{u}$ such K's; and hence $\text{card}(U_{I,\alpha+1}) \leq \max(\mathbf{u}, 2^{\text{cp}(\mathbf{u})})$.

(v) Evident from the definition.

Let us note that the proofs of the lower bounds (viz. (ii) and (iii)) appeal only to the strictly invariant equivalences $\equiv_\Gamma$. Thus the results will also hold for the strict construction $V_{I,\alpha}$ in place of $U_{I,\alpha}$. If $\mathbf{c} = \text{card}(I)$ is infinite, it follows from (i), (ii), (iv), and (v), that $\text{card}(U_{I,\alpha}) = \mathbf{u}_{c,\alpha}$. If $\mathbf{c}$ is finite, it will not be generally true that $\text{card}(U_{I,\alpha}) = \mathbf{u}_{c,a}$. However, the truth of the identity will be restored once we pass to infinite α.

Assuming the existence of the unsurpassable cardinal $\mathbf{u}_c$, we can show that the constructions will stabilize.

Lemma 11. The constructions $U_{I,\alpha}$ and $V_{I,\alpha}$ stabilize on the ordinal $\mathbf{u}_c + \mathbf{u}_c$ (given that $\mathbf{c} = \text{card}(I)$ is not an unsteady unsurpassable).

Proof. We deal with the case of $U_{I,\alpha}$, the other case being similar. It follows from lemma 10 that $\text{card}(U_{I,u/c}) = \mathbf{u}_c$. We need to show that no new cells are introduced at the ordinal stage $\mathbf{u}_c + \mathbf{u}_c$. To simplify notation, set $\mathbf{u} = \mathbf{u}_c$ and $W_\alpha = U_{I,u+\alpha}$, and note that $\text{card}(W_u) = \mathbf{u} = \text{card}(W_0)$.

We first establish:

(1) Any exp. small subset K of $W_{\mathbf{u}}$ is included in W_α for some $\alpha < \mathbf{u}$.

Proof. Suppose the condition is not true for some K. Then for each $\alpha < \mathbf{u}$, there is a β, with $\alpha < \beta < \mathbf{u}$, and a member of K that is introduced at β (in the sequence $\langle W_\xi: \xi < \mathbf{u}\rangle$). So card($K$) $\geq$ cf($\mathbf{u}$). But by lemma 2(ii), $\mathbf{u}$ is a successor cardinal. So cf($\mathbf{u}$) = $\mathbf{u}$; and therefore K is not small after all.

From (1) we establish (2) below, from which the desired result follows:

(2) Any admissible cell $\langle X,W_{\mathbf{u}}\rangle$ is indistinguishable from a cell in W_α for $\alpha < \mathbf{u}$.

Proof. It suffices to show that $\langle X,W_{\mathrm{u}}\rangle$ is indistinguishable (and hence, by lemma 8, is absolutely indistinguishable) from an admissible cell of the form $\langle Y,W_\alpha\rangle$ for $\alpha < \mathbf{u}$. For then either $\langle Y,W_\alpha\rangle$ is a member of $W_{\alpha+1}$ or it is indistinguishable (and hence absolute indistinguishable) from a member of W_α; and so, in either case, $\langle X,W_{\mathrm{u}}\rangle$ is absolutely indistinguishable from a cell in W_α.

Suppose that $\langle X,W_{\mathbf{u}}\rangle$ is K-admissible for K an exp. small subset of $W_{\mathbf{u}}$. By (1), $K \subseteq W_\alpha$ for some $\alpha < \mathbf{u}$. We distinguish two cases:

(a) X contains no exp. small set. For subsets C and D of $W_{\mathbf{u}}$, define C eq$_\mathrm{K}$ D by: $C \cap K = D \cap K$ and card($C - K$) = card($D - K$). Note that, for exp. large subsets C and D of W_{u}, $C \approx_{1,\mathrm{K}} D$ iff C eq$_\mathrm{K}$ D. So by corollary 7.16 of Part II, the cell X is closed under the relation eq$_\mathrm{K}$. Let $Y = X \cap \wp(W_\alpha)$. Then it is readily verified that $\langle Y,W_\alpha\rangle$ is admissible, that Icdstr$_K(Y)$ = Icdstr$_K(X)$, and that X and Y have a member in common. But then, from lemma 8 above, it follows that the indexed cells are indistinguishable.

(b) X contains an exp. small set C. By (1), $C \subseteq W_\beta$ for some β for which $\alpha \leq \beta < \mathbf{u}$. But now we may find an admissible cell $\langle Y,W_\beta\rangle$ that is indistinguishable from $\langle X,W_{\mathrm{u}}\rangle$. For let f be a one–one function from W_{u} onto W_β that is fixed over $C \cup K$; and let $Y = f[X]$. Then it is readily verified that $\langle Y,W_\beta\rangle$ is K-admissible, that Icdstr$_\mathrm{K}(Y)$ = Icdstr$_\mathrm{K}(X)$, and that $C \in X, Y$. So by lemma 8 again, $\langle X,W_{\mathrm{u}}\rangle$ and $\langle Y,W_\beta\rangle$ are indistinguishable.

We are now in a position to define the intended model $\boldsymbol{U}_I$. Its domain U_I is given by $U_{I,\alpha}$ for α a stabilizing ordinal. The interpretation *Abstr of* the predicate Abstr, for $x \in U_I$, $C \subseteq U_I$, and $\mathbf{R} \subseteq \wp(U_I) \times \wp(U_I)$, is given by:

> x *Abstr*$_R$ C iff $\boldsymbol{R}$ is an acceptable equivalence on $\wp(U_I)$ and $\{D: x$ admits $D\} = \{D: \boldsymbol{R}$ relates C to $D\}$.

To obtain the strict model, we define V_I in analogy to U_I and define *Abstr* by:

> x *Abstr*$_R$ C iff $\boldsymbol{R}$ is a strictly acceptable equivalence on $\wp(V_I)$ and $\{D: x$ strictly admits $D\} = \{D: \boldsymbol{R}$ relates C to $D\}$.

It should be noted that there is a sense in which the construction of $\boldsymbol{U}_I$ serves not only to generate the abstracts from stage to stage but also the means of abstraction. Of course, an equivalence over the whole domain $\boldsymbol{U}_I$ cannot be constructed at a stage at which we have not yet constructed the whole domain. However, we may at each stage take ourselves to be constructing a *representative* of such an equivalence. For an equivalence $\approx$ over M can be taken to represent the restriction of $\approx_+$ to $\wp(\boldsymbol{U}_I)$ when M is finite and to represent the restriction of $\approx_K$ to $\wp(\boldsymbol{U}_I)$ when M is infinite and $\approx$ is K-acceptable. Each abstract can then be taken to be generated from a previously given concept by means of a representative of a means of abstraction that may also be taken to be previously given.

We may at last show that the model $\boldsymbol{U}_I$ verifies the axioms of GA (the proof that $\boldsymbol{V}_I$ is a model for SGA is similar).

Theorem 12. Suppose that card(I) is not an unsteady unsurpassable. Then $\boldsymbol{U}_I$ is a model of GA$^+$ in which each cell is an abstract and each abstract a cell.

Proof. It is clear that each abstract is a cell. To show the converse, let $\langle X,U_{I,\beta}\rangle$ be a cell in $\boldsymbol{U}_I$. We deal solely with the case in which $U_{I,\beta}$ is infinite, the other case, in which it is finite, being similar. Suppose that $\langle X,U_{I,\beta}\rangle$ ε $U = U_I$. Then $\langle X,U_{I,\beta}\rangle$ is admissible. So, for some exp. small subset K of $U_{I,\beta}$, there is a K-acceptable equivalence $\approx$ on $\wp(U_{I,\beta})$ that induces X. By lemmas 1, 2(i) and 10, U is of exponential cardinality; and so, by lemma 3, K is an exp. small subset of U and the restriction $\approx'$ of $\approx_K$ to $\wp(U)$ is also an acceptable equivalence. Choose a B in X. By lemma 6, $\langle X,U_{I,\beta}\rangle$ admits a subset C of U iff $C \approx' B$; and so $\langle X,U_{I,\beta}\rangle$ stands in the relation $Abstr_{\approx'}$ to B.

Let us now verify the axioms in turn:

Identity. It suffices to establish:

(*) any two distinct cells $\langle X,U_{I,\alpha}\rangle$ and $\langle Y,U_{I,\beta}\rangle$ of U_I, $\alpha \leq \beta$, are distinguishable.

Pf. Suppose that $\alpha = \beta$. Then there is a member C of one of the cells, say X, that is not a member of the other. By lemma 7 (ii), $\langle X, U_{I,\alpha}\rangle$ admits C while $\langle Y,U_{I,\alpha}\rangle$ does not; and hence the two indexed cells are distinguishable. On the other hand, if $\alpha < \beta$, then it follows from the construction that $\langle Y,U_{I,\beta}\rangle$ is distinguishable $\langle Y,U_{I,\alpha}\rangle$.

Broad Internal Existence. Let $\approx$ be an acceptable equivalence on $\wp(U_I)$. We need to show that, for every set $C \subseteq U_I$ (which represents a concept), there is a cell $\langle X, U_{I,\alpha}\rangle$ in U_I (which represents an abstract) for which $\{D \subseteq U_I: \langle X,U_{I,\alpha}\rangle$ admits $D\} = |C|_{\approx}$. Choose an α for which $U_{I,\alpha} = U_I$. By lemma 7(ii), $\{D \subseteq U_I: \langle |C|_{\approx}, U_I\rangle$ admits $D\} = |C|_{\approx}$. By the construction, $\langle |C|_{\approx}, U_I\rangle$ is indistinguishable from a cell $\langle X,U_\beta\rangle$ in U_I; so $\{D \subseteq U_I: \langle |C|_{\approx}, U_I\rangle$ admits

$D\} = \{D \subseteq U_I$: $\langle X, U_{I,\beta}\rangle$ admits $D\}$; and so $\{D \subseteq U_I$: $\langle X, U_{I,\beta}\rangle$ admits $D\} = |C|_{\approx}$.

Exclusivity and Application. Straightforward.

Strong Minimality. Suppose that there existed a proper subset D of U_I of the sort ruled out by Strong Minimality. Let us first show that D has the same cardinality as U_I. To this end, let Γ be a non-empty set of cardinals $\leq$ cp(D). Define $\approx\, = \,\equiv_\Gamma$ as in the proof of lemma 10(ii), though relative now to the domain D; and let $\approx'$ be the restriction of $\approx_+$ to $\wp(U_I)$. Then clearly, $\approx'$ is strictly acceptable (and, given that card(D) ≥ 2, non-inflationary and not just non-inflating). Let C_0 be any subset of D whose cardinality is in Γ. Then some object x in U_I is an abstract of C_0 with respect to $\approx'$; and so there is an abstract x_Γ in D for which $\{C \subseteq D$: x_Γ is an abstract of $C\} = \{C \subseteq D$: card(C) $\in \Gamma\}$. But the x_Γ are different for different Γ. This shows that the card(D) $\geq 2^{\text{cp(card(D))}} - 1$ if D is finite and that card(D) $\geq 2^{\text{cp(card(D))}}$ is D is infinite; and from this it follows that card(D) = card(U_I).

Suppose now that $D \neq U_I$. Let β be the least ordinal for which $D \subseteq U_{I,\beta}$ fails; and let x be an indexed cell in $U_{I,\beta}$ but not in D. So β is of the form $\alpha + 1$ and x is of the form $\langle X, U_{I,\alpha}\rangle$ for $X \subseteq \wp(U_{I,\alpha})$. There is therefore an abstract x' in D that admits the same subsets of $D \subseteq U_{I,\alpha}$ as x. But then x' is indistinguishable from x; and hence $x = x'$, contrary to supposition.

Corollary 13. Let **c** be any cardinal that is not an unsteady unsurpassable. Then there is a standard model **M** of GTA containing exactly **c** individuals.

Proof. Choose I to be a set of urelements having cardinality **c**.

We state without proof:

Theorem 14. (Categoricity). Suppose that ***M*** and ***N*** are two standard models of GA$^+$ and that card(I_M) = card(I_N) is not an unsteady unsurpassable. Then ***M*** and ***N*** are isomorphic.

3. DERIVATIONS

We briefly indicate how to derive the central portions of arithmetic and analysis within the theory of abstracts and within the corresponding theory of extensions, thereby providing those branches of mathematics with a neo-Fregean foundation.

In order to carry out the derivation within the theory of abstracts, we need the premiss that there are at least two objects (i.e. $\exists x,y(x \neq y)$). For Frege, this premiss follows from the fact that sentences are the names of truth-values; and it can perhaps therefore be deemed a logical truth. But under a more orthodox approach, the premiss must be treated as an additional (and presumably non-logical) axiom. Under the approach that uses extensions instead of abstracts, this additional axiom is not required.

The only other axioms we need within the theory of abstracts are Internal Existence (in its second-order version) and Identity in its right-to-left direction (forbidding the identification of abstracts that are associated with different equivalence classes of concepts). The only axioms we need within the theory of extensions are Extensionality and Comprehension in its strict schematic formulation. We shall merely provide an informal sketch of the proofs but with enough detail, I hope, to make clear how they might be formalized. We concentrate on the more problematic case of the derivations within the theory of abstracts.

Recall that a divisor divides the universe of concepts into at most two parts (thus (Div(**R**)) is defined by $\exists$C, D$\forall$E(**R**(C,E) $\vee$ **R**(D,E))). Thus given that there are at least two objects, Internal Existence straightforwardly yields the conclusion that each internally I-invariant divisor is applicable, i.e. Eq(**R**) & Div(**R**) & I-inv(**R**) $\rightarrow$ App(**R**).

With each concept C, we may associate the strictly I-invariant divisor $\mathbf{N}_c$, where $\mathbf{N}_c$(E,F) $\leftrightarrow$ (E eq C $\leftrightarrow$ F eq C) for any concepts E and F. Thus $\mathbf{N}_c$ divides the concepts into those that are equinumerous with C and those that are not. It should be clear that $\mathbf{N}_c$ is a divisor: for if every concept is equinumerous with C then every concept bears the relation $\mathbf{N}_c$ to C; while if some concept D is not equinumerous with C then every concept bears the relation $\mathbf{N}_c$ either to C or to D. Also, it is evident from the properties of equinumerosity that $\mathbf{N}_c$ is strictly I-invariant.

So it follows from Internal Existence that each of the divisors $\mathbf{N}_c$ is legitimate; and so, for each concept C, there is a unique abstract n_c of C with respect to the equivalence $\mathbf{N}_c$. We could take n_c to be the cardinality of C. But it is also possible to derive Hume's Law and thereby see each cardinal as an abstraction from the same underlying equivalence relation.

To this end, we set up the correlation of each concept C with the abstract n_c. This correlation then establishes that equinumerosity is non-inflationary. For suppose that C and D are not equinumerous. Then n_c and n_D must be distinct; for their being the same would require that C be related by $\mathbf{N}_D$ to D. (Note that essential use is made here of the identity axiom.) Since equinumerosity is internally invariant, it follows that it is applicable; and hence Hume's Law is established.

We may also establish the legitimacy of a counterpart to sets of numbers. For let C be any non-empty concept of numbers. Then with C we may associate the divisor $\mathbf{S}_c$, where $\mathbf{S}_c$(E,F) $\leftrightarrow$ ($\mathrm{C}n_E \leftrightarrow \mathrm{C}n_F$). This divisor divides the universe of concepts into those whose number falls under C and those whose number falls outside of C. The divisor is readily shown to be internally invariant; and so for each concept C of numbers there exists the abstraction s_c of D wrt $\mathbf{S}_c$, where D is some concept whose number n_D falls under C. The abstracts s_c may then be used to represent the non-empty sets of numbers.

It should be noted that the above proof makes no appeal to a principle of extensional abstraction, of either an unrestricted or a restricted sort. Indeed, the proof depends crucially upon the fact that the concepts in question are concepts of *numbers*. For it is their being numbers which entitles us to conclude that the corresponding equivalence relations are invariant.

We therefore see that a foundation can be provided for arithmetic and analysis without making any appeal to extensional abstracts. It is indeed possible to prove a form of extensional abstraction. For let us define $C \equiv^* D$ by: C and D are both exponentially small and coextensive or neither is exponentially small. Then it may be shown that $\equiv^*$ is strictly I-acceptable and hence it may be shown that each exponentially small concept has an extension. However, this result is of no help in deriving analysis. For in order to apply it to concepts of numbers we need to know that they are exp. small; and this seems to require something like the earlier construction.

Within the theory of extensions, the derivations are much more straightforward. Hume's Law can be established directly, in much the same way as for Frege. For it is readily verified that, for each concept, the condition of being equinumerous with that concept is I-invariant and so, by Extensionality and Comprehension, it follows that there exists a unique extension whose members are the concepts that conform to the condition. We take that extension to be the number of the concept; and, with that understanding of number, it is a simple matter to derive Hume's Law.

In regard to the reals, we may, for each non-empty concept of numbers C, consider the condition of being a concept whose number falls under C. This condition is readily shown to be I-invariant and hence, for each such concept C, there will be an extension whose members are the concepts whose number falls under C. These extensions can then be used, as before, to represent the reals.

4. FURTHER WORK

Let me briefly mention three further topics that need to be considered.

(1) *Conservative Extensions*. Let us say that the §-theory $T^{\S}$ (in the language $L^{\S}$) is a *conservative extension of* the underlying second-order logic L^2 (with or without Well Ordering) if the relativization ϕ^I of an §-free formula ϕ to the predicate I (for being an individual) is a theorem of $T^{\S}$ only if ϕ is a theorem of L^2. Note that we do not require that every §-free theorem of $T^{\S}$ be a theorem of L^2 since adopting an abstraction principle

may result in our being able to prove that there are more objects than there were without the principle. It is plausible to suppose that conservativity, in this sense, is a necessary condition for what, in Section 1 of Part I, I called a 'genuine' principle of abstraction; for we do not want such a principle to have any substantive implications for the logic of individuals. For this reason, and also for its intrinsic interest, it would be desirable to prove conservativity results for particular theories, such as the theory of numbers or restricted extensions, and also for general classes of theories. However, it is not clear how the proofs should go.

(2) *Further Forms of Abstraction.* The theory of abstraction, as developed in the second two parts, is not comprehensive. For it does not deal with abstraction that is over objects, or over several concepts or relations at a time, or over the various combinations of these items. It would be desirable to extend the whole theory to these other cases. Perhaps the criterion of being non-inflationary and (predominantly) invariant could still be adopted as a means of avoiding the problem of hyperinflation. But, in other respects, the theory will be quite different. We cannot expect to have the relational analogue of Hume's Law, for example. For, as Hodes (1984, 138) and Hazen (1985, 253–40) have pointed out, the abstraction principle for relation numbers will lead to the Burali–Forti paradox. Thus the possibilities for abstraction will be much more circumscribed than they are for concepts.

(3) *Alternative Forms of Second Order Logic.* For many of our results, we have assumed that the underlying logic contains full second-order Comprehension for concepts and relations. It would be of interest to see how the theory of abstraction might be developed under various weakenings of this assumption. Some of our results also depend upon assuming that the underlying logic contains some version of the axiom of choice. But there are well-motivated theories of extension that are incompatible with such an assumption. For example, there is the Cantorian theory in which $\S C = \S D \leftrightarrow \forall x(Cx \leftrightarrow Dx)$ is taken to hold under the condition that C and D can be well ordered. By using the reasoning of the paradoxes, it can be shown, within such a theory, that the universe cannot be well ordered. Again, it would be of interest to see to what extent a theory of abstracts could be developed along these lines and whether, in particular, some of the difficulties over hyperinflation could thereby be avoided.

REFERENCES

Boolos, G. (1968), 'Degrees of Unsolvability of Constructible Sets of Integers', *Journal of Symbolic Logic*, 33/4: 497–513.

Boolos, G. (1986–7), 'Saving Frege from Contradiction', *Proceedings of the Aristotelian Society*, NS 87: 137–51.

——(1987), 'The Consistency of Frege's "Foundations of Arithmetic"', in *On Being and Saying: Essays in Honor of Richard Cartwright* (ed. J. J. Thomson), Cambridge, Mass., MIT Press, 3–20.

——(1990), 'The Standard of Equality of Numbers', in *Meaning and Method: Essays in Honor of Hilary Putnam* (ed. G. Boolos), Cambridge, Cambridge University Press, 261–77.

——(1993), *Whence the Contradiction, Proceedings of the Aristotelian Society*, suppl. vol., 67: 213–33.

Drake, F. R. (1974), *Set Theory: An Introduction to Large Cardinals*, Amsterdam, North-Holland.

Dummett, M. A. E. (1981*a*), *Frege: Philosophy of Language*, 2nd edn., London, Duckworth.

——(1981*b*), *The Interpretation of Frege's Philosophy*, London, Duckworth.

——(1991*a*), *Frege: Philosophy of Mathematics*, London, Duckworth.

——(1991*b*), *Frege and Other Philosophers*, Oxford, Clarendon Press.

Field, H. (1984), 'Critical Notice of Crispin Wright's *Frege's Conception of Numbers as Objects*', *Canadian Journal of Philosophy*, 14: 637–62, repr. as ch. 5 of Field (1989), 140–70.

——(1989), *Realism, Mathematics and Modality*, Oxford, Blackwell.

Fine, K. (with Timothy McCarthy) (1984), 'Truth without Satisfaction', *Journal of Philosophical Logic*, 13: 397–421.

——(1994), 'Essence and Modality', *Philosophical Perspectives*, 8: 1–16.

——(1995), 'Senses of Essence', in *Modality, Morality, and* Belief (ed. W. Sinnott-Armstrong), Cambridge University Press, Cambridge, 53–73.

Foreman, M., and Woodin, W. H. (1991), 'The Generalized Continuum Hypothesis Can Fail Everywhere', *Annals of Mathematics*, 133/1: 1–35.

Frege, G. (1959), *Foundations of Arithmetic* (trans. by J. L. Austin), Oxford, Blackwell.

——(1893–1903), *Grundgesetze der Arithmetik: Begriffsschriftlich abgeleitet*, H. Pohle, Jena; part trans. by M. Furth in *The Basic Laws of Arithmetic: Exposition of the System*, Berkeley, Univ. of Calif. Press.

Hale, R. (1987), *Abstract Objects*, Oxford, Blackwell.

——(1994), 'Dummett's Critique of Wright's Attempt to Resuscitate Frege', *Philosophia Mathematica*, 2/3: 122–147.

Hazen, A. (1985), Review of Crispin Wright's 'Frege's Conception of Numbers as Objects', *Australasian Journal of Philosophy*, 63/2: 251–4.

Heck, R. G. (1992), 'On the Consistency of Second-Order Contextual Definitions', *Noûs*, 64/2: 491–4.

——(1993), 'The Development of Arithmetic in Frege's *Grundgesetze der Arithmetik*', *Journal of Symbolic Logic*, 58/2: 579–601.

——(1996), 'Definition by Induction in Frege's *Grundgesetze der Arithmetik*', in *Frege. Importance and Legacy* (ed. M. Schirn), Berlin, Walter de Gruyter, 200–33.

Hodes, H. T. (1984), 'Logicism and the Ontological Commitments of Arithmetic', *Journal of Philosophy*, 81/3: 123–49.

Linsky, B. (1992), 'A Note on the "Carving up Content" Principle in Frege's Theory of Sense', *Notre Dame Journal of Formal Logic*, 33/1: 126–35.

Parsons, C. (1965), 'Frege's Theory of Number', in *Philosophy in America* (ed. M. Black), Ithaca, NY., Cornell University Press, 180–203.

Parsons, T. (1987), 'On the Consistency of the First-Order Portion of Frege's Logical System', *Notre Dame Journal of Formal Logic*, 28/1: 161–8.

Peacocke, C. (1991), 'The Metaphysics of Concepts', *Mind*, 100: 525–46.

Schroeder-Heister, P. (1987) 'A Model-Theoretic Reconstruction of Frege's Permutation Argument', *Notre Dame Journal of Formal Logic*, 28/1: 69–79.

Shapiro, S. (1991), *Foundations without Foundationalism: A Case for Second-Order Logic*, Oxford, Clarendon Press.

Tennant, N. (1997), 'On the Necessary Existence of Numbers', *Noûs*, 31: 307–36.

Wright, C. (1983), *Frege's Conception of Numbers as Objects*, Aberdeen, Aberdeen University Press.

——(1988), 'What Numbers can Believably Be', *Revue Internationale de Philosophie*, 42: 425–73.

——(1990), 'Field and Fregean Platonism', in *Physicalism in Mathematics* (ed. A. D. Irvine), Dordrecht, Kluwer, 73–93.

——(1997), 'On the Philosophical Significance of Frege's Theorem', in *Language, Thought, and Logic* (ed. R. G. Heck), Oxford, Oxford University Press, 201–44.

INDEX